James Macfarlane

An American Geological Railway Guide

James Macfarlane

An American Geological Railway Guide

ISBN/EAN: 9783744678889

Printed in Europe, USA, Canada, Australia, Japan

Cover: Foto ©berggeist007 / pixelio.de

More available books at **www.hansebooks.com**

AN ·AMERICAN·

GEOLOGICAL RAILWAY GUIDE,

GIVING THE

GEOLOGICAL FORMATION AT EVERY RAILWAY STATION,

WITH

ALTITUDES ABOVE MEAN TIDE-WATER,

NOTES ON INTERESTING PLACES ON THE ROUTES,

AND

A DESCRIPTION OF EACH OF THE FORMATIONS,

BY

JAMES MACFARLANE, Ph. D.,

AUTHOR OF "THE COAL-REGIONS OF AMERICA," AND ONE OF THE COMMISSIONERS OF
THE SECOND GEOLOGICAL SURVEY OF PENNSYLVANIA,

WITH THE CO-OPERATION OF THE STATE GEOLOGISTS, AND OTHER SCIENTIFIC GENTLEMEN.

SECOND EDITION, REVISED AND ENLARGED,

EDITED BY

JAMES R. MACFARLANE.

NEW YORK:

D. APPLETON AND COMPANY,

1, 3, AND 5 BOND STREET.

1890.

PREFACE TO THE SECOND EDITION.

The first edition of this book was published by my father, the late James Macfarla
in 1878 and, at the time of his death in October, 1885, he had prepared many of 1
chapters and collected some of the material for others for this second edition. By f
lowing the system of the work already completed, with the assistance of the gentlem
whose names appear throughout these pages, I have, after many delays, completed 1
edition.

The whole book has been carefully revised and new lines and new notes added,
that the Guide, proper, has been enlarged from 158 to 870 pages. The introducto
portion of the book has been changed only where necessary to conform its statements
the views now held by geologists. The altitudes are a new and valuable feature of tl
edition and the list is as complete as could be obtained. A few chapters were so p
pared by their authors that little work was needed before printing them, but in m
instances the labor of collecting and arranging such a mass of material into a comp
and harmonious form has been greater than would be imagined. Whatever defects a
mistakes are found in the book may be attributed to the loss of the one whose mind co
ceived its plan, and who was peculiarly fitted for its preparation.

To the contributors and my many advisors I owe a debt of gratitude that I can
express, but I know that they will feel rewarded if their work results in an increase
interest in, and knowledge of, the noble science of geology.

JAMES B. MACFARLANE.

Pittsburgh, Pa., 1890.

TABLE OF CONTENTS.

THE OBJECTS AND USES OF THIS WORK.

1. FOR THOSE WHO ARE NOT GEOLOGISTS.

The United States are intersected by numerous railroads leading in all directions, and nearly every one has occasion more or less to travel on them for considerable distances. In these railway journeys no person who has the least power of observation can fail to notice the peculiarities in the scenery and the great variety in the formations of rock to be seen in the railway cuts and cropping out on the hillsides. If we always had a professor of geology for our traveling companion, we would be glad to learn from him what these various formations of rock are, what place they occupy in the series of strata that are visible on the earth's surface, and their mineral and other productions ; also at what other localities the same rocks occur, and whether they are entirely new to us or the same we have seen elsewhere. This work is a substitute for the supposed traveling professor of geology, giving in a small space the names of the geological formations which occur along the lines of the railroads, and in another part of the book is to be found a plain but full description of each of them. There are also foot notes directing attention to interesting geological places and objects on' the routes of the railroads. One object of the work is to teach persons not versed in geology something of this science during the tedious and unprofitable hours of traveling, without study, not as in a text book, but by pointing to the things themselves as seen at railway stations and through the windows of a railway car.

No person could be so stupid as to travel all over the United States without learning the name of a single state or city through which he passes, yet how few persons know even the names of the geological formations on which they have spent their lifetimes. Every one is taught geography, and there is scarcely a child of sufficient age who cannot tell the name of the town, county and state in which he lives. But geology, which is just as well worth knowing, is neglected, and there is but little opportunity for learning any thing practically in regard to it from those about us. This is not owing to a want of a desire for knowledge, but to a want of instruction in this science, and of the practical application of what is learned by adding local geological information in a handy, cheap and accessible form, and this, which no other work affords, it is the aim of this book to furnish.

There are some kinds of knowledge too that cannot be obtained from books, but must be gathered by actual observation. The inspection of a formation in nature, which is pointed out to you, will teach you more in regard to it in a few minutes than you could learn from lectures or from reading books in as many hours, and the lesson so received will be better remembered. This book is intended as an intelligent guide to such observations. It tells you where the various formations are, and you can then see for yourself in traveling what they are.

How lonely would be a journey on which you would see not a single face that you know, and how different it would be if every one you meet were an old friend. So to the tourist new charms must be given to scenery, however attractive it may already be, if he knows something about its geology. The rocks, mountains, valleys and plains, although he sees them for the first time, are old friends in perhaps new and interesting forms. He meets them with a certain pleasure, for he understands what he sees and he is given the materials for many a happy hour of quiet and profitable reflection at home, on what he has seen on his railway journey.

2. FOR GEOLOGISTS.

But while the book is thus intended primarily as a series of object lessons for those to whom geology is yet a novelty, for the purpose of exciting an interest in, and which may ripen into a love for the science, it is believed that, being in a more convenient form than geological maps, and as no other work has attempted what is here done, all geologists, and especially students, will find it a most useful hand book on their railway journeys as well as for reference at home. It will be useful in laying down the geology in colors on any map which gives the railroads. Accurate geological maps can thus be made without expense, and there is no better exercise for students. It will also be invaluable in selecting a route of travel for geological study or for pleasure, and no geologist should make an excursion over new ground without this guide. It is a scientific catalogue of the great panorama that passes with its ever shifting scenery before the eyes of the American railway traveler, and even an artist finds a catalogue of a picture gallery very necessary. No geologist need be told that it embraces the result of a vast amount of learning, labor and research in a very small compass, and a minuteness of local geology for which he might ransack libraries in vain, and which no one man could possibly furnish. Many men for many years have devoted the finest talents in America to the study of the geology of these states, and all have contributed by their published reports, or by direct original contributions to this work, portions of the knowledge which is here indexed, otherwise it would not be becoming for the author to say so much in its praise. In order that the guide might be as accurate as possible the assistance of the state geologist of each state, or that of some scientific gentleman best acquainted with its local geology, has been invoked to revise and correct the list of formations found along the railroads. Without a single exception, and with characteristic devotion to the cause of science,* this aid has been very cheerfully and promptly rendered, and in not a few instances, where the necessary information was only in the knowledge of these gentlemen, they have filled in the geology from original sources not yet published. Due credit is given to all contributors in the notes of the proper chapter. The general accuracy of the book can be relied upon as to the formations of each locality as they were understood at the time of its publication, and it may be regarded as in harmony with the latest results of geological research. If errors are found, consider the great number of railroad stations and you will wonder there are so few.

*Scientific men freely give the results of their labors to the world, expecting only in return to enjoy the consciousness of having added by their investigations to the sum of human knowledge, and to receive the credit to which they might justly entitle them.   PROF. JOSEPH HENRY.

3. FOR USEFUL, PRACTICAL PURPOSES.

To those who take only utilitarian views and care nothing for pure science, and to all those in any way interested in the country, a means is here furnished for ascertaining the natural advantages or disadvantages of any district where there is a railroad, for it is now pretty well known to all intelligent persons that the capabilities or resources of a country, what it is and what it can become, depend chiefly on its geology.

No one in our day can doubt, that there is a definite and orderly arrangement of the rocks, that it is only in certain rocks that certain useful materials and minerals are to be obtained, and that the soil of each formation has a certain fixed value for agriculture. It was long ago shown that a geological map of England, is a map also of the distribution of its manufactures. Even the kind of people inhabiting a district, often depends on its geology. A considerable portion of the work of geologists, is devoted to tracing out the distribution of the various formations as they come out from beneath one another, and spread over the face of the country. This book is made up of a minute tabular statement or division of all places on the American railways, into classes, some of which yield useful materials or productions peculiar to them. It points out the limits to be observed in searching out new locations producing any material. Besides, if accompanied by a correct scientific knowledge of the country, it will make any man's discovery of anything useful available to his neighbors in hundreds of other places, over the whole region covered by the same formation.

The physical structure of a country being then, the means by which we can learn the range and distribution of useful materials, a strict attention to fossils is necessary, to enable us to determine the relative position of rock groups, each group, within certain limits, holding its own peculiar fossil forms, and certain economic products being confined, over wide areas, either wholly or principally to certain rocks. Many persons, ignorantly confounding the means with the end, think geologists are good authorities upon fossils, but not as to the useful properties of the formations. Sir William E. Logan, the great Canadian geologist, in answer to this objection, once said: "I am not a naturalist; I do not describe fossils, but use them. They are the geologist's friends, who direct him in the way to what is valuable. To get the necessary information from them, you must be able to recognize their aspect, and in order to state your authority, you must give their names. Some of them tell of coal—they are cosmopolites; while some give local intelligence of gypsum, or salt, or building stone. One of them helped us last year to trace out, in Canada, upwards of fifty miles of hydraulic limestone."

But it is not practicable for ordinary readers to understand the difficult science of paleontology; all they can expect to know are the results as ascertained by professional geologists, and those results are given in this little book, for every place on every railroad in America. There are many other things that might have been given, especially the structural geology of each State, geological maps, more minute lists of elevations and general physical geography, but the book contains enough for one little volume to be carried about on railway journeys.

TOWANDA, Pa., 1878.   JAMES MACFARLANE.

Prof. J. D. Dana's Table of the Geological Formations (1885),

AS NUMBERED IN THE GEOLOGICAL RAILWAY GUIDE.

Systems or Ages.	Groups or Periods.	Formations or Epochs.
20. Age of Man.	20. QUATERNARY.	20 Quaternary.
19. Age of Mammals.	19. TERTIARY.	19 c. Pliocene. 19 b. Miocene. 19 a. Eocene.
16-18. Reptilian Age.	18. CRETACEOUS.	18 c. Upper Cret. 18 b. Middle Cret. 18 a. Lower Cret.
	17. JURASSIC.	17 Jurassic.
	16. TRIASSIC.	16 Triassic.
13-15. Carboniferous.	15. PERMIAN.	15 Permian.
	14. CARBONIFEROUS.	14 c. Upp. Coal-meas. 14 b. Low. Coal-meas. 14 a. Millstone Grit.
	13. SUBCARBONIFEROUS.	13 b. Upper Subcarb. 13 a. Lower Subcarb.
8-12. Devonian, or Age of Fishes.	12. CATSKILL.	12 Catskill.
	11. CHEMUNG.	11 b. Chemung. 11 a. Portage.
	10. HAMILTON.	10 c. Genesee. 10 b. Hamilton. 10 a. Marcellus.
	9. CORNIFEROUS.	9 c. Corniferous. 9 b. Schoharie. 9 a. Cauda Galli.
	8. ORISKANY.	8 Oriskany.
2-7. Silurian, or Age of Invertebrates. — 5-7. Upper Silurian.	7. LOWER HELDERBERG.	7 Lower Helderb'g
	6. SALINA.	6 Salina.
	5. NIAGARA.	5 c. Niagara. 5 b. Clinton. 5 a. Medina.
2-4. Lower Silurian.	4. TRENTON.	4 c. Hudson River. 4 b. Utica. 4 a. Trenton.
	3. CANADIAN.	3 b. Chazy. 3 a. Calciferous.
	2. PRIMORDIAL OR CAMBRIAN.	2 b. Potsdam. 2 a. Acadian.
	1. ARCHÆAN.	1 b. Huronian. 1 a. Laurentian.

Table of the Geological Formations,

ARRANGED FOR THE SECOND EDITION OF THIS WORK BY T. STERRY HUNT, LL. D., F. R. S.

AGES.	GROUPS.	AMERICAN FORMATIONS.
Cenozoic.	20. QUATERNARY.	20. Recent.
Cenozoic.	19. TERTIARY.	19 c. Pliocene. 19 b. Miocene. 19 a. Eocene.
Mesozoic.	18. CRETACEOUS. 17. JURASSIC. 16. TRIASSIC.	18. Cretaceous. 17. New Red Sandstone. 16. New Red Sandstone.
Palæozoic.	13–15. CARBONIFEROUS.	15. Permo-Carboniferous. 14. Coal Measures. 13 b. Mississippi, (Carb. limestone.) 13 a. Waverley or Bonaventure.
Palæozoic.	8–12. ERIAN OR DEVONIAN.	12. Catskill. 11. Chemung and Portage. 10. Hamilton, (including Genesee and Marcellus.) 9. Corniferous or Upp. Helderb'g. 8. Oriskany.
Palæozoic.	5–7. SILURIAN.	7. Lower Helderberg. 6. Onondaga or Salina. 5 c. Niagara, including Guelph. 5 b. Clinton. 5 a. Medina. 5 a. Oneida.
Palæozoic.	3–4. ORDOVICIAN, (Upper Cambrian of Sedgwick or Siluro-Cambrian.)	4 c. Loraine. 4 b. Utica. 4 a. Trenton. 3 a. Chazy.
Palæozoic.	2. CAMBRIAN. (Middle and Lower Cambrian of Sedgwick.) (Keweenian.)	2 c. Calciferous. { Upper Taconic 2 b. Potsdam. { or Quebec Gr'p. 2 a. Menevian. (St. John's group.)
Eozoic.	1. PRIMARY OR CRYSTALLINE. (Primitive and Transition.)	1 f. Taconian. (Lower Taconic.) 1 e. Montalban. 1 d. Huronian. 1 c. Arvonian. 1 b. Norian. 1 a. Laurentian.

Systems or Ages.		Groups or Periods.		Formations or Epochs.
19–20. Cenozoic.	20. Age of Man.	20. Quarternary.		20. Quarternary.
	19. Age of Mammals.	19. Tertiary.		19 c. Pliocene. 19 b. Miocene. 19 a. Eocene.
16–18. Mesozoic.	16–18. Reptilian Age.	18. Cretaceous.	Rogers' Pa. and Va. No's.	18 c. Upper Cretaceous. 18 b. Middle " 18 a. Lower "
		17. Jurassic.		17. Jurassic.
		16. Triassic.		16. Triassic.
2–15. Paleozoic.	12–15. Carboniferous.	15. Permian.		15 Permo-Carboniferous.
		14. Carboniferous.	XV. XIII. XII.	14 c. Upper Coal-measures. 14 b. Lower Coal-measures. 14 a. Millstone Grit.
		13. Subcarboniferous.	XI. X.	13 b. Upper Subcarbonif'ous. 13 a. Lower "
		12. Catskill.	IX.	12 Catskill.
	8–11. Devonian, or Age of Fishes.	11. Chemung.	VIII "	11 b. Chemung. 11 a. Portage.
		10. Hamilton.	" " "	10 c. Genesee. 10 b. Hamilton. 10 a. Marcellus.
		9. Corniferous.	" " "	9 c. Corniferous. 9 b. Schoharie. 9 a. Cauda Galli.
		8. Oriskany.	VII.	8 Oriskany.
	2–7. Cambrian to Silurian, or Age of Invertebrates. — Upper Silurian.	5–7. Silurian.	VI. V. IV.	7 Lower Helderberg. 6 Salina. 5 c. Niagara. 5 b. Clinton. 5 a. Medina and Oneida.
	Lower Silurian.	3–4. Siluro-Cambrian, or Trenton.	III. II. "	4 c. Cincinnati, Hudson River or Loraine. 4 b. Utica. 4 a. Trenton. 3 b. Chazy. 3 a. Calciferous.
		2. Cambrian, or Primordial.	I.	2 b. Potsdam. 2 a. Acadian. 2 á. Georgian.
		1. Eozoic or Archæan.		1 b. Huronian. 1 a. Laurentian.

DESCRIPTIONS OF THE GEOLOGICAL FORMATIONS.

INTENDED FOR RAILWAY TRAVELERS WHO ARE NOT VERSED IN GEOLOGY.

All the rock-formations which appear on the surface of the globe, have been scientifically classified by geologists, according to the order in which they are found lying one upon another, and by the fossils they contain, and for our object may be conveniently included in twenty divisions or groups. In this work, the table of the names of the formations, groups and systems, published by Prof. J. D. Dana in his "Manual of Geology" and in his "Text Book of Geology," has been taken as the general basis, by the geologists of many of the states who have assisted in preparing the following guide, but other valuable tables and especially one arranged by Dr. T. Sterry Hunt, a general or combined table, and a list for each state at the beginning of the proper chapter, are also given. Numbers are attached to the names of the groups wherever they occur, making 20 in all. The subordinate members of each group, which are called formations, have the same number, but these sub-divisions are distinguished by the addition of small letters, a, b, c, etc., thus making in all 40 sub-divisions. By this means, the reader, although not familiar with geological tables, is at once enabled to see to what part of the general series any formation belongs, number 1 designating the oldest and number 20 the upper and last formed of all. Wherever the formations are found, they occur in the order as they are numbered, but the series in nature is never full, and in almost every locality one or more members of it are wanting.

The true method by which each of the great stratified formations is distinguished is by its own characteristic fossils, but these descriptions, having been prepared for travelers, are confined to the general aspect of the rocks as seen in passing them on the railways. They are intended to be popular rather than scientific, informing the reader what the formations are, what they look like, and their useful and valuable characters, qualities, and productions. It must also be borne in mind that this is a country of vast dimensions, and that the formations undergo important changes in their lithological character from place to place.

Paleontology, and other interesting branches constituting the purely technical portion of the subject, are omitted. That ground has been well covered by all of the excellent illustrated text-books on geology, and one object of this work is to induce persons to take up their study. Results only are here given, not the method, by which they are attained. The thicknesses of the formations are sometimes stated, but as this might mislead the unprofessional reader, it should be observed, that the width of the surface occupied by a formation depends on the amount of dip in the beds. A group less than a hundred feet thick, lying horizontally, may cover several miles, while one of several thousand feet thick, if lying at a high angle, is soon passed over.

1. EOZOIC (ARCHÆAN, AZOIC).

I. PRIMARY OR CRYSTALLINE ROCKS.

The late investigations of American geologists have enabled them to establish several divisions in the crystalline stratified rocks, which were originally called Primary or Primitive. The name Azoic, formerly given to the Primary rocks to distinguish them from the Paleozoic formations, has, since the discovery of Eozoon in the former, been exchanged for that of Eozoic. The designation Archæan or ancient rocks, is used by Professor Dana and others, and applies to the Primitive formations without distinction. Among those who have made the Primitive or crystalline rocks a special subject of study for many years, no one is more eminent than Dr. T. Sterry Hunt, whose classification of these rocks established by him in North America has since been recognized by many geologists in Europe, where the same great groups are found. The following descriptions, giving the latest conclusions as to the divisions of the Crystalline rocks, have been furnished by him for this second edition of this work.

1 a. Laurentian.—The name of Laurentian was given in 1854, by the geological survey of Canada, to the ancient crystalline terrane which forms the chief portion of the Laurentide hills, and of the Adirondacks.

Throughout these areas the prevailing rock is a strong, massive gneiss, reddish or grayish in color, sparingly micaceous, but very often hornblendic. The predominance of this mineral occasionally gives rise to a nearly pure hornblende-rock, sometimes with a little intermixed feldspar. The gneisses are, for the most part, distinctly stratified, but occasionally the evidences of stratification are not very apparent, so that these rocks have often been designated granites. This series is distinguished by the absence of chloritic, talcose, argillaceous or micaceous schists. It includes, however, crystalline limestones, of which there are supposed to exist, in the Ottawa valley, three distinct masses in the Laurentian series, each of which is, in parts, according to Logan, more than 1,000 feet in thickness. These limestones, which are generally coarsely crystalline, are often magnesian, and abound in foreign minerals, chief among which are serpentine, chondrodite, hornblende, pyroxene, magnesian mica, apatite and graphite. Most of these occur both disseminated in the beds, and, aggregated with other minerals, in veins, or endogenous masses. Associated with these limestones are often considerable beds of quartz-rock, sometimes garnetiferous. Great masses of magnetic oxide of iron are also found interstratified in this series. The measured thickness of the Laurentian gneisses, with their included limestones and other rocks, on the Ottawa, where the strata are nearly vertical in attitude, has been estimated at over 17,000 feet. Beneath these, known as the Grenville series, there is a great underlying mass of granitoid gneiss, without limestones, and of undetermined thickness, called the Ottawa gneiss, which, it is conjectured, may not be conformable with the upper portions.

In the Atlantic belt, considerable areas of Laurentian occur in Newfoundland, and probably in several parts of New England. A range of Laurentian rocks from the Western part of Connecticut extends southwestward, forming the Highlands of the Hudson, and making the South Mountain as far as the Schuylkill; while a smaller range of the same, to the southeastward, forms the Welsh Mountain, in Pennsylvania. Little is known of the distribution of the Laurentian farther southward, but gneisses near Richmond in Virginia, and at Roan Mountain, in North Carolina, are referred to this terrane.

Large areas of Laurentian occur around Lake Superior, and farther west in the Rocky Mountains, where they form the crystalline rocks of the Colorado range in the east, and those of the Wasatch in the west, and probably occur in many other parts of the region. To the Laurentian belong the gneisses of the Western Islands of Scotland, those of Scandinavia and Finland, and large portions of those of the Alps. The limestones of the Laurentian contain the remains of a foraminiferal organism known as *Eozoon Canadense* (Dawson), which has been found in several localities in Canada, and also in Bavaria, and in Finland. Accompanying it are several other small forms, regarded as organic, and referred to the protozoa.

1 b. Norian.—The upper portion of the Laurentian series on the Ottawa river, was orginally defined by the geological survey of Canada as consisting of a rock, gneissoid or granitoid in character, made up chiefly of labradorite, or related anorthic feldspars, but including also true gneisses and crystalline limestones, not unlike those already described in the Laurentian. Subsequent studies in Canada led to the conclusion that these rocks constitute a distinct terrane, resting uncomformably upon the gneisses and crystalline limestones of the preceding series, and the two were respectively designated as Lower Laurentian and Upper Laurentian or Labradorian. As the newer is very distinct from the older terrane, it has, however, been thought better to restrict the name of Laurentian to the latter. A series precisely similar to the upper one occurs in Norway, where, as in North America, it rests upon Laurentian gneisses, and where the name of norite has been given to the feldspathic rock which is its chief characteristic. Hence, the name of Norian, which has been chosen in place of Upper Laurentian, as the designation of the terrane. It is conjectured, from the fact that it has yet been found only in contact with the Laurentian, and from its including gneisses and limestones lithologically similar to those of the latter, that it is next in age.

The norites consist, for the greater part, of anorthic or plagioclase feldspars, sometimes almost without admixture, but at other times accompanied by small portions of hornblende, of pyroxene or of hypersthene, constituting what has been called hypersthenite or hyperite. Chrysolite, red garnet, green epidote, biotite, and ilmenite are often present, and these minerals are generally arranged in such a way as to give a gneissoid structure to the rock. The texture is sometimes fine-grained and compact, and at other times more coarsely granular, and even granitoid, displaying great masses of the plagioclase feldspar, frequently opalescent, and varying in composition from anorthite to andesine. The colors of the norites vary from white, pale bluish or greenish, rarely reddish, to dark lavender or smoke-blue, or nearly black.

The principal area of this terrane known in the United States is in Essex county, New York, where it covers several hundred square miles, and, although highly inclined, rests unconformably, according to Professor Hall, upon the

Laurentian. It is well displayed upon the shore of Lake Champlain, between Port Kent and Westport, and forms some of the highest hills of the interior. A second large area of Norian occurs north of Montreal, where it is similarly related to the Laurentian, and passes below the Potsdam sandstone. Other localities along the valley of the St. Lawrence are at Chateau Richer near Quebec, at Bay St. Paul, the Bay of Seven Islands, and on the River Moisie. Extensive areas of it also exist on the coast of Labrador. The same rock has been found on the east shore of Lake Huron, at the west end of Lake Superior, as at Duluth, and in Wyoming Territory.

1. c. Arvonian.—There is found in many localities a series of highly inclined stratified rocks, consisting essentially of petrosilex or halleflinta, often passing into a quartziferous porphyry. There are found with it strata of vitreous quartzite and thin layers of soft micaceous schists, besides great beds of hematite, and, more rarely, layers of crystalline limestone. This group, which has a thickness of many thousand feet, was at first included in the succeeding Huronian series, which, however, apparently overlies it unconformably.

Its relations with the preceding groups have not been clearly determined, but it appears to be identical, both in position and in character, with the group, which in Wales has, since 1878, been called Arvonian. These Arvonian rocks are well seen at many points along the coast of Massachusetts and New Brunswick and in the Atlantic belt in southern Pennsylvania. Areas of them are also seen on the north shore of Lake Superior, and rising through the paleozoic sandstones in Wisconsin. They appear under similar conditions in southeast Missouri, where they include great beds of iron-oxyd.

1 d. Huronian.—The name of Huronian was given in 1855 by the geological survey of Canada, to a series of more or less schistose crystalline rocks, shown to rest upon the Laurentian series on the north shore of Lakes Huron and Superior. A similar series is largely developed in the Atlantic belt in New-foundland, in the province of Quebec, and in New England, and farther south-westward in the Blue Ridge. The Huronian differs from the preceding series by the frequent presence of schistose rocks, and of conglomerates, which contain fragments of the underlying gneisses. The Huronian contains a considerable portion of epidote, hornblende and pyroxene, and is marked by varieties of diabasic rocks, often called gabbros, which are truly stratified, but are not to be confounded with the norites of the Norian series, to which the name of gabbro is also often given. The Huronian series moreover includes imperfect gneisses, quartzites, dolomites, serpentines, and steatite, besides large amounts of chloritic, micaceous and argillaceous schists. Its thickness has been estimated at about 18,000 feet, and it is often found resting unconformably upon the gneiss of the Laurentian. Ores of copper, nickel, chrome and iron are common in the Huronian series, which is penetrated in many localities by unstratified rocks, both granite and doleritic.

The rocks in the British Islands, which have lately been described by the name of Pebidian, are apparently identical with the Huronian; and the great series in the Alps, known to the Italians as the *pietri verdi*, or greenstone group, or at least its lower portion, has both the lithological characters and the geognostical relations of the Huronian, to which it is now generally referred. Similar crystalline schists found in California, both in the foot-hills of the Sierras and in the Coast Range, are probably Huronian. The gold veins of California traverse both these schists and the penetrating granites.

1 e. Montalban.—The name of Montalban was given in 1872 to a great series of crystalline schists which are lithologically and geognostically distinguished from the Huronian, and are well displayed in New Hampshire in the White Mountains (whence the name). It occupies large areas in New England and constitutes the gneisses and mica schists of Philadelphia, Baltimore and Washington, extending southwestward into Alabama, and, in the absense of the intermediate groups, often rests directly on the Laurentian gneiss. This is well seen on the Island of New York, on the north part of which the older gneiss, which makes up the Highlands of the Hudson, appears from beneath the Montalban, which covers the greater part of the island. The Montalban series contains fine grained white gneisses, sometimes porphyritic, but distinct from the granitoid gneisses of the Laurentian, and passing into granulites on the one hand, and very quartzose, coarse grained mica schists, chiefly muscovtic, on the other. It also includes hornblende in some parts, and the gneiss, by a predominance of this mineral, passes into a nearly black schistose hornblende-rock. Beds of granular chrysolite rock [accompanied by enstatite, and by serpentine, often with chromite, are found interstratified in this series in North Carolina and in Georgia. It also includes beds of crystalline limestone, which resemble those of the Laurentian, and moreover includes large deposits of iron pyrites and copper pyrites. The fine grained gneisses of the Montalban are sometimes called granites, but the series is penetrated by great masses of true intrusive granite. The mica schists of the series often contain garnet, staurolite, cyanite and andalusite; these species, with the exception of the first, not being, so far as known, found in the Laurentian series. The endogenous granitic veins carrying muscovite, iolite, spodumene, beryl, columbite, tinstone and apatite in the Atlantic belt, occur chiefly in the Montalban series. The Montalban is supposed to be represented by the younger gneissic and mica schist series of Scotland, which has been called Upper Pebidian, Grampian and Caledonian. It corresponds to the younger gneissic series of the Alps, where it is generally, though not everywhere, separated from the older Laurentian group by a great development of Huronian.

1 f. Taconian.—Along the great Appalachian Valley from Vermont to Alabama extends a belt of quartzite, limestone and crystalline schists with roofing-slates, which, by many geologists, have been regarded as a great development in an altered condition of the Cambrian and Ordovician (Potsdam-Loraine). These rocks, called by H. D. Rogers Primal, Auroral and Matinal, are regarded by others as older than the Potsdam, and constitute the Lower Taconic of Emmons, since called Taconian. They include the Itacolumitic series of South Carolina, and have a general thickness of 4,000 to 5,000 feet. In these are found the white marbles of the Valley, the great deposits of limonite and beds of magnetic and specular iron ores. To this series are also referred the similar series of rocks in northern Michigan and Minnesota, including what has been named the Animikie series, which have been confounded with the Huronian. A great series of similar rocks is found in the Alps between the younger gneisses and the paleozoic. T. STERRY HUNT.

2-15. PALEOZOIC.

2-4. CAMBRIAN (OR LOWER SILURIAN) AGE.

2 a. Acadian.—This series is found at Braintree, in Massachusetts, at St John, in New Brunswick, and at St. John, in Newfoundland. It includes on thousand feet or more of fossiliferous sandstone and shale, and according to Dr Hunt, corresponds to the Menevian of Great Britain. It has only been found along the north-eastern border of the Atlantic belt. It is remarkable as a fossil iferous rock below the Potsdam, which had, before its discovery, always been con sidered as the lowest formation of that description on the continent.

2 b. Potsdam.—The Potsdam sandstone, was for a long time considered as the lowest sedimentary fossiliferous rock. It is usually of a purely quartzose character, generally gray, though often striped, and sometimes partially or entirely red. In places it appears as a conglomerate, but sometimes the enclosed masses are angular, showing them to be near their source.—Hall, N. Y. R., 2'. It is a hard silicious sandstone, white, red, gray, yellowish, and frequently stripe ; Some strata of this rock are covered with the most beautifully characteriz ripple-marks as perfect as if just formed on the sand of a sea-beach, whi the rock is the most indurated kind of sandstone. Its lower portion is a granitic conglomerate, in which large masses of quartz, the size of a peck measure, are often enveloped ; they are rounded and water-worn, and held together by a finer variety of the same material. On the Canada slope, where the mass is 300 feet thick, it is wholly a conglomerate, made up of coarse materials. The part which is properly a sandstone, has two principal varieties, a close grained, sharp edged mass, with natural joints traversing it in two directions, but so closely wedged together that it is quarried with difficulty. This is the Keeseville variety, and that of Pa. and N. J. The other, the typical mass at Potsdam, is an even bedded and somewhat porous rock, at many places a distinct friable sandstone in others a yellowish-brown sandstone, the particles of which are compacte together, so as to form a firm, even-grained mass, with the planes of deposit perfectly smooth and separable from each other, the layers being from two inc' to four feet thick. At Potsdam quarries, a layer of 100 square feet may be rai and split into rails, six inches wide and ten feet long, or it may be broken ir pieces the size of a brick, with even edges of fracture, and each layer may t separated into many. The color here is yellowish-brown, and a deep red variety occurs at Chazy, resting immediately upon the primitive rock.—Mather, 102. It is nowhere charged with mineral matter, either disseminated or in veins. The native copper of Lake Superior is in an old trappean formation, and has no relation to the neighboring extensive formation of Potsdam. In an economical point of view, the Potsdam is unimportant as a depository of useful substances.

The general color of the stone at Potsdam is yellowish-brown, but the tint of each layer differs somewhat from those adjacent to it, so that the rock, upon the fractured edges, wears a slightly striped aspect.   It is the finest quarry stone in the state, being so perfectly workable and manageable.—360.   It is an excellent building material, holding mortar well, and makes a dry house.—29.   Under the Potsdam, and upon the primary rock, is the position of the specular and red oxide of iron.—V. 267.

In Minnesota, the lower portion of the formation is 400 feet thick, and is hard and often vitreous, and usually of a brick-red color, with very distinct layers, often separated into slaty layers by partings of red shale, strongly marked with fucoidal impressions, frequently ripple-marked and cracked.   The upper part of the formation, there called the St. Croix sandstone, is white or buff in color, often friable, and constitutes a heavy bedded or massive sandstone of rounded quartzose grains.—N. H. Winchell.

In Minnesota and Iowa, the Potsdam proper, omitting the St. Croix sandstone, is a friable, crumbling mass, of no value for building purposes except as sand, consisting of a pure silicious sand in minute grains, with a very slight amount of cementing matter.   Unless protected by some more resisting rock above it the Potsdam appears in steep slopes, or low, gently swelling hills and mound-like eminences.   Those portions which are hard and enduring are cemented by oxide of iron, and have a brown color.

In Wisconsin, the Potsdam is 800 to 1000 feet thick, and has a much larger surface-development than elsewhere, as will be seen by the great number of railway-stations on it.   It extends over 12,000 square miles, and contains many fossils not found in New York.   Where the Potsdam in Wisconsin is on the surface, and not covered by drift, there is usually a loose, sandy soil, with a sparse growth of small oak and pine timber.   This formation is one that has been very properly allowed to retain its original name almost undisputed all over the United States, except that Professor Owen at first called it the LOWER SANDSTONE, in the North West to distinguish it from the 3 c., St. Peters or Upper Sandstone.

In Michigan, the Potsdam is the red sandstone, which is emphatically the chief rock that appears upon the immediate coast of the whole south shore of Lake Superior, and forms the Pictured Rocks and the Falls of St. Marie.   Here it is of inconsiderable thickness, but it regularly thickens in going westward.— Houghton, 4th R., 500.   Some have referred the Lake Superior sandstone to the age of the Chazy, but the late studies of Rominger show that it is really of Potsdam age.   The Chicago Tribune office building is of this Lake Superior sandstone, and the Court House at Milwaukee is another conspicuous specimen.

In Pennsylvania, the Potsdam is a compact, fine-grained, white and yellowish vitreous sandstone, containing specks of Kaolin.

The Potsdam formation is supposed by some to be represented in the Green Pond Mountain of New Jersey by a local deposit of coarse conglomerate, 3000 feet thick, but others deny that this mountain is Potsdam.   It is less than 80 feet thick where it is seen rising from beneath the limestones of the Lehigh River, but increases in thickness westward and southward, until it comes to be represented in Tennessee by many thousand feet of alternate coarse and fine deposits.   See Safford's Geol. R. of Tenn.

3 a. Calciferous.—This group embraces in New York three distinct masses as to character and position, and these alternate and intermix with each other. The first is silicious, compact, and may probably be the continuation of the Potsdam sandstone. The second is a variable mixture of fine, yellow, silicious sand and dolomite or magnesian carbonate of lime, which, when fractured, presents a fine, sparkling grain. It is in irregular layers, which have a shattered appearance, from numerous cracks, the parts being more or less separated from each other. This is the mass from which the name Calciferous sandrock was derived. The third is a mixture of the dolomite material, which is usually yellowish, very granular when fresh broken, and of a compact limestone, which resembles the Birdseye. The action of the weather gives these layers the appearance of Gothic fret-work, and the color becomes a dark yellow-brown.—V. 21. As its name indicates, it is a sandy magnesian limestone, but it is not destitute of beds of pure limestone. The mixture of a variety of mineral matter causes the rock to weather unequally ; hence it is often rough externally, portions of the silicious part standing out in relief. There are two quite uniform characters which distinguish the Calciferous, viz: A fine crystalline structure intermixed with earthy matter and numerous small masses of calcareous spar.—E. 105. Great numbers of quartz crystals are found in the cavities of this formation, many of them very perfect as to form and transparency.—V. 30.

In the Mississippi basin this formation is called the LOWER MAGNESIAN LIMESTONE, to distinguish it from the Upper or Trenton limestone. The eastern name, Calciferous or lime-bearing sandrock, does not apply, as it is almost free from sand. As its western name indicates, it is a dolomite or magnesian limestone, and makes an excellent lime for building purposes. It usually contains about one equivalent or forty-five per cent of carbonate of magnesia. This limestone forms the summits of the bluffs of the Mississippi ; it supports high table-lands that extend back from the river, and forms prominent angles to the summits of the bluffs on either side of that river. These even and heavy layers are those usually quarried for building-stone. D. D. Owen gives descriptions of the picturesque character of the landscape in the region of the Upper Mississippi, and especially the striking similarity which the rock exposures present to ruined structures, and his report is illustrated by beautiful engravings showing the castellated appearance of the cliffs of the Lower Magnesian limestone on the Iowa river. In Pennsylvania it is a coarse, gray, calcareous sandstone, containing cavities enclosing very minute crystals of quartz and calcareous spar.

3 b. Chazy.—To the Calciferous succeeds the Chazy limestone. As a whole, it is a dark, irregular, thick-bedded limestone. At Chazy, New York, on Lake Champlain, it contains many rough, irregular, flinty or cherty masses. At Essex the beds are more regular, and form, in consequence, a better building stone. As a limestone it is purer than the Calciferous, being non-magnesian; the principal foreign matter is silica in the form of chert. It is free from the brown earthy spots, and the masses of brown calcareous spar so common in the Calciferous sandrock.

This formation is 130 feet thick on Lake Champlain, but it is less constant in the series than the others, and as it is not an important formation on the lines of the railroads, an extended description is not here necessary. It is not found in the valley of the Mohawk. Its fossils are found in Pennsylvania and Virginia, but its limits are not there defined. In the Northwestern States the St. Peter sandstone occupies the same place in the series as the Chazy in the east.

3 b. **St. Peter Sandstone** (Upper Sandstone of Owen).—This is a western formation and does not occur in the Eastern States, but Prof. Lesley thinks it may have representatives in the massive silicious members of the great lime-stone mass of from 5,000 to 6,000 feet thick, as measured along the two branches of the Juniata in Pennsylvania. It is first recognized in going west, to the south-west of Winnebago Lake. It is also seen up the Mississippi, near St. Paul and St. Anthony, and on the streams of northeast Iowa, and at La Salle, Illi-nois, where it is brought to the surface by an anticlinal axis. It is remarkable for its uniform thickness, which is from 72 to 100 feet over a space of 500 miles in length and 400 miles in width. In Central Wisconsin, however, its thick-ness is very irregular. It is also of the same character throughout, being com-posed of wonderfully uniform and exceedingly minute grains of sand, held together by the merest trace of cement, so that the mass may easily be moved with shovel and pick, as is everywhere done for the purpose of obtaining sand for mortar. This sandstone, though usually white, sometimes assumes a buff or brown color from the presence of iron, and in some localities it becomes red or is marked by bands of a bright green color. It appears like a recurrence of the Lower or Potsdam sandstone. Being composed almost entirely of pure silica, it is, when not colored by oxide of iron, one of the very best materials yet discovered in the west for the manufacture of glass. It is the same as that known in Missouri as saccharoidal sandstone, which is carried to Pitts-burgh, Pennsylvania, and used by the glass-makers in manufacturing the best kinds of glass. See note 2, Missouri.

4 a. **Trenton Limestone.**—Next in ascending order occurs the 4 a. *Trenton* limestone, which, in the Northwestern States, is divided into the Buff lime-stone and Blue limestone. In Wisconsin there are two buff and two blue beds alternating. They are undoubtedly the same as the well known Chazy, Birds-eye, Black River and Trenton limestones of New York and other Eastern States. They are known in the West wherever the exposures reach to the upper sandstone.

The upper member of the 4 a. Trenton limestone, in South Western Wis-consin and the adjoining parts of Illinois and Iowa, is the very important GALENA or lead-producing limestone, which has no exact representation in the Eastern States. It is a light gray or a yellowish-gray, heavy-bedded rock. It is compact, minutely crystalline throughout, often with small cavities lined with crystals of brown spar, and the whole thickness of the formation is 250 feet. The Galena or lead ore contains 13.4 per cent. of sulphur and 86.6 per cent. of lead, and is found in heavy bodies in crevices in this Galena dolomite or magnesian limestone. Prof. J. D. Whitney, in his admirable report on the geology of the lead region of Southwestern Wisconsin, has proved that these lead deposits must have been introduced into the fissures by precipitation from above. The lead mines of Missouri are chiefly in the Lower Magnesian lime-stone.

In Wisconsin, a very noticeable feature of the Trenton limestone is its marked division into the two parts before mentioned. One, which is the lower half, is very heavy bedded, in layers of two or three feet thick, known as the glass-rock, and the other thin bedded, in layers of two or three inches. There is always a stratum of carbonaceous shale from a quarter of an inch to a foot or more in thickness, which separates the blue or Trenton from the thin bedded Galena limestone above it.

Professor R. D. Irving describes the Galena limestone as almost invariably a very compact, hard, crystalline rock, of a yellowish-gray color, with numerous small cavities filled with a softer material, or lined with crystals of calcite. The upper portion is thick bedded and free from flints, the layers being from one to four feet thick, while the lower portion almost invariably consists of several feet of layers from one to two inches thick. Good exposures of parts of the Galena limestone are frequently to be met with. It may be seen in cliffs and ledges, on nearly all the streams in the lead region, where it weathers irregularly, leaving the surface full of small cavities, due to the removal of its softer parts. The formation contains masses of flint in layers, or in irregular pieces, which are principally confined to the middle and lower parts of the formation, although not entirely absent from any part.

In the interior valleys of Pennsylvania, as for example, in Sinking Valley, Blair Co., considerable quantities of zinc ore, and some galena, have been found in the Trenton limestone group, which is there at least 1,000 feet thick. The lead mines of Wythe Co., Virginia, are at the same, or at a somewhat lower horizon. The zinc mines near Bethlehem, Pennsylvania, and near Landisville, Lancaster Co., are nearly of the same geological age. Isolated crystals or small masses of galena occur in crevices in the limestone beds of this age throughout the entire range of the great valley from Newburgh, on the Hudson, to Chattanooga, in Tennessee. The limestones in this valley, which are the Auroral limestones of H. D. Rogers, are, by some geologists, referred to an older series.

In the State of New York the lower part of the Trenton is called the Birdseye. It is a perfectly pure limestone, and the next layer, which is the middle or Black River sub-division, is sometimes used as a marble. It is solid, hard, and easily worked, by reason of its conchoidal fracture, and is valuable for lime and for building.

The upper part of the formation, or Trenton limestone proper in New York, consists of two distinct varieties, at Trenton Falls. The first or upper part is a dark or black colored, fine grained limestone, in thin layers, separated regularly by black shale or slate, forming the great mass in which the creek has worn its channel, and in which are all the falls. See Note 62, New York.

The second, or lower part of the Trenton proper, is a gray, coarse grained limestone, in thick layers, and it is quite crystalline. This is the quarrystone at Prospect, above Trenton Falls. At Montreal, the church of Notre Dame and many other structures are constructed of the gray variety of the Trenton limestone, quarried behind the city, but the thinner layers, when not dressed, are of a more pleasing color, and make a handsomer building-stone.

The Trenton formation in all parts of the United States, is almost always a limestone. A conspicuous example of the Trenton, Utica and Hudson River formations, is seen in the long continuous and beautiful valley of the Hudson and Lake Champlain, the Kittatinny valley of New Jersey, the Cumberland valley of Pennsylvania, the Shenandoah valley of Virginia, and the valley of East Tennessee. The fertility of its limestone land is almost inexhaustible. The deposits of brown hematite iron ore, found in the soil, and occupying hollows or basins in the softer limestones below the Trenton in so many places, and in such large quantities, are supposed by some to be of aqueous origin, and not strictly a product of this formation, which is only its receptacle. But many other geologists,—R. M. S. Jackson, A. A. Henderson, Lesley, Platt, Prime and Frazer, have all agreed in advocating the oppo. site view, each from his own independent studies. They derive the limonite beds either from the solution of the ferriferous limestone layers, or from the intercalated micaceous slates, or from the pyrites-bearing slates of the neighborhood. According to Dr. Hunt, it comes from the change of masses both of iron-pyrites and of carbonate of iron, originally imbedded in the limestones and slates.

4 b. Utica Slate.—The Trenton limestone is succeeded by a dark or black carbonaceous slate, called the Utica slate. In Pennsylvania this formation is everywhere darkly colored, and the coloring matter is probably derived from abundant remains of marine plants or animals. While the black color of some of the clays in the brown hematite ore banks of the upper range (immediately beneath the Utica slate), as at the mines in Lehigh Co., Pa., and the Brandon ore mine in Vermont, seems to be derived from the black slates of the Utica, the gray color of some of the limestones, and of the carbonate ores (as at the Saucon zinc mines) is known to be due to disseminated graphite.

Within the State of New York, it is everywhere black, and usually soft and fissile. Thin beds of impure limestone are associated with it in many places, and sometimes thin layers of carbonate of iron, and it passes into the Trenton limestone by gradual interstratification. Thus bands of slate are interstratified in the limestone, and thin strata of limestone containing fossil remains in the lower part of the slate. These crumbling shales may generally be distinguished by their dark blue-black and brownish-black color, but there are some strata among the grits of the Hudson River that can scarcely be distinguished from these. The Utica slate weathers ash-gray, rapidly disintegrates, and, where it is exposed in cliffs, frost and other agents constantly break it into small fragments, which collect at the base in the form of a talus. In Pennsylvania, it outcrops, with little or no variation, as a dark blue carbonaceous slate and shale, extremely fissile in its lower beds. It forms the surface-rock along a narrow region in the Mohawk valley. In East Tennessee, the beds both of Utica and Hudson River, or Cincinnati, are of great extent, and consist of blue calcareous and sandy shales, with some layers of calcareous sandstone. Professor Hall considers the Utica slate as properly the lower member of the Hudson River group.

4. c. Hudson River (Cincinnati, Nashville, Loraine and Frankfort sandstone and shale).—The rocks of this group in New York are mostly slates,

shales and gray, slaty and thick-bedded grits. The slates and shales are
generally dark brown, blue and black, and the grits are gray, greenish and
bluish-gray. They are stratified and conformable, alternating a great number
of times, without any regular order of alternation, and in Eastern New York
are from 500 to 800 feet thick. The first New York geologists called this
formation the Greywacke, and it is still so called by the stone-cutters on the
River Hudson. Its lower portion was called the *Frankfort* slate and sandstone, and the upper part the *Pulaski* shale and sandstone, which latter were
afterwards called the *Loraine* shale. Wherever streams have passed over it
they have, in process of time, worn in the rocks a deep channel or gorge
sometimes preventing a free communication across them, as at Loraine (see
Note No. 69, New York). By decomposition, it produces a tenacious, clayey
soil, favorable for grass, forming the best dairy-land, as in Orange Co., New
York, about Goshen and Middletown. It increases in thickness southward
so rapidly that at the Delaware and Lehigh water gaps, measurements of
5,000 feet have been made through it, from its top downward, without reaching its lower limit.

In many places along its last outcrop toward the Atlantic, it has furnished many masses of a substance resembling anthracite, also beds of impure
limestone, and beds of red shale, which increase very much going south into
Virginia.

In Pennsylvania, the Hudson River slate consists of blue and greenish-
gray shale, alternating with gray calcareous and argillaceous sandstone in
thin beds. The sandstones grow more abundant as we ascend in the formation. The middle portion, where much metamorphosed and intersected
by cleavage-planes, in certain localities, produces a good roofing-slate, as at
Slatington and Delaware Water Gap, Pa.

The geologists of the Western States generally, have dropped the desig-
nation of Hudson River, at least in regard to strata west of the Alleghanies,
and have substituted for it the name, CINCINNATI, proposed by Worthen and
Meek; making this term co-extensive with the former. In this guide,
Hudson River is used in the Eastern, and Cincinnati in the Western States.
At Cincinnati the whole series is about 800 feet thick, and, according to
Dr. Newberry, by its fossils, is the equivalent of the Chazy, Trenton, Utica
and Hudson River, all blended together. In Ohio it is composed of alternating beds of limestone and shale, the latter sometimes called blue clay.
The limestone is an even-bedded, firm, durable, semi-crystalline limestone,
crowded with fossils. It is commonly called the *blue limestone*, but the
prevailing color is grayish-blue, and the weathered surface shows yellowish
or light-gray shades. In southern Illinois the lower part of the Cincinnati
is composed of brown sandy shales and sandstone, and the upper portion is
a thin-bedded, dark bluish-gray, fine grained limestone, two to six inches
thick, with shaly partings between the layers. In northern Illinois it is
bituminous, and consists of sandy shales with thin bands of limestone. In
Iowa it is the Maquoketa shales, which are bluish and brownish shales forming a stiff clay soil. In Missouri the upper shale bed only is found, with an
occasional flag-like limestone layer.

It should here be said that in the opinion of the earlier American geologists, Amos Eaton and Ebenezer Emmons, and as now maintained by Dr. Sterry Hunt, considerable portions of the strata above described, including what is called Potsdam sandstone in Pennsylvania, along the Appalachian Valley from New England to Alabama, as well as the great mass of accompanying limestones—the Auroral of Rogers—belongs to the Lower Taconic or Taconian series, and is of pre-Cambrian age. The name of Hudson River group, has hitherto been used in a very vague sense, and made to include not only the upper schistose beds, including the roofing-slate of the Taconian, and the much more recent Loraine or Cincinnati shales, but also a great intermediate series, called by Eaton the First or Transition Greywacke—the Utica, Loraine, and Oneida being his Secondary Greywacke.

This First Greywacke series, along the eastern border of the Appalachian valley in New York and New England, and thence southwest on the one hand, and northeast to the lower St. Lawrence on the other, is a great belt of disturbed strata, which were for a long time assigned by some geologists to a position above the Trenton limestone, while by others they were regarded as below that horizon, and of the age of the Potsdam and Calciferous divisions. Emmons, who for many years maintained the latter view, called these rocks the Taconic slates or Upper Taconic, a name which Logan, when he finally accepted this conclusion, changed to that of the Quebec group, divided into three parts, named by him Sillery, Lauzon, and Levis; the latter being supposed by him the oldest. It has since been shown that the Sillery is the oldest and the Levis the newest, its fauna approaching that of the Chazy; while some portions of this group (afterwards distinguished by Logan as Potsdam) contain a fauna as old, or older, than the typical Potsdam. These rocks, which have an aggregate thickness of 7,000 feet or more, are much disturbed, and include portions of strata of later date, Ordovician and Silurian. To this essentially Cambrian series, as already said, belongs a great part of what has been called Hudson River group, though this name, in paleontology, has been restricted to the Loraine shales, which belong to a higher Ordovician horizon.—T. S. H.

Keweenian.—This name has been given to the great copper-bearing series of the Lake Superior basin, which, while resting in the different parts upon various crystalline groups, is unconformably overlaid by the Cambrian sandstones of the Potsdam. It is made up chiefly of sandstones and conglomerates, with interposed layers of basic eruptive rocks of cotemporaneous origin, generally designated melaphyres. This series abounds in metallic copper, found both in veins, and in the beds, but most abundantly in certain conglomerates. The thickness of the Keweenian is not less than 20,000 feet, and perhaps much greater. Notwithstanding its great antiquity the Keweenian does not belong to the crystalline rocks. (T. Sterry Hunt.)

5-8. SILURIAN (OR UPPER SILURIAN) AGE.

5 a. Medina.—The lower member of this formation is a pebbly sandstone or grit called the Oneida conglomerate, being the same as the Shawangunk conglomerate. The upper member is called distinctively the Medina sandstone, and is usually a red or mottled argillaceous sandstone.

1. The Oneida conglomerate in New York is composed of quartz pebbles rarely exceeding three-fourths of an inch in diameter, and of white or yellowish quartz-sand. In some localities there is some interposed greenish shale. The source of its materials was to the south, the rock being 500 feet thick in the Shawangunk Mountain at Wurtsburg, on the N.Y. & Os. Mid. R. R., and 1000 feet thick in some parts of Pennsylvania and Tennessee. The greatest thickness of the Oneida in the eastern part of New York is 30 to 40 feet, but in the western part the same place is occupied by a gray quartzose sandstone, fine grained and compact. Passing upwards, the gray sandstone intermingles with the Medina sandstone, which, in its lower parts, differs chiefly in color. The red color of the Medina sandstone seems to be partially communicated to the gray below, which is often striped and spotted with red. There is, lithologically, no very strong line of demarcation between the two rocks. The oxide of iron, the red coloring matter of the upper member, has been transfused through the material of the lower as far as its particles could find admittance. The flagstones in the side-walks of Buffalo and Rochester, of a white color clouded with red, are of this formation.

In New Jersey the gray sandstone formation consists of a thick series of hard, white and whitish gray siliceous rocks, of various degrees of coarseness, from that of a fine grained, pure sandstone to that of a quartzose conglomerate with thickly-set pebbles averaging half an inch in diameter. This is the summit of the long, straight mountain ridge called the Kittatinny or North Mountain, extending from near the Hudson River into Virginia.

In Pennsylvania the Oneida conglomerate is a compact, greenish-gray, massive sandstone, containing in many places thick beds of siliceous conglomerate, and the Medina sandstone proper is a thick mass of alternating red shales and red and gray earthy sandstones. It is the North Mountain of the great Cumberland valley. At the Delaware Water-Gap the whole mass of Oneida and Medina consists of seven massive plates of coarse sand and conglomerate, separated by more argillaceous layers from each other. Going west, the number, according to Prof. Lesley, is reduced to five, and finally in Middle Pennsylvania to two, each of them very thick, and making its own mountain-crest when the dip is vertical, while the intermediate softer red mass forms a little valley between the crests. The whole formation is about 1,900 feet thick. When the dip is gentle, the Oneida makes a beautiful lofty terrace upon the flank of the mountain, the crest of which is always made by the Upper Medina. Traced southward through Virginia into Tennessee, this formation gradually thins away to 50 feet, as seen west of Knoxville.

2. The Medina sandstone proper succeeds the gray sandstone, there being no definite line of division between them. In this rock is found the *Fucoides Harlani* affording a positive character whereby to recognize it in the series. This sandstone is almost invariably of a red color, generally a brown-red, more rarely variegated light red and yellowish, and in a few rare instances of a light or whitish color,

partially greenish. It is both fine grained and coarse grained, the latter usually of the deepest color, the former more variegated. The lower falls of the Genesee, below Rochester, 110 feet in height, are formed by this rock. The deep gorge and high cliffs on both sides of the Niagara River, at Lewiston, New York, are more than one-half excavated in the Medina.

In New Jersey it is a thick formation of red and variegated sandstones and shales. Its lower beds are a dark red sandstone of a very ferruginous composition, and extreme hardness, and in the middle and upper divisions of a brownish red shale and a very argillaceous sandstone, partly calcareous.

Neither the Oneida nor Medina are found west of Ohio. Some large masses of galena and copper-pyrites with blende, have been found in the Oneida or Shawangunk grit, on the Erie R. R. east of Port Jervis and at Ellenville, but they were soon exhausted. When the Medina is a heavy coarse rock it produces a poor, barren country, but in Western New York it is more calcareous, and the soil is much better.

5 b. Clinton.—This group consists of many different kinds of rocks or masses, from which circumstance it was first called the Protean group. The name of Clinton was given to it on account of the characteristic masses being found around the village of Clinton, in Oneida County, New York. It consists of green and black-blue shale, greenish, gray and red, soft marly layers, often laminated calcareous sandstone, encrinal sandstone, and red fossiliferous iron-ore beds. The most persistent member of the group is the shale. It is bluish when fresh quarried, but when long exposed it is always of a greenish hue. The next member is the greenish sandstone, which is in thin layers, having its surface generally covered with *fucoides*. This also has a bluish tint when fresh quarried. The third persistent member consists of two iron-ore beds in New York and several in Pennsylvania.

The term Protean is still applicable to the Clinton group, which, in some places, consists of thin shaly sandstones, shales, and even conglomerates; in others, of thin bedded, impure limestones, shaly sandstones, iron-ores, etc: still again it appears as a duplicate series of shales, limestones and iron-ores, with some intermixture of sandy matter, all containing an abundance of marine shells. In the west the formation is limestone, and is of a more uniform character.

The Clinton formation produces the celebrated fossiliferous iron-ore generally known as the FOSSIL ORE, which occurs in it in every state from New York to Alabama. In all its localities this ore is red or brownish-red, very hard, and where unaltered, invariably oolitic or in larger sized concretions. In New York, where it is extensively mined, there are two beds of it, generally about 20 feet apart, and upon an average about a foot and more in thickness. The oolitic particles are usually more abundant in the lower, the larger sized concretions in the upper bed. The two beds never appear at the same locality, or in the same line of section, but where the lower one occurs the upper one is wanting, and where the upper one occurs the lower one is not found.

In Pennsylvania the Clinton is a very extensive formation, nearly 2,000 feet thick, of slate, shales, sandstones and iron-ore, with the same variety as elsewhere, and its iron ore is very rich, productive and valuable. The outcrop of the ore-beds have been traced for hundreds of miles. In Dodge County, Wisconsin, near Milwaukee, the Clinton iron-ore, at Iron Ridge, is from 15 to 18 feet thick, but this is very unusual, and it is not in the same part of the formation as the fossil ore in the east. The deposits of this ore in East Tennessee and in Alabama, called the Dye-stone ore, are still more extensive.

5 c. Niagara.—This group consists of two distinct members, a shale below and a limestone above.

The shale in New York constitutes a very uniform deposit, while the limestone, from a thin concretionary mass in the east, becomes an extensive and conspicuous rock, constantly increasing in thickness, in a western direction, even far beyond the limits of that state. The cataract of Niagara is produced by the passage of the river over this limestone and shale, and, from being a well known and extremely interesting point, as well as exhibiting the greatest natural development of these rocks in New York, this name was adopted for its designation. In this vicinity, the limestone is 164 feet thick, with the shale beneath 80 feet thick. The lower part of the Niagara group exhibits a great development of dark bluish shale, which, on exposure, gradually changes to gray or ashen color, and forms a bluish or grayish marly clay. In this state it is undistinguishable from the ordinary clays, and its outcropping edges, when long weathered, are often considered as clay beds. The Niagara is a very extensive formation, but its shales are much more persistent and wide spread than its limestone member in the east, but the limestone is more widely spread in the west. The gorge below the upper falls at Rochester is the best place to study these shales. In an agricultural point of view, this formation, like all limestones, is an admirable one. There is no better soil than that of the Niagara about Rochester, New York.

A silico-argillaceous limestone, in New York, forms the beds of passage from the soft shale below to the purer limestone above. It is of a dark or bluish color when freshly exposed, but soon changes to light gray or ashen. These beds of passage are succeeded by a dark bluish gray sub-crystalline limestone, of a rough fracture, and separated into thin courses by dark shaly matter. The third member is a coarse grained concretionary mass, in irregular layers, exhibiting a very peculiar contorted appearance, as if much disturbed while in a semi-fluid or yielding condition. The concretions often present cavities lined with crystals, or contain the remains of some organic body. This is the surface-rock in West Avenue in Rochester.

The Niagara limestone is the great limestone which, in Wisconsin, occupies the peninsula between Green Bay and Lake Michigan, and then stretches southward to the south limits of the state, and far into Illinois and Indiana. It will be noticed in looking over the Guide, how many railroad-stations in the western states, just mentioned are on the 5 c. Niagara, and how very extensive the formation must be. Its general appearance is that of a regularly bedded brown or buff dolomite, with occasional intercalations of beds of massive gray limestone. The quarries of beautiful buff limestone at Athens and Joliet, Illinois, so much used in Chicago for building-purposes, are in this formation. At Joliet there is 40 feet in thickness of this buff and gray limestone. West and northwest of Chicago the Niagara limestone is highly charged with petroleum, which oozes from the stone, blackening the face of walls built of it. On Goat Island, at Niagara Falls, the petroleum is also seen on the limestone in small quantities. In Michigan it is a grey crystalline, rather fine grained, moderately fossiliferous, dolomitic mass, 218 feet thick on Green Bay.

In Western Canada the upper part of the Niagara limestone contains peculiar fossils, and is called the Guelph, and in Wisconsin it is subdivided into the 4. Guelph, 3. Racine, 2. Waukesha and 1. Mayville beds.

This formation establishes the topographical distinction between the lower plain of Canada, in which lie Lake Ontario and Georgian Bay, and the upper plain of the United States, on which lie Lakes Erie, Huron and Michigan. Its terrace crosses Ontario, growing loftier as the thickness of the formation increases northwestward, until it becomes a range of limestone mountain-land, forming the peninsula between Lake Huron and Georgian Bay. It is there broken down in a range of islands, and reappears as a peninsula, just mentioned, cutting off Green Bay from the western shore of Lake Michigan.

The Niagara and other limestones above it, seem not to have been deposited in Pennsylvania between the Delaware and Susquehanna rivers, and in Middle Pennsylvania. While the limestones below it are well represented, the Niagara is wanting as a separate formation, and its characteristic fossils are scattered through the Clinton rocks.

6. **Salina, (Onondaga Salt Group.)**—This is an important group in the State of New York, containing all the gypsum and water-lime, and furnishing all the salt water of the salines of the city of Syracuse, which produce more salt in a small territory than any other in the world. Its soil is excellent for agricultural purposes, forming, with those south of it, including the Hamilton, the garden-region of the State of New York. The whole group is about 700 feet in thickness, and is divided into five deposits, but there are no well defined lines of division between them, except the last two.

1. The first or lowest is a red shale, showing green spots at the upper part of the mass. The great mass is of a blood red color, fine grained, earthy in fracture, with no regular lines of division, but breaking or crumbling into irregular fragments, and shows but little variation. In several localities the red shale shows numerous green spots, varying from an inch or less to several inches in diameter, which strongly contrast with the red ground on which they are placed. The green color is the result of a chemical change, the peroxide of iron being reduced to protoxide. This red shale is of great extent along the railroad, and presents a thickness of from one to five hundred feet, yet nowhere has a fossil been found in it, or a pebble, or anything extraneous, excepting a few thin layers of sandstone. The main line of the N. Y. C. & H. R. R. R. runs on the Salina formation 107 miles, from Canastota to Brighton, and nearly all of this distance on this lower or red shale portion.

2. The second deposit is the lower gypseous shales, the lower part of it alternating with the red shale, which ceases with this mass. This second deposit consists of shales and calcareous slates of a light green and drab color, with alternations of different colored masses, red, green, bluish and yellow, with a little whitish and greenish sandstone, different colors predominating in different places. In this deposit gypsum occurs in fibrous masses, either reddish or of a salmon color, which colors are peculiar to this deposit. The quantity of gypsum in this second deposit is comparatively small, and it is unimportant in an economical point of view.

Both the second and third deposits are permeable to water, which cannot be obtained in any of the hills composed of them unless the wells are sunk to the level of the water-courses, a fact which explains the absence of all brine-springs above the level of the country.

3. The third member of the Salina formation is the gypseous deposit, whicl embraces the great masses quarried for plaster or gypsum, consisting of tw ranges, between which are the hopper-shaped cavities, the vermicular lime-rock and other porous rocks. This is the most important deposit, not only on account o its plaster-beds, but because it is only in this deposit that we have positive evidenc that salt has existed in a solid state, and, therefore, the only source whence th saline springs of Syracuse could have been derived. The great mass of th deposit consists of rather soft yellowish or drab and brownish colored shal and slate, and of more compact masses which are hard, a brownish colo predominating. It is usually denominated a gypseous marl, being earthy an indurated, slaty and compact. Some of it when weathered, presents a peculia appearance, as of having been hacked by a cutting-instrument, with som regularity. The gypsum does not appear in layers or beds, but it occurs in insulate masses, and it assumes irregular not globular forms. The dark color of the gypsun is owing to carbonaceous matter. In many localities there are two ranges o these masses or plaster-beds, generally separated by the vermicular rock and th hopper-shaped cavities. There are two masses of the vermicular rock, the uppe one four feet thick, with large porous cavities, the lower one twenty feet thick with small pores. This vermicular limestone is a porous or cellular rock, resemblin lava. It is dark gray or blue in color, and perforated everywhere with curvelinca holes, but otherwise very compact. The holes or cells vary from microscopic siz to half an inch in diameter, the cells being very irregular, and communicatin with each other, some being spherical, and the resemblance in structure to porous lava is complete. Forms which are due to common salt have beer discovered in this rock, showing the presence of crystals of this substance, whicl were removed by solution.

The most interesting products of the group are the hopper-shaped cavities which must have been produced by common salt, as no other soluble minera presents similar ones. They show conclusively that salt existed in this thir deposit. When salt crystallizes, a cube first makes its appearance upon the surfac of the brine, then similar cubes form around its border, being attached to its uppe surface, near the edges, while it gradually sinks, and additional particles ar added, forming another row of cubes upon the first range. This is many time repeated, until the density of the mass formed becomes greater than the liquid when it falls to the bottom. When examined, being turned upside down, i shows a pyramid of regular steps, terminated by a cube, and when its position i reversed it presents a form like the hopper of a mill. Where two ranges of plaste beds are seen the hoppers occur between them, and between the two massses o vermicular rocks, and are from one inch to three inches and more in diameter These hopper cavities are formed in the gypseous marl, or in the more solid part of the vermicular rock. Testaceous animals cannot live in water saturated witl gypsum, hence no fossils are found in the deposit. No trace of rock-salt in New York has met the eye of any one, but the existence of it is a matter of n doubt.* The fact of the difficulty of obtaining water in the gypseous hills, ir either the second or third deposit, show there is little probability of finding sal above the level of the waters on account of its having long since been dissolved See Note 27, New York, as to the salt-wells at Syracuse.

---

*After the above was written, rock-salt was first found, in June 1878, in a boring south o Rochester.

The "Old Road," or the division of the N.Y. C. & H. R. R. R., from Syracuse to Rochester, via Auburn, runs on the gypseous portion of the formation, and the plaster-beds can be inspected at Marcellus station, close to the railroad, but the best gypsum quarries are on Cayuga Lake, just north of Union Springs, the masses being from fifteen to twenty-five feet thick. Sulphuric acid springs, and numerous sulphur springs occur in the State of New York, in the Salina formation, often rising through the crevices of the overlying Water-lime group.

4. The fourth or succeeding portion of the Salina formation, consists of those rocks which show groups of needle-form cavities, placed side by side, caused by the crystallization of sulphate of magnesia, and presenting a finely striated columnar appearance. The rock is a dark gray or drab colored, impure limestone, with cavities containing crystals and often embracing shaly beds. It appears to be a magnesian limestone, its usual color is a brownish drab, also dove color, and it breaks with an earthy fracture.

The Salina formation extends westward across Canada, and the salt-deposits of Goderich in Ontario are in it. Six large beds of rock salt have been found there in boring, measuring in all 126 feet in thickness, at from 1,027 to 1,385 feet in depth from the surface, the beds measuring from 6 feet to 35 feet each in thickness.

The salt-deposits and brine-springs of the world are by no means confined to the Salina formation; on the contrary, they are found in almost all the formations from the oldest to the youngest, and always accompanied by gypsum and red and variegated marls.

5. The fifth division of the Salina or Onondaga Salt group is the Water-lime, which has generally been considered as belonging to the Lower Helderberg, but which properly is part of the Salina. All the hydraulic cement of the State of New York, known as Rosendale Cement, and Syracuse or Manlius Water-lime, is manufactured from a portion of the stone of this Water-lime formation. It is an earthy, drab-colored limestone and usually consists of two layers of drab limestone, always separated by an intervening mass of blue; it is easily recognized by its gray or ash color when weathered. It has a thickness of not less than 30 feet, and often attains a thickness of 100 feet or more in New York. When the Water-lime is burnt the stone does not slake, if of a good quality. It is ground in a mill, and then it hardens or sets when mixed with water. and remains so under water, its goodness depending on the hardness or cohesion when set. Its peculiar quality is owing to the proportion of silica and alumina it contains. The Water-lime continues across the State of New York, the drab layers which constitute it being always found. The courses into which the layers of Water-lime are sometimes divided show a crenulated or notched surface, like the sutures of a skull, the two surfaces interlocking each other. Professor Hall says the Water-lime is a distinct member, which does not belong to the 7. Lower Helderberg group of strata, but to that below it, the 6. Salina, of which it is the upper member. It is not closely related to either, but more nearly to the Salina, and is much more widely spread than the other members of the Salina. The cement quarries of the Delaware River, between Pennsylvania and New Jersey are in this formation, but cease after passing the Lehigh River westward. The beds near Copley are Trenton or older. In Middle Pennsylvania, where the Salina group, destitute of gypsum and salt, measures 440 feet, the cement beds above measure 580 feet, and the Lewistown limestone (Lower Helderberg) 162 feet, as measured by Ashburner and Billin, in 1876.

7. Lower Helderberg.—In consequence of these rocks being so well developed on the Helderberg Mountains, near Albany, New York, they have received that name. The Lower Helderberg series consists of five limestone sub-divisions, and the Upper Helderberg of four members. They are separated by an important sandstone formation—the Oriskany. The Lower Helderberg, which is well developed in the eastern part of New York, thins out in going west, and at Syracuse disappears entirely. The sandstones also thin out and disappear, so that at Syracuse the Upper Helderberg rests on the Water-lime, the upper member of the Onondaga Salt group. The Lower Helderberg consists, in ascending order, of the 1. Tentaculite limestone, the 2. Pentamerus limestone, the 3. Delthyris shaly limestone, the 4. Encrinal limestone, and 5. Upper Pentamerus limestone.

1. The Tentaculite limestone is the lowest member of the series. Portions of it afford fine building stone, which can be procured in blocks of large size, perfectly solid, and free from cracks or flaws. They vary from ash-gray to black, and present almost every shade between these colors. The strata are intersected by two main systems of joints nearly perpendicular to each other, hence the rock can easily be quarried in large blocks. But much of it is thin-bedded, often thinly laminated, dark blue; its color, texture and composition contrasting strongly with the Water-lime below.—H. The 2. Pentamerus limestone is rarely pure, being more or less mixed with black shale, which gives a dark color to the rock, it being usually a dark gray. It is crystalline in grain, and is in layers, but the lines of division are not straight, and the surface is not even. The whole mass has a rough appearance, and it does not make a good building stone.—V. The 3. Delthyris shaly limestone, as its name implies, is a shaly mass, and consists of alternate beds of shaly and compact limestone. It is an exceedingly interesting rock from the great number of species, the abundance and perfection of its fossils.—Hall, 144. The 4. Encrinal is a compact crinoidal limestone, and the 5. Upper Pentamerus is a bluish gray limestone. In Pennsylvania, according to Rogers, the Lower Helderberg is 50 to 100 feet thick, a diversified calcareous formation, of some shade of blue, argillaceous and flaggy in its lower beds, and shaly towards the middle, with layers and nodules of chert.

8. Oriskany Sandstone.—In New York the greatest thickness of this rock is not more than thirty feet, and usually much less, but in Pennsylvania, Maryland and Virginia it is, in places, as much as 700 feet; even in New York it covers an extensive surface, and is strongly marked in its fossils, which are generally of a large size, and attract the attention of travelers. At the typical locality, Oriskany Falls, the sandstone is twenty feet thick, and is of a light yellow color, friable, and readily crumbling into pure sand ; no part of it being sufficiently solid for durable work. One characteristic of this rock is the abundance of small cavities, which have been formed by the destruction of fossils. These present themselves in all cases where the rock is well developed. The porous nature of the mass has admitted the percolation of water, which has dissolved the calcareous matter of the shells, usually leaving casts of their internal structure. As a mass the Oriskany sandstone is a coarse, rather loosely cemented, purely silicious sandstone, of a yellowish white color. Sometimes it is shaded brown or some other dark color. In Pennsylvania it forms rough ridges, with a poor sandy soil. It is used for glass-making, and contains an iron-ore too silicious to be valuable. Some of our geologists (Hall, Rogers, Dana, etc.) place the Oriskany at the top of the Silurian series, and others (Newberry, Lesley, Hunt, etc.) at the bottom of the Devonian.

9-12. DEVONIAN AGE.

9. LOWER DEVONIAN.

9 Upper Helderberg or Corniferous.—This very widely extended formation consists of four important members, the Cauda-galli, the Schoharie grit, the Onondaga limestone, and the Corniferous limestone, the upper member. But in the recent text-books on geology the whole formation is called the Corniferous, which was the name given by Eaton to the whole formation of limestone. It forms the Helderberg range, a high ridge which extends through the State of New York, forming a very rich and productive tract of country. This group of strata, as above limited, and designated the Upper Helderberg by Professor James Hall, is, in his opinion, deserving of recognition as the base of the Devonian, the Hamilton group being the middle, and the Portage, Chemung and Catskill the Upper Devonian.

9 a. Cauda-galli.—This is a fine-grained calcareous and argillaceous sandstone, usually drab and brownish, and blanching by long weathering. It readily strikes the eye by its contrast with its associated rocks, and by the singular marking of impressions strongly resembling the tail of the common barn-yard fowl, from whence its Latin name of Cauda-galli or cock's-tail. Its fossils have been found in New York and at Crab Orchard, in Kentucky. In New Jersey, northeast of the Delaware Water Gap, this and the Schoharie are three hundred feet thick.

9 b. Schoharie Grit.—This is very much like the preceding, but altogether different in its fossils. It is a fine-grained, very calcareous grit, or an arenaceous limestone, naturally brown, but weathering to a gray or drab color, containing a great number of fossils peculiar to this stratum, and is found in the mountain one and one-half miles northwest and northeast of Schoharie, New York, and extends by the Helderberg range to Kingston. The Schoharie Grit is a highly fossiliferous formation, and has a wide geographical extension. Its great number of cephalopods gives it a marked character, but it contains other fossils identical with the limestones above.—H.

The **9 c. Onondaga Limestone** in New York rarely exceeds ten to fourteen feet in thickness, but is very persistent, and is readily recognized by its light gray color, crystalline structure, toughness, and its numerous organic remains. This is one of the most valuable building stones in the Helderberg division, and has been largely quarried near Syracuse for the canal. It is an imperishable stone, having great power to resist the action of air, water and frost. It is generally the rock over which the water flows at the water-falls on the Helderberg range, as at Perryville and Chittenango Falls, and is remarkably uniform in its character. It is more extensive than the Corniferous proper, and it is very rich in beautiful and characteristic fossils. The limestones used for flagging in Syracuse are Onondaga limestone, brought from the typical localities Onondaga Valley and Split-Rock on Onondaga Hill. When wet they make a fine display of fossils of this formation. This stone is also used for building everywhere in Central New York.

9 d. Corniferous Limestone.—For all practical purposes, this and the Onondaga limestone may be regarded as one formation. It extends from the Hudson River to the Niagara River, which it crosses at Black Rock, producing there a rapid current at the International Bridge, at Buffalo, and forming a small island just above the water. It extends far into Canada, is seen at Sandusky City, Ohio, and there forms the bottom of Lake Erie. Its color varies from a light grayish-blue to a black, and is sometimes even a light gray or drab. It contains numerous nodules of flint or hornstone, from which it derives its name. But few if any of the layers afford a pure limestone. Its color varies from black to gray, brownish and light blue. It is usually in regular courses from six to eighteen inches thick, separated by layers of hornstone, and sometimes embracing flattened nodules of the same. This rock is crossed by vertical joints in two directions, giving rise to numerous copious springs of water. An upper division, called the Seneca limestone, is now included in the Corniferous. In New Jersey and Pennsylvania it is a blue and sometimes sparry limestone, including bands and nodules of chert. In Canada and the Western States it is a straw-colored and light gray rock. In its general eastern exposures it is generally bluish. Above the Corniferous are no general limestone masses in the Eastern States, but partial deposits only, the most extensive of which is the Tully limestone, found only in Central New York. There is an astonishing change from the top of the Corniferous limestone to the black shales of Marcellus. Two formations more unlike cannot anywhere be found. Both the Corniferous and Onondaga are included in the Upper Helderberg limestone of Pennsylvania, and on the Juniata they measure together only sixty feet. Immediately upon the upper surface of the Corniferous limestone, lies the valuable and extensive MARCELLUS IRON ORE. This consists of carbonate of iron, which occurs in a bed of pyritous clay, and near the outcrop is changed into limonite.

10. MIDDLE DEVONIAN.

10 a. Marcellus Shales are of a black color, usually dark brown when altered. They greatly resemble the Utica slate in mineral character, and could readily be mistaken for it. They extend in New York from the Hudson River to Lake Erie. The lower part contains some impure black limestone, not in layers or beds, but in interrupted flattened masses. The upper shales are not so highly colored as the lower ones, and are disposed to separate, when long exposed, into small, thin-edged fragments, the result of a peculiar accretionary structure. The fragments often exhibit stains, in spots, from iron rust, and also minute crystals of gypsum, the effect of the action of decomposed pyrites and limestone particles. Some portions of the lower shales are black and friable from small carbonaceous fucoids. Along the whole line of its outcrop it has been dug into in vain attempts to find coal.—Van U. 147. It has two joint planes, nearly at right angles to each other, causing projecting corners of rock, with smooth nearly vertical surfaces. These are sometimes seen in the upper members also of the Hamilton group, and the *septaria* or flattened balls of black limestone also occur in the Genesee shales.

The lower part is very black, slaty and bituminous, and contains iron pyrites in great profusion. In general character the lower part resembles the Utica slate and is not distinguishable from the 10 c. Genesee slate, in its general aspect. When long exposed, the lower part weathers to a brownish or iron-rust color, partly from the presence and decomposition of iron pyrites and partly from bituminous matter. In some situations it retains its purely black color, and scarcely separates into thin laminæ after long exposure. In many places this rock contains so much bitumen as to give out flame when thrown upon a fire of hot coals. In Western New York it is fifty feet thick, and farther east much thicker.—H.

This important formation carries its broad black outcrops across many of the Middle and Southern States, with comparatively little change, but in the South the black shale is supposed to be Genesee. In the Juniata region of Pennsylvania the Marcellus has been found to measure 875 feet thick, and is there divisible into an upper, middle and lower member, the last consisting of black and brown shales, the surface being stained with iron rust, &c., coated with bituminous matter. In Perry County, Pennsylvania, small coal beds occur in this formation, constituting the oldest known coal-measures, and significantly marking the great change in the general condition of things which either followed or was introduced by the deposit of the Oriskany sandstone.—Lesley.

In speculating upon the origin of petroleum, some geologists have sought it in a process of distillation from the black Marcellus and Genesee shales upward, and of condensation in the oil-bearing gravels and fissures of the overlying formations. Chemists, like T. Sterry Hunt, oppose this view on chemical grounds, others oppose it from other considerations of apparently equal weight. It is a curious fact, however, that at this horizon, and in the Upper Helderberg or Corniferous, occur the petroleum deposits of Upper Canada, while the Pennsylvania oil-deposits lie at successively higher and higher stages in the series.

10 b. Hamilton.—This group takes its name from the town of Hamilton, in Madison County, New York, which contains no other rock, and where the best opportunity exists of examining the members of which it is composed, and where its fossils are in great abundance. It includes all the masses between the upper shales of Marcellus, and the Tully limestone, and is from 300 to 700 feet in thickness in New York. It is important from its fine agricultural qualities, its thickness and extent, commencing at the Hudson and extending to Lake Erie. It consists of slate, shale and sandstone, with endless mixtures of these materials, or, in other words, sandy shale and shaly sandstones, and is not very easily described. There are three distinct mineral masses as to kinds, but not as to arrangement. The first, in the order of the tenuity of particles, is rather a fine grained shale, often fissile or slaty, its color some shade of blue, usually dark or blackish. The second is a coarse shale, often mixed with carbonate of lime, its color blue or dark gray when fresh, but becoming of an olive or brown color by long exposure to the weather, the color being due to manganese. It has no tendency whatever to separate into regular layers, but when a mass has been long exposed it shows numerous curved divisions, the curves very short and irregular, giving it a very peculiar appearance, which is unmistakable. The third kind, which is not so common as the two first, is a well characterized sandstone, and is generally in the upper part of the group, but more or less mixed with either of the two others. It is often in layers, though rarely straight, and usually short, interrupted, sometimes mixed with carbonate of lime. The colors of this kind are of more various shades, olive, greenish and yellowish. One thin layer produces excellent flagstones, but the group generally is deficient in building materials, the shale of the first kind readily crumbling by exposure to the air; the two latter kinds alone furnishing building stone. The best is where limestone forms the cement, and sand is in the

greatest abundance. So rare is the occurrence of regular layers in the group, that their absence is a good negative character, and its brownish or yellowish color, externally, or where weathered, a good positive one of the group generally. This applies to the central, but not to the eastern part of the State of New York. It abounds in fossils, and is admirably characterized by them, numerous species and even genera commencing with the group, and ending with it.—Van U. 150.

In the western part of the State of New York, instead of sandy shale and shaly sandstone, and even tolerably pure sandstone, as in the east, the sand has diminished and the clay increased. The group, as a whole, presents an immense development of dull olive, bluish-gray calcareous shales, which, on weathering, assume a light gray or ashen tint, some thin portions becoming brownish on exposure. The formation thins out very much in going westward, and at Lake Erie has only half the thickness found at Seneca Lake, and is so different that doubt of the identity of the two might arise, if one judged by the appearance only. The Hamilton is the New York lake formation, the following lakes being excavated in it: Otsego, Cazenovia, Skaneateles, Otisco, Owasco, Cayuga, Seneca, Canandaigua, and the north end of Hemlock Lake. The east end of Lake Erie is also cut out of the Hamilton. The upper part of the Hamilton was called the Moscow shale, from a place between Mt. Morris and Rochester, on the Genesee River.

In Pennsylvania the Hamilton shale has been measured on the Juniata, 635 feet thick. It has many hundreds of miles of outcrop, in repeated zig-zags, forming, in combination with the Genesee and Portage above it, ranges of smooth, cultivated hills, of an entirely characteristic shape, in long lines of ruffled slopes, regularly indented with short and smooth ravines. This striking topographical feature, maintains itself throughout the mountain-region into Virginia, and still farther south. The abundance of shells, without limestone beds, in Pennsylvania, furnishes a partial clue to the deposit of the (next succeeding) Tully limestone in New York.

10 b. Tully Limestone.—This is the dividing line, easy to find, between the Hamilton and Genesee, being the upper part of the former, and it is important in New York as the most southern mass of limestone in the State. It is only local, and is an impure limestone, fine-grained, usually a dark or blackish blue, often brownish. The usual thickness of the rock is about fourteen feet, and its greatest thickness twenty feet. It makes a good but not a white lime. It receives its name from the township of Tully, in Onondaga County, New York. This limestone often shows an accretionary structure, and a roughed, notched appearance, where its layers separate as in some of the layers of the water-lime. One of the lower layers is thick, the bottom one being frequently five feet in thickness, and it is owing to this circumstance, and to the softness of the shale beneath, that whenever a waterfall exists, the shale has been washed out to some depth, leaving a chamber or cavern, of which the limestone forms the roof or ceiling.—V. 169. It is a marked geological horizon in Central New York, being the termination of the Hamilton, and is succeeded by shales of a widely different character. It is often thick-bedded, but it is often divided by numerous irregular seams into small fragments. Its color, on first exposure, is blue or nearly black, but weathers to an ashen hue. It is best seen on the Cayuga Southern R. R., where it stands out in the face of the cliffs as a prominent band. It is absent west of Canandaigua Lake and in the eastern part of the state.—H. 212.

10 c. Genesee, (Black Slate of the west and south). — This is a great development of argillaceous fissile black slate. Where its edges only are exposed, it withstands the weather for a great length of time, and often presents mural banks in the ravines, river-courses, and upon the shores of lakes. When the surface of the strata is exposed it rapidly exfoliates in thin even laminæ. On disintegration it is often stained with iron, owing to decomposition of pyrites, but in many instances, and the greater number of localities, it retains a deep black color. In this it is distinguished from some beds of black slate in higher situations, which always become stained with hydrate of iron on their edges, and upon the surface of the laminæ. In color and general character it greatly resembles the Marcellus shale, and, aside from position, it would be difficult to distinguish the two, in the absence of fossils. It forms no conspicuous feature in the scenery or topography of the general surface. In ravines, and river and lake banks, it is usually seen in connection with the rocks below or above. Its greatest development, and a point where it appears more prominently alone, and the typical locality from which it was named, is at the opening of the gorge of the Genesee, at Mount Morris, where it is seen in the perpendicular cliffs for more than a mile in length. See note No. 112, New York. Another great exposure of the Genesee slate is along the Cayuga Southern Railway south of Ludlowville, where it shows from eighty to one hundred feet thick, with the Tully limestone below and the Portage shales above it. See note 83, New York. The mass decomposes much less rapidly than the soft calcareous Hamilton or Moscow shales below it, and the thin slaty laminæ resist atmospheric action a long time. In lithological character it is entirely uniform, having, from Cayuga Lake to Lake Erie, the same deep black color and laminated slaty structure, nor is there any change in its organic remains. Its fossils in Indiana are precisely identical with those of New York.—Hall 218.

There are few formations in Central New York of which the limits are so well defined as this, lying between the Tully limestone below, and the sandstone flags of the base of the Portage group, above. It may also readily be found by the black color and slaty fracture. This shale has been regarded as the main original source of the petroleum in the oil region of Ohio and Western Pennsylvania, but there is reason to believe that part, at least, of the supply of these regions has come from the Corniferous limestone below it, as maintained by Dr. Hunt.

All through the western and southwestern states there is always found a BLACK SHALE, which is often the only representative of the Devonian rocks. This is generally considered to be 10 c. Genesee. It is very remarkable that a formation of its composition, of so inconsiderable a thickness, and otherwise so unimportant, should be so widely extended, and retain throughout its character unchanged as a black shale. The researches of Dr. Newberry in Ohio tend to show its fossils to be of the Portage type. It is there 350 feet thick, and he pronounces it to be the equivalent of the Genesee and lower Portage. All the divisions of the Hamilton group, Marcellus, Hamilton and Genesee, are converted, by exposure, into a deep soil of an excellent quality for agricultural purposes, sometimes quite hilly, but forming smooth land free from stones. Some of the finest wheat-growing and hop-raising land in New York is on the Hamilton, and its rich shales have been carried south by drift and diluvial agencies, and spread over the Genesee, Portage and Chemung, greatly to their improvement.

11-12. UPPER DEVONIAN.

11 a. Portage.—This group represents an extensive development of shales and flagstones, and finally some thick-bedded sandstone towards its upper part. It is extremely variable in character at different and distant points. In New York the Portage rises sometimes in a gentle slope, and at other times abruptly from the softer shales below. Between the deep north and south valleys, in which the railroads run, the enduring sandstones of the upper part extend far northward, presenting, on the north side, a gentle slope, while on the east and west sides of the same hills, the slope is abrupt, the valleys being bounded by steep hills. *The change in the external appearance of the country indicates the commencement of these Portage rocks, although they are not seen.* Throughout the Hamilton shales, the valleys present gently sloping sides, and the country rarely rises far above the valley bottom. But on approaching the northern margin of the Portage group, the railway traveler sees a gradually increasing elevation of the hills on either side, and an abruptness in their slope, and in a short time finds himself in a deep valley bounded on either side by hills rising 400 or 500 feet, and in some instances, even 800 feet above the bed of the stream. These elevations often extend several miles unbroken, except by the deep ravines which indent their sides. The higher sandstones of the group, and in many instances the intermediate ones, produce falls in the streams which pass over them, and some of the most beautiful cascades in the State of New York, and many of the highest perpendicular falls of water, are produced by the rocks of this group, and in none others do we meet with more grand and striking scenery.—J. Hall's Report.

The pedestrian often finds his course impeded by a gorge of several hundred feet in depth, such as Watkins Glen and Havana Glen. The Portage upper, middle and lower falls are 66, 110 and 96 feet, and between the middle and lower the rocks rise in perpendicular cliffs 351 feet in height. See note No. 110, New York, as to Portage on Erie Railroad. Taghanic, Hector, and Lodi falls are also in the Portage. These points afford some of the grandest views of scenery, and admirable facilities for geological investigations. The lower division of the Portage is the 1. *Chasaqua shales,* a green shale, with thin flagstones, and sandy shale. 2. The middle portion is the *Gardeau shale* and flagstones, a great development of green and black slaty and sandy shales, with thin layers of sandstone, from which are quarried beautiful and durable flagstones. The rocks of this part of the group form high, almost perpendicular, banks on the Genesee. In a westerly direction the sandstones disappear, and the shales increase. 3. The upper part of the Portage consists of the *Portage* sandstones, thick bedded sandstones, with little shale, while below, the sandy layers become thinner, and shale beds more frequent; still it must be acknowledged that there is no abrupt change from the beginning of the Portage to the top of the Chemung. In the Portage, the sandstones and shales are less separated than above, and the sandy strata are finer grained, and contain more lime than in the Chemung. Towards the southern extremity of Cayuga and Seneca Lakes, the Portage rocks form cliffs of considerable height, which present alternating hard and soft layers, and the numerous vertical joints present the appearance of solid walls of masonry, in distinct and regular courses. The vertical joints are well seen in Havana Glen. Isolated masses, like huge columns, are often seen, standing out in bold relief from the line of the cliff, being the remains of previously exposed surfaces, which had crumbled away. On the Genesee River the group is not less than 1000 feet thick. The Portage yields less lime to the soil than the Hamilton, but for pasturage it is superior to it.—H. 224. The great dairy-country of Cortland, and other counties in Central New York, is on the Portage formation. The water of the Portage group is remarkably pure and soft. The Portage rocks have not been recognized in the eastern part of New York. In Ohio the Portage forms the upper part of the Huron shale, and the lower part of the Erie shale, of Dr. Newberry.

In Middle Pennsylvania, according to Lesley, the Portage flags are 1,450 feet thick, and the Chemung shales over them, 1,860 feet thick. It is very hard to draw a line of demarcation between them, but, as a whole, the Chemung strata are more silicious and the Portage more argillaceous. The Portage sandstones are flaggy, and, at times, very shaly, and their alternations with shale frequent, the individual beds being thin, and the shales predominant. The Chemung sandstones are more massive, ferruginous and micaceous, with fewer alternations of shale. Brachiopods and other shells are abundant in the upper Chemung shales, while the Portage rocks are almost destitute of animal forms except crinoids and fucoids. Fucoidal impressions are also very abundant in the upper Chemung, and to the decomposition of this abundant marine vegetation, Lesquereux and others ascribe the origin of the petroleum, at its various local horizons, from the Portage up to the Mahoning sandstone in the Coal Measures.

11 b. Chemung.—These rocks can everywhere be described as a series of thin-bedded sandstones and flagstones, with intervening shales, and mixtures in various proportions of these, and very rarely beds of impure limestone, resulting from the aggregation of organic remains. The whole series weathers to a brownish olive, and even the deeper green of the shales assumes that hue. The shales vary in color from a deep black to olive and green, with every grade and mixture of these. The sandstones are often brownish-gray or olive, and sometimes light gray. More generally, however, there is a tinge of green or olive pervading these strata. Towards the upper part of the group, in some localities, there is a tendency to conglomerate, and in a few places the mass becomes a well defined pudding-stone, with sometimes 150 to 200 feet of Chemung shales and sandstones above it. Towards the upper part of the group the shales are reddish, coarse and fissile, with much mica in small glimmering scales.—Hall 251. From their red color these have sometimes been mistaken for the Catskill formation.

In a few localities in Pennsylvania it contains a very excellent variety of iron ore. As a general thing, however, this formation, and all others above it, up to near the coal conglomerate, are singularly deficient in iron ore. There is little of geological interest throughout the whole extent of the Chemung group. The N. Y. L. E. & W., or Erie Railway, runs for 300 miles west of Susquehanna on this formation, and on nearly the same portion of it. In the northwestern portion of Pennsylvania the celebrated OIL REGION is in the Chemung, the oil being found stored-up in certain coarse porous sandstones, but these are merely the repository of the oil originating in lower strata. It is a very extensive formation in Southern New York, all the southern tier of counties, west of Great Bend, being covered by it, and it forms an excellent grazing and agricultural country, not quite equal to the Portage, but much superior to the Catskill. In Northern Pennsylvania this formation, as in Southern New York, consists of a vast succession of thin layers of shale, of every hue, from a deep olive and dark green to a light slaty gray, alternating with thin beds of brownish gray sandstones.

In Pennsylvania, ninety feet of strata have been carefully studied and measured on Sideling Hill, consisting of alternate beds of red and olive shales and sandstones with Chemung fossils, ripple-marks and fucoids, and a bed of iron ore long known by the name of the Larry's Creek ore, which outcrops everywhere along the face of the Allegheny Mountain. In the gaps at Blairsville and Connellsville, in Southwestern Pennsylvania, Prof. Stevenson finds Chemung fossils in what have always been called the Catskill rocks, on account of their being of a red color, and other geologists have made the same observation in Northern Pennsylvania. In Southern New York, adjacent to Pennsylvania, Professor Hall reports 150 feet of red rocks, and then thin gray rocks above with Chemung fossils.

The Erie shale of Ohio is the equivalent of the 11 b. Chemung, and the upper part of the 11 a. Portage. At Cleveland, it consists of green, gray and blue shales, soft and fine, with sheets of micaceous, silvery sandstone, from half an inch to two inches in thickness, and flattened masses of argillaceous iron ore.—Newberry. The formation also occurs in Kentucky, and Chemung fossils have been found in Utah and Nevada by Clarence King and Arnold Hague.

12. Catskill.—There is no observable line of demarcation between the Chemung and Catskill. The first sign of change is a more solid or hard rock appearing, often accompanied by red sandstone or red shale. The group consists of light colored gray sandstone, usually hard; of fine-grained red sandstone, red shale or slate; of dark colored slate and shale, of grindstone-grit, and a peculiarly accretionary and fragmentary mass, appearing like fragments of hard slate cemented by limestone, similar to what is well known in England as cornstone. The hard gray sandstone often presents a highly characteristic structure, the layers, one or more inches thick, being disposed in oblique divisions, the divisions usually overlapping each other. This peculiar angular arrangement presents altogether a singular conformation, and forms a highly picturesque rock.—V. You cán see this at Ralston, Pennsylvania.

The prevailing color of the sandstone is brick-red, though often it is lighter, and sometimes of a deeper color, from a larger proportion of iron, while the coarser parts are often gray, and the shales are green. Beds of green shaly sandstone are interstratified with the red friable sandstone, and these are succeeded by a compact kind of conglomerate rock. The formation expands, and augments in thickness, in passing eastward, till it finally rises in the high and prominent peaks of the Catskill Mountain, nearly 4,000 feet above the sea, from which the formation derives its name. See note No. 9, of New York.

The formation extends from this locality southwestward into Pennsylvania, where its outcrop, 3,000 feet thick, in combination with that of the Pocono sandstone above it, 2,000 feet thick, forms a terraced mountain, which surrounds each of the Anthracite coal fields; the red rocks of the Catskill making the terrace, and the white rocks of the Pocono forming the crest. Piled upon one another in inclined strata, they constitute the bulk of the Catskill Mountains in New York, of the Pocono plateau in Pennsylvania, and the Allegheny, Savage and Cumberland Mountains, far into Virginia and Tennessee.

Jn all the railroads approaching the anthracite coal regions of Pennsylvania one passes over these Catskill rocks, often for many miles. They contain no coal, but fossil ferns are abundant in some localities. This is the last and upper formation of the Devonian period, and is the foundation on which rests the carboniferous system. On the Delaware division of the N. Y. L. E. & W., or Erie Railway, is an opportunity of seeing the red rocks of the Catskill formation for a number of miles, and also on the N.Y. & O. Midland Railroad north of the Bloomingburgh tunnel.

In Pennsylvania it is composed of a vast succession of thin-bedded red and gray sandstones, with thin seams of red, green and mottled shales, also coarse and fine sandstones of various hues of red, brown, gray and greenish ; together with red and greenish coarse silicious conglomerate of white quartz pebbles, the whole being thick bedded, and with an oblique laminated structure. It has not much of interest, either to the scientific or practical inquirer. Its most interesting fossils are fish-remains, which, in the Catskills, extend through 100 feet in thickness of strata. It is the *Old Red sandstone* of England, lying under the coal. The English *New Red sandstone* is over the coal, being the Permian, Jurassic and Triassic formations, but these are not found directly over the coal in America.

The Catskill formation is a poor one for agricultural purposes. The fields are stony, with many projecting ledges of red rocks. Its sandstones are too hard, and too destitute of lime to produce a fertile soil, and the country covered by it is either a wilderness, or very thinly populated.

13-15 CARBONIFEROUS AGE.

13 a. Lower Sub-Carboniferous.—To a superficial observer, the remarkable substitution of great sandstone and conglomerate deposits, under the coal-measures in the east, for generally limestone deposits, under the coal-measures of the west, must seem inexplicable. But the simple explanation is, that all the sub-carboniferous sand-beds of Pennsylvania, formed near the old continent, thin away, and gradually disappear, before they reach the Mississippi; while the five great sub-carboniferous limestones of Illinois, Iowa, and Missouri, formed in a deep quiet sea, on the contrary, thin away, in going eastward, to 40 feet in Westmoreland County, and 25 feet in Somerset County, Pennsylvania; and totally disappear before reaching the Schuylkill and Lehigh Rivers. But the same limestone deposits thicken southward to 600 and 1,000 feet in Virginia, and even more in Tennessee.

In the Pennsylvania Anthracite country, the next formation above the Catskill is a gray sandstone, called by Prof. H. D. Rodgers the Vespertine. In the second geological survey, Prof. Lesley calls it the Pocono, from the name of the mountain bounding Wyoming Valley, on the south side. The miners call it the second conglomerate. It contains carboniferous fossils, but no coal of value. Invariably the Vespertine is the outside mountain surrounding the coal-basins, the inside one being the 14 a. Pottsville conglomerate, or Millstone grit, and they are separated by 13 b. Mauch Chunk red shale, of Lesley, or Umbral, of Rogers, a soft rock, which forms a valley; and all four, 12. Catskill or Ponent, 13 a. Vespertine or Pocono, 13 b. Umbral or Mauch Chunk, and 14 a. Seral or Pottsville

In Pennsylvania, the Vespertine is a white, gray and yellowish sandstone, alternating with coarse silicious conglomerates, and dark-blue, olive and black slates, and occasionally thin beds of coal.   In Michigan, it is the Marshall group, which is mostly a somewhat friable rock, with a reddish, buffish, or olive color, though in some regions becoming gray or bluish-gray.   It forms the receptacle into which the brine descends, and accumulates from the next over-lying Michigan salt group, which is 13 b., and also sub-carboniferous.   The Waverly group of Ohio is proved, by its fossils, to be of this same age.   Its sub-divisions are given at the head of the chapter on Ohio.   It produces the Berea grindstones and Waverly sandstone, the finest building-stone in Ohio, if not in the United States. In Tennessee there is a great development of the lower sub-carboniferous group, the 13 a. Barren group, and 13 b. Coral, or St. Louis limestone, formerly called by Prof. Safford the Silicious.   Its upper part is the equivalent of the St. Louis lime-stone of Missouri; the lower is a series of silico-calcareous rocks, characterized by heavy layers of chert, one inch to two feet thick.

In Illinois the series of sub-carboniferous strata consists of the 1. Kinderhook group, 2. Burlington group, 3. Keokuk group, 4. St. Louis group, the base of which was formerly called the Warsaw limestone, and the 5. Chester group; all of these are limestones and shale, with some sandstone in the first and last named. These embrace both the lower and upper sub-carboniferous, and are 1,200 to 1,500 feet thick in the south-western part of Illinois, but thin-out in going north, and entirely disappear before reaching Rock Island, where the coal-measures rest on the Devonian limestone.   In Iowa the four lower members occur, but the Chester, the thickest member, is wanting, and it is almost entirely wanting in Missouri.

In Pennsylvania a small coal-bed has been opened on the Susquehanna River, in the Pocono sandstone ; and in Huntingdon County more than a dozen small layers of coal may be traced, running through the formation.   In Montgomery County, Virginia, two similar coal-beds attain a local importance, being on Tom's Creek, respectively 4 and 8 feet thick.   These represent the lower coal of East Kentucky, Tennessee, and Alabama.

In Ohio the Subcarboniferous limestone extends through some of the south-eastern counties.   It is quite thin, and represents only the upper or Chester member of the group.   Two workable seams of coal—the Jackson and Wallston coals—are found below it.—Newberry.

13 b. Upper Sub-Carboniferous.—In Pennsylvania this is the Umbral red shale of Rogers, and the Mauch Chunk of Lesley, sometimes 3,000 feet thick, and here consists almost entirely of very soft red shales and argillaceous red sand-stone, without fossils.   It gradually becomes in Virginia a triple mass of buff, green and red shales below, a thick body of light-blue limestone, full of fossils, in the middle, and the upper part blue, olive and red calcareous shales, with massive beneath the coal-measures. It is a heavy body of limestones and shale, the latter almost one-fourth of the mass; and there is also a sandstone. See the above description of 13 a. in Illinois.

In Middle Pennsylvania, around the Broad Top coal-basin, Prof. J. P. Lesley says there appears, for the first time in this formation, going west, distinct traces of the great mountain limestone formation, which underlies all the southern and western coal-fields, and becomes one of the principal features of the geology of the Rocky Mountains, as it is also of the geology of Europe. The red shale formation is here seen, divided in two—910 feet of it above, and 141 feet of it below; a middle group of red and gray, mottled calcareous shales, and thin limestone layers, full of fossil shells—in all 49 feet thick—separating the upper and lower members of nearly pure red shale.

The narrow red shale valleys, which surround this Broad Top coal-basin, the Cumberland basin in Maryland, and the three principal groups of anthracite basins in Eastern Pennsylvania, are due to the thickness and softness of this important formation. But while it is 3,000 feet thick at Pottsville, it is but 300 feet thick along the Allegheny Mountain, and less than 100 feet thick around the coal-basins of Tioga and Bradford counties; and, therefore, instead of making valleys, only marks the top of the mountain steep slopes with a narrow terrace, over which dominates the vertical cliffs of the outcrop of the coal conglomerate.

14 a. Millstone Grit.—This is a mass of white or yellow sandstone, containing vast numbers of quartz pebbles, and forming a pudding-stone, or conglomerate. It is called the Millstone Grit, from being used for the manufacture of millstones. In Pennsylvania and Virginia the formation is 1,000 feet thick, but becomes reduced to from 10 to 175 feet in Ohio. In Kentucky it is from 50 to 500, and in Indiana from 50 to 100 feet. It is a very peculiar rock, and very wide spread, extending out beyond the coal measures proper, of which it is the base and support. There is not in the entire geological series, says Dr. Newberry, another stratum of rock so widely distributed, and presenting as strongly marked lithological characters, as this. The pebbles are generally of quartz, and well rounded. The sand, which forms the paste, and holds together the pebbles of the conglomerate, is generally coarse, and consists of rounded grains of quartz, which differ from the pebbles only in size. In the anthracite region of Pennsylvania, conglomerate rocks sometimes occur between coal-beds, but in the other coal regions they are below all the workable coal-beds. Any cases of thin beds of good coal being found in or below the conglomerate, are exceptional and rare. It does not always maintain its character as a conglomerate, being sometimes an ordinary sandstone. The great lead mines of Joplin and Granby, in Missouri, are in a ferruginous sandstone, the equivalent of the Millstone Grit, or the Chester group, and the Hot Springs of Arkansas are in the Millstone Grit, greatly metamorphosed.

14 b. and c. Lower and Upper Coal Measures.—The series of rock-strata, among which the carboniferous coal-beds are found, are called the Coal Measures, which produce all the best coal of America. They consist of repeated alternations of exceedingly diversified rocks, of every degree of coarseness, from the smoothest fire-clay to exceedingly rough, silicious conglomerates, including within those extremes a wide variety of coal-shales, or mud-rocks, of almost every color and texture—marls, argillaceous sandstones and quartzose grits, also thin bands of limestones, both pure and magnesian, and numerous seams of carbonate of iron.

The numerous coal-beds themselves, which occur among this series of strata, the most interesting and important of them all, are also found in America in all their known varieties, from the most compact anthracite to the most fusible and bituminous kinds of coal. There is no invariable order for the strata of coal measures, but usually the bed of coal has a fire-clay bed below it, and shale immediately over it. Extending our view over a considerable district, we find these rocks are coarser and more massive towards the east or southeast; that they become more fine-grained, and less sandy and earthy, and the limestones increase in size and number as we proceed westward or northwestward; that many of the strata become reduced in thickness, and some of them entirely disappear. In Pennsylvania and Ohio the middle portion of the coal measures contains no coal seams, and hence is called the Barren Measures, thus dividing the formation into Upper and Lower Productive Coal Measures. The Lower Coal Measures sometimes contain valuable beds of iron ore. Salt is produced from the Lower Coal Measures in Western Pennsylvania, Virginia, Ohio, Indiana, Illinois and Kentucky.*

15. Permian.—In the annexed Guide a large number of stations in Kansas are given as being on the Permo-Carboniferous (Permian) series, and it was for a long time supposed that these rocks occur only in Kansas. Prof. C. A. White has recently assigned a large area in Texas to the Permian, and Prof. I. C. White is inclined to refer the Permo-Carboniferous beds of Southwestern Pennsylvania and West Virginia, the No. XVI. of Rodgers, to the same age, since they are the exact counter-part of the Texas rocks in their stratigraphical relations, lithology and palæontological affinities. The Permian rocks in Europe are limestones, sandstones, red, greenish, and gray marlites or shales, gypsum beds and conglomerates, among which the limestones, in some regions, predominate. In Kansas they consist, according to Prof. Mudge, of calcareous and arenacous shales and beds of limestone. The latter are quite impure, but sometimes massive magnesian limestone, of a drab and buff color, is found, which furnishes an excellent building material. Prof. Swallow describes them as a series of limestones, marls, shales, sandstones, conglomerates and gypsums. The State capitol of Kansas, at Topeka, is built of Junction City limestone of the Permian formation. It is also used at Manhattan, and the buildings at Fort Riley are also conspicuous specimens of Permian limestone. The rocks here called Permian, are conformable to the coal measures, and contain many coal-measure fossils, with some not found below. Some geologists think there is no good reason for separating the Permian rocks from the Carboniferous system, of which they form the uppermost member (and in the Tables of Formations both Permian and Permo-Carboniferous are used.) Strata of the same age occur in Indiana, Texas and Mexico, where they contain many new and interesting reptilian remains. In most parts of the United States where the coal measures are not overlaid by the Permian beds, the latter have very probably been eroded. The Permian forms part of the New Red Sandstone of England, lying over the coal. The name is derived from Permia, a province in Russia.

---

* Having been for twenty-one years actively engaged in mining, transporting and selling coal, the author's business led him to the study of geology, particularly in its economic bearings, and he has given to the world all he knows about coal in another work entitled, "THE COAL REGIONS OF AMERICA: THEIR TOPOGRAPHY, GEOLOGY AND DEVELOPMENT," by James Macfarlane, Ph. D.

16-18. MESOZOIC.

16. Triassic.—As the railroads from Philadelphia to New York, the greatest lines of travel in this country, run·on this formation, it is the most conspicuous and well known in the State of New Jersey, and one in which geologists are now taking great interest. Every observing person must have noticed it, and its aspect and composition are so uniform and well marked, that a description of it here will answer for the whole belt through the States of Pennsylvania, Maryland, Virginia, and North Carolina, from the Hudson River to Deep River, in the latter State, and in the Connecticut Valley.

The Triassic consists of dark reddish-brown sandstone, soft, crumbly brown shales, and the upper beds are coarse conglomerates. The almost invariable dip is towards the north-west, at angles ranging from 15° to 25°. Prof. H. D. Rogers thought this uniform dip was not caused by any uplifting agency, but that the rocks were originally laid down in this manner. His theory is that the formation owes its origin to an extensive ancient river, having its source at the eastern base of the Blue Ridge, in North Carolina. Following the remnants of the Triassic formation thence north-east, it gradually, from small beginnings, becomes larger, and has throughout a descending course. At the James River, it is four, at the Potomac six, at the Susquehanna twelve, and at the Delaware, thirty miles wide—the estuary being in the region of the Raritan and the Hudson. In New Jersey, therefore, this river was at its maximum.

The uniform dip was supposed by Prof. H. D. Rogers to be the result of the oblique or slanting mode in which the sediment has been laid down by a rapid and steady current washing the material from the south-east side or shore of the river. If it were due to an upheaval, this formation, measured in the usual way, would show an unheard-of thickness. In fact, it is very thin, as is shown in the exposures of limestone in the interior of the belt. All the appearances of the formation indicate, and there is much to sustain his opinion, that it never was tilted.

But more recent study of this interesting formation, has proven two facts: (1) that it was originally extensive, far beyond its present limits; and, (2) that, in at least its middle beds, the original deposits were horizontal, and have been since upturned. The two great belts of Triassic, which cross from Virginia into North Carolina, and one of them into South Carolina, not only have their rocks dipping in opposite directions, showing a long and broad uplifted country between Raleigh and Danville; but certain groups of coal-beds, which, though now dipping in contrary directions, must of course have been originally horizontal. Traces of coal-beds have been found in the Triassic of Pennsylvania, in York county, and at Phœnixville. The intermediate country in North Carolina was, therefore, presumably once covered with the formation, and probably all Virginia, east of the Blue Ridge, and all south-eastern Pennsylvania. The formation is seen passing under the plastic clays of New Jersey, and may extend far under the bed of the Atlantic, being thus connected with the beds of the Connecticut, and even those of the Bay of Fundy.—Lesley.

Relics of vegetation are occasionally found in the Triassic, in the form of highly compact and bituminous lignite, the longitudinal sections exhibiting the fibrous structure of the wood, whence it was formed. This lignite, occurring sometimes in seams of two or three inches in thickness, amid dark shales, has been a fertile source of delusion, some persons having been induced by the hope of finding valuable coal-mines, to waste much labor in the search. Although the Richmond and North Carolina coals are Triassic, all the geological facts discountenance the notion that it contains coal in New Jersey and Pennsylvania, the detached fragments of plants, which we meet with in the form of lignite, having evidently been loosely drifted into these sediments from the land. Prof. Emmons says there is nothing which can be regarded as equivalent to the coal measures of the Chatham (N. C.) and Richmond (Va.) series in the northern beds. All this formation was produced at a period subsequent to the great Carboniferous or coal-bearing rocks. There are great numbers of fossil fish in the Trias of New Jersey and Connecticut valleys, among them twenty species of *ganoids;* also the famous bird-tracks of Dr. Hitchcock. See notes 27 and 28 Massachusetts. Fossil plants are numerous in the Trias of Virginia and North Carolina.

When a large portion of the pebbles are of limestone, in the Triassic conglomerate, and the cementing red earth which unites them, contains an adequate quantity of the same material, the rock possesses the character of a marble, as on the Potomac River. The Portland stone, or reddish-brown sandstone, so much used for building purposes in New York and other eastern cities, is from the Triassic formation.

Extensive mines for copper ore have been wrought in the Triassic, in the State of New Jersey, the ore occurring in every case adjacent to igneous traps, but not in contact with them. All these mining operations have failed, on account of the ore being diffused or disseminated through the mass of the formation, and not being found compacted in regular veins. In Europe, the upper part of the Triassic is called Keuper, or copper.

Trap-Dikes.—Numerous parallel ridges and dikes of Trap, some of them many miles in length, and with the elevation of mountains 400 feet high, and ridges of all sizes, traverse the Triassic. Indeed, nearly all the trap-dikes are confined to this formation. The material which composes these rough, rocky ridges, undoubtedly protruded in a state of fusion, slowly and gently through long narrow fissures, produced by the gaping asunder of the rocks, and not by enormous violent disruptions, like those of volcanoes, as the strata through which they passed are very little disturbed, and the dip of the strata is very little affected by them. These trap-dikes have burst through the red shale and sandstone, after they were deposited, overflowing, while in a melted and highly heated condition, the adjacent beds, and greatly altering their texture, color and mineral aspect. The finest of these trap-dikes is the Palisades, on the west side of the Hudson River, above Jersey City, and extending north of that place. (See note 5, in chapter on New York). The tunnels and deep railroad-cuts through it, in Jersey City, afford good opportunities to observe the appearance of the stone, the principal constituents of which are hornblende, feldspar, and titaniferous oxide of iron. The little mountain of iron ore at Cornwall, in Lebanon county, Pennsylvania, was thrown up by a trap-dike of the Triassic.

That the trap is not confined, however, to the Triassic rock surface, is beautifully shown by the very numerous trap-dikes which cut the Highlands of Orange county, N. Y., and of New Jersey; by the long, straight, narrow dike which issues from the South Mountain, opposite Carlisle, in Pennsylvania, and cuts across all the formations, from the Potsdam up to the Subcarboniferous, at the mouth of the Juniata, (see notes 9, 77 and 170, in chapter on Pennsylvania), and especially by the still longer trap-dike recently discovered by Prof. Frazer, in Lancaster county, Pa., which not only penetrates the Welsh hills of gneiss, but cuts across the west end of the Chester county (Pa.) Valley, near the famous nickel mine, and reaches the Susquehanna River near the roofing slates quarries at Peach Bottom.—Lesley.

The Triassic formation yields the rock-salt and brine of the greater part of Europe, especially in England, Ireland, France, and part of Germany.

17. Jurassic.—The upper portion of what is commonly called the Triassic, on the Atlantic border, may belong to the Jurassic, and is so described by Prof. P. R. Uhler, in the annexed Guide for Maryland; and by Prof. W. B. Rogers, as Juro-Triassic and Juro-Cretaceous, in Virginia. But there are beds which are undoubtedly Jurassic in several of the eastern ridges of the Rocky Mountains, and other districts of the far West. The rocks are, in general, a gray or whitish marly or arenaceous limestone, with occasional pure compact limestone beds, intercalated with laminated marls. The enormous *Dinosauri*, recently obtained by Marsh and Cope from Colorado, are from the Jurassic. It is much less important here than in England, where it is subdivided into the Liassic, Oolytic and Wealden. The name is derived from Mount Jura, in Switzerland.

18. Cretaceous.—The Cretaceous formation, along the Atlantic Coast and the lower Mississippi Valley, consists of a series of beds of strata, differing from each other; but they are all earthy in form, consisting of beds of sand and sandy-clay, except at a few points, where the strata have been cemented by oxide of iron into a kind of sandstone, or conglomerate. In Texas it contains extensive beds of gypsum. In New Jersey it produces the lower two beds of green-sand, called marl, which is extensively used in agriculture, the value of which is due to the potash and phosphates which it contains. Ninety per cent. of it is a green silicate of iron and potash, the rest being ordinary sand, and it contains no lime. But in Wyoming, Utah, and Colorado, the Cretaceous attains a thickness of 9,000 feet, and its rocks comprise beds of sand, marlite, clay, loosely aggregated shell-limestone, or rotten limestone, and compact limestone. At the middle of the Cretaceous, lie the beds of plastic-clay, outcropping across New Jersey, from Trenton to Amboy, and of great importance to the fire-brick and pottery factories, as described in the Report of Prof. Cook, of New Jersey, for 1876.

The name Cretaceous is from the Latin word for chalk, the chalk of England and Europe, being one of the rocks of this period; but in this country it contains no chalk, except in Western Kansas, 322 miles west of Kansas City, where a large bed exists. It is within one mile of Trego station on the Kansas Pacific Railroad, and is found over a tract 125 by 30 miles.

The Cretaceous formation, in the far West, passes upwards into a coal-bearing formation, several thousand feet thick, and covering on the upper Missouri River not less than 100,000 square miles in the United States, besides the portion of th' belt extending into the British possessions. The area of other lignitic basins farther south, cannot be estimated, their width being unknown. Dr. Hayden regards this coal-formation as transitional, or Lower Eocene 19. Tertiary, and in the within Guide for Colorado it is called the Lignitic Group, lying between the Cretaceous and Tertiary. Mr. Lesquereux is of the same opinion as to its Tertiary age, but nearly all other geologists regard it as Cretaceous.

In the annexed Guide for Wyoming and Utah, the formation is given at points where the coal is mined—Carbon, Separation, Black Buttes, Point of Rocks, Rock Springs, and Evanston. All the coal now mined in Wyoming is, according to the Guide, in the 18 d. Laramie Cretaceous, which corresponds with Hayden's Lignitic beds. Every division of the Cretaceous is said to be lignitic or coal-bearing, and may some day produce good coal. The Evanston beds are in the Laramie, but the Coalsville beds are probably in the 18 b. Colorado Cretaceous. The Rock Creek coal may be 18 c. Fox Hill.—A. Hague. There is no Carboniferous coal in the far west. The difference of opinion as to the age of the Lignitic or coal-bearing group, arises from the fact of its lying at the transition point from the Cretaceous to the Tertiary, where, as is not unusual, the fossils of both are mingled; and the controversy is as to precisely where the Cretaceous ends, and the Tertiary begins.

19-20. CENOZOIC.

19. Tertiary.—The Tertiary formation of the Atlantic coast is wholly of an earthy character, without solid rocks, consisting of sands and sandy blue clays, and above these yellow and brown ferruginous sand ; also clays and sands imbedding extensive layers of uncemented fossil shells. But as we trace them south and southwest through the Southern cotton-growing states, it becomes more calcareous, consisting of lead-colored sandy clays, and whitish and bluish friable limestone in North and South Carolina and Eastern Georgia. West of that, the upper member consists of two limestone strata, the middle of sand and sandy marl, and the lower part of limestone and marl. H. D. Rogers suggests that on the Atlantic slope, opposite the Appalachian Mountains, the older rocks furnished only sandy and clayey sediments, and the Tertiary deposits composed of the ruins of the former, are of that character; while farther west a wide expanse of limestones fills the upper valley of the Mississippi, and hence the Tertiary deposits bordering the Gulf of Mexico, and extending up the Mississippi River, are of a greatly more calcareous or lime-bearing character. The cotton-growing lands of the Southern States are chiefly Tertiary. In the central part of the continent, the Tertiary beds are lake sediments, or fresh-water deposits; while on the west coast they are marine. The Tertiary, in the southern part of New Jersey, furnishes great quantities of bog iron-ore, but bog iron-ore is not peculiar to the Tertiary formation. The upper bed of the green-sand of New Jersey is Tertiary. In the far-west the Tertiary strata are in a greatly more indurated or rocky condition than those of the eastern coast. The 19 a. Eocene consists of beds of clay and sand, with round ferruginous concretions and numerous seams and local deposits of lignite, according to Mr. Lesquereux. Also gray and ash-colored sandstone, with more or less argillaceous layers. The 19 b. Miocene consists of white and light drab clays, with some beds of sandstone and local layers of limestone. The 19 c. Pliocene is composed of fine, loose sand, with some layers of limestone, and contains fossil bones of animals, which are scarcely distinguishable from living species.

20. Quaternary.—The materials of the glacial drift consist of vast accumulations of sand, pebbles, and bowlders, belonging invariably to rocks lying northward of their present positions, with beds of bowlder clay of great thickness, evidently brought from a great distance from the north, by causes quite different from any now in operation, and which nearly all geologists now believe to have been glaciers. This material is spread over the whole breadth of the North American continent, down to 38° or 40° of latitude, with glacial flood-deposits farther south along the valleys; and it is also spread, in the same way, over the northern part of Europe. Nearly every recently uncovered ledge of rock in the drift-covered region has its surface marked with the characteristic striae and furrows. These scratched, polished and grooved surfaces prove the former existence, according to Agassiz's theory, of an ice sheet, many thousand feet in thickness, moving across the continent over open level plains, as well as along enclosed valleys. When softer and harder rocks alternate, they are planed off to one outline or level, as if a rigid rasp had moved over the land, leveling all before it. On the contrary, on any surface where water flows, we find the softer materials have yielded first and been worn out, while the rocks will be left standing out, and show greater resistance. Glacial surfaces are highly polished, and are marked with scratches, grooves and deeper furrows. Sometimes the smooth surfaces are like polished marble, showing that the grinding material was held steadily down in firm, permanent contact with the rocky surface against which it moved, as is the case with the glacier. There are many deep ancient channels filled by the drift.

The usual characteristic marks of glaciers extend, according to Agassiz, over the whole surface of the east half of the continent, from the Atlantic shores to the States west of the Mississippi, and from the Arctic sea to the latitude of the Ohio, about the 40th degree of north latitude. The glacier marks trend from north to south, with occasional slight inclinations to the east or west, according to the minor irregularities of the surface. The ice of the great glacial period in America, is supposed to have moved over the continent as one continuous sheet, over-riding nearly all the inequalities of the surface. The drift is spread in one vast sheet over the whole land, consisting of an indiscriminate medley of clay, sand, gravels, pebbles, bowlders of all dimensions, so uniformly mixed together, that in all parts of the country it presents a general similarity. The partial absence of stratification is one important characteristic of glacial drift. In the bowlder clays there is no arrangement of the materials according to size or weight, whereas in water the lighter materials are carried farther than the heavier ones and deposited separately. In glacial drift there are large angular fragments by which it may be distinguished from alluvium, and it retains the mud gathered during the journey, spread through its mass, while the water-rolled deposits are washed clean, and consist usually of well-rounded pebbles, and there are no scratches on the exposed surfaces of the solid rocks.

The following general description of the limit of the drift is intended to show the approximate boundary between the glaciated and non-glaciated parts of the country. Although the margins of the different drift-sheets appear to form a single margin, because the sheets overlap, it must not be inferred that they are one and the same, or that they were formed at the same time, or necessarily by the same agency. The majority of active and critical students of the drift of the interior now believe in two or more glacial epochs—not merely stages of retreat, but two or more independent ice incursions. Nor is it to be understood that the southern border is everywhere a moraine, in any special sense of the term. For more than half its extent across the country, there is no special aggregation of drift at the edge, and the precise method of its formation in certain portions is yet an open question.

In the northwestern corner of the United States, the margin of the great northern drift sheet unites or becomes confused with the local drift from the mountains, and it is impossible to say at present what is to be regarded as the margin of the great northern mantle. According to Dr. G. M. Dawson, there was a general southerly movement on the highlands of British Columbia. This appears to have penetrated to the basin of Puget Sound, but not to have reached the Columbia river. It seems also to have entered the northern edge of Washington Territory, near the northern elbow of the Columbia (Willis). It also penetrated into Idaho, as far as Lake Pend d'Oreille (Chamberlin), and also the northern border of Montana. Local mountainous glaciation was quite extensive along the Cascades, Sierra Nevada, Rocky Mountains and some minor ranges. East of the Rocky Mountains, the limit of northern drift enters the United States from Canada at the foot-hills of the mountains (G. M. Dawson), and running southward to the vicinity of Fort Shaw, curves eastward crossing the Missouri river about 40 miles above Fort Benton (Chamberlin and Salisbury). Thence it courses eastward, crossing the Yellowstone about 60 miles above its mouth, keeps north of the Northern Pacific railroad to within about 30 miles of Bismark (same authorities). Here it turns south, keeps in the vicinity of the Missouri river to Nebraska (Chamberlin, Todd), thence southerly to near the mouth of the Republican river (Todd, Mudge), thence easterly to the mouth of the Missouri river (Salisbury and Chamberlin). East of the Mississippi it forms a great loop, reaching nearly to the south end of Illinois (Worthen, Wright); swings north to the heart of Indiana (*ibid*) and south again into Kentucky (Sutton, Wright). Entering Ohio above Cincinnati it trends undulatingly northeast, and enters Pennsylvania a few miles above the mouth of the Beaver (Lewis and Wright); thence it extends northeastward into the State of New York, where, making a sharp curve, it again enters Pennsylvania in Potter county, and passes southeast to Belvidere, New Jersey (Lewis and Wright), and crosses that State with a northward arch to Perth Amboy (Cook and Smock). It traverses the whole length of Long Island (Cook, Smock, Upham) and appears on Block Island, Martha's Vineyard and Nantucket (Upham). The reader will understand that all south of the line described is unglaciated except local areas in the mountainous regions of the west, and possibly some in the Appalachians. From the Atlantic Coast to the Scioto valley, in Ohio, for the greater part, there is, on or near the margin, a well-marked terminal moraine, north of which lie other marginal moraines. From the Scioto valley westward, the margin of the drift is characterized by no sensible ridging of the nature of a terminal moraine, but terminates in a thin and often very attenuated edge. Eastward from the Atlantic shore, the edge of the glacial deposits is supposed to correspond with St. George's Bank and Sable Island Shoal, and to pass southeast of Newfoundland.

In Europe the border limit crosses the southeast corner of England, southern Holland, southern Germany, passing near Dresden, and thence onward south of Warsaw and Moscow, in a sinuous course, embracing the center of European Russia, and curving around to the northeast, runs northward to the Arctic Ocean, west of the Ural Mountains.

In no part of the United States are the phenomena of the drift displayed on a grander scale than in the Lake Superior region and on the northern borders of Wisconsin. Minnesota and Dakota are very deeply buried in drift. At the south side of Lake Superior, the drift is frequently 200 to 300 feet deep, and at the west end of that lake it is 300 or more feet thick, and it is 220 feet deep at Fargo, Dakota. The lower peninsula of Michigan is covered often from 200 to 300 feet deep.

To the southward the drift usually diminishes, and it becomes more evenly spread over the country. It is a singular fact that in the Galena lead region, at the corner of Illinois, Iowa, and Wisconsin, bounded by the Mississippi, Wisconsin, and Rock rivers, and in a considerable extent of territory north of it, no transported drift material can be found. The driftless region is 10,000 square miles in Wisconsin alone, or one fifth of the area of the State. Ohio has a very complete series of drift deposits, and they have been well studied and described by Dr. Newberry. He has classified the drift deposits as follows, in the ascending order: 1st. The Erie clay, a blue or gray unstratified bowlder clay. 2d. The forest bed, consisting of a bed of soil, with timber, the remains of an ancient forest, found in Ohio, Indiana, etc., at various depths from the present surface. 3d. Lacustrine deposits, stratified sands and clays in northern Ohio; yellow clay abounding with gravel, in southern Ohio.

The Bluff formation along the Missouri and Mississippi rivers is a very peculiar and interesting one, resting upon the drift. It is of a slightly yellowish ash color, very fine, not sandy, and yet not adhesive. It makes an excellent soil, is easily excavated by the spade alone, and yet it remains so unchanged by the atmosphere and frost, that wells dug in it require to be walled only to a point above the water line, while the remainder stands so securely without support, that the spade marks remain upon it for many years. Road embankments and excavations upon the sides of roads stand like a wall. (See general note, Mississippi chapter and note on Vicksburg, Tennessee chapter.) The peculiar outline of the bluffs along the Missouri river is very interesting. They are often naked, entirely destitute of trees, and tower up from the river bottom-land, sometimes more than a hundred feet in height, and so steep in some places that a man cannot climb them, yet they are not supported by a framework of rocks, as other bluffs are, and not a rock or pebble of any size exists in them, except a few calcareous concretions where lime-water percolates through them. It is thought to be a lacustrine deposit, a shallow lake having, during the time of the Glacial epoch, occupied the whole of the basin of the Mississippi before the great rivers had cut their valleys down to their present depths (White). In Louisiana the bluff deposit contains three distinct groups of strata, the Port Hudson below, the Loess next, and the yellow loam above, and over this the alluvium and below them all the drift (E. W. Hilgard, F. V. Hopkins).

Earthy material brought together by the ordinary action of water is said to be alluvial, and the soil or land so formed is called alluvium or alluvion. Diluvium implies the extraordinary action of water. When the drift material covers the surface, of course it forms the soil, but in driftless regions the soil is an admixture of clay, sand, lime, etc., derived from the disintegration of the rocks beneath, with decomposed animal and vegetable substances. Where neither glacial nor alluvial action has taken place—as in some parts of our Southern States—the rocks are converted into a deep and strong soil, having undergone a process of decay which has rendered them so soft, sometimes to a depth of 20 or more feet, that they may be readily cut with a spade, although retaining all the veins and layers which mark their original stratification. Without having been broken or ground up, even the hardest rocks have quietly mouldered into a soft clayey mass, which, from its peculiar structure, has a natural drainage and possesses, moreover, great fertility.

The most important of geological formations is the last of all, the soil. On this thin, superficial, earthly covering of our planet depends all the growth of all vegetation, and on that depends all terrestrial animal life. But whether the material forming the soil remains unmoved in the same spot where it was once a solid rock, or is transported bodily by a glacier, or carried from the hills into the valleys by running water, and moved from place to place by larger streams and rivers, it was originally derived from the rock formations, therefore the agricultural as well as the mineral resources of the country depend on this geology.

This completes, in brief, the description of all that can be seen of the earth, classified in geological order, from the oldest of the rocks up to the sands which are now daily washed to our feet by the currents of the rivers and the waves of the sea.

REMARKS ON THE FOREGOING DESCRIPTIONS.

Paleontologists will be disappointed in this introduction, from which that is omitted which seems to them the most important, and gives the most interest and significance to the subject, namely: the life which they find in the formations, and which serves so important a purpose in their identification and classification. But another book would have been required for that purpose, and it would have been useless without a large number of expensive engravings.* Paleontology is the province of all the text-books on geology, to which this work is a supplement, not a substitute. Its only object is to teach local geology. The descriptions were an after-thought, and they should be regarded as an attempt—to present to the unlearned a first-lesson in geology, in the vernacular tongue, in the hope that it may help on the cause of popular science. They have swollen much beyond the original design, which was definitions, rather than descriptions; but they will serve to show that paleontology is not the whole of geology, and that the formations are more than a mere cabinet of fossils.

There are some things in the descriptions that are not accepted by all geologists. But the scope of the work did not permit any account of the conflicting opinions on disputed points, or discussions of the history of geological nomenclature and classification. Whether the Oriskany sandstone should be placed at the base of the Devonian, or at the top of the Silurian; whether Hudson River, Loraine, Nashville, or Cincinnati, is the best name for that formation; and whether Cambrian should include one, or all, or none of the Lower Silurian formations, and similar questions, seem of less importance to the ordinary reader, for whom the descriptions are intended, than to the professional geologist.

All kinds of geological tables are given, for, in accepting the valuable contributions of others on local geology, it was necessary to let them have their own way, in the chapters on their own States, in regard to the names and the arrangement of the formations. A common number, attached to them throughout the book, serves to identify the formations by whatever name they are called.

The valuable part of the book is the Geological Railway Guide, the design or plan of which is original with the author, as it is believed nothing of the kind has ever appeared, in any language. It is the work of many hands, and the hearty thanks of every lover of the science are due to all those who have contributed to its pages portions of the multitude of facts, forming this index to the geology of all important places in the United States and Canada. The reader will never know the amount of time, patience, labor, and care that it has cost.

---

* See "THE ANCIENT LIFE HISTORY OF THE EARTH," a comprehensive outline of the principles and leading facts of Paleontological Science. By H. A. Nicholson. Published by D. Appleton & Co., New York. 8vo., 407 pp. $2.00. A very convenient and excellent manual of Paleontology only.

ARRANGEMENT OF THE GEOLOGICAL RAILWAY GUIDE

AND DIRECTIONS FOR USING IT.

1. The railroads are arranged by states, and the states and territories are arranged in geographical order, with reference to the great lines of travel. But to find a railroad, the reader must depend on the index. Branches are placed after the main line, which is generally first given throughout without interruption.

2. When stations are omitted for the sake of brevity, which is seldom the case, the lists being uncommonly full, their geology will be understood to be the same as that given at the stations between which they occur. If the geology of two adjacent stations is different, it is evident enough that there is a transition from one to the other formation, between the stations, but the change is often so gradual that the transition point cannot be precisely given.

3. A few feet of difference in level sometimes carries the railway track to an upper or lower formation. Railroads, too, sometimes run across narrow, projecting tails, and scalloped points of a higher or lower formation, than that given in the Guide, but which it would occupy too much space to specify. Where too, the strata are disturbed and broken-up, all the formations cannot well be specified for want of room. In such cases the Guide serves only to show nearly where you are, the prevalent formation being given.

4. The hills, bluffs and higher ground in view, are often of a different formation from that given on the railroad, but not always higher in the series. Their elevation is often due to the hardness of the strata, the softer rocks forming the valleys, in which railways generally run.

5. Keep in mind the succession of the formations, as shown on the Guide, and whether you are going from older and lower to younger or higher strata, or *vice versa*. Notice the changes in the scenery with the changes in the formations.

6. When you come to a new formation, refer to the description of it, in the beginning of the book. But it is difficult to get a clear idea of the formations from even the best description. The reader must see them for himself, and these lescriptions are intended to assist him in identifying them, and to impress their character and appearance upon his mind, or to recall them to his recollection after having seen them.

7. By a little close observation of the formations in traveling, you will find that most of them have peculiarities of their own, by which you can always know them, but which, like the features or appearances of persons, cannot be put into words, so that another who has not seen them could also recognize them. The form of the summits and slopes of the hills, and the general aspect of the country, but especially the rock-cuts on the railways, and other exposures of the formations, in quarries, and in the banks and beds of streams, should be closely observed; and if these are not visible, notice the stone used in buildings, and for the enclosures of fields, the character of the soil, and the fragments of stone mixed through its mass, which betray the nature of the solid rock formation beneath; observe also whether the rocks lie horizontally or in an inclined position.

4

The Dominion of Canada.[51]

By GEORGE M. DAWSON, D. S., F. G. S.,
Assistant Director of the Geological and Natural History Survey of Canada.

I. **Maritime Provinces.**
New Brunswick, Nova Scotia, and Prince Edward Island.

II. **Quebec and Ontario.**

III. **Manitoba and North-West Territory.**

IV. **British Columbia.**

V. **Steamboat Routes.**

---

1. The Dominion of Canada is, as a matter of convenience in this work, divided into four parts, and from a geological point of view such division is largely borne out by structural facts.
I. The Maritime Provinces includes Nova Scotia, New Brunswick, and Prince Edward Island.
II. Ontario and Quebec includes the provinces of the same names.
III. Manitoba and so much of the Northwest Territory as is traversed by railway-lines forms the third division.
IV. British Columbia, together with the eastern slopes of the Rocky Mountains (politically a part of the Northwest Territory) constitutes the fourth.
For each of these great divisions a separate table of formations is given.
For the purpose of enabling the traveler to provide himself with further information on geological points, the following notes on publications are attached:—Dominion of Canada generally: "Sketch of the Physical Geography and Geology of the Dominion of Canada," with map; Geological Survey, 1884. For economic minerals see also "Descriptive Catalogue of Exhibits at Philadelphia, 1876," and "Catalogue des Minereaux Roches, etc.," at the Exposition at Paris, 1878, by Dr. B. J. Harrington. Both published by the Geological Survey.
The "List of Publications of the Geological and Natural History Survey, 1884," enumerates all the official reports and maps to date.
I. MARITIME PROVINCES.—"Reports of Progress." Geological Survey. The whole of Cape Breton Island, part of the mainland of Nova Scotia, and nearly the whole of New Brunswick have been geologically mapped on contiguous sheets of uniform scale. Maritime Provinces generally: "Acadian Geology." Sir W. Dawson. (With supplement and map.) 1878.
The greater part of the really productive coal measures are included in the Province of Nova Scotia, the great spread of Carboniferous rocks in New Brunswick having so far been found to contain but thin, and, generally, scarcely workable, coal-seams. The deposits of the glacial period are often well shown in railway-cuttings, and extensive tracts are completely covered with these. The boulder-clay is the most persistent and universal. Peaty deposits underlying the boulder-clay have been observed locally; overlying the boulder-clay are stratified clays, sands, and gravels, and kames are frequent, particularly in New Brunswick. The stratified clays hold marine fossils in the vicinity of the coast of the southern and northern parts of New Brunswick.
The island of Cape Breton affords good coal, and a number of collieries are in operation. As it is not yet traversed by railway, it does not receive notice in the body of this work, but few places of equal area are of greater interest from a geological or picturesque point of view.
II. ONTARIO AND QUEBEC.—"Geology of Canada." Sir W. Logan. 1863. This work summarizes the main features to date, and is accompanied by an atlas of maps, sections, etc. Sir W. Logan's large map (25 miles to 1 inch, published 1866) includes, besides Ontario and Quebec, the Maritime Provinces and adjacent portions of the United States, and is much more detailed, for the region covered by it, than the map accompanying the sketch of 1884.
From 1863 reports in different portions of the provinces in annual "Reports of Progress." See also "Esquisse Géologique du Canada," etc. 1867.
III. MANITOBA AND NORTHWEST TERRITORY.—In addition to the sketch of 1884, see reports and maps in annual "Reports of Progress" of Geological Survey, "Report on Geology and Resources of 49th Parallel," by Dr. G. M. Dawson.
Much information in the possession of the Geological Survey, but yet unpublished, is incorporated in the notes on these portions of the Dominion.
IV. BRITISH COLUMBIA.—In addition to the sketch of 1884, see annual "Reports of Progress," 1871, to date. A considerable portion of the province is covered by preliminary geological maps, on a scale of 8 miles to one inch.
The greater part of the facts for the Dominion of Canada are derived from the reports and maps of the Geological Survey. Dr. G. M. Dawson also wishes to acknowledge assistance received from Dr. Selwyn, the director of the Survey, and several members of the staff, especially Messrs. R. W. Ells, R. Chalmers, and H. Fletcher. The notes on the Intercolonial Railway are chiefly due to Sir W. Dawson, as elsewhere mentioned.

I. Maritime Provinces.

Nova Scotia, New Brunswick, and Prince Edward Island.

List of Geological Formations.

Quaternary	20 c. Saxicava Sand. 20 b. Leda Clay. 20 a. Boulder Clay or Till.	**Silurian**	7. Lower Helderberg. Upper Arisaig Series 5 c. Niagara. New Canaan Series. 5 b. Clinton. Lower Arisaig Series
Triassic	16. Upper Red Sandstone, and Traps of Bay of Fundy. Upper Red Sandstones of P. E. I.	**Sil.-Camb.**	4. Cobequid Series? 4. Graptolitic Shales of New Brunswick.
Carboniferous	14 c. Upper Carb. and Permo-Carb. 14 b. Middle Carboniferous. 14 a. Millstone Grit. 13 a. Lower Carb. { Windsor Group.²⁰ (Limestone Gypsum, etc.) Horton Group.²⁰ (Lower Coal Measures.) } Bonaventure formation (in N. E. New Brunswick & E. Quebec).	**Cambrian**	2 c. Upper Cambrian. { Miré and St. Andrew Series, Cape Breton. 2 b. Middle Cambrian. Acadian Series. 2 a. Lower Cambrian. { Atlantic Coast Series, Nova Scotia }
Devonian	12. Catskill. { Scaumenac Beds (Baie des Chaleurs). 11. Chemung and Portage. 10. Hamilton. { St. John Series. (Cordiate Shale. Dadoxylon Sandstones.) 8. Oriskany, Nictau Series. Gaspé Sandstones and equivalents on Baie des Chaleurs.	**Huronian**	1 b. Felsitic, Chloritic, and Epidotic Rocks of St. John, Yarmouth, and Cape Breton, in part.
		Laurentian	1 a. Gneiss, Quartzite and Limestone of St. John and St. Anne's Mountain, Cape Breton.

Ms.	Intercolonial Railway, N. S.²		Ms.	Intercolonial Railway—*Con.*	
0	Halifax.³	2. Lower Cambrian.	90	Wentworth.⁸	5–7. Silurian.
8	Bedford.	"	96	Greenville.	13 a. Lower Carbonif.
13	Windsor Junc.⁴	"	103	Thompson.	"
30	Elmsdale.⁵	{ Contact 2 Low. Camb. and 13 a. Low. Carb.	109	Oxford.⁹	14 a. Millstone Grit.
39	Shubenacadie.	13 a. Lower Carbonif.	111	River Philip.	"
61	Truro.⁶	16. Triassic.	122	Spring Hill Jn¹⁰	"
78	Londonderry.⁷	13 a. Lower Carbonif.	126	Athol.	14 c. Upper Carbonif.

2. These notes are extracted, with little alteration, from a chapter by Sir W. Dawson, in "Hand book for the Dominion of Canada." Published by Dawson Brothers, Montreal. 1884.

3. Halifax. Quartzites and slates of the coast series, or gold series, of Nova Scotia, believed to b of Lower Cambrian age. In the vicinity of Halifax and elsewhere it contains auriferous quartz mines The nearest of these are situated at Montague and Waverly. The auriferous veins often also contain mispickel, and sometimes blend and other minerals. They run generally parallel to the strike of th inclosing rocks. The richly auriferous veins are seldom of great width, and the gold is sometime disseminated also in the contiguous slate. The age of formation, of some at least, of the veins is sub sequent to the Carboniferous, as auriferous conglomerates of Lower Carboniferous age with derive gold occur, and have actually been worked, at Gay's River. At Northwest Arm and other places ma be seen granite, which traverses these beds as thick dikes or intrusive masses, and produces con tact metamorphism. At Waverly Mine the obscure fossil named *Astropolithon* may be found in th quartzite.

4. Windsor Junction. Excellent exposures of the fossiliferous Lower Carboniferous limestones and of the great beds of gypsum characteristic of that formation in Nova Scotia.

5. Elmsdale. Beyond Gay's River, the railway enters the Carboniferous country, and in som places quarries in the Lower Carboniferous limestone may be seen near the road.

6. Truro. At and beyond Truro, the railway traverses a portion of the Triassic red sandstones o Cobequid Bay. The sandstones may be seen in the cuttings, and the red color of the soil is character istic. In approaching the Cobequid Hills, a more broken country, and beds of sandstone and con glomerate indicate the Carboniferous beds, which here reappear from under the red sandstone.

7. Londonderry. The road here enters a belt of highly-inclined slaty rocks of olive-gray and dar colors, which, at a little distance west of the railway-line, contain large and productive veins of iron

Ms.	Intercolonial Railway—Con.	Ms.	Intercolonial Railway—Con.
130 Maccan. [11]	14 b. Middle Carbonif.	275 Beaver Brook.	14 a. Millstone Grit.
138 Amherst,N.B. [12]	14 c. Upper Carbonif.	286 Bartibogue.	"
144 Aulac.	{ 14 c. " / 14 a. Millstone Grit.	296 Red Pine.	"
147 Sackville.	14 c. Upper Carbonif.	309 Bathurst. [16]	13 a. Lower Carbonif.
159 Dorchester. [13]	13 a. Lower Carbonif.	321 Petite Roche. [17]	5-7. Silurian.
167 Memramcook.	"	329 Belledune.	"
179 Painsec Junc. [14]	14 a. Millstone Grit.	338 Jacquet Riv'r. [18]	{ " and 13 a. / Lower Carboniferous.
187 Moncton. [15]	"	347 New Mills.	5-7. Silurian.
195 Berry's Mills.	"	353 Charlo.	13 a. Lower Carbonif.
206 Canaan.	"	363 Dalhousie Jn. [19]	"
215 Coal Branch.	"	372 Campbellton.	{ 8 – 12. Devonian and / Doleritic trap.
224 Weldford.	"	385 Metapedia. [20]	5-7. Silurian.
238 Kent Junction.	"	395 Mill Stream, Q.	"
244 Rogersville.	"	405 Assametquag'n.	"
255 Barnaby River.	"	420 Causapscal.	"
259 Chatham Junc.	"	433 Amqui.	"
265 Newcastle.	"		

ore, worked by the Steel Co. of Canada. This vein, or aggregation of veins, is primarily of carbonate of iron and ankerite, with some specular iron, and has been changed in many places to a great depth into limonite, which is the ore principally worked. Beyond this place the slates are seen to be pierced by great intrusive masses of red syenite and by dikes of diorite and diabase.

8. Wentworth. The rocks mentioned above are here overlain by dark-colored shaly beds, holding fossils of the age of the Clinton or older part of the Upper Silurian. The gray slates holding the iron-ore are obviously of greater age, but how much greater is uncertain. For reasons stated in "Acadian Geology," they are regarded by Sir W. Dawson as Lower Silurian. Crossing the Cobequid Hills, conglomerates are seen belonging to the southern edge of the Cumberland coal-field, on which the road now enters.

9. Oxford. Contact of Lower Carboniferous and millstone grit.

10. Springhill. Brines from Carboniferous, utilized on small scale in manufacture of salt, 2½ miles from Springhill mines. A branch road leads to the mines of the same name, the most important coal-mines on this railway. Seven coal-seams, varying in thickness from two feet to thirteen feet six inches, are known in this district. The "black seam," eleven feet thick, is that which has been most extensively worked. The mines supply the coal used on the railway.

11. Maccan. Conveyance may be taken from here to the South Joggins, on the shore of Chegnecto Bay, twelve miles distant. The section of the Carboniferous rocks on this part of the coast is one of the most instructive in existence, and has been rendered classic by the writings of Sir W. E. Logan, Sir C. Lyell, and Sir W. Dawson. The section displays over 14,000 feet in vertical thickness of strata, extending from the marine limestones of the Lower Carboniferous to the top of the coal-measures, and includes seventy coal-seams, of which, however, only two are of workable thickness. Besides numerous fossil plants (including erect sigillaria), the beds here yield reptilian remains and land-shells.

12. Amherst. Near here fine examples of the alluvial deposits of the Bay of Fundy ; more especially the great marshes of Amherst and Sackville.

13. Dorchester. Good sections of millstone grit formation. The contact between this formation and the Lower Carboniferous here. Copper-mine. Between Dorchester and Memramkook, salt-marsh.

14. Painsec Junction. On Shediac Branch, Carboniferous, chiefly or entirely millstone grit.

15. Moncton. From this point to near Bathurst the railway passes over the low Carboniferous plain of Northern New Brunswick, showing scarcely anything of the underlying rocks.

16. Bathurst. Beyond this point is the varied and interesting country of the Baie des Chaleurs, and the Restigouche and Metapedia Rivers, of which it is possible only to note some of the more striking features. Three miles beyond Bathurst, line crosses doleritel intrusion 1 mile. A short distance north of station good sections of leda clay and saxicava sand, with fossils.

17. Petite Roche. From this station to Charlo;numerous massive intrusive bodies of dolerite cutting through the Silurian rocks.

18. Jacquet River. The Lower Carboniferous here forms a narrow fringe along the shore From this station to Dalhousie, many good sections of leda clay and saxicava sand, with fossils.

19. Dalhousie. From Dalhousie the following localities may be visited : At Cape Bon Ami, near Dalhousie, a fine section of Upper Silurian shale and limestone, abounding in fossils, and alternating with very thick beds of dark-colored dolerite. Apparently resting on these are beds of red porphyry and breccia, forming the base of the Devonian. On these, a little west of Campbellton, rest agglomerate and shale, rich in remains of fishes (*Cephalaspis*, *Coccosteus*, etc.), and traversed by dikes of trap. Immediately above these, conglomerates and hard shales, the latter full of remains of *Psylophyton* and *Arthrostigma*, and at a sandstone-quarry at the opposite side of the Restigouche, are similar plants, with great silicified trunks of *Prototaxites*. All these beds are Lower Erian or Devonian. At Scaumenac Bay, opposite Dalhousie, are magnificent cliffs of red conglomerate of the Lower Carboniferous, and appearing from under these are gray sandstones and shales of Upper Erian age. These contain many fossil fishes, especially of the genus *Pterichthys*, also fossil ferns.

20. Metapedia. The rocks exposed about here are principally slates and shales with marked slaty structure, of Upper Silurian age. Fine exposures in cuttings. Fossils occur in calcareous bands. Passing Lake Metapedia, at the head of the river, the railway cuts through some limestone, probably of Hudson River age, and then passes into Lower Silurian, and probably, in part, Cambrian, shales, sandstones, and conglomerates, of which the greater part are referred to the Quebec group. At the mouth of Metapedia River leda clay and saxicava sand, with fossils.

Ms.	Intercolonial Railway—*Con.*		Intercolonial Railway—*Con.*	
			Ms.	St. John to Moncton.
441	Cedar Hall.	5–7. Silurian.	0	St. John, N.B.[55]	2. Lower Cambrian.
448	Sayabec.	"	3	Coldbrook.	"
458	Tartague.	{ 2. Cambrian, and 4. Camb. Silurian.	9	Rothsay.	1 a. Laurentian.
469	Little Metis.[21]	"	17	Nauwigewauk.	13 a. Lower Carbonif.
477	St. Flavie.	"	22	Hampton.	"
485	St. Luce.	"	26	Passekeag.	"
495	Rimouski.	"	27	Bloomfield.	"
506	Bic.[22]	"	33	Norton.	"
515	St. Fabien.	"	39	Apohaqui.	"
525	St. Simon.	"	44	Sussex.[25]	"
534	Trois Pistoles.	"	51	Penobsquis.	"
544	Isle Verte.	"	60	Anagance.	14 a. Millstone Grit. .
555	Cacouna.	"	66	Petitcodiac.	"
561	Rivière du Loup	"	76	Salisbury.	{ Contact 14 a. Millstone Grit and 13 a. L. Carb.
567	Notre Dame.	"			
573	St. Alexandre.	"	89	Moncton.[26]	14 a. Millstone Grit.
578	St. Andre.	"			
581	St. Helene.	"		Pictou Branch.	
587	St. Pascal.	"	61	Truro, N. S.	16. Triassic.
591	St. P. de Ner.	"	70	Union.	13 a. Lower Carbonif.
596	Rivière Ouelle.	"	74	Riverdale.[27]	14 a. Millstone Grit.
602	St. Anne.	"	80	West River.	5–7. Silurian.
610	St. Roche.	"	89	Glengarry.	13 a. Lower Carb., *etc.*
613	Elgin Road.	"	96	Hopewell.	"
617	St.JeanPort Joli	"	104	N. Glasgow.[28]	14 b. and c. Coal Meas.
622	Trois Saumons.	"	112	Pictou Land'g.	14 c. Up. Coal Format'n.
625	L'Islet.	"	113	Pictou.	"
629	L'Anse à Gile.	"			
632	Cap St. Ignace.	"		Shediac Branch.	
639	St. Thomas.	"	179	Painsec Jn.N.B.[14]	14. Carboniferous.
646	St. Pierre.	"	184	Dorchester Rd.	"
649	St. François.	"	188	Shediac.	"
653	St. Valier.	"	190	Pt. du Chêne.	"
657	St. Michel.	"			
663	St. Charles Jn.	"		Windsor and Annapolis Railway, N. S.	
672	Harlaka.	"	0	Halifax.[3]	2. Lower Cambrian.
677	Levis.	"	13	Windsor Junc.[4]	"
678	Point Levis[23] (op. Quebec).[24]	"	30		Intrusive Granite & 2 Lower Cambrian.

21. Little Metis. Cuttings in slates of the Quebec group. The River St. Lawrence, here thirty miles wide, suddenly breaks upon the view after passing Metis station. Beyond this point the line follows the strike of the Quebec group all the way to Point Levis, opposite Quebec.

22. Bic. Conglomerates here specially worthy of notice and well shown in cuttings.

23. Point Levis. In cuttings on a new connecting railway, about a mile from the station, beds holding *Graptolites.*

24. The rocks on which the city of Quebec stands are believed to be of Hudson River and Utica age, and fossils (*Graptolites*) lately obtained there confirm this view. The great Champlain and St. Lawrence fault cuts the north shore of the river west of Cape Rouge, and bending round, again cuts the shore immediately south of the city, and thence follows the channel of the river between Quebec and Point Levis. The falls of Montmorenci, near Quebec, are of great beauty, and show in the gorges Utica shale resting on Laurentian gneiss, which at the "natural steps" above the falls is overlain by Trenton limestone. Half way between the city and the falls, at a mill in the village of Beauport, is a bank of boulder-clay overlain by fossiliferous sand and gravel (saxicava sand), rich in *Saxicava rugosa* and other shells. Clays with a somewhat richer fauna (upper leda clay) occur in the bank of a brook a little farther from the road to the north.

25. Sussex. Brine from the Lower Carboniferous, employed to a small extent for salt-manufacture.

26. Moncton. Between this station and Salisbury, in cuttings and gravel-pits, leda clays and saxicava sands.

27. Riverdale. The millstone grit series consists of sandstones and shales, often red, and conglomerate, associated with dark-colored beds holding fossil plants and *Naiadites*, with a few under-clays and thin seams of coal ("Acadian Geology").

28. New Glasgow. In this vicinity several important coal-mines. The productive coal area, so far as yet proved, is about nine miles long by three and a half wide, with an area of twenty-two square miles. Though thus limited in extent, the seams are extremely thick. The most important of these are

Windsor and Annapolis Railway—Continued.		New Brunswick Railway—Con. St. John to Vanceboro.	
Ms.	Continued.	Ms.	
39 Newport.	13 a. Lower Carbonif.	30 Clarendon.	Granite.
45 Windsor.²⁹	" (Windsor ser.)	33 Gaspereaux.	4. Cambro-Silurian.
47 Falmouth.³⁰	" "	36 Enniskillen.	8–12. Devonian.
52 Hantsport.	" (Horton ser)	38 Hoyt.³⁷	{ 8–12. Devonian and 13 a. Low. Carbonif.
63 Wolfville.³¹	{ 13 a. Lower Carb. and 5–7. Silurian.	42 South Branch.	14 a. Millstone Grit.
65 Port William.	16. Triassic.	46 Fredericton Jn.	"
70 Kentville.³²	16. Triassic & 14. Carb.	49 Tracy.	"
82 Berwick.	"	61 Cork.	"
87 Aylesford.	"	66 Harvey.	13 a. Lower Carbonif.
98 Wilmot.	"	72 Prince William.	4. Cambro-Silurian.
101 Middleton.	"	76 Magaguadavic.	"
107 Lawrenceton.	"	85 McAdam.	"
115 Bridgetown.³³	"	91 St. Croix.	"
121 Round Hill.	"	92 Vanceboro, Me.	"
129 Annapolis.	"	118 Danforth, "	1 b. Huronian,
		160 Lincoln, "	"
New Brunswick Railway. (Formerly European and North American.) St. John to Vanceboro.		183 Old Town, "	"
		206 Bangor, "	"
0 St. John.⁵⁵	2. L. Camb. (Acadian.)	0 St. Andrews.	14 b. Middle Carbonif.
– Carleton.³⁴	"	5 Chamcook.⁴⁹	"
4 Fairville.	1 a. Laurentian.	15 Roix Road.	5–7. Silurian.
6 South Bay.	1 a. Lauren. limestones.	17 G. S. R'y Cross.	"
8 Sutton.	1 a. Laurentian.	20 Rolling Dam.	4. Cambro-Silurian.
11 Grand Bay.	{ 13 a. L. Carbonif. & Pre-Cambrian.	24 Dumbarton.	"
15 Westfield.³⁵	1. Pre-Cambrian.	28 Watt Junc.³⁸	"
20 Nerepis.³⁶	{ 1. Pre-Cambrian and 13 a. L. Carbonif.	0 St. Stephens.⁴⁴	Granite.
22 Eagle Rock.	Granite.	5 Maxwell.	4. Cambro-Silurian.
25 Wellsford.	"	8 Moore's Mills.	"
		15 Meadows.	"
		19 Watt Junc.	

the "main seam" and "deep seam." The first has a thickness of thirty-eight feet six inches, and is capable of yielding at least twenty-four feet of coal of good quality. The "deep seam" (one hundred and sixty feet below) shows seven feet eight inches of good coal with three feet six inches of shaly coal. The coals are bituminous, and yield, as a rule, a good coke. A material known as "stellar coal," which is in reality an earthy bitumen, occurs near Stellartown, but is not at present worked. It is capable of yielding from 50 to 126 gallons per ton of oil, on distillation. The New Glasgow conglomerate seen at the road-bridge and elsewhere is a peculiar deposit locally developed in the Carboniferous, possibly nearly on the horizon of the coals. On the East River, above New Glasgow, important occurrences of iron-ore, limonite, specular iron-ore, and bedded hæmatite. These have not been worked.

29. Windsor. The Windsor series, or Lower Carboniferous limestone and gypsiferous beds, is a marine formation, holding characteristic shells and corals of the Lower Carboniferous period, and containing, in addition to the limestone, thick beds of sandstone, marl, and clay, usually red, and gypsum ("Acadian Geology").

30. Falmouth. The Horton series, or Lower Carboniferous coal measures, underlies the last, and consists of hard sandstones and shales, often calcareous, associated with conglomerate and grit, and in some places with highly-bituminous shales. It holds underclays and thin coaly seams, remains of plants, fishes, and entomostracans, and footprints of batrachians, but no strictly marine remains ("Acadian Geology").

31. Wolfville. From this point to Kentville the alluviums and marshes of the Bay of Fundy shores may be seen to the north.

32. Kentville. Though marked Triassic to Annapolis, the line of the railway runs throughout near the line of junction of this formation with Silurian, Devonian (Oriskany), and intrusive granites, which form the hills to the south. To the northward is visible the continuous ridge of the North Mountain, which intervenes between the Cornwallis and Annapolis Valley and Bay of Fundy shore. This is composed of Triassic traps, which overlie the red sandstones of the same formation. Cape Blomidon (near Wolfville) is the eastern extremity of the North Mountain. In this lofty cliff (four hundred feet) columnar basaltic trap is underlain by amygdaloid, containing numerous zeolitic minerals. The base is formed of red sandstone with gypsum veins. The cliffs bordering the coast from Cape Blomidon westward afford many zeolites in fine crystals.

33. Bridgetown. At Paradise, east of this station, fine crystals of smoky quartz derived from veins in granite.

34. Carleton. This town is, like St. John, on Lower Cambrian rocks, but the railway immediately enters an area of Pre-Cambrian, and turning round northward passes into Laurentian.

35. Westfield. Immediately beyond Westfield an outlyer of Lower Carboniferous one mile wide. Pre-Cambrian rocks then extend to Nerepis, which is on (or near) a very small Lower Carboniferous outlyer.

36. Nerepis. Beyond this station Silurian 1¼ mile, followed by granite.

Ms.	New Brunswick Railway—Con.		Ms.	Between Gibson and Woodstock.
28	Watt Junc.[38]	4. Cambro-Silurian.	0	Gibson.	14 b. Middle Carbonif.
29	Lawrence.	{ 4. Cam.-Silurian and { 8–12. Devonian.	12	Keswick.	{ 4. Cambro-Silurian & { 14 b. Middle Carbonif.
43	McAdam Jun.[38]	4. Cambro-Silurian.	20	Zealand.	4. Cambro-Silurian.
49	Vanceboro, Me.	4. Cambro-Silurian.	28	Upper Keswick.	Granite.
59	Deer Lake.	Granite.	38	Millville.	4. Cambro-Silurian.
65	Canterbury.	4. Cambro-Silurian.	47	County Line.	"
75	Benton.	Syenite.	52	Woodstock Jn.	"
83	Debec. Junc.	5–7. Silurian.	57	Newberg Junc.	5–7. Silurian.
94	Woodstock.	4. Cambro-Silurian.	61	Up. Woodstock.	4. Cambro-Silurian.
			63	Woodstock.[39]	"
83	Debec Junc.	5–7. Silurian.			**Cumberland Railway.**
86	Greenville.	"	0	Springhill Jn.[10]	14 a. Millstone Grit.
90	Houlton, Me.	"	—	" Mines.	14 b. Middle Carbonif.
94	Woodstock.[39]	4. Cambro-Silurian.	—	Southampton.	14 a. Millstone Grit.
96	Up. W'dstock[40]	"	—	Half-way Lake.	13 a. Lower Carbonif.
100	Newberg Junc.	5–7. Silurian.	32	Parsboro.	"
157	Gibson.	14 b. Middle Carbonif.			**Waterloo and Magog Railway.** Province of Quebec.
107	Hartland.	5–7. Silurian.	0	Magog.[41]	5–7. Silurian.
111	Peel.	"	3	Castle Brook.	"
117	Florenceville.	"	5	Oxford L.	"
120	Kent.	"	7	Amber Brook.	1. Pre-Cambrian.
123	Bath.	"	9	Eastman.	"
135	Kilborn.	"	11	Dillonton.	"
143	Perth.	"	17	S. Stukely.[42]	"
143	Andover.	"	23	Waterloo.	"
149	Aroostook.	"			**Prince Edward Island Railway.**[43]

Ms.		Ms.	
156	F't Fairfield, Me.	"
163	East Lyndon, "	"
168	Caribou, "	"
183	Presque Isle, "	"
149	Aroostook.	"
167	Grand Falls.	"
181	St. Leonard's.	"
198	Green River.	"
201	St. Basil.	"
207	Edmundston.	"

Prince Edward Island Railway.[43]
(198 miles in operation.)
Province—Prince Edward Island.
43 The whole of this island consists of Permo-Carboniferous and Triassic rocks, with general red color, which has also been communicated to the overlying drift and soil. The surface is rolling and generally drift-covered, so that it has so far been found impossible to separate the two formations above mentioned except quite locally. The remarkably interesting Triassic reptile *Bathygnathus borealis* was found in the excavation for a well at New London. The soil of Prince Edward Island is remarkably fertile and well cultivated.

37. Hoyt. At junction Devonian and Lower Carboniferous.
38. Watt Junction to McAdam Junction. Kames and moraines frequent, and in some places cut through by the railway.
39. Woodstock to Grand Falls. Fine examples of terraces.
40. Upper Woodstock. A blast-furnace erected here, and hæmatite ores from Jacksonton at one time smelted. Bricks manufactured from drift-clays.
41. Magog. At northern or lower end of Lake Memphremagog, a very picturesque sheet of water, much frequented as a summer resort. Orford Mountain, a dioritic intrusion to the northeast.
42. South Stukely. Numerous occurrences of copper-ore in this vicinity. The Huntington copper-mine six miles distant. The ore is chiefly chloritic slate and diorite, impregnated with copper pyrites, pyrrotite, and iron pyrites. Magnesite forms enormous beds in Bolton and neighboring townships, in association with serpentine, dolomite, etc. Chromic iron also found in serpentine. (Bolton, lot 4, range 2.)
44. St. Stephen, on New Brunswick Railway: thence granite ¼ mile, Cambro-Silurian 1½ mile, granite 1 mile, Cambro-Silurian 16 miles to Watt Junction. On Grand Southern Railway: thence granite ¼ mile. Cambro-Silurian 4½ miles to Oak Bay, then Silurian.
45. Yarmouth. Highly altered rocks, consisting of chloritic and hornblendic slates, clay slates, quartz rock, etc.
46. Metegan. From this point onward the rocks differ in appearance from those previously met with, and though colored, provisionally, on the general map of the Geological Survey as Cambrian, may be Cambro-Silurian or Silurian.
47. Bloomfield. Exposures of fossiliferous Oriskany of Bear River and Clements near here.
48. Digby. Good exposures of Triassic red sandstones and trappean rocks at Digby Gut and St. Mary's Bay. Digby Gut forms the entrance to Annapolis Basin, and is passed through by steamers, connecting with railway, for St. John.
49. Chamcook. Thence Silurian 2 miles, granite 4½ miles, Silurian 1½ miles.
50. Dyers. Cambro-Silurian 2 miles. Granite 8 miles. Near Dyers, kames may be observed.

Ms.	Western Counties Railway, N. S.
0	Yarmouth.[45]	2-4. Cambrian.
5	Hebron.	"
7	Ohio.	"
10	Greencove.	"
13	Brazil Lake.	"
16	Lake Jessie.	"
18	Norwood.	"
21	Hectanooga.	"
30	Meteghan.[46]	4. Cambro-Silurian (?)
33	Saulmerville.	"
35	Little Brook.	"
37	Church Point.	"
41	Belliveau.	"
45	Weymouth.	5-7. Silurian (?)
51	Port Gilbert.	"
53	Plympton.	"
56	North Range.	"
58	Bloomfield.[47]	"
63	Jordantown.	"
67	Digby.[48]	16. Triassic.
	St. John.	
	Halifax.	

Chatham Branch Railway, N. B.
	Halifax.[3]	
0	Chatham.	14 b. Middle Carbonif.
9	Chatham Junc.	"
	Point Levis.	

Grand Southern Railway, N. B.
0	St. Stephen.[44]	Granite.
5	Oak Bay.	4. Cambro-Silurian.
14	St. Andrew's Crossing. }	5-7. Silurian.

Ms.	Grand Southern Railway—Con.
20	Dyer's.[50]	Granite.
29	Bonny River.	5-7. Silurian.
35	St. George.[51]	1. Pre-Cambrian.
44	Pennfield.[52]	"
54	New River.	
—	Lepreaux.[53]	13 a. Lower Carbonif.
58	Lancaster.[54]	1 a. Laurentian.
67	Pr. of Wales.	"
70	Spruce Lake.	"
74	Carleton.	2. Cambrian.
82	St. John.[55]	"

Albert Railway, N. B.
0	Salisbury.	14 b. Middle Carbonif.
4	Coverdale.	"
10	Turtle Creek.	"
14	Baltimore.	"
16	Dawson.	"
17	Stony Creek.	"
20	Salem.	13 a. Lower Carbonif.
22	Weldon.[56]	"
24	Hillsboro.[57]	"
29	Albert Mines.[58]	"
31	Wilson.	"
33	Curryville.[59]	14 b. Middle Carbonif.
36	Cape.	"
38	Daniels.	13 a. Lower Carbonif.
40	Shepody.[60]	"
42	The Hill.	"
44	Riverside.	"
45	Albert.	"
48	Harvey.	14 b. Middle Carbonif.

51. St. George. About three miles north of St. George, on the Magaguadavic River, a red syenite is extensively quarried. Water-power is employed to drive the polishing machinery. The stone much resembles Aberdeen "granite," and is of very fine quality and color.

52. Pennfield. Large, broad kame, or "whaleback."

53. Lepreaux. Anthracite of an impure character occurs in Devonian beds about four miles south of station. The anthracite is very impure, but is interesting, being the only known instance in America of a Devonian coal.

54. Lancaster. Between this point and next station (Prince of Wales) line passes nearly along junction of Laurentian (to north) and Devonian. At Lancaster, kames.

55. St. John. Few points are of greater geological interest than the vicinity of St. John, where within a radius of a few miles rocks occur which have been assigned to the Laurentian, Pre-Cambrian, Cambrian, Devonian, and Lower Carboniferous formations. The city stands on hard, slaty rocks of the Acadian group, which yield Primordial fossils, in some places in considerable abundance. The Devonian rocks are well exposed on the shores of Courtney Bay, and also in the vicinity of Carleton. About a mile west of the last-named place, on the shore, are the "fern ledges," which have yielded a great number of fossil plants, with some insects and crustaceans. The Devonian rests quite unconformably on the Cambrian series, and is again overlain unconformably by the conglomerates of the Lower Carboniferous.

56. Weldon. Between this point and Hillsboro the Petitcodiac salt-marsh.

57. Hillsboro. Gypsum quarries in the Lower Carboniferous rocks.

58. Albert Mines. The mineral known as Albertite, an inspissated bitumen filling veins in the black shales of the Lower Carboniferous, was at one time extensively worked here. The mines are now closed.

59. Curryville. Gray sandstone quarries.

60. Shepody. Thence to Harvey principally salt-marsh.

61. New Glasgow. (See note No. 28, under Intercolonial Railway.)

62. French River. Lower Carboniferous in valley, hills on both sides of Silurian rocks.

63. Marshy Hope. Opposite this point, on the coast, good exposures of fossiliferous Silurian rocks of Arisaig group.

64. Antigonish. Interesting display of Lower Carboniferous rocks, including beds of limestone and gypsum in this neighborhood.

65. Cape Porcupine. On the shore of the Strait of Canso, 500 feet in height. The central mass a red syenite, against which rest slaty beds, supposed by Sir W. Dawson to be Silurian. On these, conglomerates of the Lower Carboniferous.

66. Strait of Canso Wharf. Interesting exposures of Lower Carboniferous rocks at Plaster Cove and other places on north side of Strait of Canso.

Ms.	Eastern Extension Railway, N. S.		Ms.	Eastern Extension Railway—Con.	
0	New Glasgow.[51]	14. Carboniferous.	51	Pomquet.	13 a. Lower Carbonif.
5	Glenfalloch.	"	53	Heatherton.	"
10	Merigomish.	"	56	Bayfield Road.	"
13	French River.[52]	"	57	Afton.	"
18	Piedmont.	5–7. Silur. or Cam.-Sil.	61	Tracadie.	"
22	Avondale.	"	62	Girroirs.	"
24	Barney's River.	"	66	Little Tracadie.	"
27	Marshy Hope.[53]	"	70	Harb. au Bouche	"
31	James River.	13 a. Lower Carbonif.	73	C. Porcupine.[55]	{ 13 a. Lower Carb. 5–7. Silurian and Syenite.
35	Brierly Brook.	"			
41	Antigonish.[54]	"	79	Mulgrave.	13 a. Lower Carbonif.
46	South River.	"	80	S. of Canso,	
48	Taylor's Road.	"		Wh'f.[56]	"

II. Ontario and Quebec.

List of the Geological Formations in Quebec and Ontario.[223]

20. Quaternary, 20 d. Saxicava Sand.*
    20 c. Leda Clay.†
    20 a. Boulder Clay or Till.
13. Lower Carbonif., 13 a. Bonaventure
8-12. Devonian, 12. Catskill (Ont.).‡
    "   11. Chemung and Port-age.§
    "   10. Hamilton, including Marcellus and Genesee.
    "   9. Corniferous or Upper Helderberg.
    "   8. Oriskany.

*In Central Ontario. 20 d. Algoma Sand and Artemisia Gravel.
†In Central Ontario. 20 c. Saugeen Clay; 20 b. Erie Clay.
‡In Eastern Quebec. Scaumenac beds.
§8-12. Gaspé Sandstones, in eastern part of Quebec.

5-7. Silurian, 7. Lower Helderberg.
    "   6. Salina or Onondaga.
    "   5 d. Guelph.
    "   5 c. Niagara.
    "   5 b. Clinton.
    "   5 a. Medina and Oneida.
4. Siluro-Cambrian, 4 c. Hudson River
    4 b. Utica.
    4 a. Trenton.
    3 c. Chazy.
2-3. Cambrian, 3 b. Sillery and Levis.
    "   3 a. Calciferous.
    "   2 c. Upper and Lower Potsdam.
    "   2 b. Keweenian.
    "   2 a. Animikie.
1. Eozoic or Archæan, 1 c. Huronian.
    1 b. Norian or Labrador.
    1 a. Laurentian.

	Grand Trunk Railway.				Grand Trunk Railway—Con.		
Ms.	Portland to Montreal.		Alt.	Ms.	Portland to Montreal.		Alt.
0	Portland, Me.	1 c. Huronian.	14	86	Shelburne, N. H.	1 d. Montalban.	709
5	Falmouth.	1 a. Laurentian.	51	91	Gorham.	"	798
9	Cumberland.	"	85	98	Berlin Falls.	Lake Group.	1022
11	Yarmouth.	"	96	122	Groveton Junc.	1 b. Huronion.	889
27	Danville Junc.	1 d. Montalban.	203	131	Breathes.	"	876
29	Lewiston Junc.	"	248	134	North Stratford.	"	901
36	Mechanic's Falls	"	300	142	Wenlock, Vt.	"	1151
47	South Paris.	1 a. Laurentian.	392	149	Island Pond, Vt.	1 d. Montalban.	1187
70	Bethel.	"	654	165	Boundary Line.		1351
80	Gilead.	1 d. Montalban.	716		Geology in U. S. by Prof. Hitchcock.		

Ms.	Grand Trunk Railway—*Con.* Lewiston Branch.	Alt.
29	Lewiston J., Me. 1 d. Montalban.	248
33	Taylor Brook. "	205
34	Auburn. "	148
35	Lewiston, Me. "	140

Portland to Montreal.

Ms.		Alt.
165	Norton Mills, Quebec.[100] } Granite.	
169	Dixville. 5-7. Silurian.	1127
175	Coaticooke. "	1007
180	Richby. "	819
183	Compton. "	734
186	Waterville. "	646
193	Lennoxville.[101] 1. Pre-Cambrian.	600
196	Sherbrooke.[102] "	486
203	Brompton Falls. 5-7. Silurian.	471
211	Windsor Mills. "	420
221	Richmond.[103] 1. Pre-Cambrian.	391
228	Lisgar. "	629
231	Durham.[104] 2-3. Cambrian.	609
235	Danby. "	438
243	Acton Vale.[105] "	312
249	Upton. "	204
252	St. Liboire. "	
255	Britannia Mills. 4 a. Trenton.	222
257	St. Rosalie. 4 c. Hudson River.	
262	St. Hyacinthe. "	111
269	St. Madeleine. "	119
275	St. Hilare.[106] "	86
276	Beloeil. "	63
280	St. Brazile. "	
282	St. Bruno. "	98
287	St. Hubert.[107] "	91
290	St. Lambert. 4 b. Utica.	76
297	Montreal.[210] { " (Bonaventure Station).[61]	

Ms.	Grand Trunk Railway—*Con.* Montreal, Richmond, and Quebec.[108]	Alt.
0	Point Levis[82] (op. Quebec).[84] 2-3. Cambrian.	14
7	Chaudiere Curve "	229
9	Chaudiere Junc. "	
15	Craig's Road. "	335
20	St. Agapit. "	406
28	Methot's Mills. "	444
37	Lyster. "	446
41	St. Julie. "	475
49	Somerset. "	442
55	Stanfold. "	128
64	Arthabaska. "	430
71	Warwick. "	481
79½	Kingsey. "	444
84	Danville. "	
98	Richmond. 1. Pre-Cambrian.	391
137	St. Hyacinthe. 4 c. Hudson R.	111
172	Montreal.[210] { 4 b. Utica (at Bonaventure Station).[61]	

Arthabaska and Three Rivers Branch.

Ms.		Alt.
0	Arthabaska. 2-3. Cambrian.	430
4	{ Walker's Cutting. "	
11	Bulstrode. "	
18	Aston. "	
25	St. Celestin. 5 a. Medina and Oneida.	
31	St. Gregoire. 4 c. Hudson R.	
35	Three Rivers. "	

Champlain Division.

Ms.		Alt.
0	Montreal.[210] { 4 b. Utica (at Bonaventure Station).	
7	St. Lambert. "	
12	Brosseau's. "	
20	Lacadie. "	

100. The portion of the province included between the 45th parallel and Maine boundary and the St. Lawrence, generally designated the "Eastern Townships," has given rise to more discussion and difference of opinion between geologists than any other part of the Dominion. It is naturally a region of extreme geological complexity and disturbance, and can scarcely yet be considered as fully worked out. For a work like the present it is necessary, however, at least to denote the formations on one uniform system, whatever doubt may attach to the reference of some of them. For this purpose, Dr. Selwyn has kindly allowed the use of unpublished sheets, colored according to his views.

This district is the continuation northward of the Appalachian region. One of its most salient features is the great Champlain and St. Lawrence fault, which separates the undisturbed rocks of its northwestern from the plicated beds of its southeastern part. This great fracture runs from the head of Lake Champlain to Quebec and beyond. (See Note 8, New York.)

101. Lennoxville. The Hartford Mine, from which a great quantity of copper-ore has been extracted, is situated at a distance of five miles from this station. The ore is granular iron pyrites, mixed with copper pyrites.

102. Sherbrooke. Numerous occurrences of copper-ore in this vicinity and near Lennoxville. A bed of jasper in the town of Sherbrooke.

103. Richmond. The Rockland and Melbourne slate quarries are within a few miles of this station. The slates here have been somewhat extensively worked, and are unsurpassed in quality. A few miles south of Richmond, in Melbourne, fine serpentine marbles occur.

104. Durham. The line between the Pre-Cambrian and Cambrian rocks is crossed at South Durham.

105. Acton Vale. A very productive mine of variegated and vitreous copper-ore, occurring in brecciated portions of a limestone-bed, was formerly worked here, but is now abandoned. Slate quarries also in this vicinity.

106. St. Hilaire. Beloeil Mountain, one of the remarkable igneous protrusions which penetrate the flat-lying Silurian rocks of the St. Lawrence Valley, may be visited from this point. The mountain is partly composed of augite-syenite and partly of nepheline-syenite. An excellent summer hotel on the mountain. (See Note 210 on Mount Royal, Montreal.)

Grand Trunk Railway—

Ms.	Champlain Division—*Con.*	
27	St. Johns.[109]	4 b. Utica.
33	Grande Ligne.	"
39	Stottsville.	"
44	Lacolle.	"
50	Rouse's Pt.,N.Y.	"

Montreal and Province Line.

0	Montreal.[210]	{ 4 b. Utica (at Bonaventure Station).
6½	St. Lambert.	"
12	Brosseau's.	"
14	Laprairie.	"
20	St. Constant.	4 a. Trenton.
23	St. Isidore Junc.	3 a. Calciferous.
27	St. Regis.	"
33	St. Martine.	2 c. Potsdam.
38	Howick.	"
44	Bryson's.	3 a. Calciferous.
47	Ormstown.	"
56	Huntingdon.	"
64	White's.	"
74	Ft.Covington, N. Y.	"
30	St. Remi.	4 a. Trenton.
34	St. Michel.	"
37	Hughe's.	3 a. Calciferous.
39	Johnson's.	"
44	Hemmingford.	"
47	Province Line.	"
50	Moore's J., N.Y.	2 c. Potsdam.

Central Vermont Railway.
Northern Division.

0	Montreal.[210]	
0	St. Johns.[109]	4 b. Utica.
7	Verselles.	"
10	St. Brigede.	4 c. Hudson River.
14	W. Farnham.	4 a. Trenton.
21	Angeline.	2–3. Cambrian.
29	Granby.	"
37	W. Shefford.[110]	"
43	Waterloo.	1. Pre-Cambrian.
0	Montreal.[210]	
27	St. Johns.[109]	4 b. Utica.
36	St. Alexandre.	"
42	Des Rivières.	4 c. Hudson River.
45	Stanbridge.[111]	"
52	St. Armand.[112]	2–3. Cambrian.
57	Highgate Sp'gs.	3 b. Levis Limestone.
61	E. Swanton. [Vt.	2 b. Potsdam Slate.
64	Swanton Junc.	"
70	St. Albans.	"

Ms. | Quebec and Lake St. John Railway.

0	Quebec.[24]	4 c. Hudson River.
4	Junction.	"
5	Little River.	"
8	Ancine Lorette.	"
10	St. Ambrois.	1 a. Laurentian.
14	Valcartier Sta.	"
16	Jacques Cartier.	"
17	St. Gabriel.	"
23	St. Catharines.	"
24	Lake St. Joseph	"
27	Lake Sergeant.	"
30	Bourg Louis.	"
36	St. Raymond.	"
39	Côtes Road.	"
43	River Roudeau.	"
46	Lake Simon.	"
86	Lake Edward.	"

North Shore Railway.[113]

0	Quebec.[24]	
4	Lake St. John Railway Junc.	} 4 c. Hudson River.
7	Lorette.	"
13	Belair.	"
25	Point Rouge.	4 a. Trenton.
30	St. Bazile.	4 b. Utica.
34	Portneuf.	"
38	Deschambault.	" or 4 a. Trenton.
42	Lachevrotiere.	4 a. Trenton.
45	Grondines.	"
52	Ste. Anne le Perade.	} 4 b. Utica.
57	Batiscan.	4 c. Hudson River.
64	Champlain.	"
74	Piles Branch Jn.	"
77	Three Rivers[114]	"
85	Pointe du Lac.	"
92	Yamachiche.	"
97	Louiseville.	4 b. Utica.
101	Maskinonge.	"
107	St. Barthelemi.	"
111	St. Cuthbert.	"
115	Berthier Junc.	"
123	Lanoraie.	4 c. Hudson R. or Utica.
129	La Valtrie.	4 b. Utica.
132	L'Assomption.	"
136	L'Epiphanie.	"
144	St. Henri Mascouche.	} 4 a. Trenton.
148	Terrebonne.[116]	"
154	St. Vincent de Paul.	} "
159	St. Martin Jn.	3 c. Chazy.
170	Hochelaga.	4 a. Trenton.
171	Montreal.[210]	"

107. St. Hubert. Extensive peat-bogs in this vicinity, from which a considerable quantity of peat was at one time extracted and manufactured.

108. Montreal, Richmond and Quebec. This road passes for the most part over an alluvial country, in general thickly drift covered, and little is seen of the underlying rocks, except in the neighborhood of Richmond. (See Note 103.)

109. St. Johns. Pottery-works. Rough earthen-ware articles are manufactured from clay underlying the town. The clay is marine (leda clay), twenty-two feet in thickness, and covered by one foot of soil.

North Shore Railway—Con.		Ms.	The Bay of Quinte Railway.	
Ms.	**Piles Branch.**		Deseronto.	4 a. Trenton.
0 Three Rivers.	4 c. Hudson River.		East End.	"
2 Piles Branch Jn.	"		Deseronto Junc.	"
9 St. Maurice.[116]	4 b. Utica & 4 a. Trenton.		Napanee.	"
21 Lac a la Torgue.	1 a. Laurentian.		**Northern and Northwestern Railways.**	
29 Grand Piles.[117]	"		0 Port Dover.[124]	9. Cornif. and 8. Oris-
Berthier Branch.			9 Jarvis.	" [kany.
Berthierville.	4 c. Hudson River.		12 Garnett.	"
Berthier Junc.	4 b. Utica.		14 Hagersville.	"
Quebec Central Railway.			16 Ballsville.	6. Onondaga.
0 Sherbrooke.[118]	1. Pre-Cambrian.		24 Caledonia.	"
4 Lenoxville.	"		29 Glanford.	5 d. Guelph.
10 Ascot.	"		34 Rymal.	"
19 Basin.	5-7. Silurian.		40 Hamilton.[125]	5 a. Medina and Oneida.
27 Dudswell.[119]	"		48 Burlingt'n B'ch.	"
36 Weedon.	"		51 Burlington.	"
47 Garthby.[120]	"		57 St. Ann's.	5 c. Niagara (?)
57 Coleraine.	"		59 Zimmerman.	5 a. Medina and Oneida.
67 Thetf'd Min's[121]	1. Pre-Cambrian.		66 Milton.	"
78 Broughton.[122]	"		75 Stewarton.	"
91 St. Frederic.	"		77 Georgetown Jn.	"
100 Beauce.	"		77 Georgetown.	"
105 St. Joseph.[123]	2-3. Cambrian.		79 Glenwilliam.	"
110 Scotts.	"		81 Salmonville.	"
122 St. Anselme.	"		83 Cheltenham.	"
139 Levis.	"		86 Riverdale.	"
			93 Caledon East.	"

110. Shefford. The railway here passes close to Shefford Mountain, an intrusive mass described as a granitoid trachyte. A larger mass of similar trachyte forms Brome Mountain to the south.

111. Stanbridge. Bog-iron-ore in considerable quantity in this vicinity. Formerly worked.

112. St. Armand. The limestone belt between this place and Phillipsburg affords several varieties of marble of different colors. Some of these have been quarried. A black marble occurring a mile and a half southeast of Phillipsburg is particularly worthy of note.

113. The line, for the greater part of its length, is at no great distance from the north bank of the St. Lawrence, and, owing to the depth of the drift deposits and alluvium, but little of the geological structure of the county can be seen. The outlines of the formations, as represented on the geological map of Canada, are somewhat uncertain for the same reason, and must at present be considered as approximations only.

114. Three Rivers. The railway here crosses the St. Maurice, a river important from a lumbering point of view, and having a total course of about three hundred miles. The Shawanagan Falls, on the St. Maurice, twenty-one miles distant, one hundred and sixty feet in height. The falls occur over Laurentian rocks, and are very picturesque. On the river below the falls the Potsdam sandstones may be observed to overlie the Laurentian. Extensive brick-yards at Three Rivers.

115. Terrebonne. Quarries. Chazy limestone. Stone taken to Montreal in scows, and has been extensively used in enlargement of Lachine Canal.

116. St. Maurice. Iron smelting, on a small scale, has been in operation here for one hundred and fifty years. The mineral employed is bog-iron-ore.

117. Grand Piles. Navigation by steamer on the St. Maurice from this point northward, into the heart of the Laurentian country.

118. Sherbrooke. (See Note 102 under Grand Trunk, Montreal to Portland.)

119. Dudswell. About three miles northward, yellow and gray marbles capable of receiving a good polish, and highly ornamental.

120. Garthby. Deposit yielding native antimony, antimony glance, and other minerals, five miles from Garthby, in South Ham, lot 28, range 1. Lot 22, range (north) 1, Garthby; extensive deposit of iron and copper pyrites.

121. Thetford Mines. Asbestos extensively worked. The veins occur in association with serpentine rocks, which here characterize a considerable tract of country.

122. Broughton. The Harvey Hill Copper Mine, at one time extensively worked, but at present suspended, near here. Purple copper-ore, copper glance, and copper pyrites, occur in veins cutting the strata and beds conformable with the stratification.

123. St. Joseph. On the Chaudiere River. Gold occurs in placer deposits in numerous localities in this vicinity. These deposits have been worked to some extent, but are as yet imperfectly developed, as the auriferous alluviums are known to extend over an area of ten thousand square miles. The Kilgour nugget, found on the Gilbert River, weighed 51½ ounces. A handsome brecciated marble found on the Rivière Guilliaume near here.

124. Port Dover. Corniferous limestones, with pores of corals frequently filled with petroleum. Eponites occur in limestones on the lake shore.

125. Hamilton. A band of sandstone known as the "gray band," and referable to the Medina formation, is quarried here and used in building.

Northern and Northwestern Railways—

Ms.	*Continued.*	
96	Centreville.	4 c. Hudson River.
99	Palgrave.	"
105	Tottenham.	"
110	Beeton.	"
114	Thompsonville.	4 b. Utica.
116	Alliston.	"
120	Everitt.	"
123	Tioga.	4 a. Trenton.
126	Lisle.	"
129	Glencairn.	"
151	Collingwood.[126]	"
135	Allandale.	"
—	Barrie.	"

Beeton and Barrie Branch.

0	Beeton.	
—	Beeton Junc.	
9	Cookstown.	4 b. Utica.
14	Thornton.	4 a. Trenton.
19	Victoria.	"
25	Allandale.	"
—	Barrie.	"

North Simcoe Branch.

0	Allandale.	4 a. Trenton.
5	Colwell.	"
13	Minesing.	"
16	Hendrie.	"
19	Phelpston.	"
24	Elmvale.	"
26	Saurin.	"
30	Wyevale.	"
39	Penetang.	"

Allendale to Muskoka Wharf.

63	Allandale.	4 a. Trenton.
64	Barrie.	"
70	Gowan.	"
74	Oro.	"
78	Hawkstone.	"
87	Orillia.	"
90	Atherly.	"
95	Longford.	1 a. Laurentian.
100	Washago.	"
103	Severn.	"
109	Lethbridge.	"
115	Gravenhurst.	"
116	Muskoka Wharf	"

Passumpsic Railway.

Ms.	Quebec to Newport.	
	Quebec.	
	Montreal.	
	(S. E. R'y.)	
0	Sherbrooke.[102]	1. Pre-Cambrian.
3	Lenoxville.	"
8	Capleton.	1. Pre-Camb. & 2-8. Sil.
12	North Hatley.	"
19	Massawippi.	5-7. Silurian.
21	Ayer's Flats.	"
27	Libby Mills.	"
30	Smith's Mills.	"
34	Stanstead Jn[127]	Granite.
40	Newport, Vt.	5-7. Silurian.

South Eastern Railway.
Main Line.—Montreal to Richford, Vt.

0	Montreal.[210]	
0	Longueuil.	4 b. Utica.
2	St. Lambert.	"
12	Chambly Basin.	4 c. Hudson River.
13	Chamb. Canton.	"
14	Richelieu.	"
19	Marieville.	"
22	St. Angele.	"
26	St. Brigide.	"
32	Farnham.	4 a. Trenton.
37	Farndon.	2-3. Cambrian.
39	Brigham.	"
42	East Farnham.	"
45	Cowansville.	"
47	Sweetsburg.	"
50	West Brome.	1. Pre-Cambrian.
55	Sutton Junc.	"
58	Sutton.	"
63	Ambercorn.	"
66	Richford, Vt.	1 b. Huronian.

Northern Division.

0	Sorel.	4 c. Hudson River.
6	St. Robert.	"
10	Yamaska.	"
14	St. David.	"
21	St. Guillaume.	"
27	Boulogne.	"
32	St. Germain.	2-3. Cambrian.
36	Drummondville.	"
45	Wickham.	"
54	Acton.[105]	"

126. Collingwood. The Utica shales may here be observed to overlap the Trenton. These shales were at one time distilled here for oil.

127. Stanstead Junction. A considerable area of granite here, surrounded by dikes of the same material which penetrate the calcareous strata. The granite is excellent for building purposes.

128. Brome. About four miles southwest, iron-ores (specular schists) at one time worked. (See Note 110 on Brome Mountain, under Central Vermont Railway, Shefford.)

129. Sutton. Similar iron-slates to that above described in a number of places near here.

130. Abbotsford. Yamaska Mountain to the southeast, an intrusive mass about three miles in diameter, is for the most part a micaceous trachyte rock. The southeastern portion is, however, a diorite.

131. Rougemont. The intrusive mass forming the mountain of Rougemont is chiefly composed of olivine-diabase. This is one of a group of similar intrusions of which Mount Royal and Beloeil Mountain may be taken as typical.

South Eastern Railway— Northern Division—Con.		Grand Trunk Railway. Montreal to Toronto and Detroit.		
Ms.		Ms.		Alt.
60 Roxton Falls.	2–3. Cambrian.	0 Montreal.[210]	4 a. Trenton, 14 m.	61
67 South Roxton.	"	8 Lachine Jun.	"	
71 Savage's Mills.	"	14 Pointe Claire[132]	4 a. Black River.	109
77 Warden.	1. Pre-Cambrian.	21 Ste. Anne.[133]	2 b. Potsd. & Calcif.	124
80 Waterloo.	"	24 Vaudreuil.[134]	2 b. Potsdam, 12 m.	92
84 Foster.		31 St. Dominique.	"	
88 Knowlton.	"	37 Coteau Land'g.	3 a. Calc. 3 c. Chazy.	161
92 Brome Cent.[128]	"	48 Bainsville.	3 c. Chazy, 33 miles.	
96 Sutton Junc.[129]	"	54 Lancaster,Ont[135]	3 a. Calciferous.	165
		59 Summertown.	3 a. Calcif. & 3 c. Chazy.	
Champlain Division.		67 Cornwall.	3 a. Calciferous, 5 m.[162]	
0 Stanbridge.	2–3. Cambrian.	72 Mille Roches[136]	4 a. Trenton, 2 miles.	
2 Bedford.	"	77 Dickinson.	3 c. Chazy, 30 miles.	
15 Mystic.	"	81 Farran's Point.	"	242
14 Farnham.	4 a. Trenton.	92 Morrisburg.	"	
20 L'Ange Gardien.	4 a. Trenton and 4 c. Hudson River.	99 Iroquois.	3 c Chazy.	243
— Papineau.	"	104 Edwardsburg.	3 a. Calciferous.	277
26 Abbottsford.[130]	"	112 Prescott Jun.	"	303
31 St. Pie.	"	112 Prescott Jun.	3 a. Calciferous, 45 m.[303]	
39 St. Hyacinthe.	4 c. Hudson River.	164 Ottawa.[216]	3 c. Chazy, 7 miles.	
41 St. Rosalie Jn.	"	115 Gladstone.	3 a. Calciferous.	
48 St. Simon.	"	120 Maitland.	"	
53 St. Hugues.	"	125 Brockville.[137]	2 b. Potsdam.	281
61 St. Guillaume.	"	129 Lyn.[138]	"	286
		138 Mallorytown.	1 a. Laurentian.	236
St. Cesaire Branch.		147 Landsdowne.	" 34 m.	234
0 St. Cesaire.	4 c. Hudson River.	155 Gananoque.[139]	"	261
4 Rougemont.[131]	"	162 Ballantyne's.	"	361
8 Marieville.	"	169 Rideau.	3 a. Calciferous.	303
		172 Kingston.[140]	4 a. Black River.	274
St. Lambert to Longueil.		180 Collins' Bay.	4 a. Trenton, 114 miles.	
0 St. Lambert.	4 b. Utica.	194 Fredericksb'rg.	"	
2 G. T. Crossing.	"	198 Napanee.	"	
6 Longueil.	"	213 Shannonville.	"	
		223 Belleville.	"	286
Central Ontario Railway.		232 Trenton.	"	265
Trenton Junc.	4 a. Trenton.	241 Brighton.	"	304
Trenton.	"	249 Colborne.	"	322
6 Carrying Place.	"	256 Grafton.	"	
11 Consecon.	"	264 Cobourg.	"	297
16 Hillier.	"	270 Port Hope.	"	287
18 Four Corners.	"	279 Newtonville.	"	294
21 Wellington.	"	286 Newcastle.	"	296
25 Stinson's Creek.	"	290 Bowmanville[141]	4 b. Utica, 24 m.	263
28 Bloomfield.	"	294 Saxony.	"	330
32 Picton.	"	299 Oshawa.	"	333

132. Pointe Claire. Black River limestones in quarry near station. Highly fossiliferous. Much of the stone for the piers of the Victoria Bridge was quarried here.

133. St. Anne. The west point of the island of Montreal is composed of Potsdam sandstone, which is seen in the immediate vicinity of the station. Just east of this a belt of calciferous occurs, and here yields some characteristic fossils. *Scolithus Canadensis* may be found in the Potsdam. The Potsdam forms an anticlinal, and underlies the county for about eight miles westward, when it is followed by a second belt of Calciferous. On the opposite side of Lac St. Louis, at Beauharnois, six miles from St. Anne, *Protichnites* in sandstone quarries.

134. Vaudreuil. In the seigniory of Vaudreuil bog-iron-ores occur in several places, particularly at Côte St. Charles.

135. Lancaster. From this point to Cornwall the railway nearly follows the line of junction of the Calciferous and Chazy formations.

136. Mille Roches. Quarries in Trenton limestone affording good building-stone. Some beds, when polished, resemble black marble.

137. Brockville. Cliffs on the river below Brockville show good sections of the Potsdam beds, and on the river, two and a half miles above that place, an outlyer of this formation occurs, the basal conglomerate of which may be seen resting on the Laurentian. In cutting of Brockville and Ottawa

Grand Trunk Railway— Ms.	Montreal to Toronto and Detroit—*Con.*		
308 Whitby.	4 b. Utica.	266
310 Pickering.	"	287
316 Port Union.	4 c.Hudson Riv,44m.	266
324 Scarboro Jun.	"	546
333 TORONTO.	"	254
341 Weston.	"	426
354 Brampton.	5 a. Medina, 11 m.	713
362 GEORGETOWN.	"	847
365 Limehouse.[142]	5 c. Niagara.	1057
368 Acton West[143]	"	1159
374 Rockwood.[144]	"	1183
381 GUELPH.[145]	5 d. Guelph.	1066
386 Balmoral.	"	1085
391 Breslau.	"	1095
396 Berlin.	6. Onondaga, 14 m.[1101]	
403 Doon.	5 a. Guelph.	
408 Galt.[159]	"	660
402 Petersburg.	6. Onondaga.	1211
405 Baden.	7 & 8. Corn.16 m. & Oris-	
421 STRATFORD.	" " [kany.[1157]	
421 STRATFORD.	" " 33 m.	1190
432 St. Mary's.	" "	1083
444 Thorndale.	" "	936
454 LONDON.	" "	815
421 STRATFORD.	" " 26 m.	1190
432 St. Mary's.	" "	1083
447 Lucan	" "	991

Grand Trunk Railway— Ms.	Montreal to Toronto and Detroit—*Con.*		
454 Ailsa Craig.	10 b. Hamilt., 23 m.	754
461 Park Hill.	"	663
470 Widder.[147]	"	682
479 Forrest.	11b. Chemung, 91 m.	712
496 Blackwell.	"	602
501 SARNIA.	"	587
502 P. Huron, Mich.	"	586
512 Ch. & L. H. Jun.	"	623
557 Milw. Junc.	"	
561 Detroit Junc.	"	594
564 DETROIT.	10 b. Hamilton, 3 m.	581
Buffalo to Goderich and Detroit.		
0 BUFFALO.	9. Corniferous, 32 m.	588
2 Fort Erie.[146]	"	
19 Port Colborne.	"	
32 Feeder.	6. Salina, 60 miles.	
38 Dunnville.	"	
59 Caledonia.	"	
68 Onondaga.	"	
76 BRANTFORD.[148]	"	706
84 Paris.[149]	"	
82 Drumbo.	"	
97 Bright.	9. Corniferous, 68 m.	
115 STRATFORD.	"	1190
128 Mitchell.	"	
139 Seaforth.[150]	"	
148 Clinton.[151]	"	
160 GODERICH.[152]	"	730

Railway, blue boulder-clay overlaid by brownish clay. An important deposit of iron pyrites in Elizabethtown, near Brockville. Acid-works.

138. Lyn. Potsdam sandstone of good quality for building. A portion of the stone for the Parliament buildings at Ottawa was quarried here.

139. Gananoque. Quarry of red syenite on island opposite this place. The stone takes a good polish and is used for monuments, etc.

140. Kingston. Clays seen in railway cuttings near Kingston probably represent the *Saugeen clays*, a series overlying the Erie clays. These rest on a glaciated limestone surface. In one of the cuttings Silurian beds, conglomeritic, etc., and possibly Calciferous in age, are seen resting on Laurentian gneiss. The Trenton (?) here affords good building-stone. Kingston is familiarly known as "The Limestone City." A considerable quantity of apatite is brought out here from points in the vicinity of the Rideau Canal.

141. Bowmanville. Quarry in upper part of Trenton limestone.

142. Limehouse. Materials derived from the Clinton formation employed in manufacture of mineral pigments.

143. Acton West. Artemisia gravels thirty miles.

144. Rockwood. Considerable display of upper part of Niagara limestone in this vicinity. From Rockwood the slope of the country westward is at about the same rate with the dip of the beds, so that on arriving at Guelph we should be nearly on the same horizon as at the first-mentioned locality.

145. Guelph. Quarries in the Guelph formation yielding building-stone (dolomite) of a superior character. Casts of fossils.

146. The portion of this province lying between the Great Lakes, and generally designated the "Ontario Peninsula," is geologically an extension of the rock-series of the adjacent portion of the State of New York, its formations showing throughout a close correspondence to those of that State. The separation marked by the lakes and Niagara River is to be regarded rather as accidental than structural. The greater part of the surface of this portion of the province is heavily covered by deposits due to the glacial period, of which local details sufficiently precise for mention in connection with the actual lines of railways are frequently wanting.

These superficial deposits only are often seen for considerable distances along the railways.

The boulder-clay, which is thick and almost universal, is overlaid by stratified clays (Erie clays), which have not been found to hold marine fossils. The clays with marine shells, which occur in the eastern extremity of Ontario and in the Ottawa Valley, are an extension of those of the Province of Quebec, elsewhere described.

The Saugeen clays have been distinguished as an upper portion of the Erie clays, and are locally unconformable on them. They are brownish and calcareous, with beds of sand. North of Lake Huron, and between Georgian Bay and the Ottawa River, the clays are overlain by the Algoma sands, of which the Artemisia gravels, covering a considerable area in the Ontario Peninsula, are possibly a local development.

147. Widder. Near the station a cutting shows forty feet of the Hamilton formation. The rocks

Ms.	Canada Southern Railway.	Alt.
0	BUFFALO.	9. Corniferous, 2 m.	573
6	Victoria.146	6. Onondaga, 58 m.	607
8	Niagara Junc.	"	608
23	Welland.	"	599
32	Perry.	"	590
47	CANFIELD.	"	621
54	Dean's.	"	637
64	Hagersville.	9. Corniferous, 64 m.	740
72	Villa Nova.	"	732
83	Windham.	"	817
99	Tilsonburg.166	"	806
111	Springfield.	"	796
124	ST. THOMAS.	10. Hamilton, 74 m.	766
128	ST. CLAIRE JN.	"	765
137	Iona.	"	745
150	Bismarck.	"	711
162	Highgate.	"	739
187	Buxton.	"	602
198	Tilbury.	"	592
204	Comber.	9. Corniferous, 48 m.	604
213	Woodslee.	"	619
227	Colchester.	"	611
235	AMHERSTBURG.	"	600
236	Grosse Isle.	"	
239	Trenton.	"	
256	DETROIT.	10. Hamilton, 10 m.	680
0	Buffalo.	9. Corniferous.	573
8	Niagara Junc.	6. Onondaga.	608
19	Black Creek.	5 d. Guelph.	568
25	Chippewa.154	5 c. Niagara.	"
28	Clifton.155	"	
29	Susp. Bridge146	"	547
35	Queenston.	5 a. Medina.	
42	Niagara.	"	

Grand Trunk Railway.

Ms.	Great Western Division.		Alt.
	SUSP. BRIDGE.		547
0	Clifton.155	5 c. Niagara, 9 m.	
9	Thorold.169	"	
11	St.Cath'rines168	5 a. Medina, 34 m.	357
27	Grimsby.156	"	287
43	HAMILTON.	"	255
43	HAMILTON.	5 a. Medina, 32 m.	255
45	Toronto Junc.	"	305
56	Bronte.	"	
69	Port Credit.	4 c. Hud. Riv., 7 miles.	
75	Mimico.	"	
82	TORONTO.	"	
43	HAMILTON.	5 b. Clinton.	255
49	Dundas.157	{ 5 c. Niagara. / 5 b. Clinton. }	517
55	Copetown.158	5 d. Guelph.	749
59	Lynden.	"	751
62	HARRISBURG.	"	734
65	St. George.	"	
67	Dumfries.	6. Onondaga.	
72	PARIS.	" Grav. ridge.	842
79	Princeton.	"	932
84	Governor's.	9. Corniferous.	967
91	Woodstock.	"	957
110	Dorchester.	"	852
119	LONDON.	"	806
129	Komoka.	10 b. Hamilton,26 m.	811
140	Longwood.	"	752
145	Appin.	11 b. Chemung, 23 m.	743
156	Newbury.	"	702
168	Thamesville.	10 b.Hamilton, 25 m.	623
183	Chatham.	"	598
198	Prairie.	9. Corniferous, 36 m.	595

are soft marly clays with thin limestone beds, and are highly fossiliferous, yielding *Spirigera mucronata*, *Atrypa reticularis*, *Spirigera concentrica*, etc.

148. Brantford. Erie clay used in manufacture of white brick. Artemisia gravels twenty miles.

149. Paris. Gypsum quarried in a number of places in this vicinity. Two beds, each four or five feet in thickness, separated by four feet of shale.

150. Seaforth. Salt-works. Brines from the Onondaga formation employed.

151. Clinton. Salt found in boring at 1,180 feet.

152. Goderich. In cliffs on the Maitland River, near Goderich. sections of Corniferous formation —sandstones and limestones—in some places fossiliferous. In 1865 brine was discovered at Goderich, in a boring made with the hope of obtaining petroleum. In the next three years several wells were sunk here and in the vicinity, the salt being derived from the Onondaga formation. In 1867 Mr. Attrill effected a boring of 1,517 feet, for the purpose of ascertaining the amount and character of the rock-salt which had been reached in some of the wells made before that date. This boring showed a total thickness of 126 feet of rock-salt in 520 feet of strata. Dr. Hunt conducted analyses of the specimens obtained, and proved that some of the beds are extremely pure. He calculates at 880,000 bushels to the acre, the yield of salt from the best white layer of ten and a half feet in thickness. The area underlaid by these salt deposits does not extend as far north as Teeswater, but appears to have a considerable extension southward. Owing to difficulties met with in sinking a shaft to the rock-salt, the beds have not yet been worked, though a large quantity of excellent salt—particularly suitable for dairy use—is manufactured from the brines.

153. Brantford. (See Note 148 under Buffalo to G. and D.) Artemisia gravels thirty-five miles.

154. Chippewa. Base of Onondaga probably in this vicinity, but whole country covered by clays.

155. Clifton. In the slope and precipice over which the Niagara Falls occur, the whole thickness of the Niagara formation is included. On Goat Island fresh-water sands are found overlying the boulder-clay, and on the Canadian side sixteen species of fresh-water and land shells have been found in similar sands. (See Notes 39 and 42 in New York.)

156. Grimsby. Quarries in Niagara limestone and sandstone.

157. Dundas. Close to station, on north side, a fine section of Niagara and Clinton. Quarries. Great thickness of Quaternary clays in this vicinity. North of the town a gravelly ridge or shore deposit 318 feet above the lake. Brick-yards.

158. Copetown. Summit of Niagara escarpment.

159. Galt. Good exposures of Guelph formation with fossils. Quarries yielding magnesian limestone suitable for building.

160. Preston. Good sections of Guelph formation. Fossils.

Grand Trunk Railway—
Great Western Division—Con.

Ms.		
207	St. Clair.	9. Corniferous.
221	Tecumseh.	" 590
229	WINDSOR.	" 582
230	DETROIT.	10 b. Hamilton, 1 m.

Great Western Railway Air Line.

0	Buffalo.	9. Corniferous, 75 m.
16	Welland.146	"
72	Simcoe.	"
81	Delhi.	
99	Corinth.	(See Loop Line, on
102	New Sarum.	page 67.)
117	St. Thomas.	
136	Baird's.	
130	Lawrence.	
145	GLENCOE.	11b. Chemung, 2 m.
224	Windsor.	
225	Detroit.	

Northern Railway of Canada.

0	TORONTO.	4 c. Hud. Riv., 24 m.247
14	Thornhill.	" 633
18	Richmond Hill.	" 847
22	King.	" 955
30	Aurora.	4 b. Utica, 14 m.
34	Newmarket.	" 772
38	Holland.	" 743
49	Gilford.	5 d. Guelph, 34 m. 753
52	Lefroy.	" 779
57	Bramley.	" 888
63	Allandale.	" 736
74	Angus.	4 b. Utica. 627
86	Stayner.	" 717
94	COLLINGWOOD.	" 590
105	Meaford.	4 c. Hud. Riv., 16m. 674

Kingston and Pembroke Railway.

0	Mississippi.	1 a. Laurentian.
10	Oso.	"
14	Sharbot Lake.	"
18	Olden.	"
22	Parham.	"
29	Hinchinbrooke.	"
31	Bedford.	"
35	Verona.	"
39	Hartington.	Birdseye & Black River.
42	Harrowsmith.	4 a. Trenton.
47	Murvale.	"
51	Glenvale.	"
59	G. T. Junction.	Birdseye & Black River.
61	Kingston.	"

Cobourg, Peterborough, and Marmora Ry.

Cobourg.	4 a. Trenton.
Baltimore.	"
Summit.	"
Harwood.	"

International Railway.

Ms.		
0	Sherbrooke.102	1. Pre-Cambrian.
	Lennoxville.	"
	Johnville.	5–7. Silurian.
	Bulwer.	"
	Birchton.	"
	Cookshire.	"
	Robinson.	"
	Gould.	"
	Scotstown.	"
	McLeod's Cross.	"
	Marsden.	"
	Springhill.	"
	Sandy Bay.	"
69	Lake Megantic.	"

Grand Trunk Railway.
Georgian Bay and Lake Erie Division.

0	Wiarton.	5 c. Niagara, 4 m.
8	Hepworth.	5 d. Guelph, 20 m.
15	Allenford.	"
20	Tara.	"
33	Chesley.	6. Onondaga.
36	Elmwood.	"
44	Hanover.	" Artem. gr'vels.
50	Neustadt.	. "
64	Harriston.	" .
69	Palmerston.	"

0	Palmerston.	6. Onondaga.
11	Mount Forrest.	5 c. Guelph.
17	Holstein.	"
22	Varney.	"
26	Durham.	"

69	Palmerston.	6. Onondaga.
78	Listowell.	9. Cornif. & 8. Oriskany.
88	Millbank.	"
91	Milverton.	"
104	Stratford Junc.	"
105	Stratford.	"
112	Travistock Jn.	"
113	Travistock.	"
127	Woodstock.	"
136	Burgessville.	"
141	Brantford Junc.	"
144	Otterville.	"
149	Can. So. Junc.	"
160	Simcoe.	"
167	Port Dover.	"

Wellington, Grey, and Bruce (G. W. Div.).

	Brantford.153		
0	Harrisburg.	5 d. Guelph.	734
6	Branchton.	"	897
12	Galt.159	"	885
16	Preston.160	"	937
19	Hespeler.	"	943
27	Guelph.	"	1079
40	Elora.161	"	1297
43	Fergus.	"	1358
49	Alma.	"	

Grand Trunk Railway—
Wellington, Grey, and Bruce (G. W. Div.)—
Continued.

Ms.			
55	Goldstone.	6. Onondaga.	1461
58	Drayton.	"	1394
62	Moorefield.	"	1331
70	Palmerston.	"	1314
75	Harriston.	"	1264
82	Clifford.	"	1234
91	Mildmay	"	1030
97	Walkerton.[162]	"	933
101	Dunkeld.	"	
104	Cargill.	"	
105	Pinkerton.	"	861
112	Paisley.	"	776
118	Turners.	"	
125	Port Elgin.	"	675
129	Southampton.	"	616

Ms.			
0	Palmerston.		1314
5	Gowanstown.	9. Cornif. & 8.Orisk.	1385
9	Listowel.	"	1263
15	Atwood.	"	1204
19	Henfryn.	"	1166
22	Ethel.	"	1174
27	Brussells.	"	1122
34	Blue Vale.	"	1079
	Wingham Junc.	"	
38	Wingham.	"	1092
44	White Church.	"	1046
50	Lucknow.	"	910
53	Ripley.	"	807
66	Kincardine.[163]	6. Onondaga.	590

Sarnia Branch (G. W. Div.).

Ms.			
0	London.	10. Hamilton.	806
10	Komoka.	"	822
20	Strathroy.	"	747
26	Kerwood.	"	
33	Watford.	11. Chemung & Port.	787
42	Wanstead.	"	702
45	Wyoming.	"	712
51	Petrolia.[164]	"	
51	Mandaumin.	"	647
61	Sarnia.	"	589
—	Point Edward.	"	
—	Port Huron, Mic'h.	"	

Ms. | Great Western Division.—Loop Line.

Ms.			
	Buffalo.	9. Corniferous.	
	Black Rock.	"	
	Fort Erie.	"	
16	Welland Junc.	6. Onondaga.	677
23	Marshville.	"	
31	Moulton.	"	
33	Diltz.	"	
40	Canfield Junc.	"	616
48	Cayuga.[165]	"	
53	Nelles' Corners.	9. Cornif. & 8. Orisk.	715
61	Jarvis.	"	701
67	Renton.	"	
72	Simcoe.	"	719
76	Nixon.	"	
81	Delhi.	"	795
88	Courtland.	"	776
92	Tilsonburg.[166]	"	785
94	Tilsonburg Jn.	"	
99	Corinth.	"	767
107	Aylmer.	10. Hamilton.	761
102	New Sarum.	"	
117	St. Thomas.	"	767
122	Payne's.	"	
126	Baird's.	"	
129	Lawrence.	"	742
134	Middlemiss.	"	
139	Ekfrid.	"	
145	Glencoe.	11. Chem. & Portage.	728

London, Huron, and Bruce Division.

Ms.		
0	London.	9. Cornif. & 8. Oriskany.
4	Hyde Park Jn.	10. Hamilton.
8	Ettrick.	9. Cornif. & 8. Oriskany.
11	Ilderton.	"
16	Brecon.	"
20	Clandeboye.	"
26	Centralia.	"
31	Exeter.	"
37	Hensall.	"
39	Kippen.	"
43	Brucefield.	"
50	Clinton.	"
57	Londesborough.	"
61	Blyth.	"
67	Belgrave.	"
73	Wingham Junc.	"
74	Wingham.	"

161. Elora. Good sections of Guelph formation in cliffs seventy-five to eighty feet high.

162. Walkerton. Good exposure of Erie and Saugeen clays at bend of river, on 28th lot of first range north of Durham road. The Saugeen clays are deposits locally developed and overlying the Erie clay.

163. Kincardine. White and yellow bricks manufactured from drift clays.

164. Petrolia. The best petroleum wells of Ontario are in this vicinity. Surface oil had been known to exist for many years, but was first obtained by boring in 1860. The oil-producing region round Petrolia has an area of about eleven square miles. The surface is level, and consists of a bluish clay to a depth of about one hundred feet. Below this the borings penetrate about three hundred and eighty feet of dolomites, shales, and marls, to the most productive stratum, which is reached at a depth of four hundred and eighty feet. The borings at first produced flowing wells, but pumping is now necessary. Most of the oil is refined in London, Ont. It is supposed to originate in the Corniferous formation.

165. Cayuga. Extensive gypsum deposits about three miles from the town. The bed worked is about five feet in thickness.

166. Tilsonburg. Petroleum has been obtained in this vicinity.

167. Brantford. Erie clay used in manufacture of white brick. Artemisia gravel thirty-five miles.

Great Western Division.

Ms. | Brantford, Norfolk and Port Burwell R'y.

Ms.	Station	Formation	Elev.
	Harrisburg.		734
0	Brantford.[167]	6. Onondaga.	659
5	Mt. Pleasant.	"	810
7	Mt. Vernon.	"	839
10	Burford.	"	844
14	Harley.	9. Cornif. & 8. Orisk.	837
16	Hatchley.	"	
21	Norwich.	"	844
22	G.B.&L.E.Cross.	" .	
25	Middletown line,	"	
27	Springford.	"	822
32	Can.S.Ry.Cross.	"	797
34	Tilsonburg.[166]	"	785
	Tilsonburg Jun.	"	

Welland Division.
Connecting Lakes Erie and Ontario.
Port Dalhousie to Port Colborne.

Ms.	Station	Formation	Elev.
	Toronto, G. T. R.		255
	Hamilton.		255
0	Port Dalhousie.	5 a. Medina and Oneida.	
3	St.Cath'rines[168]	"	275
5	Merritton.	5 c. Niagara.	
8	Thorold.[169]	"	553
10	Allanburgh.	5 d. Guelph.	592
11	Allanburgh Jn.	"	
13	Port Robinson.	6. Onondaga.	589
17	Welland.	"	602
20	Welland Junc.	"	
24	Humberstone.	"	
25	Pt. Colborne.	9. Cornif. & 8. Orisk.	586
	Buffalo.	"	

Canada Atlantic Railway.

Ms.	Station	Formation	Elev.
0	Montreal.[210]		
38	Coteau.	3 c. Chazy.	161
42	St. Plycarpe.	"	
53	Glen Robertson.	"	
61	Alexandria, Ont.	4 a. Trenton.	
68	Kenyon.	"	
72	Maxville.	"	
70	Roxboro Grav. P	"	
87	Casselman.	"	
94	South Indian.	4 c. Hudson River.	
105	Eastman's Sp'gs	4 b. Utica.	
116	Ottawa.[216]	"	
	Chaudiere Falls	4 a. Trenton.	

Grand Trunk Railway.
Midland Division.

Ms.	Station	Formation	Elev.
0	Toronto.		
	(Union Station).	4 c. Hudson River.	255
1	Don.	"	253
9	Scarboro Junc.	"	547
14	Agincourt.	"	569

Grand Trunk Railway—

Ms. | Midland Division—Con.

Ms.	Station	Formation	Elev.
17	Millikens.	4 c. Hudson River.	551
20	Unionville.	"	577
23	Markham.	"	640
29	Stouffville.	4 b. Utica.	592
36	Ballantrae.	4 b. Utica.	
38	Vivian.	"	
42	Mt. Albert.	4 a. Trenton.	
49	Ravenshoe.	"	
54	Sutton.	"	
57	Jackson Point.	"	
34	Goodwood.	4 b. Utica.	1090
41	Uxbridge.	4 a. Trenton.	577
45	Marsh Hill.	"	
49	Wick.	"	856
50	Blackwater.	"	
53	Sunderland.	"	851
59	Cannington.	"	846
63	Woodville.	"	896
65	Lorneville Junc.	"	881
67	Argyle.	"	860
70	Eldon.	"	870
73	Portage Road.	"	911
75	Kirkfield.	"	892
78	Victoria Road.	"	837
84	Corson'sCross'g.	"	
87	Coboconk.	"	847
	Port Hope Junc.	"	
0	Port Hope.	"	
5	Quay's.	"	481
8	Perrytown.	"	652
9	Garden Hill.	"	
14	Summit.	"	910
18	Millbrook.	"	772
23	Fraserville.	"	
31	Peterborough.	"	650
24	Bethany.	"	
26	Brunswick.	"	
28	Franklin.	"	
45	Omemee.	"	
49	Reaboro.	"	
56	Lindsay.	"	865
62	Mariposa.	"	884
68	Manilla Junc.	"	955
75	Blackwater.	"	851
77	Sunderland.	"	846
83	Cannington.	"	
87	Woodville.	"	896
62	Cambray.	"	926
73	Grass Hill.	"	
65	Lorneville Junc.	"	881
73	Beaverton.	"	763
77	Gamebridge.	"	797
81	Brechin.	"	757

168. St. Catherines. Brines obtained in artesian wells here, but too impure for manufacture of salt. Mineral water.

169. Thorold. Good section of Clinton and Niagara in cutting of Welland Canal. Fossils. A band of argillacious limestone eight feet thick, in the Niagara, yields an excellent cement.

170. Madoc. Mines of magnetic iron-ore. A blast-furnace was at one time in operation in Madoc Village, but the ore is now exported. This is the typical region of the Hastings series of the Lauren-

Grand Trunk Railway—
Midland Division—*Con.*

Ms.			
84	Schepeler.	4 a. Trenton.	
88	Uptergrove.	"	
91	Atherly.	"	
93	Couchiching.	"	
94	Orillia.	"	
98	Silver Creek.	"	
102	Uhthoff.	"	
105	Foxmead.	"	
106	Alma.	"	
109	Coldwater.	"	
112	Fesserton.	"	
114	Waubaushene.	"	
116	Sturgeon Bay.	"	
120	Victoria Harbor.	"	
124	Old Fort.	"	
128	Midland.	"	

Peterborough and Lakefield Branch.

Ms.			
0	Pt. Hope.	4 a. Trenton.	
5	Quay's.	"	481
8	Perrytown.	"	652
9	Garden Hill.	"	
14	Summit.	"	910
18	Millbrook.	"	772
23	Fraserville.	"	
31	Peterborough.	"	650
33	Auburn Mills.	"	
35	Nassau Mills.	"	
40	Lakefield.	"	

Belleville Branch.

Ms.			
	Montreal. [210]		
0	Belleville.	4 a. Trenton.	286
4	Corbyville.	"	
9	Foxboro.	"	
13	Holloway.	"	
15	N. Hastings Jn.	"	516
20	Stirling.	"	415
27	Hoards.	"	
33	Cambellford.	"	507
44	Hastings.	"	635
50	Birdsall's.	"	
53	Blezard's.	"	
57	Keene.	"	
66	Peterborough.	"	650

Madoc Branch.

Ms.			
0	Belleville.	4 a. Trenton.	286
4	Corbyville.	"	
9	Foxboro.	"	
13	Holloway.	"	
15	N. Hastings.	"	516
17	W. Huntingdon.	"	
20	Ivanhoe.	"	
24	Crookston.	"	
27	Moira Lake.	" (Lake.)	519
30	Madoc. [170]	1 a. Laurentian.	584

Whitby and Haliburton Branches.

Ms.			
0	Whitby Junc.		
1	Whitby.	4 b. Utica.	288
6	Brooklin.	"	539
10	Myrtle.	"	
13	High Point.	4 a. Trenton.	
15	Manchester.	"	
17	Prince Albert.	"	539
19	Port Perry.	"	
26	Seagrave.	"	
28	Sonya.	"	
32	Manilla.	"	956
33	Manilla Junc.	"	
38	Mariposa.	"	894
42	Ops.	"	
45	Lindsay.	"	861
52	Cameron.	"	
56	Halls.	"	
59	Fenelon Falls.	"	
64	Fells.	4 a. Birdseye & Black Riv.	
69	Retties.	1 a. Laurentian.	
78	Kinmount.	"	
80	Miles R'y Junc.	"	
88	Minden.	"	
92	Ingoldsby.	"	
94	Dysart.	"	
99	Gould's.	"	
101	Haliburton.	"	

Toronto to Lindsay, Peterboro., and Port Hope.

Ms.			
0	Toronto.	4 c. Hudson River.	254
1	Don.	"	253
10	Scarboro Junc.	"	547
15	Agincourt.	"	569
18	Milliken's.	"	651
21	Unionville.	"	577
24	Markham.	"	640
29	Stouffville.	4 b. Utica.	892
35	Goodwood.	"	1092
42	Uxbridge.	4 a. Trenton.	877
46	Marsh Hill.	"	
50	Wick.	"	956
51	Blackwater.	"	
58	Manilla Junc.	"	
63	Mariposa.	"	894
67	Ops.	"	
70	Lindsay.	"	861
76	Reaboro.	"	
80	Omemee.	"	
85	Franklin.	"	
87	Brunswick.	"	
89	Bethany.	"	
94	Peterboro.	"	650
102	Fraserville.	"	
107	Millbrook.	"	772
111	Summit.	"	
116	Garden Hill.	"	
117	Perrytown.	"	652
120	Quay's.	"	481
125	Port Hope.	"	287
	Port Hope Junc.	"	

tian of the late Mr. Vennor. The rocks consist of quartzites, conglomerates, limestones, micaceous slates, and argillites, and are considered by Dr. Hunt to represent the Lower Taconic. Dr. Hunt also states that Montalban gneisses and mica schists occur in this neighborhood.

Canadian Pacific Railway.

Ms.	Ontario Division.—Main Line.		
0	Smith's Falls Jn.	3 a. Calciferous.	
6	Pike Falls.		
12	Perth.[201]	1 a. Laurentian.	431
21	Bathurst.	"	
27	Maberly.	"	
37	Sharbot Lake Jn	"	
46	Mountain Grove.	"	
51	Arden.	"	
62	Kaladar.	"	
71	Sheffield.	"	
78	Tweed.	4 a. Tren. & 1a. Laur.[571]	
87	Ivanhoe.[202]	4 a. Trenton.	
96	Cen. Ont. Jn.[203]	"	
105	Blairton.	"	
110	Havelock.	"	
116	Norwood.	"	
126	Indian River.	"	
134	Peterboro.	"	
143	Cavanville.	"	
151	Manvers.	"	
155	Pontypool.	"	1064
167	Burketon.	"	
173	Myrtle.	"	887
182	Claremont.	4 b. Utica.	865
189	Green River.	"	
197	Agincourt.	4 c. Hudson River.	571
207	North Toronto.	"	406
211	Toronto Junc.	"	394
213	Parkdale.	"	
215	Toronto.[204]	"	255
213	Lambton.	"	412
215	Islington.	"	
219	Dixie.	"	
221	Cooksville.	"	393
224	Springfield.	5 a. Medina and Oneida.	
227	Streetsville.	"	499
228	Streetsville Jun.	"	553
231	Trafalgar.	"	
234	Hornby.	"	
239	Milton.	"	663
245	Campb'lville[205]	5 c. Niagara.	929
248	McRae's.	5 d. Guelph.	
251	Schaw.	"	

Ms.	Ontario Division.—Main Line—Con.		
258	Leslie.	5 d. Guelph.	1007
264	Galt.[159]	"	935
269	Dumfries.	6. Onondaga.	
274	Ayr.	"	965
279	Wolverton.	"	962
281	Drumbo.	"	1013
285	Blandford.	9 c. Corn. and Orisk.[972]	
288	Innerkip.	"	972
294	Woodstock.	"	947
299	Beachville.	"	
303	Ingersoll.	"	
308	Putnam.	"	
313	Harrietsville.	"	
319	Belmont.	"	
327	St. Thomas.	10. Hamilton.	

Elora Branch.

Ms.			
	Toronto.[204]		255
0	Church's Falls.	5 c. Niagara.	1260
5	Erin.	"	1295
8	Hillsburg.	5 d. Guelph.	1424
12	Garafraxa.	"	1452
17	Douglas.	"	
	Spires.	"	
25	Fergus.	"	1357
27	Elora.[206]	"	1301

Orangeville Branch.

Ms.			
	Toronto.[204]		255
0	Streetsville.	5 a. Med. and Oneid.	499
1	Streetsville Jun.	"	553
3	Meadowvale.	"	566
5	Churchville.	"	
8	Brampton.	"	724
13	Edmonton.	"	
17	Campb'l's Cross.	"	
18	Cheltenham.	"	
21	Riverdale.	"	
25	Forks of Credit.[207]	"	1066
28	Church's Falls.	5 c. Niagara.	1260
31	Alton.	"	
33	Melville Junc.	"	
36	Orangeville.[208]	5 b. Clin. & 5 c. Niag.[1358]	

201. Perth. Potsdam sandstones overlapping Laurentian near here. The peculiar tracks described as *Protichnites* and *Climactichnites* in quarries in first-named formation. Dalhousie or Cowan mines twelve miles distant. Red hematite. Laurentian.

202. Ivanhoe. To Madoc iron-mines (magnetite and hæmatite) 6½ miles by road.

203. Central Ontario Junction. Branch line to Coehill Iron Mine, about 40 miles distant. Magnetite at junction of granite and crystalline limestone in Laurentian. To Delero 7½ miles by road. Marmora gold-mines. Auriferous mispickel in quartz gangue.

204. Toronto. Pleistocene clay (Erie clay), extensively wrought for the manufacture of cream-colored brick.

205. Campbellville. Escarpment of the Niagara limestone here. The outcrop of the Clinton, which is here thirty to forty feet thick, is below it, but generally concealed by talus.

206. Elora. Good sections of Guelph formation in river cliffs.

207. Forks of Credit. Extensive quarries in Medina sandstone, producing a fine reddish freestone of excellent quality.

208. Orangeville. Artemisia gravels fifty miles.

209. Owen Sound. In cliffs along the lake shore good sections, extending from Hudson River through Medina and Clinton formations, with great mass of Niagara limestone capping the plateau. Excellent yellowish-gray stone in unlimited quantity afforded by last-mentioned formation. It has been used in construction of several lighthouses on the lake. Quarries. Fossils. Deposit of yellow ocher near the town. Sections in road-cuttings exhibit relations of Erie and overlying Saugeen clays.

| Canadian Pacific Railway—*Con.* | | |
Ms.	Owen Sound Branch.		
0	Tor'nto,Union Station.	4 c. Hudson River, 255
5	Toronto Junc.	"
8	Weston.	" 429
16	Woodbridge.	" 558
21	Kleinburg.	" 715
26	Bolton.	" 838
32	Mono Road.	5 a. Medina. 976
34	Cardwell Junc.	"
41	Charleston.	5 c. Niagara. 1367
44	Alton.	" 1298
45	Melville Junc.	"
48	Orangeville.	5 c. Nia. & 5 b. Clin. 1398
52	Orangeville Jun.	5 d. Guelph. 1616
56	Laurel.	"
60	Crombies.	"
64	Shelbourne.	" 1629
68	Melancthon.	"
72	Corbettown.	"
76	Dundalk.	" 1701
81	Proton.	" 1613
86	Flesherton.	5 c. Niagara, 6 m. 1557
92	Markdale.	5 d. Guelph. 1359
98	Berkeley.	" 1329
102	Williamsford.	" 1212
106	Arnott.	"
109	Chatsworth.	5 c. Niagara, 13 m. 944
114	Rockford.	" 912
118	St.Vincent'sR'd.	"
122	Owen Sound.209	" 556

Teeswater Branch.

	Toronto.4	255
0	Orangeville.	5 b. Clin.,& 5 c.Ni., Artem. grav., 45 m. 1398
4	Orangeville Jn.	5 d. Guelph. 1616
7	Amaranth.	" 1546
10	Waldemar.	" 1495
12	Luther.	" 1544
23	Arthur.	" 1525
30	Kenilworth.	" 1486
38	Mt. Forrest.	" 1360
44	Pages.	6. Onondaga. 1282
48	Harriston.	" 1246
56	Fordwich.	9 c. Corn. and Oris. 1200
60	Gorrie.	" 1123
62	Wroxeter.	" 1122
69	Wingham Road.	"
74	Teeswater.	" 1024

Ms.	Perth and Smith's Falls.	
0	Smith's Falls.	3 a. Calciferous.
6	Pike Falls.	"
12	Perth.201	1 a. Laurentian. 431

Eastern Division.

Between Montreal, Ottawa, Pembroke, and Sudbury.

0	Montreal.210	4 a. Trenton.	
1	Hochelaga.	"	70
4	Mile End.	"	225
8	Sault aux Recollets.	"	
11	St. Martin.	3 c. Chazy.	
12	St. Martin Junc.	"	
17	Ste. Rose.	3 a. Calciferous.	85
19	Ste. Therese.	"	
27	St. Augustin.	"	227
32	Ste. Scholastique	"	238
37	St. Hermas.	"	257
43	Lachute.211	"	225
48	St. Philippe.	"	262
57	Grenville.	3 c. Chazy.	210
59	Calumet.	3 a. Calciferous.	147
64	Pointe au Chene.	1 a. Laurentian.	188
74	Montebello.	"	172
78	Papineauv'le212	"	155
83	N. Nation Mills.	"	
90	Thurso.	2 b. Potsdam.	186
93	Rockland.	1 a. Laurentian.	
99	Buckingham213	"	183
103	L'Ange Gardien.	"	
109	E.Templeton214	"	155
114	Gatineau.	"	175
118	Hull.215	4 a. Trenton.	185
120	Ottawa, Ont.216	"	
122	Skeads.217	3 c.	
125	Britannia.	"	
129	Bell's Corners.	"	
135	Stittsville.	"	
139	Cleary's.	"	
144	Ashton.	"	
146	Appleton.	3 a. Calciferous.	
149	Carleton Junc.	"	
155	Almonte.	"	
159	Snedden's.	3 c. Chazy.	
164	Pakenham.218	2 b. Potsdam.	
172	Arnprior.219	1 a. Laur. & 3 a. Calcif.	
175	Braeside.	1 a. Laurentian.	
178	Sand Point.	5 and 7. Silurian.	
184	Castleford.		

210. **Montreal.** The region about Montreal is one of much geological interest. The following formations are represented in the immediate vicinity of the city: Pleistocene, Lower Helderberg, Hudson River, Utica, Trenton, and Chazy. The Chazy is here about two hundred feet thick, and consists chiefly of limestone. Exposures may be seen north of the city, as on the St. Lawrence road, also at Caughnawaga, where there are extensive quarries. The Trenton is here about six hundred feet thick, and is composed of gray and blackish limestones for the most part. Good exposures, with numerous fossils, in quarries at the Mile End and at Pointe Claire. At the last-named locality, Black River beds occur. At the Reservoir, and at many points in Mount Royal Park, limestones, also of Trenton age, but differing in appearance from those of the above-mentioned localities, are well shown. The Chazy and Trenton formations of the vicinity supply most of the building-stone used in the city. The Utica shales may be seen at the upper end of St. Helen's Island and elsewhere, but owing to their soft character are usually concealed. The Lower Helderberg occurs in small outliers only, the most considerable being on St. Helen's Island, and consisting of a dolomitic breccia, which is trav-

Canadian Pacific Railway— Eastern Division—*Con.* Between Montreal, Ottawa, Pembroke, and Sud- Ms. | bury.		Eastern Division—*Con.* Between Montreal, Ottawa, Pembroke, and Sud- Ms. | bury.	
188	Russell's.	1 a. Laurentian.	319	Mattawa.	1 a. Laurentian.
191	Renfrew.	"	329	Renton.	"
199	Haley's.	"	342	Rutherglen.	"
206	Cobden.	"	345	Callander.	"
212	Snake River.	"	349	Nosbousing.	"
216	Graham's.	"	357	Thorncliff.	"
219	Government R'd	"	364	North Bay.	"
225	Pembroke.[220]	"	375	Beaucage.	"
236	Pettewawa.	"	381	Meadowside.	"
246	Chalk River.	"	388	Sturgeon Falls.	"
252	Weston.	"	399	Verner.	"
258	Bass Lake.	"	410	Veuve River.	"
265	Moorlake.	"	413	Veuve.	"
273	Mackey's.	"	420	Mark Stay.	"
277	Rockliffe.	"	428	Stinson.	"
287	Bissett.	"	432	Wahnapitae.	"
299	Deux Rivières.	"	438	Romford.	1 b. Huronian.
309	Klock.	"	444	Sudbury.[221]	"

ersed by dikes of nepheline-basalt. The Pleistocene is here divided into—1. Boulder clay ; 2. Leda clay ; 3. Saxicava sand. The city being built on these deposits, frequent opportunities of examining them are obtained in excavations for drains, cellars, etc. They are in some places highly fossiliferous, and are well shown in some of the quarries at Mile End, where they overlie glaciated surfaces of Trenton limestone. Near Côte des Neiges village, a Pleistocene beach with marine shells at an elevation of 470 feet.

Mount Royal is an intrusive mass, composed principally of diabase, but toward the west end is an important and more recent mass of nepheline-syenite, which is well seen at the "Corporation Quarry." Both the eruptive rock and the surrounding limestones are traversed by numerous dikes. (From "Sketch of Geology of Montreal and Environs," by Dr. B. J. Harrington, in "Hand-Book for the Dominion of Canada." Dawson Brothers, Montreal.)

In Peter Redpath Museum, McGill University, good local and general geological collections.

211. Lachute. The Palæozoic rocks here form a narrow belt of flat country bordering the Ottawa River. The Laurentian highlands may be seen to the north of this part of the railway line, and gradually approach the river.

212. Papineauville. Côte St. Pierre, one of the best localities for *Eozoon*, is reached from this station. Twelve miles by stage to St. André, thence three miles to Côte St. Pierre.

213. Buckingham is the chief point of shipment on the railway of the apatite mined at numerous places within a radius of twenty to thirty miles. Large quantities of apatite may frequently be seen piled here. Extensive deposits of plumbago near Buckingham are not at present worked.

214. East Templeton. Also an important point of shipment of apatite.

215. Hull. Within a few miles of Hull is an important deposit of magnetic iron-ore, which has been somewhat extensively mined and is exported. Also hydraulic limestone. (See note on Ottawa.)

216. Ottawa. The Laurentides, but a few miles distant, belong to the lower and middle divisions of Sir William Logan's Laurentian system. These two formations, consisting chiefly of gneisses, granites, crystalline limestones, etc., are overlain unconformably by continuous and perfectly conformable series of sedimentary strata of the Cambro-Silurian system, embracing the Potsdam (of the Ottawa and Adirondack regions), Calciferous, Chazy, Bird's Eye and Black River, Trenton, Utica, and Hudson River formations. It was in these measures that the late Mr. E. Billings made his earliest palæontological researches, and these have proved ever since, as then, to be a rich hunting-ground to the palæontologist. There are extensive and varied deposits of marine clays and sands, gravels, boulders, etc., of Pleistocene age. The Leda clay of Green's Creek, Gloucester, six or seven miles from the city, abounds in nodules holding remains of the seal, fishes, insects, shells, and plants. The total number of species representing the fossil fauna and flora of this locality does not fall far short of three hundred. Brigham's Quarries, Hull, through which the Canadian Pacific Railway runs, are undoubtedly the best Cambro-Silurian crinoid quarries in America. Deposits of magnetite, apatite, and baryta occur within a short distance of Ottawa. Both the Black River and Trenton formations yield excellent limestones for lime or building purposes, while the Chazy of Nepean afforded much of the material (sandstones) used in the erection of the Parliament buildings. A bed of hydraulic limestone occurring at the top of the Chazy has been worked and employed in the manufacture of the "Hull cement." (Note by Mr. H. M. Ami.) In Ottawa the museum and offices of the Geological Survey of Canada. Excellent collection of Canadian rocks, minerals, and fossils.

217. Skeads. Most of the sandstone used in the construction of the Parliament buildings, Ottawa, was quarried near here.

218. Pakenham. Pleistocene deposit, containing mixture of marine and fresh-water shells near Pakenham Mills, 206 feet above the sea level.

219. Arnprior. Bluish gray-banded Laurentian marble somewhat extensively quarried near here.

220. Pembroke. Excellent sections of Laurentian in railway cuttings for many miles west of this point. The rocks shown "are for the most part highly characteristic red, gray, and dark-banded gneisses ; felspathic and hornblendic, and frequently garnetiferous and micaceous. There are also some large bands of gray and white crystalline limestone ; but none of these are exposed along the line of

Canadian Pacific Railway—

Ms.	West of Sudbury Junction.	
444	Sudbury.[221]	1 b. Huronian.
455	Chelmford.	"
460	Vermilion.	"
463	Phelan's Pit.	"
478	Archer.	"
501	Pogomasing.	"
510	Spanish Forks.	1 a. Laurentian.
515	No. 23 Siding.	"
518	West Branch.	"
530	Pass Landing.	"
532	Biscotasing.	

Gap of 350 miles from Biscotasing to Port Arthur, in which no stations yet permanently located, though road for the greater part built.—Dec., 1884.

St. Eustache Branch.

0	Montreal.	4 a. Trenton.
19	Ste. Therese Jn.	3 a. Calciferous.
27	St. Eustache.	"

St. Jerome Branch.

0	Montreal.[210]	4 a. Trenton.	
1	Hochelaga.	"	
4	Mile End.	"	70
8	Sault aux Recollets.	"	225
11	St. Martin.	3 c. Chazy.	
12	St. Martin Jn.	"	
17	Ste. Rose.	3 a. Calciferous.	85
19	Ste. Therese.	"	
21	St. Lin Junc.	4 a. Trenton.	
27	St. Janvier.	3 a. Calciferous.	220

Ms.	St. Jerome Branch—Con.		
33	St. Jerome.[222]	1 c. Norian or Upper Laurentian.	311
39	New Glasgow.	"	

St. Lin Branch.

0	Montreal.[210]	4 a. Trenton.
19	Ste. Therese.	3 a. Calciferous.
21	St. Lin. Junc.	4 a. Trenton.
24	Mascouche.	"
27	Ste. Anne.	"
30	Les Plaines.	3 c. Chazy.
34	St. Lin.	3 a. Calciferous.

Aylmer Branch.

0	Aylmer.	3 c. Chazy.	222
2	Duchesne Mills.	"	
5	Belmonte.	"	
7	Hull.	4 a. Trenton.	185
9	Ottawa.	"	

Brockville Line.

0	Carleton Junc.	3 a. Calciferous.
5	Beckwith.	"
9	Franktown.	2 c. Potsdam.
15	Welsh's.	"
18	Smith's Falls.	3 a. Calciferous.
21	Story's.	"
25	Irish Creek.	"
30	Walford.	"
32	Bell's.	"
34	Jelly's.	"
36	Bellamy's.	"
39	Clark's.	"
41	Fairfield.	"
46	Brockville.	2 c. Potsdam.

the railway west of Mattawa, where it leaves the valley of the Ottawa River." (Dr. A. R. C. Selwyn, in "Descriptive Sketch of Geology, etc., of Canada.")

221. Sudbury. "After passing the Wahnapite River bridge, the Huronian rocks commence, with a series of flinty felsites or felsitic quartzites, succeeded by dark-gray quartzose conglomeritic beds; also massive crystalline diorites, red, fine-grained syenites, and a great variety of highly altered volcanic agglomerates, felspathic and dioritic." (Ibid.)

From Sudbury the Algoma Mills branch runs over Huronian rocks to the shore of the lake. The main line westward, to Port Arthur by the north shore of Lake Superior, will be in operation soon. From Sudbury it passes for about seventy miles over Huronian rocks. Thence to within about fifteen miles of the Nepigon River the Laurentian is the most widely spread formation, though intersected by belts of Huronian and with extensive granitic and dioritic intrusive masses. On both sides of the Nepigon, rocks of the Nepigon series (Cambrian) are found, and are separated by a mass of intrusive granite only from the Animikie rocks of the vicinity of Port Arthur.

222. St. Jerome. The rocks of the Norian or Upper Laurentian may be seen here, but are more typically shown at New Glasgow village, six miles distant, and the present terminus of the railway.

223. The numbers affixed to the Animikie, Keweenian, and Upper and Lower Potsdam, in the table on p. 58, are those used for convenience in this chapter, but are not intended to affirm the precise correlation of these with other formations similarly numbered in adjacent states.

III. Manitoba and North-West Territory.

Including districts of Assiniboia, Alberta, Saskatchewan, and Athabaska, to base of Rocky Mountains.

List of Geological Formations.

20. QUATERNARY.	**Alluvium.** Lake deposits of Red River Valley and Peace River, etc. **Stratified Sands and Gravels, and Moraines.** **Boulder Clay or Till.** {Upper Boulder Clay. / Interglacial Lake Deposit. / Lower Boulder Clay. / Shingle Beds.} Of Southern Alberta, etc.
19. TERTIARY.	**Miocene.** Conglomerate Sandstone and Argillite of Cypress Hills, etc.

18. CRETACEO-TERTIARY, LARAMIE.	Porcupine Hill Series. / Willow Creek Series. / St. Mary's River Series.	Of Southern Alberta	Fort Union. / Laramie.	Of Souris River, etc.	Wapite River Group.	
18. CRETACEOUS.	Fox Hill Series. / Pierre Series. / Belly River Series. / Niobrara or Benton Series.	Of Alberta	Fox Hill Ser. / Pierre Series. / Niobrara Series. / Benton Series?	Of Manitoba, etc.	Smoky River Group. / Dunvegan Group. / Ft. St. John Group.	Of Peace River.

9–12. DEVONIAN.	**Limestones of Manitoba Lake, etc.**
4. SILURO-CAMB.	**Trenton Group.** (Limestones of Winnipeg Lake, Red River Valley, etc.)
1 b. HURONIAN.	
1 a. LAURENTIAN.	

Canadian Pacific Railway.—*Con.*

Western Division.

Ms. | Winnipeg and Port Arthur Section.

Ms.			
0	Port Arthur.[224]	Animikie 2. L. Camb.	
6	Fort William.	"	602
17	Murillo.	"	944
27	Kaministiqua.	1 b. Huronian.	1010
37	Finmark.	{ 1 b. Huronian and 1 a. Laurentian.	1177
44	Buda.[225]	1 a. Laurentian.	1147
55	Nordland.	"	1550
59	Dexter.	"	
65	Linkooping.	"	1531
75	Savanne.	"	1503
86	Upsala.	1 b. Huronian.	1569
93	Carlstadt.	1 a. Laurentian.	1512
103	Bridge River.	"	1540
115	English River.	1 b. Huronian.	1514
123	Martin.	1 a. Laurentian.	1554
133	Bonheur.	"	1527
144	Falcon.	"	1504
151	Ignace.	"	1448
160	Butler.	"	1420
170	Raleigh.	1 b. Huronian.	1437
180	Taché.	"	1263
190	Brulé.	"	1352
202	Wabigoon.	"	1252
209	Barclay.	"	1248

Ms. | Winnipeg and Port Arthur Section—*Con.*

Ms.			
221	Oxdrift.	1 a. Laurentian.	1169
231	Eagle River.	"	1183
241	Vermilion Bay.	"	1216
249	Gilbert.	"	1214
256	Parrywood.[226]	"	1286
272	Hawk Lake.	"	1286
284	Beaver.	"	1183
288	Rossland.	Granite, 4 miles.	1185
297	Rat Portage.[226]	1 b. Huronian, 6 m.	1084
300	Keewatin.[227]	1 a. Laurentian.	1072
308	Ostersund.	1 a. Laurentian.	1102
313	Deception.	"	1133
320	Kalmer.	"	1214
328	Ingolf.	"	1181
	(Manitoba.)		
338	Telford.	"	1056
348	Renne.	"	1050
359	Darwin.	"	968
368	Whitemouth.	"	904
374	Shelly.	"	926
384	Monmouth.	"	876
394	Beausejour.	20. Alluvium.	811
400	Tyndall.	"	793
408	Selkirk.[228]	"	740
414	Gonor.	"	
421	Bird's Hill.	"	
428	Winnipeg Junc.	"	
429	Winnipeg.[229]	"	37

Canadian Pacific Railway—Con.				Ms. Winnipeg and Rocky Mountain Section—Con.		
**Ms.	Winnipeg and Rocky Mountain Section.**					{ 20. Glacial drift over-
0 Winnipeg.²²⁹	20. Alluvium.	737	183 Brandon.²³¹		{ lying 18. Cretaceous,	
2 Air Line Junc.	"				290 m.	1170
7 Bergen.	"		141 Kenmay.		"	1335
15 Rosser.	"	772	149 Alexander.		"	1366
29 Marquette.	"	782	158 Griswold.		"	1399
35 Reaburn.	"	781	166 Oak Lake.		"	1391
40 Poplar Point.	"	790	180 Virden.		"	1420
49 High Bluff.	"	806	197 Elkhorn.		"	1606
56 Portage la Prairie.	{ "	830	211 Fleming.		"	1760
	{		219 Moosomin.		"	1860
64 Burnside.²³⁰	"	843	226 Red Jacket.		"	1893
	{ 20. Glacial drift, prob-		235 Wapella.		"	1907
72 Bagot.	{ ably overlying Cre-		243 Burrows.		"	1924
	{ taceous.	912	249 Whitewood.		"	1939
77 McGregor.	"	937	264 Broadview.		"	1936
85 Austin.	"	981	279 Grenfell.		"	1933
93 Sidney.	"	1206	286 Summerberry.		"	1914
106 Carberry.	"	1223	294 Wolseley.		"	1926
114 Sewell.	"	1230	302 Sintaluta.		"	1960
128 Chater.	"	1165	312 Indian Head.		"	1900

224. **Port Arthur.** Good geological headquarters for examination of Nepigon, Animike, and Huronian series. Silver-mines in neighborhood and fine crystalline minerals. Attractive scenery. The formations assigned to the various stations on this line, from Port Arthur to Rat Portage, may in some cases be in error, as no geologically colored map showing the precise positions of stations is at present available. After leaving the Animike of the lake shore, the rocks are all Laurentian or Huronian, with intrusive granitic masses. Fine sections of the rocks of these series, and the dikes and veins traversing them, occur in numerous cuttings.

225. **Buda.** The reddish color of the drift deposits, characteristic of the neighborhood of Lake Superior and northeast portion of Minnesota, ends about here.

226. **Rat Portage.** On northern extremity of Lake of Woods good headquarters for excursions on lake, where Laurentian and Huronian rocks are displayed in almost continuous sections along the shores. Gold-mines. Lake extremely picturesque, with innumerable islands. Both west and east from Rat Portage, on the railway, but more particularly to east, very fine examples of perched blocks and glaciated rock surfaces. Numerous cuttings in Laurentian, Huronian, and drift deposits. From Rat Portage, in a distance of about forty miles eastward (to near Parrywood station), the succession of rocks traversed is as follows : Laurentian, Int. granite, Laurentian, Huronian, Laurentian, Huronian, Laurentian.

227. **Keewatin.** Railway twice crosses boundary between Laurentian and Huronian between Ostersund and this station. Here good opportunity of examining junction.

228. **Selkirk.** Quarries close to station in Galena limestone. Fossils.

229. **Winnipeg.** The alluvium of the Red River Valley is a deposit of a former great lake of Post-Glacial age, which Mr. Warren Upham has proposed to name Lake Agassiz. The shore lines of this body of water may still be traced, at various levels, to the east and west of the valley. The lake must have received the waters of the Saskatchewan, and had its outflow southward to the Mississippi. The alluvial deposits are of great thickness, and consist above of silty or loess-like material ; below frequently of plastic clays more or less distinctly laminated. The upper layers make excellent cream-colored brick. Alluvium completely conceals the underlying rocks in this valley ; but these are, doubtless, for the most part Silurian limestones like those of Lake Winnipeg.

230. **Burnside.** In 1874 a boring was carried out at Rat Creek, near this place, by the Geological Survey. The following section was obtained : Blue clay, 70 feet ; sand, gravel, and stones, with water, 18 feet ; white limestone (probably Devonian), 42 feet ; gray crystalline rock (Laurentian or Huronian), 77 feet. West of Burnside the country rises considerably, and this point may be assumed as the western limit. on this line, of the Red River Valley alluvium. Not far west of this the edge of the Cretaceous probably overlaps the old rocks found in the above-mentioned boring, but the whole surface is completely masked by drift deposits. (See note on Brandon.)

231. **Brandon.** From Winnipeg to Brandon, alluvium and glacial drift, the latter consisting of boulder-clay overlain by stratified sands and gravels. The western edge of the alluvial plain of the Red River Valley is indefinite on the line of the railway, which follows the wide depression of the Assiniboine. To the southeast and northwest it is marked by the escarpment of the second prairie steppe or plateau, constituting Pembina, Riding and Duck "Mountains," and the Porcupine and Basquia Hills. Sands and gravels connected with the western edge of "Lake Agassiz" may be observed in several places. The underlying rocks are completely concealed by the drift deposits, but the Cretaceous probably overlaps the Silurian and Devonian rocks of the Winnipeg basin a few miles west of Austin station. At Brandon the Assiniboine Valley itself is entered. It may be taken as typical of the wide trough-like valleys generally characterizing the rivers of the second and third prairie plateaus. Small exposures of Pierre shales (Cretaceous) in some parts of the Assiniboine Valley.

232. **Moose Jaw.** Observe the line of the Missouri Côteau in the distance, to the southwest.

233. **Mortlach.** From Brandon to Mortlach there are no exposures of the underlying rock in the vicinity of the railway, and over the second prairie plateau generally, these are seen as a rule only in the river valleys. To Mortlach, however, the whole plain is, with little doubt, based on the Pierre

Canadian Pacific Railway— Winnipeg and Rocky Mountain Section.

Ms.	*Continued.*	
324	Qu'Appelle.	{ 20. Glacial drift overlying 18. Cretaceous, 2110
332	McLean.	" 2256
341	Balgonie.	" 2164
347	Pilot Butte.	" 1993
356	Regina.	" 1862
373	Pense.*	" 1854
381	Belle Plaine.	" 1877
390	Pasquia.	" 1851
398	Moose Jaw.[232]	" 1743
406	Boharm.	" 1768
414	Caron.	" 1817
423	Mortlach.[233]	{ 20. Glacial drift overlying Ft. Union Laramie. 1935
432	Parkbeg.[259]	1958
443	Secretan.[234]	2255

* 18. Pierre Shales struck in bore-hole.

Winnipeg and Rocky Mountain Section.

Ms.	*Continued.*	
452	Chaplin.	{ 20. Alluv. overlying 18. Cretaceous. 2176
461	Ernfold.[235]	{ 20. Glacial drift overlying 18. Cretaceous. 2264
471	Morse.[235]	" 2250
480	Herbert.	" 2287
489	Rush Lake.	{ 20. Glacial drift overlying 18. Pierre shales. 2276
496	Waldec.	" 2333
510	Swift Cur'nt.[236]	{ 18. Pierre Shales, 111 miles. 2400
519	Leven.	" 2440
529	Goose Lake.	" 2441
538	Antelope.	" 2532
546	Gull Lake.[237]	" 2539
554	Cypress.	" 2632
565	Sidewood.	" 2431
575	Crane Lake.	" 2544
586	Colley.	" 2485

shales of the Cretaceous. The boulder-clay, with overlying stratified drift, and fine alluvium marking sites of former lakes or ponds, cover the entire country. At or near Mortlach the increasing elevation of the plain brings in the base of the Fort Union Laramie, but there are no exposures near the railway. No western limit is given for these beds, as their precise extent has not been determined. They do not, however, extend on the line as far as the Old Wives Lakes. They are well shown to the southeast on the Souris River, and there hold numerous seams of lignite.

234. Secretan. At Secretan the drift hills of the Missouri Côteau are well displayed. The Côteau belt, where crossed by the railway, is not so well defined as near the 49th parallel, but may be said to extend from Parkbeg station westward to a point four or five miles beyond Secretan. See Note 259.

235. Morse. Between Ernfold and Morse a second line of Côteau-like hills is crossed. The Old Wives Lakes (saline) appear to occupy an interval between this branch of the Côteau and that above described. They have evidently at one time been much more extensive, and have no outlet.

236. Swift Current. The Pierre shales (Cretaceous) are exposed on the stream a short distance north of the line, and in valleys 1½ miles northeast from station. In general the deposits of Glacial period and subsequent alluviums only are seen near the line.

237. Gull Lake. Sections of Fox Hill sandstones overlying Pierre shales in Cypress Hills, a few miles south of this station. The Cypress Hills constitute a remarkable plateau, which may be seen extending to the south of the railway for many miles east and west. It is capped by Miocene Tertiary beds, of which the most characteristic is a conglomerate [formed of well-rolled pebbles of the harder rocks of the Rocky Mountains.

238. Walsh. The dividing-line between the Pierre shales and the underlying Belly River series probably passes between Forres and Walsh stations ; but, as elsewhere in this region, the rocks are generally concealed by the later drift deposits.

239. Irvine. Half a mile south of station fine sections 'showing Pierre shales, with coaly layers near base, overlying Belly River series. Fossils.

240. Medicine Hat. Good sections of boulder-clay and drift in railway cuttings to eastward.

241. Stair. One mile southward from this station, on the banks of the Saskatchewan, lignite coal is mined in rocks of the Belly River subdivision of the Cretaceous. There are two seams, of which the lower (about five feet thick) is worked. Fine exposures of rocks all along this part of the river.

242. Langevin. In boring for water at this station, a copious flow of combustible gas has been tapped.

243. Cassels. Here also combustible gas in large quantities flows] from well. The Pierre shales must overlap the Belly River series near here, but the surface shows drift deposits only. On the river, a few miles to the south, the base of the Pierre is marked by a fine seam of coal 4' 6″ thick.

244. Bassano. Good sections showing base of Laramie and top of Pierre, four miles southwest on Bow River, where a coal-seam 4' 4″ thick occurs.

245. Crowfoot. Lignite coal 9' thick exposed on Bow River to south, and underlying Crowfoot at depth of about 100'. Shaft sunk to coal north of track, 135 feet deep.

246. Calgary. Excellent exposures of Laramie rocks along Bow River to south of line from Bassano to this point. The plain, as seen from the railway, a gently undulating drift-covered surface, showing no exposures of the underlying rocks. At bridge across the Elbow River, at Calgary, massive Laramie sandstones. Calgary is the farthest western point on this parallel to which Laurentian fragments from the northeastward have been traced. The boulders and gravel farther west appear to be entirely derived from the Rocky Mountains or of local origin.

247. Radner. For about twenty-eight miles west of Calgary the railway, following the Bow River, passes over Laramie rocks, nearly horizontal, but forming the northern extension of a wide synclinal occupied farther south by the Porcupine Hills. Between Cochrane and Radner the belt of disturbed and flexed rocks which lie along the base of the mountains, constituting the foot-hill country, is entered. Numerous fine sections of Cretaceous and Laramie in river-banks to Kananaskis.

248. Kananaskis. The Cretaceous or Laramie sandstones are here nearly flat, but appear to dip

Canadian Pacific Railway—Winnipeg and Rocky Mountain Section. Ms.	Continued.		
596 Maple Creek.	18. Pierre Shales.		2470
615 Forres.	"		2406
628 Walsh.[238]	{ 18. Belly River Series, 107 m.		2407
638 Irvine.[239]	"		2469
651 Dunmore.	"		2373
660 Medicine Hat[240]	"		2142
668 Stair.[241]	"		2403
686 Suffield.	"		2471
695 Langevin.[242]	"		2471
704 Kininvic.	"		2405
713 Tilley.	"		2438
733 Cassils.[243]	18. Pierre Shales.		2493
750 Lathom.	"		2534
757 Bassano.[244]	18. Laramie.		2563
766 Crowfoot.[245]	"		2672
776 Cluny.	"		2823
785 Gleichen.	"		2926
801 Strathmore.	"		3005
819 Langdon.	"		3265
830 Shepard.	"		3344
839 Calgary.[246]	"		3365
848 Keith.	"		3522
862 Cochrane.	"		3712
872 Radnor.[247]	{ 18. Cretaceous, and 18 Laramie.		3825
881 Morley.	"		4032
893 Kananaskis.[248]	"		4170
901 The Gap.[249]	9 & 14. Devono-Car.		4198
906 Canmore.[250]	18 Cretaceous.		4253
914 Duthil.	"		4342
919 Banff.[251]	"		4531
927 { Castle Mountain.	{ 9 and 14. Devono-Carboniferous.		4511

Winnipeg and Rocky Mountain Section. Ms.	Continued.	
938 Silver City.[252]	{ 9 and 14. Devono-Carboniferous.	4624
945 Eldon.[253]	2–4. Cambrian.	4762
955 Laggan.[254]	"	5005
962 Stephen.[255]	{ 9 & 14. Devono-Carbonif. 5296(summit).	

British Columbia boundary line.

Emerson Section.

St. Vincent.	20. Alluvium.
0 Emerson.	"
10 Dominion City.	"
18 Arnaud.	"
26 Dufrost.	"
35 Otterburne.	"
42 Niverville.	"
54 St. Norbert.	"
63 St. Boniface.	"
64 Winnipeg Junc.	"
66 Winnipeg.	"

Manitoba and Northwestern Railway of Canada.

0 { Portage la Prairie.	{ Alluvium overlying Devonian.
9 Macdonald.	"
16 Westbourne.	"
26 Woodside.	"
34 Gladstone.	"
51 Arden.	"
61 Neepawa.	Drift overlying Cretac.
66 Stony Creek.	"
78 Minnedosa.	"

below the Palæozoic limestones of the mountains, which are seen in cutting just beyond this station. Above cutting, well-marked glaciation due to former Bow Valley glacier. (The railway here enters the Rocky Mountains.) Below mouth of Kananaskis River, fine falls over Cretaceous sandstone on Bow River. The great limestone series of the mountains, characterized above as Devono-Carboniferous, is the most important constituent of the range in this part of its length. No separation, except quite locally, has yet been found possible between the Devonian and Carboniferous parts of the series.

249. The Gap. The valley beyond this point becomes quite wide, and turns to the northwest, following a belt of Cretaceous rocks.

250. Canmore. The valley here floored by the Cretaceous rocks above referred to, while limestones form the mountains on both sides. The Cretaceous is in the form of a long synclinal trough, compressed and overturned to the northeastward. Looking southeastward from this point down the valley, a section of the overturned rocks is seen in the distant hills.

251. Between Duthil and Banff, near the railway and to the north about two miles from Banff, openings have been made on anthracite coal-seams in the metamorphosed Cretaceous. Seams three to five feet. Coal of excellent quality.

252. Silver City. Castle Mountain, a remarkably bold range of Devono-Carboniferous limestone, nearly horizontal, rises immediately behind this place. Numerous discoveries of copper-ore in the vicinity.

253. Eldon. A few miles beyond Silver City the valley again turns to the northwest, following axis of anticlinal, which brings up Cambrian slates and quartzites. Mountains on both sides of valley still continue for the most part limestone.

254. Laggan. Remarkably picturesque lake, with glacier at head a few miles to the south.

255. Stephen. Near summit, between headwaters of Saskatchewan and Columbia Rivers, the general structure of the watershed range is synclinal, but complicated by minor flexures. Cambrian rocks appear a few miles down valleys both east and west of the summit. Grand peaks to north and south of valley of pass, in several cases exceeding 11,000 feet altitude. This is the only railway in North America from which actual glaciers of almost Alpine magnitude may be seen. Observe snow-field and glacier in first valley from north, west of Stephen.

256. Stonewall. Excellent exposures, in quarries, of Silurian limestones, in some beds highly fossiliferous.

257. Stone Fort. Quarries near Stone Fort and St. Andrews. Fossils.

Canadian Pacific Railway—*Con.* Pembina Mountain Section.		
Ms.		
0	Winnipeg. [229]	20. Alluvium. [737]
4	St. James.	"
18	Sa Salle.	"
30	Osborne.	"
43	Morris.	"
56	Rosenfeld.[258]	"
70	Gretna.	"
66	Plum Coulee.	"
81	Morden.	"
88	Thornhill.	"
96	Darlingford.	Pierre Shales.
102	Manitou.	"

Manitoba S. W. Colonization Railway.		
0	Winnipeg.	20. Alluvium.
7	Murray Park.	"

Manitoba S. W. Colonization Railway— *Continued.*		
Ms.		
14	Headingly.	20. Alluvium.
27	Starbuck.	"
45	Elm Creek.	"
47	Maryland.	"
51	End of Track.	"

Stonewall Section.

0	Winnipeg.	20. Alluvium.
1	Air Line Junc.	"
13	Stony Mountain.	4 c. Hudson River.
20	Stonewall.[256]	"

West Selkirk Branch.

0	Winnipeg.	20. Alluvium.
	Stone Fort.[257]	4 b. Galena Limestone.
22	W. Selkirk.	"

258. Rosenfeld. Copious flow of brine struck here in deep boring in Silurian.

259. Parkbeg. The so-called Continental moraine is represented in Dakota and the North-West Territory of Canada by the Missouri Côteau. It would appear that this and the so-called Côteau des Prairies in Minnesota and Dakota are parts of the same great feature. Their elevation is similar, and they are equally characterized by the immense profusion of erratics with which they are strewn, and by basin-like swamps and lakes. In southwestern Minnesota and eastern Dakota this elevated tract, according to Winchell, called by the earliest French explorers Côteau des Prairies, meaning highlands of the prairies, is 500 to 1,000 feet above the Minnesota River, and 1,300 to 2,000 feet above the sea. In the Côteau, then, viewed as a whole, we have a natural feature of the first magnitude, a mass of glacial *débris* and traveled blocks, with an average breadth of perhaps thirty or forty miles, and extending diagonally across the central region of the continent, from the southeastern corner of Minnesota far into northern Canada, a distance of about 800 miles. Dr. George M. Dawson, from whose writings this note is compiled, was the first to recognize the glacial origin of the Missouri Côteau. He pronounces it one of the most remarkable features of the Western plains in their northwestern extension, and as certainly the most important monument of the glacial period existing there. As to its origin, while he believes that the Côteau may possibly represent a Continental moraine, his examination of it led him to consider it as more probably due to a deposit of material from floating ice along the sloping front of the third prairie steppe. It is a question which should not be prejudged, as so many difficulties remain to be elucidated, from whatever stand-point it may be regarded. As to the similar-deposit farther south in Minnesota and Dakota, etc., T. C. Chamberlin and other geologists, who have critically studied it, are quite decided in their belief that it is a terminal moraine. The superficial deposits are to be, for geologists, the great subject of the future.   J. M.

IV. British Columbia.

List of Formations.

		COAST REGION.		INTERIOR REGION.
19. QUATERNARY.		Recent Raised Beaches. Stratified Sands, Gravels, and Clays (Marine Shells). Boulder Clay or Till.		Stratified Sands and Gravels, "White Silts" of Nechacco Basin, etc. Terrace Deposits, Moraines, Boulder Clay or Till.
20. TERTIARY.		Miocene (Volcanic). Miocene (Sedimentary, generally with Marine Shells).		Miocene (Volcanic). Miocene (Sedimentary with Lignites).

18. CRETACEOUS.		NANAIMO BASIN.	COMOX BASIN.	
	Tejon (of Cal.).	Sandst. 3,294′. Shales 960′.	Up. Cong. 820′ Up. Shales 776′ Mid. Cong. 1,100′ Mid. Shales 76′ L. Cong. 900′ L. Shales 1,000′	
	Chico (of Cal.).	1,326′ Productive	Coal Meas. 739′	
			QUEEN CHARLOTTE ISLANDS.	
	Shasta (of Cal.).	Aucella Beds of Quatsino Sd.	A. Up. Shales & Sandst. 1,500′ B. Conglomerates 2,000′ C. L. Shales & Sandst. 5,000′ D. Agglomerate 3,500′ E. L. Sandstones 1,000′	Nechacco Series. Skeena R. Sandstones with Coal. Iltasyouco Beds 10,000′; Skeena Volcanic Series; Porphyrite Series (?). Aucella Beds of Tatlayoco, Jackass Mt., and Skagit 7,000′ or more; Porphyrite Series (?).

		COAST REGION.		INTERIOR REGION.
16. TRIASSIC.		Monotis Beds and Contemporaneous Volcanic Rocks of Queen Charlotte and Northern Vancouver Islands. Volcanic Rocks of Sooke R. (?)		Monotis Beds of Northern Rocky Mts.; Red Beds of Southern Rocky Mts.; Nicola Series (Volcanic) of S. Interior Plateau. Auriferous Schists (in part?).
14. CARBONIFEROUS (possibly in part Devonian).		Crystalline and Metamorphic Rocks of Vancouver and Coast Range (largely altered Volcanic, but include Limestones, etc.).		Cache Creek Series. (Fusuline Limestone, Quartzites, Volcanic Materials, etc.)
9–12. DEVONIAN.				Limestones of Rocky Mts.
2–4. CAMBRIAN.				Basal Series of South. Rocky Mts.; also largely in Purcell and Selkirk's Ranges (Auriferous Schists in part?).
1. ARCHÆAN.		Basal Rocks of Coast Range (?).		Gneissic Rocks and Crystalline Schists of Shuswap and Okanagan Lakes and Gold Range.

Ms.	Canadian Pacific Railway.			Ms.	Canadian Pacific Railway—*Con.*	
0	Port Moody.	19. Tertiary overlain by drift.		117	North Bend.	Metamorphic rocks o: Coast Ranges. ⁴⁸¹
12	Port Hammond.	"	*¹⁷	127	Keefers.³⁰⁶	16. Triassic (?) ⁵⁵⁴
20	Whannock.	"	¹⁵	137	Fraser R.	18. Cretaceous,
30	"St. Mary Msn."	"	³²		Bridge.³⁰⁷	" Shasta Group."⁵³¹
40	Nacomin.	"	²⁴	143	Lytton.³⁰⁸	Metamorphic rocks o Coast Ranges. ⁶⁸¹
49	Harrison River.	"	⁴²			
58	Agassiz.³⁰²	18. Cretaceous overlain by drift. ⁵¹		149	Section House.	" ⁷⁶¹
				153	Section Ho.³⁰⁹	19 b. Mio.(Volcanic).⁶⁸¹
68	Ruby Creek³⁰³	Metamorphic rocks of Coast Ranges. ⁹⁶		160	Drynok.	" ⁷⁶⁵
				166	Spence's Bridge.	" ⁷⁸⁵
76	Hope.	"	²⁰⁹	177	Chinaman's Ranch.³¹⁰	13. Carboniferous.
82	Texas Lake.³⁰⁴	"	¹⁹⁵			⁸⁷'
85	Emory.	"	¹⁸²	194	Ashcroft.³¹¹	18. Cretaceous. ¹⁰⁸'
90	Yale.³⁰⁵	"	²¹⁶	206	Penny's Ranch.³¹²	18. Miocene (Vol canic). ¹²⁷'
100	Spuzzum.	"	³⁶⁶			

* Reduced levels above ordinary high water of Pacific Ocean.

301. The rocks forming the south side of Burrard Inlet, and underlying the flat or gently undu lating tract about the mouth of the Fraser, are, so far as known, Tertiary, and, at least in part, of Mio cene age. The covering of drift being, however, thick, and the region as yet but partially explored, it i difficult precisely to fix the limits of these rocks. Cretaceous rocks of the Shasta group, and possibl; of the overlying series to which the coals of Vancouver Island belong. also occur.

302. The Cretaceous rocks above referred to are supposed to cross the Fraser about here. The; are somewhat extensively developed on Harrison Lake, and hold abundance of *Aucella Piochii,* whicl may be considered as the most characteristic fossil of the Cretaceous of the mainland of British Co lumbia.

303. The metamorphic rocks of the Coast Ranges, named the "Cascade Crystalline series " in th preliminary classification, consist of a great variety of gneissic and schistose materials. Orthoclas felspars are seldom developed, and dioritic rocks are abundant. The series also includes limestones It is, with little doubt, of the same age with the similar rocks of the vicinity of Victoria, and thes are known to be Palæozoic, and probably, in part at least, Carboniferous. The series has been largel; built up of contemporaneous volcanic rocks which have since been extremely metamorphosed. Larg granitic and syenitic intrusive masses are frequent.

304. At Silver Peak, near Hope, at a height of about seven thousand feet, exceptionally rich silver ores occur. These exist in veins traversing a small outlier of the Shasta Cretaceous which occupie the summit of the mountain. Litigation has so far prevented the development of these mines.

305. At this point the line enters the Cañon of the Fraser, and the scenery becomes grand in th extreme, the river breaking through the axial portion of the Coast Range. From the mouth of the An derson River (Boston Bar) the valley becomes again comparatively wide, and the mountains retreat t a greater distance.

306. The immediate valley of the river is excavated, in this part of its course, in dark slaty o schistose rocks, which have been referred to as the "Anderson River series " in preliminary reports The age of these is uncertain, but they are very possibly Triassic. They underlie the lowest Creta ceous, and rest between it and the older crystalline rocks, and have evidently been the source of th gold which is found on this part of the Fraser. The bar and bench diggings of the Fraser were at on time very remunerative, and were the first in British Columbia to attract attention and lead to an in flux of miners. Subsequently the mines of the Cariboo country and rich gold finds in other districts drew away the mining population.

307. A trough of Shasta Cretaceous here crosses the river obliquely. It forms the hills and mount ains which rise above the valley on the east, for many miles to the southward. The rocks consist o hard, greenish sandstones or quartzites, with beds of conglomerate, and evidently represent, for th most part, the deposit of a shore-line. At Jackass Mountain, on the wagon-road, they are well shown and have yielded specimens of *Aucella Piochii* and other fossils.

308. The line here leaves the Fraser to follow the Thompson River. Immediately north of Lyt ton the Cretaceous trough above referred to—which appears in the intervening distance to be inter rupted—resumes, and characterizes the Fraser Valley for a long way to the north.

309. The Tertiary rocks of this part of the province are all provisionally classified as Miocene, an are probably of the age of the "Truckee Miocene" of the 40th Parallel Report. They consist gener ally of sandstones, shales, etc., capped by a great thickness of volcanic materials which are largel; basaltic. The sedimentary part of the formation frequently holds lignites or coals, and a number o fossil plants have been obtained from it.

310. The rocks provisionally classed as Carboniferous are, at least in great part, of that age, an hold limestones characterized by *Fusulina.* They consist, however, for the most part, of quartzite and hard shales, and contain great beds of contemporaneous volcanic matter, in association witl which serpentines occur. These rocks are well displayed on the wagon-road from Ashcroft north ward to Clinton. The serpentines, with associated conglomerates, etc., are best seen on this road be tween Hat Creek and Mundorf's.

311. The rocks in this vicinity are much altered, but those in the valley appear to belong to an iso lated Cretaceous area.

312. *General Note on Unfinished Portions of Line east of Kamloops Lake.*—The line may no (December, 1884) be said to be practically completed to Kamloops Lake, leaving, under construction, length of about one hundred and eighty miles eastward from this point to the mouth of the Kicking Horse River, on the Columbia. The lower end of Kamloops Lake lies on rocks of the Câche Creel

series, which have been characterized in a previous note ; the greater part of the lake is, however, bordered by volcanic rocks of Tertiary age. Cherry and Battle Bluffs, on opposite sides of the lakes, are believed to represent the core of an ancient Tertiary volcano. In the former considerable veins of magnetite occur. Remunerative gold placers have been worked for many years on the Tranquille River, which flows into the lake. Near the town of Kamloops the rocks of the Câche Creek series reappear and characterize the banks of the South Thompson River to the lower end of Little Shuswap Lake, though the higher portion of the plateau to the south is composed of volcanic Tertiary rocks. White silty deposits, due to the last stage of the glacial period, are cut into terraces along the banks of the river. Little and Great Shuswap Lakes, with Adam's Lake, are fjord-like bodies of water occupying deep, mountain-bordered valleys in the western portion of the Gold Range. The lakes are bordered by gneissic rocks and crystalline schists, which have been referred to collectively, in the reports of the Geological Survey, as the *Shuswap series*, and are now believed to be Archæan. These rocks probably exceed thirty-two thousand feet in thickness, and are divisible into several subordinate series. For further information on the country from the mouth of the Fraser to this point, see "Descriptive Sketch of Physical Geography, and Geology of Canada, 1884," and "Report of Progress, 1877-1878." Leaving Shuswap Lake, the line follows up the valley of Eagle Creek and traverses the Gold Range by the Eagle Pass to the west crossing of the Columbia River. Thence it crosses the Selkirk range to the east crossing of the Columbia, and follows that river up (southward) to the mouth of the Kicking Horse. This portion of British Columbia may be said to be geologically unknown, but consists, so far as ascertained, of rocks similar to those of the Shuswap Lakes, with quartzites and schists which are probably Cambrian.

V. Steamboat Routes.

I. Montreal to Quebec. Little of geological interest is to be seen on this route, the river-banks being generally low, or where higher usually showing only drift deposits. Near Quebec, sections of Cambrian and Cambro-Silurian rocks.

Quebec and Gulf Ports. Quebec to Picton, Nova Scotia, with calls at intermediate ports. A picturesque and geologically interesting route.

Quebec. (See Note 24, under Intercolonial Railway.) Soon after leaving Quebec, a fine distant view of the Montmorenci Falls. Beyond the east end of the Island of Orleans, Laurentian rocks form the north shore. At St. Paul's Bay, Little Mal Bay, and Murray Bay, small outliers of Cambro-Silurian. Beyond these the north shore is entirely Laurentian. Behind Murray Bay the mountains are particularly bold. The south shore to beyond St. Anne des Monts is composed of Cambrian rocks, which form picturesque hills near Bic.

Father Point. Pilot station. Cambrian.

Metis. Cambrian. A sea-side resort.

Beyond Matanne the Shickshock Mountains to the south. The higher portions composed of Pre-Cambrian rocks with extensive granitic intrusions. Beyond St. Anne des Monts the south shore is fringed with Cambro-Silurian rocks to Gaspé Bay.

Gaspé. Ship Head, at northern entrance to Gaspé Bay, a bold promontory. Lower Helderberg limestone. The shores of Gaspé Bay are generally characterized by Devonian rocks. Excellent sections. Fossil plants. The south point of Gaspé Bay is composed of rocks of the Bonaventure (Lower Carboniferous) series. This occupies the coast to the Baie des Chaleurs.

Percé Silurian limestones here appear below the Bonaventure, and form the remarkable pierced rock, two hundred and ninety feet high, which gives the place its name.

Baie des Chaleurs. (See notes under Intercolonial Railway.) The northern shore of the eastern part is principally composed of Silurian and Bonaventure rocks ; the southern, at Bathurst, Bonaventure formation ; eastward, to Point Miscou, Middle Carboniferous.

Miramichi Bay. Shores all Middle Carboniferous. Carboniferous rocks constitute the whole New Brunswick shore to Pictou. Prince Edward Island, Permo-Carboniferous and Triassic.

Quebec to Saguenay River.

Quebec. (See notes under Intercolonial Railway and Quebec and Gulf Port steamers.)

Murray Bay. An outlier of Cambro-Silurian rocks here occupies the coast for a distance of six miles, and runs up the Murray River for a similar distance, gradually narrowing out. The rocks are well displayed in White Point at the wharf and at Les Ecorchés on the east side of the bay. They consist of limestones and calcareous sandstones, Black River, and Trenton, and are highly fossiliferous in some places. Fossiliferous glacial clays on some parts of the beach at low tide. Ancient sea-margin terraces with marine shells to height of over 600 feet in this vicinity.

Rivière du Loup. Cambrian. Marine shells in glacial clays of beach on east side of bay at mouth of river.

Tadousac. At mouth of Saguenay River. Laurentian. Fine examples of terraces at several levels. The Saguenay River, from this point to Ha Ha Bay, is the finest example of a fjord on the eastern coast of North America, and is celebrated for its grand and gloomy scenery. It possesses all the characters of a true fjord—bold rocky shores without beaches, uniformity in width, great depth in its upper part, and comparatively shallow water at its mouth. From Tadousac to Ha Ha Bay is a distance of about sixty miles. Near this point the valley bifurcates, one branch reaching to Lake St. John—forty miles—by Chicoutimi, while the other is occupied in part by Lake Kenogami. The rocks to Ha Ha Bay and Chicoutimi are all Laurentian, and generally heavily glaciated. Near the wharf at Ha Ha Bay an intrusive mass characterized by anorthosite felspar. Round Lake St. John extensive area of Norian rocks, with overlying Cambro-Silurian, and glacial clays with marine shells. The existence of this great fjord is probably due to the greater drainage area tributary to it as compared with other rivers on the north shore, and it was probably in the first instance excavated by the river at a period of greater continental elevation than the present.

Port Mulgrave to Sydney, C. B. (Steamers connecting with Eastern Extension Railway at Port Mulgrave and running through the Bras d'Or Lakes to Sydney, C. B.)

Port Mulgrave. (See Notes 65 and 66, under Eastern Extension Railway.)

The Bras d'Or Lakes are celebrated for their picturesque scenery. They are almost altogether surrounded by a fringe, of varying width, of Lower Carboniferous rocks, behind which rise hills of Pre-Cambrian rocks. The formations met with in Cape Breton generally are, however, very varied.

Sydney. Coal-formation rocks, with the most important coal deposits of Cape Breton. The principal workings are in the Sydney main seam, averaging about six feet thick, and these already extend in some places to a considerable distance beneath the sea. Fine section on northwest side of Sydney Harbor, described by Mr. Brown as including thirty-four seams of coal and forty-one underclays with *Stigmaria*. Erect trees and *Calamites* at eighteen distinct levels. Sydney mines afford good coal for gas-making and steam purposes, yielding a strong coke.

II. Toronto or Kingston to Montreal by Steamer. This is a favorite route with tourists. After leaving Toronto, the north shore of Lake Ontario is composed of Hudson River rocks for twenty miles. Thence Utica twenty miles, Trenton one hundred miles. The rocks are generally heavily covered with drift, which often forms steep banks. Both shores, and the islands at the eastern extremity of the lake, are based on Black River limestones. The north shore is then occupied by Laurentian for about thirty miles, the river cutting through a narrow neck of these rocks, which connects the great Laurentian area to the north with that occurring in New York State. This produces the well-known scenery of the Thousand Islands. For ten miles above Brockville the rocks on the north shore, Potsdam ; south shore, Laurentian and Potsdam. Thence Calciferous on both shores twenty-five miles. Thence to Mill Roches (twenty-seven miles), north shore, Chazy ; south shore, Calciferous. Thence Calciferous on both shores, twenty-four miles. Thence to Coteau (fifteen miles), north shore, Chazy ; south shore, Calciferous. Thence, four eight miles, both shores and Grand Island, Calciferous. Thence, in twenty-six miles, Potsdam, Calciferous, Black River, Trenton, Utica, in regular succession to Montreal. (See notes on Grand Trunk Railway, which runs parallel to north shore of lake and river.)

THE RAPIDS OF THE ST. LAWRENCE.—Throughout that portion of the river characterized by rapids, the rocks are those of the Cambro-Silurian system. The Lachine Rapids occur over the outcrop of the Trenton limestone, the wide basin occupied by the river below being excavated in the softer Utica shales. With this exception, no very marked connection between the geological structure and the existence of the rapids is evident. The rapids may be said to begin below Prescott, but are unimportant till the Upper Long Sault is reached, thirty miles below that place. Four and a half miles below these are the Longue Sault Rapids, which are twelve miles in length, with a fall of forty-eight feet. Farther down, at Côteau, the rapids recommence, and are known as the Côteau Rapids. Below these is calm water for about five miles, when the Cedar Rapids, a mile and a half long, occur. After three miles of calm water are the Cascade Rapids, below which Lake St. Louis, at the mouth of the Ottawa River, is entered. The Lachine Rapids, between this lake and Montreal, are the last, with a descent of forty-five feet. Above the Lachine Rapids the descent of the river is one hundred and seventy-five feet, making the total descent, from Lake Ontario to the head of ocean navigation in the harbor of Montreal, two hundred and twenty feet. The average fall of the river is about eighteen inches to the mile, but a large part of this descent is accomplished in the various rapids. These are surmounted by vessels ascending the river by a series of canals, aggregating forty-two miles in length.

III. Routes from Sarnia, Owen Sound, Collingwood, etc., to Port Arthur (connecting there with C. P. Railway).

Two main routes are followed—one to the south of Manitoulin Islands to Sault St. Marie, the other to the north of the islands to the same point. The boats leaving the last-mentioned ports frequently take the north shore route, which, from a geological or picturesque point of view, is to be preferred.

The south shore of the Manitoulin Islands is throughout composed of Niagara limestones, with outlying patches of Guelph in some places.

After clearing Notawasaga Bay, the northeast shore of Georgian Bay is Laurentian to and at Killarney. Thence the shore of the mainland is for seventy-five miles Huronian, the off-lying islands consisting of Cambro-Silurian rocks, from the Black River series to the Niagara. The north shore is then for twenty miles Laurentian, this formation forming a narrow band with Huronian behind. Then twenty miles Huronian to Bruce Mines.

Bruce Mines. Good locality for studying the Huronian rocks. Copper-mines at one time extensively worked; at present closed. The veins traverse a mass of interstratified diorite. The ore is chiefly copper pyrites. From Bruce Mines for ten miles, north shore, Huronian; south shore, Cambro-Silurian. Thence to Lake Superior, both Sugar Island and the southwest main shore of peculiar red and spotted sandstone of Potsdam or Chazy age. Thence to Port Arthur steamers generally run far from land. The north shore is principally Laurentian and Huronian to Nipigon Bay, whence Lower Cambrian rocks characterize the shore and form all the off-lying islands to Thunder Bay.

Thunder Bay. (See Note 224, under C. P. Railway.)

IV. Victoria to Nanaimo and Comox and Northward.

Victoria. Highly altered rocks dioritic, felspathic, and micaceous, in a few places becoming almost gneissic, with interbedded black argillites and crystalline limestones. The latter in a few places hold obscure fossils, which are Palæozoic and very probably Carboniferous. Many intrusive syenitic, etc., masses; one of which characterizes both sides of Victoria Harbor at the entrance. The rocks of this vicinity may be taken as typical of those forming the axial portions of Vancouver Island, and are largely altered volcanic products. Limestone may be observed near entrance to Beacon Hill Park, and at the shore at the west end of the town. Fossils in limestone on road near east side of Esquimalt Bay. Very fine glaciated rocks everywhere along the shore. These are overlain by boulder-clay, and this again by stratified clays and sands which in some places yield marine shells. Good sections of all these deposits in shore cliffs. (See papers in "Quart. Jour. Geol. Soc.," Vol. XXXIV., p. 89, and ibid., 1881.)

From Victoria, northward along coast, similar rocks to Saanich Point, the end of which is fringed by Cretaceous.

Cowichan Harbor. South side, Cretaceous. North side, metamorphic rocks (Carboniferous?).

Maple Bay. South side, Cretaceous; north side and at wharf, similar metamorphic rocks. From Maple Bay, for eight miles, coast metamorphic, off-lying islands Cretaceous. Thence to Dodd Narrows, coast and island Cretaceous. (Productive coal measures.) Just north of Dodd Narrows, high cliffs of these rocks.

Nanaimo and Departure Bay. Productive coal measures (Cretaceous). Extensive coal-mines. Seams worked five to fifteen feet. These are true bituminous coals, yielding a good coke, and suitable for gas manufacture. From Departure Bay, for fourteen miles, the coast chiefly of metamorphic rocks like those above described. Thence to Comox, forty-two miles, Cretaceous.

Comox. An extensive coal-field, but by reason of the more accessible position of Nanaimo the mines here are not at present worked. On Texada Island, to the northeast, fine deposit of magnetic iron-ore.

N. B.—The route above described is that taken by coasting steamers. Steamers bound northward to Port Simpson and Alaska generally pass farther out near the off-lying islands. These are almost altogether composed of Cretaceous rocks, and, in consequence of their general northeastward dip, the outer tier of islands displays the higher members of the formation as here developed. The southwestern sides of the islands generally form low sandstone cliffs.

Route Northward from abreast Comox to Port Simpson and Alaska. From Comox the Cretaceous rocks probably extend in a wide belt along the shore nearly to Seymour Narrows, but are heavily covered by drift deposits, which form white cliffs. High mountains in the interior of Vancouver Island composed, so far as known, of crystalline rocks, with extensive granite intrusions.

Seymour Narrows and northward to Alert Bay. Metamorphic and crystalline rocks. (See Note 303, Can. Pacific Railway, W. Coast portion.) Near Port McNeil, Cretaceous rocks again form a strip of low country, extending back from the shore, and continue to Beaver Harbor. Thomas Point and north shore of Beaver Harbor, and thence to north end of Vancouver Island, all rocks of the older series. Similar metamorphic and crystalline rocks, with interbedded slaty argillites and limestones, and granitic intrusions northward to Wrangel, in Alaska. In vicinity of Port Simpson, slaty argillites and mica schists with limestones extensively developed. Near Wrangel similar mica schists yield very fine garnet crystals. Wrangel is at the mouth of the Stickeen River, by which the gold-mines of Cassiar are reached.

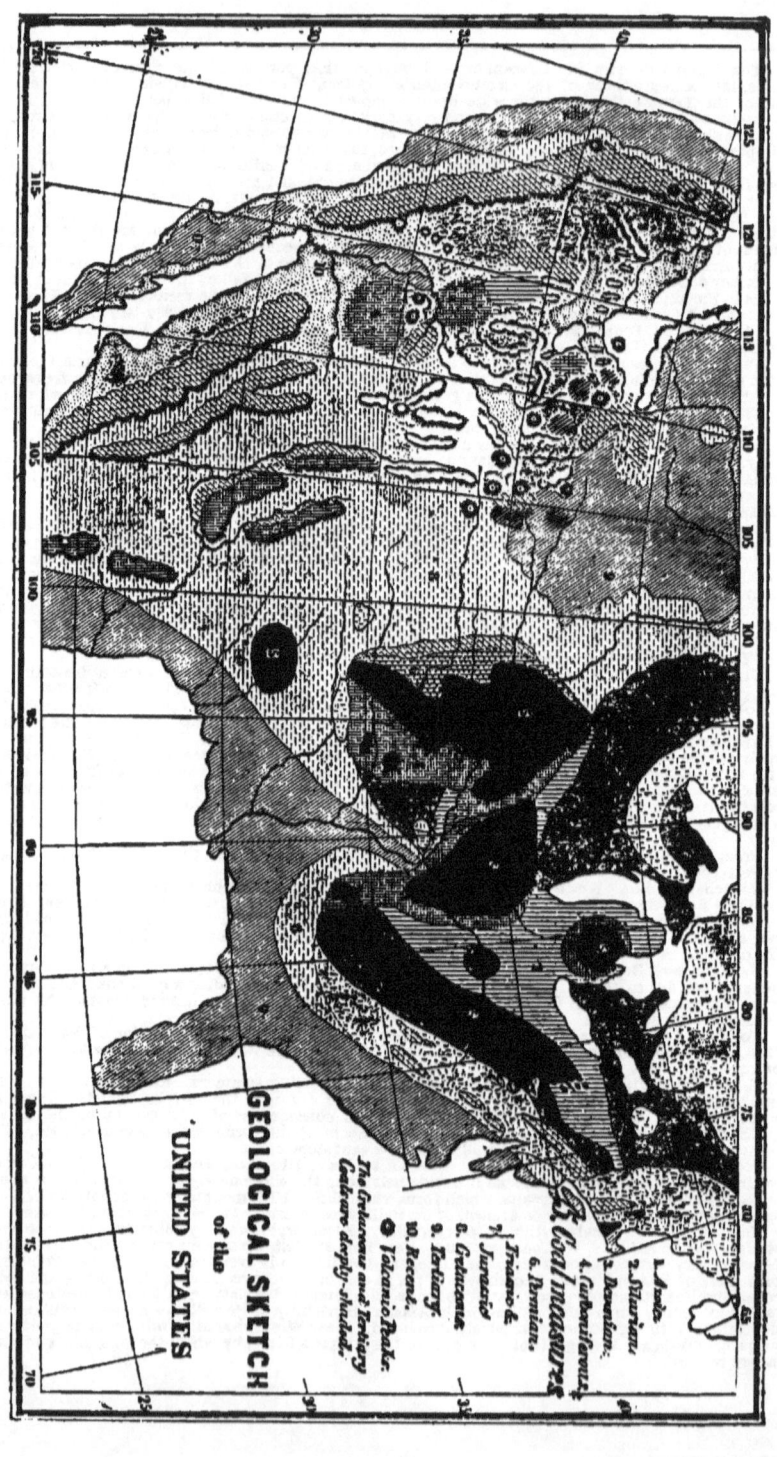

GEOLOGICAL SKETCH
of the
UNITED STATES

1. Azoic
2. Silurian
3. Devonian
4. Carboniferous
5. R. Coal measures
6. Permian
7. Triassic
8. Jurassic
9. Cretaceous
10. Tertiary
11. Recent.
● Volcanic Peaks.

The Cretaceous and Tertiary
Coal are deeply shaded.

The New England States.

THE geology of the New England States is much more difficult than that of the country west of the Hudson River and Lake Champlain. The rocks are very largely crystalline, besides being greatly contorted and folded. Both Archæan and metamorphic Paleozoic groups are represented, and geologists have disagreed as to the extent occupied by each of these two series. A quarter of a century since (before 1885) the opinion was commonly entertained that these crystallines consisted entirely of Paleozoic rocks in an altered condition; now it is generally conceded that many of the older areas are to be found. Different views are also entertained as to the value of lithological distinctions for chronological purposes. Fortunately, a few fossiliferous areas have escaped the ravages of upheaval and denudation, and it is only by a study of the relations of these to the underlying or overlying crystallines, that any attempt at correlation is possible. The principal localities where fossils are found are (1) the region of the Taconic schists and Stockbridge limestones; (2) that of probably Devonian limestone in the Connecticut Valley at Bernardston; and Niagara limestones at Littleton, N. H.; and (3) that of carboniferous rocks in Rhode Island and their continuation northeastward into Massachusetts. Devonian fossils have been found in the northern part of Maine, and Silurian and Devonian in the eastern part of Maine. The 16. Triassic of Connecticut Valley need not be named as one of these doubtful areas.

The scheme of classification proposed by Professor C. H. Hitchcock for the whole of New England is printed on an introductory page, while his determinations as to the formation at each railroad station are given in this "Guide" for Maine, New Hampshire, Vermont, and Connecticut. In the chapter on Massachusetts, the determinations for each railway station are given by Professor W. O. Crosby, representing a class of geologists holding widely different views, who recognize the Taconic system and believe that the white crystalline marble, 3,000 feet thick, in Berkshire County, Mass., lies below the Cambrian, and is a distinct and much older formation; and claim that the fossils referred to occur in outliers of the newer, resting on these older formations, just as they often do elsewhere. They also claim that the highly crystalline Taconic schists can not be correlated successfully with the Cambrian or with the Hudson River group.

The following scheme of classification of the New England crystallines, by Professor Hitchcock, is also very different from that given by Professor W. O. Crosby for Massachusetts. The differences are occasioned chiefly by the views entertained concerning the igneous rocks, syenites, granite, and porphyry. In Dr. Hitchcock's scheme these are regarded as of later origin than the gneisses, which have been disturbed by their eruption; but Professor Crosby seems to regard many of the syenites, felsites, and diorites as older than the gneisses; because the latter appear to rest or lean upon the unstratified rocks. The difference is so radical that the schemes can not be harmonized. But, in a work of this character, it is right that the different views should be represented.

Professor Hitchcock also thinks that the word Montalban is misleading, and, as restricted by him in New Hampshire, it would not embrace over one sixth part of the rocks so named by Professor Crosby. The typical area of Montalban in the White Mountains is said by the former to be either overlaid or cut by the rock called Norian by Dr. T. Sterry Hunt and Professor Crosby. Hence, it is claimed, the Norian is the newer of the two, and the scheme proposed for Massachusetts is by him considered erroneous.

However the reader may differ with either party, he will find much positive knowledge which all will accept in these pages, where the kinds of rock along the railroads are given, i. e., gneiss, mica schists, granite, etc., and we can leave it to time to give to these formations of doubtful age their true place in the series, for it is believed that the discovery of fossils here and there about New England may, after a while, settle the geology of a large portion of that difficult country, and that even an accepted classification of the crystalline rocks may be accomplished.   J. M.

Maine, New Hampshire, Vermont, Massachusetts, Rhode Island, and Connecticut.

Table of the Geological Formations of the New England States.
By Professor C. H. Hitchcock.

Cenozoic.

20. Quaternary.	20 c. Terraces.
"	20 b. Champlain Clays.
"	20 a. Till, drumlins, Terminal Moraine.
19. Tertiary.	19 c. Pliocene.
"	19 b. Miocene.
"	19 a. Eocene.

Mesozoic.

16. Triassic.	16. Triassic.

Paleozoic.

14. Carbonifer's.	14 b. Coal Measures.
"	14 a. Lower Carboniferous.
8–10. Devonian.	10 s. s. Probably Hamilton. Slates of St. Croix River.
"	9. Upper Helderberg l. s.
"	8. Oriskany Group.
5–7. Silurian.	7. Lower Helderberg.
"	5. Niagara.
3–4. Cambro-Silurian.	4 d. Magnesian Slate (Emmons), possibly Cambrian.
"	4 c. Lorraine Shales.
"	4 b. Utica Slate.
"	4 a. Trenton Limestone. Black River and Birdseye l. s.
"	3 c. Chazy l. s.
"	3 b. Levis Limestone.
"	3 a. Calciferous Sandrock.
2. Cambrian.	2 b. Potsdam ss. sl. qu. Georgia Group, Clay Slate.
"	2 a. Acadian. Clay Slates unfossiliferous. Taconic Slate (in part).

Foliated Crystalline Series.

E. Groups of debatable age, probably pre-Cambrian.
Rockingham Group, Slates and Quartzites.

Coös Group.	Calciferous Mica Schist.
	Staurolite Slates and Schists.
	Quartzites.
	Kearsarge Group.

Foliated Crystalline Series—Con.

D. Huronian.	Hydromica (talcose) Schists and Grits.
Subdivided in Connecticut Valley into Auriferous conglomerate, Lyman and Lisbon groups	Volcanic Group of Selwyn Hornblende Schist. Merrimack Group and Schists. Rockingham Group (in part). Ferruginous Slates (N. H).
C. Upper Laurentian	Montalban.
B. Middle Laurentian	Green Mountain Gneiss. Lake Winnipiseogee Gneiss. Bethlehem Gneiss.
A. Lower Laurentian	Porphyritic Gneiss. Adirondack Gneiss. K. 2. and K. 3. of Conn.

Eruptive Crystalline Rocks.

Basic.	Mesozoic Diabase or Dolerite. Older Diabase. Diorite. Melaphyr. Gabbro.
Acidic.	Felsite. Porphyry. Granite. Syenite. Protogene.

Cambrian and Cambro-Silurian Rocks of the Champlain Valley, with their thickness in feet.

4 c. Lorraine Slate	400
Hydromica Schist, Taconic Range	2,000
4 b. Utica Slate	300
4 a. Trenton Limestone	400–600
Black River, or La Motte and Birdseye Limestone	40
3 c. Chazy Limestone	400
3 b. Levis Limestone	600
3 a. Upper Calciferous Sandrock	200
Lower " "	400
Fucoidal Layer	200
Potsdam Sandstone, red	500
" " gray	310
" quartzite	1,200
Georgia Slates	3,000
Cambrian Slates and Schists	4,000
Total thickness	14,150

Eruptive Crystalline Rocks of New Hampshire, with local names.

Basic.	Mica Diabase. Porphyritic Diabase. Anorthite Diabase. Olivine Diabase. Ordinary Diorite. Porphyritic Diorite. Mica Diorite. Labradorite Diorite. Gabbro.	Acidic.	Felsite. Porphyry. Quartz Porphyry. Orthoclase Porphyry. Pequawket Breccia. Muscovite Granite. Muscovite Biotite or Concord Gr. Franconia Breccia Granite. Biotite or Conway Granite. Mica Hornblende or Chocorua Gr. Hornblende or Albany Granite. Protogene. Granitell. Granite of Veins. Augite Syenite. Hornblende Syenite.

Maine.[1]

Ms.	Maine Central Railroad.		
0	Portland.	D. Huronian.	13
8	Falmouth.	B. Laurentian.	49
15	Yarmouth.	"	88
20	Freeport.	"	127
25	Oak Hill.	"	125
29	Brunswick.	"	64
37	Bowdoinham.	"	10
44	Richmond.	"	77
56	Gardner.	"	33
60	Hallowell.	" Granite.	56
62	Augusta.	" "	48
70	Riverside.	2. Cambrian.	
81	Waterville.	"	117
89	Clinton.	D. Huronian.	133
94	Burnham.	"	167
101	Pittsfield.	"	210
108	Newport.	"	200
117	Etna.	"	
125	Herman Pond.	"	
135	Bangor.	"	13

Ms.	Lewiston Division.		
0	Portland.	D. Huronian.	13
8	Falmouth.	B. Laurentian.	49
19	Gray.	"	106
29	Danville Junc'n.	C. Montalban.	200
36	Lewiston.	"	200
46	Leeds Junction.	"	271
55	Winthrop.	"	220
61	Readfield.	"	
74	North Belgrade.	"	
84	Waterville.	2. Cambrian.	117

Belfast Division.

0	Burnham.	D. Huronian.	167
8	Unity.	"	222
12	Thorndike.	"	267
22	Brooks.	B. Laurentian.	376
32	City Point.	E. Pre-Cambrian.	29
34	Belfast.	"	29

Skowhegan Division.

0	Waterville.	2. Cambrian.	117
11	Pishon Ferry.	D. Huronian.	
19	Skowhegan.	"	

Dexter Division.

0	Newport.	D. Huronian.	
7	Corinna.	"	
14	Dexter.	"	

1. The eruptive rocks of Maine have not been studied yet. The "traps" along the sea-shore are of at least four different ages. The oldest is porphyritic; the second metalliferous; the third was ejected earlier than the Devonian; while the fourth has cut Hamilton sandstones. In the northern part of the State is a trappean conglomerate, with pebbles more than a yard in diameter. A light-colored, coarse diorite forms a mountain mass in Rangely, and the same material is commingled with serpentine farther north, nearer the Canada line. The granites and syenites are as varied as those of New Hampshire. The granite of Biddeford is the same as the Conway granite of New Hampshire, but with fewer cavities to produce disintegration. A drab-colored porphyry occurs in mountain masses upon Moosehead Lake and near Mount Katahdin. Siliceous slates and jaspers abound on the coast of Washington County.

The Lower Helderberg is also cut by trap dikes in several localities.

Maine Central Railroad—*Con.*	
Ms. |	**Androscoggin Division.**

Ms.			
0	Bath.	B. Laurentian.	
9	Brunswick.	"	64
20	Lisbon.	C. Montalban.	
27	Lewiston.	"	100
34	Leeds Junction.	"	171
44	North Leeds.	"	180
54	Livermore Falls.	D. Huronian.	
67	Wilton.	B. Lake Gneiss.	
74	Farmington.	E. Pre-Cambrian.	

Bangor to Vanceboro.

0	Bangor.	D. Huronian.	11
4	Veazie.	"	110
7	Buson Mills.	"	56
9	Orono.	"	
10	Webster.	"	
12	Great Works.	"	
13	Old Town.[4]	"	88
14	Milford.	"	
19	Costigan.	"	111
23	Greenbush.	"	
27	Olamon.	"	191
31	Passadumkeag.	"	131
36	Enfield.	" and granite.	190
45	Lincoln.	"	105
56	Winn.	"	201
58	Mattawamkeag.	"	
66	Kingman.	"	325
79	Bancroft.	"	333
88	Danforth.	"	379
93	Eaton.	"	400
98	Forrest.	"	435
102	Toma.	"	
114	Vanceboro.[5]	3–4. Camb. Silurian.	394

Bangor to Mt. Desert.

137	Bangor.[6]	D. Huronian.	13
148	Holden.	Granite.	
164	Ellsworth Falls.	D. Huronian.	
166	Ellsworth.	D. Huronian.	
176	Hancock.	"	
179	{ Mt. Desert Ferry.	"	

Ms. |	**Knox and Lincoln Railroad.**

0	Bath.	B. Laurentian.
11	Wiscasset.	"
18	New Castle.	"
30	Waterloo.	"
37	Warren.	"
45	Thomaston.[3]	{ 3–4. Limestone. Cambro-Silurian.
49	Rockland.	" and Quartzite.

Bangor and Piscataquis Railroad.

0	Bangor.	D. Huronian.	11
12	Old Town.[4]	"	88
21	Alton.	"	
31	Lagrange.	"	
40	Milo.	"	
53	Dover.	"	
61	Guilford.	2. Cambrian.	
64	Abbot.	"	
65	Blanchard.	"	
81	Shirley.	"	
88	Greenville and Moosehead.	"	

Portland and Rochester Railroad.

0	Portland, Me.	D. Huronian.	10
3	Westbrook.	C. Montalban.	19
5	Cumberland Ms.	E. Pre-Cambrian.	56
6	Saccarappa.	"	
10	Gorham.	"	
15	Buxton Centre.	"	
18	Saco River.	"	
21	Hollis Centre.	"	
25	Cen. Waterboro.	"	
28	S. Waterboro.	"	
32	Alfred.	Syenite.	
36	Springvale.	C. Montalban.	
43	E. Lebanon.	E. Kearsarge Group.	
49	E. Rochest., N.H.	" "	
52	Rochester.	" "	

Somerset Railroad.

0	North Anson.	D. Huronian.
4	Anson.	"
12	Norridgewock.	"
25	Oakland.	"

2. **Livermore.** Station at gorge in Pemigewasset River, and shows finely several dikes of igneous rocks of different ages. As carefully studied by Dr. Hawes, they are diabase, olivine diabase, diorite, syenite, and granite.

3. **Thomaston.** The location of the limestone-quarries furnishing the famous Rockland or Maine lime.

4. **Oldtown.** Most of the ancient valleys of New England have an escar or ridge of coarse gravel and sand following the channel of the current as the ice of the glacier period began to melt. These ridges are more common in Maine than elsewhere.

5. **Vanceboro.** The pale argillites along the St. Croix River, near and below Vanceboro, are called Devonian by Messrs. Bailey and Matthew, provincial geologists of New Brunswick, because of the discovery of the remains of Lepidodendron in it in the Magaguadavic Valley.

6. **Eastport.** These same authors regard the red sandstones near Eastport as of Lower Carboniferous age, instead of the Hamilton Devonian, as they have been heretofore referred. St. Andrews, N. B., or Calais, Me., is the nearest railroad station to Eastport.

New Hampshire.[7]

Ms.	Grand Trunk Railway.		
0	Portland, Me.	D. Huronian.	
5	Falmouth.	B. Laurentian.	49
11	Yarmouth.	"	94
18	Pownal.	C. Montalban.	143
27	Danville Junc'n.	"	200
36	Mechanic Falls.	"	298
41	Oxford.	"	331
47	South Paris.[8]	B. Laurentian.	389
55	West Paris.	"	483
65	Locke's Mills.	"	718
70	Bethel.	"	646
80	Gilead.	C. Montalban.	711
86	Shelburne, N. H.	"	704
91	Gorham.	"	794
98	Berlin Falls.	B. Lake Group.	1016
103	Milan.	"	1060
122	Groveton.	D. Huronian.	884
134	North Stratford.	"	902
142	Wenlock.	Granite.	1152
149	Island Pond.	"	1197
166	Norton Mills.	"	1357
175	Coaticooke.	E. Calcife's Mica Schist.	

(Continued in Canada.)

Portland and Ogdensburg Railroad.

Ms.			
0	Portland, Me.	D. Huronian.	16
5	Westbrook.	C. Montalban.	19
11	So. Windham.	"	
17	Sebago Lake.	"	274
24	Steep Falls.	"	305
32	Baldwin.	"	
36	Hiram.	"	
43	Brownfield.	"	396
49	Fryeburg.	"	420
55	Conway C., N.H.	Conway Granite.	455

Ms.	Portland & Ogdensburg R. R.—Con.		
60	North Conway.[9]	Conway Granite.	527
66	Glen Station.	Albany Granite.	530
72	Upper Bartlett.	Conway Granite.	660
78	Bemis.	C. Montalban.	996
87	Crawford's.[10]	"	1903
91	Fabyan's.	"	1571
96	Twin Mount'n.[11]	B. Bethlehem Gr.	1375
100	Bethlehem Junc.	"	1187
104	Wing Road.	A. Laurentian.	1019
114	Lunenburg, Vt.	D. Huronian.	

Boston and Lowell Railroad.

Ms.			
0	Concord.[19]	Concord Granite.	852
10	Canterbury.	E. Rockingham Schist.	
18	Tilton.	B. Lake Gneiss.	458
27	Laconia.	C. Montalban.	
33	Weirs.[14]	A. Porphyritic Gneiss.	
48	Ashland.[15]	"	
51	Plymouth.	C. Montalban.	474
59	Rumney.	"	520
67	Wentworth.	B. Lake Gneiss.	
71	Warren.	"	736
84	Haverhill.	D. Huronian.	
93	Wells River.	" Lyman.	443
103	Lisbon.	" Lisbon.	577
	North Lisbon.	5. Niagara.	667
113	Littleton.[16]	E. Coös and 8. Niag.	817
120	Wing Road.	A. Porphyritic Gn.	1019
124	Bethlehem.	B. Bethlehem Gn.	1187
129	Twin Mountain.	" (Loc. Glacier)[13]	1375
134	Fabyan's.	C. Montalban.	1571
120	Wing Road.	A. Porphyritic Gn.	1019
128	Dalton.	D. Huronian.	856
135	Lancaster.	"	870
145	Groveton Junc.	"	901

7. The New Hampshire formations are believed to possess thickness as follows: Niagara, 500 feet; Calciferous mica schists, 4,800 feet; Coös group, 7,300 feet; Cambrian slates of Connecticut Valley, 3,000 feet; Kearsarge group, 1,300 feet; Rockingham mica schists, 6,000 feet; Merrimack group, 4,300 feet; Huronian, 12,000 feet; Montalban, 10,000 feet; Lake Winnipiseogee gneiss, 18,000 feet; Bethlehem gneiss, 5,000 feet; porphyritic gneiss, 5,000 feet.

8. Paris. Locality of the famous red and green tourmalines. At least one hundred remarkably fine specimens of tourmaline have been taken from this vein and placed in museums or cut as gems. Forty varieties of minerals occur in a coarse granite, one of which is mica in large plates.

9. North Conway. Mount Kiarsarge, in full view from the station, is a conical mass of Albany granite which has broken through both the Conway granite and a slate, and contains numerous fragments of both these rocks in its igneous embrace.

10. Crawford House. The railroad passes from here through the well-known notch of the White Mountains and around the base of Mount Willard, a region as famous for its varieties of granite as for scenery. The cut at the summit is through typical Montalban schists. Opposite Dismal Pool it is traversed by an enormous vein of fine-grained granite, which has also cemented together immense fragments of the Montalban schists. The junction between this Franconia breccia and the succeeding Conway granite, may be followed up a cliff for one thousand feet higher than the railroad, the latter rock having been erupted last. Between this Conway granite and a dark slate often filled with large pencils of andalusite is the interesting vein, three hundred feet wide, of Albany granite, which illustrates the action of a melted rock upon slates, giving rise to "contact phenomena." The slates have been rendered more crystalline; have been altered into hornstone; the broken pieces have been cemented by a siliceous paste full of microscopic tourmalines; and Carlsbad twin crystals of orthoclase, with dihexagonal pyramids of quartz, are developed in the lower part of the Albany granite. All these and other interesting phenomena may be seen along the railroad in a walk of half a mile.

11. Twin Mountain. The large boulders of granite east of the hotel are part of the moraine of a local glacier which has moved in a northwest direction. The boulders have certainly been transported from some ledge nearer Mount Washington than Fabyans's.

Boston and Lowell Railroad—Con.

Ms. | Concord to Nashua.

0	Concord.	Concord Granite.	
5	Suncook.	C. Montalban.	281
9	Hooksett.[17]	"	206
13	Martin's.	B. Lake Gneiss.	199
18	Manchester.[18]	"	181
26	Reed's.	"	137
29	Thornton's.	"	125
35	Nashua.	D. Merrimack Gr'up.	120

Suncook Valley Branch.

0	Hooksett.[17]	C. Montalban.	206
20	Pittsfield.	E. Rockingham Sch.	493

Northern Division.

0	Concord.	Concord Granite.	252
7	Penacook.	C. Montalban.	268
14	Nor. Boscawen.	"	290
17	Franklin.	"	362
25	East Andover.	"	561
31	Potter Place.[12]	E. Kearsarge Gr.	653
44	Grafton.[13]	A. Porphyr. Gneiss.	848
52	Canaan.	D. Hornblende Schist.	965
59	Enfield.	B. Bethlehem Gneiss.	768
65	Lebanon.	"	510
69	W. R. Junction.	D. Hornblende Sch.	369

Concord and Claremont Division.

0	Concord.[19]		252
8	Mast Yard.	D. Ferrug. Schists.	375
12	Contoocook.	Concord Granite.	373
18	Warner.	B. Lake Gneiss.	422
23	Roby's Corners.	A. Porphyritic Gneiss.	
27	Bradford.	"	679
34	Newbury.	"	1130
43	Newport.	B. Lake Gneiss.	892
48	Kelleysville.	"	707
54	Claremont.	E. Calc. Mica Sch.	543

12	Contoocook.	Concord Granite.	373
20	Henniker.	A. Porphyr. Gneiss.	439
27	Hillsboro.	B. Lake Gneiss.	674
33	Antrim.	"	

Ms. | Concord and Claremont Division.—Con.

35	Bennington.	A. Laurentian.	
37	Hancock Junct.	"	
44	Peterboro.	B. Lake Gneiss.	744

Nashua to Keene.

0	Boston.		135
40	Nashua.	D. Merrimack Gro'p.	120
45	S. Merrimack.	"	
48	Amherst.	B. Laurentian.	
51	Milford.	" and granite.	
55	East Wilton.	C. Montalban.	328
59	S. Lyndeboro.	E. Rockingham.	624
66	Greenfield.	C. Montalban.	500
71	Hancock Junc'n.	A. Laurentian.	
75	Hancock.	"	
82	Harrisville.	"	1334
89	Marlboro.	C. Montalban.	739
96	Keene.	B. Bethlehem Gr'up.	466

Mt. Washington to Wing Road.

0	Mt. Washington.	C. Montalban.	6291
3	Base Mt. W'n.[20]	"	2668
9	Fabyan's.	"	1571
10	Wh. M't'n. House	Conway Granite.	
14	Twin Mt. H'se.[11]	B. Bethlehem Gr.	1375
19	Bethlehem Jun.	"	1187
23	Wing Road.	A. Laurentian.	1019

Pemigewasset Valley Branch.

0	Plymouth.	C. Montalban.	474
2	Livermore F'ls.[2]	"	531
4	Campton.	"	539
7	Campton Vill.	"	583
9	Thornton.	A. Laurentian.	555
13	W. Thornton.	"	580
16	Woodstock.	B. Laurentian.	642
20	N. Woodstock.	"	734

Profile and Franconia Notch Railroad.

0	Bethlehem.	B. Bethlehem Gr.	1187
10	Profile House.	A. Laurentian.	1937

12. Potter Place. Mount Kearsarge may be reached from this station, or from Warner upon the Concord and Claremont Railroad. The rock is an andalusite mica schist, the same with that of Mount Monadnock in Jaffrey and the base of Mt. Kiarsarge near North Conway. (Please notice the spelling of Ki and Kearsarge.)

13. Grafton. Locality of the largest beryl known, weighing two and one half tons. This was formerly preserved beneath a rude shed built to protect the mineral, but the shed and crystal have now fallen into decay. Very large crystals of the same mineral are now found occasionally in one of the mica-quarries.

14. Weir's. About half a mile from the station is a thick bed of clay lying between the lower and upper till.

15. Ashland. Between Weir's and Ashland many excellent exposures of porphyritic or oldes gneiss may be seen along the railroad. Over twenty of these areas have been described in the State and are supposed to represent the earliest known ejections of igneous matter, in which foliation ha been superinduced in concentric layers resembling strata.

16. Littleton. The fossiliferous limestone, here first called Lower Helderberg, is regarded by Pro fessor R. P. Whitfield as Niagara, because of the presence of the chain coral and of *Pentamerus ny sius.*

17. Hooksett. The railroad-bridge over the Merrimack River rests upon islands of a white quartz which are the outcrops of a remarkable vein, traced for over 125 miles, from Royalston, Mass. to Bridgeton in Maine. A second vein, parallel to this, crosses the river just north of Manchester, te miles distant.

18. Manchester. The prevailing rock is a coarse saccharoidal gneiss, believed to correspond ver closely in lithological aspect with the typical Laurentian of New York and Canada.

19. Concord. The traveler will do well to visit the State-House, with its large relief map of th State, and the large quarries of Concord granite two miles toward West Concord.

Ms. | Monadnock Railroad.

Ms.			
0	Peterboro.	B. Lake Gneiss.	744
7	Jaffrey.	C. Montalban.	1032
11	Rindge.	"	1003
17	{ Winchen-don, Mass. }	Gneiss.	993

Concord and Portsmouth Railroad.

0	Manchester.	B. Lake Gneiss.	161
8	Auburn.	"	269
18	Raymond.	D. Huronian.	172
24	Epping.	E. Rockingham.	154
31	New Market.	Exeter Syenite.	52
41	Portsmouth.	E. Rockingham.	

Manchester and Lawrence R. R.

0	Manchester.	B. Lake Gneiss.	161
8	Wilson's.	D. Merrimack Group.	
14	Windham.	"	324
22	Messers.	"	
26	Lawrence.	"	65

Manchester and North Weare Railroad.

0	Manchester.	B. Lake Gneiss.	161
11	Oil Mills.	" and A.	
19	North Weare.	"	489

Cheshire Railroad.

0	Bellows Falls.[24]	C. Montalban.	303
4	Walpole.	E. Coos Sch. & Qu.	217
10	Westmoreland.	D. Hornblende Sch.	512
22	Keene.	B. Bethlehem Group.	466
32	Troy.	C. Montalban.	1002
37	Fitzwilliam.	Concord Granite.	1062
43	State Line.	C. Montalban.	898
46	Winchendon.	"	448
54	S. Ashburnham.	"	1014
64	Fitchburg.	"	430

Ashuelot Railroad.

0	Keene.	B. Bethlehem Group.	466
8	Westport.	"	
15	Ashuelot.	A. Porphyr. Gneiss.	434
24	South Vernon.	E. Coös Quartz.	

Whitefield and Jefferson Railroad.

0	Whitefield Jun.	D. Huronian.	931
1	Whitefield Vill.	"	
3	Hazen's Mills.	B. Laurentian.	

Ms. | Whitefield & Jefferson R. R.—Con.

7	Cherry Pond.	B. Laurentian.
10	Jefferson.*	"

Montpelier and Wells River R. R., Vt.

0	Montpelier.	Clay Slate.	484
6	E. Montpelier.	E. Calcife's Mica Schist.	
10	Plainfield.	"	752
15	Marshfield.	"	1140
21	Summit.	Granite.	
28	Groton.	E. Calcif. Mica Sch.	778
34	Boltonville.	"	624
38	Wells River.	D. Huronian.	448

This railroad is in Vermont.

Saratoga and Champlain Railroad.

0	Rutland.	Calcif. Sandrock.	519
11	Castleton.	2. Cambrian Slates.	475
8	Granville, N. Y.	"	
19	Rupert.	"	
26	Salem.	"	
34	Eagle Bridge.	"	

Worcester, Nashua and Rochester R. R.

0	Worcest'r, Ms.[22]	Mica Schist.	473
9	W. Boylston.	"	448
10	Oakdale.	"	382
12	Sterling Junc'n.	"	436
17	Clinton.	E. Pre-Cambrian.	309
19	Lancaster.	"	259
25	Harvard.	"	288
28	Ayer Junction.	D. Merrimack Group.	230
32	Groton.	"	303
36	Pepperell.	"	205
40	Hollis, N. H.	"	195
46	Nashua.	"	120
49	Hudson.	"	
52	W. Windham.	"	
56	Windham.	" & B. Lau'n.	
63	Hampstead.	"	256
65	Sandown.	"	
70	Fremont.	"	
74	Epping.	"	154
79	Lee.	"	
88	Barrington.	" & B. Lau'n.	
93	Gonic.	E. Kearsarge Group.	
95	Rochester.	"	226

☞ Railroads not found under New Hampshire heading will be found in Massachusetts.

20. Mt. Washington. Boulders that have been transported as much as twelve miles, and up-hill nearly four thousand feet, by the ice sheet, occur upon the top of this mountain. Striæ occur here and upon all the Presidential summits, running southeasterly.

22. Worcester. Mr. Joseph H. Perry announces the discovery of a Lepidodendron in the plumbago of Worcester. Lesquereux, after examination of photographs, pronounces it to be like the L. acuminatum of the Carboniferous limestone of Siberia. If there is no mistake about this discovery, it will prove the existence of an outlier of the Lower Carboniferous in Central Massachusetts. The schists have been supposed by us to belong rather to the Huronian or Cambrian.

* Upon July 10, 1885, a new slide scarred the north side of Cherry Mountain. It originated in the giving way of a ledge near the top of the mountain, when the ground was exceedingly wet. The earth slid one and a half miles in about four minutes' time, killing cattle in the field and fatally wounding one man. The lower end is very near this station.

Vermont.[23]

Central Vermont Railroad.			Ms.	Central Division—Con.	
Ms.	Southern Division.		292 Milton.	2 Potsdam Limes.	351
127 Brattleboro.	2. Cambrian.	225	296 Georgia.[30]	Potsdam Slate.	355
130 Putney.	E. Coös Schist.	257	306 St. Albans.	2 Potsdam Slate.	390
141 Westminster.	2. Cambrian.	264	Rutland Division.		
145 Bellows Falls.[24]	C. Montalban.	276			
153 Ch'rlest'wn,N.H.	E. Coös Group.	375	0 Bellows Falls.[24]	C. Montalban.	305
163 Claremont, N.H.	E. Calcife's Mica Schist.		5 Rockingham.	E. Calcif's Mica Sch.	333
171 Windsor.[25]	"	331	10 Chester.	B. Lake Gneiss.	501
179 North Hartland.	2. Camb.& D.Huro'n.	387	22 Cavendish.	"	921
185 White River Jn.	D. Hornbl. Sch.	" 369	27 Ludlow.[27]	D. Huronian.	1061
Central Division.			34 Summit.	B. Green Mt. Gneiss.	
			39 E. Wallingford.	"	1195
171 Hartford.	2. Cambrian.	485	46 E. Clarendon.	3 b. Camb. Sil. Limest.	
198 Sharon.	E. Calcif's Mica Sch.	507	52 Rutland.[28]	2 e. Calcifer's Sandrock (Stockbridge).	519
205 Royalston.	"	517	59 SutherlandFalls.	3 c. Chazy Marble.	
216 Bethel.	D. Huro'an Soapst.	576	69 Brandon.	19 a. Eocene Tert'y.	353
217 Randolph.	"	698	74 Leicester Junc.	3 c. Chazy Marble.	351
223 Braintree.	"	784	79 Salisbury.	3 b. Levis Limest.	346
232 Roxbury.	" VerdeAnt.	1016	85 Middlebury.	"	341
239 Northfield.	D. Huro'an Soapst.	739	89 Brooksville.	3 c. Chazy Limest.	301
249 Montpelier.	" & ClaySlate.	629	93 New Haven.	4 a. Trenton Limest.	291
258 Waterbury.	"	434	99 Vergennes.	3 c. Chazy Limest.	201
266 Bolton.[26]	B. Green Mt. Gneiss.	346	104 Nor. Ferrisburg.	"	131
272 Richmond.	D. Huronian.	328	108 Charlotte.[29]	"	161
281 Essex Junc'n.	Clay Slate.	360	113 Shelburne.	2 j. Potsdam Sand.	151
286 Winooski.	3 b. Camb.Sil.Limes.	190	120 Burlington.	"	109
289 Burlington.	2 Potsdam Sandst.	109			

23. LIST OF ERUPTIVE ROCKS OF VERMONT. — Diabase, diorite, trachytic porphyry, muscovite granite, mica hornblende granite, protogene, granitell, concretionary granite, granite of veins, syenite, brecciated syenite. The trachytic porphyry is supposed to have been erupted at the close of the Silurian.

24. Bellows Falls. The finest exhibition of terraces along the Connecticut River north of Massachusetts is just south of the village of Bellows Falls.

25. Windsor. An interesting escar has been traced from Lyme, N. H., to Windsor, Vt., about thirty miles long. Portions of it have been removed by the wearing action of the Connecticut. It appears to have been deposited by a powerful current derived from the melting of the glacial sheet prior to the accumulation of terraces. Mt. Ascutney, 3,186 feet high, is proved to be an eruptive mass of syenite and granite which has been protruded through a narrow orifice and poured out over a floor of the calciferous mica schist about one thousand feet above the sea, very much as lava accumulates around a volcanic vent. The melted material penetrated cracks in the underlying calciferous mica schist, forming veins indurating the clayey layers, calcining and glazing the limestones, but where it flowed over gneiss the floor remained unaffected. Many other granite mountains in Northern New England show similar proofs of protrusion at the surface.

26. The center of the anticlinal axis of the Green Mountains. At least eight of the general sections of the Vermont survey show this feature of structure, proving this formation to be older than the Huronian adjacent upon both sides. This structure was denied by Logan for the continuation of the Vermont rocks in Canada in his generalizations, but his descriptions of the rocks confirm the views of the Vermont geologist. Dr. Selwyn, the successor of Logan in office, accepts the Vermont view.

27. Ludlow. In Plymouth, ten miles north, gold is now (1885) being profitably milled from quartz. It is in the Huronian, which may be followed continuously to Zoar and Chester, Mass., upon the Fitchburg Railroad.

28. Rutland. The Rutland Railroad follows the Champlain Valley, noted for the presence of the entire series of Lower Silurian groups. The valley itself is a part of the great Appalachian Valley, extending from the St. Lawrence to Alabama, and constituting a natural and well-marked boundary between the crystalline groups on the east, known as the Green Mountains, Highlands of New York and New Jersey, Blue Ridge of Virginia, and the true Appalachian Mountains on the west from the Catskills to the Cumberland plateau, in Tennessee.

29. Charlotte. Champlain clays. The bones of a *Beluga*, a species of white whale, were found near here while excavating a railroad cut in 1849, one hundred and fifty feet above the ocean. The subdivision proposed by C. B. Adams in 1846 was that of the lower "Blue clay," containing a deep-sea fauna, and an upper "Brown clay," carrying littoral species. Several years later, Dawson proposed the names of "Leda clay" and "Saxicava sand" for the synchronous deposits in the St. Lawrence Valley.

30. Georgia. This town has furnished thirty or forty species of trilobites and other fossils of the Middle Cambrian, or a horizon between the Potsdam sandstone of New York and the St. Johns or Acadian group of New Brunswick and Eastern Massachusetts.

Ms.	Central Vermont Railroad.	
	Western Division.	
0	St. Albans.	2 j. Potsdam Slate. 390
9	Swanton.	" 160
	Northern Division.	
0	St. Albans.	2 j. Potsdam Slate. 390
	Georgia.[30]	"
9	East Swanton.	"
17	Province Line.	3 b. Levis Limestone.
	Eastern Division.	
0	St. Albans.	390
10	Sheldon.	D. Huronian. 374
18	Enosburg Falls.	" 436
28	Richford.	" 473
	Addison Division.	
0	Leicester Junc.	3 c. Chazy. 361
3	Whiting.	"
7	Shoreham.	" and 3 a.
9	Orwell.	2 c. Calcifer's Sandrock.
15	Larabee's Point.	4 a. Trent. & La Motte.
16	Ticonderoga.	3 a. Calciferous s. s.

	Woodstock Railroad.	
0	White River Jn.	369
1	Hartford.	D. Huronian. 485
6	Dewey's Mills.	Calcif. Mica Schist.
7	Queechee.	" 650
11	Taftsville.	" 657
14	Woodstock.	" 697

	Bennington and Rutland Railroad.	
0	Rutland.	2 a. Calcif's Sandr'k.[619]
6	Clarendon.	" 639
9	Wallingford.	"
13	S. Wallingford.	3 c. Chazy Marble.
18	Danby and Mt. Tabor.	3 a. Calcif's Sandstone.
25	East Dorset.	" & Chazy Marble

Ms.	Bennington and Rutland R. R.—*Con.*	
30	Manchester.[18]	3 b. Camb. Sil. Limest.
39	Arlington.[31]	" 471
44	Shaftsbury.	"
51	N. Bennington.	"
55	Bennington.	"
61	T. & B. Junc'n.	2. Cambrian(Taconic) sL

	Boston and Lowell Railroad.	
	Vermont Division.	
0	Lunenburg.	Lyman Gp. and D. Hur.
7	Miles Pond.	C. Montalban. 851
13	West Concord.	E. Coös Group. 857
21	St. Johnsbury.[32]	E. Calcif's Mica Sch.[691]
33	Danville.	" 1375
41	Walden.	" 1673
49	Greensboro.	" 1166
57	Hardwick.[36]	" 891
62	Wolcott.	D. Huronian. 705
70	Morrisville.	" 659
73	Hyde Park.	" 586
78	Johnson.	" 541
86	Cambridge Jun.	" 473
104	Sheldon.	" 374
118	Swanton.	" 160
120	Maquam Bay.	"

	Passumpsic Railroad.	
0	Sherbrooke, P.Q.	1. Pre-Cambrian. 486
3	Lennoxville.	" 500
12	North Hatley.	" & 2–7. Silur'n.
30	Smith's Mills.	5–7. Silurian.
34	Stanstead Junc.	Granite.
40	Newport, Vt.	E. Calc. Mica Schist.[703]
45	Coventry.	"
55	Barton.	" 959
68	West Burke.	" 1040
76	Lyndonville.	" 741
84	St. Johnsbury.[32]	" 691
87	Passumpsic.	"
94	Barnet.	" 466
105	Wells River.	D. Huronian. 443

31. **Arlington.** A few miles east, in the edge of Sunderland, is the best-known exposure of the junction of the Potsdam quartzite with the unconformably underlying gneiss of the Green Mountains. The blue quartz of the granite veins crossing the gneiss is recognized as the source of the grains of sand in the quartzite. Also an excellent locality for the *Scolithus.*

32. **St. Johnsbury.** Eastern Vermont is largely underlaid by a mica schist having a micaceous limestone interstratified with it, to which the name of " calciferous mica schist " is applied in the State reports. It is called " Silurian " when it passes into Canada, and " Montalban mica schist " in Massachusetts. Protracted studies show the strata to be disposed in a synclinal attitude, overlying clay slate. Numerous areas of granite have been erupted through it, both in Vermont and Canada. There is an excellent development of this rock at St. Johnsbury Center and at Danville.

33. **Fairlee.** A few miles west of this station is the famous Ely copper-mine, for many years the greatest producer of the metal from the yellow sulphuret of any mine in the United States. Six miles west of Pompanoosuc are other copper-mines, and an establishment producing copperas.

34. **Norwich and Hanover.** A few rods east of the station, on the east side of the Connecticut, the escar has been cut through by erosion, showing an anticlinal ridge of gravel underlying the terraces of Hanover Plain. The same ridge has been cut by White River at White River Junction, where the same structure is observable.

35. **Hanover.** The collections of the Geological Survey of the State are placed in the Museum of the State Agricultural College. A marked feature is the arrangement of over three thousand lithological specimens in geographical order, taken along thirteen parallel sectional lines across New Hampshire and Vermont. Colored geological profiles accompany the specimens, with the locations and dips indicated, so that one can discover the mutual relations of the rocks without the labor of traveling over the country. In the same room is a large relief map of the same States, colored geologically, upon the horizontal scale of one mile to the inch.

Ms.	Passumpsic Railroad.—Con.			Ms.	Passumpsic Railroad.—Con.	
110	Newbury.	D. Huronian.	436	129	North Thetford. D. Huronian.	40?
113	S. Newbury & Haver-hill, N.H.	"	412	131	Thetford & Lyme, N.H. } E. Coös Group.	413
117	Bradford.	"	410	141	Norwich[34] & Hano-ver,[35] N. H. } D. Hornblende Sch.	406
124	Fairlee & Orford, N. H	"	438	145	White River Jn.	" · 369

Connecticut.[37]

New York, New Haven and Hartford Railroad. New York and New Haven Division.				Hartford Division.—Con.		
0	New York.	C. Montalban.	52	86	Wallingford.	16. Triassic.
11	W'ms Bridge.	Crystalline Limestone.		89	Yalesville.	"
14	Mount Vernon.	"		92	Meriden.	" 131
17	New Rochelle.	B. Mid. Lau'n Gneiss.[32]		99	Berlin.	" 63
21	Mamaroneck.	"		105	Newington.	"
22	Harrison.	"		110	Hartford.	" 39
24	Rye.	"		116	Windsor.	"
26	Port Chester.	"		121	Windsor Locks.	" 40
29	Greenwich.	"		122	Warehouse Pt.	"
30	Cos Cob.	"		124	Enfield Bridge.	"
34	Stamford, Conn.	"	12	127	Thompsonville.	"
37	Noroton.	"		136	Springfield.	"
38	Darien.	"			Shore Line Division.	
42	South Norwalk.	"			New York.	C. Montalban.
45	Westport.	"		0	New Haven.	16. Triassic. 10
50	Southport.	"		2	Fair Haven.	"
51	Fairfield.	"		8	Branford.	Laurentian Gneiss.
56	Bridgeport.	"	9	11	Stony Creek.	"
60	Stratford.	E. Calcif's Mica Schist.		16	Guilford.	Anthophyllitic Gneiss.
61	Naugatuck Jun.	"		20	Madison.	"
64	Milford.	D. Huronian.		23	Clinton.	"
74	New Haven.	16. Triassic.	10	28	Westbrook.	Gneiss.
	Hartford Division.			31	Saybrook.	" light colored.
				33	Conn. River.	
74	New Haven.	16. Triassic.	10	34	Lyme.	Laurentian Gneiss.
80	North Haven.	"		39	South Lyme.	"
				43	East Lyme.	"

36. Hardwick. A few miles north, in Craftesbury, is the celebrated concretionary granite, in which concentric balls of mica are numerously interspersed, to which the local name of "petrified butternuts" has been applied.

37. NOTE.—The very minute description of the foliated crystalline rocks of Connecticut by J. G. Percival furnishes the basis for the following attempted correlation of them with similar groups elsewhere. The Trias divides the crystalline into an eastern and western "Primary"—and Roman letters were used by Percival for the subdivisions of the western primary group. A. is undoubtedly the Huronian of the upper Connecticut. B. is the range of clay slate to the west, the same with that in Bernardston, near Guilford, Vt., and the Ammonoosuc gold-field, N. H. C. is the calciferous mica schist. D. is probably Middle Laurentian. E., F., G., H., and I. belong to the Green Mountain gneiss, perhaps partly Montalban. K. is Lower or typical Laurentian. L., M., N., O., and P. are the Cambro-Silurian lime-stones and schists called Taconic by Emmons. The A. and B. of the eastern Primary comprise both Lower and Middle Laurentian. C. is probably Montalban. D. and E. are the southward extension of the ancient Laurentian gneiss of Worcester County, and F. is closely allied to the Montalban.

Percival did not determine the nature of the "traps" of Connecticut, but showed their arrangement in curves; Professor Dana determined the constituent minerals to be pyroxene and labradorite with magnetite. Dr. G. W. Hawes confirmed this determination, but uses the name diabase instead of dolerite; Percival found, in both the eastern and western primary, systems of dikes parallel to the borders of the Trias entirely through the State; these are anhydrous, while those in the sandstones are mostly hydrous and amygdaloidal.

Ms.	Shore Line Division.—Con.	
47	Waterford.	Laurentian Gneiss.
50	New London.	" 9
112	Providence.	14. Coal Measures.
156	Boston.	2. Cambrian.

New Canaan Railroad.

0	New Canaan.	B. Middle Laurentian.
9	Stamford.	"

Danbury and Norwalk Railroad.

	Wilson Point.	B. Middle Laurentian.
0	South Norwalk.	"
18	Sanford.	"
24	Bethel.	"
27	Danbury.	Limestone. 397

Ridgefield Branch.

0	Ridgefield.	B. Middle Laurentian.
	South Norwalk.	"

Housatonic Railroad.

	New Haven.	16. Triassic.
0	Bridgeport.	B. Middle Laurentian.
10	Stepney.	"
15	Botsford.	"
19	Newtown.	"
23	Hawleyville.	" 206
27	Brookfield Jun.	B. Mid. Laurentian.
29	Brookfield.	" 238
35	New Milford.	Limestone abundant. 224
42	Merwinsville.	"
48	Kent.	"
57	Cornwall Bridge.	A. Lower Laurentian.
61	West Cornwall.	"
65	Lime Rock.	3–4. Camb. Sil. Limest.
67	Falls Village.	"
73	Canaan.	" 627
75	Ashley Falls.	"
79	Sheffield.	"
85	Gt. Barrington.	"
87	Van Deusenville.	"
89	Housatonic.	"
91	Glendale.	"
93	Stockbridge.	"
95	South Lee.	"
99	Lee.	"
101	Lenox Furnace.	"
102	Lenox.	"
106	Dewey's.	"
110	Pittsfield.	"
	North Adams.	"
87	Van Deusenville.	"
95	W. Stockbridge.	"
98	State Line.	3–4. Camb. Sil. Schists.

Shepaug Railroad.

0	Litchfield.	B. Middle Laurentian.
6	Morris.	"
8	Romford.	"
12	New Preston.	Limestone.
13	Washington.	B. Middle Laurentian.
20	Roxbury.	"

Ms.	Shepaug Railroad—Con.	
24	Roxbury Falls.	B. Middle Laurentian.
27	Shepaug.	"
32	Hawleyville.	"
38	Bethel.	"

Naugatuck Railroad.

	New Haven.	16. Triassic.
0	Bridgeport.	B. Middle Laurentian.
3	Stratford.	E. Calcifer's Mica Schist.
5	Junction.	"
14	Derby.	"
16	Ansonia.	B. Middle Laurentian.
20	Seymour.	"
23	Beacon Falls.	"
27	Naugatuck.	"
28	Union City.	"
32	Waterbury.	"
35	Oakville.	"
38	Watertown.	"
35	Waterville.	"
42	Thomaston.	"
47	Campville.	"
49	Litchfield.	"
52	Torrington.	A. Lower Laurentian.
57	Burrville.	"
61	Winsted.	"

Hartford & Conn. Western R. R.

0	Hartford.	16. Triassic.
6	Bloomfield.	"
10	Scotland.	"
12	Tariffville.	Diabase Range.
15	Simsbury.	16. Triassic.
22	Canton.	B. Middle Laurentian.
24	Collinsville.	"
28	Pine Meadow.	"
29	New Hartford.	" 369
35	Winsted.	A. Lower Laurentian.
	Naugatuck Dep.	"
36	West Winsted.	"
	Colebrook.	"
45	Norfolk.	" 1220
48	West Norfolk.	B. Middle Laurentian.
52	East Canaan.	2 b. Potsdam Quartzite.
55	Canaan.	3–4. Camb. Sil. Limest.
60	Chapinsville.	Cambro-Silurian.
62	Salisbury.	Camb. Sil. Limestone.
64	Lakeville.	" 670
66	Ore Hill.	4 c. Lorraine Group.
67	State Line Junc.	3–4. Camb. Sil. Limest.
70	Mount Riga.	"
74	Boston Corners.	"
78	Copake.	"
84	Ancram.	2–4. Camb. Sil. Schists.
86	Gallatinville.	"
91	Jackson Corners.	"
96	Ellerslie.	"
103	Red Hook.	"
107	Rhinebeck.	"
	Rhinecliff.	"
110	Rhinebeck Junc.	"

Ms.	Central Vermont Railroad.	
256	Stafford.	B. Middle Laurentian.
262	Tolland.	"
266	Merrow.	"
268	Mansfield.	"
270	Eagleville.	"
276	Willimantic.	"
280	S. Windham.	C. Montalban.
283	Lebanon.	"
286	Franklin.	"
289	Yantic.	"
293	Norwich.	"
296	Mohegan.	"
298	Massapeag.	A. Older Laurentian.
300	Montville.	"
303	Waterford.	"
306	New London.	"

Providence and Worcester Railroad.

0	Providence.	14. Coal Measures.
4	Pawtucket.	"
6	Valley Falls.	"
7	Lonsdale.	3–4. Camb. Silurian.
9	Ashton.	"
11	Albion.	"
13	Manville.	"
16	Woonsocket.	"
18	Waterford.	A. Laurentian.
	Blackstone.	"
20	Millville.	"
25	Uxbridge.	"
26	Whitin's.	"
31	Northbridge.	"
33	Farnum's.	"
34	Saundersville.	"
35	Sutton.	"
38	Millbury.	"
43	S. Worcester.	Mica Schist.
44	Worcester.	"

Stonington and Providence Railroad.

0	New London.	A. Laurentian.
9	Mystic.	"
12	Stonington.	"
18	Westerly.	"
26	Wood Riv. Jun.	"
35	Kingston.	"
42	Wickford Junc.	"
48	Greenwich.	14. Carboniferous.
53	Hill Grove.	"
57	Auburn.	"
62	Providence.	"

New York and New England Railroad.

0	Boston.	3–4. Cambrian.
46	East Douglass.	Quartzite.
53	E.Thompson,Ms.	C. Montalban.
57	Thompson, Ct.	"
61	Putnam.	"
66	Pomfret.	B. Middle Laurentian.
68	Abington.	"

Ms.	N. Y. & New England R. R.—Con.		
74	Hampton.	B. Middle Laurentian.	
86	Willimantic.	A. Laurentian.	233
95	Andover.	"	
105	Vernon.	"	242
109	Manchester.	C. Montalban.	
115	E. Hartford.	16. Triassic.	
117	Hartford.	"	39
121	Elmwood.	"	
123	Newington.	"	
127	New Britain.	"	179
132	Plainville.	"	191
133	Forrestville.	"	
136	Bristol.	B. Middle Laurentian.	
140	Terryville.	"	
148	Waterville.	"	
150	Waterbury.	"	260
158	Towantic.	"	
161	Southford.	"	
164	Pomperaug Val.	16. Triassic.	
169	Sandy Hook.	B. Middle Laurentian.	
171	Newtown.	"	
174	Hawleyville.	"	306
180	Danbury.	"	397
185	Mill Plain, N. Y.	"	
191	Brewster.	"	406
196	Towner's.	A. Older Laurentian.	433
198	Patterson.	"	
204	Pawling.	A. Older Laurentian.	
207	Poughquag.	3–4. Camb. Sil. Limest.	
210	Stormville.	"	
215	Hopewell.	"	
219	Brinkerhoff.	"	223
221	Fishkill, N. Y.	"	213
225	Matteawan.	2 b. Potsdam.	
228	Fishkill Land'g.	4 c. Lorraine.	
229	Newburgh.	"	

Norwich Division.

0	Worcester.	Mica Schist.
1	S. Worcester.	"
5	Auburn.	"
9	North Oxford.	"
11	Oxford.	"
15	North Village.	B. Middle Laurentian.
16	Webster, Mass.	"
20	{ N. Grosven-	
	ord'le, Ct. }	"
21	Grosvenordale.	"
24	Mechanicsville.	"
26	Putnam.	C. Montalban.
31	Dayville.	"
34	Danielsonville.	"
39	Wauregan.	"
40	Central Village.	"
44	Plainfield.	"
50	Jewett City.	"
58	Greeneville.	"
60	Norwich.	"
73	New London.	Laurentian.

N. Y. & New England R. R.—Con.

Ms.	Hartford Division.	
0	Springfield.	16. Triassic.
3	Armory Station.	"
4	Water-Shops.	"
7	E. Longmeadow.	"
10	Shaker Station.	"
12	Hazardville.	"
16	Melrose.	"
17	Broad Brook.	"
19	Osborn.	"
23	E. Windsor Hills.	"
26	South Windsor.	"
27	Burnham's.	"
29	East Hartford.	"
31	Hartford.	"

Melrose Branch.

16	Melrose.	16. Triassic.
17	Sadd's Mills.	"
19	Ellington.	"
21	Windermere.	C. Montalban.
23	West Street.	"
24	Rockville.	"

Providence Division.

0	Providence.	14. Coal Measures.
4	Cranston.	"
7	Oak Lawn.	"
	Pontiac.	"
9	Natick.	Laurentian.
11	River Point.	"
	Arctic.	"
	Centerville.	"
13	Quidnick.	"
14	Anthony.	"
15	Washington.	"
18	Coventry.	"
	Summit.	A. Laurentian.
24	Greene.	"
27	Oneco.	"
29	Sterling.	"
32	Moosup.	"
35	Plainfield.	C. Montalban.
	Packerville.	"
40	Canterbury.	"
	Jewett City.	"
46	Versailles.	"
48	Baltic.	"
51	Scotland.	"
55	S. Windham.	"
58	Willimantic.	B. Middle Laurentian.

New Haven and Northampton R. R.

0	New Haven.	16. Triassic.	
6	Centreville.	"	
9	Mount Carmel.	"	114
15	Cheshire.	"	166
20	Hitchcock.	"	
	Plantsville.	"	
22	Southington.	"	152
27	Plainville.	"	191

New Haven & N'thampton R. R.—Con.

Ms.			
31	Farmington.	16. Triassic.	204
37	Avon.	"	242
39	Weatogue.	"	
42	Simsbury.	"	167
47	Granby.	"	204
	Congamond.	"	227
55	Southwick, Mass.	"	242
61	Westfield.	"	
68	Southampton.	"	195
71	Easthampton.	"	169
76	Northampton.	16. Triassic and Syenite.	
80	Hatfield.	"	
85	Whately.	16. Triassic.	
88	South Deerfield.	"	
93	Conway.	E. Calcif's Mica Schist.	
95	Conway Junc.	"	
99	Shelburne Falls.	Middle Laurentian.	
108	Charlemont.	D. Huronian.	
111	Zoar.	"	
116	Hoosac Tunnel.	B. Middle Laurentian.	
123	North Adams.	3-4. Camb. Sil. Limest.	

Boston and New York Air Line.

0	New Haven.	16. Triassic.	
5	Montowee.	"	
8	Northford.	"	
12	Wallingford.	"	
18	Middlefield.	"	
19	" Centre.	"	
20	Rockfall.	"	
24	Middletown.	"	83
25	Portland.	"	
30	Cobalt.	C. Montalban.	
33	East Hampton.	"	
36	Lyman Viad.	"	
39	West Chester.	B. Middle Laurentian.	
44	Turnerville.	"	
49	Liberty Hall.	"	
54	Willimantic.	"	

ADDITIONAL RAILROADS IN MAINE.

St. Croix and Penobscot Railroad.

0	Calais.	Granite and Syenite.
2	Milltown.	" "
5	Baring, N. B.	Syenite.
	Princeton, Me.	Calciferous Mica Schist.

Sandy River Railroad.

0	Farmington.	E. Pre-Cambrian.
3	N. Farmington.	" Mica Schist.
11	Strong.	" "
18	Phillips.	" with Limestone.

Bangor and Katahdin Iron Works R. R.

0	Bangor.	Huronian.
39	Milo Junction.	"
45	Brownville.	Cambrian slate quarries.
	Katahdin I. W.	Bog ore making char-coal-iron.

The RAILROADS OF RHODE ISLAND are given in the chapters on Massachusetts and Connecticut.

This blank space is intended for additional geological notes in pencil by the traveler.

Massachusetts.

By Professor W. O. Crosby, of the Massachusetts Institute of Technology, Boston, Mass.

Table of the Geological Formations of Massachusetts.

Cenozoic.		Eozoic.[14]	
20. Quaternary.	20 b. Champlain Clay and Gravel.	4. Taconian.	4 c. Taconian Schist.
"	20 a. Glacial Drift.	"	4 b. Stockbridge Limestone.
19. Tertiary.	19 b. Miocene.	"	4 a. Quartzite.
"	19 a. Eocene.	3. Montalban.	3 f. Serpentine and Chlorite Schist.
Mesozoic.		"	3 e. Hornblende Rock and Schist, and Hydro.-Mica Schist.
16. Triassic.	16. Triassic.	"	3 d. Argillite and Quartzite.
		"	3 c. Mica Schist (many varieties)
Paleozoic.		"	3 b. Gneiss (many varieties)
		"	3 a. Granite.
14. Carbonifer's	14 b. Coal Measures.	2. Huronian.	2 e. Limestone and Serpentine.
"	14 a. Millstone Grit.	"	2 d. Stratified Diorite, Slate, Quartzite, etc.
6. Silurian.	6. Lower Helderberg.	"	2 c. Eruptive Diorite, etc.
5. Cambrian.	5. Acadian.	"	2 b. Petrosilex and Felsite.
		"	2 a. Granite.
		1. Norian.	1. Syenite, etc.

Ms.	Eastern Railroad.	Alt.	Ms.	Eastern Railroad—Con.	Alt.
0	Boston.[1]	20 a. Glacial Drift. [10]	37	Newburyport.	2 a. Gran. & 2 c. Dio.[124]
2	Somerville.	5. Acadian Slate. [8]	39	Salisbury.	"
3	Everett.	20 b. Clay and Gravel.	43	Seabrook.	3 c. Mica Schist.
5	Chelsea.[2]	20 a. Glacial Drift.	47	Hampton.	"
6	Revere.	"	51	Greenland.	"
11	Lynn.[3]	{ 2 b. Petrosilex and Felsite.	57	Portsmouth.	"
			58	Kittery.	" [17]
13	Swampscott.	{ 2 c. Eruptive Diorite, etc.	63	Elliott.	" [21]
			67	Conway Junc.	"
16	Salem.	1. Syenite.	70	S. Berwick Jn.	"
18	Beverly.	2 a. Granite.	75	North Berwick.	"
21	North Beverly.	{ 2 c. Eruptive Diorite, etc.	80	Wells.	2 a. Granite.
			89	Kennebunk.	5. Cambrian.
23	Wenham.	"	94	Biddeford.	" and Granite.
28	Ipswich.	2 a. Gran. & 2 c. Diorite.	95	Saco.	5. Cambrian.
31	Rowley.	"	103	Scarboro.	2. Huronian.
34	Knight's Cross.[4]	" and 2 b. Felsite.	108	Portland.	" [13]

1. The central portion of Boston, embracing the termini of all the railroads entering the city, rests on an unbroken drift formation; but numerous excavations and borings have shown that the underlying rock is the Acadian or Braintree slate. Artesian wells on Causeway and Providence Streets have penetrated the slate to depths of 1,700 and 2,500 feet.
2. The hills in Chelsea and vicinity are fine examples of lenticular drift hills or drumlins.
3. The adjacent rocky peninsula of Nahant consists chiefly of coarse diabase, which intersects Acadian slate and limestone at East Point.
4. This is an interesting locality. South of the station is the Parker River basin, which is a closed synclinal of Acadian slate and conglomerate, resting on banded petrosilex, and including contemporaneous beds of melaphyre. Within half a mile of the station, toward the northwest, are the Devil's Den and Devil's Basin, abandoned quarries of limestone and serpentine, which have afforded specimens of

Eastern Railroad—*Con.*

Saugus Branch.

Ms.		
3	West Everett.	20 b. Clay and Gravel.
5	Malden.	5. Acadian Slate.
7	Maplewood.	"
8	Linden.	20 a. Glacial Drift.
9	Cliftondale.	2 b. Petrosilex & Felsite.
10	Saugus.	"
11	East Saugus.	"
12	Raddins.	"

Swampscott Branch.

13	Swampscott.	2 c. Eruptive Dior., etc.
15	Phillip's Beach.	"
16	Clifton.	"
17	Marblehead.[5]	"

South Reading Branch.

18	Peabody.	2 c. Erupt. Diorite, etc.
22	Lynnfield.	2 a. Granite.
23	Montrose.	2 c. Erupt. Diorite, etc.
25	Wakefield.	"

Salem and Lawrence Branch.

18	Peabody.	2 c. Erupt. Diorite, etc.
20	Danversport.	"
21	Danvers.	"
22	Beaver Brook.	2 d. Stratified Dior., etc.
25	Middleton.	2 c. Erupt. Diorite, etc.
29	Boxford.	3 b. Gneiss.
34	North Andover.	"
36	Lawrence.	3 c. Mica Schist. 6 5

Gloucester Branch.

18	Beverly.	2 a. Granite.
22	Beverly Farms.	"
25	Manchester.[6]	"
27	Magnolia.	"
31	Gloucester.	"
35	Rockport.[7]	"

Essex Branch.

23	Wenham.	2 c. Erupt. Diorite, etc.
24	Hamilton.	"
28	Essex.	2 a. Granite.

Amesbury Branch.

39	Salisbury.	2 a. Granite.
43	Amesbury.	20 a. Glacial Drift.

Dover Branch.

57	Portsmouth.	3 c. Mica Schist.
61	Newington.	"
65	Cushing's.	"
68	Dover.	3 a. Granite.

Conway Branch.

Ms.		
67	Conway Junc.	3 c. Mica S(
69	Salmon Falls.	3 d. Argillit
73	Great Falls.	"
79	Rochester.	3 c. Mica S(
87	Milton.	"
97	Wolfboro Junc.	3 b. Gneiss.
104	N. Wakefield.	"
114	Ossippee.	"
124	Madison.	"
138	Conway.	3 a. Granite

Wolfboro Branch.

97	Wolfboro Jn.	3 b. Gneiss.
109	Wolfboro.	"

Boston and Maine Railr

0	Boston.[1]	20 a. Glacia
2	Somerville.	5. Acadian
4	Edgeworth.	
5	Malden.	{ 2 b. Pet Felsite
6	Wyoming.	2 b. Petrosil
7	Melrose.	2 d. Strat. I
8	Stoneham.	2 b. Pet. & I
9	Greenwood.	2 b. Petrosil
10	Wakefield.	2 c. Erupt.
12	Reading.	2 a. Granite
16	Wilmington.	3 b. Gneiss.
18	Wilmington Jn.	"
20	Lowell Junc.	"
23	Andover.	"
27	Lawrence.	3 c. Mica S(
32	Bradford.	
33	Haverhill.	
36	Atkinson.	
38	Plaistow.	
41	Newton.	
46	East Kingston.	
51	Exeter.	
54	S. Newmarket.	3 a. Granite
57	Newmarket.	"
62	Durham.	"
64	Madbury.	3 d. Argillit
67	Dover.	3 a. Granite
72	Salmon Falls.	3 d. Argillit
78	N. Berwick.	3 c. Mica S(
85	Wells.	3 a. Granite
90	Kennebunk.	5. Cambriar
100	Saco.	"
109	Scarboro.	2. Huronian
116	Portland.	"

Medford Branch.

2	Somerville.	5. Acadian
4	Glenwood.	20 b. Cham
6	Medford.	5. Acadian S

5. The rocky peninsula of Marblehead Neck, lying opposite the town, across the h
posed chiefly of granite (2 a) and many varieties of petrosilex and felsite (2 b). On the
the town are fine exposures of the Norian syenite (1), both stratified and eruptive.
  6. The celebrated singing beach is not far from the station.
  7. The most important of the Cape Ann granite-quarries are in the town of Rockpor

Boston and Maine Railroad—Con.

Ms. | Georgetown and Newburyport Branch.

Ms.	Station	Rock	
10	Wakefield.	2 c. Erupt. Diorite, etc.	
13	Lynnfield.	2 e. Limest. & Serpent'ne	
15	W. Peabody.	2 c. Eruptive Diorite.	
19	Danvers.	"	
25	Topsfield.	2 a. Granite.	
28	Boxford.	2 d. Strat. Diorite, etc.	
31	Georgetown.	2 c. Erupt. "	
34	Byfield.	"	
40	Newburyport.	2 a. Granite.	124

Georgetown and Bradford Branch.

Ms.	Station	Rock	
31	Georgetown.	2 c. Erupt. Diorite, etc.	
34	Groveland.	3 c. Mica Schist, Argil.	
38	Bradford.	"	

Lowell and Andover Branch.

Ms.	Station	Rock	
20	Lowell Junc.	3 b. Gneiss.	103
22	Tewksbury.	"	124
27	Lowell.	3 c. Mica Schist.	99

Dover and Alton Bay Branch.

Ms.	Station	Rock
67	Dover.	3 a. Granite.
75	Gonic.	3 d. Argillite, etc.
77	Rochester.	3 c. Mica Schist.
85	Farmington.	"
91	New Durham.	"
94	Alton.	"
95	Alton Bay.	3 b. Gneiss.

Boston and Lowell Railroad.

Ms.	Station	Rock	
0	Boston.[1]	20 a. Glacial Drift.	12
3	Somerville.	5. Acadian Slate.	8
4	College Hill.	"	21
5	West Medford.	"	21
8	Winchester.	2 c. Erupt. Dior., etc.	27
10	Woburn.	"	
11	Stoneham.	"	
15	Wilmington.	3 b. Gneiss.	97
19	Billerica.	"	110
22	North Billerica.	"	120
26	Lowell.	3 c. Mica Schist	99
28	No. Chelmsford.	"	105
32	Tyngsboro.	3 a. Granite.	
40	Nashua.	3 c. Mica Schist.	124
45	Merrimack.	3 d. Argillite, etc.	
48	Amherst.	3 b. Gneiss.	258
51	Milford.	"	244
55	Wilton.	3 c. Mica Schist.	228
59	So. Lyndeboro.	3 b. Gneiss.	
66	Greenfield.	"	235
71	Hancock Junc.	"	
75	Hancock.	"	
82	Harrisville.	"	
89	Marlboro.	"	278
96	Keene.	"	

Boston and Lowell Railroad—Con.

Ms. | Middlesex Central Branch.

Ms.	Station	Rock	
3	Somerville.	5. Acadian Slate.	
4	W. Somerville.	"	
5	Arlington.	2 a. Granite.	
6	Arlingt'n H'ghts.	2 c. Erupt. Diorite, etc.	
9	East Lexington.	"	
11	Lexington.	"	
15	Bedford.	3 b. Gneiss.	
19	Concord.	"	135
21	Prison Station.	"	

Salem and Lawrence Branches.

Ms.	Station	Rock	
26	Lowell.	3 c. Mica Schist.	99
31	Tewksbury Jn.	3 b. Gneiss.	124
33	Hagget's.	"	
38	Lawrence.	3 b. Mica Schist.	65
34	Wilmington Jn.	3 b. Gneiss.	88
38	North Reading.	2 d. Strat. Diorite, etc.	
43	West Peabody.	"	
46	Peabody.	2 c. Erupt. Diorite, etc.	
48	Salem.	1. Syenite, etc.	

Stony Brook Branch.

Ms.	Station	Rock	
26	Lowell.	3 c. Mica Schist.	99
29	N. Chelmsford.	"	
31	W. Chelmsford.	3 a. Granite.	
33	Westford.	"	102
35	Graniteville.[8]	"	
36	Forge Village.	"	
42	Ayer Junction.	3 c. Mica Schist.	230

Nashua and Acton Branch.

Ms.	Station	Rock	
0	Nashua.	3 c. Mica Schist.	
6	Dunstable.	3 b. Gneiss.	51
9	East Groton.	3 c. Mica Schist.	
15	Westford.	3 a. Granite.	
16	East Littleton.	3 b. Gneiss.	
20	North Acton.	"	
22	Acton.	"	44
23	Prison Station.	"	

Boston, Revere Beach, and Lynn Railroad.

Ms.	Station	Rock	
0	Boston.[1]	20 a. Glacial Drift.	10
1	East Boston.	"	
3	Winthrop Junc.	"	
4	Beachmont.[9]	"	
6	Atlantic.	20 b. Beach Gravel.	
7	Point of Pines.	"	
9	West Lynn.	2 b. Petrosil. and Felsite	
10	Lynn.	"	

8. The Chelmsford granite, so called, is extensively quarried near this station.

9. This railroad runs from Beachmont to Point of Pines on the crest of Revere Beach, a remarkable barrier thrown up by the surf between the sea and the marshes of Revere and Saugus.

10. The celebrated Trilobite quarry, a quarry in the Acadian slate, which has afforded large and fine specimens of Paradoxides Harlani, is on the banks of Hayward's Creek and Weymouth Fore River, two miles southeast of Quincy station, and one mile north of East Braintree station.

11. Fall River is on the boundary between the Carboniferous conglomerate and the Montalban

Ms.	Old Colony Railroad.		Ms.	Plymouth and Sou
0	Boston.[1]	20 a. Glacial Drift. [10]	10	Braintree. [2]
3	Savin Hill.	5. Acadian Conglom.	11	E. Braintree.[10] [5]
4	Harrison Square.	"	12	Weymouth.
5	Neponset.	"	13	N. Weymouth. [2]
6	Atlantic.	"	15	East Weymouth.
7	Wollaston.	20 a. Glacial Drift.	16	West Hingham. [5]
8	Quincy.[10]	5. Acadian Slate.	17	Hingham. [2]
9	Quincy Adams.	2 a. Granite.	19	Nantasket.
10	Braintree.	"	22	Cohasset.
11	South Braintree.	"	25	Egypt.
14	Randolph.	"	27	Scituate.[15] [2]
17	Stoughton.	2 c. Eruptive Diorite.	30	E. Marshfield.
22	North Easton.	2 a. Granite.	34	Marshfield.
24	Easton.	14 b. Coal Measures.	36	Webster Place.
30	Raynham.	"	38	Duxbury.
35	Taunton.	"	39	South Duxbury.
37	North Dighton.	14 a. Millstone Grit.	42	Kingston.
39	Dighton.	"	46	Plymouth.[32]
42	Somerset.	"		
48	Fall River.[11]	"	11	South Braintree. [2]
54	Tiverton.	"	15	S Weymouth.
56	Bristol Ferry.	"	18	N. Abington.
58	Portsmouth.[12]	14 b. Coal Measures.	21	S. Abington.[16] [1]
68	Newport.[13]	"	24	South Hanson.
			30	Plympton. [2]
	Bridgewater and Myrick's Division.		33	Kingston.
11	South Braintree.	2 a. Granite.	18	N. Abington. [2]
15	Holbrook.	"	20	Rockland.
17	East Stoughton.	"	25	Hanover.[16] [1]
20	Brockton.	"		
21	Campello.	14. Carboniferous.		**Cape Cod**
26	Bridgewater.	"	34	Middleboro.[17] [2]
34	Middleboro.	" 96	39	Rock. [3]
42	Myrick's.	"	45	Tremont. [2]
45	Assonet.	3 a. Granite.	49	Wareham.
50	Fall River.[11]	14 a. Millstone Grit.	54	Buzzard Bay.
			62	Sandwich.
	Shawmut and Milton Branches.		69	W. Barnstable.
4	Harrison Square.	5. Acadian Conglom.	73	Barnstable.
5	Shawmut.	5. Acadian Slate.	75	Yarmouth.
6	Cedar Grove.	5. Acadian Conglom.	80	So. Yarmouth.
7	Milton L. Mills.	"	84	Harwich.
8	Mattapan.	"	89	Brewster.
			94	Orleans.
	Granite Branch.		97	Eastham.
6	Atlantic.	5. Acadian Conglomer.	103	Wellfleet.
8	E. Milton.	5. Acadian Slate.	111	Truro.
9	West Quincy.[14]	2 a. Granite.	120	Provincetown.

granite (3 a). There are important quarries in the granite, and the quartzite erate contain Primordial forms of *Lingula.*

12. The most extensive coal-mines in New England are at the Coal Mine

13. The shore east and south of the city gives a very good section of The chasm called Purgatory is on the shore two miles from Newport. New posed of granite and metamorphic slates.

14. The important granite-quarries of Quincy are chiefly in the immedia

15. Outcrops are almost unknown between Scituate and Plymouth, but most points on Huronian granite (2 a).

16. The drift of this region is thick and unbroken, and there is much dou aries of the underlying formations.

17. South and east of Middleboro the rocks are very rarely exposed, ar which the greater part of this division lies, does not include a single outerc land Light, in Truro, on the extremity of Cape Cod, afford fine sections of tl include fragments of calcareous sandstone, filled with characteristic Eoce occurrence of Eocene strata under this part of Massachusetts Bay.

Old Colony Railroad—*Con.*

Fair Haven Branch.

Ms.		
45	Tremont.	20 a. Glacial Drift.
50	Marion.	3 b. Gneiss.
55	Mattapoisett.	"
60	Fairhaven.	"

Wood's Holl Branch.

54	Buzzard Bay.	20 a. Glacial Drift.
58	Pocasset.	"
62	N. Falmouth.	"
65	West Falmouth.	"
71	Wood's Holl.[33]	"

Middleboro and Taunton Branch.

34	Middleboro.	20 a. Glacial Drift. [96]
39	East Taunton.	14. Carboniferous.
44	Taunton.	"

Fall River, Warren, and Providence Division.

49	Fall River.[11]	14. Carboniferous.
52	Swansea.	"
56	Warren.	" [593]
60	Bristol.	"
68	Providence.	"

Fall River Branch.

49	Fall River.[11]	14. Carboniferous.
52	Hemlock.	3 a. Granite.
57	N. Dartmouth.	3 b. Gneiss.
62	New Bedford.	"

New Bedford Branch.

35	Taunton.	14. Carboniferous.
42	Myrick's.	
49	Braley's.	3 a. Granite.
53	Acushnet.	3 b. Gneiss.
56	New Bedford.	"

Attleboro and Taunton Branch.

35	Taunton.	14. Carboniferous.
40	Barrowsville.	"
45	Attleboro.	14 b. Coal Measures.

Fitchburg and Taunton Division.

0	Fitchburg.[24]	3 c. Mica Schist and 3 a. and b. [430]
3	W. Leominster.	3 c. Mica Schist.
5	Leominster.	" [373]
9	Pratt's Junction.	" [429]
12	Sterling.	"
13	Clinton.	3 d. Argillite, etc. [309]
16	Bolton.	3 a. Granite.
18	West Berlin.	3 c. Mica Schist.
20	Berlin.	3 b. Gneiss.
23	Northboro.	"
30	Marlboro.	2 d. Stratif. Diorite.[278]
31	Southboro.	3 b Gneiss. [307]
32	Fayville.	"
35	Framingham.	" [188]
37	S. Framingham.	" [163]
40	Sherborn.	2 d. Strat. Dior., etc.[177]

Fitchburg and Taunton Division—*Con.*

Ms.		
46	Medfield.	2 d. Strat. Dior., etc.
50	Walpole.	14 a. Millstone Grit.[157]
53	South Walpole.	" [227]
55	Foxboro.	2 a. Granite. [284]
58	Mansfield.	14 b. Coal Measures.[172]
63	Norton.	"
65	Crane's.	14. Carboniferous.
69	Taunton.	"

Lowell and Framingham Division.

0	Lowell.	3 c. Mica Schist. [99]
4	Chelmsford.	3 b. Gneiss.
6	S. Chelmsford.	"
9	Carlisle.	"
13	Acton.	" [44]
15	Concord Junct.	" [135]
18	North Sudbury.	2 d. Strat. Diorite.
20	Sudbury.	" [127]
22	South Sudbury.	3 b. Gneiss.
26	Framingham.	" [188]

Boston and Providence Railroad.

0	Boston.[1]	20 a. Glacial Drift. [6]
2	Roxbury.	5. Acadian Conglom. [20]
4	Jamaica Plain.	" [33]
5	Forest Hills.	" [36]
6	Mount Hope.	5. Acadian Slate.
7	Clarendon Hills.	2 b. Petrosil. & Fels. [50]
8	Hyde Park.	5. Acadian Conglom. [51]
9	Readville.	" [61]
14	Canton Junct.	2 a. Granite.
15	Canton.	2 c. Erupt. Diorite. [101]
18	Stoughton.	" [220]
18	Sharon.	" [220]
22	East Foxboro.	2 a. Granite. [211]
24	Mansfield.	14 b. Coal Meas. [169]
26	West Mansfield.	"
31	Attleboro.	" [129]
35	North Attleboro.	"
38	Hebronville.	"
39	Pawtucket.	"
40	Providence.	14. Carboniferous.

Dedham Branch.

5	Forest Hill.	5. Acadian Conglom. [36]
6	Roslindale.	5. Acadian Slate.
8	West Roxbury.	"
10	Dedham.	2 a. Granite.

New York and New England Railroad.

0	Boston.	20 a. Glacial Drift. [10]
3	Dudley St.	5. Acadian Conglom.
4	Mount Bowdoin.	"
5	Dorchester.	5. Acadian Slate.
6	Mattapan.	2 b. Petrosil. & Felsite.
8	Hyde Park.	5. Acadian Conglom. [51]
10	Readville.	" [61]
11	Elmwood.	2 a. Granite.
13	Ellis.	"
15	Norwood.	"

New York and New England Railroad—Continued.

Ms.		
19	Walpole.	14 a. Millstone Grit.
23	Norfolk.	2 c. Eruptive Diorite.
27	Franklin.	"
30	Wadsworth's.	"
36	Blackstone.	3 c. Mica Schist. 197
40	Ironstone.	3 b. Gneiss.
46	East Douglas.	" 517
48	Douglas.	"
52	East Thompson.	"

Southbridge Extension.

Ms.		
52	East Thompson.	3 b. Gneiss.
58	East Webster.	3 c. Mica Schist.
59	Webster.	"
64	Quinnebaug.	"
67	West Dudley.	3 b. Gneiss.
70	Southbridge.	"

Woonsocket Division.

Ms.		
0	Boston.[1]	20 a. Glacial Drift. 10
10	Newton Upper Falls.	5. Acadian Congl.
12	Needham.	2 b. Petrosil. & Felsite.
14	Charles River.	2 a. Granite.
16	Dover.	"
20	Medfield.	2 c. Eruptive Diorite.
25	Medway.	"
29	N. Bellingham.	3 c. Mica Schist.
35	E. Blackstone.	"
38	Woonsocket.	"

Norwich Division.

Ms.		
0	Worcester.[18]	3 c. and d. Argillite and 3 a. and b. 475
4	Auburn.	3 c. Mica Schist.
9	North Oxford.	"
11	Oxford.	"
16	Webster.	3 b. Gneiss.

Hartford Division.

Ms.		
0	Springfield.	16. Triassic. 175
7	E. Longmeadow.	"

Providence Extension.

Ms.		
27	Franklin.	2 c. Erupt. Dio., etc.292
31	W. Wrentham.	2 a. Granite.
33	Diamond Hill.	3 b. Gneiss.

Providence and Worcester Railroad.

Ms.		
16	Woonsocket.	3 c. Mica Schist.
18	Blackstone.	" 197
25	Uxbridge.	3 b. Gneiss. 231
31	Northbridge.	" 269
35	Sutton.	" 331
38	Millbury.	" 392
44	Worcester.[18]	3 c. and d. Argillite, and 3 a. and b. 475

Boston and Albany Railroad.

Ms.		
0	Boston.[1]	20 a. Glacial Drift. 10
5	Brighton.	5. Acad. Sl. & Congl. 24
7	Newton.	" 46
10	Auburndale.	" 63
12	Newton Lower Falls.	20 a. Glacial Drift.
13	Wellesley Hills.	2 a. Granite.
15	Wellesley.	" 140
18	Natick.	2 a. and d. Granite & Strat. Diorite. 170
21	S. Framingham.	3 b. Gneiss. 163
24	Ashland.	" 184
28	Southville.	" 263
32	Westborough.	" 300
38	Grafton.	" 368
44	Worcester.[18]	3 c. & d. Schist & Argillite, also 3 a. & b. Gran. & Gneiss. 473
53	Rochdale.	3 b. Gneiss. 731
57	Charlton.	" 888
62	South Spencer.	" 704
67	Brookfield.	" 606
69	West Brookfield.	" 604
73	Warren.	" 593
79	West Brimfield.	" 391
84	Palmer.	" 336
89	N. Wilbraham.	" 264
92	Indian Orchard.	16. Triassic. 241
99	Springfield.	" 70
108	Westfield.	" . 147
116	Russell.	3 c. Mica Schist. 273
120	Huntington.	" 373
126	Chester.[19]	3 c. Mica Schist and 3 e. and f.
131	Middlefield.	3 b. Gneiss. 595
135	Becket.	" 1207
138	Washington.	" 1437
142	Hinsdale.	" 1431
146	Dalton.	4 a. Quartzite. 1198
151	Pittsfield.	4 b. Limestone. 1013
159	Richmond.[20]	" 1047
162	State Line.	4 c. Taconic Schists. 914

Brookline and Newton Highlands Branch.

Ms.		
0	Boston.[1]	20 a. Glacial Drift. 10
4	Brookline.	5. Acad. Sl. & Congl. 16
6	Reservoir.	"
8	Newton Centre.	" 46
9	Newton Highl'ds	"

Milford Branch.

Ms.		
21	S. Framingham.	3 b. Gneiss. 163
25	East Holliston.	" 169
26	Holliston.	" 191
30	Braggville.	"
12	Milford.	" 244

18. The Worcester slates include a bed of anthracite one mile east of the city. It was mined fifty years ago, and granite is now quarried in that vicinity, on Millstone Hill.

19. The emery-mine, one half mile from the station, is an important mineral locality. One mile west of the station the railroad crosses an immense bed of serpentine (3 f).

20. The Taconian limonite deposits are extensively mined in Richmond, and the celebrated boulder trains are in the western part of the town.

Boston and Albany Railroad—*Con.*
Webster Branch.

Ms.		
44	Worcester.[18]	3 c. & d. & 3 a. & b.[473]
48	Jamesville.	3 b. Gneiss. [564]
54	N. Oxford Mills.	"
56	Howarth's.	"
60	Webster Mills.	"

Ware River Branch.

Ms.		
0	Winchendon.	3 b. Gneiss. [993]
6	Baldwinville.	" [901]
10	Templeton.	" [964]
16	Williamsville.	" [633]
22	Cold Brook.	" [672]
25	Barre Plains.	" [588]
33	Gilbertville.	" [546]
37	Ware.	" [489]
45	Thorndike.	" [345]
49	Palmer.	" [336]

Athol Branch.

Ms.		
0	Springfield.	16. Triassic. [70]
7	Indian Orchard.	" [241]
11	Red Bridge.	3 b. Gneiss.
17	Three Rivers.	"
19	Bondsville.	" [350]
23	West Ware.	" [387]
27	Enfield.	" [415]
31	Greenwich.	" [445]
38	North Dana.	" [462]
40	New Salem.	" [522]
43	South Athol.	" [561]
49	Athol.	" [546]

Pittsfield and North Adams Branch.

Ms.		
0	Pittsfield.	4 b. Limestone. [1013]
3	Coltsville.	"
6	Berkshire.	"
9	Cheshire.[21]	"
12	Cheshire Harb'r.	"
14	Adams.	"
20	North Adams.[22]	" [686]

Worcester, Nashua, and Rochester Railroad.

Ms.		
0	Worcester.[18]	3 c. & d. and 3 a. & b.[476]
9	West Boylston.	3 c. Mica Schist. [442]
12	Sterling Junc.	"
17	Clinton.	3 d. Argillite and c. [309]
19	Lancaster.[23]	" [259]
25	Harvard.	" and 3 a.[288]
28	Ayer Junc.	" and 3 c.[230]
31	Groton.	3 c. Mica Schist. [303]
36	Pepperell.	" [205]
41	Hollis.	"

Worcester, Nashua, and Rochester Railroad—*Con.*

Ms.		
46	Nashua.	3 c. Mica Schist.
49	Hudson.	" [221]
57	Windham.	"
63	Hampstead.	"
70	Fremont.	" [60]
74	Epping.	"
80	Lee.	3 b. Gneiss.
88	Barrington.	3 c. Mica Schist.
93	Gonic.	"
95	Rochester.	"

Boston, Barre, and Gardner Railroad.

Ms.		
0	Worcester.[18]	3 c. & d. and 3 a. & b.[476]
3	Barber's.	3 c. Mica Schist.
6	Chaffin's.	"
8	Holden.	" [755]
10	Jefferson's.	3 b. Gneiss.
13	Brooks.	" [30]
16	Princeton.	"
20	Hubbardston.	"
27	Gardner.	" [1009]
38	Winchendon.	" [993]

Fitchburg Railroad.
Hoosac Tunnel Route.

Ms.		
0	Boston.[1]	20 a. Glacial Drift. [11]
3	Somerville.	5. Acadian Slate. [8]
4	Cambridge.	"
6	Belmont.	" and 2 c.[73]
7	Waverly.	" " [133]
10	Waltham.	" "
12	Stony Brook.	2 c. Erupt. Dior., etc.[91]
13	Weston.	" [95]
17	Lincoln.	2 d. Strat. Dior., etc.[205]
20	Concord.	3 b. Gneiss. [135]
22	Concord Junc.	"
25	South Acton.	" [199]
32	Littleton.	" and 3 c.[228]
36	Ayer Junction.	3 c. and 3 d. [230]
40	Shirley.	3 d. Argillite. [283]
42	Lunenburg.	"
45	Leominster.	3 c. Mica Schist. [373]
50	Fitchburg.[24]	" & 3 a. & b.[430]
54	Wachusett.	3 b. Gneiss.
60	Ashburnham.	" [1106]
65	Gardner.	" [1009]
71	Baldwinville.	" [891]
77	Royalston.	"
83	Athol.	" [546]
87	Orange.	" and 3 a.
90	Wendell.	3 a. Granite and 3 b.
92	Erving.	3 b. Gneiss.

21. The celebrated Berkshire sand, used in glass-making, results from the disintegration of the Taconic quartzite, and is most extensively quarried in the town of Cheshire.

22. At the Natural Bridge, one and a half miles northeast of the station, is a fine gorge cut out of the Taconic limestone, and a large marble-quarry.

23. The micaceous argillite of Lancaster is noted for the numerous and fine crystals of chiastolite which it contains.

24. Rollstone Hill, immediately south of the city, and Pearl Hill, two miles north, are interesting localities for minerals and rocks. Rollstone Hill is a boss of micaceous granite (3 a.) which is extensively quarried.

Fitchburg Railroad—
Hoosac Tunnel Route—Con.

Ms.			
98	Miller's Falls.	3 b. Gneiss and 8.	292
102	Montague.	16. Triassic.	129
106	Greenfield.	" Sandst. & Trap.	
110	West Deerfield.	" and 3 c.	181
114	Bardwell's.	3 c. Mica Schist.	238
119	{ Shelburne Falls.²⁵	} 3 b. Gneiss.	430
122	Buckland.	3 c. Mica Schist.	
128	Charlemont.	"	
132	Zoar.	"	
136	Hoosac Tun'l.²⁶	3 e. and 3 f.	
	Hoosac Mount.	.	2610
	Do., E. Summit, over Tunnel.		2269
	Hoosac Tunnel, East Portal.		759
	Do., Cent. Shaft.		819
	Do., West Portal.		759
143	North Adams.²²	4 b. Limestone.	686
148	Williamstown.	"	560
152	Pownal.	" and 4 c.	

Watertown Branch.

5	Fresh Pond.	20 b. Champlain Clay.	
6	Mount Auburn.	5. Acadian Slate.	
8	Watertown.	"	
10	Waltham.	"	

Marlborough and Hudson Branch.

25	South Acton.	3 b. Gneiss.	199
28	Maynard.	"	
31	Whitman's Cros.	"	
32	Rockbottom.	"	
34	Hudson.	"	221
38	Marlboro.	"	278

Ms.	Peterboro and
36	Ayer Junction.
40	West Groton.
44	Townsend Harb.
46	Townsend Cent'r
48	W. Townsend.
52	Mason Centre.
55	Pratt's.
60	Greenville.

Turner's F

0	Greenfield.
3	Montague City.
5	Turner's Falls.²⁷

New London No

50	Stafford.
61	Monson.
65	Palmer.
68	Three Rivers.
70	Barrett's Junc.
75	Belchertown.
80	Dwight's.
85	Amherst.
88	North Amherst.
91	Leverett.
94	Mount Toby.
96	Montague.
100	Miller's Falls.
103	Northfield F"ms.
109	Northfield.
111	South Vernon.
116	Vernon.
121	Brattleboro.

25. The falls of the Deerfield River are near the station, and are int
numerous large pot-holes exposed, and the contortions and metamorphism
marks an important anticlinal axis. One mile west of the station ancient
the railroad cut, fifty feet above the present bed of the river.

26. The rocks traversed by the tunnel are well shown in the vast depos
station and the eastern portal. The side of the mountain above the portal i
that crosses the Boston and Albany Railroad near Chester. One half mil
quarry in soapstone and chlorite schist, affording green foliated talc.

Travelers on the Boston and Albany, and Fitchburg Railroads, have a go
the stratigraphy of the mountainous district between the Berkshire and Con

The main Hoosac range is probably an overturned or broken anticlinal,
all dipping to the east. A synclinal axis is reached at Chester, on the Bo
near Zoar, on the Fitchburg.

Beyond this the strata dip to the west until we reach the anticlinal axis
Fitchburg, beyond which they dip to the east again for about eight miles, o
assic beds.

The second anticlinal is not exposed on the Boston and Albany road, p
before it reaches that line.

27. The noted locality of fossil footmarks is on the west bank of the ri
above the village. W. W. Draper was the first person to observe them, in
they were "turkey tracks made two thousand years ago." His impressio
Colonel Wilson, who called the attention of Dexter Marsh to them. Mr. 1
slabs, and showed them to Dr. James Dean, who requested Professor E. Hitc
scientifically. This was done, and the results accumulated in the Hitchcoc
Amherst, where are over twenty thousand separate ichnites, illustrating abo
species, all from the Connecticut Valley.

28. This is the locality furnishing for the Amherst Museum the large row
Giganteum, the largest of the Triassic birds. Across the river, in South Ha
ity of *Otozoum Moodii*, so named for Pliny Moody, who was the first person
known to have observed any of the footmarks. A specimen is preserved v
saying that "the tracks were made by Noah's raven."

29. This is the town where the celebrated Helderberg limestone crops
remnant of a once extensive deposit, preserved accidentally from erosion, a
beneath the Coos quartzite.

Ms.	Connecticut River Railroad.		
0	Springfield.	16. Triassic.	70
4	Chicopee.	"	79
6	Chicopee Falls.	"	
8	Holyoke.	"	94
13	Smith's Ferry.	"	122
15	Mount Tom.[28]	"	
17	Northampton.	" and 3 a.	125
21	Hatfield.	16. Triassic.	
14	North Hatfield.	"	172
26	Whateley.	"	186
28	South Deerfield.	"	207
33	Deerfield.	"	221
36	Greenfield.	"	181
43	Bernardston.[29]	{ " and 3 c. / and d.	359
50	South Vernon.	3 c. and 3 d.	

Ms.	New Haven and Northampton Railroad.		
47	Granby.	16. Triassic.	
55	Southwick.	"	242
61	Westfield.	"	147
68	Southampton.	"	195
72	Easthampton.	"	169
77	Northampton.	" and 3 a.	125

Ms.	New Haven and Northampton Railroad—Continued.		
80	Florence.	3 a. Granite.	273
82	Leeds.	"	336
84	Haydenville.	"	432
85	Williamsburg.	" and 3 c.	492
88	South Deerfield.	16. Triassic.	207
93	Conway.	3 c. Mica Schist. '	
99	Shelb'rne F'ls.[25]	3 b. Gneiss.	430

Ms.	Housatonic Railroad.		
75	Ashley Falls.	4 b. Limestone.	
79	Sheffield.	"	
85	Gt. Barrington.	"	
87	Van Deusenville.	"	
89	Housatonic.	4 a. Quartzite.	
91	Glendale.	" and 4 b.	
93	Stockbridge.	4 b. Limestone.	
99	Lee.[30]	"	
102	Lenox.	"	
106	Deweys.	"	
110	Pittsfield.	"	1013
87	Van Deusenville.	4 b. Limestone.	
95	W. Stockbridge.	"	
98	State Line.	4 c. Taconian Schists.	

30. The Taconic limestone is here a beautiful white marble, and it is extensively quarried. Less important quarries, worked for lime or marble, occur the entire length of the Berkshire Valley.

31. Amesbury. This and the adjoining towns, also the immediate city of Boston, are chiefly occupied by a profusion of lenticular-shaped drift hills, believed to be moraines of ancient glaciers, and different from the usual ground moraine of glacial drift. The hills may be two hundred feet high, and their longer axes run southeasterly, being parallel with the course of the striæ in the neighborhood. They consist of till, and resemble the drumlins of Scotland. They also occur conspicuously in southern New Hampshire, and other parts of New England, and in western New York. In the Merrimack and Connecticut Valleys a few have been found having a direction to the south and west of south, but agreeing with the course of adjoining striæ.

32. Plymouth. This township is said to contain three hundred and fifty-six ponds. These lie in hollows of the drift.

33. Wood's Holl. The extreme terminal moraine of the ice-sheet, which constitutes the "backbone" of Long Island, also Block Island, and the hilly part of Martha's Vineyard, from Gay Head to Vineyard Haven. It also appears at Chappaquiddick and Tuckernuck Islands, and forms Saul's Hills and Sankaty Head on Nantucket. A *second terminal moraine*, five to fifteen miles north from the foregoing, extends on the north shore of Long Island, from Port Jefferson to Orient Point, forms Plum and Fisher's Islands, reaches along the south shore of Rhode Island, from Watch Hill nearly to Point Judith, forms the chain of Elizabeth Islands, and continues on the peninsula of Cape Cod, from Wood's Holl to North Sandwich, and thence east to Orleans.

The portions of Martha's Vineyard, Nantucket, and Cape Cod, south of these moraines, and also Eastham, Wellfleet, and Truro, are modified drift.

Manomet Hill, east of Plymouth, is a moraine connected with that of Cape Cod and the Elizabeth Islands.

34. The numbers attached to the Norian, Huronian, Montalban, and Taconian, and their subdivisions, are used for convenience in this chapter; they only apply to Massachusetts, and are not intended to indicate correlation with formations similarly numbered in other parts of the book.

Notes 31, 32, and 33 are by Prof. Warren Upham; and 28 and 29 are by Prof. C. H. Hitchcock, from the first edition.

This blank space is intended for additional geological notes in pencil by the traveler.

New York.

By James Macfarlane.[1]

GEOLOGICAL FORMATIONS OF THE STATE OF NEW YORK.[2]

FORMATIONS AND SUB-DIVISIONS.	FORMATIONS AND SUB-DIVISIONS.
20. Quaternary.	**7. Lower Helderberg.*** *(Upper Silurian.)*
16. Triassic.	**6. Waterlime.**
	6. Salina or Onondaga Salt group.
12. Catskill.	**5 c. Niagara.**
11 b. Chemung.	**5 b. Clinton.**
11 a. Portage, { 3. Portage s. s. / 2. Cardeau shales. / 1. Chasaqua shales.	**5 a. Medina,** { 2. Medina Sandstone. / 1. Oneida Conglom.
10 c. Genesee.	**4 c. Hudson River,** { 3. Lor. sha. / 2. Frankfort. sh. & s. s.
10 b. Hamilton, { 3. Tully Limestone. / 2. Moscow shales. / 1. Hamilton shales.	**4 b. Utica.**
10 a. Marcellus.	**4 a. Trenton,** { 3. Trenton l. s. / 2. Black River l. s. / 1. Birdseye l. s.
9 c. U. Held'berg { 4. Seneca l. s. / 3. Corniferous l. s. / **or Corniferous,** { 2. Onond'a l. s. / 1. Schoharie.	**3 b. Chazy.** / **3 a. Calciferous.**
9 a. Cauda Galli. / **3 Oriskany.**	**2 b. Potsdam** = dicellocephalus beds. / **2 a. Acadian** = paradoxides beds. [Note 2] / **2 á. Georgian** = olenellus beds.
*Consisting in the ascending order of : 1, the Tentaculite limestone ; 2 Pentamerus limestone ; 3, Delthyrus shaly limestone ; 4, Encrinal limestone ; and 5 Upper Pentamerus limestone.	**1 d. Montalban.** / **1 c. Norian.** / **1 a. Laurentian.**

Left-column group labels: Devonian. Right-column group labels: Upper Silurian; Lower Silurian or Ordovician; Cambrian; Archaean.

GENERAL NOTE. The State of New York is to the geologist what the Holy Land is to the Christian, and the works of her Palæontologist are the Old Testament Scriptures of the science. It is a Laurentian, Cambrian, Silurian and Devonian State, containing all the groups and all the formations of these long ages, beautifully developed in belts running nearly across the State in an east and west direction, lying undisturbed as originally laid down. Railroads running north and south pass over a number of the formations in short distances, while those running east and west run for long distances on the same formation, as for example the N. Y. C. & H. R. R. R. on the 6. Salina, and the Erie Railway on the 11 b. Chemung. In the eastern part of the State the formations are more irregularly disposed. New York localities are those to which we must always go back as the standard by which any disputed formation of these ages is to be tested.

1. The author has bestowed more of his own labor and research on the local geology of this State, than any other, having besides diligent study of all the official reports, made personal observations of the exposures of the formations in traveling for many years on all the railroads. It was from making geological notes on the margin of railroad time tables that he conceived the idea of this geological railway guide book for the State, and by calling in the aid of scientific gentlemen of other States, he has been enabled to extend it over the whole United States and Canada. To Prof. James Hall, of Albany, the State Geologist, he is indebted for much information as to some of the localities in this State. [Note to first edition.] In revising this chapter the editor has made changes in the first edition only where recent investigations have rendered them necessary. In the revision he has been advised by the gentlemen whose names appear as authority for new lines and new notes and especially by Prof. W. B. Dwight of Vassar College. When no authority is given for any portion of the chapter, it will be understood that it has been taken from the first edition. J. R. M.

2. The table here given is not satisfactory to all of the contributors to this chapter, but, where terms are used by them in a different sense, the change is indicated by the number or otherwise. The Cambrian, as given in the table, is also divided into Lower (2 á), Middle (2 a.) and Upper (2 b.). In the first edition "Cambrian" included 2 b.—4 c. and was divided into Lower (2 b.), Middle (3 a., 3 b. (Quebec), and 3 c. (Chazy)), and Upper (4 a., 4 b., and 4 c.) J. R. M.

3. N. Y. C. & H. R. R. R. GRADES CAUSED BY GEOLOGICAL STRUCTURE.—This railroad undoubtedly occupies the finest locality for an east and west railroad in the United States. It owes this to geological structure, the outcrop of the formations running east and west, and the Salina or Onondaga, Utica and Hudson River soft shales are cut into low valleys through which the railroad and Erie Canal are built. If the formations had run north and south, as they do in Pennsylvania, Maryland, etc.,

New York Central and Hudson River Railroad.[3]		Alt.	New York Central Railroad.
Ms.			Ms.
0	New York.[178] See Note 4.	22	34 Croton.
11	Spuyten Duyvil. 1 a. Laurentian.		37 Crugers.
12	Riverdale.[5] "		38 Montrose.[6]
13	Mt. St. Vincent. "		41 Peekskill.
15	Yonkers. "		45 { Ft. Montgomery.
19	Hastings. "		{ Highlands.
20	Dobb's Ferry. "	12	
22	Irvington. "		49 { Garrison's.
25	Tarrytown. "		{ (West Point.)
29	Scarborough. "		52 Cold Spring.
30	Sing Sing.[5] "	9	54 Cornwall.[6]

and been turned up edgewise, the hard sandstones would have been high r
tains to overcome, as they are everywhere from the Mohawk Valley to Ala
stone ridge of the Helderberg range, which bounds this valley on the sou
direction, as the 2-4. formations do, a tunnel would probably hav
the western part of the State these Helderberg limestones continue, but no
The road via Geneva, runs on them at Auburn, Clifton Springs, etc., but w
than the direct road, and at Buffalo they are level with the plain. It shoul
Laurentian mountains at Little Falls and at Peekskill have been cloven f
opening the gateways for the traffic and travel of the West. The popular in
is a level plain like the prairies of the West, derived from traveling on t
is altogether erroneous. There is only a narrow trough through the centr
the railroad and canal are located, that is of this level character.

4. New York island is 12 miles long and nearly two miles wide. The v
one-quarter miles at 14th St. Below Grand street it gradually becomes r
north end. The lower part of the city, below Wall street, is half a mile '
island is gneiss, except a portion of the north end, which is limestone. The
with deep alluvial deposits, which in some places are more than 100 feet in
croping of the gneiss appeared on the surface about 16th street, on the east
diagonally across to 31st street on 10th Avenue. North of this much of the
It contains a large portion of mica, a small proportion of quartz and still les
an abundance of iron pyrites in very minute crystals, which, on exposur
consequence of these ingredients it soon disintegrates on exposure, render
poses of building. The erection of a great city, for which this island fu
very greatly changed its natural condition.

Dr. Hunt claims that the New York gneiss is in great part of Montalba
with that of Philadelphia, Baltimore and Washington, and that it rests upo
of the Highlands, which he says is the surface rock in the northern part of
Dana thinks it extremely probable that the limestone and conformably asso
ter County and New York Island, as well as those of the Green Mountain
New York Island, are metamorphosed Lower Silurian (including Cambrian

5. On the opposite side of the river may here be seen for many miles th
mountain ridge close to the water's edge. Its upper half is a perpendicular
columnar structure from 100 to 200 feet in height, the whole height of the r
from 400 to 600 feet, and the highest point in the range opposite Sing Sing l
son, known as the High Torn. The width of the mountain is from a half mile
western slope being quite gentle. In length it extends from Bergen Point
erstraw, and then westward in all 48 miles, the southern portion being r
lower half of the ridge on the river side, is a sloping mound of detritus, c
accumulated at the base of the cliff, being derived from its weathered ai
talus and the summit of the mountain are covered with trees, with the ba
the Palisades between. Viewed from the railroad or from a steamboat on t
precipice with its huge weathered masses of upright columns of bare
straight, unbroken ridge overlooking the beautiful Hudson River, is cert
esque. Thousands of travelers gaze at it daily without knowing what it is.

This ridge consists of a great sheet of basalt lying upon 16. Triassic sar
glomerates, which are often exposed along the river bank extending up the
for a considerable distance to an irregular contact with the igneous rock.
the trap has come from below as a dike through a long rent'or fissure and
by intrusion between the layers of sedimentary rock. Subsequent erosion l
ing strata near the crest line and for some distance back but at many p
side of the ridge, the dike structure and relations to the overlying st
See Notes 145 and 134.

(See description of the 16. Triassic formation and its Trap Dikes.) Her
uncommon instance of a great geological blank. On the east side of this
long either to the Archaean and oldest rocks, or to the Cambro-Lower S
while on the west side they are No. 16. all the intermediate Silurian, Dev
formations being wanting. This state of things continues all along the J
the 18. Cretaceous or 17. Jurassic taking the place of the 16. Triassic farther

6. 38 Montrose to 54 Cornwall. This celebrated passage of the Hudson
is a gorge nearly 20 miles long from 3 miles south of Peekskill to Fishkill, a
Laurentian rocks far below mean tide water. The hills on its sides
much as 2,600 feet, and in many places the walls are very precipitous. The
that is not easily disintegrated or eroded, nor is there any evidence of ar

New York Central & Hudson River Railroad.—Continued.[7]		Alt.
57 { Dutchess and Columbia Junction.[7]	4 c. Hud. Riv. Group.	
58 Fishkill.	"	213
62 Low Point.	"	
64 New Hamb'g.[118]	Calciferous-Trenton.	
69 Camelot.	4 c. Hud. Riv. Gr'p.[139]	
73 Poughke'psie.[119]	"	
78 Hyde Park.	"	
83 Staatsburg.	"	
88 Rhinebeck.	4c.&H.R. 2á.&2b.Cam.	
94 Barrytown.	"	
98 Tivoli.	"	
104 Germantown.	"	
107 Livington.	"	
109 Catskill.	"	
114 Hudson.[9]	4 b. Utica.	
118 Stockport.	2 á. Cambrian.	
121 Coxsackie.	"	
123 Stuyvesant.	4 c. H'd. R. & 2 á.Cam.	
129 Schodack.[8],[120]	"	
133 Castleton.	4 c. Hudson River.	19
142 East Albany.	"	28
142 Albany.[10],[121]	"	30
148 Troy.[7],[10]	4 c.Hud. R. & 2 á. Cam.	

New York Central & Hudson River Railroad.—Continued.		Alt.
142 Albany.[10],[131]	4 c. Hudson Riv., 27 m.	
145 West Albany.[11]	"	196
160 Schenectady.[122]	4 b. Utica,	246
169 Hoffman's Ferry.	4 b. Utica, 7 miles.[266]	
174 Crane's Village.	"	270
176 Amsterdam.[13]	4 a. Trent. 10 ms.	279
182 Tribes Hill.[117]	" quar. 1 m.[308]	
187 Fonda.[13]	4 b. Utica, 5 miles.[299]	
192 Yost's.[14]	{ Two bluffs or noses { of Calc. on Laur.[300]	
195 Spraker's.[14]	{ 3 a. Calc. hill. Laur'n { at R. R. track.[301]	
198 Palatine Bridge. [15,180]	{ 4 a. Trent. 3 ms.[304] { Hills to north Calcif.	
200 Fort Plain.[16]	{ 4 a. Trenton, 18 ms. { and Huds'n Riv.[305]	
206 St. Johnsville.[180]	"	319
209 East Creek.	"	334
216 Little Falls.[17]	1 a.Lauren'an, 1 m.[376]	
223 Herkimer.[180]	4 b. Utica, 28 miles.[398]	
225 Ilion.	"	400
227 Frankfort.	"	402
237 Utica.[18]	"	410
241 Whitesboro.[19]	"	415
244 Oriskany.[20]	4. c. Hud. Riv. 8 m.[423]	

It is clearly a case of erosion, but not by the present river, which has but very slight fall in crossing them to join tide water near Peekskill. This therefore was probably a work mainly performed in some past period when the continent was at a higher level. Most likely it is a valley of great antiquity. Also see note 17.

7. From Dutchess Junction to Troy, revised by Prof. W. B. Dwight, from Rhinebeck to Troy the stratigraphy being given on the authority of Mr. S. W. Ford, except that his nomenclature has been modified so as to harmonize with that adopted in this chapter.

8. *Schodack.* A series of great dislocations with upthrows on the east side traverse eastern North America from Canada to Alabama. One of these great faults has been traced from near the mouth of the St. Lawrence River, keeping mostly under the water up to Quebec just north of the fortress, thence by a gently curving line to Lake Champlain or through Western Vermont across Washington and Rensselaer Counties into Columbia County. The line of faulting has been recently traced southward to Schodack Landing and to the south of Poughkeepsie and is supposed to run in to another series of faults, probably of a later date, which extend as far as Alabama. It brings up the rocks of the 2 b. Potsdam group in Vermont and New York on the east side of the fracture to the level of the 4 c. Hudson River and 4 a. Trenton l. s. on the west. In some places the Trenton appears on the east. J. M.

This fault is met with, a little more than half a mile east of Troy along the line of Jacob street. The rocks upon its eastern side (Potsdam) there hold an interesting fauna. From that point the fault takes a somewhat irregular course, being nearly two miles inland from the Hudson at Greenbush, and comes out upon the Hudson about a mile and a half south of Schodack landing. 8. W. F.

9. *Catskill Mountains.* For many miles on this railroad are beautiful views of the Catskill Mountains, 3,000 feet high, (12. Catskill,) several miles distant on the opposite or west side of the river and which furnish the name for the Catskill formation. The wide valley between them and the river is composed of 11 b. Chemung, 10. Hamilton, 7 Lower Helderberg and 4 c. Hudson River. The geology on the east or railroad side is entirely different.

10. *Albany.* The clay beds at Albany are more than 100 feet thick, and between that city and Schenectady they are underlaid by a bed of sand that is in some places more than 50 feet thick. There is an old glacial clay and boulder drift below the gravel at Albany, but Professor Hall says 't is not the estuary stratified clay. At the south end of the city of Troy the gravel and sand beds are subject to dangerous land slides. See also Note 121.

11. The distant mountain to the southwest is the Helderberg range. See notes 24 and 41.

12. *Amsterdam.* Precipice of 4 a. Trenton limestone back of the town, and quarries at the track. For 40 miles to Little Falls the railroad runs on Trenton limestone 3 a. Calciferous, 4 b. Utica and c. Hudson River irregularly alternating. See also Note 180.

13. Branch railroad north to Johnstown and Gloversville, in a valley of Utica slate.

14 Between Fonda and Palatine Bridge are fine bluffs of 3 a. Calciferous. The talus of fragments of rock at the foot of the precipice whiten out in weathering like the stones about an old lime-kiln. It is from the cavities of the Calciferous that the beautiful quartz crystals are produced, of which great quantities have been found. A similar bluff on south side of river. No Potsdam here.

15. The railroad skirts along the base of a ridge of Trenton limestone here and at Fort Plain.

16. At Fort Plain village the transition from the Birdseye to the Trenton limestone is to be seen, the first layers of the latter being of a drab color.

17. At Little Falls for one mile is a rare opportunity of seeing the 1 a. Laurentian formation, being a gorge cut by the Mohawk River through a spur of the Adirondack Mountain, which here crosses the railroad. You are now on the bottom rocks of the geological series, for nothing older

New York Central & Hudson River Railroad.—*Continued.*			
Ms.			Alt.
251	Rome.[21]	445	4 c. Hudson River.
255	Green's Cors.[22]		5 a. Medina, 2 ms.[466]
259	Verona.[23]	467	4 b. Clinton 9 miles.
264	Oneida.[24]	440	4 c. Niagara. 3 miles.
266	Wampsville.[25]		"
269	Canastota.[26]	426	{ 6. Salina or Onondaga Salt group. 23 miles.

New York Central & Hudson River Railroad.—*Continued.*			
Ms.			Alt.
273	Canaseraga.	418	{ 6. Salina or Onondaga Salt group.
275	Chittenango.		" 417
279	Kirkville.		" 423
282	Manlius.		" 415
289	Syracuse.[27,181]		" 403

The railroad *via* Auburn is better than the Direct road to Rochester for geological observation.

has ever been found beneath them. The scenery has suddenly changed, and nothing is seen but bare, weatherworn precipices of crystalline rocks, from which all the elements through all the ages, have failed to produce a soil, yet a certain strange interest is attached to them. The oldest picture in the world, the oldest statue or other work of art, would excite the greatest attention, yet what are these in antiquity compared with these grand old Laurentian rocks, the oldest formation and the oldest dry land on the face of the earth, dating far back of the first appearance of either animal or vegetable life of any kind on our planet. The river channel through these rocks is an unequivocal example of river erosion, as pot-holes are found at various heights. See also notes 6 and 56.

18. *Utica.* The 4 b. Utica slate was named from this city. To study the Trenton, Black River and Birdseye limestones at their original, historical localities, change cars at Utica and go up the Utica and Back River Railroad to Trenton Falls. (See the within guide for that railroad.) You can then go on to Watertown on these limestones. Return by the Rome, Watertown & Ogdensburg Railroad to Rome or Syracuse, examining the Loraine shales at Adams and Pulaski.

19. From here to Syracuse there is no lock in the canal. This long level is 427 feet above tide.

20. *Oriskany.* The formation of this name, is not exposed here, but at Oriskany Falls on the D. L. & W. R. R. from which the name is derived. The best fossils of it are found east of Union Springs in Cayuga County. Along the part of the road east of Oriskany, the Utica shale forms the bottom of the valley. The south wall of the valley consists of the outcrops of the 4 c. Hudson River, 5 a. Oneida Conglomerate, 5 b. Clinton, the 6 Waterlime and 9. Upper Helderberg. See 191.

21. *Rome.* No more 2–4 formations west of this in New York. From Rome to Buffalo and from Lake Ontario south to the Pennsylvania line all the formations are 5–11 Silurian and Devonian, and they are finely displayed in numerous gorges, ravines, canons and precipices, very regularly disposed in belts of outcrop running east and west. The typical localities from which most of the formations were named, are situated in this district. It is all historical geological ground, and you can scarcely go amiss in looking for fossils.

22. West of Little Falls the lower formations pass abruptly to the north and cross under Lake Ontario into Canada. The 4 c. Hudson River first crosses the valley, and then the Oneida conglomerate. Other rock formations now appear between Rome and Oneida, which had no existence in the basin east of Little Falls. These are the 5 a. Medina and Clinton, which overlie the Oneida, and form all the south shore of Lake Ontario, and extend across Canada West. Also 5 c. Niagara and the 6. Salina or Onondaga salt group, on which the N. Y. C. & H. R. R. R. runs from Oneida nearly to Rochester. The non-existence of these extensive formations east of Little Falls (the 5 a. Medina, 5 b. Clinton, 5 c. Niagara and 6. Salina), which cover the best part of Western New York, must be owing to the two parts of the State being separated in these early ages by the old Laurentine ridge at Little Falls into separate basins, in which the rock-forming conditions were different.

23. *Verona.* The Clinton fossil iron ore crops out on the railroad, but not of a good quality.

24. *Oneida.* The prominent ridge bounding the valley on the south of Utica, Oneida and Syracuse, called Stockbridge Hill, Pompey Hill, Cazenovia Hill and Onondaga Hill, is the Helderberg range, a continuous mountain 800 feet high, forming the back-bone of the State, and composed at its base of the 6 Waterlime, of the Salina group, all the members of the 7. Lower Helderberg being wanting as well as the 8. Oriskany sandstone and other sandstones that separate the Lower and Upper Helderberg, except a mere trace. On the Waterlime rests the Onondaga limestone, the most valuable building stone, and above this the Corniferous. Over these three great limestone formations is always found the 10 a. Marcellus shales, the 10 b. Hamilton and the 10 c. Genesee, forming the fine fertile country extending south from this ridge. Still farther south is the 11 a. Portage with its glens, gorges and precipices, and 11 b. Chemung, extending to the Pennsylvania State line. The Oneida conglomerate, which is 30 or more feet thick in Herkimer and Oneida, gradually attenuates in going west, being a grey band, from 4 to 5 ft. thick at Rochester. It was named from Oneida County.

25. *Wampsville.* Numerous fragments of Niagara limestones are seen mixed with the soil, showing its existence underneath. The Niagara limestone and shales which, at Niagara, Lockport and Rochester are 150 ft. thick, thin out in going eastward, being only two or three ft. thick at Saquoit Creek near Utica.

26. *Canastota.* Stop off and take the branch railroad to Cazenovia, rising 750 feet in 15 miles. Fine geological sections of 6. Salina with gypsum beds, 9 Upper Helderberg and 10 b. Hamilton. Magnificent view across Oneida Lake and a beautiful village and lake at Cazenovia. ••

27. *Syracuse.* Onondaga Lake, which is in sight and on the north side of the railroad at the west end of Syracuse City, is 5 miles long, 1 mile wide; its greatest depth is 60 feet, and its surface is 363 feet above tide water. It is excavated in the red shale of the (6.)Salina formation. The lake is what remains of an ancient much more extensive and deeper excavation, all of which has been filled in with sand, gravel and rolled stones, except the part occupied by the lake. The bottom and sides of the lake are covered with lake marl six feet thick. The ancient excavation underneath answers an excellent purpose as a reservoir into which the salt waters are received and retained, and the marl of the bottom of the lake serves an equally good purpose by separating the fresh water of the lake from the salt water stored away in the basin or reservoir of sand and gravel beneath. There could be no better material for the purpose. Into this basin the various borings of the salt wells are made, not through

Ms.	New York Central & Hudson River Railroad.—Continued. Old Road, via Auburn.	Alt.	Ms.	New York Central & Hudson River Railroad. Old Road, via Auburn—Continued.	Alt.
289	Syracuse.[27]	6. Salina, 9 miles. [403]	346	Oaks Corners.[31]	9 c. Cornif. l. s., 18 m.
298	Camillus.	"	349	Phelps.	"
300	Marcellus.[28]	" Gypsum beds.	353	Clifton Spri'gs.[40]	" [618]
303	Half Way.	9 c. Upp. Helderberg,	358	Shortsville.	"
307	Skaneateles.[29]	or Cornifer. 14 m. [610]	364	Canandaigua.[157]	10 Hamilton 6 ms. [740]
310	Sennett.	"	368	Paddleford.	"
316	Auburn.[30]	" [715] { Quar. of Corn. l. s.	369	Farmington.	
321	Aurelius.	6. Salina, 10 miles.	370	W. Farmington.	{ 9 c. Cornifer's l. s. and Salina.
326	Cayuga.[78]	" (Lake.[388])	374	Victor. [182]	"
331	Seneca Falls.	9 c. Corn. l. s. 8 miles.	379	Fisher's.[182]	9 c. Salina 11 miles.
334	Waterloo.	9 c. Seneca limestone.	384	Pittsford.	
341	Geneva.[31]	{ Deep drift overlying 6. Salina and 9 c. Cornifer. l. s. [452]	388	Brighton.	5 c. Niagara, 4 miles.
			392	Rochester.[36],[187]	" [508]

or into rock, but only through the lake marl and other loose material mentioned, to a depth of 150 to 450 feet. No rock salt or bed of salt has ever been discovered in this State, although it has been in Canada; but in this Salina formation are two porous or Vermicular masses of limestone, looking as if perforated by little worms, and hence the name; and between them are certain hopper shaped cavities in the shale in which, as well as in the perforations of these limestones, salt in a crytalline and solid state, it has been conjectured, formerly existed, the saline materials of which have been dissolved in water which percolated through the formation and passed into the basin where it is now found, the bed of marl on which is Onondaga Lake, being afterwards formed over it. But the origin of the salt water may be said to be at present unknown. Forty gallons of the brine produce a bushel of salt, weighing 60 pounds. These are the most productive salt wells in the world in so small a territory—two miles long and one-fourth of a mile wide.

28. *Marcellus*, from which the formation is named, is three miles south of this station.

29. *Skaneateles*. From the Junction with the N. Y. C. & H. R. R. R. the Skaneateles railroad runs south up the outlet of the lake of that name over the Corniferous limestone. The lake outlet with its falls, amounting to 463 feet to Jordan, affording excellent mill sites and many exposures of the rock. Before reaching Skaneateles Village the railroad passes over the Marcellus shales. Skaneateles Lake, where the railroad terminates, is 14 miles long, from a half to a mile and a half wide; its greatest depth south of Borodino is 320 feet and its surface 879 feet above tide. The sides of the northern end of this lake, at the beautiful village of Skaneateles, gradually slope to the water, corresponding in inclination to each other and adding greatly to the beauty of the lake. The water line, with the exception of the south part, is excavated in the Hamilton group. The south part of the lake is more narrow, and the banks rise abruptly to a considerable height above the water. The Tully limestone, at the top of the Hamilton, and over that of the Genesee slate, appear to the south of Borodino, rising, when first seen, 150 feet above the lake, and the south end or head of the lake is surrounded by the Portage group. Fossils along the lake. *Cyathophylloid* corals.

30. *Auburn*. The Corniferous member of the 9. Upper Helderberg limestone and the Onondaga limestone, which is its lower member, are extensively quarried at Auburn. The State Prison and the facings of many of the buildings of this handsome little city are entirely made of this limestone, and several fine churches are built of it. The formation ends at the main street where the 10 a. Marcellus shale begins, and it extends in the stream up to the outlet of the lake. Beginning below the city and following up the stream to the State Prison, the outlet exposes the following section: eight feet of the upper part of 6. the Waterlime of the Salina formation, one foot of 8. Oriskany sandstone, over eight feet of 9 c. Onondaga limestone and twenty-seven feet of the Corniferous exclusive of its upper member the Seneca limestone.

31. *Geneva*. The Seneca limestone of the upper part of the 9. Upper Helderberg limestone disappears near Waterloo and reappears at a distance of six or seven miles west near Oaks Corners. The whole mass of limestone, and all the rocks north of it to Lake Ontario, have been removed from all the intermediate space, and along the shore of that lake the great depth of alluvium conceals the rock if any be present. Near Oaks Corners the limestone suddenly terminates as if broken off and removed, leaving an abrupt descent to the east which bears evidence of the erosive action of water. Seneca Lake and Lake Ontario probably originally communicated by this deep old channel. Ontario is 196 feet lower than Seneca. The same state of things seems to exist north of Cayuga Lake, where the drift material causes the Montezuma marshes and the shallowness of that lake at that end. Seneca Lake is 40 miles long, 3 miles wide, 530 feet deep, and its surface is 441 feet above tide water.

32. *Jordan*. Between Skaneateles Junction and Elbridge the Oriskany sandstone is over 30 feet thick, being at its maximum. At Auburn it is from six inches to two and a half feet thick.

33. *Weedsport*. At many points between Syracuse and Rochester, and on the Southern Central and other cross roads, are seen numerous hills or short ridges running from north to south, from fifty to one hundred feet high, with steep slopes and very sharp crests. These are not of drift or alluvium, as they appear to be, but are in reality outliers of the marly deposits of the Salina or Onondaga salt group, with only a thin covering of loose materials. Mount Hope at Rochester, the hills south of Brighton, Fort Hill Cemetery in Auburn, James street hill and University hill in Syracuse, and numerous hog-back ridges about Jordan and other places, are of this character, being Salina shales in place, spared when the adjoining valleys were eroded. There are, however, some hills composed of gravel, or a mixture of gravel and sand, but very little glacial drift on this R. R.

34. Great crops of peppermint are raised here, and this place supplies the world with peppermint oil. There seems to be some peculiarity in the soil which adapts it for the production of

New York Central & Hudson River Rail-road.—*Continued.*				New York Central &	
Ms.	Direct Road.		Alt.	ro. Ms.	Niagara Falls Div
289	Syracuse.[27],[161]	{ 6. Salina or Ononda. Salt gr'p, 71 ms.[403]		426	Lockport.[38] 600
299	Warner's.	"	427	430	Lockport Junc.
302	Memphis.	"	410	436	Hall's.
307	Jordan.[32]	"	406	441	Tonawanda.
311	Weedsport.[33]	"	404	448	Black Rock.[40]
314	Port Byron.	"	406	449	Intern'l Bridge.
324	Savannah.[31],[78] 407	" Marshes.		452	Buffalo.[40]
328	Clyde.	"	396		
335	Lyons.	"	407		Direct
340	Newark.	"	413	370	Rochester.[36],[137]
348	Palmyra.[34]	"	433	377	Coldwater.
353	Macedon.	"	471	381	Chili.
360	Fairport.	"	456	385	Churchville. 570
366	Brighton.[35]	5 c.Niagara l. s. 10 ms.		388	Bergen.
370	Rochester.[36],[137]	"	505	391	West Bergen.
				395	Byron.
	Niagara Falls Division.			402	Batavia.[41] 395
				408	Crofts. 563
370	Rochester.[36],[137]	5 c. Niaga., 10 ms. 505		414	Corfu.
380	Spencerport. 580	5 b. Clinton, 12 miles.		418	Crittenden. 848
383	Adams Basin.	{ Railroad runs between Clinton and Medina.		421	Wende.
				423	Town Line.
389	Brockport.	"	546	428	Lancaster.
392	Holley. 532	5 a. Medina, 23 miles.		438	Buffalo.[40]
396	Murray.	"	568		
481	Albion.	"	547		Buffalo and Niag
407	Knowlesville.	"		0	Buffalo. 584
411	Medina.[37]	"	545	3	Intern'l Bridge.
415	Middleport.	5 b. Clinton, 4 miles.		5	Black Rock.[40]
420	Gasport.	"	521	11	Tonawanda.
426	Lockport.[38]	5 c. Niaga., 21 ms. 600		17	La Salle.
437	Sanborn.	"		22	Niagara Falls.[39]
446	Suspens. Bridge	"	580	24	Suspens. Bridge.
447	Niagara Falls.[39]	"	574	30	Lewiston.[42] 388

35. *Irondequoit.* A few miles east of the mouth of the Genesee River, empties into the lake, ;flowing in a deeper channel than the Genesee, but and gravel. Professor Hall suggests with much probability that the Genes the Irondequoit, but when that was filled with gravel and the region ele turned westward and compelled to cut its present rocky bed like the Niag is not rare, is not many times repeated in this State. See notes 31, 38, 39 ar

36. *Rochester.* See Genesee Falls out of the car windows on the north the station house. The gulf of the Genesee River, from Rochester to Char the striking example of erosion which it exhibits. The distance is seven forms three cataracts over three distinct formations, the Medina sandstone the Clinton 25 feet one and three-fourth miles below, and the Niagara grou railroad bridge. It is evidently the different hardness of the groups or the composition that have produced these falls. These three falls at first were bu lower ones are gaining probably on the upper one and the time may come wh

37. The 5 a. Medina formation is named after this place. Layers filled wit

38. At Lockport is a repetition of the Rochester and Niagara Falls ravi stone and shales here crossed by the railroad on a high bridge. Here too, you can see on the north side of the railroad an old, dry channel from whicl ted by the drift, corresponding to the Irondequoit at Rochester and St. D There is another of these dry, old channels at Oak Orchard. Niagara fossil

39. Niagara Falls are six and a half miles south from Lake Ontario at I distance the river runs in a gulf, which, at the falls, is 160 feet, and at Lewi generally about twice as wide at the top as at the bottom. The rocks passed falls are the Medina sandstone, the Clinton group of limestone and shale, ar and shale. These rocks have a slight southerly dip, and all except the Ni peared beneath the bed of the river, the falls being now in the Niagara g lying beneath the limestone. At the whirlpool, a little more than three mil west bank of the river, the continuity of the rock forming the bank is inter filled with drift material. This ravine many be traced two miles in a north thence another depression can be followed to Lake Ontario at St. David's fo town. When the ravine to St. David's was blocked up by drift materials the

New York Central & Hudson River Rail-road. —*Continued.*		
Ms. Canandaigua and Tonawanda Division.		Alt.
0 Canandaigua.[157]	10 b. Hami n, 16 m.	740
8 East Bloomfield.	"	853
12 Miller's Cor's.[163]	"	896
15 West Bloomfield.	"	
18 Honeoye Falls.	9 c. Cornifer. 2 ms.	777
25 West Rush.	6. Salina, 22 miles.	
26 Erie R. R. Junc.	"	
28 Maxwell's	"	
33 Caledonia.[125]	"	655
40 Le Roy.[125]	9 c. Cornif., 25 ms.	872
44 Stafford.	"	894
50 Batavia.[41]	10 b. Hamilton.	895
57 East Pembroke.	9 c. Corniferous.	885
63 Richville.	"	828
65 Falkirk.	"	843
67 Akron.[125]	"	765
74 Clarence Centre.	6. Salina, 21 miles.	643
77 Transit.	"	
80 Gettzville.	"	
86 Tonawanda.	"	880

New York Central & Hudson River Rail-road.—*Continued.*		
Ms. Charlotte Branch.		Alt.
370 Rochester.[26,137]	{ 5 c. Niagara.	508
	5 b. Clinton.	
379 Charlotte.[35]	5 a. Med., (Lake, 245)	

Troy & Schenectady.

148 Troy.	Hud. Riv. & 2 b. Pots.
151 Cohoes.	" Falls, 70 Feet.
154 Crescent.	"
160 Niskayuna.	"
166 Aqueduct.	4 b. Utica.
170 Schnectady.	"

Skaneateles Railroad.[29]

Syracuse,	(As before.)	403
0 Skaneateles Jc.	9 c. Corniferous.	810
3 Mottville.	10 a. Marcellus.	
4 Kellogg's Mills.	"	
5 Skaneateles.[29]	10 b. Hamilton.	890

to find its present rocky channel. Even though the drift rose only a foot higher than the rocks it would as effectually force the water over the rocks as if it formed a mountain. Could the river have once surmounted the drift, its work would have been comparatively easy in wearing out a bed through the old ravine, but till it was able to flow over the barrier it would have no power over it, and must commence its slow work of wearing away the solid rock. The present gulf shows us what it has done since the drift period. J. HALL and SIR CHARLES LYELL.

40. At Black Rock there is only from 6 to 14 inches of the Onondaga limestone which is of a grayish color, crystalline and contains few fossils. The Corniferous limestone above it is 25 to 30 feet containing abundance of hornstone. It is dark colored, fine grained, and in its fresh fracture, and particularly when wet, it presents an almost black appearance, which has given the name of Black Rock to the place. It affords good quarries of excellent building stone. From the occurrence of the Corniferous along the south end of Lake Erie and its dip southward, it seems probable that the bed of this lake has never been excavated below it, and that it now forms the floor beneath the deposit of alluvium. It seems that there are others of the lake bottoms composed of limestone, especially Lake Ontario. See note 71. This is probably for the reason that it received a polish from the action of glaciers which then passed over it, while the resistance of the grit of the sandstones and shales was more favorable for deeper excavation. Lake Erie is 230 miles long, 50 miles wide, 140 feet deep and its surface is 569 feet above tide.

41. Batavia is the highest point on the N. Y. C. & H. R. R. R., and one of the highest in Western New York, being 895 feet above tide. This is caused by there crossing the 9 c. Helderberg formation, which maintains its elevation although not observable as a mountain range, being overcome by easy grades. Notice the elevations of the railroad crossings of the Helderberg and Hamilton range, although the railroad seeks the lowest points; Buffalo, 584; Batavia, 895; Le Roy, 872; Canandaigua, 740; Auburn, 715; Skaneateles, 890; Tully, 1249; Cazenovia, 1249; Cooperstown, 1193. When the valleys cut through the limestone, the summit is farther south on the Hamilton or Portage.

42. *Lewiston.* Tourists should not fail to go down to Lewiston, the terminus of the Buffalo and Niagara Falls division. This railroad ride, although little known, is one of the finest in the United States. It follows the bank of the Niagara River, affording admirable views of the rapids and the formations displayed in the gulf. Nowhere in the State are there better geological sections. On the Canada side, also the Canada Southern Railway, running to the mouth of the Niagara River at Niagara City, affords one good view of the falls, but no such remarkable sections of the rocks as on the American side, where the railroad overhangs the fearful torrent of the river for several miles.

43. *Knowersville.* The Helderberg mountain shows finely on the left or southwest side of the railroad opposite Guilderland and Knowersville. The railroad passes through it between that place and Duanesburgh. The mountain is capped by the 7. Lower Helderberg limestone forming a steep precipice along its summit, and this rests on the 4 c. Hudson River slates. Back of Knowersville two notches are cut out of the mountain by two streams, leaving a picturesque, fortress-like bluff of the limestone. The Helderberg formations are named from this mountain. See Note 158.

44. At Howe's Cave large quarries on the railroad track. Good place to examine Lower Helderberg limestone and to collect fossils. The cave is an old underground water channel, and it is several miles long. Notice that the limestone at Cobleskill is *Upper* Helderberg and that at Howe's Cave *Lower* Helderberg. On no other railroad can you see them both.

45. Cooperstown is seated at the south end of Otsego Lake on a dike of alluvium. This lake is a handsome sheet of water seven miles long, one and a half wide, 1193 feet above the ocean. It has a high ridge of the Hamilton group on the east side, a low and interrupted range of the same on the west side, and an elevated projection on the northeast end. This lake is one of the head waters of the Susquehanna, the valley spreading out to the southwest. See also 186.

46. *Sharon Springs.* All the large sulphur springs of the State, Avon, Clifton, Richfield, etc., and many small ones, rise from the waterline. Glacial Striae here and at Cherry Valley.

47. *Cherry Valley.* The railroad is on Corniferous, but the cliffs and gorge are Waterlime, Lower Helderberg, Cauda Galli, and, slightly, Oriskany. Marcellus and Hamilton form the hills on the south.

Delaware & Hudson Canal Co's Railroads.

Albany and Susquehanna Railroad.

Ms.	Station		Formation	Alt.
0	Albany.[10],[121]		4 c. Hudson River.	30
6	Adamsville.		"	212
7	Slingerlands.		"	214
11	New Scotland.		"	327
14	Guilderland.[158]		"	329
17	Knowersville.[43]		"	459
24	Duanesburg.	793	" and Utica.	
27	Quaker Street.		"	
31	Esperance.		"	769
36	Central Bridge.		7. L. Helderberg.	
39	Howe's Cave.[44]		"	782
			8. Oriskany. (Portage Hills.)	
45	Cobleskill.	908	9 c. U. Helderb'g l. s.	
			10 a. Marcellus.	
50	Richmondville.	1173	" 10b. Ham.	
57	East Worcester.		10 b. Hamilton.	
62	Worcester.	1310	"	
67	Schenevus.	1272	11 a. Portage. (Kame-like Hills.)	
70	Maryland.	1220	"	
75	Cooperstown Junction.[45]		"	
76	Colliers.		11 b. Chemung.	1118
79	Emmons.		"	1127
82	Oneonta.		"	1087
90	Otego.		"	1054
95	Wells Bridge.		"	1049
99	Unadilla.[184]		"	1022
103	Sidney.	990	12. Catskill, synclinal.	
108	Bainbridge.		"	994
114	Afton.		11 b. Chemung.	979
119	Nineveh.		"	1032
127	Tunnel.		"	
132	Osborn Hollow.		"	1115
134	Port Crane.		"	1041
142	Binghamton.[185]		"	859
	Saratoga.	265	3 a. Calciferous and 4 a. Trenton.	304
0	Ballston.	310	4 c. Hudson River.	
15	Schenectady.		"	246
29	Quaker Street.		"	
45	Cobleskill.	908	9 c. Upper Helderberg.	
50	Hyndsville.		"	1112
54	Seward.		"	1177
59	Sharon Spr'gs.[46]		7. Low. Helderb.	1853
68	Cherry Valley.[47]		9 c. Corn. & Marc.	1821

Cooperstown and Susquehanna Valley R. R.

Ms.	Station		Formation	Alt.
75	Junction.		11 a. Portage.	
91	Cooperstown.[45]		10 b. Hamilton.	1193

Delaware and Huds Railroads.

Middleburg and Schoha ley Rail

Ms.	Station
0	Central Bridge or Schoharie Junction.
3	Hollenbeck's.[48]
6	Schoh'e C. H.[49]
9	Borst's.
12	Middleburg.

Nineveh

Ms.	Station
119	Nineveh.
122	Centre Village.
127	Ouaquaga.
130	Windsor.
133	Comstock.
140	Jefferson Junc.

Saratoga and Ch

Ms.	Station	Alt.
0	Albany.[10],[121]	
6	West Troy.	
9	Cohoes.[50]	
12	Albany Junction.	
0	Troy.	
6	Albany Junc.	
12	Mechanicsville.	
25	Ballston.	
32	Saratoga.	265
43	Gansevoorts.	
49	Fort Edward.	
57	Smith's Basin.	
60	Fort Ann.	
64	Comstock's.	
71	White Hall.[179]	
0	White Hall [51]	
7	Chubb'sDock.	
10	Dresden.[52]	
14	Putnam.	
20	Pattuiwa. (Mt. Defiance.)	
22	Ft. Ticonderoga. (Ticon'ga Creek, (Tunnel.)	
24	Addison Junc.	

48. On either side of the valley, according to Prof. Hall, is the followi shales, (Clinton group); Coralline limestone, (Niagara); Waterlime, (Salina); Delthyris shaly limestone; Upper Pentamerus, (Lower Helderberg); Oris harie grit; Onondaga limestone, (Upper Helderberg). At Hollenbeck's are man's Nose."

49. The Schoharie grit formation was named from this place. The f found in the mountain one and a half miles northwest and northeast of Sch

50. See from car windows the great falls of Mohawk, 70 feet high, over

51. White Hall is usually called the head of Lake Champlain, but the more than 100 to 150 yards wide. It is in fact a mere channel between mud Lake Champlain is 112 miles long, 600 feet deep, and the surface being on

Delaware and Hudson Canal Company's Railroads.—Con. Ms. Saratoga and Champlain Division.—Con.		Alt.
	1 a. Laurentian bluff.	
	4 a. Trenton.	
32 Crown Point.	1 a. Laurentian bluff.	
	4 a. Trenton, 7 miles.	
	Val'y chiefly 1 a. Laur.	
40 Port Henry.[58]	1 a. Laurentian.	
(Tunnel.)	"	
51 Westport.[54]	"	
54 Wadham's Mills.	"	
57 Whallonsb'gh.[55]	{ For 13 miles deep cuts through bluffs, 1 a. Laur'n. Beautiful sections.	
64 Willsborough.[55]		
77 Port Kent.[56]	1 a. Laurentian ends.	
(Ausable R.)[57]	2 b. Potsdam.	
84 Valcour.	{ 2 b. Pots'm. Heavy beds of sand & clay.	
90 Plattsburg.	" [119]	
95 Beekmantown.	{ 4 a. Trenton and 3 b. Chazy.	
99 West Chazy.	"	No rock exposure.
100 Chazy.[58]	"	
105 Sciota.	"	
111 Mooer's Junc.	"	
118 Champlain.	{ 3 a. Calciferous & 3 b. Chazy.	
99 West Chazy.	"	
122 Rouse's P'nt.[179]	"	
(Con. in Canada, see Grand Trk. R'y.)		

Delaware and Hudson Canal Company's Railroads.—Con. Ms. Ausable Branch.		Alt.
0 Plattsburg.	2 b. Potsdam.	[119]
5 Salmon River.	3 a. Calciferous.	
8 Laphams Mills.	1 a. Laurentian.	
10 Peru.	"	
14 Harkness.	"	
17 Ferronia.	"	
20 Ausable.[57]	"	
Glens Falls Branch.		
49 Fort Edward.	4 a. Trenton.	[141]
53 Sandy Hill.	"	
55 Glens Falls.	" Utica sl. above.	
Lake George Branch.		
22 Ticonderoga.	1 a. Laurentian.	
26 Baldwin on Lake George.[59]	"	
Rutland and Washington Division. 164		
0 Rutland, Vt.	Calciferous-Trenton.	
4 W. "	" & 4 c. H. R.	
10 Castleton, Vt.	2 Lower Cambrian.	
14 Poultney, Vt.	" "	
21 Middle Granville	" & 4 c. H. R.	
26 Granvi'e, N.Y.[140]	4 c. Hudson River.	
30 W. Pawlet.	L. Camb. & 4 c. Hud. R.	
37 Rupert, Vt.	2 Lower Cambrian.	
45 Salem, N. Y.	" "	
52 Shushan.	2 L. Camb. & Hud. Riv.	
56 Cambridge.	4 c. Hudson River.	
62 Eagle Bridge.[140]	" "	

extends 500 feet below the level of the ocean. Its bed is a deep chasm in the Laurentian or Primitive rocks. On the west side, where the mountain ranges reach it, the slope is abrupt, but on the east side it is longer and more gradual. At many places the lake is bordered by steep banks of blue and yellowish brown clay and yellowish brown sand, rarely over 15 feet thick, but its greatest height is 100 feet at Burlington. It contains marine fossils in the mixture of clay and sand, but none in the clay beneath. This drift formation extends north to the mouth of the St. Lawrence River. In Albany County it is an immense mass and is known as the Albany clay.

52. From Dresden to Port Kent, 67 miles the Laurentian hills are the western boundary of the valley of Lake Champlain. But at many points this mountain ridge recedes from the lake, leaving nooks and valleys, in which are patches of 3 b. Chazy and 4 a. Trenton limestone along the railroad.

53. The magnetic iron ore mines back of Port Henry are worth a visit, the bed of the ore being more than 100 feet thick. The mining of these heavy beds is on a grand scale.

54. From 51 Westport to 77 Port Kent, the formation, according to Dr. Hunt, is 1 c. Norian or Upper Laurentian.

55. At the village of Essex, on the lake and between Whallonsburgh and Willsborough stations, is a bold bluff, 100 to 200 feet high above the lake, of 3 b. Chazy limestone.

56. The Adirondack Mountains commence at Little Falls, rising suddenly from the Mohawk Valley, and run northeast to Port Kent on Lake Champlain. The most elevated peak, Mount Marcy, is 5,467 feet high, the summit being just upon the region of perpetual frost. There are four other peaks 5,000 feet high, each distant about 6 miles from the other. This group of Adirondack Mountains is the culminating point of the State around the sources of the Hudson, Ausable, Racket and Black Rivers, and dividing the north half of the State into two separate geological basins. They are directly west of Westport, several miles to the west of the railroad. Only a glimpse of one of them can be had from the railroad. In the Adirondack pass in Essex County, is a perpendicular precipice or naked wall of rock 1,000 feet high and more than half a mile long. There is not probably in the Eastern States an object of the kind so vast and imposing as this. Emmons, 218.

57. Stop at Port Kent and visit the Ausable valley, which is interesting for the Ausable chasm, where for at least two miles the Ausable River, a large and rapid stream, is compelled to flow through a rocky gorge in the 2 b. Potsdam sandstone with perpendicular walls of 100 feet with a width only varying from 20 to 40 feet. Here the *lingula antiqua* is found in great abundance, and there is here a better development of the Lower Silurian or Cambrian rocks than in any other part of the State. Emmons, 267. *Lingula* and *trilobites* near foot of Cathedral rocks.

58. 3 b. Chazy formation was named from this locality. Off line of R. R. are abundant Chazy fossils, *Maclura Rhynchonella*, etc. See Note 55. Also as to Isle La Motte see Note 67.

59. The rock which forms Diamond Island in Lake George is a good example of 3 a. Calciferous. Lake George is 30 miles long, 1½ miles wide, and its surface is about 80 feet above tide water.

Ms.	Adirondack Railroad.	Alt.
0	Saratoga. ³⁰⁴	4 a. Trenton & 3 a. Cal.
6	Greenfield.	2 b. Potsdam. ⁵⁶⁴
10	King's.⁶⁰	" ⁵³⁵
13	South Corinth.	" ⁶⁰⁶
17	Jessup'sLanding.	" ⁶⁰⁶
22	Hadley.⁶⁰	1 a. Laurentian. ⁶⁰⁶
30	Stony Creek.	" ⁵⁶⁹
36	Thurman.	" ⁵³⁵
44	The Glen.	" ⁷¹²
47	Washbu'n'sEddy.	"
50	Riverside.	" ⁸¹⁵
58	North Creek.	" ⁹⁷⁶

Chateaugay Railroad.164

Ms.		Alt.
0	Plattsburg.¹⁶¹	4 a. Trenton.
8	Morrisonville.	2 b. Cambrian. (?)
12	Cadyville.	"
17	Dannemora.	1. Laurent.& 2. b. Cam.
22	Saranac.	"
34	Lyon Mt.	"

Crown Point Iron Co's R. R.

Ms.		
0	Crown Point.	1. Laurt. & 4 a. Trent.
13	Hammondville.	1. Laurentian.

Utica and Black River R. R.

Ms.		Alt.
0	Utica.	4 b. Utica, 12 ms. ⁴⁴⁶
6	Marcy.	" ⁵⁸⁷
10	Stittville.	" ⁵⁶⁰
12	Holland Patent.	4 a. Trenton, 32 ms.⁶³⁰
16	Trenton.	" ⁸⁴⁰
18	Trenton Falls.⁶²	" ⁸⁴⁰
19	Prospect.⁶²	" ¹⁰¹⁰
21	Remsen.	" ¹¹⁸⁵

Ms.	Utica and Black	
25	East Steuben.	4
28	Alder Creek.	
35	Boonville.⁶³	
38	Leyden.	
42	Port Leyden.	
45	Lyons Falls.⁶⁴	1
51	Glendale.	4
54	Martinsburg.⁶⁵	
58	Lowville.	
66	Castor Land.	
70	Deer River.	
74	Carthage.⁶⁶	1
81	Great Bend.	4
83	Felt's Mills.	
85	Black River.	
92	Watertown.⁶⁷	1
104	Sacket's Harbor.	
74	Carthage.⁶⁶	1
92	Theresa Junc.	2
98	Orleans Corners.	8
101	Lafargeville.	
108	Clayton.	2
74	Carthage.⁶⁶	1
83	Sterlingsville.	8
87	Philadelphia.	2
90	Shurtliff's. ⁴¹⁶	
93	Theresa Junct.	
95	Theresa.	·
101	Redwood.	
108	Rossie. ⁸²⁶	
113	Hammond.	2
118	Briar Hill.	
123	Morristown. ²⁵¹	

60. This railroad cuts through Trenton, Calciferous and Potsdam withi[n]
Saratoga. Fine sections of ripple marked Potsdam in railroad cut in Greenfi[eld]
is repeated at the High Falls of the Hudson at Luzerne or Hadley station on
road, in Warren County, where the river flows for a mile through a gorge [of]
Potsdam sandstone and the gneiss. The walls rise in some places to a heigl[t]
61. *Potsdam.* This is the locality which gave the name to the Potsda[m]
description of that formation in another part of this volume.
62. *Trenton Falls.* For about three miles between Trenton Falls station [and]
a mile or two east of the railroad, the East Canada Creek has cut a passa[ge]
limestone, the sides of the excavation rising vertically with an average hei[ght]
this passage are the Trenton Falls or Cascades which have given so much cel[ebr]
ly meriting by their number, beauty and position, the admiration they rece[ive]
at Prospect Village there are six falls, five of which are placed at intervals [that]
occupy the middle part of the excavation. The rock is in thin layers of from[?]
ness, separated by thin layers of shale, and contains trilobites in prodigious [?]
tion derives its name from this place. It is 500 feet thick and about seven m[iles]
east or south it grows thinner and is about 30 feet thick in the Mohawk Vall[ey]
at Prospect and used at Utica, is the upper part of the Trenton, which is he[re]
a more solid and crystalline structure and appearance. Going on north by t[he]
for many miles on a terrace of the limestones of this group, forming th[e]
which has its rocky channel in this formation all the way to Watertown, wit[h]
at Lyons, Carthage and Watertown and many cascades. Very picturesque [?]
ing geology, with an abundance of fossils.
63. *Boonville.* The first range or cliff of limestone on Black River, exte[nds]
river from opposite Boonville to Watertown, is the Birdseye limestone. It [?]
which by long exposure to the weather becomes of a light ash gray or white. [?]
layers, with straight, vertical joints, giving the rock when quarried the appe[arance]
has a compact grain and smooth fracture.
64. At Lyons Falls, Black river falls 63½ feet over gneiss or 1 a. Lauren[tian]
Carthage it falls but 9 feet and there is another fall over gneiss rock.
65. The high hills west of Martinsburg are of the Hudson River group.

Rome, Watertown and Ogdensburg Railroad.

Ms.		Alt.	
0	Rome.	445	4 c. Hudson River.
11	Taberg.		" 11 miles.
14	McConnellsville.		{ 5 a. Medina and Oneida Conglomerate, 31 miles.
18	Camden.		" 520
23	West Camden.		"
28	Williamstown.		" 604
31	Kasoag.		" 636
37	Albion.		" 547
42	Richland.[68]		"
47	Sandy Creek.[659]		4 c. Hudson R. 12 ms.
52	Mannsville. [725]		" Lora. shales.
54	Pierrep't Manor.		" deep gulfs.
59	Adams.[69]	599	4 a. Trenton limestone.
63	Adams Centre.		" 619
72	Watertown Junc.		Tren., Birdseye
73	Watertown.[67]	403	and Black Riv.
78	Sanford'sCorners	435	}(Sandy drift)
83	Evan's Mills.		3 a. Calciferous.
90	Philadelphia.		2 b. Potsdam. 485
96	Antwerp.		1 a. Laure'n, Iron ore.
101	Keene's.		" "
108	Gouverneur.		2 b. Potsdam.
115	Richville. [828]		1 a. Laurentian.
123	De Kalb Junc.		" Iron ore.
129	Rensselaer Falls.		2 b. Potsdam.
134	Heuvelton.		
142	Ogdensburg.		3. a. Calciferous. 248

Ms.		Alt.	
42	Richland.[68]		5 a. Medina.
47	Pulaski.[70]		4 c. Hudson River.[377]
50	Sandhill.		5 a. Medina. 813
55	Mexico.		" 375
60	New Haven.		" 306
63	Scriba.		"
71	Oswego.[71]	280	" Lake, 245
73	Watertown.[67]		4 a. Trenton. 455
72	Watertown Junc.		" 403
76	Brownville.[72]		"
86	Chaumont.		" 294
89	Three-Mile Bay.		"
93	Rosiere.		"
97	Cape Vincent.		" 253

Ms.		Alt.	
123	De Kalb Junc.		1 a. Laurentian.
131	Canton.		2 b. Potsdam.
142	Potsdam.[61]		
148	Potsdam Junc.		3 a. Calciferous.

Rome, Watertown & Ogdensburg R. R.—Con. Syracuse Division.

Ms.		Alt.	
0	Syracuse.[27]		{ 6. Salina or Ononda-ga Salt group.[403]
5	Liverpool.		"
8	Woodward.		5 c. Niagara.
11	Clay.		5 b. Clinton.
15	Brewerton.[102]		" 384
18	Central Square.		5 a. Medina.
22	Mallory.		"
24	Hastings.		"
27	Parish.		" 474
31	Union Square.		4 c. Hudson River.
34	Holmesville.		" 320
39	Pulaski.[70]		" 377
45	Sandy Creek Ju.		" 569

Lake Ontario Division, West.

Ms.		Alt.	
0	Oswego.[71]	280	5 a. Medina. Lake, 245.
4	Furniss.		"
7	Wheeler's.		"
10	Hannibal.		5 b. Clinton.
13	Sterling Valley.		"
16	Sterling.		"
20	Red Creek.		" 525
26	Wolcott. [360]		"Fossil iron ore.
31	Rose.		"
36	Alton.		"
38	Wallington.		"
41	Sodus.		" 430
47	Williamson.		" 604
52	Ontario. [415]		"Fossil iron ore.
56	Union Hill.		" "
59	Webster.		"
64	Pierce's.		"
66	Sea Breeze.[35]		5 a. Medina.
70	Charlotte.[35]		" 255
76	Greece.		"
80	North Parma.		"
83	East Hamlin.		"
86	Hamlin.		" 310
90	East Kendall.		"
92	Kendall.		"
97	East Carlton.		"
100	Carlton.		"
103	Waterport.		" 349
106	Carlyon.		"
110	Lyndonville.		"
114	County Line.		"
118	Somerset.		" 382
123	Hess Road.		"
127	Newfane.		"
128	Coomer Road.		"
132	Wilson.		" 300
147	Rawsonville.		"
156	Lewiston.[42] [858]		" Lake, 245.

66. The Laurentian rocks cover the whole of the country east of the Black River and the later formations west of the river, the opposite sides forming the strongest contrast imaginable as to rocks, soil, vegetation and population.

67. At Watertown the banks of the Black River present fine sections of the limestone visible from the car windows, showing the Trenton limestone, Black River limestone and the Birdseye limestone. There is a mass forming the Black River sub-division, known to quarrymen as the seven feet tier, lying between the Birdseye and Trenton limestone. At the Isle LaMotte, near Chazy, in Lake Champlain, it is a black marble, but at Watertown it is only suitable for ordinary purposes.

Delaware, Lackawanna and Western Railroad.			Delaware, Lackaw Railroa
Ms.		Alt.	Ms.
0 Binghamton.[185]	11 b. Chemung.	846	60 Poolville. 1099 1
7 Chenago.[190]	"		64 Hubbardsville.
11 Chenango Forks.[901]	" Moraine.		68 Nor. Brookfield.
21 Whitney's Point.	"		72 Sangerfield Cen.
23 Lisle.	"		73 Waterville.[188]
30 Marathon,	"	1026	78 Paris. 1422
35 State Bridge.	" Moraine.		81 Richfield Ju.
44 Cortland.[191]	11 a. Portage "	1116	84 Clayville.[191]
47 Homer.	" "	1131	86 Sauquoit.
54 Preble. 1183	10 a. Genesee,"		87 Chadwick's.
59 Tully.[78] 1200	10 b. Hamil'n, "		98 Washing'n Mills.
61 Apulia.	" "	1227	91 New Hartford.
66 Onativia.	10 c. Marcellus.		95 Utica.[18]
73 Jamesville.[74]	9 c. Corniferous.	585	
80 Syracuse.[27]	6. Salina.	403	81 Richfield Junc'n.
			85 Bridgewater.[190]
80 Syracuse.[27]	6. Salina.	403	86 Unadilla Forks.
	5 c. Niagara.		88 West Winfield.
92 Baldwinsville.	5 b. Clinton.	390	90 Cedarville.[193]
98 Lamson's.	5 a. Medina.		92 Miller's Mills.
104 Fulton.[75]	"	887	99 South Columbia.
115 Oswego.[71] 280	" Lake, 245.		102 Richfield Spgs.[46]

Cayuga Division.

0 Owego.[188]	11 b. Chemung.	822		0 Utica.[18]
4 Cattatonk.	"			4 New Hartford.
10 Candor.	"	822		9 Clinton.[76]
14 Wilseyville.	11 a. Portage.	940		11 Franklin I. W.
33 Ithaca on hill.	" Striae. 840			14 Deansville.
33 Ithaca on Lake.	189 "	392		18 Oriskany Falls.[20]
				21 Solsville.[191]
0 Binghamton. [185]	11 b. Chemung.	846		24 Bouckville.
11 Chenango Forks.[901]	" Moraine.			26 Peaksport.
19 Greene.[188]	"	916		29 Hamilton.[198]
25 Brisbin.[188]	"			31 Smith's Valley.
29 Coventry.[188]	"			
33 Oxford.[188] 980	10 a. Portage.		Heavy drift	0 Clinton.[76]
41 Norwich. 1001	10 b. Hamilton.			2 Kirkland.
47 North Norwich.	"			3 Clark's Mills.
52 Sherburne.	"	1042		5 Westmoreland.
57 Earlville.[94,191]	"	1071		7 Bartlett.
				13 Rome.

The Falls of Black River in Watertown are 35 feet perpendicular over the sion Bridge, and 112 feet within the city limits in six separate falls. Good

68. There are two miles of rapids in Salmon River, which termina high water the sheet of water is 250 feet wide, and at low water about hal over the grey sandstone of the 5 a. Medina, and is seven miles northeast fr

69. *Adams.* The Gulf of Loraine, on South Sandy Creek, is a genuine c flowing through the Loraine or Hudson River slates, Utica slate and Trent of Loraine, from which some geologists prefer that name for the formation dicular and vary in height from 100 to 300 feet, and the gulf varies in width several of these gulfs in Jefferson County, some of them 12 miles in length. points of the streams. A convenient place to study the Loraine shales, a l is the pleasant village of Adams. There are two of these gulfs within two town of Loraine, but not on the stream in the village, which is on Trenton observe a remarkable moraine of naked Laurentian boulders, some of then crosses the railroad just south of Adams, where are many boulders in the fi tend from Lake Ontario south of Woodford northeast into Canada. The ri along Lake Ontario, also occurs here a little nearer the lake than the ridge

70. The shales and sandstones at Pulaski are the upper part of the 4 c were at first called Pulaski Shales, or the Shales of Salmon River, and Lora rock at Pulaski village and is full of fossils, while the lower or Frankfort di

71. *Oswego.* Lake Ontario, like all other New York lakes, is a lake c northeast shore, in Canada, is the 4 a. Trenton limestone. On its south or the 5 a. Medina sandstone extending from Oswego, the whole length of Canada. The lake is excavated 50 feet in the red and 100 feet in the gra 230 feet in the Hudson River and 120 feet in the 4 b. Utica slate, the whol 500 feet or the real depth of the lake, the surface of the 4 a. Trenton lime It is 180 miles long, 40 miles wide, 492 feet deep and its surface is 245 feet a

Delaware, Lackawanna and Western Railroad.—Con.		
Ms.	Binghamton to Buffalo.	Alt.
207 Binghamton.[90]	11 b. Chemung.	863
215 Vestal.	"	828
221 Apalachin.	"	819
228 Owego.[188]	"	818
233 Lounsberry.	"	
236 Nichols.	"	789
242 Litchfield.	"	
246 Waverly.[188]	"	828
250 Williwanna.	"	801
Lowmansville.	"	828
263 Elmira.	"	855
267 Horseheads.	"	911
272 Big Flats.	"	906
Gibson.	"	
278 Corning.[188]	"	929
281 Painted Post.	Fossils. "	945
284 Coopers.	"	
287 Curtis.	"	
289 Campbells.	"	1013
293 Savonia.	"	
298 Bath.[205]	"	1101
302 Kanona.	"	
306 Avoca.	"	1193
Wallace.	"	1232
314 Cohocton.	"	1287
319 Bloods.	"	1317
327 Perkinsville.	"	
Wayland.	"	1359
332 Dansville.	11 a. Portage.	1035
332 Groveland.	"	598
846 Mt.Morris.	10 c. Genesee.	574
849 Leichester.	"	850
858 York.	"	929
863 Roch. & Pitts. Ju.	"	
867 East Bethany.	"	938
374 Alexander.	10 b. Hamilton.	890

Del., Lack. & Western R. R.—Con.		
Ms.	Binghamton to Buffalo.—Con.	Alt.
380 Darien.	10 b. Hamilton.	875
387 Alden.	10 b. Ham. & 9 c. Corn.	800
396 Lancaster.	9 c. Corniferous.	853
403 East Buffalo.	"	577
409 Buffalo.[90]	"	538

Northern Central Railroad.		
0 Elmira.[104]	11 b. Chemung.	863
6 Horse Heads.	865 " Valley drift.	
10 Pine Valley.	" "	865
13 Millport.	11 a. Portage.	
19 Havana.[85,191]	"	447
22 Watkins.[86,194]	473 " Lake,441	
29 Rock Stream.	"	
31 Big Stream.	10 c. Genesee, Gulf.	
33 Starkey.	"	810
37 Himrod's.	"	799
41 Milo.	"	857
45 Penn Yan.[67]	756 " & Portage.	
49 Benton.	" "	
51 Bellona.	10 b. Hamilton.	868
55 Hall's.	"	
58 Stanley.	"	904
61 Lewis.	"	
68 Hopewell.	"	850
69 Canandaigua.[88]	Lake,668 "	740
0 Sodus Point.	5 a. Medina, Lake 245.	
4 Wallington.	"	
6 Sodus Centre.	5 b. Clinton.	
10 Zurich.	"	
13 Fairville.	5 c. Niagara.	
16 Newark.	6. Salina.	418
20 Marbleton.	"	
22 Outlet.	"	
23 Phelps.	9 c. Corniferous.	
27 Orleans.	"	
31 Flint.	"	
34 Stanley.	10 b. Hamilton.	904

72. Midway between Watertown and Brownville the whole river falls 60 feet in less than half a mile, running in a gorge with high banks.

73. *Tully.* The Tully limestone, separating the Hamilton from the Genesee, which is named from this place, is not seen on the railroad, but is found further to the west. Outcrop in grove 8. E. of the village. The swamp near Preble is supposed to be underlaid by the Tully limestone.

74. Between Syracuse and Jamesville are good natural sections of the 6. Waterlime and 9. Onondaga and Corniferous limestones, many quarries and natural cliffs. Beyond Jamesville observe the transition into the Hamilton group where the high hills begin, the Marcellus shales being deeply excavated. Visit Green Lake, near Jamesville.

75. The red sandstone of the 5 a. Medina formation is well displayed at Fulton, in Oswego County, where it causes the Oswego Falls and forms the banks and bed of the river above and for half a mile below. The upper layers are covered with *Fucoides Harlani*, some of them of gigantic size.

76. The 5 b. Clinton formation is named from this place.

77. This is one of the best railroads in the State for geological observations. There are many points on the Cayuga Railroad where the junction of the Hamilton with the Tully limestone and of the latter rock with the Genesee shale, and of the Genesee with the Portage group are perfectly seen in juxtaposition. The lake affords every evidence and facility for geological sections, with fossils.

78. Cayuga Lake is 40 miles long, 3½ miles wide, 390 ft. deep, and its surface is 376 ft. above tide.

79. The gypsum beds are finely displayed just north of Union Springs, and large quantities are produced for market. South of the town the 9.Upper Helderberg range crosses, and causes an islet in the lake. Its lower layers, the Onondaga limestone, make beautiful quarries.

80. The low clayey land extending nearly to Levanna is on the 10 a. Marcellus shale. The first rock south of this is the dividing line between the Marcellus and Hamilton.

81. The 10 b. Hamilton presents its first bluff south of Aurora, 20 to 50 feet high, containing numerous fossils. Further south are many others, some of them 100 feet high, extending for miles. Nothing could be finer than these geological sections of the Hamilton.

82. The Tully limestone first appears at Lake Ridge, from which the station is named. It is the dividing line between the 10 b. Hamilton and the 10 c. Genesee. It dips as you go south and rises again. This looks like a flexure of the formations, but it is caused by the change in the course of

Lehigh Valley Railroad.			Ms.	Pa. & N. Y. Ca[nal]
Ms.	Cayuga Branch.[77]	Alt.	0	Freeville.
0 Cayuga.[78] 888	6 Salina. Lake, 376.		4	West Dryden.
6 Union Springs.[79] 394	6. Salina, with Gypsum beds. 9 c. Corniferous quarries.		7	Asbury Road.
			10	South Lansing.
			14	North Lansing.
10 Levanna.[80]	10 a. Marcellus. 10 b. Hamilton.		17	Genoa.
			23	Venice Centre.
13 Aurora.[81]	"	925	27	Scipio.[197]
16 Willett's.	"	405		
20 King's Ferry. 394	" Bluffs 100 ft.			**Geneva, Ithaca**
22 Atwater's.	"	394		
25 Lake Ridge.[82] 401	" Tully limes.		0	Sayre.[109]
27 Taughannock. 411	" "		2	West Waverly.
32 Ludlowville.[83] 898	10 c. Genesee and Portage.		9	Bingham's
			16	Van Ettenville.
38 Ithaca.[84]	11 b. Portage.	392	19	Spencer.[188]

Pa. & N. Y. Canal & R. R.				
0 Sayre.[109]	11 b. Chemung.	774	23	North Spencer.
7 Barton.	"	803	27	West Danby.
10 Smithboro.	"	799	31	Newfield.[191]
14 Tioga.	"	805	38	Ithaca.[84]
20 Owego.[188]	"	822	44	Willow Creek.
24 Flemingville.	"	907	46	Taghanic Falls.
29 Newark Valley.	"	966	48	Trumansburg.
35 Berkshire.	"	1048	51	Covert. 858
39 Richford.	"	1097	54	Farmer.
43 Hartford Mills.	"		57	Ovid Centre.
45 Hartford.[198]	"	1186	61	Hayt's Corners.
51 Dryden.[196] 1079	" Sum'it,1215		65	Romulus.
54 Freeville.	11 a. Portage.	1049	70	West Fayette.
56 Peruville.	"			
59 Groton.[196]	"	997	77	Geneva.[81] 459
65 Locke.[197]	799 " on 10 c. Gen.			
69 Moravia.[98]	"	732		**Syracuse, Geneva** [
73 Cascade.[99]	10 b. Hamilton.	724	0	Geneva.[81]
76 Scipio.[197]	730 " (Glen.)		9	Earle.[89]
70 Wyckoff's.[99]	"	726	14	Dresden.[87]
(Foot of Lake.)			21	Himrod's.
86 Auburn.[30]	9 c. Corniferous. 666		26	Dundee.
90 Throop.	6. Salina, 13 miles.		30	Rock Stream.
95 Weedsport.[33]	. "	429	33	Reading Centre.
99 Brick Church.	"		36	Watkins Glen.
104 Cato.	"	423	37	Glen Bridge.[86]
108 Ira.	5 c. Niagara.		45	Beaver Dam.
112 Martville.	5 c. Clinton.	867	49	Post Creek.
115 Sterling.	"		52	Ferrenburg.
116 Fair Haven.	5 a. Medina, 3 miles.		58	Corning.[188]
118 N. Fair Haven.[71]	" Lake, 245			

(vertical labels in central column: "20. Valley Drift." and "Owasco Lake.")

the lake. After rising again it forms a beautiful coping of the Hamilton Taughannock. See the description of the 10 b. Tully limestone.

83. This is one of the best localities of the Hamilton group which we k ville the 10 c. Genesee shale appears above the Tully limestone. It is unifc structure, fine grained, a hard and brittle mud rock, its edges resisting the when exposed falling into pieces. You get a good section of the base of the is a well marked dividing line here between the Genesee and Portage, bein thick, very compact and solid, with its under surface filled with fucoids rais inches long with their ends depressed. The eye readily follows it as it dips

84. Every part of the Portage group can be inspected in the ravines and v of Itha..n.

85. There is a glen here, one mile southeast from the station, quite equ is also in the Portage. See Note 86.

86. Watkins Glen is in the 11 a. Portage. It is a great wonder and very grand view of the chasm in crossing the bridge over it at Glen Bridge on t Corning Railroad. The gulfs on that road are perfectly characteristic of the

Elmira, Cortland & Northern, formerly		
Ms.	**Utica, Ithaca and Elmira Railroad.**	**Alt.**
0	Elmira.	11 b. Chemung.	862
5	Horse Heads.	"	899
10	Breesport.	"	1097
14	Erin. 1249	"	
17	Park. 1513	"	
21	Swartwood. 1059	"	
25	Van Etten.[198]	1012 "	
28	Spencer.[188] 996	"	
32	West Candor.	"	
34	North Candor.	"	
37	Wilseyville.[188]	940 "	
42	White Church.	958 "	
44	Mott's Corners.	11 a. Portage. 945	
46	Besemer's.	" 949	
50	Ithaca.[84,189]	Striae. " 840	
53	Varna.	"	
54	Snyder's.	" 995	
57	Etna.	" 1010	
60	Freeville.	" 1049	
62	Malloryville.	" 1059	
63	McLean.	" 1090	
67	Sou. Cortland.[100]	" 1151	
70	Cortland.	" 1116	
71	D. L. & W. Dep't.	" 1116	

(Hills of Portage.)

0	Cortland.	11 a. Portage. 1116
12	Truxton. 1135	" V'y drift.1135
16	Cuyler.	" 1225
20	De Ruyter.[190]	10 c. Genesee. 1276

0	De Ruyter.[190]	10 c. Genesee. 1276
10	Otselic.	11 a. Portage.
20	Plymouth.	11 b. Chemung.
28	Norwich.	" 1001

Elmira, Cortland & Northern R. R.[26]

0	Canastota.[26]	6. Salina. 426
3	Clockville.[195]	" 637
4	Colton. [·95]	"
5	Oak Hill.	" Gypsum in cuts.
6	Quarries.[95]	9. Onondaga limest'ne.
8	Perryville.[96]	" 1041
9	Hyatt's.	"
11	Chitt'go Falls.[97]	10 c. Marcellus. 1051
12	Bingley.[191]	" 1041
13	Shelter Valley.	"
14	Firndell.	10 a. Hamilton.
15	Cazenovia.[93,191]	" 1176
17	Syr. & Chen. Ju.	" 1245
22	New Woodstock.	" 1293
26	Shedd's Corners.	" 1388
30	De Ruyter.[190]	10 c. Genesee. 1276

Ms.	**New York, Ontario & Western R. R.**	**Alt.**
	New York, (Erie Railroad), N. W.	
0	Middletown.	4 c. Hudson River.[550]
5	Fair Oaks.	"
10	Bloomingb'g. 198 101	{ 5 a. Oneida. 757 / Tunnel, 3,840 feet.
12	Wurtzboro.	
15	Summitville.[198]	{ 10. Hamilton, 11 a.[545] / Portage & Chemung.
30	Fallsburg.	12. Catskill. Tunnel,
39	Liberty Falls.	Striae. " 1,017 ft.
40	Liberty.	"
46	Parksville.	" 1798
51	Morseton.	11. Chemung.
63	Cook's Falls.	"
73	East Branch.	"
82	Hancock.[188]	12. Cat'l. Tun'l, 1,100 ft
89	Codosia Summit.	" 1462
93	Rock Rift.[188]	" 1152
101	Walton.[188]	June'n of the 11. 1220
108	Zig Zag.[180]	Chem. &12. Catsk.[1665]
117	Sidney Centre.	12. Catskill, synclinal.
125	Sidney Plains.	11 b. Chemung. 967
127	New Berlin Jun.	"
134	Guilford.	" 1399
143	Oxford.	"
148	Norwich.[190]	11 a. Portage. 768
163	Earlville.[188]	10 c. Genesee.
167	Smith's Valley.	10 b. Hamilton.
172	Eaton.	10 a. Marcellus.
174	Morrisville.[191]	9 c. Cornifer. l. s. in
181	Munnsville.[191]	" hills.
183	Cook's Corners.	6. Salina.
187	Oneida Comm'ty.	5 c. Niagara.
190	Oneida.	5 b. Clinton. 412
192	Durhamville.	"
200	North Bay.[102]	" "Lake, 367 (Deep drift glass sand.)
209	Cleveland.	"
216	Constantia.[102]	"
223	Central Square.	"
230	Pennellville.	"
238	Fulton.[75]	5 a. Medina. 335
250	Oswego.[71]	" Lake, 245.

101	Walton.[188]	(As before.)
105	Colchester.	12. Catskill.
109	Hawley's.	"
112	De Lancey's.	"
118	Delhi.	"
127	New Berlin Jun.	11 b. Chemung.
134	Mount Upton.	"
140	Holmesville.	"
145	New Berlin Cen.	10. Hamilton.
149	New Berlin.	"

87. The outlet of Crooked Lake from Penn Yan to Dresden is through the Genesee slate, Tully limestone, and the upper part of the Hamilton—all finely displayed. Crooked Lake is 20 miles long, one mile wide, 100 feet deep, and its surface is 718 feet above tide water. Its northern half is divided by a bluff of Portage (800 feet high) into two branches—one of them 12 and the other 8 miles long.

88. Canandaigua Lake is 14 miles long, from one to two miles wide, its surface is 668 feet above tide, and its greatest depth is 100 feet, but it is very shallow at both ends. It is excavated from the Hamilton and Portage groups.

89. The drift described in note 31 extends nearly to Dresden.

90. The D., L. & W. From Binghampton to Buffalo is by Prof. H. S. Williams of Cornell University. Compare formations and notes on N. Y., L. E. & W.

Ms. New York, Ontario & Western.—Con.	Alt.
0 Middletown.	4 c. Hudson River.
15 Summitville.	"
17 Phillipsport.	"
19 Homowack.	"
23 Ellenville.	" and Trenton.

Cornwall to Middletown.[123]

0 Cornwall.[116],[142]	4 c.Hudson River.
3 Montana.	"
6 Meadow Br'k.[124]	Red Grits and Cong.
7 Dennistons.[142]	4 c. Hudson River.
12 Rock Tavern.	"
14 Burnside.	"
16 Campbell Hall.	"
18 Stony Fork.	"
21 Ireland.	"
23 Mechanicstown.	"
25 Middletown.	"

New York, Lake Erie and Western R. R.
(Late Erie Railway.)

New York.	See Note 4.	
0 Jersey City.[103]	16. Triasic. Tunnel in intrusive basalt sheet.	
(Tide Marshes.)[104]	16. Triasic.	
9 Rutherford P'rk.	16. Triasic.	50
11 Passaic.[127]	"	55
16 Paterson.	"	89
21 Ridgewood	"	137
23 Hohokus.	"	190
25 Allendale.	"	270
27 Ramsey's.	20. Quaternary.	345
31 Suffern, N. J.[105]	1. Archaean.	298
33 Ramapo, N. Y.	"	510
34 Sterling Junc.	"	
35 Sloatsburg.	"	350
41 Southfield.	"	491
48 Greenwood.[105]	"	520

Ms. New York, Lake Erie & West'n—Con.	Alt.
47 Turner's.[128]	3? Low. Silur'n l. s.[555]
49 Monroe.[129]	4 c. Hudson River.
50 Schunemunk Mt.	10? Middle Devonian.
51 Oxford.	3? Low. Silur'u l. s.[540]
53 Greycourt.[130]	4 c. Hudson River.
59 Goshen.	" 431
66 Middletown.	" 552
70 Howell's.	" 599
75 Otisville.[106]	" 570
Kittatiny, Blue, or Shawangunk Mountain.	5 a. Oneida, or Shawangunk and Medina.
87 Port Jervis.[101] 188	7. Low'r Helderberg. 8. Oriskany. [442] 9. Cauda Galli & Up. Heldg. & 10. Hamilt.
Sparrowbush.	11 a. Portage.
99 Pond Eddy, Pa.	11 b. Chemung. 571
106 Shohola.	" 648
110 Lackawaxen.[107]	" 648
116 Pine Grove.	" 668
122 Narrowsburg.[107]	" 720
131 Cochecton, N. Y.	12. Catskill ridge. 748 11 b. Chemung.
135 Callicoon.	" 781
136	12. Catskill,(bluffs).
143 Hawkins.	"
147 Basket.	"
154 Lordville.	"
159 Stockport.	11 b. Chemung.
163 Hancock.	12. Catskill. 926
172 Hale's Eddy.	11 b. Chemung. 950
176 Deposit.	" 1008
184 Summit.[199]	1873 "Mt.toN.Cats
192 Susquehan'a.[108]	" 914
200 Great Bend.[200]	" 834

91. Just south of the Erie Canal there is a deep cut in a bluff of Waterlime Group.

92. Picturesque view of Pompey Valley.

93. Cazenovia Lake is a beautiful lake, 4½ miles long, ¾ mile wide, and 70 feet deep, 1,189 feet above tide water, and is excavated in the Hamilton group. It discharges its waters into Chittenango Creek, which runs northward.

94. Lebanon and Earlville are both good localities for Hamilton fossils.

95. Extensive and beautiful view extending over Oneida Lake.

96. Canaseraga Falls similar to Chittenango Falls. Note 97.

97. The Falls are in sight in the valley to the west. Here Chittenango Creek falls 120 feet perpendicularly into a canon over the 9. Onondaga limestone, with the Corniferous bed over it, which forms the sides of the creek at the top of or above the Falls. Under the Onondaga limestone is the Oriskany sandstone, only six inches thick. Above the Falls the creek flows through a small, handsome valley, its lower sides formed of Marcellus, and the tops of the hills Hamilton.

98. Moravia is an excellent locality for Hamilton fossils. The Tully limestone, the dividing line between the Hamilton and Genesee, is half way up the hill sides, and appears to dip below the valley north of Locke. It is met with at the falls of Dry Creek, south of Moravia.

99. Owasco Lake is 10 miles long, a mile and a half wide at the north at Auburn, and a half mile at the south end, and 750 feet above tide water. The whole of the lake is in the Hamilton group.

100. Marl is here taken from the bottom of ponds; dried like bricks, and burnt into lime.

101. From Bloomingburg tunnel to Sidney, the geology is the same as from Port Jervis to Susquehanna on the Erie Railway. In the hills at Port Jervis, fossils of L. H., Oriskany and Hamilton.

102. Oneida Lake is 19 miles long, 6 miles wide, its greatest depth not over 40 feet, and in general al it is quite shoal. Its surface is 367 feet above tide water. It is excavated in the 5 b. Clinton group the rocks of which appear on its south shore and west end. Its north shore is covered with sandy alluvium which is 100 feet deep at the east end and furnishes glass sand used in the glass factories in this vicinity.

103. The Erie railway tunnel at Jersey City is through Bergen Hill, which is the southern end of the mountain ridge of basalt or trap rock of the 16. Triassic age, 48 miles long, known farther north as the Palisade Mountain. See note 5.

104. The railroads out of New York through New Jersey pass over very extensive tide marshes, covered with reeds and coarse sedge grass, growing in soft mud, which is in some places forty feet deep, and all overflowed in high tide. These vast salt marshes so near New York City, which excite

Ms. N. Y., Lake Erie & Western.—Con.

Ms.	Station	Formation	Alt.
295	Kirkwood.	11 b. Chemung.	876
214	Binghamton.[105]	"	868
223	Union.	"	840
229	Campville.	"	830
236	Owego.[188]	"	822
246	Smithboro.	"	799
248	Barton.	"	803
255	Waverly.[109]	"	886
260	Chemung.	"	817
266	Wellsburg.	"	831
273	Elmira.[108]	"	868
290	Corning.[188]	"	942
301	Addison.	"	998
331	Hornellsville.	"	1161
343	Canaseraga.	Mor.? "	
355	Nunda.[191]	11 a. Portage.	1836
361	Portage.[110,191]	"	1314
365	Castile.[191]	"	1401
374	Warsaw.	"	1326
380	Dale.	"	1190
391	Attica.	"	998
395	Griswold's.	10 b. Hamilton.	1044
397	Darien.[160]	"	1024
403	Alden.	10 a. Marcellus.	864
408	Town Line.	9 c. Corniferous.	742
412	Lancaster.	"	663
420	East Buffalo.	"	607
422	Buffalo.[40,197] 568	" Lake.	569.

(Valley drift. Kame-like knolls.)

Ms.	Station	Formation	Alt.
0	Corning.	11 b. Chemung.	942
1	Painted Post.	"	945
5	Coopers'.	"	970
7	Curtis'.	"	997
9	Campbell's.	"	1014
14	Savonia.	"	1053
20	Bath.[205]	Mor.? "	1105
23	Kanona.	"	
27	Avoca.	"	1195
30	Wallace's.	"	1235
35	Liberty.	Mor.? "	1293
39	Blood's.	"	1825
45	Wayland.	"	1889
50	Springwater.[191]	11 a. Portage.	1870
53	Webster.[191]	"	1848
57	Conesus.	"	1280
60	South Livonia.	11 b. Hamilton.	1167
64	Livonia.	"	1030
67	Hamilton.	"	920
76	Avon.[111] 568	9 c. Cornif. and Water-	
80	Rush. 541	6. Salina.	lime.
82	Scottsville.	"	558
86	Henrietta.	"	564
90	Red Creek.	"	525
94	Rochester.[32] 527	5 c. Niagara, 8 miles.	

(Valley drift. Kame-like knolls.)

Ms. N. Y., Lake Erie & Western.—Con.

Ms.	Station	Formation	Alt.
331	Hornellsville.	11 b. Chemung.	1161
340	Alfred.[201]	Fossils. "	1660
349	Andover.	"	1640
357	Genesee.	"	1526
365	Phillipsville.	"	1390
369	Belvidere.	"	1584
373	Friendship.	"	1539
382	Cuba. 1542	"Sum't, 1698.	
389	Hindsdale.	"	1501
394	Olean.[201]	"	1488
398	Allegany.	"	1422
407	Carrollton.	"	1399
410	Great Valley.	"	1398
413	Salamanca.	"	1384
421	Little Valley.	" Mor.	1594
428	Cattaraugus.[208]	"	1411
437	Dayton.	" Mor.	1846
440	Perrysburg.	"	1260
447	Smith's Mills.	"	1010
451	Forestville.	"	883
454	Sheridan.	11 a. Portage.	760
459	Dunkirk.	"	598

(Valley drift.)

Ms.	Station	Formation	Alt.
76	Avon.[111] 585	9 c. Cor. & 6. Water Li.	
83	Caledonia. 658	"	
90	Le Roy. 872	"	
94	Stafford. 910	"	
100	Batavia.[41] 895	"	
107	Alexander.	10 b. Hamilton.	928
110	Attica.	11 a. Portage.	998

(Good exposures of the rocks.)

Ms.	Station	Formation	Alt.
76	Avon.[111]	9 c. Corniferous.	585
80	South Avon.	" and Marcell.	
85	Geneseo.	10 b. Hamilton.	600
89	Cuylerville.	"	528
90	Shaker's. 574	11 a. Chasaqua shale.	
91	Mt. Morris.[112]	10 c. Genesee.	595
94	Sonyea.	"	592
98	McNair.	"	576
102	West Sparta.	11 a. Portage.	
106	Dansville.[118]	"	691

New York, Pennsylvania and Ohio R. R.[186]

Ms.	Station	Formation	Alt.
0	Salamanca.	11 b. Chemung.	1393
12	Steamburg.	"	
18	Randolph.	"	1313
25	Kennedy.	"	1264
34	Jamestown.[118]	"	1321
39	Lakewood.[115]	"	
41	Ashville.	"	1356
51	Bear Lake, Pa.	"	1550
58	Columbus.	"	1427
61	Corry, Pa. 1428	" Carbonif.	

the wonder of strangers, contain from 250,000 to 300,000 acres or from 400 to 470 square miles. Future generations may build dikes and reclaim them, but at present they are dismal swamps without a single tree or shrub, and wholly impassable to either man or beast. The two hills which rise abruptly in the salt meadow south of the Erie Railway and north of the Pennsylvania Railroad, are called Big Snake Hill and Little Snake Hill. The large one is half a mile long and 200 feet high. Both of these hills are outbursts of trap from between the underlying sandstone strata, similar to the Palisade Mountain.

105. *Suffern to Greenwood.* Here is a long natural gap through the Laurentian Highland range or Ramapo Mountains.

New York, Lake Erie & Western.—Con.
Ms. Suspen'n Bridge & Niagara Falls Branch. Alt.

420	Buffalo.	9 c. Corniferous 5 5 5
420	East Buffalo.	" 6 0 7
425	Main Street.	" 6 8 0
431	Tonawanda.	6. Salina. 5 8 0
437	La Salle.	" 5 7 2
442	Niagara Falls.³⁹	5 c. Niagara. 5 7 4
443	Susp. Bridge.⁴²	" 5 8 0
444	Clifton, Ont.	"

Lockport Branch.¹³⁶

0	Buffalo.	9 c. Corniferous. 5 8 8
8	Tonawanda.	6. Salina.
18	Hodgeville.	
22	Lockport ³⁸	5 c. Niagara.

Piermont Branch.

0	Suffern.¹³¹	16. Triassic. 2 9 5
9	Nanuet.	" 2 8 4
17	Piermont.¹³²	" Trap. 6

Northern Railroad of New Jersey.

0	Jersey City.¹⁰²⁄¹⁰⁴	16. Triassic.	Trap.
4	Homestead.¹³³	"	
6	New Durham.¹³⁴	"	
7	Granton.¹³⁵	"	Trap.
9	Ridgefield.	"	
12	Leonia.	"	
14	Englewood.	"	
15	Highland.	"	
16	Tenafly.	"	
17	Cresskill.	"	
19	Closter.	"	
21	Norwood.	"	
23	Tappan.	"	
24	Sparkill.¹³²	"	20 Quat.
25	Piermont.	"	Trap.
29	Nyack.	"	

New York, Lake Erie & Western.—Con.
Mɴ.   Walkill Valley Railroad.   Alt.

0	Jersey City.	(See Main Line ErieR.)
59	Goshen.¹⁰⁵	4 c. Hudson Riv. 4 3 1
61	Ripp's.	"
64	Campbell Hall.	" 3 9 6
66	Neely Town.	3 a. L. Sil l. s.(fos.)³⁸⁰
68	Beaver Dam.	" 4 0 5
69	Montgomery.	" 3 8 6
73	Walden.	3 5 1 " Fossils.
76	Shawangunk.	{ 5 a. On'da or Shaw'k Grit and Medi. 2 7 7
79	New Hurley.	{ 7. Lower Helderberg and 9. Upper Held'g, mainly Upper.
82	Gardner.	" 3 1 1
85	Forest Glen.	"
87	New Platz.	" 2 6 6
91	Springtown.	"
94	Rosendale.¹¹⁴	4 c. Hudson River.¹⁶⁷
96	Katson's Cave.	"
98	Whiteport.	" 1 8 9
102	Kingston.¹¹⁴	1 9 6 " & Waterli

Monticello and Port Jervis Railroad.

0	Port Jervis.¹⁰¹	10. Hamilton. 4 4 2
6	Huguenot.²⁰⁶	"
8	Rose Point.	11 b. Chemung.
12	Paradise.	"
13	Oakland.	"
16	Hartwood.	"
18	Gillman's.	"
20	Barnum's.	"
24	Monticello.²⁰⁷	12. Catskill.

106. *Otisville.* A short distance west of Otisville the Hudson River Slates are seen in contact with the Shawangunk Grits along a fault line. This is the dividing line between two of the great geological groups or periods, the Lower Silurian and Upper Silurian. In a moment the whole character of the country is changed from cultivated grazing land on the Hudson River slates, the Orange County milk country to the east of this line, to a poor, barren, rocky region on the Oneida or Shawangunk and Medina formations, showing in a striking manner how the character of the country depends on its geology. In descending the Shawangunk Mountain towards Port Jervis there is an alternation of beds of the Oneida conglomerate, which is of a light gray color, and the Medina sandstone, which is of a high red color. Some pockets of galena were discovered and mined here, but were soon exhausted. At Port Jervis we are in the Hamilton, a formation producing a country capable of supporting a population. The intermediate formations are very thin and compressed together.

107. *Lackawaxen.* From Port Jervis to Narrowsburg, the Delaware River and Erie Railway pass through a deep and crooked gorge about 25 miles long, exhibiting some of the wildest scenery in the country. The railroad is cut out of rock in many places and overhung as it were by ragged precipices.

108. *Binghampton.* West of Susquehanna the Erie Railway and its branches run for more than 300 miles on the 11 b. Chemung formation. Most of it is a fine fertile country with some handsome towns, the largest of which are Elmira and Binghampton, in valleys filled with gravel alluvium, and the higher country formed of the calcareous Chemung shales, is quite productive, much of it being a good grazing country; but there is no variety in its geology. East of Susquehanna the Chemung formation is composed of harder sandstone. It contains less calcareous shale, and the soil is poor. The country improves rapidly going westward from Susquehanna. See also 185.

109. Just west of Waverly are the Chemung Narrows, where 100 feet of rock are exposed. The quarries have produced an abundance of characteristic fossils of the Chemung group in their greatest beauty and perfection, the formation having been named from this locality. Five miles south of Waverly the opening of the Susquehanna Valley may be seen, where the Chemung River from the west and the Susquehanna from the east unite and traverse the State of Pennsylvania to Chesapeake Bay. At the west end of Waverly Village is a curious flat-topped hill, about 60 feet high, called "Spanish Hill." It is an eddy hill of gravel formed in the drift period; but it can be seen to better advantage on the south side, at Sayre on the Pa. & N. Y. R. R. and the G. I. & S. R. R. There is a similar eddy hill in the village of Union. The plain at Sayre is "Valley Drift."

110. *Portage.* Here the railroad crosses the very deep gorge of the Genesee River on a high iron bridge 820 feet long and 235 feet high. There are three falls within a distance of two miles which

Ms. Station	Group	Alt.
0 Carrolton.	11 b. Chemung.	1899
6 Limestone.	"	1416
11 Bradford's, Pa.	"	1464

Buffalo and Southwestern.

Ms. Station	Group	Alt.
0 Buffalo.[40]	9 c. Corniferous.	555
3 Junction.	"	
5 Limestone Ridge.	"	
10 Abbott Road.	"	
13 Hamburg.	10. Hamilton.	635
16 Eden Valley.	11 a. Portage.	
19 Eden Center.	"	
23 North Collins.	"	846
27 Lawton's.	11 b. Chemung.	
30 Collins.	"	
33 Gowanda.	"	776
39 Dayton.	" Moraine	
43 Pine Valley.	"	885
48 Cherry Creek.	"	
53 Clear Creek.	"	
56 Randolph.	" Moraine	
60 Kennedy.	"	
69 Jamestown.[115]	" Moraine	

Tioga, Elmira & State Line Railroad.

Ms. Station	Group	Alt.
0 Elmira.[108]	11 b. Chemung.	868
1 Erie Junction.	"	
3 State Line Junc.	"	909
7 Wells.	"	995
9 Seeley Creek.	"	1041
10 State Line.	"	
12 Millerton, Pa.	"	1246
15 Trowbridge.	12. Catskill.	1440

Middletown & Crawford Branch.

Ms. Station	Group	Alt.
0 Middletown.	4 c. Hudson River.	562
3 Crawford Junc.	"	
5 Circlesville.	"	
8 Bellville.	"	
10 Thompson Ridge.	"	
13 Pine Bush.	"	

Newburg Branch.[123] (Short Cut.)

Ms. Station	Group	Alt.
0 Greenwood.	1 Archæan.	520
2 Junction.[128]	3? Lower Silurian, l.s	
Central Valley.	"	
5 Highl'd Mills.[126]	Silurian Grits.	480
7 Woodbury,	{ 10? Green Pond Mt. S'rs, Mid. Dev'n.[442]	
Mountainville.	3? Lower Silurian, l.s.	
13 Cornwall.[126]	4 c. Hud. Riv. 280,[142]	
15 Vails Gate Junc.	"	280
17 New Windsor.	"	192
20 Newburg.[133]	"	25

Ms. Station	Group	Alt.
0 Greycourt.[130]	4 c. Hudson River.	
2 Craigville.[142]	"	
7 Washingtonville.	"	
9 Salisbury.	"	
13 Vails Gate.	"	250
16 New Windsor.	"	192
20 Newburg.[133]	"	25

Pine Island Branch.[123]

Ms. Station	Group	Alt.
0 Goshen. 142	4 c. Hudson River.[431]	
3 Orange Farm.	3? Lower Silurian.	
6 Florida.	"	
12 Pine Island.	"	

Syracuse, Ontario & New York Railroad.

Ms. Station	Group	Alt.
0 Syracuse.[27]	6. Salina.	405
8 Manlius Cen.[91]	7. L. Held., Waterli.[435]	
10 Fayetteville.	" & 9. Onon. l. s.[538]	
12 Manlius.	{ 9. Onondaga limest. Heavy beds. 742	
15 Oran.[92]	9. Onondaga l. s. 897	
Tunnel. 1318	{ 10 a. Marcellus. 10b. Tunnel in Hamilton sandstone.	
20 Cazenovia.[93]	10. Hamilton.	1191
23 Webster's.	"	
29 Erieville.	"	1577
32 Georgetown.	"	1450
38 Lebanon.[94]	{ 10 c. Genesee. 1338 11 a. Portage, cliffs.	
45 Earlville.[135]	10 c. Genesee.	1071

New Jersey and New York R. R.[123]

Ms. Station	Group	Alt.
0 Spring Valley.	16. Triassic.	
Pomona.	"	
Mt. Joy.[139]	"	
Thials.	"	
9 Haverstraw.	"	
11 Stony Point.	"	

Dunkirk, Allegheny Val'y & Pitts. R. R.[136]

Ms. Station	Group	Alt.
0 Dunkirk.	11 a. Por. & 11b. Che.[595]	
3 Fredonia.	11 a. Portage.	765
5 Laona.	"	810
13 Lily Dale.	"	
14 Cassadaga.	11 b. Chemung.	1309
18 Moons.	"	1303
22 Sinclairville.	"	1330
26 Gerry.	"	
29 Ross' Mill.	"	1262
32 Falconer.	"	1258
33 Junction.	"	1262
38 Frewsburg.	"	1261
Con. in Pa.		

are 60, 90 and 110 feet high, besides the intervening rapids. Two of them are visible from the car windows on the north side. The bridge crosses the upper falls. The river pursues a meandering course through this deep gorge and over these three successive cascades, descending more than 500 feet, and passes out into the Valley of the Genesee at Mount Morris. The gorge is 20 miles long by the river, or 14 by the public road, and its depth in some places is not less than 350 feet, its width only about 600 feet, and the banks nearly perpendicular. The place is well worth a visit. It is cut out of the 11 a. Portage group, except the lower end, which is in the 10 c. Genesee shale. The Portage group was named from this place. See note 112, Mount Morris. There is an ancient channel from Portage to Nunda, filled up by drift, compelling the river to cut its present deep, torturous channel. For other examples of this see notes 31, 35, 38 and 39.

111. *Avon.* You have 9. Upper Helderberg, and 10 a. Marcellus shale in the creek.

112. To study the Genesee shales stop at Mount Morris. Go through the village one mile

Ms. Lake Shore & Mich. Southern R. R. Alt.

Ms.			Alt
0	Buffalo.⁴⁰	9 c Corniferous	588
10	Hamburg.¹⁴⁸	10 Hamilton.	685
21	Angola.	"	887
26	Farnham.	"	628
29	Irving.	"	586
31	Silver Creek.	10 c. Genesee.	683
40	Dunkirk. 598	11 a. Port. & Chemung.	
49	Brocton Junct'n.	689 " "	
57	Westfield.	697 " "	
65	Ripley, Pa.	" "	
73	North East.	805 "	
80	Harbor Creek.	781 "	
84	Wesleyville.	"	
88	Erie.	686 "	
98	Fairview.	"	
103	Girard, Pa	717 "	
115	Conneaut, Ohio.	11. Erie Shale.	
123	Kingsville.	672 "	
128	Ashtabula.¹⁴⁸	646 "	
	(Continued	in Ohio.)	

Portage along the lake. Chemung to the E. in the hills.

New York, Chicago & St. Louis Ry.

Ms.		
0	Buffalo.	9 c. Corniferous.
2	Erie Junction.	"
9	Bay View.	10. Hamilton.
15	Lake View.	"
28	Irving	"
32	Silver Creek.	10 c. Genesee.
42	Dunkirk.	11 a. Port. 11 b. Chem.
50	Brocton Ju.	" "
58	Westfield.	" "
66	Ripley, Pa.	" "
88	Erie.	" "
103	Girard.	" "
116	Conneaut, Ohio.	11. Erie Shale.

Bath and Hammondsport R. R.

Ms.		
0	Bath.²⁰⁵	11 b. Chemung. 1105
5	Cold Spring.	
9	Ham'ndsport.¹⁹⁷	"

Ms. Buffalo, Rochester & Pittsb'h R. R. Al?

Ms.			Alt
0	Rochester.	5 c. Niagara.	48
5	Maplewood.	"	
7	Brookdale.	6. Salina.	
11	Scoftsville.	"	55
14	Garbuttsville.	6. Waterlime.	
15	Wheatland.	"	59
17	Mumford.	"	61
21	Lime Rock.	9 c. U. Helderberg.⁷⁷	
25	Le Roy.	"	87
30	Pavilion Center.	10. Hamilton.	94
33	Pavilion.	"	94
38	Wyoming.	10 c. Genesee.	96
43	Warsaw.	11 a. Portage.	112
48	Rock Glen.	"	
54	Gainesville.	" Mor.	140
62	Bliss Corners.	"	
65	Eagle Village.	Moraine." Sum't. 190?	
83	Machias.	1646 " & 11 b. Che	
93	Ashford.	Mor. " "	
97	Ellicottsville.	Moraine.	156
102	Great Valley.²¹⁰	"	139
108	Salamanca.	Valley drift.	139

Buffalo Division.¹⁸⁶

Ms.		
0	Buffalo.	9 c. Corniferous.
2	Buffalo Creek.	10. Hamilton.
5	W. Seneca.	"
10	Hamburg.	"
11	Orchard Park.	"
16	West Falls.	"
21	Colden.	11 a. Portage.
23	Glenwood.	"
28	E. Concord.	"
31	Springville.	"
38	Riceville.	"
41	W. Valley.	"
48	Ashford.	11 a. Por. 11 b. Chem
57	Gt. Valley Cent.	11 b. Chemung.
62	Bradford Ju.	"
63	Kilbuck.	"
66	Carrolton.	"
72	Limestone.	"

northwest to the mouth of the gorge, where the Genesee River, after running 20 miles through the deep canon from Portage, breaks out into the beautiful broad and fertile Genesee Valley. There is a good section close to the bridge over the river. Get a boat and row one mile up the pool of the State dam, which flows to the foot of the precipices all that distance. This is the finest exposure of the 10 c. Genesee in the State, the typical locality from which it was named, and the scenery is in itself remarkably good. The cliffs are 100 to 200 feet perpendicular, full of *Septaria*, like flattened cannon balls sticking in the walls. It is curious that so soft a shale rock should stand the weather so well and not form sloping banks when the edges only are exposed. See note No. 110, Portage.

113. Dansville is in a beautiful ampitheatre of Portage hills with very picturesque views from the Water Cure and other elevated points. Moranic Kame-like hills of glacial origin.

114. The Rosendale Cement, manufactured near Rondout, is from the 6. Waterlime rock, which is here between the Medina sandstone and the Lower Helderberg limestone, the intermediate formations being wanting. It is a light blue, fine grained limestone, with smooth conchoidal fracture. The same formation furnishes the Hydraulic Cement, made at Syracuse, N. Y., and elsewhere.

115. *Jamestown.* Chautauqua Lake is 18 miles long, 2 miles wide, 1291 feet above tide water and 726 above Lake Erie. Its northern extremity is only 8 miles from Lake Erie, and yet it empties its waters by the Conewango, Alleghany, Ohio and Mississippi into the Atlantic. It is a beautiful sheet of water, bounded on its eastern side by gravelly sloping banks, and on the west by more level and in some places marshy shores. It is excavated in the Chemung group, the Portage being along its outlet and on the shores of Lake Erie below, but of much less thickness than further east.

116. *Cornwall.* Just south of this station contact of the Trenton slates (See Note 142.) and the Archæan rocks of the highlands; the former overturned and dipping beneath the latter. See also Notes 130 and 126.   N. H. Darton.

Buffalo, New York and Philadelphia, now, Ms. Western N. Y. & Penna. R. R.		Alt.	B., N. Y. & P., now W'n. N. Y. & Pa. R. R., Ms. Rochester Division.—Con.		Alt.
0 Buffalo.[40]	9 c. Corniferous.	558	47 Tuscarora.	11 a. Portage.	
13 Elma.	10. Hamilton.	527	50 Nunda Ju.	, "	
17 Aurora. 525	" & 11 a. Portg.		53 Nunda.	"	
22 Wales.	"		62 Swains.	"	
26 Holland.	"	1176	52 W. Nunda.	"	
29 Protection.	Moraine. "	1385	55 Lewis.	"	
36 Arcade.[185] 1457	11 a. Por.& 11 b. Chem.		59 Portage.	"	
39 Yorkshire.	Moraine. "	1456	64 Wiscoy.	11 a. Por. & 11 b.Chem.	
43 Machias.	" "		68 Filmore.	11 b. Chemung.	
50 Franklinville.	11 b. Chemung.	1593	72 Houghton.	"	
57 Ischua.	Vall'y drift. "	1541	75 Caneadea.	"	
63 Hinsdale.	" "	1501	91 Cuba.	"	
69 Olean.[201]	Moraine. "	1433	99 Hinsdale.	"	
76 Portville.	"	1442	106 Olean.[201]	" to Conglomer.	
84 Eldred, Pa.	12. Catskill.	1448	0 Olean.	" "	
89 Larabees.	"	1451	8 Alleghany.	11 b. Chemung.	
97 Port Allegeny.	"	1452	9 S. Vandalia.	"	
107 Keating Summit.	"	1551	13 S. Carrolton.	"	
121 Emporium. 1024	{ 14 a. Carboniferous, summit of hills.		19 Salamanca.	"	
			25 Red House.	"	
Pittsburgh Division.186			33 Wolf Run.	"	
			39 Corydon, (Pa.)	"	
0 Buffalo.[40]	See Lake Shore R. R.				
10 Hamburg.[148]	"		**Michigan Central Railway.** 136		
40 Dunkirk.	"		Buffalo.	9 c. Corniferous.	
49 Brocton.	11 b. Chemung.	672	0 Fort Erie.	"	
56 Prospect.	"	1221	13 Chippawa.	·	
63 Mayville.	"	1300	16 Niagara.	5 c. Niagara.	
69 Summerdale.	"	1629	17 Clifton. (Can'da).	"	
73 Sherman.	"	1565			
79 Panama.	"	1545	**Tonawanda Valley & Cuba Ry.** 136		
83 Clymer.	"	1146	0 Attica.	11 a. Portage.	
(Continued in Pennsylvania.			9 Johnsonburg.	"	
			13 N. Java.	"	
Rochester Division. 136			19 Curriers.	"	
0 Rochester.[137]	5 c. Niag. 5 b. Clinton.		26 Arcade.	11 a. Por. 11 b. Chem.	
6 Genesee Ju.	" "		36 Fairview.	11 b. Chemung.	
12 Scottsville.	6. Salina.		59 Cuba.	"	
20 Avon.[111]	9 c. Cornif. 6. Waterli.		30 Sandusky.	"	
26 York.	10 b. Hamilton.				
29 Pifford.	"		**Rochester and Lake Ontario Railroad.**136		
33 Cuylerville.	"		0 Rochester.	5 c. Niag. 5 b. Clinton.	
35 D., L. & W. Cros.	"		Lake Beach.	5 a. Medina.	
39 Mt. Morris.	10 c. Genesee.				
41 Sonyea.	"				

117. *Tribes Hill.* Good Trenton fossils at quarries and along outcrop. Canastota, Cazenovia and surrounding country excellent ground for Lamellibrachiati of Hamilton group, and there and at Hamilton best locality for *Homolonotus Dekayi.* R. P. WHITFIELD, Curator of Museum of Nat. Hist. of N. Y.

118. *New Hamburgh.* Wappinger Creek, entering the River here is bordered for nearly its entire course of thirty miles from Stissing Mountain, mostly on west, by ridges of limestone. This belt of limestone, like another one lying further east along the Harlem Railway, traverses the Hudson River shales of the County from N. E. to S. W.; like the shales, it consists of denuded folds, dipping mainly eastward, often forced over so as to overlie the younger slates. These limestones have lately been proved, on the evidence of fossils, to comprise at least the following formations:
1. Strata of associated limestone and quartzose rock, of the Lower Cambrian, containing Olenellus trilobites. These are best seen at the bases of Stissing and Fishkill Mountains.
2. Limestones and calcareous shales of Middle Cambrian or Paradoxides horizon.
3. The Upper Cambrian, or Potsdam, arenaceous limestones interstratified with calcareous shales and sandstones.
4. A prominent stratum, probably Calciferous, but containing mostly a new and unique fauna. Its most characteristic locality is Rochdale, four miles northeast of Poughkeepsie.
5. Trenton limestone, with a fauna of Canadian type, shown at Rochdale and Pleasant Valley.

Fonda, Johnstown and Gloversville Railroad.

Ms.	Station		Alt.
0	Fonda.[13]	4 b. Utica.	299
6	Johnstown.	" Striæ.	
8	Gloversville.	{ 4 b. Utica and	800
		{ 4 a. Trenton.	
22	Northfie.d.[180]	{ 4 b. Utica and	
		{ 1 a. Laurentian.	

Lackawanna & Pittsburg R. R.136
Olean Division.

Ms.	Station	
0	Olean.	11 b. Chemung.
4	Gordons.	"
6	Postville.	" & Conglom.
7	White House.	"
10	Ceres.	"
15	Little Genesee.	Chemung to Conglom.
18	Bolivar.	11 b. Chemung.
20	Richburg.	"
29	Friendship.	"
38	Narrow Gage Ju.	"
44	Angelica.	"

Lackawanna Division. 136

Ms.	Station	
0	Nar'w Guage Ju.	11 b. Chemung.
6	Angelica.	"
16	Birdsall.	"
24	Swains.	"
29	Canaserago.	"
37	Rogersville.	"
41	Wayland.	"
0	Swains.	"
10	Nunda.	11 a. Portage.
12	Junction.	"

Ulster and Delaware Railroad.

Ms.	Station		Alt.
0	Rondout.[114]	{ 4 c. Hudson Riv.[6]	
		{ 6. Water Lime.	
4	Kingston. 159	7. Lower Helderberg.	
9	West Hurley.	10. Hamilton.	534
12	Olive Branch.	11 b. Chemung.	504
15	Brook's Crossing.	11 a. Portage.	
17	Broadhead Bra.	"	504
18	Shokan. 887	11. Chem. & 11. Cats.	
21	Boiceville.	12.Catskill.	604
24	Mount Pleasant.	"	
27	Phœnicia.[206]	"	795
32	Fox Hollow.	"	1004
33	Shandaken.	"	1072
36	Big Indian.	"	1213
39	Pine Hill. 1879	{ " Lowest Pass	
		{ of the Catskill Mts.	
44	Griffin's Corners.	12. Catskill	1504
48	Dean's Corners.	11. Chemung.	
51	Kelly's Corners.	208 "	1878
53	Halcottville. 205	"	1408
57	Straton's Falls.	12. Catskill.	
59	Roxbury.[206]	"	1801
65	Moresville.	"and Chemung.	
74	Stamford.[209]	"	1771

Lehigh and Hudson River R. R.

Ms.	Station		Alt.
0	Greycourt.[180]	4 c. Hudson River.	
1	East Chester.	"	
3	Sugar Loaf.	"	
4	Lake.	4 a. Trenton.	542
9	Warwick. 141	"	502
12	New Milford	"	

New York, Susquehanna & West'n R. R. 123

	Station	
71	Quarryville, N. J.	4 c. Hudson River.[142]
72	Van Sickles.	"
75	Unionville.	"
78	West Town.	"
81	Johnsons.	"
83	Slate Hill.	"
85	Spring Side.	"
88	Middletown.	"

West Shore R. R. 143

Ms.	Station		Alt.
0	Weehawken,N. J.	144Trias.; Trap dike.[5]	
2	New Durham.	16 Triassic.	4
6	Little Ferry.	"	8
7	Ridgefield Park.	"	8
8	E. Hackensack.	"	50
9	Teaneck.	"	95
10	W. Englewood.	"	74
12	Bergen Fields.	"	57
13	Schraalenburgh.	"	82
16	Randalls.	"	46
18	West Norwood.	"	52
19	Tappan, N. Y.144	"	74
21	Orangeburgh.	"	93
22	Blauveltville.	"	122
24	Nyack T'pike.[145]	"	Trap.[56]
26	Valley Cottage.	"	125
29	Congers.	"	178
33	Haverstraw.[146]	"	75
37	Tompkin's Cove.	147?Slates & limest's.[5]	
39	Jones' Point.	1 a. Laurentian.	6
41	Iona Island.	"	7
43	FortMontgomery.	"	8
47	Cranston's.	"	8
48	West Point.		8
52	Cornwall.[116]	4 c. Hud. Riv.[142]	10
57	Newburgh.[138]	{ Hudson Riv. and 28	
		{ Cambro-Silu. limest.	
61	Clark's Dock.[149]	{ 3. Lower Silurian	
		{ limestones.	10
65	Marlborough.[150]	4 c. Hudson River.	10
68	Milton.	9 4 c. Hud. Riv. Group.[9]	
72	Highland.	"	9
78	West Park.[151]	"	108
80	Esopus.[152]	"	113
83	Ulster Park.	"	145
88	Kingston.[153]	9 c. Corniferous.	182
95	Mt. Marion.[154]	"	159
99	Saugerties.[154]	9 a. Cauda Galli.	156
103	West Camp.[154]	4 c. Hudson River.[118]	

This limestone crosses the Hudson River obliquely in two strips, between Hampton, (just south of Marlborough), and Danskammer Point. At the north end of the New Hamburgh tunnel, the limestone is well shown overlying, by inversion, the Hudson River shale.

The shales throughout this County are mainly of the Hudson River Group, with here and there Graptolitic layers, which are by some geologists assigned to the Utica slates. W. B. D.

Ms.	West Shore.—Con.	Alt.	Ms.	West Shore.—Con.	Alt.
110 Catskill.[155]	4 c. Hudson Riv. ?	93	255 Wampsville.	5 c. Niagara.	456
115 West Athens.	"	127	257 Canastota.	6. Salina.	432
120 Coxsackie.	"	137	261 Canaseraga.	"	417
125 New Baltimore.	"	185	264 Chittenango.	"	410
128 Coeyman's Ju.	"	177	268 Kirksville.	"	420
133 Selkirk.	"	145	270 Manlius Centre.	"	412
141 Albany.	"	15	274 Dewitt.	"	410
128 Coeyman's Ju.	"		278 Syracuse.	"	399
132 S. Bethlehem.	"	202	285 Amboy.	"	402
136 Feura Bush.	"	235	288 Warners.	"	428
New Scotland.	"	297	290 Memphis.	"	405
142 Voorheesville.	"	327	295 Jordan.	"	393
146 Guilderland.	"	312	300 Weedsport.	"	423
147 Fullers.	"	285	303 Port Byron.	"	399
152 S. Schenectady.	"	346	307 Montezuma.	"	389
Saratoga.	4 a. Trent. & 3 a. Calc.		309 Seneca River.	"	
160 Rotterdam Ju.	4 b. Utica.	287	311 Savannah.	"	405
161 Pattersonville.	"	270	317 Clyde.	"	389
168 Port Jackson.	4 a. Trenton.	351	324 Lyons.	"	403
173 Fort Hunter.	"	394	329 Newark.	"	433
174 Auriesville.	"	303	333 Port Gibson.	"	430
178 Fultonville.	4 b. Utica.	302	338 Palmyra.	"	436
183 Downing.	"	296	341 Macedon.	"	472
187 Sprakers. 309	{ 1 a. Laur. capped by / 3 a. Calcifer. hills.		349 Fairport.	"	449
			353 Pittsford.	"	470
193 Canajoharie.	4 a. Trenton.	302	356 Edgewood.	"	500
194 Fort Plain. 306	4 a.Birdseye, 4 a.Tren.		360 Red Creek.	"	542
199 St. Johnsville.	4 c. Hudson River.	327	362 Genesee Ju.	5 c. Niagara.	523
200 Mindenville.	"	331	367 Rochester.	"	
204 Indian Castle.	"	339	363 Maplewood.	"	535
209 Little Falls.	1 a. Laurentian.	382	365 Chili.	"	549
212 Jacksonburgh.	"	388	368 Buckbees.	"	563
217 Mohawk.	4 b. Utica.	396	372 Churchville.	6. Salina.	567
219 Ilion.	"	390	374 Bergen.	"	580
221 Frankfort.	"	395	381 Byron.	"	615
225 W. Frankfort.	"	403	387 Elba.	'	760
229 E. Utica.	"	497	392 Oakfield.	"	765
231 Utica.	"	515	398 Alabama.	"	710
238 Clark's Mills.	4 c. Hudson River.	516	404 Akron.	9 c. Corniferous.	678
242 Heckla.	5 a. Medina.	527	410 Clarence.	"	708
247 Vernon.	5 b. Clinton.	595	415 Bowmansville.	"	695
252 Oneida Castle.	5 c. Niagara.	485	423 E. Buffalo Ju.	"	620
			426 Buffalo.[145]	"	579

119. *Poughkeepsie.* From the north end of the New Hamburgh tunnel, with the exception of a short strip of Potsdam limestone a little south of Camelot, Hudson River shales and grits occupy continuously the east bank of the River as far as Rhinecliff and beyond, passing under the city of Poughkeepsie. Also they form the west bank from Hampton to Rondout. At several points there appear, without any definite divisional lines, layers of graptolitic shales which some geologists consider characteristic of the Utica Slate. Such layers occur in the R. R. cuts at the dock opposite the N. Y. State Hospital for the Insane, and at West Park on the west bank above the City.

At a point immediately south of the Driving Park, and on the Spackenkill road are localities of fossiliferous Potsdam. At the first point there is a conspicuous fault between the Potsdam and Hudson River Groups, which continues three miles southeasterly, striking the river in a bold bluff south of Camelot. Here are extensive and valuable beds of moulding sand, which are evidently in part at least derived from the disintegration of the Potsdam arenaceous limestone. This fault is a part of the great system of faults described in Note 8. W. B. D.

120. *Schodack Landing.* The Hudson River shales in the neighborhood abound in graptolites and about a mile and a half south are overlaid in apparent conformity by schists and limestones, containing fossils of the Lower Cambrian group, the latter rocks making the third promontory along the R. R. track south of the station. When the foliage is absent, the line of contact of the two groups can be seen from the cars. S. W. FORD.

121. *Albany.* Two miles below Albany at Kenwood in ravine near Knitting Mill is the famous locality for the Norman's Kill graptolites in Utica Slate. Beds nearly covered by buildings at present. The bed is seen near the middle of D. & L. R. R. cut. R. P. W.
Champlain deposits here. T. C CHAMBERLIN.

Ms.	New York City & Northern R. R.156 Alt.			Ms.	N. Y. Central and Hudson River R. R. Harlem Division. 162. 174, 175, 176.		Alt.
0	155 Street.[178]			0	New York.	See Note 4.	
1	High Bridge.	Limestone.	8	9	Fordham.	Middle Laurentian.	
8	South Yonkers.	Middle Lauren.	145	11	Williams Bridge.	Limestone.	
11	N. Yonkers.	"	164	14	W. Mt. Vernon.	"	
13	Odells.	"	119	16	Bronxville.	"	
15	Ashford.	"		17	Tuckahoe.	" Marble.	
18	Elmsford.	"		20	Scarsdale.	"	
20	E. Tarrytown.	"		22	White Plains.	Middle Laurent'n. 202	
21	Tarrytown.	"		31	Pleasantville.	Limestone. Marble.	
23	Tarrytown Hts.	"	387	33	Chappaqua.	"	
27	Whetson's.	"		37	Mount Kisco.	Middle Laurentian.	
30	Merritts Cors.	"	346	40	Bedford. 291	" Feldspar pro-	
32	Croton Lake.	"		45	Golden's Bridge.	" duced for pot-	
37	Yorktown.	"	459	47	Purdy's.	" teries,	
38	Amawalk.	"	384	48	Croton Falls.	" 356	
39	West Somers.	"	517	53	Brewster's. 414	L. Laure. Iron ore W.	
42	Baldwin Place.	"	621	56	Dykeman's.	" on summit.	
44	Mahopac.	Lower Laurentian.	641	61	Patterson.	Camb. Silurian l. s.	
47	Crafts.	"	482	64	Pawling.	"	
49	Carmel.	"	519	71	South Dover. 415	" Iron ore W.	
52	Tilly Foster Mines	"	401	76	Dover Plains.	" Limest. on E.	
54	Brewster.	"	406				

(column 37–48 on the right bracketed as **Highlands**)

122. The limestones and sandstones used for flagging and building in the various cities along the line of the N. Y. C. & H. R. R. R., are as follows: At Albany and Schenectady, 4 c. Hudson River; Utica and Rome, 4 a. Trenton limestone, generally of the Birdseye portion, which produces the thickest stone; at Syracuse, Auburn and Geneva, the 9. Upper Helderberg, generally the Onondaga or lower portion of it; from Rochester to Buffalo the 5 a. Medina sandstone is the favorite for these purposes. Some 5. Niagara limestone are used at Rochester and 9 Upper Helderberg or Corniferous at Buffalo, especially for lime burning. But the best flagstones are from the Hamilton and Chemung formations, and generally come from the shores of Cayuga Lake. Large quantities of flagstones are also brought from the upper part of the Hamilton group in the higher parts of the Helderberg, and from the same geological position along the west side of the River Hudson from below Catskill as far as Kingston.

123. By Mr. Nelson H. Darton, of the U. S. Geological Survey. Mr. Darton prefers to use the term 4 a. Trenton rather than Hudson River for the wide areas of slates in Orange and adjacent counties, which contain a mixed Hudson River and Trenton limestone fauna, but for the sake of uniformity Hudson River is used throughout the chapter.

124. *Meadow Brook.* About three-fourths of a mile east, the railroad crosses the ridge described in note 126. The red grits near this station are the same as those in the ridge there described, brought up by a synclinal.   N. H. D.

125. Caledonia and Stafford, two of the best places in the State for silicified Upper Helderberg corals. Akron also. Excellent corals at Le Roy.   R. P. W.

126. *Cornwall.* Just west of this station is a ridge composed of red and grey conglomerates similar to those near Highland Mills and probably near Oneida in age. It is flanked on the western side by Lower Helderberg limestone from the Waterlime to the Delthyris shaly limestone, the latter holding a bed of Limonite and plentiful fine casts of about a hundred varieties of fossils. The occurrence of this fossiliferous rock so far from the main mass of the formation is very interesting. See also Note 124.   N. H. D.

127. *Passaic.* South of this station the palisadal front of the First Watchung or Orange Mountain is in sight. This canoe-shaped ridge and some others behind it to the west and south are capped by the outcropping edges of great sheets of basalt lavas, which were outpoured at intervals on the floor of the Triassic sea during the deposition of the formation. The upper surfaces of these sheets, when not too deeply eroded, are deeply vesicular and at some points they are exposed in contact with unaltered shaly sediments. The more or less vesicular and altered bases of these sheets lie with perfect conformity on the shales, which often extend for some distance up the steep sides of the ridges and dip at low angles westward. Basal contacts in the quarries on the ridge slopes southeast of Paterson may be seen from the cars and are fine exposures in the deep gorge, into which the Passaic River falls in crossing the First Watchung ridge in Paterson.   N. H. Darton.

128. *Turner's.* On emerging from the highlands north of Greenwood the line of the road passes over a broad valley encircling and extending northeastward from Turner's, and is in greater part underlaid by limestones of undetermined, but probably Lower Silurian age, and by slates of Trenton age.   N. H. D.

129. *Monroe.* A mile west of this station a synclinal holding Middle Devonian is crossed, but no outcrops are visible from the cars. These rocks extend for many miles southward into New Jersey. In New York they form Bellvale Mountain to the Erie R. R. and thence extend northward in the high, rough, double crested ridge known as Schunemunk Mountain. The lower members are flagstones and slates, the upper a coarse pebble conglomerate. In a flagstone quarry, two miles N. N. W. of Monroe, the remains of Devonian plants are quite abundant. In the valley westward the series is underlaid by a white Quartzite succeeded by limestone holding an Upper Silurian fauna and an unfossiliferous limestone lying on Gneiss. The two last are exposed in the railroad cut a mile east of Oxford. This gneiss is flanked on the west by an inconsiderable thickness of limestone which is overlaid by the slates which are thence exposed nearly to Oxford.   N. H. D.

N. Y. Central & Hudson River R. R.—*Con.*
Harlem Division.—*Con.*

Ms.		Alt.	
82	Wassaic.		Cam.-Sil. Schists.
84	Amenia.		" " l. s.
87	Sharon.		" " "Burd'n's gun
93	Millerton.	702	" "bar'l iron ore W
97	Mount Riga.		" " l. s. (Summit).
100	Boston Corners.		" " " Iron ore W.
106	Copake.		" " " Iron Works.
109	Hillsdale.	671	Cambro-Silurian.
116	Martinsdale.		" "
120	Philmont.		" "
126	Ghent.		" "
127	Chatham.		" "

All the iron ore is produced on the west side—none on the east side of railroad.

N. Y., Rutland & Montreal Ry.

0	Chatham 4 cor.	4 c. Hud. Riv. Group.
5	Chatham.	"
11	Rider's Mill.	"
18	New Lebanon.	"
27	Lebanon Springs.	"
31	N. Stephentown.	"
34	Centre Berlin.	"
39	Berlin.	"
44	Petersburg.	"
45	N. Petersburg.	"
47	T. & B. Junction.	2. Cambrian sl.
53	Bennington, Vt.	3. Lower Silurian l. s.

See Notes 174-75-76.

Ms. N. Y., New Haven & Hartford R. R. Alt.

0	New York.[178]	See Note 4.	
12	Williams Bridge.	"	
15	Mount Vernon.	1 d. Montalban, probably.	
18	New Rochelle.	"	70
22	Mamaroneck.	"	
25	Rye.	"	
27	Port Chester.	"	
30	Greenwich.	"	
31	Cos Cob Bridge.	"	
35	Stamford, Conn.	"	

Harlem River Branch.

0	Harlem River.	Montalban or Metamorphic. See Note 4.
1	Port Morris.	
5	West Chester.	"
12	New Rochelle.	"

Middletown Branch. 164

0	New Britain.	16 Triassic.
3	Berlin.	"
13	Middletown.	"

130. *Greycourt.* West of the Oxford limestone to the Blue, or Shawangunk Mountain, at Otisville there is a rolling country underlaid by Slates, which have been recently found to be Trenton in age. (See Note 142.) They extend northeastward to the Hudson River and south across part of New Jersey. They are underlaid by limestones, which hold Lower Silurian faunas. N. H. D.

131. *Suffern.* A short distance east is Union Hill composed of a thin sheet of trap lying upon heavy beds of Conglomerate. N. H. D.

132. *Sparkill.* At many points south of here overlying stata are found in contact with Palisade trap sheet, as stated in Note 5. North of this station the R. R. crosses the sheet and skirts the east side of the ridge at a considerable altitude. The under contact of trap and sandstone maybe found near Piermont-on-the-Hill, and near Grandview, above the R. R. N. H. D.

133. *Homestead.* See Note 5. This road crosses the Palisade trap ridge in the Erie tunnel and skirts its western base to Sparkill where it recrosses to Piermont. A few hundred yards S. E. of the station, and in sight from the cars, contact of trap and overlying shales is exposed in a small quarry. N. H. D.

134. *New Durham.* Three-fourths of a mile east in a cut at entrance to W. S. R. R. tunnel the dike structure of Palisade trap is exposed at unconformable contact with overlying sandstones. N. H. D.

135. *Granton.* A short distance north is a small dike and sheet of trap separated from the Palisade sheet by a slight thickness of sandstone. N. H. D.

136. By Prof. H. S. Williams, of Cornell University.

137. *Rochester.* Shales below falls filled with corals and *Brachiopods* of Niagara group. Entire Clinton exposed and many layers filled with excellent fossils. Several beds of graptolites known by the black color of the seam. Lower fall gives limestone filled with *Pentamerous Elongatus* and below Medina sandstone with fucoides, etc. R. P. Whitfield.

See Note 36 and Glacial Note 181.

138. *Newburgh.* The city rests upon strata which are evidently similar to those identified in Duchess County. The entire water-front is composed of Hudson River shale, while that part of the city west of West street is on the belt of limestone which crosses the river from New Hamburg in Duchess County. On the river road three miles north of the city, there are highly fossiliferous ledges of the Trenton group, containing the Coral Solenopora Compacta, and very large Crinoid columns. With this exception this great belt of limestone from Hampton to Long Pond appears to be entirely without fossils. A comparison with the more northern extension of the belt makes it probable that besides the Trenton, Calciferous and Cambrian strata are present. Snake Hill to the south and Cronomer's Hill to the west, are Archæan gneiss. W. B. D.

139. *Mt. Joy.* Road crosses Palisade trap sheet.

140. *Eagle Bridge.* At Eagle Bridge, Cambridge and Granville, the railroad passes over a narrow strip of Hudson River Shales flanked on either side by broad masses of Lower Cambrian or "Georgia" shales and limestones, which are not more than a mile distant, or less. At Salem a broad belt of Hudson River shale lies a short distance to the west. Fossiliferous localities of the Lower Cambrian have been found near Shushan, Salem, Rupert and Granville. (Some of the chief localities described are one mile south of Shushan one and one-half miles east and west, and one mile south of N. Greenwich (near Salem) two miles south of North Granville, and at Low Hampton, just west at the crossing of Poultney River.) W. B. D.

Ms.	Boston and Albany Railroad.	Alt.
0	Albany. 4 c. Hudson River.	32
1	Greenbush. "	24
9	Schodack. 208 Doubtful, 174,175&176	
17	Kinderhook. "	318
20	Chatham Centre. "	315
24	Chatham.163 4 c.Hud. Riv. Gr'p.	462
29	East Chatham. "	691
34	Canaan.173 "	869
39	State Line. "	914
	(Continued in Massachusetts).	

Ms.	Hudson & Chatham Branch.	
0	Hudson.	4 b. Utica.
4	Claverack.	Doubtful.
9	Millerville.	"
11	Pulver's.	"
15	Ghent.	"
17	Chatham.	4 c. Hud. Riv. Group.

Ms.	New York & Massachusetts R. R.164	
0	Poughke'psie.119	4 c. Hud. Riv. G'p.179
6	Pleasant Val.165	4 a. Trenton.
11	Salt Point.166	4 c. Hud. Riv. Group.
13	Clinton Cors.167	4 c. Hud. Riv. Shale.
16	Willow Bro'k.168	Cambri.(?) limestones.
18	Standfordville.	4 c. Hu. Riv. Shale.328
20	McIntyre.	Calciferous limestone.
21	Stissing.169	2 á and 2 a Cambrian.
27	Pine Plains. 470	2á and 2 (?) Cambrian.
31	Ancram L'd. Ms.	" 570
37	Boston Corners.	" 738

Ms.	Hartford & Conn. Western R. R.	Alt.
0	Rhinecliff.	4 c. Hudson River.
3	Rhinebeck.	"
7	Red Hook.	2-4 Camb. Sil. Schists.
11	Spring Lake.	"
17	Jackson Corners.	"
25	Ancram.	"
35	Boston Corners.	3-4 Camb. Sil. Limest.
42	State Line.	"
	See Connecticut.	

Ms.	Newburgh, Dutchess & Conn. Railroad.164	
0	Dutchess Junc.	4 c. Hud. Riv. Group.
2	Matteawan.170	" 119
4	Glenham.170	" 213
6	Fishkill.	Calcif.-Trent.(?)I's.213
11	Hopewell.	" 252
13	Clove Branch Ju.	" 289
17	Sylvan Lake.	"
19	Billings.	4 c. Hudson River.891
25	Verbank.	" 553
30	Millbrook.	" 566
37	Bangall.171	"
40	Stissing Junc.	" 437
45	Pine Plains.	Cambrian(Upper?)470
47	Bethel.	3 a. Calciferous.
50	Shekomeko.172	{ Calciferous and 505 / Upper Cambrian.
52	Husted.	Cambrian (Upper?)
54	Winchell's.	4 c. Hudson River.667
59	Millerton.	Calciferous-Trent.?702

141. *Warwick.* At Edenville, four miles west, compare the "blue limestone" of Primordial or Lower Silurian age with the "white limestone" of the Archæan, which there crop out in parallel and almost contiguous ridges. The Archæan limestone is highly crystallized and contains many crystals of foreign matter. W. B. D.

142. This series of slates, occupying large areas in Orange County, New York, and extending southward into New Jersey, contains a mixed Hudson River and Trenton limestone fauna, and should perhaps be designated Trenton. (See Note 123.) N. H. D.

143. West Shore R. R. Stations from Weehawken to Nyack Turnpike are by Prof. W. B. Dwight of Vassar College, thence to Cornwall by Mr. Nelson H. Darton, U. S. Geologist, thence to Esopus by Prof. Dwight, and thence to Albany by Prof. Dwight and Hon. James G. Lindsey of Rondout. From Albany to Buffalo the tables are by Prof. H. S. Williams of Cornell. On this portion see notes on New York Central, running nearly parallel.

144. For stations in N. J. see also New Jersey Chapter.

145. *Nyack Turnpike.* From some distance southward of this station and thence northward, this road skirts the western side of the palisade trap sheet, and crossing it in a tunnel north of Congers, follows its eastern side to Haverstraw, where the high ridge formed by the trap, curves westward to the highlands. In the cut at the southern end of the tunnel the highly altered sedimentary beds are exposed, abutting against the steep trap dike, while on the east side of the ridge, they are exposed dipping gently beneath the trap, indicating the dike and sheet structure described in Note 5. N. H. D.

146. *Haverstraw.* One mile north of the station there is a cut through 16. Triassic calcareous conglomerate. A few hundred feet farther, on Stony Point, the deep cut gives fine exposures of some members of the Cortland series of intrusives and metamorphics. N. H. D.

147. *Tompkin's Cove.* Extensive quarries of blue and grey limestones near station. Age of the beds uncertain but probably Lower Silurian. They are separated from the Archæan rocks of the highlands by black slates of unknown age, which are exposed at many points in this vicinity and southward to Pompton, N. J. N. H. D.

148. *Hamburg.* Eighteen Mile Creek and vicinity are most excellent localities for Hamilton fossils, along lake shore and up stream a short distance and also at Hamburg in cutting on R. R. (R. P. W.)

Sub-aqueous drift; lake terraces along the lake shore to Ashtabula. (Chamberlin.)

149. *Clark's Dock.* Interesting clay beds of the Champlain Period deposited in the form of three inverted, truncated cones, instead of horizontally, as is usual in the beds lining both banks of the Hudson. W. B. D.

150. *Marlborough.* Hampton Point, three quarters of a mile south is the northern edge of the limestone belt crossing from Duchess County, (See Note 118.) and passing to the west of Newburgh. Here Kerr's Hydraulic Cement Works are now in successful operation. The limestone is apparently Cambrian with perhaps Lower Silurian. See Note 138. W. B. D.

151. *West Park.* On the north side of a railroad cut just south of Hazen's (or Adam's Dock), and between one and two miles south of the railroad station, slabs of slate covered with excellent graptolites, may be obtained. These are referred by Prof. Whitfield to the Utica slate; by some other geologists to the Hudson River Group. W. B. D.

Ms.	New York & New England R. R.164	Alt.
0	Newburgh.[188]	4 c. Hudson River.
1	Fishkill.[118]	"
4	Matteawan.[170]	"
8	Fishkill Village.	Calcif.-Trent. l's. 218
10	Brinkerhoff.	" 225
14	Hopewell.	"
19	Stormville.	"
22	Poughquag.	"
25	Pawling.	"
31	Patterson.	Laurentian.
33	Towners.	" 432
38	Brewster.	" 406
44	Mill Plain.	"

Troy and Boston Railroad.164
(Fitchburg Railroad.)163

0	Troy.	Hud.Riv. and Georgia.
4	Lansingburgh.	"
9	Melrose.	"
18	Schaghticoke.	" Trenton?
14	Valley Falls.	4 c. Hudson River.
17	Johnsonville.	"
21	Buskirk's.	4 c. H. Riv.& Georgia.
24	Eagle Bridge.	"

Ms.	Troy and Boston.—Con.	Alt.
26	Hoosic Junction.	4 c. H. Riv. & Georgia.
	State Line.	{ 4 c. Hud. Riv. and Calcif.-Chazy-Tren.
27	Hoosic Falls.	4 c. Hudson River.
30	Hoosac.	{ 4 c. Hud. Riv. and Calcif.-Chazy-Tren.
32	Petersburg.	Calcif.-Chazy-Trent.
36	North Pawnal.	" " "
43	Willi'mstown.[163]	" " "
45	Blackinton.	{ Hudson River and Calcif.-Chazy-Tren.
48	North Adams.	Calcif.-Chazy-Trenton.

Greenwich and Johnsonville Railroad.
Washington Co. 164

9	Johnsonville.	4 c. Hudson River.
5	Lee's.	"
6	S. Cambridge.	"
8	W. Cambridge.	"
10	Summit.	"
13	Easton.	Lower Cambrian.
16	Greenwich.	

152. *Esopus.* On leaving the river in Esopus, before crossing Rondout Creek, going north, the road crosses the ends of a synclinal arch; the first rock is nearly vertical section of Niagara, then Waterlime-Pentamerus, Catskill Shaly, Upper-Pentamerus, Catskill-Shaly, Pentamerus, Upper Pentamerus. After crossing the creek, the road enters a tunnel the south end of which is Catskill Shaly, the middle section Upper Pentamerus and the north end Oriskany, all nearly vertical. After the tunnel is passed the Cauda Galli is entered and perhaps Schoharie Grit, and then Corniferous and it may be the Onondaga. J. G. L.

153. *Kingston.* Unconformability of Lower and Upper Silurian well shown here. Remarkable contortions of strata. Fossils abundant. At Rondout, now included in the city of Kingston, are seen Hudson River Group; Oneida; Coralline limestone of Niagara Group; all the divisions of Lower Helderberg; Oriskany; Cauda Galli and Corniferous; all but the last two quite fossiliferous. At old Kingston, on Esopus Creek, Marcellus and Hamilton. Immense Cement quarries in Helderberg limestones.
See "Non-conformity at Rondout" by W. M. Davis, Am. Journ. Science, November, 1883.
W. B. D.
Station is on terrace of Alluvium and Drift overlying Corniferous, which crops out in a high ridge to the eastward, dipping to the northwest. To the west bluff of Marcellus overlying Corniferous. J. G. L.

154. *Mount Marion.* The road (going north) continues on Corniferous nearly to Saugerties, where it comes again to the Cauda Galli and, before it reaches West Camp, it passes back over all the intervening layers to the Hudson River which it does not leave, except a few cuts into the Waterlime between West Camp and Catskill. J. G. L.
At Glenerie a little over a mile southeast from Mount Marion station along the east bank of Saugerties Creek, are abundant exposures of Oriskany, crowded with finely weathered fossils. W. B. D.

155. *Catskill.* The Helderberg rises sharply to the west nearly all the way to Coeyman's.
156. By Prof. C. H. Hitchcock.
157. *Canandaigua.* Go up the lake six miles to Monteith's Pt. up ravine, most excellent Hamilton fossils, all classes. Also all along lake shore to Black Pt. Heads of Monteith's ravine, Genesee slate with plants, and gas springs. R. P. W.

158. *Knowersville and Guilderland.* Go up mountain to first plateau, rocks filled with Lower Helderberg fossils. *Tentaculites* and *Leperditia* at base of vertical layers. Thompson's Lake one and a half miles back from top of bluff at Indian Ladder road, Schoharie grit and Upper Helderberg fossils. Also Clarksville 12 miles southwest of Albany has yielded immense numbers of Lower Helderberg Bryozoans and Corals. R. P. W.

159. *Schoharie.* In the hill east and west from the village the entire Helderberg series occurs, and fossils are numerous in the Coralline limestone. Lower Helderberg, Oriskany sand, Schoharie grit and Upper Helderberg. R. P. W.

160. *Darien.* Best locality in the state for Hamilton in streams at Darien City, and also two miles west of Darien Centre in small stream at Milldam, and for one mile below slate road Corals and Shells. R. P. W.

161. The formations are given on this road approximately, no definite information having been published. From Dannamora to Lynn Mt. both the Laurentian and the Potsdam are given, implying that both strata are in the neighborhood. W. B. D.

162. Revised by Prof. C. H. Hitchcock. From Pawling to Chatham Prof. Dwight prefers "Calciferous" or "Calciferous-Trenton." This limestone, he says, is the eastern fork of the Copake-Hillsdale belt of which the Wappinger Valley limestones are the western fork. Calciferous fossils occur in it. Cambrian strata may be present. At North East Center, one and one-half miles south of Millerton, Calciferous fossils occur on Edward Clark's farm.

Ms. Ogdensburg & Lake Champlain R.R.	Alt.	Ms.	Catskill Mt. & C
0	Ogdensburg.	3 a. Calcif. 20 ms. ²⁴⁸	0	Catskill Landing.	4
9	Lisbon.	"	1	Catskill.	7
17	Madrid.	"	8	S. Cairo.	
25	Norwood.	"	14	Mountain House.	
28	Knapps.	2 b. Potsdam, 53 ms.	16	Palenville.	
36	Brasher Falls.	"	**Stony Clove and Cats**	
41	Lawrence.	"	0	Hunter.	1
47	Moira.	"	2	Kaatersville Ju.	
55	Bangor.	"	4	Stony Clove.	
61	Malone.	"	6	Edgewood.	
73	Chateaugay.	1 a. Laurentian, 5 ms.	9	Lanesville.	
81	Cherubusco.	2 b. Potsdam, 36 ms.	12	Chichesters.	
89	Ellenburg.	"	14	Phœnecia.	
90	Dannemora.	" 1356		
97	Altona.	"		**Kaatersvill**
103	Mooer's Forks.	"	0	Kaatersville Ju.	1
106	Mooer's Junction.	3 b. Chazy.	8	Kaatersville.	
114	Champlain.	3a. Cal.&3b.Chazy,4ms		**Long Island**
118	Rouse's Point.	3 b. Chazy, 2 miles.	0	Hunter's Point.	2
122	Alburgh.	4 b. Utica, 13 miles.	10	Jamaica.	
126	Alburgh Springs.	"	19	Mineola.	
133	Swanton.	4 c. Hudson River.	25	Hicksville.	
136	Swanton Junc.	"	29	Syosset
142	St. Albans, Vt.	2 b. Potsdam, 6 miles.		

163. *Williamstown.* An important point in the typical area of the o Recent researches of laborious stratigraphic and paleontological field-wo ed in securing, in general, a well-assured stratigraphy for this entire ing the great synclinals of limestones, shales, schists and quartzytes tain ridges and the adjacent rolling country on the east and west flanks. extensive discoveries of fossils were made by Mr. C. D. Walcott in 1 years previous. Stratigraphic maps have been lately published by Pro Mr. Walcott. These show beyond question that the main central ridge sist of Potsdam, Calciferous, Chazy, Trenton and Hudson River strata, a belt of Potsdam and pre-Cambrian rock, and on the west by a wid brian somewhat intermixed with Hudson River Shales.

Some of the principal localities of fossils are at Pownal, and thr nington, Vt., north side of Graylock Mt., Mass. near Hoosac, and Hoo points for which see Note 140.

164. By Prof. W. B. Dwight, of Vassar Collge.

165. *Pleasant Valley.* Fossiliferous Trenton in cut near north of depo mile south. Calciferous limestone in ridges west of the Trenton, at qu Potsdam limestone a little northwesterly from railroad station. Hudso side of the belt of these limestones. About half way between this an ous Potsdam mainly composes hill on east side of the railroad near the sc

166. *Salt Point.* Limestone belt passes to east of depot through H Clinton Corners passes west of station. Exposure of Trenton and Calc a little Potsdam at Wallace's quarry one mile south of Salt Point.

167. *Clinton Corners.* Limestone of Potsdam and Calciferous grou station.

168. *Willow Brook.* A ledge of quartzite of Lower Cambrian occur the southwest and some of the limestone may belong to the same ho

169. *Stissing.* Station stands on one of the Wappinger limestones, in a little gully near track and in cuts to the north and south. Bein is uncertain, but probably either Potsdam, Rochdale or Trenton. Betw the base of Stissing Mountain (Archæan gneiss) is a strip of red group. On ascending the southern slopes of the Mountain, the r by an underlying stratum of limestone of the "Olenellus" group, *Micaus;* underlying this a little higher up the declivity is quartzose nellus" group and immediately overlying the gneiss. In some spots ginous and highly fossiliferous containing *Olenellus asaphoides* and other fos

170. *Matteawan and Glenham.* The stations (Newburg, Dutches shales of the Hudson River Group, which near Glenham become in s and also bright purplish red. Ledges of an impure irregular granite near Fishkill Creek surrounded by shales or limestones. On the so in Matteawan and Glenham are conspicuous ridges of limestone belo ger Valley series, but not yet exactly determined by fossils. On farm cott, southwest from Matteawan and three miles from the Hudson Lower Cambrian crops out, immediately overlying the gneiss rock

171. *Bangall.* A broad belt of Calciferous and Cambrian limeston from Bangall for about a mile and a half along the Hull's Mills road fossiliferous at some points. In this vicinity there are numerous faul River Group, and the two stratigraphic components of the limestone.

Ms.	Long Island Railroad.—*Con.*	Alt.	Ms.	Long Island Railroad.—*Con.*	Alt.
34	Huntington.	20. Quartenary, with	10	Jamaica.	20. Quartenary.
40	Northport.	Tertiary or Cretaceous.	16	Valley Stream.	"
59	Port Jefferson.	"	19	Ocean Point.	"
30	Farmingdale.	"	21	Far Rockaway.	··
65	Manor.	"	25	Sea Side House.	"
94	Greenport.	"	22	Freeport.	"
0	Hunter's Point.	"	36	Babylon.	"
3	Woodside.	"	47	Oakdale.	"
4	Winfield.	"	54	Patchogue.	"
5	Newtown.	"		**Staten Island Railroad.**	
8	Flushing.	"			
9	College Point.	"	0	Stapleton.	18 c. Cretaceous. (Plastic clay formation.)
11	Whitestone.	"			
14	Brookdale.	"		Richmond.	"
0	Brooklyn.	20. Quartenary.	11	Pleasant Plains.	"
8	Richmond Hill.	"	13	Tottenville.	"

172. *Shekomeko.* An independent strip of limestone about six miles long extends from "The Square" two mile south of Shekomeko, up the valley to Pulver's Corners. It consists of Calciferous, and probably the Potsdam, which runs frequently into calcareous shales. At Husted Station, the latter formation skirts the west flank of Winchell's Mountain, and is well shown in a deep cut just north of the station. In a cut south of the Shekomeko Station is a conspicuous fault between the Calciferous and Hudson River Group, and a little further south, the Calciferous contains fossils.   W. B. D.

173. *Canaan 4 Corners.* The limestone belt between Canaan 4 Corners and State Line Station, which with the overlaying argillaceous and arenaceous rocks, formed a portion of the original "Taconic Series" of Emmons, have recently been shown by indisputable paleontological evidence to belong, in part at least, to Lower Silurian formations. Fossils have been recently discovered at the railroad tunnel (No. 290) and south of it, also on Drowne's farm one mile east of Canaan 4 Corners. These fossils indicate certainly Lower Silurian strata, probably of the Trenton and Calciferous groups. See note 163.   W. B. D.

Geology of Eastern New York.

174. The geology of the country between the Hudson River and the Connecticut and Massachusetts State Line was involved in almost entire obscurity until within a few years. In the State geological survey of forty-eight years ago, the slates were assigned, for stratigraphic reasons, to the Hudson River Group, and the limestones without any evidence of any value derived from fossils, was assigned to the Calciferous and Trenton groups. Afterwards, the entire mass of rocks was indefinitely assigned to the Quebec Group and was so designated in the first edition of this GUIDE. The difficulty of ascertaining the true order was much increased from the fact that the strata are much metamorphosed, flexed and faulted.

It is now known, on abundant paleontological evidence, that the shales and schists with some attendant "grits" are of the Hudson River Group, and perhaps of the Utica Slate; and that the limestones and some quartzytes are Cambrian or Silurian, that is, comprising strata either of the "Georgia" ["Olenellus"], Paradoxides, Potsdam, Calciferous, or Trenton.

It is certain that tne three latter formations are largely represented. The fossils are unique and important, but they are in general altered, fragmentary, difficult to obtain and difficult to study.   W. B. DWIGHT.

A general sketch of the geology of this region is given in Notes 175 and 176 by Drs. Hunt and Dana, who represent diverse views on some of the important questions connected with the stratigraphy, and much information will be found in the tables and notes on tations in this region, especially in Notes 118, 119, 138, 163 and 173.

175. To the east of the Hudson River in New York we find besides the Laurentian rocks of the Highlands, a great development of the gneiss and mica-schists of the Montalban and of two other and very unlike series. The first of these is the Lower Taconic, consisting of the Stockbridge limestone with quartzites and peculiar slates. This series together with the Primary crystaline schists, stretches up northward, passing along the southeast side of the Highlands, and occupying portions of Eastern New York and Western New England. On the northwest side of the Highlands, extending northward along the valley of the Hudson, and as far as Lake Champlain, is found another series, variously designated as the Hudson River Group, the Taconic Slates or Upper Taconic series of Emmons, and the Quebec group of Logan. These rocks have been supposed to be Upper Cambrian or Silurian, Utica, Loraine and Oneida) but are now believed to be chiefly of Lower and Middle Cambrian ages. They are generally disturbed and often inverted, and include small outliers and involved portions of Upper Cambrian and occasionally of Silurian strata. This Upper Taconic or Cambrian group is distinct from and superior to the Lower Taconic. It is impossible in the present state of our knowledge of their distribution to define the limits of these various groups of strata to the east of the Hudson, or to say at what stations the Upper Taconic, the Lower Taconic (Taconian) or the Primary rocks are met with.   T. S. HUNT.

NOTE.—Dr. Hunt here uses the terms Cambrian, etc. as given in the first edition. See Note 2, also Dr. Hunt's table in the Introduction.

176. To the north of Putnam County, N. Y., whose rocks are with small exceptions archæan, there is a large development along the boundary between New York and New England of the "Lower Taconic Series" of Emmons, consisting of limestone, called in part the Stockbridge limestone, with hydromica and mica-schists and quartzite. These rocks

extend northward over a portion of Eastern New York and neighboring portion of Connecticut, Massachusetts and the southern half of Vermont. The limestones have afforded Lower Silurian fossils in Canaan, (see Note 173), Columbia County, New York and in West Rutland and elsewhere in Central Vermont. The rocks near Poughkeepsie were made part of the "Lower Taconic" and have recently afforded Lower Silurian and some Cambrian fossils. The slates were formerly all referred to the Hudson River Group. In Rensselaer Co., N. Y., occur slates and other rocks made "Upper Taconic" by Emmons, containing Cambrian fossils and similar rocks occur in parts of western and northern Vermont.   J. D. DANA.

Note on the Glacial Drift on Long Island
by Mr. Warren Upham, Assistant U. S. Geologist.

177. On Long Island the terminal moraine of the continental ice-sheet extends from Fort Hamilton twenty-four miles in a nearly northeast course to Roslyn; thence it runs nearly due east sixty miles to Canoe Place and the Shinnecock Hills; next it turns northeast about eight miles to near Sag Harbor; and thence its course is east and east-northeast about twenty-five to Montauk Point. This range of hills long ago was called "The backbone of the island."

From the Narrows to Roslyn, this moraine varies from 100 to 250 feet in height, is mainly composed of unmodified drift, upper till on the surface, with glaciated pebbles and boulders in deep excavations. Its irregular contour is well seen in Greenwood Cemetery and Prospect Park and at Ridgewood Reservoir.

East of Roslyn it is almost wholly composed of modified drift, being waterworn gravel and sand with few or no boulders. These deposits are stratified, but often with oblique bedding and seem to constitute the entire mass of hills from 200 to nearly 400 feet high. Harbor Hill, a half mile east from Roslyn is the highest, 384 feet above sea, and is of this kind. In the same class are Jane's Hill, 354 feet; Rutland's, 340 feet; Osborn's or Bald Hill, a few miles southwest from Riverhead, 293 feet, The portion of this moraine forming the peninsula of Montauk, ten miles long and 150 to 200 feet high, is stratified, but contains frequent embedded boulders, which are also spread over the surface.

Long Island, south of this series of hills, consists of plains of fine gravel and sand 5 to 10 miles wide and 100] long. The north portion at the foot of the moraine is 50 to 150 feet above sea, from which height they slope southward. Numerous ancient water courses 10 to 25 feet deep and 100 to 300 feet wide cross from north to south. In some cases these channels continue beneath the the sea level of the southern bays to the beach ridge, by which they are divided from the ocean.

A later terminal moraine 100 to 200 feet high, formed during a halt in the final retreat of the ice-sheet, of modified drift, except near Greenport and Orient, forms the north shore from Port Jefferson to Orient Point. It is separated from the extreme moraine by plains, also crossed by old channels of drainage.

Glacial Notes,
BY PROF. T. C. CHAMBERLIN,
Of the United States Geological Survey and State Geologist of Wisconsin.

178. Roches Moutonnees at New York and for several stations east on the N. Y. & N. R. R.
179. Champlain.
180. Striæ.
181. Between Syracuse and Rochester drumlins have very fine development.
182. Between Victor and Fisher's, kame-like, semi-morainic hills are well developed.
183. Kame-like, semi-morainic hills.
184. Kame-like gravel hills.
185. Glacial flood deposits.
186. Gravel hills and terraces.
187. Moraine.
188. Valley drift, kame-like knolls.
189. Sub-aqueous drift.
190. Valley drift.
191. Morainic and glacial flood gravels.
192. Moraine and sub-aqueous drift.
193. Morainic(?) hills.
194. Sub-aqueous till; striæ.
195. Morainic(?) knolls.
196. Morainic glacial flood gravels.
197. Sub-aqueous till.
198. Kame-like knolls.
199. Kame-like knolls; Moraine(?).
200. Valley drift; Kame-like knolls; Moraines(?)
201. Kame-like and morainic hills.
202. Valley drift; moraine.
203. Morainic knolls.
204. Morainic kame-like hills.
205. Kame-like knolls and glacial flood gravels; moraine(?).
206. Valley drift; gravel knolls.
207. Striæ; moraine(?) in vicinity.
208. Valley drift; gravel knolls; moraine(?)
209. Moraine; gravel knoll.
210. Glacial flood gravels.
211. Morainic terrace.

New Jersey.

By Professor Jno. C. Smock, Assistant State Geologist, New Brunswick, N. J.

Geological Formations or Epochs found in New Jersey.

20. Quaternary and Recent	{ 20 b. Champlain. { 20 a. Glacial Drift.	

Tertiary.

19. Tertiary.	19 c. Pliocene.
"	19 b. Miocene.
"	19 a. Eocene (Upper Marl in part).

Cretaceous.

18. Cretaceous.	18 g. Upper Marl (in part).
"	18 f. Yellow Sand.
"	18 e. Middle Marl.
"	18 d. Red Sand.
"	18 c. Lower Marl.
"	18 b. Clay Marls.
"	18 a. Raritan Clays or Plastic Clays.
16. Triassic, or New Red Sandstone.	

Devonian.

	Green Pond Mountain Rocks.
10. Hamilton.	10 a. Marcellus Shale.
9. Upper Helderberg or Corniferous	{ 9 d. Corniferous. { 9 c. Onondaga. { 9 a. Cauda Galli.
8. Oriskany.	8. Oriskany Sandstone.

Upper Silurian.

7. L. Helderb'g	Upper Pentamerus Limest.
"	Encrinal "
"	Delthyris Shale "
"	Lower Pentamerus "
"	Tentaculite "
6. Salina.	6. Water Lime.

Lower Silurian.

5. Niagara.	5 a. { Medina Sandstone. { Oneida Conglomerate
4. Hudson.	4 c. Hudson River Slate.
"	4 b. Utica Slate.
4. Trenton.	4 a. Trenton Limestone.
3. Canadian.	3 a. Magnesian Limestone.
2. Primordial or Cambrian.	2 b. Potsdam Sandstone.
1. Archæan.	1 b. Huronian.
"	1 a. Laurentian.

Notes on the Table of Formations.—No. 21, Recent, includes the tidal meadows, the alluvial, upland necks of the southern part of the State, the sand-beaches of the Atlantic coast, and some of the peat-deposits of the interior.

Under 20 b., Champlain, are placed the modified drift bordering some of the rivers; and deposits of the ancient lake basins.

No. 20 a., Glacial, represents the glacial drift north of the terminal moraine.

The yellow sand and gravel of the southern part of the State is represented as Pliocene, 19 c.

The Miocene, 19 b., is identified by its characteristic fossils in Cumberland County, but it is not on any railroad line.

The Eocene, 19 a., is recognized in the upper layers of the upper green-sand marl-bed.

The Cretaceous, 18, includes the green-sand marls of the southern part of the State and the plastic clays here designated as the Raritan clays.

Under 16, Triassic, the trap-rock outcrops are included with the red sandstone.

The Green-Pond Mountain series of shales, sandstones, and conglomerates are of Devonian age, but there is some uncertainty as to their true position. They are provisionally assigned to the Upper Devonian.

The Marcellus Shale, the Corniferous and Onondaga Limestones, the Cauda Galli Grit, the Oriskany Sandstone, the Lower Helderberg Series, and the Water Lime group occur in the Upper Delaware Valley, west of the Kittatinny Mountain. No railway line runs nearer to them than the New York, Lake Erie and Western Railway, at Carpenter's Point, and Port Jervis.

The 8 a. e. c., Magnesian Limestone, is the equivalent of the calciferous sandstone of New York.

The 4 b. e. c., Utica Slate, has not been outlined on any of the State maps, as it is almost impossible to separate it from the Hudson River slate.

In No. 1, Archæan, the subdivision is based on lithology alone. The gneissic, granitic, syenitic, and other associated crystalline rocks are assigned to the Laurentian, and the fine crystalline, hornblendic, schistose rocks to the Huronian.

The reference to the newer and superficial formations is not made in all cases; and the more characteristic and typical localities only of the Recent and Quaternary ages are given.

Some of the stations are on the boundaries of formations and cover two outcrops. The aim is to give the most conspicuous and well-developed one in such localities.

Northern Railroad of New Jersey.*

Ms.		
0 Jersey City.[1] [2]	1. Archæan, 16. Trias.[6]	
7 New Durham.[3]	{ 16. Triassic, 20. Qua- { ternary, 21. Recent.[4]	
8 Granton.	"	4
10 Ridgefield.	"	5
13 Leonia.	"	4
15 Englewood.	"	15
16 Highland.	"	55
17 Tenafly.	"	45
18 Cresskill.	"	40
20 Closter.[4]	"	35
22 Norwood.	"	40

New York, West Shore, and Buffalo Railway.

Jersey City.	1. Archæan, 16. Trias.[10]	
Weehawken.[5]	16. Triassic.	10
1 New Durb'm.[5] [7]	{ 16. Trias., 20. Quater- { nary, 21. Recent.	4
5 Little Ferry.	"	4
6 Ridgefield Park.	"	10
7 Hackensack.	"	40
9 Teaneck.	"	80
10 W. Englewood.	"	75
12 Bergen Fields.	"	70
12 Schraalenburgh.	"	90
16 Randall's.	"	60
17 West Norwood.	"	50
19 Tappan, N. Y.	"	85

New York, Susquehanna, and Western Railroad.

0 New York.		
1 Jersey City.	1. Archæan, 16. Trias.[10]	
7 Schuetzen Park.	16. Triassic.	4
7 New Durham.[8]	16. Trias., 21. Recent.[4]	4
12 Little Ferry.	"	
12 Ridgefield Park.	"	10
14 Bogota.	"	5
14 Hackensack.	"	10
16 Maywood.	"	65
17 Rochelle Park.	"	45
19 Dundee Lake.	"	40
21 Paterson.[9]	"	100
24 Van Winkle's.[10]	"	125

New York, Susquehanna, and Western Railroad—Con.

Ms.		
26 Midland Park.	16. Trias., 21. Recent[225]	
27 Wortendyke.	"	275
28 Wyckoff.	"	345
30 Campgaw.	"	390
31 Crystal Lake.[11]	"	340
32 Oakland.[12]	"	275
35 Pompton.[13]	{ 1 a. Laurentian, 20 b. { Champlain.	220
38 Butler.	"	360
44 Charlotteb'gh.[14]	"	725
45 Newfo'ndland.[15]	12. Catskill Devon.	770
47 Oak Ridge.	{ 4 c. Hudson River (?) { 20. Quaternary.[830]	
51 Stockholm.[16]	1 a. Laurentian.	980
53 Summit.	"	1032
54 Two Bridges.	"	960
57 Ogdensburgh.[17]	{ 1 a. Laurentian, 20 a. { 20 a. Glacial.	660
60 Franklin.[18]	{ 1 a. Laurentian, 2 b. { 2 b. Potsdam.	530
63 Hamburgh.	3 a. Magnes. Limest.[425]	
67 Deckertown.	4 c. Hudson River.	465
71 Quarryville.[19]	"	560
75 Unionville, N. Y.	"	520
54 Two Bridges.	1 a. Laurentian.	960
57 S. Ogdensb'gh.[20]	{ 1 a. Laurentian, 20 a. { Glacial.	615
61 Sparta.	3 a. Magnes. Limest.660	
63 Sparta Junc.[21]	{ 3 a. Mag. Limest., 20 { b. Champlain.	580
69 Washingt'nv.[22]	4 c. Hudson River.	
72 Swartswood.	"	
76 Stillwater.	"	460
80 Marksboro.[23]	"	390
82 Paulina.	"	360
83 Blairstown.	3 a. Magnesian.	350
85 Kalarama.	"	370
89 Hainesburg.	"	320
91 Warrington.	"	310
92 Columbia.[24]	{ 3 a. Magnesian, 20 b. { Champlain.	305
96 Dunnfield.[25]	5 a. On'da & Medina.[280]	
98 Dela. Wat. Gap.	5 a. Medina.	325

* The altitudes are from the topographical sheets of "Atlas of New Jersey," prepared by the Geological Survey of New Jersey, Professor George H. Cook, State Geologist, and compiled by C. C. Vermeule, C. E., topographer.

1. The Archæan rocks are now all covered by improvements, and there are no outcrops; but a large part of the city has this formation as its underlying rock.

2. The Palisade range of Bergen Hill trap-rock in the western part of the cut, as seen at the tunnel.

3. The trap-rock of the Palisade range is seen on the east side, the whole length of this road to the New York line. (See Note 5, under New York.) On the left are the recent formations of the Hackensack meadows.

4. The sandstone lying upon the trap-rock can be seen on the mountain southeast of the station and near its crest.

5. At the east entrance to the tunnel the indurated shale, and above it the trap-rock, can be seen. One mile to the south there are good exposures of the latter rock cutting across the sandstone and shaly rocks. And sandstone was met with in the tunnel-cutting.

6. The sandstone on the west of the trap-rock is beautifully exposed in the west entrance to the tunnel. There are good sections showing glacial drift also.

7. The recent formations of the meadows along the Hackensack are seen on the left or west side from here to Hackensack.

8. (See Notes 3 and 6.)

9. The Garret Rock ridge of trap-rock is prominent in the southwest and south of the city. Passaic Falls, where the Passaic River falls seventy feet over ledges and through fissures of trap-rock.

Ms.	Green Pond Mine Railroad.		
0	Charlotteburgh.	1 a. Laurentian.	725
5	Green P'd Mines|	"	940

New York, Lake Erie, and Western Railroad.			
	New York.		
1	Jersey City.	1. Archæan, 16. Trias.	6
6	Secaucus.[26]	16. Trias., 21. Recent.	5
9	Rutherford.	.¹	55
12	Passaic.	"	55
14	Clifton.	"	60
15	Lakeview.	"	100
17	Paterson.[27]	" 20 b. Champ.[77]	
22	Ridgewood.	"	137
24	Hohokus.[28]	"	197
26	Allendale.	"	330
28	Ramsey's.	"	345
30	Mahwah.	"	275
10	Rutherford Jn.	"	150
13	Garfield.	"	60
20	Ridgewood Jn.	"	110

Ms.	Newark and Paterson Railroad.		
	New York.		
1	Jersey City.	1. Arch., 16. Trias.	6
9	Newark.	16. Triassic.	10
11	Belleville.	"	35
12	Avondale.[29]	"	100
13	Franklin.	"	70
16	Peru.	"	135
17	Athenia.	"	130
20	Paterson.	"	77

New Jersey and New York Railroad.			
1	Carlstadt.[30]	16. Trias., 21. Recent.	5
2	Woodridge.	"	15
6	Hackensack.	"	10
7	Cherryville.	"	10
9	New Milford.	"	10
10	Oradell.	"	10
13	Westwood.	"	75
14	Hillsdale.	"	65
15	Pascack.	"	115
16	Park Ridge.	"	155

In Morris Hill, near the falls, fine section of sandstone and conglomerate, bedded trap-rock capped by the columnar trap.

10. Columnar trap-rock seen on west of road in the second mountain range.

11. Morainic drift surface is noticeable on north of road, from here to Oakland, where the modified or terrace drift can be seen, thence to Pompton on the left side of car.

12. Here the train approaches the gneissic rocks (1 a. Laurentian) in the eastern face of the Highlands.

13. South of Pompton Junction ¼ mile, and in the left bank of the Pequannock River, there is an isolated outcrop of black, slaty rock, which is probably Huronian. The locality is in sight from the railroad track. Graphite mine ¼ mile south of Bloomingdale, a flag-station between Pompton and Butler. From Pompton to Charlotteburgh the road follows the Pequannock River, and excellent views of the Highland ranges are to be had from the car-window.

14. The bold escarpment of the Copperas Mountain here comes in view, and west of this station the road passes through a gap in the range. It belongs to the Green-Pond Mountain series of Devonian age.

15. Green Pond Mountain is seen to the southwest of the station. Green Pond, a beautiful, natural lake, 1,048 feet high, is three miles south of Newfoundland.

16. East of Stockholm the line re-enters the outcrop of the Laurentian rocks, and runs thence over them to Franklin Furnace.

17. The railroad line here runs on a remarkable moraine, which, excepting the narrow passage for the Wallkill, stretches across the valley and is one hundred or more feet high, affording pretty views on each side. West of the station there are cuts in the white, crystalline limestone. The Sterling Hill zinc-mines are southwest of the station.

18. The noted Mine Hill is northeast of and in sight from the station. The zinc-mines of *franklinite* ore are here. Famous mineral locality. The Potsdam sandstone is cut a few rods northwest of the depot.

19. The extensive meadows of the Drowned Lands are on the east of the road. Quarries of flagging-stone on Flagstone Hill west of the station.

20. The valley of the Wallkill River is on the west.

21. Modified drift of Germany Flats conceals the limestone.

22. The road here runs near the line between the slate and the magnesian limestone of the Paulinskill Valley. The ridge bordering the valley on the southeast from Washingtonville to the Delaware River is slate.

23. Near Marksboro, White Pond is noted for its shell marl deposits of *Recent* age.

24. The station is on the river terrace. Northward two miles, the road enters the slate belt. Quarries of roofing-slate a little way east of the road.

25. The railroad line follows the river through the gap in the conglomerate of the main southeast ridge, and then across the Medina red, gray, and olive-colored shales and sandstones. Grand scenery.

26. The road here crosses a low, upland strip of sandstone. To the southwest are to be seen the Snake Hill and Little Snake Hill—trap-rock hills. The meadows to the southeast and to the northwest are RECENT.

27. (See Note 9.) The modified drift is beautifully exposed in hills east of the depot and in the city.

28. The red sandstone is cut down deeply by the gorge east of the road. Northward to the State line the rock is covered by drift, and several side-cuttings show this drift.

29. The Belleville quarries, southeast of the station, yield annually a great amount of very excellent brownstone.

30. Tidal meadows to right. Sandstone ridge on left. The line follows the Hackensack and then the Pascack Rivers. Very few exposures of the rock; drift surface generally.

31. This railway west of the Erie line runs westerly, and cuts into the sandstone at the south side of Snake Hill, which is trap-rock mainly. West of Arlington it cuts deeply across the sandstone ridge.

New York and Greenwood Lake Railroad.

Ms.		road.
0	New York.	
1	Jersey City.	1. Archæan, 16. Trias. 6
7	Arlington.[31]	16. Triassic. 120
8	Newark.	" 60
11	Bloomfield.	" 140
13	Montclair.[32]	" 280
16	Montclair H'ghts	" 360
17	Great Notch.[33]	" 305
18	Cedar Grove.	16. Trias., 20 a. Glac.250
19	Little Falls.[34]	16. Triassic. 200
20	Singac.	{ 16. Triassic, 20 b. Champlain. 170
22	Mount'n View.[35]	" 185
24	Pequannock.	" 180
26	Pompton Plains.	" 190
27	Pompton.	" 225
32	Midvale.[36]	{ 1 a. Laurentian, 20 b. Champlain. 255
34	Ringwood Junc.	" 280
36	Ringwood.[37]	1 a. Laurentian. 340
38	Hewitt.	" 450
41	Cooper.[38]	" 621
	Surface of Greenwood Lake.	
44	State Line.	" 630

Orange Branch.

11	Watsessing Jn.	16. Triassic.	145
14	Orange.[39]	"	160

Delaware, Lackawanna, and Western Railroad.

Morris and Essex Division.

0	New York.		
1	Hoboken.[40]	16. Triassic.	
9	Newark.	"	35
12	Orange.[41]	"	185
15	South Orange.	"	140
19	Milburn.	"	147

Delaware, Lackawanna, and Western Railroad—Con.

Morris and Essex Division.

Ms.		
20	Short Hills.[42]	{ 16. Triassic, 20 a. Glacial. 210
21	Summit.	" " 381
24	Chatham.	" " 232
27	Madison.	" " 245
29	Convent.[43]	" " 385
31	Morristown.	{ 1 a. Laurentian ; 16. Triassic. 326
33	Morris Plains.[44]	{ 16. Triassic ; 20 b. Champlain. 405
37	Denville.	1 a. Laurentian. 523
39	Rockaway.	" 557
43	Dover.[45]	{ " 20 a. Glacial. 575
48	Drakesville.	1 a. Laurentian. 797
52	Stanhope.	" 873
56	Waterloo.[46]	" 717
61	Hackettstown.[47]	3 a. Mag. Limestone.567
67	Port Murray.	4 c. Hudson River. 600
71	Washington.[48]	{ 1 a. Laurentian ; 2 b. Potsdam. 500
76	Broadway.	1 a. Laurentian. 380
80	Stewartsville.	3 a. Magnesian. 360
84	Phillipsburg.[50]	" 220

2 Newark and Bloomfield Branch R. R.

	Newark.	16. Triassic.	35
4	Bloomfield.	"	115
5	Montclair.	"	260

3 Passaic and Delaware R. R.

	Summit.	16. Triassic.	381
2	N. Providence.[51]	"	230
5	Berkel'y H'ights.	"	215
8	Sterling.	"	230
10	Millington.	"	280
12	Lyons.	"	315
15	Bernardsville.[52]	"	360

A slight fault is seen in this cut. The historic Schuyler mine (copper) is one mile northeast of this station.

32. The road here approaches the trap-rock range (First Mountain).

33. The railroad line crosses the First Mountain range part way through a gap. Good exposures of trap-rock in cuts. Going toward Cedar Grove, beautifully glaciated surfaces and good sections of glacial drift on the side of track.

34. Falls of Passaic River over trap-rock ledges in village northeast of station. Quarries in brown sandstone. Fine examples of trap-rock columns on shale one mile northeast of village and near the river.

35. The road here passes through a gap in the Towakow-Packanack range of trap-rock and enters the Pompton Plains basin, a part of the old glacial Lake Passaic. The southern portion is still wet, peaty meadow. Northward a gravelly plain. The Archæan highlands are seen on the left—or west side of the plains.

36. The isolated crests of gneissic ridges, nearly buried in the drift gravel, characterize this valley.

37. The long-worked and celebrated iron-mines of Cooper and Hewitt are here reached by this branch railway.

38. The largest lake in the State, lying between the Laurentian ridges on the east and the rough Bearfort and Bellvale Mountains on the west. The latter are of the Green-Pond Mountain series of rocks. At the south end and west side of the lake there are small outcrops of 4 c. Hudson River, 5 a. Oneida, and Medina.

39. Famous basaltic columns at O'Rourke's quarry, west of the town.

40. At Castle Point, north of ferry, serpentine outcrops.

41. (See Note 39.)

42. Hills of glacial drift here are prominent ; and the terminal moraine crosses the Second Mountain range south of Summit. Thence to Morristown the southern edge of the drift is, on the average, a half mile south of the railroad.

43. West of the station deep sink-holes appear near the line of road.

Ms.	4 Chester Branch R. R.	
Dover.	1 a. Laurentian.	575
6 Succasunna.[53]	. "	20 b.
	Champlain.	705
8 Ironia	1 a. Laurentian;	20 b.
	Champlain.	710
13 Chester.	1 a. Laurentian;	20 b.
	Champlain.	685

Ms.	5 Boonton Branch R. R.	
0 New York.		
1 Hoboken.	16. Triassic.	10
4 Secaucus.	"	5
8 Kingsland.	"	40
9 Lyndhurst.	"	20
12 Passaic.	"	70
16 Paterson.[54]	"	180
19 Little Falls.	"	185
22 M'ntain View.[56]	"	185
24 Lincoln Park.[57]	"	170
26 Whitehall.[58]	"	225
29 Montville.[59]	"	360
31 Boonton.[60]	1 a. Laurentian.	400
35 Denville.	"	522

Ms.	6 Warren R. R., or Main Line.—Con.	
66 Washington.[61]	1 a. Laurentian; 2 b. Potsdam.	480
71 Oxford Furnace. [62]	3 a. Magnesian; 2 b. Potsdam.	492
75 Bridgeville.	3 a. Magnesian.	395
77 Manunka Chunk. [63]	4 c. Hudson.	320
80 Delaware	"	295

Ms.	Central R. R. of New Jersey.	
0 New York.		
1 Jersey City.	1. Arch'n; 16. Trias.	10
4 Greenville.	16. Triassic.	20
6 Bayonne.	"	20
7 Bergen Point.[64]	"	15
10 Elizabethport.	"	10
12 Elizabeth.	"	29
15 Roselle.	"	70
17 Cranford.	"	66
19 Westfield.[65]	"	130
21 Fanwood.	" 20 a. Glac'l	160
24 Plainfield.[66]	"	105
26 Dunellen.	"	60
31 Bound Brook.	"	36
35 Somerville.	"	69
36 Raritan.	"	75
40 North Branch.	"	93
45 White House.[67]	"	181
49 Lebanon.[68]	"	298
51 Annandale.	1. Archæan.	349
53 High Bridge.[69]	"	335
56 Glen Gardner.	"	471
57 { Junction, Summit of N. J. C. R. R. }	"	513
61 Asbury.[70]	3 a. Magnesian.	436
63 Valley.[71]	"	398
65 Bloomsbury.	"	324
68 Springtown.	"	312
74 Phillipsburg.[72]	"	223

Ms.	2 Newark and New York R. R.	
1 Jersey City.	1. Archæan.	10
8 Newark.	16. Triassic.	35

44. The Archæan rocks are west of the plains. The drift is thick and the plains are a part of the old glacial Lake Passaic. The road enters the Highlands north of this station.

45. Dover is the center of the iron-mine district of Morris County.

46. The Musconetcong Valley is here entered, the road passing through the terminal moraine a half mile north of Hackettstown.

47. The beautiful and fertile valley is here spread out before the traveler. Going south to Port Murray, deep cuts show slate. The Schooley's Mountain table-land is seen on the east.

48. The railroad cut exposes Potsdam sandstone and Laurentian gneiss. The Pohatcong Valley is here entered, and hence to Broadway the line follows at the side of the valley.

50. The railroad cut near Phillipsburg cuts a slaty rock, which may be Utica slate.

51. The railroad line runs down from Summit into the valley of the Passaic and along the south-east foot of Long Hill.

52. Bernardsville is at the border of the Laurentian Highlands.

53. Modified drift forms the surface of these plains.

54. The road runs close under Garret Rock. Quarries of sandstone on the east side of this mountain, where the trap-rock can be seen upon the sandstone. On the left side of the track there are side cuts in trap-rock and sandstone. On the right one sees the same rocks exposed in the bluff west of the mills. Fine view of the city is here also had.

56. (See Note 35.)

57. Here the road follows on northern foot of Hook Mountain and south of the Pompton Plains.

58. Between Whitehall and Montville there are very fine sections of high terrace hills at the right of the track. Footprints in red sandstone at quarry one mile southeast of the station.

59. Famous locality for serpentine and chrysolite at Gordon's quarry two miles north of this station. Fossil fish locality is about two miles southeast.

60. To the east and southeast the passenger looks over the red sandstone plain—to the distant Second Mountain range of trap-rock.

61. (See Note 48.)

62. Extensive iron-works and iron-mines. Tunnel through the gneissic rocks east of the station.

63. Tunnel in slate. Beautiful view of the Delaware and of Water Gap.

64. Railroad cut west of the station, near Newark Bay, shows old sand-dune upon sandstone drift.

65. Beyond this station, and on to Netherwood, railroad cuts show good sections of glacial drift where the terminal moraine is crossed.

66. The plain country southwest of the moraine is here reached. First Mountain (of trap-rock) is on the north.

Ms. | 3 Delaware and Bound Brook R. R.

Ms.		
0	New York.	
1	Jersey City.	1. Arch'n; 16. Trias. 10
31	Bound Brook.	16. Triassic. 36
35	Weston.	"
41	Van Akon.	"
45	Skillman.[73]	"
48	Hopewell.	"
53	Pennington.	"
57	Ewing.	"
61	Trenton.	1 Archæan.

4 South Branch R. R.

Ms.		
0	New York.	
1	Jersey City.	1. Arch'n; 16. Trias. 10
35	Somerville.	16. Triassic. 69
	Roycefield.	" 109
	Flaggtown.	" 135
	Neshanic.	" 94
	Three Bridges.	" 114
52	Flemington.	" 195

5 High Bridge Branch R. R.

Ms.		
0	New York.	
1	Jersey City.	1. Arch'n; 16. Trias. 10
53	High Bridge.	" 335
58	Califon.[74]	2 b. Potsdam. 485
61	Middle Valley.	3 a. Mag. limestone. 505
64	German Valley.	" " 545
66	Naughright.	" " 575
68	Bartley.[75]	{ 1. Archæan (?); 20 b. Champlain. 630
70	Flanders.	" 587
75	Kenvil.[76]	" 727
78	Port Oram.	1. Arch.; 20 a. Gla'l. 670
79	Dover.[77]	" " 570
83	Rockaway.	" " 540

Hibernia Mine R. R.

Ms.		
4	Hibernia.[78]	1. Arch.; 20 a. Gla'l. 540

6 Ogden Mine R. R.

Ms.		
75	Kenvil.	{ 1. Archæan; 20 h. Champlain. 727
80	Hopatcong.[79]	" 925
	Surface of lake	
83	Hurdtown.[80]	" 950
90	Ogden Mines.	" 1225

Ms. | 7 Chester Branch R. R.

Ms.		
64	German Valley.	3 b. Mag. limestone. 545
70	Chester.[80a]	1. Archæan. 845

Euston and Amboy R. R.
Lehigh Valley R. R.

Ms.		
0	New York.	
1	Jersey City.	1. Arch'n; 16. Trias. 10
26	Metuchen.[81]	16. Trias.; 20 a. Glac. 100
33	Perth Amboy.	18 a. Raritan clays. 20
32	New Market.	16. Triassic. 52
36	Bound Brook.	" 39
47	Neshanic.	" 113
54	Flemingt'n Junc.	" 116
63	Clinton.	3 a. Mag. limestone. 200
61	Landsdown.	16. Triassic. 200
64	Midvale.	" 350
66	Pattenburg.[82]	" 445
69	West End.	1. Arch'n; 3 b. Mag. 450
71	Bloomsbury.	3 b. Magnesian. 395
75	Phillipsburg.	" 322

Pennsylvania R. R.
1. United Railroads of New Jersey.

Ms.		
	New York.	
1	Jersey City.[83]	1. Arch'n; 16. Trias. 10
3	Marion.	16. Trias.; 20 a. Glac. 4
4	Meadows.[84]	21. Recent; 16. Trias. 4
8	East Newark.	" " 10
9	Newark.	16. Triassic. 10
11	Waverly.	" 10
14	Elizabeth.	" 29
17	Linden.	" 25
19	Rahway.	" 25
21	Houtenville.	" 35
23	Iselin.	" 55
24	Menlo Park.[85]	" 90
26	Metuchen.	" 110
29	Stelton.	" 90
31	N. Brunswick.[86]	" 50
35	Adams.	" 110
38	Deans.	" 83
41	Monmouth Junction.[87]	{ 18 a. Cretaceous, Plastic clay. 92
45	Plainsboro.	18 a. Cretaceous. 81
47	Princeton Junc.	" 83
50	Princeton.	16. Triassic. 230
51	Lawrence.	{ 18 a. Cretaceous, Plastic clay. 90
56	Trenton.[88]	{ 1. Archæan; 20 b. Champlain. 33

67. Round Valley Mountain to the southwest, a peculiar, horse-shoe shaped ridge of trap-rock. The railroad line is at north side of it.
68. About half a mile west of Lebanon the Archæan territory is entered.
69. Here the deep valley of the north branch of Raritan is crossed.
70. Limestone dipping under the gneiss of mountain is noticeable in the railroad cut northeast of the station. Hence to Bloomsbury the line runs near foot of the Musconetcong Mountain.
71. Large iron-mines one mile southwest.
72. (See Note 50.)
73. Sourland Mountain (trap-rock) appears on right side of the car, to northwest. Beyond the next station (Hopewell) the road cuts across the end of the Mount Rose or Rocky Hill range.
74. Here the road enters the German Valley, shut in by Archæan ranges of mountains.
75. The underlying formation (presumably Archæan) is here concealed by drift. The same is true at the succeeding stations of Drakesville and Kenvil. The low ridges on the east of the line are of sandstone (Green Pond Mountain series).

2 Woodbridge and Perth Amboy R. R.

Ms.			
	New York.		
19	Rahway,	16. Triassic.	25
20	Perth Amboy Jn.	"	20
22	Edgar's	{ 18 a. Cretaceous, Raritan clays.	40
23	Woodbridge.[89]	18 a. Cretaceous.	15
24	Spa Spring.	"	10
26	Perth Amboy.[90]	"	40

2 a. Belvidere Delaware R. R.

0	Trenton.[91]	1. Arch'n; 2 b. Potsd.[33]	
4	Asylum.[92]	16. Triassic.	51
8	Somerset.	"	64
9	Wash'ton Cross.	"	65
10	Titusville.[93]	"	67
12	Moore's.	"	68
16	Lambertville.[94]	"	72
19	Stockton.[95]	"	82
23	Bull's Island.	"	93
26	Tumble.	"	96
31	Frenchtown.	"	125
35	Milford.[96]	"	137
38	Holland.	"	135
42	Riegelsville.[98]	3 b. Mag. limestone.	163
45	Carp'nterville.[99]	" "	175
50	Phillipsburg.[100]	" "	195
53	Harmony.	" "	220
57	Martin's Creek.	" "	231
64	Belvidere.	" "	265
68	Manunka Chunk.	4 c. Hudson.	320

Ms. | Lehigh and Hudson River R. R.

0	Philadelphia.			
50	Phillipsburg.	3 a. Mag. limestone.		195
64	Belvidere.	"	"	265
69	Buttsville.	"	"	391
73	Townsbury.[101]	"	"	500
75	Gt. Meadows.[102]	} 20 b. Champlain.		522
81	Allamuchy.	"		536
83	Andover.[103]	"		590
89	Sparta Junction.	"		580
96	Franklin Junc.	"		520
98	Hamburgh.[104]	{ 3 a. Mag. limestone. 20 a Glacial.		460
103	McAfee.[105]	1. Archæan.		440
106	Vernon.	3 a. Mag. limestone.		416
124	Greycourt, N.Y.			

Flemington Branch R. R.

16	Lambertville.	16. Triassic.	72
19	Mt. Airy.	"	147
23	Ringoes.	"	242
26	Copper Hill.	"	169
28	Flemington.[106]	"	182

3. Millstone Branch R. R.

	New York.		
	New Brunswick.	16. Triassic.	50
33	Millstone Junc.	"	90
34	Voorhees.	"	110
35	Clyde.	"	125
37	Middlebush.	"	115
39	East Millstone.	"	55

76. Northeast of Kenvil, about one mile, the terminal moraine is entered, and the railroad cuts afford good sections of the glacial drift, thence to Port Oram.

77. (See Note 45.)

78. Large mines of magnetic iron-ore, for which this road is the outlet.

79. Largest lake wholly in the State.

80. Iron-mines. Apatite locality. This railroad line has its terminus at large Ogden Mines.

80 a. Iron-mines in and near the village.

81. The terminal moraine is crossed by this road southeast of the station.

82. Here the road leaves the red sandstone territory and enters the gneiss in the Musconetcong tunnel. A fold of the magnesian limestone in it. At the west end entrance of the tunnel the deep cut exposes disintegrated gneisses, and to west the magnesian limestone and hydro-mica slates. West End iron-mines.

83. Bergen Cut, in trap-rock, between Jersey City and Marion.

84. The road here crosses the Newark Meadows. Much buried cedar timber in the black earth; and the stumps and fallen trunks may be seen from the car-windows.

85. The terminal moraine is crossed between this station and Metuchen.

86. The red sandstone forms bluffs in right bank of the Raritan, which are seen crossing the bridge.

87. Low cuts here and hence to Trenton in drift sand and gravel. They conceal the underlying formations.

88. The gneissic rocks are to be seen in the Delaware River above the railroad bridge. Northeast of the station a long cut exposes a gravel formation, which belongs to the Trenton terrace level. Mastodon tusk has been found in it. Rude flint implements found by Dr. Abbott in this formation, south of station, in the river bluff.

89. Center of fire-clay digging and fire-brick works. Very large banks west and south of the village.

90. Southern limit of glacial drift at mouth of the Raritan River.

91. A micaceous sandstone (Potsdam) near the Warren Street station.

92. Coarse, pebbly beds of the Triassic are noticeable near Asylum station. Thence, up the river, many cuts in the red sandstone. Near Greensburg there are large quarries of sandstone.

93. Trap-rock of Smith's Hill, north of Titusville.

94. Goat Hill (trap-rock) south of this station. North of it, and east of the town, remarkable examples of indurated shales. Tourmaline locality.

95. Sandstone quarries.

96. Flagstone quarries north and northeast of village. Pebble bluff, a huge wall of red conglomerate northwest of the village, at foot of which is the road. Nockamixon Cliffs on opposite (Pennsylvania) side.

98. Musconetcong Mountain range of gneiss south of station.

Ms.	4. Rocky Hill Branch R. R.	
	New York.	
41	Monmouth Junction.	{ 18 a. Cretaceous, Raritan clay. 92
45	Kingston.	16. Triassic. 60
47	Rocky Hill.[107]	" 60

5. Amboy Division.

Ms.		
	New York.	
	So. Amboy.[108]	{ 18. Cretaceous ; a. Raritan clays. 20
8	Old Bridge.	" 10
10	Spotswood.	" 29
14	Jamesburg.	" 73
16	Prospect Plains.	" 140
18	Cranbury.	{ 18. Cretaceous ; b. Clay marls. 110
21	Hightstown.	" 99
24	Windsor.	" 85
27	Newtown.	" 122
31	Yardville.	" 53
34	Bordentown.	" 10
	Trenton.[109]	1. Archæan. 33
35	White Hill.[110]	{ 18. Cretaceous ; a. Plastic clays ; b. Clay marls. 10
37	Kinkora.	"
39	Florence.	{ 18. Cretaceous ; a. Plastic clays.
43	Burlington.	" 10
46	Edgewater.	"
47	Beverly.	"
49	Delanco.	"
50	Riverside.	"
53	Riverton.	"
54	Palmyra.[111]	"
57	Fish House.[112]	"
61	Camden.	"
62	Philadelphia.	"

6. Freehold and Jamesburg Agricultural R. R.

Ms.		
41	Monmouth Junction.	{ 18. Cretaceous ; a. Raritan clay. 92
43	Dayton.	" 90
49	Jamesburg	" 73
54	Englishtown.[113]	18. Cret. ; a. b. Clay m'ls.
58	Freehold.	{ " d. Red sand.[188]
		" c. Lower marl.
61	Howell's.	" e. Middle marl.
66	Farmingdale.[114]	{ " f. Yellow sand.
		" g. Upper marl.
		{ Eocene.
69	Allaire.	"
73	Manasquan.	19. Tertiary.
74	Sea Girt.	"

7. Pemberton and Hightstown R. R.

0	Hightstown.	18. Cret's ; b. Clay marls.
5	Sharon.	" "
7	Imlaystown.	" "
10	Cream Ridge.[115]	{ " d. Red sand bed.
		" c. Lower mrl bed.
12	Hornerstown.	" e. Middle marl.
15	New Egypt.[116]	{ " f. Yellow sand.
		" g. Upper marl.
20	Wrightstown.	" f. Yellow sand.
		" " "
23	Lewistown.	" e. Middle marl.
25	Pemberton.[117]	{ " f. Yellow sand.
		" g. Upper marl.

9. Burlington R. R.

	Burlington.	{ 18 Cretaceous ; a. Plastic clay. 10
	Mount Holly.[118]	{ 18. Cret'ous ; b. Clay marl ; c. Lower mrl ; d. Red sand.

99. Pohatcong range of gneiss north of this place.
100. Two miles to north the railroad line runs at river foot of Marble Mountain. Hornblendic schists, crystalline limestone, steatite (quarries) and gneisses. Some of these may be Huronian. River terraces at Belvidere.
101. The line skirts mountain on west, Pequest Valley on east. Terminal moraine lies across valley near Townsbury.
102. Great Meadows is an old glacial lake-basin filled by drift and recent alluvial deposits.
103. The once famous Andover iron-mine is northeast of station and near the track. To northeast a chain of natural lakes in a modified drift, valley underlain by limestone.
104. A remarkable cut in glacial drift south of the station.
105. Large quarries in white, crystalline limestone in this vicinity and near Hamburgh. On east the high Wawayanda Mountain ; on the west, Pochuck Mountain ; both ranges of gneissic rocks.
106. Copper-mine west of town.
107. Trap-rock quarries south of station.
108. Fossil-leaf locality in clay-pits near shore.
109. (See Notes 88 and 91.)
110. Fine sections of clay-marls, and the clays in the bluff, and at clay-banks near Kinkora. Northwest of Florence station and in the river bluff the yellow gravel covers thirty or more feet of Cretaceous clays and sands.
111. Fine section of gravel, sands, and Cretaceous clay in south bank of the Pensauken Creek.
112. Clay-pits. Locality of fossil antos in clay.
113. Marl-pits north of railroad line—as near Freehold. Red sand forms surface at Freehold.
114. Extensive marl-pits in vicinity. Lower layer of upper bed mostly opened. Upper layer is Eocene. Many fossils.
115. Lower marl is opened in this neighborhood for marls.
116. Good section along Crosswicks Creek, showing all the marl-beds and their layers. Upper marl-bed is worked in vicinity of New Egypt. Many fossils.
117. Large pits near the village, in the middle bed.

8. Kinkora Branch R. R.

Ms.		
0	Kinkora.[119]	{ 18. Cretaceous. / a. Plastic clay. / b. Clay marls.
4	Columbus.[120]	{ 18. Cret's b. Clay mrl.
7	Jobstown.	" c. Lower marl.
9	Juliustown.	" d. Red sand.
10	Lewistown.	" e. Middle marl.
		" f. Yellow sand.

10. Camden and Burlington County R. R.

Ms.		
0	Philadelphia.	
1	Camden.	18. Cret's; a. Plas. clay.
6	Merchantville.	" "
11	Moorestown.	" b. Clay marl
14	Hartford.	" "
15	Masonville.	" c. Lower marl.
18	Hainesport.	" "
20	Mt. Holly.[121]	" d. Red sand.
22	Smithville.	" d. Red sand.
24	Birmingham.[122]	" e. Middle marl.
25	Pemberton.	" "

11. Pemberton and Sea-Shore R. R.

Ms.		
25	So. Pemberton.	{ 18. Cretac's; g. Upper marl; f. yellow sand.
29	New Lisbon.	19. Tertiary; c. Pliocene.
43	Whitings.	" "
52	Toms River.	" "
55	Island Heights.	" " 10
58	Barnegat Pier.	21. Recent. 10
	Seaside Park.[123]	" 5
60	Berkeley.[124]	" 5
64	Chadwick.	" 5
70	Bay Head.	"
	Bay Head Junc.	"
71	Point Pleasant.	"
72	Brielle.	"
73	Manasquan.	" 135
74	Sea Girt.	"

12. Medford Branch R. R.

Ms.		
0	Mount Holly.	{ 18. Cret's; b. Cl'y mrls. / " c. Lower marl. / " d. Red sand.
3	Lumberton.	" d. Red sand.
7	Medford.[126]	{ " e. Middle marl. / " f. Yellow sand. / " g. Upper marl.

New York and Long Branch R. R.

0	New York.	
1	Jersey City.	1. Arch'n; 16. Trias. 10
13	Elizabethport.	16. Triassic. 10
14	Elizabeth.	" 29
21	Sewaren.	" 25
25	Perth Amboy.	{ 18. Cretaceous; a. Raritan clays. 30
27	South Amboy.	" 10
28	Morgan.[127]	" 10
30	Cliffwood.	18. Cret's; b. Clay marls.
32	Matawan.[128]	" "
34	Hazlet.	" "
38	Middletown.[129]	{ " c. Lower marl. / " d. Red sand.
42	Red Bank.	" "
44	Little Silver.	" d. Red sand.
47	Branchport.	{ " e. Middle marl.
48	Long Branch.[130]	" "
50	Elberon.	" "
52	Deal Beach.[131]	{ " f. Yellow sand. / " g. Upper marl.
55	Asbury Park.[132]	19. Tertiary; c. Pliocene.
	Key East.	" "
56	Ocean Beach.	" "
58	Spring Lake.	" "
60	Sea Girt.	" "

Freehold and New York R. R.

0	New York.	
1	Jersey City.	1. Arch'n; 16. Trias. 10
12	Matawan.	18. Cret's; b. Clay marls.
14	Keyport.	" "

118. Holly Mount consists of red-sand bed capping lower marl rising above the clay-marl plain.
119. (See Note 110.)
120. Here, as at many localities in West Jersey, the strata are concealed ; and the dip of beds is so slight that there is some uncertainty in some localities what are the underlying strata.
121. (See Note 118.)
122. (See Note 117.)
123. Sea-beaches (Recent).
124. Artesian well here strikes the marl-beds after penetrating overlying gravels, sands, and clays.
125. Or, possibly, Pliocene.
126. Marl-pits in both the middle and upper beds in the vicinity of village.
127. The railroad line here cuts into the stoneware clay-bed, going toward South Amboy. Southward the dark-colored clays and the clay-marls are exposed in the cuts.
128. Matavan Creek cuts into clay-marls.
129. Railroad cut through lower bed, at station. Deep cut in red sand south, one mile.
130. Surface clays and gravels may be Pliocene.
131. Pits in upper marl-bed—west of railroad line—at Poplar, also near Deal Beach.
132. The superficial beds are probably Pliocene. Artesian-well borings pass through these and reach the Cretaceous marl series.
133. Mount Pleasant Hills (red-sand bed and lower marl) to southeast.
134. Numerous marl-pits in vicinity, and many fossils. Red-sand bed forms hills generally.
135. A sandy strip of beach-sand and Recent.
136. Navesink Highlands to west of river—of red-sand bed, capping lower marl. Latter is seen in north or Raritan Bay side of Highlands.

Ms.	Freehold and New York R. R.—*Con.*	
	Morganville.[133]	18. Cret's; b. Clay marls.
	Wickatunk.	{ " c. Lower marl,
		" d. Red sand.
	Marlboro'gh.[134]	" "
22	Freehold.	" "

New Jersey Southern R. R.

	New York.	
0	Sandy Hook.[135]	21. Recent.
4	Highlands.[136]	"
6	Seabright.	"
8	Monmo'th Be'ch.	"
10	E. Long Branch.	19. Tertiary.
11	Branchport.	18. Cretaceous.
13	Oceanport.	"
15	Eatontown.	{ " d. Red sand.
		" e. Middle marl.
16	Red Bank.	" "
17	Shrewsbury.	" " 54
15	Eatontown.	" "
	Eatontown.	
21	Shark River.[137]	{ " f. Yellow sand.
		" g. Upper marl.
25	Farmingdale.	" "
26	Squankum.	" "
32	Lakewood.	19. Tert.; c. Pliocene.[53]
40	Manchester.	" " 45
45	Whitings.	" " 187
50	Wheatland.[138]	" " 143
53	Woodmansie.	" " 136
58	Shamong.	" " 98
69	Atsion.	" "
	Atsion.	" "
78	Atco.	" "
78	Winslow Junc.	" "
79	Winslow.[139]	" "
84	Cedar Lake.	" "
89	Landisville.	" "
94	Vineland.	" "
97	Bradway.	" "
100	Rosenhayn.	" "
106	Bridgeton.	" "
108	Bowentown.	" "
113	Greenwich.[140]	21. Recent.
115	Bayside.	"

2. Atlantic Highlands Branch R. R.

	Red Bank.	{ 18. Cret's; d. Red s'nd.
0		" e. Middle marl.
	Chapel Hill.	" d. Red sand.
6	Hopping.	" b. Clay marls.
8	AtlanticHighlds.	" d. Red sand.
6	Port Monmouth.	21. Recent; 18 a. Cl. mrl.

Ms.	3. Toms River and Waretown R. R.	
	New York.	
0	Sandy Hook.	21. Recent.
40	Manchester.	19. Tert'ry; c. Pliocene.
47	Toms River.	" "
51	Bayville.	" "
53	Cedar Creek.	" "
55	Forked River.	" "
59	Waretown.	" "
62	Barnegat.	" "

Tuckerton R. R.

0	Whitings.	19. Tert'ry; c. Pliocene.
5	Bamber.	" "
7	Lacy.	" "
11	Middle Branch.	" "
15	Waretown Junc.	" "
17	Barnegat.[141]	" "
21	Manahawken.	" "
26	West Creek.	" "
29	Tuckerton.	Recent.

Camden and Atlantic R. R.

0	Philadelphia.	
1	Camden.	18. Cret's; a. Plas. cl'ys.[6]
7	Haddonfield.	{ " b. Clay marls. 72
10	Ashland.	" c. Lower marl.
		" d. Red sand.
12	Kirkwood.[142]	" e. Middle marl.[69]
17	Berlin.	19. Tert.; c. Plioc'ne.[176]
19	Atco.	" "
23	Waterford.	" "
27	Winslow.[139]	" "
30	Hammonton.	" "
33	Da Costa.	" "
36	Elwood.	" "
41	Egg Harbor.	" "
47	Pomona.	" "
52	Absecon.	" and 21. Recent.
59	Atlantic City.	21. Recent. 5

Philadelphia, Marlton and Medford R. R.

0	Philadelphia.	
1	Camden.	18. Cret's; a. Plas. cl'ys.[6]
7	Haddonfield.	" b. Clay marls. 72
13	Marlton.	" e. Middle marl.
		" "
18	Medford.[126]	{ " f. Yellow sand.
		" g. Upper marl.

Williamstown R. R.

0	Atco.	19. Tert'ry; c. Pliocene.
7	Williamstown.	" "

137. Much sandy gravel on hills in vicinity, which may be Pliocene. Shark River marl-pits near village and southeast of station. Noted Eocene fossil locality.
138. Clay-pits near station.
139. Glass-sand pits. Glass-works. Artesian well reached Cretaceous marls three hundred and sixty feet deep.
140. A very fertile alluvial upland neck.
141. The lower upland points are probably Recent, as are the tidal marshes along this coast.
142. Pits in middle marl-bed at side of track.

May's Landing and Egg Harbor R. R. Ms.			Ms.	West Jersey R. R.—Con.	

	May's Landing and Egg Harbor R. R. Ms.			West Jersey R. R.—Con.	
	Egg Harbor.	19. Tert'ry; c. Pliocene.	46	Manumuskin.	19. Tert.; c. Pliocene.
	May's Landing.	" "	53	Belleplain.	" "
			56	Woodbine.	" "
Philadelphia and Atlantic City R. R.				Sea Island City.	21. Recent, Sea-beach.
0	Camden.	18. Cret's; a. Plas. clays.	62	Seaville.	19. Tert'ry; c. Pliocene.
3	Oakland.	"	69	Cape May, C. H.	" "
4	Linden.	" b. Clay marls.		Anglesea.	21. Recent, Sea-beach.
5	Dentdale.	" "	75	Rio Grande.	19. Tert'ry; c. Pliocene.
		" "	78	Bennett.	" "
7	Magnolia.	" c. Lower marl.	81	Cape May.[147]	21. Recent.
8	Somerville.	" d. Red sand.	0	Camden.	18. Cret's; a. Plas. cl'ys.[6]
9	Laurel.	" e. Middle marl.	18	Glassboro.	19. Tert.; c. Pliocene.[144]
11	Clementon.	" g. Upper marl.	20	Union.	" "
14	Albion.	19. Tert'ry; c. Pliocene.	24	Monroe.	" "
15	Lansborough.	" "	26	Elmer.	" " 112
16	Willi'mst'wn Jn.	" "	29	Palatine.	" " 116
19	Cedar Brook.	" "	31	Husted.	" " 96
21	Blue Anchor.	" "	38	Bridgeton.[148]	" " 51
23	Winslow.	" "	0	Camden.	18. Cret's; a. Plas. cl'ys.[6]
27	Hammonton.	" "	26	Elmer.	19. Tert.; c. Pliocene.[117]
30	Da Costa.	" "	31	Daretown.	" "
33	Elwood.	" "	34	Yorketown.	" "
38	Egg Harbor.	" "	37	Riddleton.	" "
43	Pomona.	" "	38	Alloway.	" "
49	Pleasantville.	" "	43	Salem.	{ " e. Middle marl.
53	Atlantic City.	21. Recent.			{ " 21. Recent.
West Jersey R. R.			0	Camden.	18. Cret's; a. Plas. cl'ys.[6]
0	Camden.	18. Cret's; a. Plas. cl'ys.[6]	8	Woodbury.	" b. Clay marls. 34
30	Newfield.	19. Tert.; c. Plioc'ne.[114]	13	Clarksboro.	" "
33	Forest Grove.	" "	19	Swedesboro.[149]	{ " c. Lower marl.
36	Buena Vista.	" "			{ " d. Red sand.
47	May's Landing.	" " 10	26	Woodstown.[150]	" e. Mid. marl.
59	Pleasantville.	" "	30	Riddleton.	19. Tert'ry; c. Pliocene.
66	Somers Point.	" " 10	**Delaware River R. R.**		
64	Atlantic City.	21. Recent. 5	0	Camden.	18. Cret's; a. Plas. cl'ys.[6]
0	Camden.	18. Cret's; a. Plas. cl'ys.[6]	8	Woodbury.	" b. Clay marls. 34
4	Gloucester.	" b. Clay marls. 16	13	Paulsboro.	" "
5	Westville.	18. Cret's; b. Clay m'rls.[9]	20	Bridgeport.	" "
8	Woodbury.	" " 34	24	Pedricktown.	21. Recent.
11	Wenonah.	{ " d. Red sand. 36	28	Penn's Grove.	"
13	Barnsboro.[143]	{ " e. Middle marl.[63]	**Cumberland and Maurice River R. R.**		
18	Glassboro.[144]	19. Tert.; c. Pliocene.[148]	0	Bridgeton.[148]	19. Tert.; c. Pliocene. 51
21	Clayton.	" " 143		Fairton.	" "
24	Franklinville.	" " 123		Newport.	" "
28	Malaga.	" " 106		Dividing Creek.	" "
30	Newfield.	" " 114	20	Port Morris.	" "
34	Vineland.[145]	" " 110			
40	Millville.[146]	" " 36			

143. Large marl-pits, and branch railroad line to them.
144. Glass-sand pits between this place and Williamstown.
145. The gravel well exposed in railroad cut at station.
146. Glass-sand pits along Maurice River below the town.
147. On an upland island.
148. Glass-sand bed opened south of town in river-bank.
149. Lower marl-bed along Raccoon Creek.
150. Middle marl-bed here opened for marl digging.

This blank space is intended for additional geological notes in pencil by the traveler.

Pennsylvania.

By J. P. Lesley, State Geologist.

LIST OF THE GEOLOGICAL FORMATIONS OF PENNSYLVANIA.

Prof. Dana's Table of the Formations.	Names Provisionally adopted in the Second Geological Survey of Pennsylvania, by Prof. J. P. Lesley.		Old Penn. Nos. of 1st Geo. Sur.
20. Quaternary.	20. Quaternary.		
16. Triassic.	16. Triassic.		
14 c. Upper Coal Measures.	14 c. { Green Co. Group.		XVII.
	{ Washington Co. Group.		XVI.
" "	" Monongahela River Series.		XV.
14 b. Lower Coal Measures.	14 b. Barren Measures.		XIV.
" "	" Allegheny River Series.		XIII.
14 a. Millstone Grit.	14 a. Pottsville Conglomerate.		XII.
13 b. Upper Sub-Carboniferous.	13 b. Mauch Chunk Red Shale.		XI.
13 a. Lower Sub-Carboniferous.	13 a. Pocono Gray Sandstone.		X.
12. Catskill.	12. Catskill Red Sandstone. ⎤		IX.
11 b. Chemung.	11 b. Chemung.		VIII f.
11 a. Portage.	11 a. Portage.		VIII e.
	10 c. Genesee.		VIII d.
{ Genesee.		Devonian.	
10. Hamilton, { Hamilton.	10 b. Hamilton.		VIII c.
{ Marcellus.	10 a. Marcellus.		VIII b.
9. Corniferous.	9. Upper Helderberg.		VIII a.
8. Oriskany.	8. Oriskany. ⎦		VII.
7. Lower Helderberg.	7. Lower Helderberg. ⎤		VI.
6. Salina.	6. Salina.		V c.
5 c. Niagara.	5 c. Niagara.	Silurian.	V b.
5 b. Clinton.	5 b. Clinton.		V a.
5 a. Medina.	5 a. Medina.		IV b.
	" Oneida. ⎦		IV a.
4 c. Hudson River.	4 c. Hudson River. ⎤		III b.
4 b. Utica.	4 b. Utica.	Siluro- Cambrian.	III a.
4 a. Trenton.	4 a. Trenton.		II b.
3. Canadian.	3 a. Calciferous.		II a.
2. Primordial or Cambrian.	2 b. Potsdam. ⎦		I.
1. Archæan.	1. Azoic.		

NOTES ON THE TABLE OF FORMATIONS. All beneath the Potsdam is styled Azoic, because no survey has yet sufficiently differentiated the mass into its several systems. The term Eozoic is rejected, partly because both too vague and too shifting, and partly because it would suit the Cambrian system better than the Huronian and Laurentian, both of which remain to all intents and purposes Azoic. The terms Huronian and Laurentian are known to apply lithologically to rock masses in Pennsylvania, but their geographical relationships in the State are but imperfectly made out.

Much uncertainty still exists about the lines of demarcation between some of the formations in Pennsylvania, such as between the Catskill and Chemung; the Lower Helderberg and Clinton; the Hudson River and Utica; the Calciferous and Potsdam.

Niagara, Onondaga or Salina, Corniferous and other names were omitted, in the first edition, because of their uncertain presence in many districts of the State; and because of the narrowness of their upturned outcrops where they do exist.

Some of the places named in the following lists occupy positions covering the width of two or more steeply outcropping formations, to any one of which, therefore, they might be assigned.

In the northern and western counties it is often impossible to say precisely whether places stand upon Chemung, Catskill, Pocono or Mauch Chunk rocks. In such cases, Chemung has been preferred, because the others might be studied in the surrounding hills on account of the general horizontality of the bedding.

The last column in the table gives the numbers assigned to the Paleozoic formations in 1837, and their modifications since 1874. All above XII are additions.

J. P. L.

Pennsylvania.*

Ms.	Pennsylvania Railroad. New York Division.		Alt.
0	W. Philadelphia.	1. Azoic.	32
6	Kensington.[1]	20. Quaternary.	27
13	Holmesburg.	"	
23	Bristol.	"	21
26	Tullytown.	"	20
32	Morrisville.	1. Azoic.	34
33	Trenton, N. J.	(See New Jersey.)	63

Ms.	Pennsylvania Division—Main Line.		Alt.
0	W. Philadelphia.	1. Azoic.	32
5	Merion.	"	247
9	Bryn Mawr.	"	416
20	Paoli.	"	534
22	Malvern.	"	546
28	Oakland.[2] 266	{ 2 – 4. Siluro-Cambrian. (Calcif'ous?)	
33	Downingtown.	{ 3 a. & 4 a. Magnesian Limesto's & Marbles	
39	Coatesville.	"	330
44	Parkersburg.	2 b. Potsdam s. s.	537
47	Penningtonville.	"	500
51	Gap.[3]	1. Azoic.	559
57	Lemon Place.[4]	{ 2 – 4. Siluro-Cambrian Limesto's.	552

Ms.	Pennsylvania Railroad. Pennsylvania Div.—Main Line—Con.		Alt.
61	Bird-in-Hand.	{ 2–4. Siluro-Cambrian Limestones.	359
69	Lancaster.	"	359
76	Landisville.[5]	"	405
81	Mount Joy.	"	366
87	Elizabethtown.[6]	16. Triassic.	457
95	Branch Inter.[7]	"	
96	Middletown.	"	314
106	Harrisburg.	{ 4 a. Trenton Limestone and edge of 4 b. Utica Slate.	320
111	Rockville.[8] 350	4 c. Hudson Riv. Slate.	
113	Marysville.	5 a. Oneida Conglom'e.	
120	Duncannon.[9]	12 Catskill s. s.	356
133	Newport.	11 b. Chemung.	395
138	Millerstown.[10]	{ 5 b. Clinton fossil iron ore beds.	408
143	Thompsontown.	7. L. Helderberg.	419
148	Tuscarora.	10. Hamilton.	429
152	Perrysville.[11]	"	441
155	Mifflin.	5 b. Clinton.	441
162	Narrows.[12]	"	
167	Lewistown.	7. L. Helderberg.	495
178	McVeytown.[13]	"	522

1. *Kensington.* This line runs along the Delaware river over alluvion and modified glacial drift, based upon Azoic rocks, upon which lie the bottom layers of the Cretaceous of New Jersey.

2. *Oakland.* Here the line finally leaves the Azoic rocks, across a fault, and passes white marble quarries to the Westchester Valley, rocks vertical, and probably identical with those of western Vermont.

3. *Gap.* Beds of quicksand. Wharton's famous nickel mine not far off.

4. *Lemon Place.* From here to Elizabethtown, over the garden of Pennsylvania, the great limestone plain of Lancaster; steep dips; plications and faults innumerable; structure difficult.

5. *Landisville.* Zinc mines recently worked one mile to the east.

6. *Elizabethtown.* Road runs for a mile or two along part of a greenstone trap dike, twenty miles long, extending from the Cornwall iron mines near Lebanon, to the Susquehanna river at Falmouth, and into the trap region of York County. Good place to study the action of the trap rock in metamorphosing the beds of New Red.

7. *Branch Inter.* South edge of the limestones of the Great Valley.

8. *Rockville.* Finest section in the State here. Seven miles thickness of rock, nearly vertical, slightly overturned, so that the upper formations seem to plunge beneath the lower, may here be measured, viz: From the Hudson River slates (Siluro-Cambrian), up to the Coal Measures on the summit of the Third Mountain.

9. *Duncannon.* Here a greenstone trap dike only 4 feet thick, crosses the road and river. It carries iron ore. One mile west, a coal bed is opened in the Pocono Sandstone, the representative of the New River Coal System of Montgomery County in Virginia. Five miles east is a curious notch in the summit of Peter's (Fourth) Mountain, where the Dauphin-Halifax Turnpike crosses its crest. The vertical wall is scored horizontally with *glacial striæ* (?). Notice the terrace which the Catskill makes on the north flank of Peter's Mountain opposite Duncannon; it is the finest exhibition of Catskill terrace erosion in the State. See Notes 77 and 170.

10. *Millerstown.* Clinton fossil ore bed extensively worked here and at Mifflin.

11. *Perrysville.* Best place to study the little coal beds in Hamilton (Lower Devonian) rocks.

12. *Narrows.* Long Narrows. River flows in a narrow synclinal between anticlinals of Medina.

13. *McVeytown.* Good place to study Oriskany glass sand quarries, one mile back of McVeytown on the opposite (north) side of river.

* The altitudes in this chapter are taken from Report N, by Charles Allen, Assistant Geologist, and from other reports of the survey. The datum is high water in the Schuylkill and seven feet have been added to reduce to mean surface of the Ocean.

Ms.	Pennsylvania Railroad. Pennsylvania Div.—Main Line—Con.	Alt.	Ms.	Pennsylvania Railroad. Pennsylvania Div.—Main Line.—Con.	Alt.
188	Newton Hamil'n. 10. Hamilton.	599	308	Derry. 14 b. Barren Mres.	1172
191	Mount Union. 5 b. Clinton.	597	313	Latrobe.[24] 1006 { 14 c. Monongahela	
195	Mapleton.[14] 7. L. Helderberg.	593		{ Riv. Series of C. M.	
203	Huntingdon.[15] 10 b. Hamilton.	632	323	Greensburg. "	1091
210	Petersburg. 6. Salina.	678	328	Penn. .:	974
216	Spruce Creek.[16] 4 a. Trenton L. s.	777	333	Irwin's. "	884
220	Birmingham.[17] 3 a. Calciferous.	866	343	Brinton's. " ,	787
223	Tyrone. 5 b. Clinton.	907	347	Wilkinsburg. 14 b. Barren Mres.	923
227	Tipton.[46] 10. Hamilton.	990	354	Pittsburgh.[25] ' "	745
231	Bell's Mills.[18] "	1060			
237	Altoona. "	1173		**Philadelphia and Erie Division.**	
242	Kittaning Pt.[19] 12. Catskill.	1594	0	Sunbury.[26] 11 b. Chemung.	447
	{ 14 b. Coal Meas-	2161	2	Northumberland. 12 Catskill.	467
249	Gallitzin. { ures of the Alle-		9	Montandon. 6. Salina.	464
	{ gheny Riv. Series. "		13	Milton.[27] "	476
252	Cresson. "	2017	17	Watsontown. "	482
255	Lilly.[20] "	1867	19	Dewart. { 10.Hamilton and	468
262	Wilmore. "	1557	24	Montgomery. { 7. L. Helderberg.	491
265	South Fork.[21] "	1485	28	Muncy.[28] 5 b. Clinton.	520
269	Mineral Point. "	1414	40	Williamsport.[29] 10. Hamilton.	526
274	Conemaugh. "	1225	45	Linden. 11 a. Portage.	543
276	Johnstown. "	1184	52	Jersey Shore.[30] 11 b. Chemung.	595
285	Ninevah. "	1121	57	Pine. "	566
290	New Florence. "	1076	60	Wayne. "	573
295	Bolivar.[22] "	1033	65	Lock Haven.[31] "	559
301	Blairsville Int.[23] "	1113			

14. *Mapleton.* Vertical Oriskany glass sand quarry on the opposite (east) bluff.

15. *Huntingdon.* Plenty of middle Devonian fossils to the south of the town, across the flat. One mile further on, high and picturesque pulpit rocks of Oriskany crown the bluffs on both sides of the river. Best view to be got by crossing the turnpike bridge at Huntingdon and riding a mile towards Petersburg. Fine pulpit rocks stud the crest of Warrior's ridge to the north and far to the north-east.

16. *Spruce Creek.* To the south are the Springfield Furnace mines. To the north-east, up Spruce Creek a dozen miles, are the largest limonite mines of the interior of the State.

17. *Birmingham.* Here Potsdam comes up in the center of the overturned anticlinal.

18. *Bell's Mills.* Blair's mine, between Bell's Mills and Altoona. An open quarry in limonite on Oriskany and Helderberg outcrops; very curious. Unique exposure of *celestine* in the bank of the creek below Bell's Mills.

19. *Kittaning Pt.* Horseshoe Bend, on 1° gradient, cuts off the point of a spur of horizontal Devonian measures, between two ravines; coal mines at the head of each ravine; curious scenery.

20. *Lilly.* Coal mines and coke ovens for miles.

21. *South Fork.* The anticlinal at the Viaduct brings up the Mauch Chunk Red Shale 20 feet above grade, and produces the three-mile loop in the river. A very curious place. Notice the boulders of false bedded Pocono sandstone lying in the bed of the valley below, under the viaduct.

22. *Bolivar.* A vast bed of fire-brick clay half a mile back.

23. *Blairsville Int.* Notice the arch of Pocono and Catskill opposite. On the opposite mountain top lies a small patch of the lowest coal bed of the Allegheny River series. See also note 73.

24. *Latrobe.* Here the Pittsburgh Coal Bed is first met—the lowest bed of the upper productive (Monongahela River) Coal Series. Down the Loyalhanna, left bank, six miles, the hill slope is covered with cubic blocks of sand rock 20 feet high and 100 feet on a side, moved several hundred feet down a gentle slope from their original sites.

25. · *Pittsburgh.* The Pittsburgh Coal Bed is seen mined at the hill tops south of the city, 350 feet above the Monongahela River level. At the south end of the hill behind the city, stands an oil well derrick 70 feet high, 100 feet above the streets. It has been bored to a depth of 2,300 feet, through the Butler Oil Rocks, but yields nothing but a stream of strong brine.

26. *Sunbury.* Fine cliffs opposite, west side of the river. Superb landscape from hill ¼ mile back of station.

27. *Milton.* In the centre of a rolling plain of Salina anticlinals and synclinals crossing the river from east to west, bounded on the west by anticlinal Oneida and Medina Mountains called the "Buffalo," "Seven Mountain," "Jacks," etc., around the bases of which run the outcrops of the fossil ore.

28. *Muncy.* Plenty of fossils; fine cliffs of Chemung and Portage facing the river on the east side. Last appearance of Silurian Mountains of Middle Pennsylvania towards the north-east—the end of the Bald Eagle Mountain (5 a. Medina) close along the railroad. Facing the spectator, in the north, appears the wall of the Allegheny Mountain with patches of the lowest coal on the broken forest plateau above.

29. *Williamsport.* Five miles south, through a gap, lies the little secluded Musquito Valley of Siluro-Cambrian limestone, with black marble quarries of Trenton limestone.

Pennsylvania Railroad.				Pennsylvania Railroad.		
Ms.	Philadelphia and Erie Division—*Con.*		Alt.	Ms.	Philadelphia and Erie Division—*Con.*	Alt.
69	Queen's Run.[32]	11 b. Chemung.	564	234	Pittsfield.	11 b. Chemung.	1241
75	Ferney.	"	595	238	Garland.[43]	"	1309
80	Whitham.	"	519	244	Spring Creek.	"	1395
86	Hyner.	"	644	249	Columbus.	"	1407
89	North Point.	"	657	251	Corry.[44]	"	1445
92	Renovo.[33]	"	672	256	Concord.	"	1384
98	Westport.	"	691	262	Union.	"	1270
102	Cook's Run.	"	709	269	Waterford.	"	1192
106	Keating.	"	719	275	Jackson.	"	1227
110	Round Island.	"	755	281	Belle Valley.[45]	11 a. Portage.	1006
117	Sinnemahoning.	"	794	288	Erie.[189]	"	585
120	Driftwood.[34]	12. Catskill.	815				
129	Sterling.	"	914			Sunbury Branch.	
133	Cameron.[35]	"	982	0	Sunbury.[36]	12. Catskill.	451
139	Emporium.[36]	"	1081	11	Danville.[47]	5 b. Clinton.	471
148	Beechwood.	"	1252	20	Catawissa.	Catskill–Chemung.[478]
160	St. Mary's. 1667	{ 14 b. Allegheny Riv. Series of Coal Mres.		54	Conyngham.	"
165	Daguscahonda.[37]	12. Catskill.	1478		Cranberry.	14 b. Anth. Coal Mres.
170	Ridgeway.[38]	11 b. Chemung.	1593		Hazleton.[48]	" " " 1835
178	Wilmarth.	12. Catskill.	1447	36	Nescopec.[49]	10 b. Hamilton.
184	Wilcox.[39]	"	1526	58	Nanticoke.[50]	14 Coal Measures.
189	Sergeant.	"	1716	63	Wilkesbarre. 132	"
193	Kane.[40] 2020	14 a. Pottsville Conglo.		26	Mainville.[51]	Pocono–Catskill. 597
199	Wetmore.	"	1808	35	Mt. Grove.[52]	13 b. Mauch Chunk.
202	Ludlow.	"	1604	37	Rock Glen.[53]	Conglomerate. 929
209	Sheffield.[41]	"	1839	39	Gowen.	14 Coal Mres. 1017
212	Tiona.	13 a. Pocono?	1862	43	Tomhicken.	" 1256
217	Stoneham.	12. Catskill.	1857				
222	Warren.[42] 1195	11 b. Oil Sand Group.					
228	Irvineton.	"	1158				

30. *Jersey Shore.* Gap into secluded Nippenose or Oval Valley (anticlinal Trenton limestone, fossils) four miles south, and across the river in the gap stands a remarkable conical hill.

31. *Lock Haven.* Five miles south gap into Nippenose Valley; limestone; limonite mines; Trenton fossils, etc.

32. *Queen's Run.* Here the road enters the gate of the long gorge of the West Branch Susquehanna, and continues in it 51 miles to Driftwood; the floor of the gorge being sometimes Chemung and sometimes Catskill. Steep walls of Catskill and Pocono rocks, a thousand feet high, hem in the river, with its innumerable bends. Side gorges of the same nature open on both sides. On the hogback mountain tops between, covered with broken rocks and forest, lie patches of coal measures. The strata gently rise and fall in successive undulations, crossing the river at right angles. Old iron furnace of cut stone at Farrandsville. Total failure to work sub-conglomerate carbonate iron ore. Similar failure in same ore at head of Tangascowtac Creek, opposite, to the west.

33. *Renovo.* Good hotel; machine shops of the company; coal mines on the top of the mountain, back of the town.

34. *Driftwood.* Low grade road to the great Jefferson county coal field, up Bennett's Branch.

35. *Cameron.* Coal mines on top of the mountain.

36. *Emporium.* Valley of erosion in Chemung rocks straight north into New York State. From here, the road (and river) rises fast, and reaches the general level of the upland at St. Mary's.

37. *Daguscahonda.* The lowest coal beds are mined all about here, and south of Daguscahonda. The road descends rapidly into the winding gorge or trench of the Clarion River to Ridgeway.

38. *Ridgeway.* Down the Clarion are coal mines and salt and oil borings (no oil).

39. *Wilcox.* Deep gas wells (no oil). The Bishop Summit coal mines, 10 miles to the northeast; Johnson's Run coal basin to the east.

40. *Kane.* Summit of the country. Lowest coal bed. Road northeast, through forest, 15 miles, to Alton coal mines; thence railroad down Tuniangwant to the Bradford oil wells.

41. *Sheffield.* Here the Olean conglomerate may be well studied in connection with the lowest coal bed.

42. *Warren.* Capital centre point for the geological student. Fossils in the hills around. Fine cliffs of Olean conglomerate crown the hill tops. Butler-Venango oil sands crop out in the foot-hills. Oil wells sunk in the valley bottom reach Warren oil sand group at 500 to 600 feet. Railroads down the river; and across to Titusville. Good hill-roads to Pleasantville and Oil City, along the great original oil belt.

43. *Garland.* Olean conglomerate quarries on the peak of the hill, one mile northwest. Top of oil sand crops out in the valley bed.

44. *Corry.* Oil refineries; very high land.

45. *Belle Valley* descends rapidly through a ravine, in Chemung and Portage rocks, to the lake shore.

Ms.	Pennsylvania Railroad—*Continued.* Columbia Branch.	Alt.
0	Lancaster. { 2 – 4. Siluro-Cambrian Limesto's.	359
7	Mountville. "	404
12	Columbia.54 "	251
16	Marietta. "	260
23	Bainbridge.55 "	271
27	Falmouth. 16. Triassic.	
30	Highspire. "	300
33	Baldwin. 2-4. Siluro-Cambrian.	
37	Harrisburg. 4 b. Utica Slate.	320

Pomeroy and Newark Railroad.

Ms.		Alt.
0	Pomeroy. { 2 – 4. Siluro-Cambrian.	453
3	Newlin. 1. Azoic.	
6	Doe Run. "	374
12	Chatham. " Serpentine.	
15	Avondale.56 "	282
18	Landenberg. "	
22	Thompson. "	
38	Delaware City. Del.	16

Frederick Division.

Ms.		Alt.
0	Columbia.54 2-4. Siluro-Camb.	251
5	Stoner. '	
14	York.57 "	365
19	Graybill. "	426
25	Minges Mill. "	435
32	Hanover. "	599
39	Littlestown. "	619
47	Taneytown, Md. "	493
70	Frederick, " 4. a. Trenton.	280

Ms.	Pennsylvania Railroad—*Continued.* East Brandywine and Waynesboro.	Alt.
0	Downingtown. 4 a. Trenton.	256
6	Brooklyn. 1. Azoic.	331
12	Barneston. "	456
18	Honeybrook. "	
22	Beartown. "	
28	New Holland. "	

Williamsburg Branch.

Ms.		Alt.
0	Williamsburg.58 4 a. Trenton.	847
6	Reese's. 10. Hamilton.	903
11	Frankstown.59 "	915
14	Hollidaysburg. 5 b. Clinton.	942

Ebensburg and Cresson Branch.

Ms.		Alt.
0	Cresson. { 14 b. Coal Mrs.	2028
	Allegheny Riv. Ser.	
6	Kaylor's. "	
11	Ebensburg. "	2022

Bedford Division.
(See Huntingdon and Broad Top Railroad.)

Ms.		Alt.
0	Mount Dallas.60 5 b. Clinton.	1058
8	Bedford.61 1062 7. Lower Helderberg.	
13	Napier. 5 b. Clinton.	1108
18	Sulphur Springs. "	
22	Bard's. 10. Hamilton.	
31	Hyndman.62 7. Low. Held.	930
36	Cook's Mills. "	774
39	State Line, Md. "	728
41	Mt. Savage, Jn." "	837
45	Cumberland, " "	638

46. *Tipton.* Branch railroad to mines recently opened in Pocono coal measures. Very important geological locality.

47. *Danville.* Famous and extensive fossil ore (Clinton) iron mines, sunk deep. Iron works here and at Bloomsburg. Ore crops along both sides of mountain ridge for 15 miles. May be studied on the anticlinal arch in the gaps at both places. Medina arch in the gap through Montour's Ridge. Fine cliffs of Portage and Chemung along the river. Fine collecting ground for fossils at the limestone quarries.

48. *Hazleton.* Mammoth and other anthracite beds mined extensively along this road; remarkable open cut mines.

49. *Nescopec.* Fine gap through the Nescopec mountain to the south.

50. *Nanticoke.* A remarkable mining accident occurred in the vicinity of Nanticoke, December 18, 1885. The roof of a coal mine which was only three feet thick, but which was overlaid by 257 feet of glacial drift, caved in. The glacial gravel filled the mine and entrapped 26 miners. Exposure of red beds of No. XI, 500 feet thick on south side of river extending from Nanticoke gap to Shickshinny. The mountain on the north side of the river is made of No. X. No. XII caps the mountain on the south side of the river. The thickening of the red shale between Pittston and Nanticoke is gradual. See Note 122.

51. *Mainville.* Fine gap and section of Upper Devonian and Lower Carboniferous rocks here.

52. *Mt. Grove.* Pass the isolated synclinal McCauley's mountain and coal basin between here and next station.

53. *Rock Glen.* Enter here the northern basin of the Eastern Middle Anthracite coal field. Fine views down upon the red shale. Cunningham valley northward.

54. *Columbia.* Five miles back toward Lancaster, famous limonite iron mines. Road runs up the east bank of the river, six miles, under cliffs, to Chicques. Chicques rock, 800 feet high, Potsdam. Geology still obscure and very interesting.

55. *Bainbridge.* One mile after passing this, enter Trias (dipping N. W.) and continue on it to Highspire.

56. *Avondale.* Serpentine belt crossed here, and before reaching here.

57. *York.* This road follows the York county belt of the Cadorus (S.-C.) limestones, with the south-east edge of the Trias, not far off on the right, and the north-west edge of the Azoic country on the left. Pigeon Hills (Azoic or perhaps Potsdam?) to the right before reaching Hanover. Trap dikes just west of Hanover, and at Littlestown.

58. *Williamsburg.* The great Springfield furnace limonite mines are (by Mine Railroad) five miles to the south.

59. *Frankstown.* Old and extensive Clinton (fossil) ore mines here.

Pennsylvania Railroad—Continued. Bald Eagle Valley Division.			Alt.
Ms.			
0	Tyrone.	5 b. Clinton.	907
5	Bald Eagle.⁶³	10. Hamilton.	1058
10	Hannah.	"	1057
14	Port Mathilde.	"	1007
21	Julian.	"	851
26	Unionville.	"	782
29	Snow Shoe Junc.	"	722
31	Milesburg.⁶⁴	"	700
34	Curtin.	"	
40	Howard.	"	679
44	Eagleville.	"	635
51	Mill Hall.	"	573
55	Lock Haven.	"	555

31	Milesburg.⁶⁴	"	700
33	Bellefonte.⁶⁵	4 a. Trenton.	744

Tyrone and Clearfield Division.			
0	Tyrone.	5 b. Clinton.	907
6	Vanscoyoc.	12. Catskill.	1427
13	Summit.⁶⁶ 2043	14 a. Pottsville Conglo.	
19	Osceola.⁶⁷	14 b. Coal Mrs.	1488
24	Phillipsburg.	"	1425
29	Wallaceton.	"	1727
34	Woodland.	"	1472
41	Clearfield.	"	1103
47	Curwinsville.	"	1141

Pennsylvania Railroad—Continued. Phillipsburg and Moshannon Branch.			Alt.
Ms.			
0	Morrisdale.	14 b. Coal Measures.	
8	Osceola.⁶⁷	"	1468
13	Sterling.	"	
17	Ramey.	"	

Hollidaysburg and Morrison's Cove Branch.			
0	Altoona.	10. Hamilton.	
4	Canaan.	"	
8	Hollidaysburg.	5 b. Clinton.	942
11	Reservoir.	"	987
17	Roaring Spr's⁶⁸	4 a. Trenton.	1196
22	Martinsburg.	"	1366
28	Henrietta.⁶⁹	"	1409

Southwest Pennsylvania Branch.			
0	Fairchance	14 c. U. Coal Mres.	
2	Oliphant.	"	
7	Uniontown.	"	983
11	Lamont Furn.⁷⁰	"	1023
16	Dunbar.⁷¹	"	995
20	Connellsville.⁷²	14 b. Barren Mrs.	915
24	Pennville.	"	1054
	Tarr's.	"	1099
39	Youngwood.	"	957
45	Greensburg.	14 c. U. Coal Mrs.	1091

60. *Mt. Dallas.* Extensive fossil ore mines at Everett, east of Mount Dallas; and in the gap of the mountain approaching Bedford.

61. *Bedford.* Mineral waters. Abundance of Helderberg and Oriskany fossils; interesting and varied geology; iron mines around. Dunning mountain, fossil iron ore mines, north-east.

62. *Hyndman.* At north end of, but outside of the Cumberland coal basin.

63. *Bald Eagle.* This and the following stations are at old iron furnaces, not able to use their fossil ore close by, and therefore hauling Sil.-Cambrian limonites from the Warrior Mark Valley, over the Bald Eagle mountain.

64. *Milesburg.* Entrance gap to the Nittany Limestone Valley, which is full of iron ore banks.

65. *Bellefonte.* Trenton fossils abundant here. To the south-east, seven miles, Nittany Mountain, in the centre of the valley; fine views; curious geology; synclinal ships-keel mountain; turnpike road. Fine section of limestone beds on the great anticlinal of Nittany Valley.

66. *Summit.* Summit of Allegheny Mountain and east edge of the bituminous coal fields. Here Powell's semi-bituminous coal mines.

67. *Osceola.* Many coal mines along the Moshannon and below this in the 1st sub-division of First Basin. Road gets into 2d sub-division over a low anticlinal. All the mines along this road are on beds of the Allegheny River series.

68. *Roaring Springs.* Here enter Morrison's Cove by a gap in the nearly vertical Medina and Oneida rocks of Dunning's Ridge. Fossil ore outside (W.); Bloomfield limonite mine (very famous) inside (E.) U. S. cannon made at Pittsburgh from pig metal from the furnace in the gap. Sinking springs up the run.

69. *Henrietta.* Old limonite mines (very rich), Schoenberger's. A few miles further on are the large, recent, and curious Leathercracker Cove limonite mines of the Cambria Company. Remarkable faults.

70. *Lamont Furnace.* Important outcrop of the iron ore beds underlying the Pittsburgh Coal bed.

71. *Dunbar.* Mauch Chunk red shale iron ore beds in the ravines of the mountain.

72. *Connellsville.* Centre of the coke region. Miles of coke ovens along the road from here toward Greensburg and toward Mount Pleasant. (See Coke Report, L. 1877, Second Geological Survey of Pa.) Pittsburgh bed 12 feet thick in this narrow basin.

73. *Blairsville Int.* Occupies the same position on the Kiskaminitas that Connellsville (72) does on the Youghioghany, in the center of the narrow first gas coal basin west of Chestnut ridge. Pittsburgh coal bed on the hills opposite, south side river. See also Note 23.

74. *Saltsburg.* Two miles further the Pittsburgh bed occupies the central hills of the third gas coal basin. Old salt wells along the river bringing up brine from the Pocono sandstone.

75. *Leechburg.* Famous gas well 1,250 feet deep, on south side of river. Gas from first (?) oil sand (of Butler and Venango) brought across the river on bridge, to rolling mill. Gas furnaces for puddling iron here first successfully used. See Report L. Geological Survey. Some miles to the south are the famous Murraysville gas wells.

76. *Tarentum.* Group of great gas wells; gas piped to Pittsburgh.

77. *Millersburg.* End of the long trap dike is just back of this. See Notes 9 and 170.

78. *Allegheny City.* Remark the typical Eddy Hill in the centre of plain, on which the Observatory stands.

Pennsylvania Railroad—*Continued*.
Western Pennsylvania Division.

Ms.		Alt.
0	Blairsville Int.[73] 14 b. L. Coal Mrs.[1114]	
8	Livermore.[74] 14 b. Barren Mrs.	945
17	Saltsburg.[74] "	891
24	Roaring Run. "	830
32	Leechburg.[75] 14 b. L. Coal Mrs.	
37	Allegheny Junc. "	785
38	Freeport. "	772
45	Tarentum.[76] "	757
51	Springdale. 14 b. Barren Mrs.	749
57	Montrose. " "	
62	Sharpsburg.[102] " "	739
67	Allegh'y City.[78] " "	743
0	Butler.[79] 14 b. L. Coal Mrs.[1009]	
10	Delano. " "	1283
21	Butler Junction. " "	768

Lewistown Branch.

1	Lewistown.	7. Lower Heldbrg.	699
6	Mann's.[80]	4 a. Trenton.	573
13	Milroy.	4 and 3 a. Calcif.	748

Indiana Branch.

0	Blairsville Int.[33]	14 b. L. Coal Mrs.[1113]	
3	Blairsville.	14 c. U. Coal Mrs.	1011
13	Homer.	14 b. Barren Mrs.	
19	Indiana.[81]	"	1311

Lewistown Division.

0	Sunbury.[26]	12. Catskill.	444
5	Selinsgrove.	10. Hamilton.	
17	Middleburg.	5. b. Clinton.	
25	Beavertown.	"	
50	Lewiston.	7. L. Helderberg.	498

{ Allison ore So. of R. R. }

Pennsylvania Railroad—*Continued*.
Lewisburg and Tyrone Railroad.

Ms.		Alt.
0	Montandon. 5 b. Clinton.	
2	Lewisburg. "	482
11	Mifflinburg. "	585
19	Laurelton.[32] "	607
37	Coburn.[82] 4 a. Trenton.	1028
43	Rising Springs[83] "	
57	Oak Hall.[84] '	
58	Lemont. "	1002

Lewisburg and Tyrone Branch.

0	Scotia.[85]	3 a. Calciferous.	
9	Penn. Furnace.[86]	"	1074
12	Marengo.	"	
18	Warriors Mark.	"	
21	Pennington.	"	
25	L. & T. Junc.[87]	5 a. Oneida.	
26	Tyrone.	5 b. Clinton.	

Bellefonte and Snow Shoe Branch.

0	Bellefonte.[63]	4 a. Trenton.	744
3	Milesburg.[64]	10 a. Marcellus.	722
4	Snow Shoe Int.[68]	"	
6	School Hse. Cross.	12. Catskill.	
22	Snow Shoe City.	14 b. Low. Cl. Mrs.[1572]	

Newry Branch.

0	Newry.	12. Catskill.	
2	Duncansville.	7. L. Helderberg.	990
3	Y Switches.	6. Salina.	
4	Hollidaysburg.	"	953

Springfield Branch.

0	Springfield Junc.	4 c. Hudson Riv.	876
8	Mines.[89]	3 a. Calciferous.	1374

79. *Butler.* To get to the first productive deep oil wells one must go several miles north-east from Butler toward St. Jo., Petrolia, etc. The road descends to the Allegheny River over lower coal measures.

80. *Manns.* In the gap of Jack's Mountain is the spring and former residence of "Logan the Indian." Trenton rocks form cliffs. The Kishacoquillas Valley is shut in east of Milroy by two remarkable "ships keel" (synclinal) mountains of Medina and Oneida. The hull is Oneida, the keel Medina. The valley and its three arms are all surrounded by terraces *of erosion.* Taylor thought it was a terrace of deposit, and that the valley had been a lake. A turnpike drive across the valley from Logan's Gap, north-west, by the old iron mines, and over the Standing Stone mountain, to Greenwood furnace, with its fossil ore mines and fine scenery will repay. A fault cuts the mountain. The Clinton shales are curiously crumpled in the cuttings descending to the furnace.

81. *Indiana.* The barren coal measures cover most of Indiana County; underneath lie the Allegheny River coal series.

82. *Laurelton, Coburn.* Between Laurelton and Coburn the road gets through the Seven Mountains by following the deep tranverse gorge of Penn Creek, crossing the anticlinals, which make the Buffalo Mountains in Union County; the last two being those of Poe Valley and Lick Valley. It issues at Coburn upon the wide limestone valley, full of sink holes and caves, with beds of limonite iron ore. Roundhead (synclinal) splits the east end. Brush Mountain forms the north wall.

83. *Rising Springs.* Egghill to the west, a synclinal knob of Medina left standing in the valley. Notice Long's cave at west end of Brush Mountain, at the opening of Brush Valley. Notice sink hole two miles west of Old Fort, which communicates, under Nittany Mountain, with the great spring one mile west of Pleasant Gap. Curious eddy hill in pleasant gap.

84. *Oak Hall.* Here Nittany Mountain ends, the Hudson River slates swinging round it. Oneida rocks on top; fine view toward Bellefonte, northward, and toward Tyrone, westward. Remarkable uncovered cavern, with more recent cavern under it along Big Hollow, four miles west. (See Report T. 4, p. 422.)

85. *Scotia.* Brown hematite (limonite) iron mines.

86. *Penn. Furnace.* The greatest old brown hematite mine in middle Pennsylvania. Excellent place to study the origin of such deposits. Other mines near the next three stations.

87. *L. and T. Junction.* In the Bald Eagle Gap.

88. *Snow Shoe Int.* Rocks all vertical. Oriskany outcrop continuous from here eastward to Lockhaven; none seen westward toward Tyrone.

Pennsylvania Railroad.—Continued.		
Ms.	**Bloomfield Branch.**	**Alt.**
0	Roaring Sprg.[88] 4 a. Trenton.	1196
3	Orehill. 3 a. Calciferous.	

Pittsburgh, Virginia and Charleston Ry.
Now Monongahela Div. P. R. R.

0	Pittsburgh.[25] 14 b. & c. Bar. Mrs.	766
15	McKeesport.[90] "	737
32	Mo'gahela City. 14 c. Upper Cl. Mrs.	748
55	Brownsville. "	767
59	Tippecanoe. 14. Coal Measures.	854
63	Wolf Run. "	895
65	Upp. Middletown "	911
70	Redstone Junc. "	951
77	Uniontown. "	990

Westchester Branch.

0	Philadelphia. 1. Azoic.	32
24	Frazer.[91] "	490
26	Woodland. "	581
28	Greene Hill. "	
29	Fern Hill.[92] "	
31	Westchester.[93] "	420

Schuylkill Division.

0	Philadelphia. 1. Azoic.	60
4	Park. "	165
7	W. Laurel Hill. "	158
8	Manayunk.[139] "	89
9	Shawmont.[94] "	101
13	Conshohock'n[140] 3 a. Calciferous.	68
17	Norristown. 16. Trias.	85
28	Phœnixville.[143] "	131
40	Pottstown.[144] "	140
48	Birdsboro. "	198
58	Reading.[146] 3 a. Calciferous.	209

Pennsylvania Railroad.—Continued.		
Ms.	**Columbia and Port Deposit Branch.**	**Alt.**
0	Columbia.[54] 1 Azoic.	231
3	Washington. "	232
5	Cresswell. "	
11	Safe Harbor.[95] "	198
14	Pequea.[95] "	
10	McCall's Ferry[96] "	169
24	Fishing Creek. "	109
27	Peachbottom. 4 c. Hudson Riv.	99
32	Conowingo. 1 Azoic.	71
35	Octoraro. "	
38	Rock Run. "	
40	Port Deposit, Md. "	9
44	Perryville. "	21

Phila., Germantown & Chestnut Hill Branch.

0	Philadelphia. 1 Azoic.	32
12	Chestnut Hill.[97] "	

Northern Central Railway.

0	Baltimore, Md. (See Maryland.)	
47	Hanover Jun.[98] 2–4. Siluro-Camb.	422
57	York. "	386
67	Conewago.[99] 16. Triassic.	289
73	Goldsboro.[100] "	804
79	Red Bank. "	
84	Bridgeport.[101] 4 a. Trenton.	855
88	Harrisburg. 4 b. Utica.	
91	Marysville. 5 a. Oneida.	850
93	Dauphin.[8] [349] 13 b. Mh. Ck. Red sh.	
99	Clark's Ferry. 12. Catskill.	366
106	Halifax. 12. Catskill.	880
111	Millersburg.[77] { 13 b. Mauch Chunk / Red Shale.	396
118	Mahantango. 12 Catskill.	404
127	Trevorton.[108] "	430
133	Selinsgrove.[104] { 10. Hamilton & 7 / Lewiston limestone.	438
138	Sunbury.[26] { 12. Catskill or / 11 b. Chemung.	444

(Philadelphia and Erie to Williamsport.)

89. *Mines.* One of best and largest brown hematite iron mines in Pennsylvania on the sharp anticlinal axis of Canoe Valley, five miles east of Hollidaysburg.

90. *Port Perry, McKeesport.* Mines in the Pittsburgh coal bed line the river on both sides in a continuous series; the bed descending slowly from 360 feet above water level at Pittsburgh to within 30 or 40 feet in the neighborhood of Monongahela City. The bed rises again and goes into the air, ascending the Youghiogheny River; the banks becoming hillslopes of the Barren measures.

91. *Frazer.* From here to Fern Hill, study the belt of South Valley Hill talcose mica slate.

92. *Fern Hill.* Cross the serpentine belt.

93. *West Chester.* Supposed Laurentian gneiss belt.

94. *Shawmont.* Fine fresh rock cuttings of gneiss all along this part of the line; contortions; steatite quarry.

95. *Safe Harbor, Pequea.* Iron works.

96. *McCall's Ferry.* At Toquan Creek the great anticlinal crosses the river, which runs on north-eastward by Quarryville and Christiania into Chester County, north of the Chester Valley.

97. *Chestnut Hill.* The Valley of the Wissahiccon Creek on the west gives a fine section of the Chestnut Hill sub-division of the gneisses of the Philadelphia Azoic belt.

98. *Hanover Junc.* Magnetic and limonite iron ores from one to five miles west of this and in the ridges to the north and south.

99. *Conewago.* Cliffs of greenstone trap overhang the road and river.

100. *Goldsboro.* More trap cliffs from here to Red Bank. Magnetic iron ore bed above, back from the river.

101. *Bridgeport.* Fine long cuttings through Calciferous limestone opposite Harrisburg.

102. *Sharpsburg.* Iron works here were fired by natural gas brought in a pipe, 40 miles long, from the great gas wells in northern Butler County long before its introduction into general use in or near Pittsburgh.

Ms.	Northern Central Railway.—Con.		Alt.
178	Williamsport.[29]	10. Hamilton.	540
187	Cogan Valley.	12. Catskill.	
192	Trout Run.[105]	"	694
198	Bodine's.	"	
202	Ralston.	14 b. Coal Meas.	560
203	McIntyre.[106]	"	
207	Roaring Run.	12. Catskill.	940
212	Carpenter's.	11 b. Chemung.	
218	Canton.	"	1201
220	Minnequa Sprgs.	"	1261
222	Alba.[107]	12. Catskill.	1230
231	Troy.	"	1340
236	Columbia X R'ds	11 b. Chemung.	1148
241	Snediker's.	"	1148
247	State Line.	"	1106
256	Elmira, N. Y.	"	863

Shamokin Division.

138	Sunbury.[26]	12. Catskill.	442
156	Shamokin.[108]	{ 14 b. Anthracite Coal Measures.	738
164	Mt. Carmel.[109]	"	1054

Summit Branch Railroad.

0	Millersburg.[186]	{ 13 b. Mauch Chunk Red Shale.	397
8	Elizabethville.	"	
14	Lykens.[110]	"	677
17	Dayton.	"	
20	Williamstown.	"	1127

New York, Lake Erie & Western R. R.
Jefferson Branch.

0	Susquehanna.	11 b. Chemung.	914
11	Starrucca.	12. Catskill.	
14	Thompson's.	"	1703
25	Herrick Centre.	"	1803
33	Forest City.	13 a. Pocono.	1481
38	Carbondale.	{ 14b. Anthracite[1079] Coal Measures.	

, N. Y., Lake Erie & Western R. R.—Con

Ms.	Honesdale Branch.		Alt.
0	Lackawaxen.	12. Catskill.	650
4	Rowland's.	"	700
8	Millville.	"	780
12	Kimble's.	"	849
16	Hawley.	"	899
20	White Mills.	"	923
25	Honesdale.[111]	"	966

Tioga Railroad.

0	Corning.	(See C.C. & A.R.R.)	942
15	Lawrenceville.	"	1006
23	Tioga.	11 b. Chemung.	1043
31	Mansfield.	{ 11 b. Chemung Iron ore.	1140
36	Covington.	11 b. Chemung.	1208
41	Blossburg.	{ 14b. Semi-Bitumin's Coal Measures.[1343]	

F. B. C. Co. R. R.

48	Fall Brook.	"	1642
41	Blossburg.	"	1348
45	Morris Run.	"	1678
41	Blossburg.	"	1348
45	Arnot.	"	1682
0	Elmira, N. Y.	11 b. Chemung.	865
10	State Line.	"	1092
12	Millerton.	"	1246
15	Trowbridge.	"	1440
17	Summit.	"	1593
23	Tioga Junction.	"	1021

Bradford Branch.

0	Carrolton, N. Y.	(See Erie Railw'y)	1399
11	Bradford.[112]	11 b. Chemung.	1444
19	Big Shanty.	"	1666
26	Gilesville.	14 b. Coal Mres.	2055
14	Custer City.	Catskill & Chemung.	
27	Kinzua B'dge[113]	{ Carboniferous Con. and 13a. Pocono s.s.	
32	Mt. Jewett.	14. Coal Measures.	
42	Midmont.	"	
53	Johnsonburgh.	13a. Pocono Sandstone.	

103. *Trevorton.* West end of the anthracite coal field. No anthracite west of this. Fine study of the lowest beds in the gap of the Conglomerate mountain.

104. *Selinsgrove.* Easternmost limit of the fossil ore outcrops of the Lewistown belt. Good anticlinal sections of 10. Genesee, Hamilton, Marcellus and 7. Lower Helderberg l. s. between here and Sunbury.

105. *Trout Run.* Entrance to the long gorge of the Lycoming Creek through the Allegheny Mountain plateau; similarly situated to Queens Run (32). Gorge exactly like that of the West Branch Susquehanna (32). Coal patches 1,000 feet above road level, up Trout Run.

106. *McIntyre.* Old iron mines under the cliffs of Pottsville conglomerate forming the cornice of the mountain walls. Great incline plain up mountain to McIntyre coal mines.

107. *Alba.* The Armenia Mountain of Catskill and Pocono dominates this on the west. On its top is the east end of the Blossburg-Antrim semi-bituminous coal basin.

108. *Shamokin.* In the gap opposite the town five ribs of Pottsville conglomerate enclose the four lowest anthracite coal beds. A cross section of the coal measures up to the 12th bed can be made here.

109. *Mt. Carmel.* In the center of the Shamokin group of three anthracite sub-basins.

110. *Lykens.* Here is a range of collieries on the southern outcrop of the famous Lykens Valley anthracite coal bed, which lies 50 or 100 feet above the Mauch Chunk red shale formation No. XI, and is, therefore, worked from the outside conglomerate wall of the Bear Creek coal basin. The bed seems to correspond to the famous block or iron furnace coal bed of Sharon in Mercer County, and of Nelsonville in Ohio. It is the lowest workable bed in the anthracite region.

N. Y., Lake Erie & Western R. R.—Con.

Ms.	Toby Branch.	Alt.
0	Brockwayville.	14 b. Lower Coal Mres.
4	Brockport.	"
6	Hellen Mills.	"
10	Kyler's Corners.	"
12	Dagus Mines.[114]	"

New York, Pennsylvania & Ohio R. R.

Ms.			Alt.
0	Salamanca.	(See New York.)	1393
61	Corry.[44]	Oil Sand Group.	1431
72	Union City.	"	1301
79	Mill Village.	"	1216
88	Cambridge.	"	1163
92	Venango.	"	1163
96	Seagertown.	Sub-Conglomerate	1116
102	Meadville.	"	1080
110	Geneva.	"	1069
116	Evansburg.	14. Conglomerate.	1284
121	Atlantic.	"	
129	Greenville.	Sub-Conglomerate.	984
131	Shenango.	"	936
135	Transfer.	"	993

(Continued in Ohio.)

Franklin Branch.

Ms.			Alt.
0	Meadville.	Sub-Conglomerate	1089
6	Shaw's.	"	1092
11	Cochranton.	"	1064
19	Utica.	"	1035
28	Franklin.[115]	"	987
36	Oil City.	"	1006

Delaware, Lackawanna & Western Railroad.

Ms.			Alt.
0	New York.	(Cont. from N. Jersey.)	
84	Delaware.	4 c. Hudson River.	
92	Water Gap.[116]	5 a. Oneida.	319
96	Stroudsburg.[117]	10. Hamilton.	403
100	Spragueville.	Catskill–Chemung.	490
104	Henryville.	" "	596
109	Oakland.	12. Upp. Catskill.	1011
115	Forks.	"	
122	Tobyhanna.	"	1932
128	Goldsboro.[118]	"	
136	Moscow.	"	1558
139	Dunning's.[119]	"	1400
149	Scranton.	{ 14b. & c. Anthra-[745] cite Coal Measures.	
159	Abington.	12. Catskill.	1058
164	Factoryville.[120]	"	920
174	Nicholson.	"	769
176	Foster.	Catskill–Chemung.	
183	Montrose.	" "	1053
190	New Milford.	" "	1087
196	Great Bend.	11 b. Chemung.	879
210	Binghamton.	(Cont'd in N. Y.)	846

Bloomsburg Division.[121]

Ms.			Alt.
0	Scranton. 743	{ 14 b. and c. Anth'e Coal Measures.	
6	Lackawanna.	" 576	
9	Pittston.[124]	" 576	
12	Wyoming.	" 563	
20	Plymouth.	" 542	
24	Nanticoke.[50]	" 535	
33	Shickshinny.[122]	14a. Pottsville Con.[520]	
41	Beach Haven.	10 b. Hamilton. 530	
47	Briar Creek.	10. Hamilton. 501	
54	Espy.[123]	7. Low. Helderberg. 490	
58	Rupert.	11 b. Chemung. 482	
68	Danville.[47]	5 b. Clinton. 457	
80	Northumberland.	12. Catskill. 452	

(Right margin, rotated:) Over the great Lack'na and Wyo-ming coal basin.

111. *Honesdale.* Head of the Delaware and Hudson Canal supplied with Carbondale and Scranton anthracite coal of the third great basin by railroads coming out of the basin over the Wyoming mountains.

112. *Bradford.* Petroleum was first found in the Bradford (Chemung) black oil sand in 1871. The area of productive oil territory in the Bradford district up to January, 1885, was 121 square miles, and during 14 years had produced on an average 820,000 barrels of crude oil per square mile (C. A. Ashburner). The most productive oil region in the State, and, until the discovery of oil at Smethport and Kane, the lowest of the Pennsylvania oil horizons, 1,775 feet below the Olean conglomerate. (J. P. L.)

113. *Kinzua Bridge.* Highest bridge structure in the world; 301 feet high, 2,052 feet long; contains 3,500,000 pounds iron; cost $275,000.

114. *Dagus Mines.* Extensive workings in the Lower Kittaning coal bed by the New York, Lake Erie and Western R. R. Co.

115. *Franklin.* Lubricating oil from the first sand. At Stoneboro and Mercer, on the road to Newcastle, local glacial moraines are reported by Prof. T. C. Chamberlin of the U. S. Survey.

116. *Water Gap.* Celebrated for its scenery. Large hotels. Indian staircase in the gap made by massive north dipping outcrops of Medina and Oneida. One mile before reaching these rocks are quarries of Hudson River roofing slate on both sides of the Delaware River. Best headquarters for studying the great Terminal Glacial Moraine, which crosses the river at Belvedere and the mountain at Fox Gap, and runs past Lake Poponoming, northward, to the top of Penobscot Knob and so west by Long Pond to the Lehigh. See descriptions, pictures and maps in Report Z, Geological Survey.

117. *Stroudsburg.* Excellent geological headquarters. Fine exposures of Oriskany, Waterline, etc., etc., in the ravine of Broadhead's Creek between the gap and Stroudsburg. Fossils abundant around Stroudsburg. Buttermilk and other cascades to the right of the road (east). Noble carriage drive and exquisite scenery, for 30 miles from Stroudsburg to Milford. Lake on top of the Blue (Kittatinny) Mountain, 10 miles east of S. Fine drive south-west through Red Valley (Clinton) and over outcrops of Helderberg to the Wind Gap. Ascent of the Pocono Knob (Catskill) to the north-west.

Ms.	Lehigh Valley Railroad.	Alt.
0	Perth Amboy.	(See New Jersey.)
61	Easton.[125]	3 a. Calciferous. 210
73	Bethlehem.[126]	" 235
88	Allentown.	" 254
81	Catasauqua.[127]	4 a. Trenton. 282
87	Laury's.	4c. Hudson Riv. Sh.[329]
94	Slatington.[128]	" 365
103	Lehighton.[129]	11 b. Chemung. 465
107	Mauch Chunk[130]	13b. M'ch Ch'k r.s.[544]
114	Penn Haven.	" 705
120	Drake's Creek.	12. Catskill.
130	Tannery.	"
132	Whitehaven.	13 b. Mauch Ch'k.[1143]
142	Summit Siding.	13 a. Pocono. 1723
146	Fair View.[131]	" 1673
152	Newport. [1023]	13b. Mc'h Ch'k r.s. }
		14 a. Potts. Cong. } Wyoming Valley.
158	Sugar Notch. [566]	14b. An. Cl. Mres.
162	Wilkesbarre.[132]	" 549
168	Fort Blanchard.	"
	Pa. & N. Y. R. R.	
170	Pittston.	" 571
172	L. & B. Junction.	" 569
183	Falls.[133]	12. Catskill. 567
186	McKunes.[134]	" 597
194	Tunkhannock.	" 610
199	Vosburg.	" 615
206	Mehoopany.	" 624
209	Meshoppen.	" 643
217	Laceyville.	Catskill-Chemung. 657
227	Wyalusing.	" 674
233	Frenchtown.	11 b. Chemung. 659
237	Rummerfield.	" 696

Ms.	Pa. & N. Y. R. R.—Continued.	Alt.
244	Wysauking.[135]	11 b. Chemung. 718
248	Towanda.[136]	" 737
255	Ulster.	" 742
259	Milan.	"
263	Athens.	" 779
265	Sayre.	" 774
268	Waverly, N. Y.	" 830

Mahanoy, Hazelton & Beaver Meadow Branches.

Ms.		Alt.
0	Penn Haven Jc.	13b. M'ch Ch'k r. s.[705]
4	Black Creek Jc.	" 1015
5	Weatherly.	" 1090
11	Beaver Meadow.	14b. An. Cl. Mres.[1355]
15	Audenreid.	" 1733
10	Lumber Yard.	"
14	Jeddo.	"
16	Ebervale.	"
16	Freeland.	Carbonif. Conglom.
15	Hazelton.[48]	14 b. Anth. Cl. Mres.
23	Tomhicken.	"
18	Quakake Junct.	13 b. Mauch Ch'k.[1315]
22	Delano.	14b. An. Cl. Mres.[1565]
27	Mahanoy City.	" 1230
30	Shenandoah.[137]	"
35	Girardville.	"
38	Ashland.	" 856
36	Raven Run.	"
40	Centralia.	• " 1484
45	Mt. Carmel.[109]	" 1058
59	Shamokin.[108]	" 730

118. *Goldsboro.* Head waters of Lehigh, on the extreme highland, "shades of death," "beach woods," a plate of Pocono rocks covered here and there by synclinal outstretches of Mauch Chunk red shale.

119. *Dunnings.* Commence descent into third anthracite coal field by a ravine through the Pottsville conglomerate. Under it the iron ore of XI has been opened.

120. *Factoryville.* Now over the Elk Mountain synclinal range of Pocono in the first bituminous coal basin; but no coal.

121. *Scranton to Pittston.* Terraces and drift hills along railroad, also glacial striae at Pittston and Taylorville.

122. *Shickshinny.* River cuts across the coal field, leaving a small ridge of coal measures isolated on the west side. Here all the measures from No. X to No. XIII, inclusive, can be seen from the station. The Susquehanna's course through the synclinal at right angles to its axis is interesting here. See Note 50.

123. *Espy.* Square across to the north, six miles, is seen the high end of the Shickshinny (Pocono) Mountain, reached by a good road from Bloomsburg, seven miles, and affording one of the finest panoramic views in Pennsylvania. The glacial moraine crosses that mountain from Berwick northward.

124. *Pittston.* In the gap north of the station the red shale beds of No. XI are missing.

125. *Easton.* Famous collecting ground for rare minerals. Azoic ridge to the north, with serpentine belt. Remarkable outcrops, natural and artificial, of the calciferous limestones along the river north bank to Bethlehem. Many iron works. Laurentian rocks south of the river all the way up.

126. *Bethlehem.* Zinc works. Zinc mine in Saucon Valley to the south, easily reached by N. P. Railroad.

127. *Catasauqua.* Perhaps the best limonite open mine in America for study, lies four miles west (Ironton). Best reached on wheels; also by rail, over a long, high iron bridge. Manganese, kaolin, lignite, with the ore. Mine very large and old.

128. *Slatington.* Extensive roofing slate quarries here where the roofing slate belt from the Delaware river crosses the Lehigh river on its course west into Berks County. Note the duplication of the slate bands by anticlinals and synclinals, as described in Report D. 3, Vol. I, Geological Survey. Two miles further enter the Lehigh Water Gap between sloping walls of Oneida and Medina. Issue upon Clinton red shale. Notice a fine Eddy Hill opposite. Behind it is a local moraine,? which a glacier, formerly descending the Lehigh, left across the mouth of the Aquashicola Creek, forcing that stream to excavate a new channel in the solid Medina rocks of the mountain. Two miles farther, at the bend of the river, north bank, the ice has crushed over the slates, polished the surface and loaded it with till. From the Gap Hotel ride to the top of Stone Hill (Oriskany outcrop) for the view through the Gap. Hydraulic lime quarries on the way up.

Ms.	Barclay Railroad.	Alt.
0	Towanda.¹³⁶	11 b. Chemung. 725
7	Greenwood.	12. Catskill. 823
16	Barclay.¹³⁸	14 b. Coal Mres. 1756

State Line and Sullivan Railroad.

Ms.		Alt.
0	Towanda.¹³⁶	11 b. Chemung. 725
4	Monroeton.	" 762
24	Dushore.	12. Catskill. 1593
29	Bernice.	{ 14 b. Loyalsock Coal Measures,semi- Anthracite. 1858

Montrose Railroad.

Ms.		Alt.
0	Montrose.	12. Catskill. 1656
8	Hunter's.	" 1347
14	Springville.	" 1257
22	Lobeck.	"
28	Tunkhannock.	" 811

Ms.	Philadelphia and Readir	
0	Philadelphia.	1. Azoic.
4	Belmont.	"
8	W. Manay'k.¹³⁹	"
14	W. Consho'n.¹⁴⁰	"
17	Bridgeport.¹⁴¹	3 a. Calci
22	Port Kennedy.	2 b. Potsd
24	Valley Forge.¹⁴²	"
28	Phœnixville.¹⁴³	16. Triass
32	Royer's Ford.	"
40	Pottstown.¹⁴⁴	"
45	Douglasville.	"
47	Monocacy.	"
52	Exeter.¹⁴⁵	"
58	Reading.¹⁴⁶	3 a. Calci:
66	Leesport.	4 b. Utica
70	Shoemakersville.	4c. Huds'i
75	Hamburg.	'
78	Pt. Clinton.¹⁴⁷	5 b. Clint
83	Auburn.¹⁴⁸	7. Low. H
86	Landingville.	11 b. Che
93	Pottsville¹⁴⁹ ⁶¹⁴	14 b. & c. .

129. *Lehighton.* On the crest of one of the grandest anticlinals in the State. Th₁ dipping Chemung and Hamilton here turn over and descend vertically. From here to the vertical Devonian and Bernician systems are crossed at right angles, so as to give a of 10,000 feet, up to the coal measures.

130. *Mauch Chunk.* Fine geological headquarters. The gap in the Second moun whole Pocono and Catskill. The river above gives the Mauch Chunk red shale. ₁ Pottsville conglomerate. Nine miles up the "passenger tourist's gravity road " lie Summit Mine, mammoth coal bed, 60 feet thick, open quarry. In the gap notice the the very earliest anthracite iron furnace once stood. Good specimens of dendrites t the plates in the mountain opposite the hotel. From here to Penn Haven, the fin₁ Lehigh, with its ox bow bend and walls of Catskill rocks. Glacial Moraine at Sand Ru

131. *Fair View.* Ascend 400 feet higher to the summit of Penobscot Knob, afford₁ view in the State. Notice the glacial scratches on the rock on the highest summit From here all the colleries are visible below, and the whole structure of the third anthr can be made out. Down Solomon's Gap by three incline planes, notice the erosion of under the conglomerate rocks.

132. *Wilkesbarre.* Anthracite coal was first mined and used at Wilkesbarre in 17₁ two blacksmiths named Gore. First shipment made to government arsenal at Carlisl₁

133. *Falls.* Buttermilk Falls, not the falls of that name near Stroudsburg, but same rocks, with the hollows filled with gravel.

134. *McKune's.* Enter the long gorge of the North branch of the Susquehann₁ Allegheny mountain plateau, capped (further west) by the Mehoopany coal basin.

135. *Wysauking.* A small but remarkable fault in the 11 b. Chemung rocks ₁ Narrows. It slants up the hillside and may be studied on the R. R. and on the com feet above. The centre line of the Towanda anticlinal crosses the river at the northe cliff, 1,050 feet above the fault.

136. *Towanda.* Fine cliffs, "The Red Rocks," just north of the fault and east fro station. Chemung fossils. Also another cliff directly opposite Towanda on east sid₁ Going north no such precipices are seen, the Chemung shales forming hills with roun Good view of Towanda village from the railroad. Boulders of white limestone from York found in the river were formerly burnt for lime. Picturesque view at Ulster Na₁

137. *Shenandoah.* The greatest overlap in the mammoth coal bed in the Antl occurs in the Shenandoah City colliery. See Atlas of Geological Survey, where it is fu

138 *Barclay.* Barclay or Towanda C. Co.'s, Long Valley and Shraeder Mines on Towanda Mountain, 1,300 feet above the river at Towanda. Incline planes. High f₁ gorges splitting the mountain. Laurel swamps. Semi-bituminous coal.

139. *W. Manayunk.* Beautiful ravine of the Wissahiccon to the east, deeply trenc: belt. Serpentine and soapstone quarries at Lafayette above Manayunk.

140. *W. Conshohocken.* Picturesque vertical trap dyke left standing in the limes quarries east and west of here.

141. *Bridgeport.* On south edge of the Trias country. Bone cavern in limeston Port Kennedy studied by Dr. Leidy and Prof. Cope. Great limestone quarries south c one of which the trias beds are seen lying on the upturned edge of the old limestone b

142. *Valley Forge.* Ditto. The hill back of it is the east end of the ridge of Potsd forming the north wall of the Chester Valley far to the south-west. Under its north the Azoic.

143. *Phœnixville.* In the tunnel here Mr. Wheatley found his coal plants (Tria₁ bones. Two miles south-west runs the edge of the Trias, with breccias, copper veins, Azoic. Trias continues hence to near Reading.

144. *Pottstown.* Trap hills to the north.

Philadelphia & Reading R. R.—*Continued.* Ms. Lehigh and Susquehanna Division.		Alt.
75 Easton.[125]	3 a. Calciferous.	215
86 Bethlehem.[126]	"	235
95 Catasauqua.[127]	4 a. Trenton.	238
109 Lehigh Gap.[128]	11 b. Chemung.	392
120 Mauch Chnk.[129]	13 b. Mch. Chk. r. s.[532]	
127 Penn Haven Ju.	"	703
145 White Haven.	12 Catskill.	1130
158 Penobscot.[131]	"	
171 Ashley. 634	14 b. Anth'e Coal Mres.	
174 Wilkesbarre. 550	"	
183 Pittston. 571	"	
187 Spring Brook.	"	
193 Scranton. 740	"	
195 Green Ridge.	"	

(bracketed to the right of the above rows: "Wyoming & Lackawanna Valleys & Coal field.")

East Penna and Lebanon Valley Branch.		
0 Allentown.[150]	3 a. Calciferous.	431
6 Emaus.	"	434
10 Millerstown.	"	383
15 Shamrock.	"	433
18 Topton.	"	435
25 Fleetwood.	"	449
31 Temple.	"	387
36 Reading.[146]	"	266
45 Wernersville.	"	388
51 Womelsdorf.	"	436
58 Myerstown.	"	474
64 Lebanon.[151]	"	466
69 Annville.	"	442
74 Palmyra.	"	455
81 Hummelston.[152]	"	376
90 Harrisburg.	4 b. Utica Slate.	321

Philadelphia & Reading R. R.—*Continued.* Little Schuylkill, East Mahanoy, Mine Hill and Ms. Mahanoy & Shamokin Branches.		Alt.
0 Herndon.	12 Catskill.	431
14 Trevorton.[768]	14 b. & c. An. Cl. Mres.	
21 Shamokin.[108]	"	738
25 Excelsior.	"	
30 Mount Carmel.	"	
43 Ashland.[153]	"	859
45 Girardville.	"	1021
47 Mahanoy.[154]	"	1343
98 Tamaqua.[155]	"	803
102 Ringgold.[156]	5 b. Clinton.	558

Chester Valley Branch.		
0 Bridgeport.	3 a. Califerous.	76
6 Centreville.	"	202
10 Cedar Hollow.	"	246
16 Exton.	"	324
22 Downington.	"	267

Schuylkill & Susquehanna Branch.		
0 Auburn.[148]	9. Up. Helderberg. 466	
5 Hannon.	10. Hamilton.	
12 Rock.	"	
18 Pine Grove.	11 b. Chemung. 520	
24 Ellwood. 673	13 b. Mauch Chu'k r. s.	
30 Rausch Gap.	"	909
35 Yellow Spring.	"	777
38 Rattling Run.	"	692
46 Forge.	"	435
51 Dauphin.	"	349
54 Rockville.[3] 349	4 c. Hudson Riv. Slate.	
59 Harrisburg.	4 b. Utica Slate.	321

145. *Exeter.* Trap dikes to the south and west, across the river. Remarkable horseshoe ridge of trap to the east. See map of the South Mountains in Report D 3, Vol. II, Part 1, Atlas Geological Survey.

146. *Reading.* The "White Spot" high on the mountain to the east is a remnant of Potsdam sandstone left lying unconformably on Laurentian.

147. *Port Clinton.* A noble fault crosses the river three times in the gap; once at the canal locks, again at the rock at the west mouth of the old tunnel, and then runs vertically up the steep. Hudson River slates dipping 10° south abut against the bottom plate of Oneida standing vertical. Between this and Auburn very fine exposures of Clinton red shales. No fossil ore.

148. *Auburn.* Back of this, on the south side of Summer Hill, multitudes of Hamilton and Chemung fossils.

149. *Pottsville.* Center of the soft anthracite collieries. Fine geological headquarters. For four miles before reaching this place the whole Devonian and Bernician systems stand vertical, affording a section of 20,000 feet of rock up to the top of the lower productive coal series in the fold of the great synclinal in the lower part of the town. View from the top of Sharp Mountain, 800 feet high, instructive. Hotel at Mount Carbon close to where Dr. Isaac Lea found fossil footprints. See Note 169.

150. *Allentown.* Road runs along the base of the Laurentian Mountains over Calciferous limestone holding limonite beds.

151. *Lebanon.* Cornwall Magnetic Iron Mines six miles to the south; holds copper, trap and marble.

152. *Hummelton.* Iron mines, limonite, south of the town.

153. *Ashland.* Remarkable large fossil tree stems visible in the coal measures here. Glacial stria (?) cross white pebbles in the conglomerate crest of mountain west of the Ashland Gap, opposite Mt. Carmel.

154. *Mahanoy.* Large collieries. Shaft sunk by diamond drill.

155. *Tamaqua.* Little Schuylkill here makes a cross section of the Pottsville coal basin. Mr. C. A. Ashburner estimates that the center of the mammoth coal bed basin south of Tamaqua is 1800 feet deep.

156. *Ringgold.* From here down to Port Clinton the Little Schuylkill cuts through ten anticlinals.

157. *Union.* All along here the thinness of the Trias upon the Cambro-Silurian is revealed by erosion.

158. *Ironville.* Famous old and large limonite iron ore mine.

159. *Tremont.* View from the mountain to the southwest of it down the fish tail double red shale valley, split by the great mass of the Pocono rocks, is fine and instructive.

Philadelphia & Reading R. R.—Continued.

Ms. Schuylkill Valley Branch. Alt.

0	Pottsville.[149]	14b.&c.An.Cl.Mres	614
4	Port Carbon.	"	639
7	New Philadelp'a.	"	690
13	Tuscarora.	"	909
18	Tamaqua.[153]	"	802

Pickering Valley Branch.

| 0 | Phœnixville.[145] | 16. Triassic. | 110 |
| 11 | Byers. | 1. Azoic. | 428 |

Reading and Columbia Branch.

0	Reading.[166]	3 a. Calciferous.	265
6	Sinking Springs.	"	348
13	Reinholds.	16. Triassic.	449
16	Union.[157]	"	399
20	Ephrata.	3 a. Calciferous.	384
27	Litiz.	"	375
32	Manheim.	"	402
37	Landisville.[5]	"	404
41	Ironville.[158]	2 b. Potsdam.	
46	Columbia.[54]	3 a. Calciferous.	250

Lancaster and Quarryville Branch.

0	Lancaster Jun.	3 a. Calciferous.	371
8	Lancaster.	"	312
14	West Willow.	"	449
20	New Providence.	1. Azoic.	401
23	Quarryville.	"	488

Lebanon and Tremont Branch.

0	Brookside.	14 b. Anth. Coal Mres.	
13	Tremont.[159]	14 b. Coal Mres.	766
20	Pine Grove.	11 b. Chemung.	520
24	Irving.	10. Hamilton.	499
29	Murray.[160]	"	456
37	Jonestown.	4 c. Hudson River.	422
44	Lebanon.[151]	3 a. Calciferous.	466

Mine Hill and Schuylkill Haven Branch.

0	Schuylkill Hav.	11 b. Chemung.	629
9	Minersville.[161]	14 b. and Cl. Mres.	700
14	Glen Dower.	"	

Philadelphia & Reading R. R.—Continued.

Ms. Catawissa and Williamsport Branch. Alt.

0	Philadelphia.	(See Main Line.)	
78	Port Clinton.[147]	5 b. Clinton.	410
98	Tamaqua[155]	14 b.& c. Cl. Mres.	803
107	Tamanend. [1305]	13 b. Mh. Ck. r.s.& s.s.	
114	Girard.	"	1407
118	Brand'nville.[162]	13 b. Mh.Ck. r. s.	1285
124	Ringtown.	"	1129
132	Beaver Valley.	"	924
136	McAuley.[163]	"	759
139	Mainville.[164]	12 Catskill.	672
146	Catawissa.	Catskill-Chemung.	476
154	Danville[47]	5 b. Clinton.	494
162	Mooresburg.	10 Hamilton.	616
167	Pottsgrove.	"	489
170	Milton.[37]	6 Salina.	465
175	White Deer.	"	476
182	Montgomery.	11 a. Portage.	486
187	Muncy.[28]	5 b. Clinton.	494
190	Hall's. [512]	7 Lower Helderberg.	
195	Montoursville.	10 Hamilton.	524
199	Williamsport.[29]	11 a. Portage.	519

Mill Creek and Mount Carbon Branch.

0	Pottsville.[149]	14 b. An.Cl. Mres.	614
4	Dormer's.	"	647
7	New Castle.	"	876
12	Frackville.	"	1479

Colebrookdale Branch.

0	Pottstown.[144]	16 Triassic.	150
6	Colebrookdale.	1. Azoic.	816
13	Mt. Pleasant.	"	

Philadelphia and Chester Branch.

| 0 | Eddystone. | 1. Azoic. | |
| 4 | Thurlow. | " | |

Chestnut Hill Branch.

| 0 | Philadelphia. | 1. Azoic. | 47 |
| 11 | Chestnut Hill. | " | 410 |

160. *Murray.* Passing out of the gap Hole Mountain stands on the left (east) a curious synclinal outlier of Oneida capping a ridge of Hudson River, proving that no non-conformability exists.

161. *Minersville.* A line of great collieries on the mammoth vein extend westward. The gap of the west branch Schuylkill above Minersville, shows a superb arch of the conglomerate. Back of Mine Hill is the mine which burned for thirty years.

162. *Brandonville.* Making down grade from the conglomerate along the southern and western sides of the red shale valley of the Catawissa Creek crossed by numerous anticlinals from between the Beaver Meadow, Hazleton and Black Creek basins, to the east, and zigzagging the (Pocono) Catawissa Mountain to the west.

163. *McAuley.* A curious little oval mountain basin of anthracite lower coal beds (McCauley) stands out on the red shale plain to the right. Notice the rift in its southern side, and its fortress like outline.

164. *Mainville.* Fine gap through the Nescopic Mountain and section of white Pocono rocks with terraces of Red Catskill on its northern flank.

165. *Gwynedd.* Plants in the Trias as at Phœnixville. Trap ridge pierced by the tunnel.

166. *Coopersburg.* Saucon valley zinc mines.

167. *Steelton.* Bessemer steel works, Pennsylvania Steel Co.

168. *Cornwall.* Cornwall magnetic iron mines located here; this is the largest deposit of iron ore in Pennsylvania.

169. *Pottsville Ju.* The deepest shaft (1575 ft.) in Pa. is located here. The carboniferous conglomerate is boldly and beautifully exposed in the gap south of the town. The dip of the conglomerate is overturned and is toward the south, although the coal beds above the conglomerate lie in the synclinal to the north. See Note 149.

Philadelphia & Reading R. R.—Continued.
Ms. Schuylkill and Lehigh Branch. Alt.

Ms.	Station		Alt.
0	Reading.[166]	3 a. Calciferous.	265
43	Slatington.[128]	4 c.HudsonRiv. s.l.	166

North Pennsylvania and Bound Brook Div.

Ms.	Station		Alt.
0	Philadelphia.	1. Azoic.	25
10	Abington.	"	254
14	Ft. Washington.	16. Triassic.	170
18	Gwynedd.[166]	"	271
22	Landsdale.	"	365
25	Hatfield.	"	311
31	Sellersville.	" and Trap.[331]	
38	Quakertown.	"	496
44	Coopersburg.[166]	"	549
51	Hellertown.	3 a. Calciferous.	276
54	Bethlehem.[126]	"	237

Bound Brook Route.

Ms.	Station		Alt.
0	Philadelphia.	1. Azoic.	25
8	Jenkintown.	"	203
15	Somerton.	"	156
21	Langhorn.	16. Triassic.	96
29	Yardley.	"	79
88	Jersey City.	(See New Jersey.)	

Steelton Branch.

Ms.	Station		Alt.
0	Harrisburg.	4 a. Trenton.	821
3	Steelton.[167]	"	

Germantown and Norristown Branches.

Ms.	Station		Alt.
1	Philadelphia.	1 Azoic.	47
7	Germantown.	"	215
	School Lane.	"	103
	Wissahickon.	"	39
	Schurz.	"	71
	Shawmont.	"	69
	Princeton.	"	62
	Lafayette.	"	53
	Spring Mill.	3 a. Calciferous.	58
	Potts.	"	63
	Magee's.	"	64
	Norristown.	16 Trias.	75

Stony Creek R. R.

Ms.	Station		Alt.
0	Norristown.	16 Trias.	63
10	Lansdale.	"	362

North East Penna. R. R.

Ms.	Station		Alt.
0	Abington Ju.	1 Azoic.	259
	Hillside.	2 b. Potsdam.	
4	Willow Grove.	3 a. Calciferous.	259
	Heaton.	16 Trias.	
7	Hatboro.	"	329
10	Hartsville.	"	262

Philadelphia & Reading R. R.—Continued.
Ms. Cornwall and Mt. Hope R. R. Alt.

Ms.	Station		Alt.
0	Lebanon.[151]	3 a. Calciferous.	
1	Donaghmore.	"	
4	Midway.	"	
5	N. Cornwall.	"	
6	Cornwall.[155]	"	
7	Miners Village.	16 Trias.	
8	Overlook.	"	
9	Penryn.	"	
12	Mt. Hope.	"	

People's Railway.

Ms.	Station		Alt.
0	Pottsville.[149]	14 b. Coal Mres.	614
5	Pottsville Ju.	"	
15	Tremont.[159]	"	

Coudersport and Port Allegheny R. R.

Ms.	Station		Alt.
0	Coudersport.	12 Catskill.	1661
3	Olmstead.	"	
9	Pomery Bridge.	"	
13	Silver Spring.	"	
17	Port Allegheny.	"	1481

Warren and Farnsworth Vy. R. R.

Ms.	Station		Alt.
0	Clarendon.	13 a. Pocono s. s.	1396
3	Underwood's.	"	
6	McCalmont.	"	
8	East Branch.	"	
10	Garfield.	Carbonif. Cong.	

Nanticoke Branch.

Ms.	Station		Alt.
0	Wilkes Barre.[133]	14 Coal Mres.	550
3	Ashley.	"	634
5	Sugar Not^h.	"	659
8	Hanover.	"	654
12	Nanticoke. 50	"	540
13	Wanamie.	"	644

Nescopeo Branch.

Ms.	Station		Alt.
0	White Haven.	13 b. Mauch Ch'k.[1120]	
8	Upper Lehigh.	14 Coal Mres.	1602

Drifton Branch.

Ms.	Station		Alt.
0	Drifton Ju.	13 b. Mauch Ch'k r. s.	
7	Council Ridge.	Carbonif. Conglomert.	
8	Eckley.	14 Coal Mres.	
10	Jeddo.	"	
11	Drifton.[203]	"	

Tamaqua Branch.

Ms.	Station		Alt.
0	Mauch Ch'nk.[140]	13 b. MauchC'k.r.s.[633]	
5	Nesquehoning.	"	501
9	Hanto.	"	1005
10	Lansford.[171]	14 Coal Mres.	
11	Coledale.	"	962
15	Tamaqua.[155]	'	787

170 *Carlisle.* Trap dike 3 miles before reaching Carlisle; visible a long way off as a low mound across the great valley covered with trees, while all around is cultivation. West of Carlisle notice "Wagner's Gap" and "Doubling Gap" in the North or Blue Mountain. They are really not gaps but folds, caused by anticlinals passing through the mountain and elevating the vertical 5 a. Medina strata. The mode in which this was done may be understood by holding up the edge of a sheet of paper in a perpendicular manner and then elevating it in one spot from beneath, which will cause the upper edge to fold in an S shape, similar to these so-called gaps.

Ms	Gettysburg & Harrisburg R. R.	Alt.
0	Carlilse Junct'n.	4 a. Trenton	477
8	Upper Mill.[172]	1. Azoic.	
10	Hunter's Run.	1. Azoic.	
15	Laurel.	3 a. Calciferous.	412
18	Pine Grove,[173]	"	1221
10	Hunter's Run.	1 Azoic.	
15	Starner's.	"	
16	Idaville.	16 Trias.	
17	Gardener's.	"	
19	Bendersville.	"	
22	Sunnyside.	"	
23	Biglersville.	"	
26	Goldenville.	"	
32	Gettysburg.[206]	"	

Perkiomen Railroad.

0	Perkiomen.	16 Triassic.	109
6	Collegeville.	"	155
11	Schwenksville.	"	152
14	Salford.	"	
18	Green Lane.	"	246
22	Hanover.	"	
43	Allentown.[150]	3 a. Calciferous.	257

Wilmington and Northern Railroad.

0	Reading.[146]	3 a. Calciferous.	
9	Birdsboro.	16. Triassic.	173
21	Springfield.[174]	1 Azoic.	645
27	Waynesburg Ju.	"	
36	Brandywine.	"	556
39	Coatesville.	4 a. Trenton.	315
45	Laurel Iron W'ks.	1. Azoic.	341
57	Chadd's Ford.	"	175
72	Wilmington, Del.	(See Del. and Md.)12	

Phila. Wilmington and Baltimore R. R. Central Division.

0	West Philadel'a.	1. Azoic.	14
7	Clifton.	"	109
14	Media.	"	210
18	Linni.	"	136
27	West Chester.	"	406

Philadelphia & Bal[Ms. Phila. Wilming[
0 Philadephia.
14 Lamokin Junc.
20 Rockdale.
25 Concord.
33 Fairville.
40 Avondale.
46 Penn.[175]
52 Oxford.
112 Baltimore.

Phila., Wilmingto[

0 Philadelphia.
2 Gray's Ferry.[176]
13 Chester.[177]
14 Lamokin.
16 Thurlow.
18 Linwood.
20 Claymont.
22 Holly Oak.
23 Belleview.
26 Edge Moor.
28 Wilmington.
(Continued i[

Chester C[

0 Lamokin.
4 Knowlton.
5 Rockdale.
6 Lenni.
7 Wawa.

Peachbotto[

0 Oxford.
20 Dorsey.[178]

Buffalo, New York Western New[

0 Buffalo.
78 State Line.
88 Larrabees.
96 Port Allegany.
107 Keating.
114 Shippen.[8]
121 Emporium.[86]

171. *Lansford.* The Mauch Chunk red shale and Pottsville conglom[between Hanto and Lansford.

172. *Upper Mill.* Passes into the Papertown Gap of the South Mount[(S. W.), up the Mountain Creek Valley, with its range of old and exten[quarries; ore heavily charged with manganese. Ride to the left (E.)over the [ler's mine, and down to the Big bank. Very instructive. Over Strickler'[died with a 30-foot plate of Potsdam(?). In the Papertown gap beginnin[Holly Springs Village are 3,000 feet (horizontal distance) of upturned qua[perhaps to the Huronian system of Canada. These make the Mountai[Reports C and C2.

173. *Pine Grove.* Extensive, well arranged, limonite mine, planned [

174. *Springfield.* Warwick iron mine three miles to the east, on the [copper, etc. Jones' mine 1½ to the north at the east extremity of the C[caster Co. limestone. French Creek copper mines further east than Warw[

175. *Penn.* Line of serpentine to the left. Road runs along the bel[several miles. Great serpentine quarries at Avondale.

176. *Gray's Ferry.* Azoic Rocks here decomposed into kaolin.

177. *Chester.* The road runs on the edge of the Azoic, masked by drift a[

178. *Dorsey.* Roofing slate quarries at Peach Bottom on the Susqueha[able fossil locality, the only one in the southern Azoic belt; apparently sea[of the Hudson River slate formation.

B., N. Y. & P.—Continued.

Ms.	Buffalo and McKean Railroad.		Alt.
0	Larrabees.	11 b. Chemung.	1476
9	Smethport.	"	1493
15	Colegrove.	12. Catskill.	1543
22	Clermont.[179]	14 b. Coal Mres.	2074

Pittsburgh Division.

0	Irvineton.	Oil Sand Group	1168
9	Thompson.	"	1143
15	Tidioute.[180]	"	1113
23	Hickory.	"	1091
30	Tionesta.	"	1060
41	Oleopolis.	"	1032
50	Oil City.	"	1008
54	Rouseville.	"	1037
55	Rynd Farm.	Sub-conglomerate	1043
57	Columbia.	"	1067
58	PetroleumCentre.	"	1089
60	Pioneer.	"	1099
63	Miller Farm.	"	1130
68	Titusville. [181]	"	1194
79	Centreville.	"	1296
86	Spartansburg.	"	1455
95	Corry.[44]	Oil Sand Group.	1433

Oil City and Ridgeway Railroad.

	Oil City.	11 b. Chemung.	1008
	Sidney's.	14 b. Coal Measures.	

Union and Titusville Branch.

0	Titusville.[182]	13 Sub-conglomer.	1194
8	Tryonville.	"	1320
16	Lincolnville.	"	1861
25	Union City.	Oil Sand Group.	1270

New Castle and Franklin Railroad.

0	New Castle[182]	14 a.Conglomerate.	783
9	Wilmington.	"	923
16	Leesburg.	"	1045
22	Mercer[115]	"	1097
30	Garvin's.	"	1327
36	Stoneboro.[115]	"	1171
57	Franklin.[115]	Sub-Conglomer.	1017

B., N. Y. & P.—Concluded.

Ms.	Buffalo Division.		Alt.
0	Olean, N. Y.	11 b. Chemung.	1432
11	Knapp's Creek.	"	
17	Red Rock, Pa.	12 Catskill.	
22	Tarport.	11 b. Chemung.	
23	Bradford.[112]	"	
51	Kinzua.	"	
76	Portville, N. Y.	"	
79	Bullis Mills, Pa.	"	
84	Eldred.	"	1440
0	Eldred.	"	1440
6	Duke Centre.	Chemung and Catskill.	
11	Summit City.	13 a. Pocono.	
16	Sawyer.	11 b. Chemung.	
18	Tarport.	"	
19	Bradford.[112]	"	
7	Larrabees.	"	1478

Dunkirk, Allegheny Valley and Pittsburg Railroad.

0	Dunkirk.	(See New York.)	598
47	Russellsburg.	11 b. Chemung.	1233
55	Warren.[42]	Oil Sand Group.	1200
61	Irvineton.	"	1164
67	Pittsfield.	"	1245
71	Garland.[43]	"	1293
79	Newton.	"	1411
90	Titusville.[181]	Sub-carbonife'us.	1181

Lake Shore & Michigan Southern R. R.

436	Girard.	11 a. Portage.	717
441	Fairview.	"	725
451	Erie.	"	686
459	Harbor Creek.	"	730
466	North East.	"	504

(Continued in Ohio.)

Franklin Division.

36	Jamestown.	Sub-conglomerate.	990
45	Salem.	14 a.Conglomerate.	998
52	Clark.	"	1164
57	Stoneboro.[115]	"	1171
65	Raymilton.	"	1138
71	Summit.	"	1165
78	Franklin.[115]	Sub-conglom'rate	1017
86	Oil City.	"	1010

179. *Clermont.* Coal mines on the highest land at the only practicable north and south pass over the great water shed between the Pennsylvania and New York waters.
180. *Tidioute.* The valley of the Allegheny River is full of derricks from here to Oil City; and the valley of Oil Creek up to Titusville.
181. *Titusville.* Here is the deepest of all oil wells, but unproductive.
182. *New Castle.* Old iron making centre. Banks of the river faced with terraces of Ferriferous limestone supporting large deposits of limonite ("buhr stone") iron ore, of the lower productive coal series.
183. *Kittanning.* Two Kittanning coal beds in the river hills low down; two Freeport coal beds high up. These constitute the chief beds of the Lower Coal Measures.
184. *Red Bank.* Between the mouth of the Mahoning and the mouth of the Redbank, the westermost of the great anticlinals, brings up the conglomerate 100 feet above water level. The anticlinal sinks 500 feet in 40 miles before reaching and crossing the Ohio River 4 miles below Pittsburgh.
185. *Brady's Bend.* Great iron works and iron and coal mines. Wells strike oil here 1,100 feet beneath the river bed in the third oil sand of the Venango oil group.
186. *Parkers.* High cliffs of conglomerate back of the town. A forest of oil well derricks on both river banks and on top of the cliffs. Here the Butler Co. oil belt crosses the river into Clarion County. Oil wells numerous at intervals all the way up to Franklin and Oil City.
187. *Stigo.* Deep old oil wells. Very old iron furnace, centre of a former region of 50 charcoal blast furnaces.

Ms.	Shenango and Allegheny R. R.	Alt.
0	Greenville.	Sub.conglomerate. 961
2	Shenango.	"
6	North Hamburg.	14aConglomerate.1158
12	Cool Spring.	" 1127
17	Mercer.[115]	" 1108
33	Harrisville.	14b.Allegh'yR.Cl.1340
35	Centreville.	"
37	Branchton.	Conglomerate.
38	Bovard.	"
43	Anandale.	"
47	Hilliard.	14 b. Allegheny R. Cl.
37	Branchton.	Conglomerate.
	Coaltown.	14 Coal Measures.
38	Keisters.	"
41	Hallston.	"
46	Euclid.	"
49	Jamisonville.	"
52	Oneida.	"
58	Butler.	"

Allegheny Valley Railroad.

0	Pittsburgh.[25]	14b. BarrenMres. 745
4	Sharpsburg.	" 745
10	Verona.	" 746
17	Parnassus.	" 768
21	Tarentum.	14b.Allegh'y R.Cls.778
29	West Pa. Junct.	" 791
35	Kelly's.	" 780
44	Kittanning.[188]	14b.Lower Cl Mres.810
48	Cowanesha'ock.	" 808
55	Mahoning.	14a.Pottsv.Conglo. 824
64	Red Bank[184]	" 850
68	Brady's Bend[185]	" 856
71	Catfish. 859	14 b. Lower Cl. Mres.
82	Parker's.[186] 889	14a. Pottsville Conglo.
85	Foxburg.	" 897
89	Emlenton.	" 905
106	Scrubgrass.	" 944
115	Foster.	10 Sub-conglomer. 869
123	Franklin.[115]	" 988
132	Oil City.	" 1009
149	Titusville.[181]	"
188	Corry.[44]	Oil Sand Group.

Low Grade Division.

0	Red Bank.[184]	14 b.Coal Mres. 851
15	Leathwood.	" 1027
20	New Bethlehem.	" 1080
40	Brookville.	" 1235
55	Reynoldsville.	" 1377
70	West Summit.	"
77	Pennfield.	"
87	Tyler's.	"
98	Grant.	12. Catskill. 995
110	Driftwood.	" 814

Allegheny Valley Railroad.—Continued.

Ms.	Plum Creek Branch.	Alt.
0	Pittsb'rgh.[25] 745	14 b. & c. Barren Mres.
12	Ink Works.	14 b. Lower Coal Mres.
17	Coal Works.	"

Sligo Branch.

0	Sligo Junction.	14 b. Lower Coal Mres.
10	Sligo.[187]	" 1115

Pittsburgh, Ft. Wayne & Chicago Railway.

0	Pittsburg.[25] 745	14 b. & c. Barren Mres.
13	Sewickley.	" 738
21	Baden. 706	14 b.Lower Coal Mres.
26	Rochester,	" 710
29	New Brighton.	" 750
35	Homewood.[188]	" 949
46	Enon.	" 994

(Continued in Ohio.)

New Brighton and New Castle R. R.

0	Kenwood.	14 Coal Measurers.
2	Fetterman.	"
5	Thompson.	Conglomerate.
9	Rock Point.	"
11	Chenton.	" 900
12	Wampum.	" 801
13	Wampum Ju.	"

Erie and Pittsburgh R. R.

0	Erie.[189]	11 a. Portage. 685
11	Fairview.	" 735
15	Girard.	" 697
20	Crosses.	11 b. Chemung. 785
26	Albion.	" 857
35	Conneautville.	" 1066
39	Summit.	" 1141
43	Linesville.	Sub-conglomerat. 1033
47	Espyville.	" 1088
56	Jamestown.	" 979
63	Greenville.	" 961
71	Clarksville.	" 894
77	Sharon.[190]	" 853
83	Middlesex.	" 833
87	Pulaski.	" 826
94	Harbor Bridge.	" 816
98	New Castle.[182]	14 a. Conglomerat.809
150	Mahonington.	Sub-conglomerate. 789
151	Lawrence Junct.	" 774
154	Moravia.	Conglomerate. 806
156	Newport.	" 813
157	Wampum.	" 801
160	Clinton.	" 900
168	Homewood.	" 950

188. *Homewood.* Immense sandstone cliffs (at the base of the coal measures) wall in the valley of the Beaver. Homewood Furnace. Ferriferous limestone and ore all around.

189. *Erie.* Numerous gas wells used for lighting the city, heating, rolling iron, etc.

190. *Sharon.* The Sharon bed as a "block coal" raw fuel for iron furnaces becomes the great bed of Ohio; it is the lowest workable coal bed; overlies the Olean conglomerate, which is the lowest of the three divisions of the Pottsville conglomerate formation, No. XII. The coal bed is in the hill tops.

Ms.	Ashtabula and Pittsburgh R. R.	Alt
0	Pittsburgh.25	14b.& c.Bar'nMres.745
47	Lawrence Junc.	14a. Potts. Conglo.774
57	Lowell.	826

(Continued in Ohio.)

Cleveland and Pittsburgh Railroad.

0	Pittsburgh.25	14b.&c. Bar'nMres745
26	Rochester.	14b. Lower Cl.Mres710
34	Industry.	" 701
40	Smith'sFer'y.191	" 699

(Continued in Ohio.)

Pittsburgh, Cincinnati and St. Louis Railroad.

0	Pittsb'rgh25 745	14 b. & c. Barren Mres.
8	Mansfield.	14c. Up. Cl. Mres. 776
15	Noblestown.	" 926
23	Bulger.192	" 1156
32	Hanlon's.	" 942

(Continued in Ohio.)

Chartiers Division.

0	Pittsb'rgh.25 745	14c. Upper Coal Mres.
8	Mansfield.	" 773
22	Canonsburg.	" 985
31	Washington.231	" 1051

Baltimore and Ohio Railroad.
Pittsburgh Division.

0	Pittsb'rgh.25 751	14b. & c. Bar. Cl.Mres.
11	Port Perry.90	" 765
15	McKeesport.	" 765
22	Coultersville.766	14c. Upper Coal Mres.
33	West Newton.	" 782
40	Jacob's Cr'k. 797	14 b.& c. Bar. Cl. Mres.
49	Oakdale.	" 849
57	Connellsville.72	" 894
65	Indian Creek.193	12. Catskill. 990
74	Ohio Pyle.194	14 b. Coal Mres. 1287
84	Confluence.195	" 1348
92	Pinkerton.196	" 1649
101	Mineral Pt.197	" 1825
109	Yoder's.	"
116	Sand Patch198	14 a.Pottsv.Congl.2255
126	Glencoe.	12. Catskill. 1623
135	Hyndman.62	10 Hamilton. 941
141	Cook's Mills.	" 774
146	Mt. Savage Jun.	" 687
150	Cumberland, Md.	7.LowerHelderb'g. 688

	Baltimore and Ohio R. R.—Continued.	
Ms.	Wheeling and Pittsburgh Branch.	Alt.
0	Pittsburgh.25	14 b. Barren Mres.
5	Glenwood.	" 760
11	White Hall.	14c.Up. Cl. Mres.1148
19	Gastonville.	" 895
21	Finleysville.	" 914
24	Crouches.	" 958
84	Zediker.	" 1006
38	Washington.199	" 1022
45	Taylorstown.	" 1027
54	W. Alexander.	14 c. Coal Mres. 1161
70	Wheeling, W. Va.	" 629

Somerset and Cambria Branch.

0	Johnstown.	14 b.Low.Cl.Mres.1154
7	Ingleside.	"
9	Border.	"
13	Bethel.	"
19	Hooversville.	14 b. Barren Mres.1669
23	Stoyestown.	14 b. L. Coal Mres.
33	Geiger's.	"
36	Somerset.	14 b. Barren Mres.
38	Roberts.	14 b. L. Coal Mres.
40	Millford.	"
42	Shamrock.	"
45	Rockwood.	Conglomerate.

Fayette County Branch.

0	Connelsville.72	14 c. U. Coal Mres.894
1	Gibson.	14 b. Barren Mres.
2	Fayette.	14 b. L. Coal Mres.921
3	Watts.	" 991
4	Dunbar.71	" 1011
6	Mt. Braddock.	" 1175
12	Lemont.	14 b. Barren Mres.1084
13	Uniontown.	14c. Up. Cl. Mres. 962

Pittsburgh Southern Division.

0	W. Pittsburgh.	14 b. Barren Mres.
3	Banksville.	"
6	Mt. Lebanon.	14 c. U. Coal Mres.
12	Castle Shannon.	"
17	Upper St. Clair.	"
21	Library.	"
25	Finleyville.	"

Mt. Pleasant Branch.

0	Mt. Pleasant.	14 b. Bar'n Mres.1057
1	Stauffer.	" 105?
3	Iron Bridge.	" 1051
4	W. Overton.	14c. U. Coal Mres.1045
5	Everson.	"
7	Tinstman's.	" 1076
9	Morgan.	" 944
10	Broadford.	" 8.-
12	Connellsville.72	" 494

191. *Smith's Ferry.* Numerous old oil wells producing a little from the conglomerate and subconglomerate.

192. *Bulger.* Prof. Stevenson's "Bulger anticlinal" crosses here. The Pittsburgh coal bed dwindles through to a small bed in Ohio, but grows thicker southwestward through Washington county into Greene county, as the new wells testify.

193. *Indian Creek.* Fine gorge of the Youghiogheny through Chestnut Ridge, walls 1,300 feet high. Pulpit rocks of Piedmont sandstone (top member of Pottsville conglomerate) left standing like stranded ships on the broad summit of the mountain. Dry oil wells and old salt wells in the floor of the gorge on the river bank. Cow rock on the southern brow of the gorge covered with the sculptures of the aborigines.

Huntingdon and Broad Top Mountain Railroad.

Ms.		Alt.
0	Huntingdon.[15]	10 b. Hamilton. · [621]
7	Grafton.	10 a. Marcellus.
15	Coffee Run.	10 b. Hamilton. [872]
24	Saxton.[200]	12. Catskill. [849]
31	Hopewell.[201]	13 b. Mch. Ck. r. s. [898]
43	Everett.[202]	10 b. Hamilton. [1118]
53	Bedford.[61]	7. Lo. Helderberg.[1062]

Cumberland Valley Railroad.

Ms.		Alt.
0	Harrisburg.	14 b. Utica Slate. [322]
8	Mechanicsburg.	9. Corniferous. [436]
19	Carlisle.[170]	4 a. Trenton. [477]
30	Newville.	" [533]
41	Shippensb'g.[204]	" [654]
52	Chambers'g.[205]	" [618]
63	Greencastle.	" [585]
74	Hagerstown, Md.	" [572]
94	Martinsburg.	(See Maryland.) [634]

South Penn. Branch.

Ms.		Alt.
0	Chambersb'g.[205]	4 a. Trenton. l. s. [616]
7	Marion.	8 a. Calciferous.
9	So. Penn Junct.	" [632]
15	Williamson.	4 c. Hudson River.
19	Lehmaster's.	8 a. Calciferous.
20	Mercersburg Ju.	4 c. Hudson River.
22	Mercersburg.	4 a. Trenton. l. s.
25	London.	"
28	Richmond.	"

Dillsburg Branch.

Ms.		Alt.
0	Harrisburg.	4 a. Trenton. l. s. [322]
8	Mechanicsburg.	8 a. Calciferous. [427]
9	Dillsburg.	16 Trias. [542]

Hanover Junction, Hanover and Gettysburg Railroad.

Ms.		Alt.
0	Gettysburg.[206]	16. Triassic.
4	Granite.	" Trap dike.
5	Gulden's.	"
10	Oxford.	"
13	Valley.[57]	9. Corniferous.
17	Hanover.	" Trap dike.
20	Smith's,	1. Azoic.
22	Porter's.	"
26	Jefferson.	"
27	Cold Spring.	2—4 Siluro-Cambrian.
28	Strickhauser's.	"
30	Hanover Junc.[98]	"

East Broad Top Railroad.[207]

Ms.		Alt.
0	Mt. Union.[208]	{ 5 a. Medina. [597] 8. Oriskany. 10 a. Marcellus. 10 b. Hamilton.
4	Aughwick. [560]	" { Oriskany Ridge on east. Hamilton on w.
7	Shirley.	10 a. Marcellus.
11	Rockhill.[209]	" [624]
14	Beersville.	{ 11 a. Portage. 11 b. Chemung. [658] 10 a. Marcellus.
18	Three Springs.	{ 8. Oriskany, cut. 7 L. Helderberg l. s. 5 b. Clinton anticlin.
20	Saltillo. [781]	{ 6 Salina & Wat'lime. 7. L. Helderberg l. s. 8. Oriskany. 10 a. Marcellus. 11 b. Chemung gap. 12. Catskill. 13 a. Pocono tunnel. 13 b. Mauch Ck. r. s. 14 a. Pott. con. on top
25	Coles. [1359]	{ 13 b. Mh. Ck. r. s. E. " tunnel. 14 a. & 14 b. on west.
28	Cook's. [1541]	{ 13 b. Mauch Ck. r. s. 14 a. Conglomerate.
31	Robertsdale.[210]	14 b. L. Cl. Series.[1785]

Shade Gap Branch.

Rockhill.[209]	7 L. Helderberg. [624]
Shade Gap.	5 b. Clinton.

Corning, Cowanesque and Antrim R. R.

Ms.		Alt.
0	Corning.	11 b. Chemung. [942]
15	Lawrenceville.	" [1006]
23	Tioga.	" [1052]
39	Wellsboro.	" [1819]
51	Antrim.	{ 14 b. Semi-Bitumi's Coal Mres. [1672]
15	Lawrenceville.	11 b. Chemung. [1006]
27	Elkland.	" [1142]

194. *Ohio Pyle.* Fine Cascade. The whole river falls over a horizontal plate of coal measure sandstone. Wild scenery all around. Coal bed 4 feet thick under the falls.

195. *Confluence.* The Turkey Foot. Junction of the three great branches of the Youghiogheny. Fort Hill, a very remarkable oval hill of coal measures terraced by coal bed outcrops all around as if artificially, several hundred feet high; its flat top, a field from which many Indian skeletons have been ploughed up ever since the first settlement of the country.

196. *Pinkerton.* Fine mountain nose full of coal beds and terraced by sandstone of the barren measures.

197. *Mineral Point.* The fine isolated Pittsburgh coal basin of the Salisbury Ridge, to the south, capped with fossiliferous limestones of the upper coal measures. Romantic falls on Elk Lick Creek not far up from its mouth.

198. *Sand Patch.* Summit of the Allegheny Mountain.

199. *Washington.* Great gas and oil wells recently struck in this neighborhood.

200. *Saxton.* Turn in here to the Broad Top Coal Mines up Shoup's Run. Hotel at Broad Top City, as high as the top of the Allegheny Mountain. Fine scenery. Curious geology.

Corning, Cowenesque & Antrim R. R.—Con.			Delaware & Hudson Canal Co.—Con.		
Ms.	Pine Creek Division.	Alt.	Ms.	Gravity R. R.	Alt.
58 Corning, N. Y.			Carbondale.	14 b. An. Cl. Mres.	1015
93 Stokesville Ju.	12 Catskill.	1170	Head Plane, 1 ⎫	Carboniferous,	1385
97 Matson's.	"		" " 2	Conglonmerate,	1293
101 Ansonia.	"	1134	" " 3	Mauch Chunk,	1494
110 Tiadaghton.	11 b. Chemung.	995	" " 4 ⎬	and Pocono.	1777
118 Blackwells.[211]	12 Catskill.	875	" " 5		1938
123 Cedar Run.	"	802	" " 6		1921
128 Slate Run.	"		" " 7 ⎭		1587
133 Ross.	"		Honesdale.	12 Catskill.	1003
134 Cammal.[212]	"	693	Bangor and Portland Ry.		
136 Miller's.	"		0 Portland.	4 c. Hudson River.	
139 Jersey Mills.	"	655	2 Mt. Bethel.	"	
143 Waterville.[213]	"	624	5 Johnsonville.	"	
146 Ramsey's.	"	605	9 Bangor.	"	
151 Safe Harbor.	"		10 Flicksville.	"	
155 Jersey Shore.[30]	7 L. Helderberg.	595	13 Ackermanville.	"	
157 CementHol'w.[214]	"	567	16 Pen Argyl.	"	
164 Linden.	"	511	19 Miller.	"	
168 Newberry Ju.	"	506	23 Stockertown.	"	
171 Williamsport.[29]	"		24 Tatamy.	"	
			26 Nazareth.	4 a Trenton.	

Addison & Northern Penna. Ry.			Beech Creek, Clearfield and South Western Railroad.		
0 Addison.	11 b. Chemung.	993	0 Philipsburg.	14 b. Bar'n Mres.[1425]	
5 Freeman's.	"		15 Peale.	14 b Low Coal Mres.	
11 Nelson.	"		18 Gorton Heights.	"	
14 Elkland.	"		24 SnowShoe Sum'it.	"	1617
16 Osceola.	"		27 Snow Shoe.	"	
21 Knoxville.	"		31 South Fork.	Conglomerate.	
25 Cowenesque.	12 Catskill.		37 Panther Run.		
27 Westfield.	"		41 Hayes.	Sub-Conglomerate.	
31 Sabinesville.	11 b. Chemung.		46 Monument.	12 Catskill.	
32 Summit.	"		49 Mapes.	11 b Chemung.	
35 Davis.	12 Catskill.		53 Beech Creek.	7 L. Helderberg.	616
41 Gaines.	"		59 Mill Hall.	"	
46 Galeton.	"		62 Lock Haven.[31]	"	576
			66 Wayne.		
Delaware and Hudson Canal Co.			73 Jersey Shore.[30]	"	597
0 Carbondale.	{ 14 b. Anthra. Coal Measures.	1079	76 Larry's Creek.	10 b. Hamilton.	
7 Jermyn.	"	968	81 Linden.		
13 Dickson.	"		85 Newberry Juc.	7 L. Helderberg.	
16 Scranton.	"	739	Newberry.	"	
			89 Williamsport.[29]	"	

201. *Hopewell.* Juniata flows in the red shale under cliffs of conglomerate on one side and a Pocono sandstone (terrace) mountain on the other. Iron works. Fine section up Yellow Creek into Morrison's Cove. Great outcrop of Hamilton limonite.

202. *Everett.* Long outcrop of Clinton fossil ore. Beautiful turnpike carriage drive, south, along the river, and over Wray's Hill, with wonderful sections of contorted Catskill all the way.

203. *Drifton.* The extensive coal mines of Hon. Eckley B. Cox, are clustered around Drifton.

204. *Shippensburg.* Five miles due east is a great spring rising at the south end of the limestone, and foot of the mountain; the head of Yellow Breeches Creek.

205. *Chambersburg.* Back-set of the mountains to the east and cross fault along the turnpike to Gettysburg. A mile or so south of the turnpike immense old limonite ore banks (Pond Bank, etc.) in which kaolin and lignite deposits occur like those of Brandon in Vermont. Five miles further south, in the foot slope of the mountain, are the Mont Alto ore banks. Back of Mont Alto in the mountains are magnetic ore beds, porphory rocks, copper ores.

206. *Gettysburg.* "Round Top," "Cemetery Hill," "Macfarlane's Hill" and "Culp's Hill," forming the ridge on which the Union Army fought the great battle of Gettysburg, July 2d and 3d, 1863, are all trap dikes. Good place to study trap dikes. Scenery beautiful and full of historical interest. (See description of Triassic formation in Report C and C2.)

Ms.	Williamsport & North Branch R. R.	Alt.
	Williamsport.[39]	7 Lower Helderberg.	
0	Halls.	"	512
2	Pennsville.	10 a Marcellus.	
3	Lime Ridge.	7 Lower Helderberg.	
4	Opp's Cross.	"	
6	Hughsville.	10 b. Hamilton.	579
8	Bryan.	11 b. Chemung	
9	Picture Rocks.	12 Catskill.	667
10	Lyon Saw Mill.	11 b. Chemung	
11	Tivoli.	"	
13	Corson.	12 Catskill.	
14	Glen Mawr.	"	
16	Edkins.	"	
17	Strawbridge.	"	
19	Stroups.	"	
20	Muncy Vy.	"	
22	Sonestown.	"	945

Bells Gap R. R.

Ms.		Alt.
0	Bells Mills.[16]	10 a Marcellus.	1050
2	Root's.	11 b. Chemung.	1322
4	Collier Siding.	12 Catskill.	1642
5	Shaw Run.	13 a Pocono.	
6	Look Out.	Conglomerate.	1915
7	RhododendronPk		
8	Lloydsville.	14 b. L. Cl. Mres.	2180
13	Mountaindale.	"	1965
16	Glascow.	"	1772
25	Irvona.	"	

Bradford, Eldred and Cuba and Bradford, Bordell and Kinzua Railroads.

Ms.		
0	Bradford.[112]	11 b Chemung.
	Taylor.	12 Catskill.
9	Kinzua Jc.	13 a Pocono.
	Van Vlicks.	"
	Simpsons.	"
	Ormsbys.	Carbonif. Cong.
	Smethport.	Catskill and Chemung.
24	Eldred.	11 b Chemung.
40	Bolivar.	"
56	Wellsville.	"
0	Cuba.	11 b Chemung.
21	Bolivar.	"
42	Richburg.	14 b L. Coal Mres.

Ms.	Catasauqua and Foglesville R. R.	Alt.
0	Catasauqua.[127]	3 a Calciferous.	282
3	Seiples.	"	465
5	Guth's.	"	491
6	Walbert.	"	550
9	Chapman.	"	541
12	Trexlertown.	"	411
14	Breinigsville.	"	
17	Lichty.	"	
13	Spring Creek.	"	383
15	Alburtis.	"	455
20	Rittenh'se Gp.[215]	Azoic.	940

Cornwall & Lebanon & Colebrook Valley Railroads.

Ms.		
0	Conewago.	16 Trias.	
1	Mt. Vernon.	"	
2	Aberdeen.	"	
3	Beverly.	"	
5	Bellair.	"	
7	Flag.	"	
8	Roseland.	"	
10	Colebrook.	"	
12	Mt. Gretna.	"	
15	Cold Spring.	"	
16	Cornwall.	3 a Calciferous.	608
19	Midway.	"	
22	Lebanon.[151]	"	466

Ligonier Valley Railroad.

Ms.		
0	Latrobe.[24]	14 c. U. Cl. Mres.[1006]
3	Kingston.	14 b. Barren Mres.
11	Ligonier.	14 b. L. Cl. Mres.[1148]

Meadville & Linesville R. R.

Ms.		
0	Meadville.	Oil Sand Group.	
1	Kerrtown.	Sub Conglomerate.	
3	Mercer Pike.	"	
7	Watson Run.	"	
9	West Vernon.	"	
12	Conneaut Lake.	"	1082
15	Harmonsburg.	"	
16	Gehrton.	"	
17	Shermansville.	"	
21	Linesville.	"	1083

207 See Report F. of the second geological survey.

208. *Mt. Union.* Jack's Mountain on the west, 5 a. Medina, with 5 b. Clinton fossil ore on its flanks. Blue Ridge, 5 a. Medina in the distance on the east. End of Chestnut Ridge, southeast from station, composed of Lewiston on 9 Upper Helderberg limestone and 8 Oriskany sandstone.

209. *Rock Hill.* On the east, Blacklog Mountain, 5 a. Medina. Shade Mountain also Medina. Blacklog valley between them, is anticlinal Chazy and Trenton limestone.

210. *Robertsdale.* Coal openings on both sides of the railroad. The two upper seams worked, the lower seam not worked.

211. *Blackwells.* Third Basin crosses about one and a half miles north. Flagstone quarry. The Terminal Moraine crosses this road near the station. A quarter of a mile below the mouth of Babb's Creek. A hill covered with boulders on the west side of Pine Creek, rises 100 feet above the creek. No similar accumulation occurs below this point. The creek flows in a deep gorge between nearly vertical cliffs of Catskill sandstone.   H. C. Lewis.

212. *Cammal.* Second Basin crosses near this station.   A. Hardt, C. E.
213. *Watervelle.* First Basin crosses near here.   A. H.
214. *Cement Hollow.* Cement was produced here years ago.   A. H.

Phila., Newtown & N. Y., R. R.

Ms.		Alt.
0	Philadelphia. 1 Azoic.	
8	Fox Chase. "	190
12	Huntington V'y. "	117
15	County Line. "	
16	Southamton. "	289
18	Churchville. "	184
19	Holland. "	
23	Newtown. 16 Trias.	144

York & Peachbottom R. R.

Ms.		
0	York.	3 c. Calciferous 381
7	Dallastown.	Chlorite Schists. 637
9	Red Lion.	1 Azoic. 900
14	Felton.	" 536
18	Laurel.	" 411
21	Muddy C'k F'ks.	" 366
27	Woodbine.	" 294
40	Peachbottom.	4 c. Hudson Riv.(?)[116]

Harrisburg & Potomac R. R.

Ms.		
0	Shippensb'g.[204]	3 a. Calciferous Lime.
5	Leesburg.	"
7	Jacksonville.	"
9	Haysgrove.	"
11	Doner's.	"
12	Huntzdale.	"
14	Moore's Mill.	"
17	Barnitz.	"
19	Mt.Holly Springs.	"
20	Gt. & Har. Cros'g.	"
24	Boiling Springs.	"
25	Leidigh's.	"
27	Brandtsville.	"
29	Mech. & Dill's Jc.	"
32	Bowmandale.	"

Mont Alto R. R.

Ms.		
0	Waynesboro.	3 a. Calciferous. 1200
1	Price's Church.	"
2	Nunnery.	"
3	Quincy.	"
5	Zion.	"
6	Altodale.	"
7	Intersection.	"
9	Mt. Alto.	" 968
11	Fayetteville.	"
13	Font Hill.	"
14	Woodstock.	" 715
15	Brookside.	"
16	Junction.	" 714
20	Chambersb' g.[205]	4 a. Trenton Lime.

Lehigh & Lackawanna R. R.

Ms.		Alt.
0	Bethlehem.[116]	3 a. Calciferous.
4	Shimer.	"
5	Ritter.	"
7	Broadhead.	"
8	Steuben.	"
10	Clyde.	"
12	Bath.	4 a. Trenton Lime.
15	Chapman.	4 c. Hudson Riv. Slate.
17	Point Phillips.	"
20	Katellen.	"
22	Horn's Springs.	"
25	Wind Gap.	"
27	Pen Argyle.	"
28	Hulls.	"
29	Bangor Junction.	"
30	Bangor.	"

New York, Susquehanna & Western R. R.

Ms.		
98	Del. Wat'r G'p.[116]	5. b. Clinton
102	Stroudsburg.[117]	10 a. Marcellus.
105	Gravel Place.	"

Buffalo, Rochester and Pittsburgh R. R.

Ms.		
107	Bradford Junc.	11 b. Chemung.
120	Limestone.	"
122	Babcock.	"
123	Kendall.	"
124	Bradford.[112]	"
127	Custer City.	"
129	Howard Jc.	"
	Clarion Junction.	Sub-Conglomerate.
	Whistletown.	"
174	Ridgway.	"
182	Carmon.	"
	Short's Mill.	"
189	Forestville.	Conglomerate.
192	Brockwayville.	14 b. L. Coal Mres.
	Lane's Mills.	"
195	Beech Tree Ju.	"
200	Grove Summit.	"
204	Falls Creek.	14 b. Barren Mres.
206	Du Bois.	"
	Carlisle.	"
214	Sykes.	"
	Cramer.	"
	Bells Mills.	"
228	Punxsutawney.	"
229	Clayville.	"
231	Walston.	"

215. *Rittenhouse Gap.* Magnetic iron is mined along the terminus of this road. The ore is used by the Crane and Thomas iron companies.

216. *Sheffield.* The Hague gas well is located one and a half miles east of the town and is one of the most remarkable gas wells in Pa. (See Carll's report on Warren County, I 4.)

217. *Chewton.* Good geological headquarters for studying XIII in hills and XII along wild gorge of Connoquenessing River.   I. C. W.

218. *Youngstown.* In vicinity of Youngstown the Sharon coal which comes near the base of XII may be studied.

219. *Renfrew.* Near this is the celebrated Thorn Creek oil district, which has furnished the largest wells in America, one, the Boyd and Semple putting out 9,000 barrels the first 24 hours.

     I. C. WHITE.

Sharpsville R. R.

Ms.	Station	Alt.
0	Sharpsville.	Sub-conglomerate.
3	Mt. Hickory.	Conglomerate.
4	Hermitage.	"
5	Oakland.	"
6	Summit.	"
7	Neshannock.	"
9	Lackawan'ck Jc.	"
12	Lyle.	"
15	New Wilmington.	"
17	Wilmington Jc.	"

Tionesta Valley R. R.

Ms.	Station	Alt.
0	Sheffield Junct.	13 a. Pocono.
6	Brookston.	"
10	Donaldson.	"
13	Sheffield.[216]	"
19	Garfield.	Carbonif. Conglom.

New York, Pittsburgh & Chicago R. R.

Ms.	Station	Alt.
0	New Galilee.	14 b. Low. Cl. Mres.
3	Darlington.	"
6	Cannelton.	"
9	Negley.	"
12	Mill Rock.	"
14	Rogersville.	"

Pittsburgh & Castle Shannon R. R.

Ms.	Station	Alt.
0	Pittsburgh.[25]	14 b. Barren Mres.
9	Castle Shannon.	14 c. U. Coal Mres.

Pittsburgh & Lake Erie R. R. *

Ms.	Station	Alt.
0	Pittsburgh.[25]	14 b. Barren Mres. [730]
5	Chartiers.	" [726]
6	McKee's Rocks.	14 c. Mahoning s. s.[726]
7	Davis Island.	" [725]
11	Moon Run.	" [718]
12	Montour Jc.	" [718]
13	Middletown. ·	" [722]
14	Lashell.	" [716]
15	Stoop's Ferry.	" [719]
17	Shousetown.	" [761]
18	Shannopin.	14 b. L. Cl. Mres. [777]
19	West Economy.	" [765]
21	Woodlawn.	" [742]
22	Alliquippa.	" [756]
23	Logstown.	" [752]
24	Stobe.	" [752]
25	Kiasola.	" [752]
26	Monaca.	" [751]
27	Phillipsburg.	" [752]
	Beaver.	" [752]
28	Bridgewater.	" [780]
29	Fallston.	" [719]
31	Brighton.	14 a. Conglomer. [722]
32	Beaver Falls.	14 a. Top of XII. [740]
34	College.	Middle of XII. [750]
36	Homewood.[188]	Lower half of XII.[749]
40	Clinton.	" [754]
	Rock Point.	" [754]

Pittsburgh & Lake Erie R. R.—Con.

Ms.	Station	Alt.
43	Wampum.	Lower half of XII. [766]
44	Newport.	Basal portion XII. [772]
46	Moravia.	" [786]
49	New Castle Jc.	Base of XII. [795]
52	New Castle.[182]	"
50	Mahoningtown.	" [800]
54	Edenburg. [793]	13 d. Cuyahoga Shale.
57	Carbon.	" [808]
59	Lowellsville, O.	" [822]
62	Struthers.	" [827]
68	Youngstown. [218]	"

Pittsburgh, McKeesport & Youghiogheny Railroad.

Ms.	Station	Alt.
0	Pittsburgh.[25]	14 b. Barren Mres.[730]
5	Hayes.	"
7	Homestead.	" [755]
8	City Farm.	" [759]
9	Rankin.	" [742]
10	Braddock.	" [735]
	Bessemer.	" [739]
11	Port Perry.[90]	" [734]
12	Saltsburg.	" [748]
13	Demmler.	" [742]
15	McKeesport. [90]	" [754]
19	Boston.	" [742]
22	Greenock.	" [756]
25	Stringtown.	" [756]
28	Scott Haven.	" [762]
33	West Newton.	" [765]
38	Port Royal.	" [780]
40	Jacob's Creek.	" [785]
46	Layton.	" [811]
54	Dickerson Run.	" [853]
56	Broad Ford Jc.	14 c. U. Cl. Mres. [873]
57	Broad Ford.	"
58	New Haven.	" [804]

Montour Railroad.

Ms.	Station	Alt.
0	Montour Junc.	14 b. Barren Mres.[718]
11	Imperial.	"

Pittsburgh, Chartiers & Youghiogheny Railroad.

Ms.	Station	Alt.
0	Pittsburgh.[25]	14 b. Barren Mres.
5	Chartiers.	"
12	Mansfield.	"
15	Bower Hill.	"
20	Beechmont.	14 c. U. Coal Mres.

Pittsburgh & Western R. R.

Ms.	Station	Alt.
0	Allegheny.[78]	14 b. Barren Mres.
3	Bennett.	"
5	Sharpsburg.	"
9	Elfinwild.	14 b. L. Coal Mres.
14	Wildwood.	"
16	Gibsonia.	14 b. Barren Mres.

*By Prof. I. C. White, U. S. Geologist.

Ms. Pittsburgh & Western R. R.—Con.	Alt.	Ms. Pittsburgh & Western R. R.—Con.	Alt.
18 Bakerstown.	14 b. Barren Mres.	101 Lucinda.	14 b. Low. Coal Mres.
20 Valencia.	"	107 Tylersburg.	Conglomerate.
25 Callery Jc.		120 Warrensville.	"
28 Evans City.	14 b. Low. Coal Mres.	135 Sheffield Jc.	"
32 Harmony.	"	153 Kane.	Coal Measures
33 Zelienople.	"	157 Kanesholm.	"
43 North Sewickley.	"	164 Mt. Jewett.	"
45 Wurtemburg.	14 a. Comglom.	**Waynesburg & Washington R. R.**	
51 Chewton.[217]	"		
54 Moravia.	"	0 Waynesburg.	14. c. Greene Co. Group.
57 New Castle Jc.	"	5 Sycamore.	14 c. U. Coal Mres.
60 New Castle.[182]	"	7 Swart.	. "
58 Mahoningtown.		9 Deer Lick.	"
62 Edenburg.	Sub-conglomerate.	11 West Union.	"
67 Lowellville, O.	"	12 Dunn.	"
75 Youngstown.[218]	"	14 Lindley's Mills.	"
25 Callery Jc.	14 b. Barren Mres.	15 Hackney.	"
33 Renfrew.	"	16 Johnson.	"
40 Butler.[79]	14 b. L. Coal Mres.	18 Luellen.	"
48 St. Joe.	"	19 Baker.	"
53 Millerstown.	"	21 McCracken.	"
57 Karns.	"	23 Vankirk.	"
58 Petrolia.	"	26 Braddock.	"
62 Bruin.	"	29 Washington.[199]	"
67 Parker.	Conglomerate.	**Youghiogheny R. R.**	
70 Foxburg.	14 b. L. Coal Mres.		
74 St. Petersburg.	"	Irwins.	14 c. U. Coal Mres.[884]
78 Turkey.	"	Shaft No. 2.	" 993
86 Knox.	"	Chambers.	" 1032
91 Shippenville.	"	McGrew's.	" 981
95 Clarion Jc.	"	Millville.	" 867
100 Clarion.	"	Cowans.	"
98 Arthurs.	"	Marchands.	" 788
		Sewickley.	" 780

Mineral Localities.

The following notes are taken from a list of Mineral Localities sent to the editor by Mr. Joseph Wilcox, of Media, Pennsylvania, one of the Commissioners of the Second Geological Survey.

P. W. & B. R. R. Swarthmore. At Avondale quarries, one mile south, Garnets and Tourmaline; one mile north, Andalusite.

Media. At Blue Hill, two miles north, Green Quartz, Chrysotile. In Upper Providence, Andalusite, Stellate, Antophyllite, Amethyst, Asbestos, Actinolite.

Elwyn. In Middletown, Actinolite, Green Feldspar, Corundum, Chromic Iron, Moonstone, Sunstone.

Bridgewater. Sphene.

Morgan. Amethyst, Corundum.

Rockdale. Amethyst, Asbestos.

Concord. Two miles south, in Green's Creek, Garnet (so-called Pyrope). Garnet mined as a substitute for emery.

Fairville. Mica in large crystals.

Rising Sun Station. Near New Texas in Lancaster Co., Chromic Iron has been largly mined. Brucite, Ripidolite, Picrolite, Emerald, Nickel, Williamsite, Genthite.

Brandywine Summit. Two miles southwest, Kaolin mines. Near Elam, Garnet, Mica, Feldspar.

Moore's. Near Moore's Ferry, Kyanite.

Chester Station. In Leiperville quarries, Garnet, Beryl, Feldspar, Tourmaline, Pink Zoisite, Mica.

Newport. At Brandywine Springs, Fibrolite.

West Chester. Two miles south at Brinton's quarry, Clinochlore, Jefferisite, Oligoclase. Serpentine is largely quarried there.

Wilmington and Northern R. R. Hall's. One mile and a half southwest, Corundum mines, Diaspore, Margarite, Garnet, Feldspar, Tourmaline.

P. R. R. Gap Station, Lancaster Co. Gap mine four miles, Millerite, Siderite, Chalcopyrite, Pyrrolite (niccoliferous.)

This blank space is intended for additional geological notes in pencil by the traveler,

Ohio.*

GEOLOGICAL FORMATIONS FOUND IN OHIO.

Groups.	Ohio Sub-Divisions.	Equivalents in other States.
20. Quaternary.	20 c. Stratified Drift. Terraces, &c., Valley Drift, Kames, Osars, &c. 20 b. Forest Bed (local). 20 a. Boulder Clay, Till., Erie Clay.	
14. Coal Measures and Conglomerate Coals.	14 c. Upper Barren Measures. 14 c. Upper Productive " 14 b. Lower Barren " 14 a. and b. Lower Productive and Conglomerate Coal Measures.	Coal Measures of Pennsylvania, and Conglomerate Coals.
14. Conglomerate (in part).	14 a. Sharon Conglomerate.	Sharon Conglomerate of Pennsylvania.
13. Sub-Carboniferous Limestone.	13 f. Maxville Limestone.	Chester Limestone, Illinois.
13. Waverly.	13 e. Logan Group, Olive Shales, Logan Sandstone, Waverly Conglomerate. 13 d. Cuyahoga Shale. 13 c. Berea (or Waverly) Black Shale. 13 b. Berea Grit. 13 a. Bedford Shale.	Shenango Sandstone in part, Pennsylvania. Marshall Group, Michigan. Crawford Shales, Pa. Orangeville Shale in part, Pennsylvania. Pithole Grit, or Third Mountain Sand, Pennsylvania.
11. Ohio (Black) Shale.	11 c. Cleveland Shale. 11 a. and b. Erie Shale. 10 c. and 11 a. Huron Shale.	Chemung, Portage, and Genesee, of New York.
10. Hamilton.	10 b. Hamilton Shale. Olentangy Shale.	Hamilton Group, New York (in part).
9. Corniferous.	9 b. Delaware Limestone. 9 a. Columbus Limestone.	Marcellus Shale, Corniferous and Onondaga Limestones of New York.
6 & 7. Waterlime.	6 and 7. Waterlime.	Waterlime and L. Helderberg, New York.
6. Salina.	6. Salina Shales & Plaster Beds.	Salina Group, New York.
5. Niagara.	5 h. Hillsboro' Sandstone. 5 g. Cedarville Limestone. 5 f. Springfield Limestone. 5 e. West Union Limestone. 5 d. Niagara Shale. 5 c. Dayton Limestone. 5 b. Clinton Limestone. 5 a. Medina Shale.	Guelph, Canada. Niagara Group, New York. Clinton Group, New York. Medina Sandstone, New York.
4. Hudson River or Cincinnati.	4 c. Lebanon Beds. 4 b. Cincinnati Beds. 4 a. Pt. Pleasant Beds.	Hudson River and Utica Shale of New York.

* In the first edition this chapter was furnished by Dr. J. S. Newberry, the State Geologist at that time. It has been very much enlarged for this edition, the new railroads added, the whole care-

Ms. | Ashtabula and Pittsburg Railroad.

0	L. S. & M. S. R. R.
1	Ashtabula.	11. Erie Shale. 650
8	Austinburg.	"
12	Eagleville.	"
16	Rock Creek.	"
24	Orwell.	" & 13. Waver.
29	Bloomfield.	13 e. Waverly.
34	Bristolville.	"
40	Champion.	"
45	Warren.	13 d. " 862
50	Niles.	14 a. Conglomerate. 911
55	Girard.	{ 13 Wav., 14 a. Congl., 14 b. Coal Meas. 885
60	Youngstown.	14 a.Con. & Cl. Meas. 865
65	Struthers.	14 b. Coal Measures.
68	Lowell.	"

Bellaire, Zanesville and Cincinnati R. R.
Ms. | In driftless region.

0	Bellaire.	{ 14 c. Upper Prod. Meas. Pittsburg Seam, No. 8. 657
12	Bethel.	14 c. Up. Barren Meas.
33	Jerusalem.	"
42	Woodsfield.	"
49	Lewisville.	"
59	Summerfield.	"
77	Caldwell.	"
88	Cumberland.	14 b. Low. Barr. Meas.
110	Zanesville.	{ The Sewickly coal mined near known as Cumberland Seam. 14 b. Low. Prod. Meas., Kittan. Coals, Nos. 5 and 6. 711

Baltimore and Ohio and Chicago Railroad (B. & O. R. R.).

0	Chicago Junc.	
8	Attica.	9. Cornif. & 10. Huron.
16	Republic.	9. Corniferous.
24	Tiffin.	5. Niag. & 7. Held. 758
30	Bascom.	5. Niagara.
37	Fostoria.	"
44	Bloomdale.	5. Niag. & 7. Helderb'g.
50	New Baltimore.	
62	Deshler.	7. Helderberg.
74	Holgate.	
88	Defiance.	10 c. Huron Shale. 700
94	Delaware.	"

Central Ohio Railroad (B. & O. R. R.).

0	Baltimore, Md.	
376	Bellaire.	{ 14 c. C'l Meas. Pittsburg S'm, No. 8. 657
385	Glencoe.	"
395	Belmont.	{ 14 c. Coal Meas. Up. Barren Measures.
403	Barnesville.	{ 14 c. Coal Meas., Sewickly Seam, No. 86.
413	Salesville.	14 c. Coal Measure.
428	Cambridge.	{ 14 c. Coal Meas., Up. Freeport S'm, No. 7.
437	Concord.	"
447	Sonora.	"
454	Zanesville.	{ 14 c. Coal Meas. Kit. S'ms, Nos. 5 & 6. 711
468	Pleasant Valley.	13 c. "
470	Black Hand.	13 e. Waverly.
480	Newark.[1]	" 821
486	Union.	13 d. "
495	Pataskala.	
504	Taylor's.	{ 11 c. Hur. & 13 a. & b. Waverly.
513	Columbus.	{ 9. Cornif., 10. Ham., 11. Ohio Shale. 746

Straitsville, Somerset and Newark R. R.

0	Newark.	13 e. Waverly. 821
9	Avondale.	14 b. Coal Measures.
17	Glenford.	{ 13 s. and c. Limestone and 14 a. Congl. "
27	Wellans.	
38	Bristol.	{ 14 b. Coal Meas., Kittanning Seams, Nos. 5 and 6. 965 "
43	Shawnee.	

fully revised, and about fifty foot-notes appended by Professor Edward Orton, the present State Geologist. Several additional glacial notes are by Rev. G. Frederick Wright, of Oberlin, one of the United States Geologists, who has been engaged under Professor T. C. Chamberlain in making a glacial survey of the terminal moraine through Ohio, Indiana, Kentucky, and Illinois. His notes are signed G. F. W., and all the other notes are by Professor Orton except No. 62.   J. M.

1. Newark. Glacial boundary at Newark. G. F. W.
2. Chicago and Atlantic Railway. Route heavily covered with drift.
3. Marion. Fine exposures of limestone in Marion quarries. Fossils abundant.
4. Lima. Waterlime quarried here. Strong building-stone. Some beds fossiliferous.
5. Winchester. Near margin of glacial drift.
6. Mineral Springs. Springs derived from black shale.
7. Miamisburg. Cedar trees and peat 100 feet beneath glacial deposits at Germantown, three miles southwest from Miamisburg. G. F. W.
8. Amanda. Glacial boundary three miles east of Amanda. G. F. W.
9. Lancaster. On the glacial boundary. Granite boulder two miles northeast, 18 x 11 x 6 feet out of ground. G. F. W.
10. Bremen. Glacial boundary two miles northwest. G. F. W.
11. Cecil. Region heavily covered with drift. Very few outcrops of strata to be found. These mainly in beds of streams.
12. Greenville. At Greenville an interesting outcrop of Guelph division of the Niagara occurs, rich in fossils. A number of new species have been obtained here. The rock is dolomitic, but contains more carbonate of magnesia than carbonate of lime.

Ms. | Chicago and Atlantic Railroad.

0 Marion, Ohio.[2]	9. Corniferous.	970
6 Espyville.	7. Waterlime.	956
7 Moran's.	"	
11 Clifton's.	"	971
16 Hepburn.	"	956
19 Dudley.	"	971
25 Kenton.	"	990
29 Sage.	"	998
33 Oakland.	"	994
35 Scioto.	"	999
38 Preston.	"	999
42 Harrod's.	"	1009
45 Westminster.	"	995
49 Townsend.	"	
52 Lima.[4]	"	899
55 Shawnee.	"	862
58 Kemp.	"	855
61 Conant.	"	845
65 Spencerville.	"	848
72 Yorkville.	"	837
80 Enterprise.	9. Corniferous.	840
84 Glenmoore.	"	835
88 Greenwood.	"	836
92 Rivare, Ind.		847
96 Decatur, Ind.		820

Chicago, St. Louis and Pittsburg R. R.

0 Columbus.	9 Cor., 10. Ham., & 10. Huron.	846
18 Pleasant Valley.	7. Helderberg.	
28 Milford Centre.	"	
38 Cable.	"	
47 Urbana.	7. Held. & 5 g. Niag.[1033]	
58 St. Paris.	5. Niagara.	
73 Piqua.	" & 5 c. Niag.[985]	
83 Bradford Junc.	5. Niagara.	
95 Greenville.	5 g.	1055
108 New Madison.	"	
114 New Paris.	5 f. Niagara.	
0 Bradford Junc.	"	
10 Pikeville.	"	
21 Union.	"	

(Continued in Indiana.)

Cincinnati and Eastern Railway.

0 Cincinnati.[62]	4 b. Cincin. Group. 507
14 Batavia.	"
27 New Richm'd.[62]	"
32 Williamsburg.	4 c. "
40 Mt. Oreb.	"
47 Sardinia.	"
57 Winchester.[5]	" & 5 a. & b. Niagara.
62 Irvington.	4. Cincinnati Group.
75 Mineral Spr'gs.[6]	11. Ohio Shale & 13 a. and b. Waverly.
90 Henley.	13 d. Waverly.
106 Portsmouth.	"

Ms. | Cincinnati, Hamilton & Dayton R. R.

0 Cincinnati.[62]	4 b. Cincinn. Group.[507]
5 Cumminsville.	"
15 Glendale.	"
19 Jones.	"
25 Hamilton.	" 604
37 Middletown.	4 c. "
49 Miamisburg.[7]	4.
60 Dayton.	4 c. & 5 a. b. c. Niag.[754]

Cincinnati, Hamilton and Indianapolis Railroad.

0 Cincinnati.	4 b. Cincin. Group. 507
25 Hamilton.	" 604
32 McGonigle.	"
39 Oxford.	4 c. "
44 College Corn'rs.	"

Cincinnati & Muskingum Valley Railroad.

0 Cincinnati.[62]	4 b. Cincin. Group. 507
36 Morrow.	4 b. & c. " 642
46 Clarksville.	4 c. "
56 Wilmington.	5 b. & c. Ni. & 5 c. Ni.
66 Sabina.	5. Niagara.
77 Washington.	7. Helderberg. 957
87 New Holland.	10 c. Huron Shale.
95 Williamsport.	10 c. Hur. Shale and 9 a. Corniferous.
104 Circleville.	"
116 Amanda.[8]	13. Waverly.
125 Lancaster.[9]	13 e. " 826
130 Bremen.[10]	"
134 New Lexington.	14 b. Coal Meas., Kit. Coals, Nos. 5 & 6.
152 Roseville.	" 711
157 Zanesville.	"
168 Ellis.	737
176 Dresden Junc.	14 b. c. m. Mercer Horiz.

Cincinnati, Richmond & Chicago R. R.

0 Cincinnati.[62]	4 b. Cincin. Group. 507
25 Hamilton.	" 604
36 Collinsville.	"
44 Camden.	4 c. " 839
53 Eaton.	5 d. & e. f. Niagar.[1044]
60 Florence.	
70 Richmond, Ind.	See Indiana.

Cincinnati, Van Wert & Michigan R. R.

0 Cecil.[11]	
7 Paulding.	9. Cornif. & 10. Ham.
19 Van Wert.	9. Corniferous. 788
43 Celina.	5 g. Niagara. 850
76 Greenville.[12]	" 1055

Cleveland, Columbus, Cincinnati and Indianapolis Railroad.

0 Cleveland.	11. Erie Shale. 599
13 Berea.[63]	13 b. & c. Waverly. 796
25 Grafton.	" 803

13. Malvern. Glacial boundary five miles north. Glacial terrace extensive along Big Sandy Creek.
G. F. W.

Cleveland, Columbus, Cincinnati and Indianapolis Railroad—Con.

Ms.		
36	Wellington.	13 b. & c. Waverly. 861
47	New London.	" 936
55	Greenwich.	" 1050
67	Shelby.	13 c. " 1119
70	Vernon.	"
76	Crestline.	" 1186
80	Galion.	13 b. " 1170
93	Gilead.	11 c. Cleve. Shale. 1041
97	Cardington.	10 c. Huron Shale. 1012
104	Ashley.	" 987
114	Delaware.	{ 9. Cornif., 10. Ham., & 10 c. Huron. 953
122	Lewis Centre.	10. a. & c. Hu. Shale. 962
129	Worthington.	" 915
138	Columbus.	{ 9. Cornif., 10. Hamil., & 11. Ohio Sh. 746

Indianapolis Division.

80	Galion.	13. Waverly. 1170
92	Caledonia.	9. Corniferous.
101	Marion.	" 977
111	N. Bloomington.	7. Helderberg.
122	Mt. Victory.	"
132	Rushsylvania.	"
141	Bellefontaine.	{ 7. Held., 9. Cornif., & 10 c. Huron. 1115
150	De Graff.	5. Niagara.
157	Pemberton.	"
164	Sidney.	" 958
182	Versailles.	"
190	Ansonia.	"
197	Union.	"

Cincinnati Division.

0	Delaware.	{ 9. Cornif., 10. Ham., & 10 c. Huron. 953
9	Ostrander.	9. Corniferous.
17	Marysville.	7. Helderberg.
22	Milford.	"
32	Mechanicsburg.	5. Niag. & 7. Helderb.
43	Moorfield.	5. Niagara.
50	Springfield.	5 d. e. f. g. Niagara.
63	Osborn.	Cincinnati Group.
74	Dayton.	{ 4 c. Cin. Group & 5 a. b. c. Niagara. 754
81	Carrollton.	4 c. Cincinnati Group.
90	Franklin.	"
99	Henderson.	"
108	Maud's.	4 b. "
120	Carthage.	"
130	Cincinnati.	" 507

Cleveland, Loraine & Wheeling Railroad.

0	Uhrichsville.	{ 14 b. Coal Meas., Kit. Seam, 5 and 6.
12	Dover.	"
23	Barr's Mills.	{ 14 b. Coal Meas., Mercer Horizon.
35	Massillon.	{ 14 b. C. Meas., Sharon Seam No. 1.
48	Warwick.	"
59	Russell.	13 a. Waverly.

Cleveland, Loraine and Wheeling Railroad—Con.

Ms.		
72	Medina.	13 d. & e. Waverly.
85	Grafton.	13 b. & c. "
16	Black River.	11. Ohio Shale.

Cleveland, Akron and Columbus R. R.

0	Hudson.	14 a. Conglomerate.
7	Cuyahoga Falls.	"
14	Akron.	"
27	Clinton.	{ 14 b. C. Meas., Sharon Seam No. 1.
38	Orrville.	13 e. Waverly. 1074
52	Fredericksburg.	{ 13 e. Waverly, 14 a. Con. Coal Meas.
61	Millersburg.	"
81	Gann.	13 e. Wav., 14 a. Cong.
90	Howard.	13 e. Waverly.
100	Mt. Vernon.	" 991
109	Mt. Liberty.	"
124	Sunbury.	13 a. & b. Waverly.
133	Westerville.	{ 10 c. 11 a. b. c. Ohio Shale. 931
145	Columbus.	{ 9. Cornif., 10. Ham., & 11. Ohio Sh. 476

Cleveland and Pittsburg Railroad.

0	Cleveland.	11. Erie Shale. 599
8	Newburg.	13 b. Waverly. 802
14	Bedford.	" 954
26	Hudson.	14 a. Conglomerate.
38	Ravenna.	14 b. Coal Measure.
52	Limaville.	"
57	Alliance.	" 1099
63	Homeworth.	"
69	Bayard.[14]	{ 14 b. Coal Meas., Kit. Seam, 5 and 6. 1076
81	Millport.	{ 14 b. C'l Meas., Freeport Seams, 6 a. & 7. " 861
87	Salineville.	"
94	Irondale.	" 6 a.
102	Wellsville.	{ 14 b. Coal Meas., Kit. Seam, 5 and 6. 690

River Division.

0	Bellaire.	14 c. Coal Measures. 667
6	Martin's Ferry.	"
13	Portland.	"
20	La Grange.	14 b. "
26	Steubenville.	{ 14 b. Coal Meas., L. Freeport Seam. 665
35	Sloan's.	" 700
46	Wellsville.	{ 14 b. Coal Meas., Kit. Seams.

Tuscarawas Branch.

0	Bayard.[14]	{ 14 b. Coal Meas., Kit. Seams, 5 & 6. 1088
8	Malvern.[13]	{ 14 b. Coal Meas., Kit. Seams. 1001
12	Waynesburg.	" 1001
23	Zoar.	{ 14 b. Coal Meas., Mercer S'ms, 3 & 5 a. 889
32	New Philad'a.	{ 14 b. Coal Meas., Kittanning Seams. 906

Cleveland, Youngstown and Pittsburg Railroad.

Ms.	Railroad.
0 Mt. Union.	14 b. Lower Coal Meas.
15 Palmyra.[15]	{ 14 a. Cong. and 14 b. Cong. Coal Meas.
22 Newton Falls.[16]	44 a. Conglomerate. 96 e
27 Phalanx.	"

Columbus & Cincinnati Midland R. R.

0 Columbus.	9. Cor. & 11. O. Sh. 746
Mt. Sterling.	7. Waterlime.
Bloomingsburg.	"
Washington C.H.	" 957
Sabina.	5 g. Niagara.
Wilmington.[17]	5 c. d. e. f. Niagara. 992
Clinton Valley.	4 c. Cincinnati Group.

Columbus and Eastern Railway.

0 Hadley Junc.	{ 13 d. Wav. Drift, deposits heavy.
8 Thornport.[18]	{ 13 c. Wav. Drift, near boundary of drift.
14 Glenford.[19]	{ 13 f. Sub Carb. Lime. & 14 a. Conglom.
20 Mt. Perry.	{ 14 b. Low. Coal Meas., Mercer Horizon.
26 Fultonham.	"
35 Redfield.	{ 14 b. Low. Coal Meas., Kit. Coals, 5 & 6.

Columbus, Hocking Valley and Toledo Railroad.

0 Columbus.	{ 9. Corn. & 11. Ohio Sh., Drift heavy. 746
12 Groveport.	{ 11. Ohio Shale, Drift beds heavy.
23 Carroll.	13 d. Waverly. 815
32 Lancaster.[20]	{ 13 d. & e. Wav., conglom. prominent. 821
42 Millville.	{ 13 e. Wav., conglom. quarried largely.
50 Logan.	{ 13 e. Wav., type locality of Log. gr'p. 730
60 Lick Run.	{ 14 b. L. Coal Meas., Kit. Coals, Nos.5 & 6
62 Nelsonville.[21]	" 683
70 Salina.[22]	{ 14 b. L. Coal Meas., Up. Freeport C'l. 659
76 Athens.	{ 14 b. L. Barren Meas., Crinord'l Limest. 656

Ohio River Division.

50 Logan.	13 e. Waverly. 730
58 Union Furnace.	{ 14 b. Con. Coal Meas., Mercer Horizon.
71 Creola.	{ 14 b. L. Coal Meas., Mer. Hor., Block ores
76 McArthur.	{ 14 b. L. Coal Meas., Ferrif. Limes & Hor.
84 Eagle Furnace.	"
93 Minerton.[23]	"
115 Gallipolis.	14 b. L. Barren Meas.
130 Middleport.[24]	{ 14 c. Up. Prod. Meas., Pittsburg Coal.
132 Pomeroy.[25]	"

Straitsville Branch.

0 Logan.	13 e. Waverly. 730
5 { Webb's Summit.[26]	{ 13 f. Sub-Carboniferous Limestone.
9 Oreville.	{ 14 b. L. Coal Meas., Ferrif. Limestone.
11 Straitsville.	{ 14 b. L. Coal Meas., Kit. Coal, No. 6. 756
Greendale.	{ 14 b. L. Coal Meas., Mercer Horizon.
Carbon Hill.	{ 14 b. L. Coal Meas., Kittanning Coal.
Snow Fork Junc.	"
Nelsonville.	" 683

Toledo Division.

0 Columbus.	9. Cor. & 11. O. Sh. 746
14 Powell's.	9. Corniferous.
24 Delaware.	9. Cor. & 11. O. Sh. 953
41 Owen's.	9. Corniferous.
46 Marion.	" 977
64 Up. Sandusky.	7. Waterlime, drift he y
74 Carey.	5 g. Ni. & 7. Waterl. 820
88 Fostoria.	5 g. Niagara.
96 Rising Sun.	"
106 Pemberville.	"
124 Toledo.	7. Waterlime. 587

Columbus and Xenia Railroad.

0 Columbus.	{ 9. Cor., 10. Ham., & 11. Ohio Shale. 746
9 Alton.	9. Corniferous.
25 London.	" 1015
41 Selma.	5. Niagara.
55 Xenia.	{ 4 c. Cin., 5 a. b. and c. Niagara.

14. Bayard. Glacial boundary passes through Bayard. G. F. W.
15. Palmyra. Sharon coal in valuable basins.
16. Newton Falls. Fine development of conglomerate.
17. Wilmington. Fine exposures of Clinton limestone in Todd's Fork, near Wilmington.
18. Thornport. Near boundary of drift.
19. Glenford. Fine quality of S. C. limestone quarried here. Carboniferous conglomerate ground for glass-sand near by.
20. Lancaster. Glacial boundary passes through Lancaster. G. F. W.
21. Nelsonville. Fine sections of lower coal measures.
22. Salina. Salt manufacture; the Logan group furnishes the brine.
23. Minerton. The Clarion or Ferriferous limestone coal is mined here.
24. Middleport. Brown or paper coal found in the Pittsburg seam at one point.
25. Pomeroy. Extensive mining of coal (Pittsburg seam) and manufacture of salt. Brine derived from Waverly conglomerate, Logan group.
26. Webb's Summit. Typical locality of Sub-Carboniferous limestone for Ohio. Maxville is ad-

Ms. | Connotton Valley Railroad.

Ms.	Station	
0	Cleveland.	11. Ohio Shale. 599
		12 a. and b. Waverly.
12	Bedford.	Typical locality for Bedford shale. 964
32	Kent.	14 a. Con. Massive. 1049
40	Mogadore.27	14 b. L. Coal Meas., Mercer Horizon.
60	Canton.28	" 1049
76	Minerva Junc.	14 b. L. C'l Meas., Kit. C'ls, Nos. 5 & 6. 1011
87	Carrollton.	14 b. L. Coal Meas., Up.Freep't C'l, No.7.
95	Dell Roy.29	"
102	Sherrodsville.	"

Dayton and Michigan Railroad.

Ms.	Station	
0	Cincinnati.	507
60	Dayton.	4 c. Cincin. Group, & 5 a. b. & c.Niag. 754
74	Tippecanoe.	Cincinnati Group.
87	Troy.	" 845
88	Piqua.	4. Cin. Group, 5 a. Clin., & 5. Niag. 935
100	Sidney.	903
119	Wapakoneta.	5 g. Niagara.
131	Lima.	7. Helderberg. 893
144	Columbus Grove	" 877
151	Ottawa.	" 769
165	Deshler.	" 730
176	Weston.	8. Orisk. & 9. Corn. 683
182	Tontogany.	7. Helderberg.
193	Perrysburg.	" 689
202	Toledo.	" 589

Dayton and Union Railroad.

Ms.	Station	
0	Dayton.	4 c. Cin. Group and 5 a. b. c. Niag. 754
12	Brookville.	5 a. b. and c. Niagara.
21	Baltimore.	5 f. Niagara.
28	Arcanum.	"
35	Greenville.	5 g. " 1055
47	Union.	"

Indiana, Bloomington & Western R. R.

Ms.	Station	
0	Springfield.	5 d. and e. Niagara.
11	Plattsburg.	5. Niag. and 7. Helder.
20	London.	7. Helderberg.
32	Georgesville.	9. Corn. and 7. Helderb.
45	Columbus.	9. Corn., 10. Ham., & 11. Ohio Shale.

Ms. | Lake Erie and Western Railroad.

Ms.	Station	
0	Sandusky.	9. Corniferous. 600
6	Castalia.31	" 600
23	Fremont.	7. Waterlime. 637
44	Fostoria.	5 g. Niagara.
60	Findlay.	5 g. Niag. & 7. Helder.
75	Bluffton.32	7. Waterlime.
91	Lima.	7. Waterlime, drift heavy. 874
112	St. Mary's.	" 853
123	Celina.	" 850
138	Fort Recovery.	"

Lake Shore and Michigan Southern R. R.

Ms.	Station	
0	Buffalo, N. Y.	See New York.
116	Conneaut.	11 a. and b. Erie Sh. 553
129	Ashtabula.	" 650
138	Geneva.	" 669
144	Madison.	" 717
155	Painesville.	" 681
174	Nottingham.	"
183	Cleveland.	" 594
196	Berea.63	13 b. & c. Waverly. 795
209	Elyria.	" 730
217	Oberlin.	" 827
227	Wakeman.	"
239	Norwalk.	" 730
243	Monroeville.	11. Ohio Shale. 736
251	Bellevue.	" & 9. Cor. 766
258	Clyde.	7. Helderberg. 702
267	Fremont.	" 637
279	Elmore.	5. Niagara.
296	Toledo.	7. Helderberg. 589
338	Wauseon.	11. Ohio Shale. 775
353	Stryker.	" 721
360	Bryan.	" 773
370	Edgerton.	" 845

Ms.	Station	
0	Elyria.	13 b. Waverly.
10	Brownhelm.	
14	Vermilion.	11. Ohio Shale.
21	Ceylon.	"
34	Sandusky.	9. Corniferous. 600
46	Port Clinton.	7. Helderberg.
58	Oak Harbor.	5. Niagara.
65	Graytown.	5 g. Niagara.

Franklin Division.

Ms.	Station	
0	Ashtabula.	11. Erie Shale. 650
11	Jefferson.	"
24	Andover.	13. Waverly.
30	Simon.	"
36	Jamestown.	See Penna.

27. Mogadore. Coal measures clays worked on a large scale in potteries.
28. Canton. Road here passes out of drift-covered territory. The old moraine in great force near Canton.
29. Dell Roy. One of the best fields of Upper Freeport coal in State.
30. Nickel Plate. Much of the line is in a heavily drift-covered country. In the western part of Ohio particularly few exposures of the rocks are found.
31. Castalia. One of the strongest springs of Ohio.
32. Bluffton. Stone quarried extensively for railroad ballast.
33. Chillicothe. Glacial boundary two miles north. Glacial terraces extensive all along the river. Immense kames on Paint Creek, five miles west. (See Note 48.) G. F. W.
34. New Lisbon. Extensive glacial terraces containing kidney iron-ore. The glacial boundary is on the highlands just south. G. F. W.

Ms. | Little Miami R. R. (P. Cin. & St. L.).

Ms.	Station	Geology
0	Cincinnati.⁶²	4 b. Cincin. Group. ⁵⁰⁷
9	Plainville.	"
17	Miamiville.	"
23	Loveland.	"
36	Morrow.	4 b. & c. " ⁶⁴²
45	Freeport.	4 c. "
56	Claysville.	"
65	Xenia.	4 b.Cin., 5 a.b.&c.Ni.⁸⁵⁰

Marietta & Cincinnati R.R.(B. & O. R.R.).

Ms.	Station	Geology
0	Cincinnati.⁶²	4 b. Cincin. Group. ⁵⁰⁷
5	Cummingsville.	"
20	Remington.	"
31	Cozaddale.	"
41	Blanchester.	4 c. " ⁹⁷⁹
50	Martinsville.	5 b. Niagara. ¹⁰⁴⁵
62	Lexington.	7. Helderberg.
74	Greenfield.	" ⁸⁹⁸
85	Frankfort.	11. Ohio Shale. ⁷⁶⁵
98	Chillicothe.³³	{ 11. Ohio Shale, and 13 a. and b. Wav. ⁶³⁷
105	Schooley's.	13 d. Waverly. ⁶⁶⁸
117	Raysville.	{ 14 a. Cong. & Cornif. Coal Meas. ⁶³⁸
127	Hamden.	14 b. Cong. C'l Meas.⁷²³
139	Zaleski.	{ Coal Meas., Mercer & Kit., Nos. 3 to 6.⁷²³
152	Marshfield.	Camb. Limestone. ⁸²⁸
159	Athens.	Cam. & Crin. Limest.⁶⁵⁶
	New England.	14 c. Coal Measure.
	Cutler.	" ⁷⁷⁹
	Moore's Junct.	"
	Marietta.	" ⁶²⁵

Ms.	Station	Geology
0	Blanchester.	4 c. Cincin. Group. ⁹¹⁹
11	Lynchburg.	"
21	Hillsboro.	5 c. d. e. f. g. h. Ni.¹¹³⁵

Ms.	Station	Geology
0	Hamden.	{ 13 s. c. Limest., 14 Coal Meas., Sharon Coal Horiz.
12	Jackson.	{ 14 a. Cong. and Cong. Coal Measure.
19	Vaughan's.	14 b. Coal Measure.
28	Washington.	Coal Meas., Fer. Limest.
38	Webster.	{ 14 b. Coal Meas., Mercer Horizon.
50	Sciotoville.	13 e. Waverly.
56	Portsmouth.	13 d. "

Ms.	Station	Geology
0	Athens.	{ 14 b. Coal Measure, Crin. Limest. ⁶⁵⁶
11	Guysville.	14 c. Coal Measure.
23	Coolville.	"
28	Little Hocking.	" ⁷⁵⁷
36	Parkersburg.	"

Marietta, Pittsburg and Cleveland R. R.

Ms.	Station	Geology
0	Marietta.	14 c. Coal Measure. ⁶²⁵
7	Caywood.	"
18	Warner.	"
27	Dexter.	"Crin. Limest.
36	Caldwell.	"
45	Glenwood.	"

Marietta, Pittsburg and Cleveland Railroad—Con.

Ms.	Station	Geology
59	Cambridge.	{ 4 b. Coal Meas., Up. Freep't Sm., No. 7.
70	Kimbolton.	{ 4 b. Coal Meas., Kit. Seam, Nos. 5 & 6.
80	New Comerst'wn	" ⁷⁹⁸
90	Phillipsburg.	"
100	Dover.	" ⁸⁶⁰

"Nickel Plate." ⁹⁰

New York, Chicago and St. Louis R. R.

Ms.	Station	Geology
0	Buffalo.	
116	Conneaut.	11. Ohio Shale. ⁶⁵⁰
129	Ashtabula.	" ⁶⁵³
138	Geneva.	"
154	Painesville.	" ⁶⁵¹
160	Mentor.	" ⁶⁸⁴
165	Willoughby.	"
173	Euclid.	"
183	Cleveland.	" ⁵⁹⁹
192	Rocky River.	"
202	Avon.	"
210	Lorain.	"
221	Vermilion.	13 a. and b. Waverly.
229	Berlin Heights.	"
236	Milan.	11. Ohio Shale.
248	Bellevue.	" ⁷⁶⁶
260	Green Springs.	7. Waterlime.
280	Fostoria.	5 g. Niagara.
300	Mt. Comb.	7. Waterlime.
310	Leipsic.	"
325	Continental.	"
341	Latty.	9. Corniferous.
353	Smiley's Station.	"

New York, Pennsylvania & Ohio R. R.

Ms.	Station	Geology
0	Cincinnati.⁶²	⁵⁰⁷
59	Dayton.	{ 4. Cincin. Group, & 5 a. b. & c. Niag.⁷⁵⁴
70	Osborne.	4. Cincinnati Group.
76	Enon.	5 d. and e. Niagara.
80	Springfield.	5 d. e. f. g. " ⁹¹⁰
89	Bowlinsville.	Niagara.
95	Urbana.	5 g. Ni. & 7. Held. ¹⁰²⁹
105	Mingo.	7. Helderberg.
114	Pottersburg.	"
121	Broadway.	"
129	Richwood.	" ⁸⁴⁴
138	Green Camp.	"
144	Marion.	9 a. and b. Cornif. ⁹⁶¹
153	Caledonia.	" ¹⁰⁶⁸
164	Galion.	13 b. Waverly. ¹¹⁷¹
172	Ontario.	13 c. " ¹³⁷⁷
179	Mansfield.	13 e. Waverly. ¹¹⁵⁶
187	Windsor.	" ¹⁰⁶⁹
196	Ashland.	" ¹⁰⁸⁶
207	Polk.	" ¹²⁴²
213	West Salem.	" ¹⁰⁸⁸
216	Burbank.	"
221	Pike.	"
225	Russell.	"

Ms.	New York, Pennsylvania and Ohio Railroad—Con.		Ms.	North-Western Ohio Railway.	
232	Wadsworth.	14 b. Coal Meas. [1117]	0	Toledo.	7. Helderberg. [569]
240	New Portage.	14 a. Conglomerate. [967]	6	Walbridge.	"
246	Akron.	" [1005]	18	Woodville.	5. Niagara.
250	Tallmadge.	{ 14 b. Coal Measure, Sharon Seam. [1108]	26	Helena.	"
			31	Burgoon.	"
256	Kent.	14 a. Conglomerate. [1049]	42	Tiffin.	" & 7. Held. [758]
263	Ravenna	14 a. & b. C'l Meas. [1095]	52	Bloomville.	9. Corniferous.
269	Freedom.	" [1150]	62	New Washingt'n	10 c. Hur. & 10. Ham.
279	Braceville.	13 d. and e. Wav. [901]	75	Vernon.	13 d. Waverly.
283	Leavittsburg.	13 d. & e. Waverly. [892]	86	Mansfield.	13 e. " [1167]
286	Warren.	3 d. Waverly. [902]			
294	Cortland.	"		**Ohio Central Railway.**	
307	Orangeville.	13 c. and d. Wav. [945]	0	Toledo.	7. Lower Helderb. [567]
	Mahoning Division.		10	Stony Ridge.	5 g. Niagara.
			35	Fostoria.	"
0	Sharon.	{ 14 a. & b. C'l Meas., Sharon C'l, No. 1.	69	Bucyrus.	11. Ohio Shale. [1009]
7	Hubbard.	14 a. & b. Coal Meas.	89	Mt. Gilead.	13 a. and b. Wav. [1100]
15	Youngstown.	{ 14 a. Cong. & 14 a. & b. Sharon Coal No. 1. [865]	108	Centerburg.	13 d. Waverly.
			124	Granville.	13 e. "
			142	Lakeside. [35]	13 d. "
23	Niles.	" [911]	156	Rushville. [36]	13 e. "
31	Leavittsburg.	" [897]	167	Junction City.	14 b. Low. Mer. Horiz.
40	Mahoning.	14 a. Conglomerate.	172	New Lexington.	14 b. Kit. C'ls, 5 & 6. [871]
51	Mantua.	" [1111]	179	Moxahala. [37]	"
57	Aurora.	" [1090]	184	Corning. [38]	"
65	Solon.	" [1032]			
75	Newburg.	13 a. Waverly. [815]		**Ohio and Mississippi Railroad.**	
80	Cleveland.	11. Erie Shale. [599]	0	Cincinnati.	14 b. Cincin. Group. [507]
	Niles and New Lisbon Branch.		9	Delhi.	"
			13	North Bend. [39]	"
0	Niles.	{ 13 d. Waverly and 14 a. Conglom. [911]		**Ohio Southern Railway.**	
6	Austintown.	{ 14 a & b. C'l Meas., Low. Merc. Horiz.	0	Springfield. [40]	5 f. and g. Niagara. [953]
12	Canfield.	{ Coal Meas., Ferrif. Limest. Horiz. [1100]	12	S. Charleston.	{ 5 f. & g. Ni. Drift heavy, no rock visible.
18	Green.	{ Coal Meas., Low. Kittanning Coal.	36	Washingt'n C.H.	{ 7. Waterlime. No rock visible. [957]
23	Loetonia.	" [1036]	43	Good Hope.	7. Waterlime.
25	Franklin.	"	50	Greenfield. [41]	" [898]
33	New Lisbon. [34]	{ Coal Meas., Ferrifer. Limest. to Mahoning Sandstone. [968]	62	Bainbridge. [42]	{ 7. Waterl., 11. Ohio Sh., 13 a. & b. Wav.
			84	Waverly.	{ 11 c. Ohio Sh., 13 a. b. and c. Waverly.
	Liberty and Vienna Branch.		97	Beaverton.	13 e. Wav. & 14 a. Con.
			109	Jackson. [43]	14 a.& b.Con.& C'l Meas.
0	Vienna.	14 b. Coal Meas.	113	Coalton. [44]	"
8	Vienna Junct.	"	119	Wellston. [45]	"

35. Lakeside. Lake produced by glacial accumulations near margin of glacial area.

36. Rushville. The upper beds of the Waverly here yield an abundant series of fossils, part of them agreeing with the Sub-Carboniferous limestone forms of Illinois.

37. Moxahala. Between Moxahala and Corning the change occurs which converts the middle Kittanning coal seam (No. 6) from a 3½ foot seam into a 10–12 foot seam. The Mid. Kittanning coal, and also the Lower Freeport seam, are both mined at Moxahala. In the tunnel south of the town the Upper Freeport horizon is well shown except the coal.

38. Corning. The Upper Freeport coal (No. 7) is also worked near Corning. It is known here as the "upper vein," or Norris coal.

39. North Bend. Extensive glacial deposits at North Bend railroad-tunnel, on the I. C. & L. R. R., passes through a glacial deposit 150 feet deep. G. F. W.

40. Springfield. Fine exposures of Niagara. Worked on large scale for building-stone and lime.

41. Greenfield. Best showing of Lower Helderberg in Ohio. Stone of great value. Quarried on large scale for building-stone. All fragments and spalls burned for lime ; stone remarkably even bedded.

Ms.	Painesville & Youngstown R. R.	
0	Youngstown.	14 a. and b. Cong. & Cong. Coals. 865
9	Niles.	14 d. Conglomerate. 911
15	Warren.	13 d. Waverly. 892
25	Southington.	"
31	Bundysburg.	14 a. Conglomerate.
38	Burton.	"
48	Chardon.	"
59	Painesville.	11. Erie Shale. 695

Ms.	Pittsburg, Cincinnati and St. Louis R. R.	
0	Columbus.	9. Corn., 10. Ham., & 11. Ohio Shale. 746
10	Black Lick.	13 b. Waverly.
17	Pataskala.	13 d.
33	Newark.[46]	13 e. " 821
41	Hanover.	" 882
49	Frazeysburg.	14 b. Coal Meas., Mercer Horizon. 753
53	Dresden Junc.	" 737
62	Conesville.	14 b. Coal Meas., Kit. Seams, 5 and 6. 740
69	Coshocton.	" 773
75	West Lafayette.	"
83	N. Comerston.	" 798
89	Pt. Washington.	" 815
97	Trenton.	" 835
100	Uhrichsville.	Coal Measures. 865
110	Bowerston.	C'l Meas., Freep't S'ms.
121	Fairview.	Coal Measures. 1011
130	Unionport.	" 948
138	Smithfield.	" 775
150	Steubenville.	C'l M., L. Free. Sms. 730

Ms.	Pittsburg, Fort Wayne & Chicago R. R.	
0	Chicago.	See Indiana.
168	Dixon.	7. Helderberg. 800
173	Convoy.	" 793
181	Van Wert.	" 788
193	Delphos.	" 786
201	Elida.	" 800
208	Lima.	" 884
216	Lafayette.	" 938
222	Ada.	"
232	Dunkirk.	" 951
239	Forrest.	5. Niagara. 940
251	Upp. Sandusky.	7. Helderberg. 862

Ms.	Pittsburg, Fort Wayne & Chicago Railroad—Con.	
259	Nevada.	9. Corniferous. 934
267	Bucyrus.	9. Cor., 10. Ham., & 11. Ohio Sh. 1009
280	Crestline.	13 d. Waverly. 1169
293	Mansfield.	13 e. " 1167
307	Perrysville.	" 1008
318	Lakeville.	13. Wav., 14 c. Con., & 14 b. C'l M. 956
333	Wooster.	13 e. Waverly. 91-
344	Orrville.	13 e. Wav., 14 c. Con., & 14 b. C'l M. 1074
359	Massillon.	14 a. & b. Coal Mea. 967
367	Canton.	Coal M., Mer. Hor. 1059
379	Strasburg.	Coal Measure. 1101
385	Alliance.	" 1099
392	Damascus.	" 1190
405	Leetonia.	Coal Meas., L. Kit. Seam, No. 5. 1036
414	N. Waterford.[47]	Freeport Seams. 1078
	(Continued in Pennsylvania.)	

Ms.	Sandusky, Mansfield and Newark Railroad (B. & O. R. R.).	
0	Sandusky.	9. Corniferous. 600
8	Prout's.	11. Ohio Shale.
15	Monroeville.	11 c. Ohio Shale. 736
23	Havana.	13 b. Waverly.
28	Chicago Junc.	13 c. "
35	Plymouth.	"
42	Shelby Junc.	" 1119
49	Spring Mill.	"
54	Mansfield.	13 e. " 1167
63	Lexington.	"
74	Independence.	"
84	Frederick.	"
91	Mt. Vernon.	" 991
103	Utica.	"
116	Newark.[46]	" 821

Ms.	Scioto Valley Railroad.	
0	Columbus.	9. Cor., 10. Ham., 11. Ohio Shale. 746 11. Ohio Sh. Whole
30	Circleville.	region heavily covered with drift.
39	Kingston.	13 d. Waverly.

42. Bainbridge. Sections from Helderberg limestone to Berea grit found in steep hills. The Ohio shale is fossiliferous here to small extent. The valley of Paint Creek has unusual geological interest.

43. Jackson. The lowest coal of the series is mined largely here. It has great excellence as an iron-making fuel. Four furnaces depend upon it.

44. Coalton and Wellston. At these places is the only field of the State in which the second seam of the coal series is worked. The coal has great excellence and value. It is also an iron-making fuel in the raw state.

45. Barr's Mills. Glacial boundary passes through Barr's Mills. G. F. W.

46. Newark. Glacial boundary passes through Newark, running north and south. G. F. W.

47. North Waterford. Glacial boundary five miles south. Glacial deposits extensive at East Palestine. G. F. W.

48. Chillicothe. The road here passes out of the glacial area. At Chillicothe all divisions of Waverly well shown. (Also see No. Note 33.)

49. County Bridge. At this point fine exposures of Waverly black slate.

50. Waverly. From Waverly the division of rocks received its name, the main element being the quarry-stone, which is the southern extension of the Berea grit.

51. Sciotoville. At Sciotoville the famous Sub-Carboniferous fire-clay that accompanies the limestone is largely worked and manufactured.

Ms.	Scioto Valley Railroad—Con.		Toledo, Cincinnati and St. Louis Railroad—Con. Ms.	
50	Chillicothe.⁴⁸	{ 11 c. Ohio Sh., 13 a. b. c. d. e. Wav. ⁶³⁷	30 Jamestown.	{ 5. Niagara. Drift beds heavy.
61	County Bridge⁴⁹	13 b. c. & d. Waverly.	66 Frankfort.	11. Ohio Shale. ⁷⁵⁵
70	Waverly.⁵⁰	{ 11 c. Ohio Sh., & 13 a. b. c. Waverly.	80 Chillicothe.	{ 11. Ohio Sh. & 13 a.b. c. d. e. Wav. ⁶³⁷
76	Piketon.	" ⁵⁷⁸	93 Richmondale.	14 a. Con. & 13 e. Wav.
90	Lucasville.	13 c. d. e. Waverly.	104 Byers' Station.	"
100	Portsmouth.	13 e. - " ⁴⁸⁹	110 Coalton.	14 a. & b. Con. & C'l M.
105	Sciotoville.⁵¹	{ 13 e. Wav., 13 f. Sub-Carb. Limestone.	115 Wellston.	"
114	{ Franklin Furnace.	{ 14 a. and b. Coal Measures.	115 Wellston.	"
			136 Centerton.	14 b. Coal Measures.
124	Hanging Rock.	{ 14 b. Coal Meas. and Ferrif. Limestone.	152 Mt. Vernon.	{ 14 b. Coal Meas., Fer. Limestone.
127	Ironton.⁵²	{ 14 b. Coal Meas., Kit. Coals, 5 and 6.	159 Etna.	"
131	Ashland.	"	168 Ironton.	"

Toledo, Cincinnati & St. Louis Railroad.

0	Toledo.	7. Waterlime. ⁵⁸⁷
24	Grand Rapids.	9. Corniferous.
42	Holgate.	"
74	Delphos.	{ 7. Waterlime. Drift heavy. ⁷⁸⁶
108	Decatur.	9. Corniferous.
74	Delphos.	7. Waterlime. ⁷⁸⁶
92	Mendon.	"
104	Celina.	5 g. Niagara. ⁸⁵⁰
139	Covington.	5 f. & g. "
150	West Milton.	5 b. "
156	Harrisburgh.⁵³	5 b. "
169	Dayton.⁵⁴	{ 4 c. Cin. & 5 a. b. c. d Niagara. ⁷⁵⁴
183	Centerville.	"
199	Lebanon.⁵⁵	4 c. Cincinnati. ⁷⁴⁰
207	Mason.	4 b. & c. " ⁷⁰⁰
229	Cincinnati.⁶²	4 b. " ⁶⁰⁷
0	Dayton.	{ 4 c. Cincin. and 5 a. b. c. d. Niag. ⁷⁵⁴
17	Xenia.	{ 4 c. Cincin. and 5 a. and b. Niagara.

Valley Railway.

Cleveland.	11. Ohio Shale. ⁵⁹⁹
Independence.⁵⁶	13 a. b. c. Waverly.
Peninsula.⁵⁷	"
Akron.	{ 14 a. Cong. and 14 b. Coal Measure. ¹⁰⁰⁵
Greentown.	{ 14 b. Brookville or Gray Limest. Coal.
Canton.⁵⁸	14 b. Merc. Horiz. ¹⁰⁴⁹
No. Industry.	14 b. Kit. Cls., No. 5 & 6.
Mineral Point.⁵⁹	"
Valley Junc.	14 b. Mercer Horiz. ⁹⁰⁰

Wabash, St. Louis and Pacific Railroad.

0	Toledo.	7. Helderberg. ⁵⁸⁷
0	South Toledo.	"
17	White House.	9. Corniferous. ⁶⁵⁴
29	Liberty.	10 c. Huron. ⁶⁸⁴
35	Napoleon.	10. Ham. & 11. O. Sh.⁶⁸²
52	Defiance.	" ⁷⁰⁰
61	Emerald.	10. Hamilton.
71	Antwerp.	9. Corniferous. ⁷³⁸
94	Ft. Wayne.	See Indiana.

52. Ironton. The charcoal iron manufacture of Ohio is centered here.
53. Harrisburgh. Clinton limestone, white and marble-like here.
54. Dayton. Junction of Lower and Upper Silurian well shown at Soldiers' Home. Valuable quarries in Dayton stone at many points. The Clinton limestone highly fossiliferous in this region.
55. Lebanon. One of the typical localities for fossils of the Upper Cincinnati beds.
56. Independence. Valuable quarries in Berea stone. Grit especially valuable for millstones for grinding wood pulp, pearl barley, etc.
57. Peninsula. Large quarries in Berea grit.
59. Mineral Point. Valuable bed of Kittanning clay. Best fire-clay in the State.
60. Lodi. Excellent locality for Upper Waverly fossils.
61. Massillon. Lowest coal (Sharon) mined largely here.
62. The Cincinnati Glacial Dam. The survey of the terminal moraine in Ohio, made by Rev. G. F. Wright in 1882, proved that the southern boundary of the great ice-sheet crossed the Ohio River near New Richmond, twenty-two miles by the river above Cincinnati, and extended across the northern counties of Kentucky, four or five miles south of the river, recrossing the Ohio near Aurora, Indiana. Mr. Wright inferred that one effect of this glacier was to form an immense dam of ice and moraine débris, 500 to 600 feet high, which effectually closed the old channel of the Ohio for forty-nine miles by the windings of the river, and set back the water of the river and its tributaries until, as shown by Mr. I. C. White, it probably occupied the channel between the Kanawha and the Ohio Valleys, through West Virginia, now the line of the Chesapeake and Ohio Railroad. The site of Pittsburg, Pa., was submerged to the depth of 300 feet, the remarkable terraces in the valleys of the Ohio, Allegheny, Monongahela, and other branches, for the origin of which no satisfactory explanations had before been given, being then formed, according to White and Lesley, around the shores of this great inland lake. (See Note No. 62, in West Virginia.) J. M.

Ms.	Wheeling and Lake Erie Railway.		Ms.	Wheel'g & Lake Erie Railway—*Con.*	
0	Toledo.	7. Waterlime. [587]	133	Sippo.	{ 14 a. Congl. & 14 b. Lower Coal Meas.
36	Fremont.	7. Waterlime. [637]	137	Massillon.[61]	" [967]
59	Monroeville.	11. Ohio Shale. [736]	148	Navarre.	{ 14 b. Con. Coal Meas., Mercer Horizon.
64	Norwalk.	13 a. & b. Waverly.	154	Zoar.	" [891]
85	Wellington.	13 d. Wav. D'ft h'vy. [861]	157	Valley Junction.	"
100	Lodi.[60]	13 d. & e. Waverly.			
121	Orrville.	13 e. Waverly. [1074]			

63. The Berea Grit, the most important member of the Sub-Carboniferous formation in Ohio, is quarried here on a very large scale. The Berea Shale that makes the roofs of the quarries is highly fossiliferous.

This blank space is intended for additional geological notes in pencil by the traveler.

Michigan.[1]

LIST OF THE GEOLOGICAL FORMATIONS OF MICHIGAN.

PROBABLE EQUIVALENTS OF DANA.	LOCAL DESIGNATIONS.
20. Quaternary.[2]	20. Quaternary, Lacustrine Drift.[4]
14 c. Upper Coal Measures.	14 c. Coal Measures.
14 a. Millstone Grit.	14 a. Parma Sandstone.
13 b. Upper Sub-Carboniferous.	13 b. Carboniferous Limestone.
"	13 b. Michigan Salt Group.
13 a. Lower Sub-Carboniferous.	13 a. Marshall Group.
11 b. Chemung.	11. Huron Group, Chemung Shale.
11 a. Portage.	11. Huron Group, Portage Shale.
10 c. Genesee.	11. Huron Group, Black Shale.
10 b. Hamilton.	10 b. Little Traverse Group.
9 c. Corniferous and 9 b. Schoharie.	9. Corniferous Group.
7. Lower Helderberg.	7. Lower Helderberg.
6. Salina.	6. Salina Group.
5 c. Niagara.	5. Niagara Group.
5 b. Clinton.	"
4 c. Cincinnati.	4 c. Cincinnati.
4 a. Trenton.	4 a. Trenton.
3. Canadian.	3 c. and 3 a. Chazy and Calciferous.
2 b. Potsdam.	2 b. Lake Superior Sandstone.
1 c. Keweenian.	1 c. Cupriferous Rocks, Sandstones,
1 b. Huronian.	Conglomerates and Traps.
1 a. Laurentian.	1 b. Huronian.
	1 a. Laurentian.

Sketch of the Geology of Michigan.*

The State of Michigan is divided, geographically, into two parts by Lake Michigan and the Straits of Mackinaw, but geologically there is no such division, the upper and lower peninsula, as they are called, being, with the portion now covered by water, one uniform series of formations succeeding each other in their proper order. For the clear understanding of its geological structure we should imagine the water of the lakes removed, or the strata extending under it. The city of Cincinnati, in Ohio, stands upon a dome or ridge of upraised older strata which have been uncovered by the planing off of their higher beds, until on both sides of it the outcrop of several of the formations appear. The strata dip from this ridge towards the east and towards the west, and the line of it extends towards the common corner of Ohio, Indiana and Michigan. It bifurcates, however, before reaching that point, the east branch running up to the west end of Lake Erie, causing several islands there, and subsides in Canada near the River Thames; while the west branch passes across the northern part of Indiana and Illinois to the head of Lake Michigan, and thence northwest through Wisconsin.

On the north another ridge of still older rocks, the 1. Laurentian, extends through Canada around the north shores of Lakes Huron and Superior. It also appears in the upper peninsula. This, the oldest of the formations, is the lowest and foundation of all, the later formations resting upon it, dipping south and southwest away from the Laurentian. The whole State of Michigan, including the parts covered by the lakes, is therefore surrounded on all sides by ancient axes of elevation, which isolated her rock formations from the adjoining regions. It may be considered as one great basin, for even if the surrounding regions do not in all cases actually occupy a higher level, yet we find the strata dip from all sides towards the centre. The upper peninsula, or that portion of the State north of Lake Michigan, is bounded around the entire south shore of Lake Superior by the 2 b. Potsdam red sandstone, of which the Pictured Rocks are composed, and reposing upon it are the south-dipping Lower Silurian series in regular belts, in a general east and west course, and extending up to 5 c. Niagara limestone, which extends between Green Bay and Lake Michigan, and forms the shores of Lake Michigan and Lake Huron. The Upper Helderberg also appears on Mackinaw and other islands.

1. This chapter was prepared for this work by Prof. Alexander Winchell, LL. D., of the University of Michigan, former Director of the Geological Survey of Michigan.
2. The rocky formations of the lower peninsula are deeply and generally covered by drift. In all the western half of the State, south of Little Traverse Bay, no good characteristic exposures exist, save in Kent county and near Holland in Ottawa county. Hence in most cases our knowledge of the underlying rocks is only a matter of inference. A. W.

* Derived chiefly from Prof. A. Winchell's Geological Reports of this State.

Michigan Central Railroad.		
Ms.		Alt.
0 Detroit.	{ 10 b. Little Traverse, / ben. Lacustrine.	581
3 Grand Trunk Jun	11. Hu. ben. Lacus.	
10 Dearborn.	"	614
17 Wayne.	"	662
30 Ypsilanti.	{ 13 a. Mashall (?) / Lower Ridge.	714
38 Ann Arbor.	{ 13 b. Mich. salt, / Terminal Moraine.*	771
43 Delhi.	{ 13 b. Mich. salt, / Deep Drift.	
47 Dexter.	{ 13 b. Carbon. lime s. / Deep Drift.	858
55 Chelsea.	13 b. Carb. lime s.	913
62 Francisco.	"	1016
66 Grass Lake.	"	986
69 Leoni.	"	980
76 Jackson.	14 c. C. Mes. Mines	927

(Air Line Division.)

Ms.		Alt.
76 Jackson.	927 14 c. Coal Mes. Mines.	
83 Snyder's.	13 b. Carb. l. s.	971
90 Concord.	"	987
99 Homer.	13 a. Marshall	972
108 Clarendon.	"	966
109 Tekonsha.	"	937
117 Union City.	{ 11. Huron. Kid'y / Iron Ore.	900
124 Sherwood.	"	872
129 Colon.	"	838
136 Wasepi.	"	842
140 Centreville.	"	848
145 Three Rivers.	"	805
152 Corey's.	"	871
160 Vandalia.	10 b. L. Trv.(?)	878
165 Cassopolis.	"	881
170 Dailey.	9. Corniferous	871
174 Baron Lake.	"	768
179 Niles.	"	861

Michigan Central Railroad—*Con*		
Ms.	(Kalamazoo Division.)	Alt.
76 Jackson.		927
81 Trumbull's.	14 c. Coal Meas.	
87 Parma.	956 14 a. Parma s. s. outc'p	
92 Bath Mills.	13 b. Carb. limestone.	
96 Albion.	"	943
101 Marengo,	13 a. Marshall.	921
108 Marshall.	" outcrops.	898
113 Ceresco.	"	802
115 White's.	"	900
121 Battle Creek.	" "	819
126 Bedford.	"	809
130 Augusta.	"	789
135 Galesburg.	" (?)	788
140 Comstock.	11. Huron.	782
144 Kalamazoo.	"	777
149 Ostemo.	"	962
156 Mattawan.	"	860
160 Lawton.	"	778
162 White Oaks.	"	842
168 Decatur.	"	781
172 Glenwood.	10 b. L. Tra.(?)	751
179 Dowagiac.	9. Cornifer.(?)	760
185 Pokagon.	"	733
191 Niles.	"	681
197 Buchanan.	"	733
202 Dayton.	"	718
205 Galien.	"	682
209 Avery's.	"	655
211 Three Oaks.	"	669
218 New Buffalo.	" Sand Dunes.	602

(Deep Drift bracket: 162–211)

(Continued in Indiana.)

(Grand Rapids Division.)		
0 Jackson.	14 c. Coal Measures.	
10 Rives Junction.	"	904
17 Onondaga.	"	895
24 Eaton Rapids.	"	876
35 Charlotte.	14 a. Parma Sand.	906
40 Chester.	"	883
46 Vermontville.	13 b. Carb. Lime.	817

The lake is excavated chiefly in the 6. Salina formation, Prof. James Hall estimating that two-thirds of it is from that formation. The geological strata were first laid down extending across where the lakes now are, so that eastern Wisconsin is a part of this basin. The lakes rest in troughs which have been excavated subsequently nearly along the strike or outcropping edges of some of the softer formations. In the lower peninsula, or the main portion of the State between Lake Michigan and Lake Erie, all the Michigan series above the Niagara and up to the Carboniferous appear on the surface, but all of them much thinner than in the States farther east.

To make it still more clear we might begin at the highest formation, the 14 b. Coal Measures, which extends, in an oval form, from Jackson to Saginaw Bay. This is the upper layer of rocks, and the other formations crop out in successive layers below it on all sides. The annexed Railway Guide shows their exposures on the lines of the railroads, as they have been carefully made out by Prof. Alex. Winchell. Each rocky stratum, therefore, may be considered as dish-shaped, and taken together they form a nest of dishes or basins, the highest being the coal field near the centre of the lower peninsula, and passing from this in any direction we travel successively over the outcropping edges of older and older strata.

The Lake Superior iron ore is found in the 1 b. Huronian formation, directly west of Marquette. The copper is found chiefly in a great trap-dyke, which extends for many miles along Keweenaw Point. These iron ore and copper producing mines are the richest and most productive in America.

Michigan is therefore a distinct and independent geological area. Its topmost formation is a coal basin, underlaid by the Devonian formations, very much thinned out it is true, and below that the Silurian largely developed and extending out to the oldest Laurentian rocks on the north, and all this within the bounds of the State, with small portions only of this separate geological world extending into adjoining States on the west side. The whole of the peninsula is covered with drift, from one hundred to three hundred feet deep, and rock exposures are very rare

* Drift 164 feet on Main Street and 292 in Observatory Hill contains fossil wood at depth of 60 feet.

Michigan Central Railroad.
(Grand Rapids Division.)—Continued.

Ms.			Alt.
50	Nashville.	13 b. Carb. l. s.	807
55	Sheridan.	"	856
62	Hastings.	"	791
73	Middleville.	"	717
79	Caledonia.	"	799
85	Hammond.	"	754
94	Grand Rapids.	" Ext. exposures.	605

(South Haven Division.)

Ms.			Alt.
0	Kalamazoo.	11. Huron.	777
8	Alamo.	"	705
14	Kendell's.	"	792
17	Pine Grove.	"	777
18	Gobles.	"	803
22	Bloomingdale.	"	781
24	Beaver Lake.	"	
27	Columbia.	"	682
29	Grand Junction.	"	676
31	Geneva.	"	695
39	South Haven.	"	583

(Concealed by Drift.)

(South Bend Division.)

Ms.			Alt.
0	Niles.	9. Corniferous.	681
5	Bertrand.	"	939
9	Notre Dame.	"	
11	South Bend.	"	

(Saginaw Division.)

Ms.			Alt.
0	Jackson.	14 c. Cl. Mr. Mines	842
11	Rives Junction.	"	
15	Leslie.	"	883
25	Mason.	"	
37	Lansing. \	"	852
53	Laingsburg.	"	806
65	Owosso.	"	745
87	St. Charles.	{ 14 c. Coal Measures Lacustrine.	591
101	Saginaw City.	"	591
103	East Saginaw.	"	
105	Carrollton.	"	
116	Wenona.	"	589
121	Bay City. 3	"	592

(Mackinaw Division.)

Ms.			Alt.
0	Bay City.4	14 c. C Mes., Lacus	597
6	Kawkawlin.	"	627
29	Standish.	"	774
41	Wells.	"	997
54	West Branch.	13 b. Carb limestone.	
67	St. Helenas.	" (?)	1158
78	Roscommon	" (?)	1126
93	Grayling.	13 b. Mich. Salt.	1148
102	Forrest.	13 a. Marshall.	1226
113	Otsego Lake.	"	
121	Gaylord	" (?)	1849

Michigan Central Railroad.—Con.
(Bay City Division.,)

Ms.			Alt.
0	Detroit.	11 b. L. Trav.	581
10	Norris.	11. Hu. Lac.	631
14	Warren.	"	641
17	Oakwood,	"	650
24	Utica.	"	
29	Yates.	"	
31	Rochester.	13 a. Mars'll	747
35	Goodison's.	"	842
41	Orion.	"	995
44	Oxford.	13 b. Mich. St	1058
52	Metamora.	"	1053
60	Lapeer.	"	830
61	Junction.	"	
64	Millville.	13 b. Carb. limestone.	
65	Carpenter's.	"	801
70	Columbiaville.	"	77
74	Otter Lake.	13 b. Mich. Salt.	860
80	Millington.	14 a. Parma s. s.	737
87	Vassar.	14 c. Coal Meas.	643
95	Reese.	"	629
110	Bay City 4	"	592

(Quat'ry Deposits overlying.)

Lake Shore & Michigan Southern R. R.
(Michigan Division.)

Ms.			Alt.
0	Cleveland.		
113	Toledo.	9. Corniferous.	
123	Sylvania.	"	
130	Ottawa Lake.*	"	688
133	Riga.	"	692
135	Blissfield.	10 b. Lit. Traverse.	684
139	Palmyra.	11. Huron.	707
141	Lenawee Junc.	"	714
145	Adrian.	"	810
155	Clayton.	"	905
162	Hudson.	13 a. Marshall,	945
168	Pittsford.	"	1109
172	Osseo.	"	1126
178	Hillsdale.	" Ext. Quarries	1095
182	Jonesville.	"	1097
187	Allen's.	"	1064
194	Quincy.	1L Huron.	1027
200	Coldwater.	" worked for Brick	983
215	Bronson.	"	927
218	Burr Oak.	"	896
224	Sturgis.	"	934
229	Douglas.	"	
236	White Pigeon.	"	824

(Detroit Division.)

Ms.			Alt.
0	Toledo,	9. Corniferous	
7	West Toledo.	"	
10	Alexis.	"	
15	Vienna.	"	
20	La Salle.	"	
25	Monroe Junction.	" & L. Held'g.	579

* Sunken in the limestone, and has underground communication with Lake Erie

3 Lacustrine deposits of Saginaw Valley 100 feet deep
4 The shallow salt wells here are supplied from the base of the Coal Measures

Lake Shore & Michigan Southern R. R.
Ms. (Detroit Division.)—Con.   Alt.

Ms.	Station	Geology	Alt.
25	Monroe Junc.[5]	9. Cornifer.	
32	Newport.	"	584
38	Rockwood.	"	582
44	Trenton.	" exposu.	584
48	Wyandotte.	10 b. L. Trv.	580
51	Ecorces.		
57	Grand Trunk Jun	11. Huron.	586
62	Det. & Mil. Junc.	"	
65	Detroit.	10 b. L. Trv.	581

(bracketed note: Generally beneath Lacustrine deposits.)

Ms.	Station	Geology	Alt.
0	Monroe Junction.	9. Corniferous.	579
10	Ida.	6. Salina, expos'es	632
17	Petersburg.	9. Corniferous.	670
20	Deerfield.	"	670
26	Wellsville.	10 b, Lit. Traverse.	690
29	Lewanee Junc.	11. Huron.	714
33	Adrian.	"	810

(Jackson Division.)

Ms.	Station	Geology	Alt.
0	Adrian.	11. Huron.	810
4	Lenawee Junc.	"	714
8	Chase's.	"	
13	Tecumseh.	"	807
18	Clinton.	13 a. Marshall.	832
25	Manchester.	"	907
32	Norvell.	"	
36	Napoleon.	{ " exposures extensively quarried.	964
40	Eldred.	13 b. Carb. l. s. (?)	
46	Jackson.	14 c. Cl. Measures	928

(Kalamazoo Division.)

Ms.	Station	Geology	Alt.
0	White Pigeon.	11. Huron.	834
4	Constantine.	"	808
12	Three Rivers.	"	805
17	Moore Park.	"	842
20	Flowerfield.	"	864
24	Schoolcraft.	"	854
30	Portage.	"	860
37	Kalamazoo.	"	777
43	Cooper.	13 a. Marshall.	749
46	Argenta.	"	772
49	Plainwell.	"	774
52	Otsego.	"	710
62	Allegan.	"	705
70	Hopkins.	"	703
73	Hilliards.	"	710
77	Dorr.	13 b. Mich. Salt(?)	696
83	Byron Center.	"	740
89	Grandville.	"	628
93	Eagle Mills.	13 b. Carb. l. s.	601
95	Grand Rapids.	" exposures.	605

(bracketed note: Generally concealed by drift.)

(Lansing Division.)

Ms.	Station	Geology	Alt.
0	Jonesville.	13 a. Mars'll expo.	1097
7	Litchfield.	"	1009
14	Homer.	"	972
22	Albion.	13 b. Carb. l. s.	948
29	Devereux.	14 a. Parma s. s.	990

Lake Shore & Michigan Southern R. R.
Ms. (Lansing Division.)—Con.   Alt.

Ms.	Station	Geology	Alt.
33	Springport.	14 a. Parma s. s.	986
38	Charlesworth.	14 c. Coal Meas.	916
42	Eaton Rapids.	"	864
52	Diamondale.	"	
59	South Lansing.	"	807
60	Lansing.	"	827

Grand Rapids & Indiana Railroad.

Ms.	Station	Geology	Alt.
0	Cincinnati, O.	(See Indiana.)	
143	Lima.	11. Huron.	
147	Sturgis.	"	934
157	Nottawa.	"	852
159	Wasepi.	"	842
163	Mendon.	"	842
168	Portage Lake.	"	834
173	Vicksburg.	"	852
178	Austin.	"	862
185	Kalamazoo.	"	777
194	Travis.	13 a. Marshall.	742
197	Plainwell.	"	744
202	Monteith.	"	828
203	Martin.	"	827
207	Shelby.	"	832
210	Bradley.	" (?)	757
213	Wayland.	13 b. Mich. Salt.	747
221	Ross.	"	777
227	Fisher.	13 b. Carb. l. s.	682
234	Grand Rapids.	"	605
237	D. & M. Crossing.	"	
244	Belmont.	"	661
248	Rockford.	"	689
251	Edgerton.	14 c. Parma s. s.	755
255	Cedar Springs.	14 c. Cl. Measure.	846
257	Lockwood.	"	852
260	Sand Lake.	"	912
262	Pierson.	"	906
266	Maple Hill.	"	872
268	Howard City.	"	
274	Morley.	"	887
281	Stanwood.	"	934
290	Low. Big Rapids.	"	916
291	Up. Big Rapids.	"	
295	Paris.	"	927
302	Reed City.	" (?)	1027
309	Ashton.	" (?)	1152
314	Le Roy.	" (?)	1232
319	Tustin.	13 b. Mich. Salt(?)	1212
331	Clam Lake.	"	
334	Linden.	13 b. Carb. l. s.	874
343	Manton.	"	1142
352	Walton.	13 a. Marshall.	1047
352	Walton.	"	1047
356	Fife Lake.	"	1019
362	South Boardman.	"	1005
371	Kalkaska.	"	1022
375	Leetsville.	"	1050
380	Havana.	"	

5. Extensive quarries, exposing in places the waterlime of Lower Helderberg.

Grand Rapids & Indiana Railroad—Continued.

Ms.		Alt.
384	Mancelona.	13 a. Marshall. 1119
390	Cascade.	11. Huron.
394	Simons.	"
399	Elmira.	" 1284
408	Boyne Falls.	10 b. Lit. Trav.(?) 712
415	Melrose.	" 677
424	Petoskey.	" ext. cliffs. 658

(Traverse City Railroad.)

Ms.		Alt.
852	Walton.	13 a. Marshall. 1047
361	Kingsley.	" 786
364	Mayfield.	11. Huron.
378	Traverse City.	" Lacustrine.

Detroit, Grand Haven & Milwaukee R. R.

Ms.		Alt.
0	Detroit.	10 b. Lit. Traverse. 581
3	L. S. & M. S. Jun.	11. Huron.
4	Gd. Trunk Jun.	" 586
13	Royal Oak.	" 663
18	Birmingham.	13 a. Marshall. 779
26	Pontiac.	" 934
31	Drayton Plains.	13 b. Mich. Salt. 967
33	Waterford.	13 b. Carb. l. s. 988
35	Clarkston.	" 1006
41	Davisburg.	" 959
47	Holly.	14 a. Parma s. s. 938
50	Fenton.	14 c. Coal Meas. 909
55	Linden.	" 874
63	Gaines.	" 859
70	Vernon.	" 770
75	Corunna.	" Mines. 776
78	Owosso.	" 745
88	Ovid.	" 735
92	Shepardsville.	" 749
98	St. Johns.	" 767
107	Fowler.	" 748
112	Pewamo.	" 744
117	Muir.	" 657
124	Ionia.	{ " Quarries in / upper sandstone. 659
132	Saranac.	14 c. Coal Meas. 643
139	Lowell.	14 a. Parma s. s. 641
148	Ada.	13 b. Carb. l. s. 666
158	Grand Rapids.	" ext. quarries. 639
167	Berlin.	13 b. Mich. Salt. 667
173	Coopersville.	13 a. Marshall. 646
180	Nunica.	" 631
186	Spring Lake.	" 596
187	Ferrysburg.	11. Huron. 596
189	Grand Haven.	{ " Remarkable / Sand Dunes. 594

Flint & Pere Marquette Railroad.

Ms.		Alt.
0	Toledo.	9. Corniferous. 579
25	Monroe.	" & 7. Low. Held'g.
34	Grafton.	9. Corniferous.
36	Carlton.	"
39	Waltz.	10 b. Little Traverse.
40	Belden.	11. Huron.

Flint & Pere Marquette Railroad—Continued.

Ms.		Alt.
43	New Boston.	11. Huron.
51	Wayne.	" 662
58	Plymouth.	" 747

(D., L. & L. M. Crossing.)

Ms.		Alt.
62	Northville.	13 a. Marshall.
66	Novi.	"
70	Wixom.	13 b. Mich. Salt.
76	Milford.	"
80	Highland.	13 b. Carb. limestone.
83	Clyde.	"
91	Holly.	14 a. Parma s. s. 938
100	Grand Blanc.	14 c. Coal Meas.
108	Flint.	" 715
115	Mount Morris.	"
119	Pine Run.	"
123	County Line.	"
125	Birch Run.	"
134	Bridgeport.	"
138	S. & M. C. Jun.	"
142	E. Saginaw.6	"
142	E. Saginaw.	{ 14 c. Cl. Mres. buried / 100 ft. ben. Lacus. dp.

(J., L. & S. Crossing.)

Ms.		Alt.
152	Freeland.	14 c. Cl. Mes.
162	Midland.	"
167	Averill.	"
169	Sanford.	"
175	North Bradley.	"
181	Coleman.	"
186	Loomis.	"
191	Clare.	"
196	Farwell.	"
200	Remick.	"
203	Lake.	934 "
209	Chippewa.	"
213	Sears.	"
217	Evart.	"
226	Hersey.	"
230	Reed City. 1037	" (?)
237	Chase.	" (?)
239	Summitville.	" (?)
241	Nirvana.	" (?)
248	Baldwin. 1011	13 b. Carb. l. s.
264	Weldon Creek.	"
272	Amber.	"
278	Ludington.	"

Rocks totally concealed beneath heavy beds of Quaternary deposits. No rock exposures. Drift 200 to 300 feet.

(Flint River Division.)

Ms.		Alt.
0	Flint.	14 c. Coal Meas. 715
4	Junction.	"
8	Genesee.	"
14	Otisville.	14 a. Parma sandstone
19	Otter Lake.	13 b. Mich. Salt.
124	E. Saginaw.6	14 c. Coal Meas. 1441
153	Portsmouth.	"
155	Bay City.	" 592

6. Salt wells 850 feet deep to Marshall sandstone; supplied from overlying Michigan salt group.

Detroit, Lansing & Northern R. R.

Ms.			Alt.
0	Detroit.	10 b. Lit. Traverse.	581
8	Gd. Trunk Junc.	11. Huron.	566
13	Redford.	"	681
15	Fisher's.	"	681
16	Elmwood.	"	688
19	Livonia.	13 a. Mashall.	669
23	Plymouth.	"	747
29	Salem.	"	958
34	South Lyon.	13 b. Carb. l. s.	953
43	Brighton.	14 a. Parma s. s.	929
46	Genoa.	14 c. Coal Meas.	978
52	Howell.	"	
57	Fleming.	"	984
60	Fowlerville.	"	902
65	Le Roy.	"	1282
71	Williamston.	outcrops. "	891
76	Meridan.	"	850
79	Okemos.	"	874
85	Lansing.	"	
86	North Lansing.	"	
92	Delta.	" .	867
94	Ingersoll's.	"	861
97	Grand Ledge.	outcrops."	850
102	Eagle.	"	851
106	Danby.	"	782
109	Portland.	"	730
114	Collins.	"	777
118	Lyons.	"	786
122	Ionia.	{ " Quarries in upper sandstone.	659

Ms.			Alt.
0	Ionia.	14 c. Coal Meas.	659
5	Stanton Junc.	"	821
9	Wood's Corners.	"	881
14	Fenwick.	"	848
19	Sheridan.	"	856
24	Stanton.	"	904

(Cone'd)

Ms.			Alt.
122	Ionia.	14 c. Cl. Me.	659
130	Palmer's.	"	668
133	Chadwick.	"	655
135	Kiddville.	"	802
141	Greenville.	"	819
146	Gowen.	"	848
151	Trufant's.	"	884
153	Maple Valley.	"	928
156	Coral.	"	897
160	Howard.	"	

(Beneath drift, from 100 to 200 feet deep.)

Chicago & West Michigan Railroad.

Ms.			Alt.
.....	Chicago.		
0	New Buffalo.	9. Cornif. S. Dunes.	602
7	Chickaming.	"	
10	Troy.	"	
15	Bridgeman.	"	
16	Morris.	"	
20	Stevensville.	"	
28	St. Joseph.	"	
30	Benton Harbor.	"	

Chicago & West Michigan Railroad. *Continued.*

Ms.			Alt.
39	Coloma.	9. Corf. (?) Sand Dunes	
42	Watervliet.	10 b. Lit. Traverse.(?)	
47	Hartford.	11. Huron.	
54	Bangor.	"	
58	Breedsville.	"	
62	Grand Junction.	"	
75	Rennsville.	"	678
79	Richmond.	" [fossils.	
90	Holland.	13 a.Marshall,outcrops	
90	Holland.	13 a. Marshall.	
95	Zeeland.	"	
104	Hudsonville.	"	
110	Grandville.	13 a. Michigan Salt.	
115	Grand Rapids.	13 b. Carb. limestone.	
90	Holland.	13 a. Marshall.	
99	Olive.	"	
109	Robinson.	"	
110	Nunica.	"	651
116	Fruitport.	"	
126	Muskegon.	"	
126	Muskegon.	"	594
130	B. R. Junction.	"	
136	Twin Lake.	"	
142	Holton.	"	
150	Fremont Centre.	"	
160	Allyton.	13 b. Carb. limestone.	
161	Morgan.	"	
170	Traverse Road.	"	
181	Big Rapids.	14 c. Cl. Measure.	916
126	Muskegon.	13 a. Marshall.	594
142	Whitehall.	"	
143	Montague.	13 b. Mich. Salt.	887
157	Shelby.	{ 13 b. Car. l. s., extensive deta'ed tab.	508
163	Mears.	13 b. Carb. limestone.	
170	Pentwater.	"	595

Grand Rapids, Newaygo & Lake Shore Railroad.

Ms.			Alt.
0	Grand Rapids.	13 b. Carb. l. s.	608
7	Alpine.	"	609
14	Sparta.	"	
19	Tyrone.	"	
21	Casinovia.	"	
25	County Line.	"	
27	Ashland.	"	
80	Grant.	"	
36	Newaygo.	"	
39	Croton.	"	
46	Morgan.	"	
67	Big Rapids.	4 c. Coal Measure.	918

Detroit, Hillsdale & S. W. Railroad.

Ms.			Alt.
0	Ypsilanti.	13 a. Marshall	714
11	Saline.	"	889
17	Bridgewater.	"	
28	Manchester.	"	907
36	Brooklyn.	"	
41	Woodstock.	"	1195

Detroit, Hillsdale & Southwestern R. R.—Continued.		
Ms.		Alt.
44	Somerset.	18 a. Marshall.
49	Jerome.	"
53	North Adams.	"
61	Hillsdale.	"Outcrops foss. 1095
65	Banker's.	" 1067
.....	Reading.	11. Huron. 1200
.....	Camden.	"

Chicago & Canada Southern Railroad.		
0	Fayette.	11. Huron.
7	Morenci.	"
13	Weston.	"
17	Fairfield.	10 b. Lit. Traverse. 799
20	Ogden.	"
25	Blissfield.	" 684
32	Deerfield.	" 870
36	Petersburg.	" 870
40	Dundee.	9. Corniferous. 661
42	North Rainsville.	" ext. quarries.
47	Maybee.	"
50	Exeter.	"
55	Carlton.	"
57	Bryar Hill.	"
61	Flat Rock.	"
67	Slocum Junction.	"

Toledo, Canada Southern & Detroit R. R.		
0	Detroit,	10 b. Lit. Traverse. 581
2	M. C. Junction.	11. Huron.
9	Ecorces.	10 b. Little Traverse.
12	Wyandotte.	" 580
16	Trenton.	9. Corniferous. 684
17	Slocum Junction.	"
15	Stony Creek.	{ "and 7. L. Held. ext. expos. & quar.
20	Monroe.	6. Corn. & 7. Heldberg
25	La Salle.	9. Corniferous.
30	Vienna.	"
34	Alexis.	"
40	Toledo.	"

Deep La. deposits.

Grand Trunk Railroad.		
196	Port Huron.	11. Huron 633
207	Smith's Creek.	"
217	Ridgeway.	"
223	New Haven.	"
237	Mount Clemens.	" 617
250	Milwaukee Junc.	" 602
255	Detroit Junction.	"
258	Detroit.	10 b. L. Trav. Driftover 100 feet deep.

Deep Quaternary deposits. Many surface signs of Petroleum.

Chicago & Grand Trunk Railroad.		
0	Port Huron.	11. Huron. 633
4	Gd. Trunk Junct.	" 636
10	Thornton.	"
19	Emmet.	" 779
27	Capac.	18 a. Marshall. 617

Chicago & Grand Trunk Railroad.—Continued.		
Ms.		Alt.
34	Imlay City	13 a. Marshall. 888
39	Attica.	" 898
46	Lapeer.	13 b. Mich. salt. 880
53	Elba.	13 b. Carb. l. s. 850
57	Davison.	14 a. Parma s. s. 701
66	Flint.	{ 14 c. Coal Measures. Not worked. 715
88	Durand.	601
87	Bancroft.	{ 14 c. Coal Meas. 656 Some exposures, 662 but not worked. 679
96	Perry.	
100	Shaftsburg.	
112	Trowbridge.	{ 14 c. Coal Meas. 651 Slightly worked. 646
115	Lansing.	
120	Millett's.	14 c. Coal Measures.
125	Sevastopol.	"
127	Potterville.	"
134	Charlotte.	" 906
142	Olivet.	14 a. Parma sand s.
147	Bellevue.	13 b. Car. l. s., quar. fos.
152	Madison.	13 b. Michigan salt.
160	Battle Creek.	{ 13 a. Marshall, outcrop fossil. 919
170	Climax.	13 a. Marshall.
175	Scott's.	11. Huron.
179	Indian Lake.	"
183	Vicksburg.	" 852
189	Schoolcraft.	"
200	Marcellus.	"
204	Volinia.	"
209	Jamestown.	10 b. Little Traverse.
213	Cassopolis.	9. Corniferous.
222	Edwardsburg.	" 661
	(Continued in Indiana.)	

Saginaw Valley & St. Louis Railway.		
0	East Saginaw. 6	
2	Saginaw.	14 c. Coal Measures.
6	Tittabawassee Jc	"
9	Swan Creek.	"
11	Graham's.	"
12	Sand Ridge.	"
16	Hemlock.	"
19	Porter's.	"
22	West Mill.	"
26	Wheeler's.	"
28	Breckenridge.	"
35	St. Louis.	"
....	Elm Hall.	"

Chicago & Northwestern Railroad. Green Bay & Lake Superior Line.		
0	Chicago, Ill.	(See Wisconsin.)
264	Menomonee.	4 a. Trenton.
273	Little River.	"
279	Wallace.	"
285	Stephenson.	"
291	Gravel Pit.	"
295	Bagley.	"
302	Kloman.	"
305	Spaulding.	3 a. Calciferous.

Chicago & Northwestern Railroad.
Ms. Green Bay & Lake Superior Line.—*Con.* Alt.

Ms.		Alt.
316	Bark River.	3 a. Calciferous.
321	Ford River.	4 a. Trenton.
328	Escanaba.	"
331	Flat Rock.	"
333	Bay Siding.	"
337	Mason.	" 838
340	Day's River.	"
345	Beaver.	"
352	Maple Ridge.	"
357	Centreville.	" 313
362	Helena.	3 a. Calc., 3 c. Chazy.
369	Little Lake.	2 b. Lake Superior s. s.
370	Smith Mine Junc.	1 a. Laurentian.
382	Cascade Junction	1 b. Huronian.
384	Goose Lake.	"
389	Negaunee.	{ 1 b. Huron, Iron Mines. 1879
393	Ishpeming.	" 1443
401	Marquette.	" 849
441	L'Anse.	2 b. L. Superior s. s. 606

Marquette, Houghton & Ontonagon R. R.

Ms.		Alt.
0	Marquette.	1 b. Huronian. 649
3	Bancroft.	" 936
7	Morgan.	" 1280
8	Eagle Mills.	" 1379
12	Negaunee.	" Iron Mines. 1443
15	Ishpeming.	" Exten. Min. 824
21	Greenwood.	" 1544
25	Clarksburg.	" 1535
26	Humboldt.	"
35	Republic.	" 1510
31	Champion.	" Iron Mines. 1597
38	Michigamme.	" 1864
47	Sturgeon.	1 a. Laurentian. 1643
56	Palmer.	1 b. Huronian. 868
63	L'Anse.	2 b. L. Super. s. s. 608
93	Houghton.	{ 2–4. Eruptive rocks, with Native Copper Mines. 607
93	Hancock.	

Michigan & Ohio Railroad.

Ms.		Alt.
0	Toledo.	{ Deep Lacustrine deposits over 9. Cornif.
23	Dundee.	9. Corniferous.
33	Britton.	11. Huron. No expos.
34	Ridgeway.	" "
38	Tecumseh.	" " 807
51	Cambridge.	" "
60	Addison.	13 a. Marshall.
67	Jerome.	"
70	Moscow.	{ 13 a. Marshall, many expo., fossil casts.
75	Hanover.	13 a. Mar. Quarry 1114
79	Pulaski.	" Expos. 1048
88	Homer.	13 a. Marshall. 1114
100	Marshall.	{ " Old quarry filled. 898

Michigan & Ohio Railroad.
Ms. *Continued.* Alt.

Ms.		Alt.
105	Ceresco.	13 a. Marshall. 892
114	Battle Creek.	{ " Outcrops fossils. 819
123	Augusta.	13 a. Marshall. 789
127	Yorkville.	{ " Rare exposures.
129	Richland.	{ 13 a. Marshall.(?) No exposures.
145	Monteith.	13 a. Marshall.(?) 836
149	Fisk.	" (?)
151	Kellogg.	" (?)
156	Allegan.	{ 11. Huron. No convenient exposurs 708

Port Huron & Northwestern Railroad.
(East Saginaw Division.)

Ms.		Alt.
0	Port Huron.	{ 11. Huron, under Lacustrine. Buried trees.
.....	Gratiot Centre.	11. Huron. 618
11	Kingsley.	" 736
12	Saginaw Junct'n.	"
20	Green's Corners.	"
25	Brockway Centre	"
33	Yorks.	13 a. Marshall.
37	Brown's City.	"
45	Marlette.	"
50	Clifford.	13 b. Mich. Salt Group
59	Mayville.	13 b. Carbon. l. s.
65	Juniata.	14 a. Parma s. s.
71	D. & B. C. Junct.	14 c. Coal Measures(?)
72	Vassar.	" (?) 643
83	Fraukenmuth.	{ 14 c. Coal Measures. Lacustrine.
91	East Saginaw.	{ 14 c. Coal Measures. Lacustrine, 100 feet. Many brine wells.

(Sand Beach Division.)

Ms.			Alt.
0	Port Huron.	Road runs along the strike of the formation.	11. Huron.
15	Grant Centre.		" 745
26	Croswell.		" 730
32	Anderson.		" 742
45	Downing.		
52	Palms.		
70	Sand Beach.		

(Almont Division.)

Ms.			Alt.
0	Port Huron.	11. Huron. No outcrops. Some surface indications of petroleum and asphalt. Gas escapes.	
4	G. T. Junct'n. 886		
11	Burn's.		
16	Lamb's.		
20	Memphis.		
26	Berville.		13 a. Marshall.
34	Almont.		"

Port Huron & Northwestern R. R.—Con. Ms. (Port Austin Division.) Alt.	
0 Port Huron.	11. Huron.
52 Palms.	"
60 Tyre.	13 a. Marshall.
70 Bad Axe.	"
77 Filion.	"
87 Port Austin.	" Salt wells.

Grand Rapids & Indiana Railroad.	
425 Petosky. ***	10 b. Lit. Trav. Fine expo., many fossils.
426 Bay View. ***	
436 Alanson.	10 b. Little Traverse.
460 Mackinaw City.	9. Corniferous. Fine exposures across the Straits.

Michigan Central Railroad. (Mackinaw Division.)	
119 Gaylord.	13 a. Marshall.(?) ¹⁶⁴⁹
127 Vanderbilt.	11. Huron. (?)
138 Wolverine.	" (?)
160 Mullet Lake.	10 b. Little Traverse.
166 Cheboygan.	9. Corniferous.
182 Mackinaw City.	" Outcrops.

Detroit, Mackinaw & Marquette Railroad.	
0 Point St. Ignace.	9. Corniferous. Fine exposures Salina Gypsum near.
.... St. Ignace.	"
9 Allenville.	5. Niagara lime.
11 Moran.	Crossing Niag., Cin., and Calcif. forma-
20 Palms.	tions. Country mostly
23 Johnson.	covered by Peat, Bog, Iron
27 Trout Lake.	Ore, and Drift. At Au
37 Hendrie.	Train is outlet of a de-
55 Newberry.	pressed passage to White
64 McMillan.	Fish River and Little
76 Seney.	Bay de Noquet.
84 Driggs.	
91 Creighton.	
101 Jerome.	
109 Munising.	2 b. L. Superior s. s. Cliffs. Fine expos- ures on Grand Is.
122 Au Train.	2 b. L. Superior s. s.
127 Rock River.	"
132 Deerton.	"
134 White Fish.	"
136 Sand River.	" 627
147 Chocolay.	" 617
151 Marquette.	1 b. Huronian. 649 Glaciated rocks.

Grand Trunk Railway. (Michigan Air Line Branch.)	
0 Ridgeway.	11. Huron.
25 D. & B. C. Cross.	13 a. Marshall.
35 Pontiac.	"
39 Orchard Lake.	"

Grand Trunk Railroad. Ms. (Michigan Air Line Branch.)—Con. Alt.	
59 South Lyon.	13 b. Mich. Salt Gr.
67 Hamburg.	14 a. Parma s. s.(?)
106 Jackson.	14 c. Coal Measures.

Michigan Central Railroad. (South Haven Division.)	
0 Kalamazoo.	11 Huron. 777
9 Alamo. 705	
15 Kendall's. 792	
18 Pine Grove. 777	Whole dist. over Huron group. Only very scant outcrops. Surface level. Some scattered blocks of hard purple sand-stone not identified.
23 Bloomingdale 781	
25 Berlamont. 700	
28 Columbia	
29 Grand Junc. 678	
32 Lacota.	
40 South Haven. 588	

Chicago & Northwestern Railroad. (Menominee River Railroad.)	
0 Chicago.	5. Niagara l. s.
305 Powers.	2 b. L. Superior s. s.
313 Cedar.	"
216 Wauceda.	1 b. Huronian.
319 Sturgeon.	Menominee Iron
323 Vulcan.	Ranges. Many outcrops of Diorites, Quartzites, Gran- "
.... Curry.	ites, and vast beds of Sla- "
326 Norway.	tes and Marbles, besides the "
.... Indiana.	ores of Iron which are now extensively worked. "
330 Quinnesec.	"
334 Iron Mountain, M	"
336 Lake Antoine Jc.	"
339 River Siding.	"
343 Spread Eagle, Ws.	"
.... Commonwealth J.	"
349 Florence, Wis.	"
349 Florence, Wis.	"
356 Stager, Mich.	These roads pass through the "
358 Mastodon.	"
361 Panola.	"
364 Crystal Falls.	"
353 Brule.	"
356 Stager.	"
361 Armstrong.	"
371 Palatka.	"
373 Stambaugh, Mich	"
374 Iron River.	"

Toledo, Ann Arbor & Grand Trunk R. R.	
0 Toledo.	Deep Lacustrine, over 9. Corniferous.
18 Monroe Junction.	9. Corniferous.
22 Dundee.	9. Cornif. Quarries nr.
32 Milan.	13 a. Marshall.
40 Pittsfield.	13 b. Mich. Salt Gp.
46 Ann Arbor.	Deep (204 ft.) Drift, over 13 b. Michigan Salt Group.
55 Worden's.	13 b. Michigan salt.
61 South Lyon.	13 b. Carbon. l. s.

Indiana.

BY PROF. JOHN COLLETT, STATE GEOLOGIST.

LIST OF THE GEOLOGICAL FORMATIONS FOUND IN INDIANA.[1]

20. Quaternary.*	13 b. Upper Sub-Carbonifer's.	5 c. Niagara.
14 c. Upper Coal Measures.	13 a. Lower Sub-Carbonifer's.	5 b. Clinton.
14 b. Middle Coal Measures.	9-12. Devonian.	4 c. Cincinnati.
14 a. Millstone Grit and Lower Coal Measures.		

Michigan Central Railroad.

Ms.			Alt.
0	Chicago.	(See Illinois.)	589
23	Gibson's.	5 c. Niagara.	600
29	Tolleston.	"	607
35	Lake.	"	617
44	Porter.	"	647
50	Furnessville.	"	609
56	New Buffalo.	"	602
	(Continued in Michigan.)		

(Joliet Division.)

Ms.			Alt.
0	Lake.	5 c. Niagara.	617
7	Ross.	"	636
14	Dyer.	"	635
45	Joliet, Ill.	(See Illinois.)	543

Lake Shore & Michigan Southern R. R.
(Western Division.)

Ms.			Alt.
0	Chicago.		589
14	Colehour.	5 c. Niagara.	
30	Miller's.	"	635
41	Chesterton.	"	589
45	Burdick.	"	
49	Otis.	"	765
51	Holmesville.	"	800
59	Laporte.	9-12. Devonian.	811
66	Rolling Prairie.	"	821
73	New Carlisle.	"	772
75	Terre Coupee.	"	760
80	Warren.	"	781
86	South Bend.	"	735
90	Mishawaka.	"	722
96	Osceola.	"	737
101	Elkhart.	"	755

(Air Line Division.)

Ms.			Alt.
0	Elkhart.	9-12. Devonian.	755
10	Goshen.	"	739
18	Millersburg.	"	635
25	Ligonier.	"	666
30	Wawaka.	"	696
34	Brimfield.	"	945
41	Kendallville.	"	974

Lake Shore & Michigan Southern R. R.
(Air Line Division)—Continued.

Ms.			Alt.
47	Corunna.	9-12. Devonian.	957
50	Sedan.	"	923
54	Waterloo.	"	897
62	Butler.	"	863
69	Edgerton.	"	830
	(Continued in Ohio.)		

Baltimore & Ohio Railroad.
(Chicago Division.)

Ms.			Alt.
0	Chicago.	(See Illinois.)	589
34	Mich. Cen. Junc.	5 c. Niagara.	
50	L. N. A. & C. Junc.	"	
58	Wellsboro.	"	
72	Walkerton Junc.	9-12. Devonian.	
89	Bremen.	"	
106	Milford Junction.	"	841
110	Syracuse.	"	870
118	Cromwell.	"	
128	Albion.	"	927
138	Avilla.	"	989
143	Garrett.	"	893
146	Auburn Junc.	"	868
147	Auburn.	"	872
163	Hicksville.	"	

Pittsburg, Fort Wayne & Chicago R. R.

Ms.			Alt.
0	Chicago.	(See Illinois.)	589
16	Sheffield.	5 c. Niagara.	
20	Cassello.	"	
24	Clarke.	"	
31	Liverpool.	"	
37	Wheeler.	"	666
44	Valparaiso.	"	736
53	Wanatah.	"	781
59	Hanna.	9-12. Devonian.	
78	Donelson.	"	
84	Plymouth.	"	1781
95	Bourbon.	"	
99	Etna Green.	"	
104	Selby.	"	
109	Warsaw.	"	824

* Four-fifths of the State of Indiana is covered with drift. It is 90 feet to the rock in Indianapolis. At some points north of Wabash River the drift has been bored into 400 to 600 feet. It thins out as you go toward Ohio River, does not reach it at some points, and is sparingly found south of that stream. (See Notes No. 62 Ohio and No. 62 West Virginia.)

Ms	Pittsburg, Fort Wayne & Chicago R. R.—Continued.	Alt.
115	Kosciusko.	9–12. Devonian.
117	Pierceton.	"
122	Larwill.	"
129	Columbia.	" 886
140	Arcola.	" 833
148	Fort Wayne.³⁴	" 775
158	Maples.	"
	(Continued in Ohio.)	

Pittsburg, Cincinnati & St. Louis R. R.
(First Division.)

Ms		Alt.
0	Indianapolis.	9–12. Devonian. 709
11	Cumberland.	"
17	Philadelphia.	"
21	Greenfield.	"
28	Cleveland.	"
30	Charlottsville.	"
84	Knightstown.	"
85	Raysville.	"
38	Ogden's.	5 c. Niagara.
39	Dunreith.	"
44	Lewisville.	"
51	Dublin.	"
53	Cambridge City	" 941
58	Germantown.	"
63	Centerville.¹ ⁴ ³⁵	4 c. Cincinnati.
68	Richmond.²	" 969
74	New Paris.²	" 825
79	Wiley's ²	"
	(Continued in Ohio.)	

(Second Division.)

Ms		Alt.
0	Chicago.	589
20	Dalton.	5 c. Niagara.
27	Lansing.	"
84	Shereville.	"
41	Crown Point.	" 714
47	Cassville.	" 654
51	Hebron.	" 714
61	Koutt's.	" 685
67	La Crosse.	9–12. Devonian. 675
77	North Judson.	" 702
91	Winamac.⁴⁶	" 715
97	Star City.	" 706
101	Rosedale.	"
105	Royal Centre	" 735
111	Gebhardt.	" 762
117	Logansport.	" 606
121	Anoka.	9–12. Devonian. 696
127	Onward.	" 763
132	Bunker Hill.	" 800
140	North Grove.	" 817
142	Amboy.	" 810
145	Converse.	" 815
148	Mier	" 816
157	Marion.	5 c. Niagara. 811

Ms	Pittsburg, Cincinnati & St. Louis R. R.—(Second Division.)—Continued.	Alt.
162	Jonesboro.	5 c. Niagara. 846
169	Upland.	"
175	Hartford.	"
185	Dunkirk.	"
189	Red Key.	"
193	Power's.	"
197	Ridgeville.	" 994
200	Deerfield.	"
203	Warren.	" 731
210	Union.	" 1108
	(Continued in Ohio.)	

(Columbus, Chicago & Indiana Central Division.)

Ms		Alt.
0	Chicago.	589
117	Logansport.⁸	9–12. Devonian. 606
122	Anoka.	" 696
127	Walton.	"
130	Lincoln.	"
133	Galveston.	"
139	Kokomo.	"
145	Tampico.	5 c. Niagara.
149	Nevada.	"
152	Windfall.	"
157	Curtisville.	"
161	Elwood.	" 858
166	Frankton.	"
171	Florida.	"
175	Anderson.³	" 850
.....	Bellefontaine Crossing.	
184	Middletown.	5 c. Niagara.
187	Honey Creek.	"
190	Sulphur Springs.	"
195	Junction.	"
197	New Castle.	" 1075
201	Ashland.	"
204	Millville.	"
208	Hagerstown.	"
215	Washington.	" 884
....	Centreville Pike.	
224	Richmond.⁸	4 c. Cincinnati. 885

(Indianapolis & Vincennes Division.)

Ms		Alt.
0	Indianapolis.	9–12. Devonian. 709
4	Maywood.	" 695
8	Valley Mill.	". 759
11	West Newton.	13 a. L. Sub-Carb. 779
12	Friendswood.	" 742
16	Mooresville.	" 625
18	Mathews'.	" 691
20	Brooklyn.	" 659
23	Centerton.³⁷	" 681
26	Hastings.	" 607
30	Martinsville.⁷⁸	" 598
33	Hynds.	" 600
87	Paragan.	" 577
44	Gosport.³⁷	13 b U Sub-Carb.⁵⁹⁸

---

1. Glacial markings.	4. Pre-historic mounds.
2. Crowded with fossils of Lower Silurian age.	5. Coal fossils.
3. Rich in fossils, Devonian and Up. Silurian.	6. Devonian fossils.

Pittsburg, Cincinnati & St. Louis R. R. Mᵘ. (Indianapolis & Vincennes Div.)—Con. Alt.	
53 Spencer.³⁷⁴³⁸	13 b. U. Sub.Carb. ⁵⁵⁷
62 Freedom.	" 538
65 Farmer's.	14 a.Millstone Grit.⁵²⁸
71 Worthing'n.⁴⁴⁸⁷	{ 14 a.Mills.Gr. & 14 b. / L. Coal Meas. 522
78 Switz City.⁸⁹	" 526
82 Lyons.	" 509
87 Marco.⁷⁴	" 482
97 Edwardsp't.⁵⁴⁸⁷	14 c. U. Coal Meas.⁴⁶⁰
108 Bruceville.	" 515
117 Vincennes.³⁷	" 417

Detroit & Eel River Railroad.

0 Logansport.⁶	9–12. Devonian.
18 Denver.	"
21 Chili.	" 725
27 Roann.	" 750
33 Laketon.	" 762
37 N. Manchester.	" 775
45 Collamer.	" 795
47 South Whitley.	" 805
51 Taylor's.	" 854
56 Columbia City.	" 836
62 Collin's.	" 870
66 Cherubusco.	" 895
70 Potter's.	" 881
74 C. R. Crossing.	"
76 Cedar Creek.	" 861
81 Auburn Junction.	" 868
82 Auburn.	" 872
88 Mooresville.	" 877
93 Butler.	" 863

Wabash, St. Louis & Pacific Railroad.
(Late Toledo, Wabash & Western R. R.)

0 Toledo.	9–12. Devonian.
88 New Haven.	" 753
94 Fort Wayne.	5 c. Niagara. 775
109 Roanoke.	"
118 Huntington.	" 784
131 Lagro. {41	" 698
136 Wabash.⁷ {43	" 740
150 Peru.⁸	" 655
157 Waverly.	"
166 Logansport.⁶	{ 9–12. Devonian,10 b. / Hamilton. 606
180 Rockfield.	"
186 Delphi.⁹	"
195 Buck Creek.	"
203 Lafayette.	13 a. L Sub-Carb. 59?
213 West Point.	"
225 Attica.⁴¹	14 a Mills. Grit. 540
233 West Lebanon.	" 720
242 State Line.	14 c. Mid. Coal Meas.
(Continued in Illinois.)	

Wabash, St. Louis & Pacific, R. R.—Con. Ms. (L. M. & B Division.) Alt.	
0 Lafayette Junc.	18 a. L. Sub-Carb. 595
8 Porter's.	" 647
10 Montmorency.	" 672
21 Templeton.	14 b. L. Coal Meas. 675
23 Oxford.	" 703
29 Boswell.	" 734
37 Ambia.	" 710

Cincinnati, Lafayette & Chicago R. R.

..... Cincinnati.	
..... Indianapolis.	9–12. Devonian. 709
0 Lafayette.	13 a. L Sub-Carb. 505
7 Porter's.	" 647
9 Montmorency.	" 672
13 Otterbien.	13 b. L. Sub-Carb. 685
18 Templeton.	14 b. L. Coal Meas.675
23 Atkinson.	"
28 Fowler.	"
35 Earl Park.10	"
41 Raub.	"
46 Sheldon.	"

Indianapolis, Bloomington & Western R.R.

0 Indiana.	9–12. Devonian.
14 Brownsburg.	"
18 Pittsboro.	18 a. Lower Sub-Carb.
22 Lizton.⁴⁴	"
27 Jamestown.⁸⁷	"
33 New Ross.	18 b. Upper Sub-Carb.
44 Crawfordsville 11	" 741
54 Wayneto'n.¹²⁴⁴⁵	14 a. Millstone Grit
65 Veedersburg.	14 a. Mills.Gt. & 14 b.L.
72 Covington.¹³⁴³⁹	14 c. " Coal Meas.
85 Danville, Ill.¹³	14 c.
(Continued in Illinois.)	

Cleveland, Columbus, Cincinnati & Indianapolis Railroad.
(Indianapolis Division.)

0 Indianapolis.	9–12. Devonian. 709
9 Lawrence.	" 872
14 Oakland.	" 846
16 McCord's.	" 854
21 Fortville.	" 857
28 Pendleton.¹⁴⁴⁴⁶	" 847
35 Anderson.⁴⁷	5 c. Niagara. 880
41 Chesterfield.	" 907
43 Daleville.	" 910
48 Yorktown.	" 924
54 Muncie.	" 948
60 Selma.	" 1005
67 Farmland.	" 1037
75 Winchester.	" 1069
84 Union.	" 1108
(Continued in Ohio.)	

7. Upper Silurian cephalipodes.	11. Keokuk crinoids.
8. Upper Silurian and Devonian fossils.	12. Glacial markings.
9. Pentamerous and black slate.	13. Coal measures fossils.
10. Drift and knolls.	14. Devonian fossils.

Indianapolis & St. Louis Railroad.

Ms.			Alt.
0	Indianapolis.	9–12. Devonian.	709
2	Asylum.	"	
6	Sunnyside.	13 a. Lower Sub-Carb.	
8	Spray.	"	
12	Avon.	"	
16	Easton.	"	
19	Danville.	"	813
23	Hadley.	"	
27	Reno.	"	
31	Malta.	13 b. Upper Sub-Carb.	
32	Darwin.	"	
38	Greencastle.	{ 13 b. U. Sub-Carb. & 14 a. Mills. Grit.	
44	Fern.	"	
48	Lena.	14 a. Millstone Grit.	
53	Carbon.	14 b. Low. Coal Meas.	
56	Perth.	"	
61	Fountain.	"	
64	Grant.	14 c. Mid. Coal Meas.	
67	Markle.	"	
69	Gravel Pit.	"	879
72	Terre Haute.	"	498

St. Louis, Vandalia, Terre Haute & Indianapolis Railroad.

Ms.			Alt.
0	Indianapolis.	9–12. Devonian.	709
4	Fairview.	"	
9	Bridgeport.	13 a. L. Sub-Carb	748
14	Plainfield.	"	742
17	Cartersburg.	"	
19	Belleville.	"	
20	Clayton.	"	859
25	Amo	"	820
28	Coatsville.	"	878
33	Fillmore.	13 b. U. Sub-Carb.	844
39	Gr'ncastle.[15][16]	13 b.&14a.Mills.Gt.	834
43	Hamrick's.	14 a. Mills. Grit.	703
47	Reelsville.	"	618
50	Eagle's.	"	
53	Harmony.	14 b. L. Coal Meas.[17]	672
54	Knightsville.[16]	[16]	
57	Brazil.[16][49]	"	645
60	Williams.	14 c. M. Coal Meas.[16][16]	
62	Staunton.	"	643
65	Seeleyville. } [50]	"	885
73	Terre Haute.	"	492

Cincinnati, Hamilton & Indianapolis R. R.

Ms.			Alt.
0	Cincinnati.	(See Ohio.)	
25	Hamilton.	4 c. Cincinnati.	
32	McGonigle's.	"	
39	Oxford.	"	703
44	College Corner.	"	
52	Liberty.	"	979
58	Brownsville.	"	793
66	Connersville.	5 c. Niagara.	832

Cincinnati, Hamilton & Indianapolis R. R. Continued.

Ms.			Alt.
76	Glenwood.	5 c. Niagara.	
84	Rushville.	"	972
91	Arlington.		
98	Morristown.	9–12. Dev. 9 c. Cor.[842]	
103	Fountaintown.	"	
123	Indianapolis.	"	709

Indianapolis, Cincinnati & Lafayette R. R.

Ms.			Alt.
0	Cincinnati.	(See Ohio.)	
18	Valley Junc.[76]	"	
20	Elizabethtown.	"	446
25	Lawrenceburg.	4 c. Cincinnati.	479
26	Newton.[18]	"	
33	Guilford.	"	508
34	Hansell's.	"	
40	Harman's.[18]	"	747
42	Weisburg.	"	929
46	Sunman's.	"	1015
48	Spades.[51]	5 c. Niagara.	1015
51	Morris.	"	982
54	Batesville.	"	968
60	New Point.	"	
62	Smith's Crossing.	"	1008
65	McCoy's.	"	1027
68	Greensburg.	"	942
74	Adams.	"	880
78	St. Paul.[17]	"	852
81	Waldron.[17]	"	819
84	Prescott.	"	
88	Shelbyville.	9–12 Devonian.	769
95	Fairland.	"	774
99	London.	"	775
100	Brookfield.	"	
102	Acton,	"	792
106	Gallaudet.[19]	"	852
115	Indianapolis.	"	709
125	Augusta.	13 b. Up. Sub-Carbon.	
130	Zionsville.	"	
135	Whitestown.	"	
138	Holmes.	"	800
143	Lebanon.	"	925
148	Hazelrigg.	"	
152	Thorntown.	"	818
157	Colfax.	"	825
163	Clark's Hill.	"	782
166	Stockwell.	"	
171	Culver's.	"	
179	Lafayette.	"	895

Jefferson, Madison & Indianapolis R. R.

Ms.			Alt.
0	Indianapolis.	9–12 Devonian.	722
7	Southport.	"	761
11	Greenwood.	"	858
13	Worthsville.	"	
15	Whiteland.	"	805
20	Franklin.[20]	"	732

15. Good fossils.
16. Block coal.
17. Rich in Upper Silurian fossils; good quarries.
18. Lower Silurian fossils.
19. Healthy summit
20. Collette Glacial River bed.
21. Lower Silurian fossils.
22. Geodes.

Jefferson, Madison & Indianapolis R. R.— Continued.

Ms.		Alt.
25	Amity.	9-12. Devonian.	693
31	Edinburg.	13 a. L. Sub-Carb.	674
35	Taylorsville.	"	656
38	Lowell.	"	636
41	Columbus.	"	630
46	Walesboro.	"	613
48	Waynesville	"	607
52	Jonesville.	"	594
57	Rockford.[52]	"	585
59	Seymour.	"	605
64	Chestn't R'ge }20	"	553
66	Langdon's.	9-12. Devonian.	529
69	Retreat.	"	540
71	Crothersv'le.	"	562
75	Austin.	"	549
77	Marshfield.	"	543
82	Vienna.	13 a. L. Sub-Carb.	566
89	Henryville.	9-12. Devonian.	479
93	Memphis.[50]	"	490
100	Sellersburg.	"	478
108	Jeffersonv'le	"	455

Ohio & Mississippi Railroad— Continued.

Ms.		Alt.
165	Montgomery's.[57]	14 b. L. Coal Mrs.	
178	Washington.[57]	14 c. Mid. Cl. Mrs.	484
180	Wheatland.[78]	"	
185	Richland.	"	
191	Vincennes.[58]	14 d. Up. Coal Mrs.	
	(Continued in Illinois.)		

Fort Wayne & Jackson Railroad.

(L. S. & M. S.)

Ms.		Alt.
0	Fort Wayne.	9-12. Devonian.	762
16	New Era.	"	859
23	Auburn.	"	872
28	Waterloo.	"	914
33	Summit.	"	1001
37	Pleasant Lake.	"	975
42	Angola.	"	1052
50	Fremont.	"	1055
54	State Line.	"	
	(Continued in Michigan.)		

Ohio & Mississippi Railroad.

Ms.		Alt.
0	Cincinnati.	(See Ohio.)	
20	Lawrenceburg.	4 c. Cincinnati.	479
24	Aurora.[76]	"	493
26	Cochran.	"	493
33	Dillsboro. }21	"	
37	Cold Springs	"	
40	Moore's Hill.	"	
42	Milan.	"	955
45	Pierceville.	"	1010
47	Delaware.	"	
52	Osgood.	5 c. Niagara.	950
56	Poston.	"	
58	Holton.	"	
62	Nebraska.	"	
66	Butlerville.	"	
78	North Vernon.[54]	9-12. Devonian.	727
79	Hardenburg.	"	
83	Fleming's.	"	
87	Seymour.	"	605
92	Shields' Mill. .	13 a. L. Sub-Carbon.	
98	Brownstown.[77]	"	
101	Velonia.	"	
106	Medora.	"	
111	Sparksville.	"	
114	Ft. Ritner.[23,55]	"	
117	Tunnelton.[55]	13 a. and 13 b.	
121	Scotville.	13 b. Up. Sub-Carbon.	
127	Mitchell.	"	676
133	Georgia.	"	
139	Huron.[23,56]	13 b. & 14 a. Mills. Gt.	
150	Shoals.[24]	14a. & 14b. L.C.Mr.[450]	
158	Loogootee.	14 b. L. Coal Mrs.	532
162	Clark's.[57]	"	

Grand Rapids & Indiana Railroad.

Ms.		Alt.
275	Sturgis.	(See Michigan.)	
286	La Grange.	9-12. Devonian.	915
290	Valentine.	"	952
295	Wolcottville.	"	938
297	Rome City.	"	920
304	Kendallville.	"	974
310	Avilla.	"	969
314	La Otto.	"	
320	Huntertown.	"	827
333	Fort Wayne.	"	752

Cincinnati, Richmond & Fort Wayne Railroad.

Ms.		Alt.
333	Fort Wayne.	9-12. Devonian.	
338	Adams.	5 c. Niagara.	796
354	Decatur.	"	807
360	Monroe.	"	
366	Berne.	"	
370	Geneva.	"	
374	Briant.	"	
381	Portland.	"	904
392	Ridgeville.	"	993
400	Winchester.	"	1088
406	Snow Hill.	"	
409	Lynn.	"	1174
416	Newport.	"	
418	Haley.	"	
422	Parry.	"	
424	Richmond.	4 c. Cincinnati.	969
(Co	ntinued in Ohio,	Cinn. Rich. & Ch. R.R.)	

23. Kaolin and caves.
24. Pentremites.

25. Glass sand.
26. Good Sub-Carbonif. fossils and Oolitic stone.

Fort Wayne, Muncie & Cincinnati R. R. Ms.		Alt.
0 Fort Wayne.	9-12. Devonian.	775
3 Wabash Junc'n.	"	730
7 Ferguson's.	"	806
11 Sheldon.	"	
14 Ossian.	"	831
19 Eagleville.	"	
24 Bluffton.	5 c. Niagara.	857
35 Keystone.	"	871
38 Montpelier.	"	867
47 Hartford.	"	895
54 Eaton.	"	
65 Muncie.	"	943
71 McGowan's.	"	
75 Springport.	"	1015
78 Summit.	"	818
80 N. C. Junction.	"	
83 New Castle.	"	1075
90 New Lisbon.	"	1096
96 Cambridge City.	"	941
98 Milton.	4 c. Cincinnati.	
103 Beeson's.	"	875
108 Connersville.	"	832

Cincinnati, Wabash & Michigan R. R.		
0 Anderson Junc.	8. Orisk. & 9 c. Cor.	894
13 Alexandria.	5 c. Niagara.	872
34 Marion.	"	811
54 Wabash.	"	742
69 N. Manchester.	9-12. Devonian.	774
90 Warren.	"	731
103 Milford.	"	850
115 Goshen.	"	759
125 Elkhart.	"	741

Wabash, St. Louis & Pacific Railway.		
0 Indianapolis.	19-12. Devonian.	709
6 Malott Park.	"	
11 Castleton.	"	
15 Fisher's.	"	
17 Britton's.	"	
22 Noblesville.	"	
28 Cicero.	"	
31 Arcadia.	"	
34 Buena Vista.	"	
40 Tipton.	5 c. Niagara.	607
42 Jackson's.	"	
46 Sharpsville.	"	
49 Fairfield.	"	
54 Kokomo.	"	
59 Cassville.	"	884
61 Bennett's.	"	
63 Miami.	"	
67 Bunker Hill Cr'g.	"	800
75 Peru.	"	655
81 Courter.	9-12. Devonian.	
83 Denver.	"	

Wabash, St. Louis & Pacific Railway—Continued. Ms.		Alt.
85 Deed's.	9-12. Devonian.	
88 Birmingham.	"	
90 Lincoln.	"	
93 Wagner's.	"	
98 Rochester.	"	
102 Sturgeon.	"	
103 Tiosa.	"	
105 Walnut.	"	
108 Railsback's.	"	
110 Argos.	"	
118 Plymouth.	"	769
125 Tyner.	"	
128 Knott's.	"	
132 Walkerton.	"	
136 Kankakee.	"	622
141 Stillwell.	"	
148 La Porte.	"	811
155 Webbers.	5 c. Niagara.	
161 Michigan City.	"	603

Louisville, Evansville & St. Louis R. R.		
0 Princeton.	14 c. U. Coal Mrs.	433
5 Lyle's.	"	
10 Mount Carmel.	(See Illinois.)	
11 C. & V. Junction.	"	
15 Brown's.	"	
19 Bellmont.	"	
27 Crackle's.	"	
29 Albion, Ill.	"	

Louisville, New Albany & Chicago R. R.		
0 New Albany.[59]	9-12. Devonian & 13 a. L. Sub-Carb.	438
6 Smith's Mills.	"	
12 Wilson's.	"	
18 Providence [36][60]	13 a. Lower Sub-Carb.	
23 Pekin.	"	
27 Farabee's.	"	
30 Harristo'n.[26][61]	13 b. U. Sub-Carb.	872
35 Salem.[26][61]	"	714
40 Hitchcock's.	"	
45 Campbellsburg.	"	
47 Saltillo.	"	
52 Lancaster.	"	
56 Orleans.[26][63]	"	633
61 Mitchell.[26]	"	676
65 Juliet.	"	
71 Bedford.[32][63]	"	679
78 Salt Creek.	"	
82 Guthrie.[27]	"	
85 Harrodsburg.	"	506
89 Smithville.	"	717
92 Clear Creek.	"	
97 Bloomington.[26]	"	743
101 Wood Yard.	"	

27. Geodes.
28. Cave and brook.

29. Rich in Keokuk crinoides.
30. Ferns.

Louisville, New Albany & Chicago R. R.—Continued.

Ms.		Alt.
104	Ellettsville [26][62] 13 b. U. Sub-Carb.[682]	
109	Stinesville.[63] "	
113	Gosport. "	595
117	Spring Cave.[96] "	
122	Quincy.[79] "	749
125	Oakland. "	846
128	Cloverdale. "	782
134	Putnamville. "	687
139	Greencastle. 13b.&14a.U.C.M.[884]	
143	Maple Grove. 13 b. Up. Sub-Carbon.	
148	Bainbridge. "	936
152	Carpentersville. "	
156	Ashby's. "	
159	Ladoga. "	
163	Whitesville. "	874
170	Crawfordsville [29] "	741
175	Cherry Grove. "	
180	Linden. "	
184	Corwin. "	
187	Raub's. "	
190	Taylor's. "	854
198	Lafayette. 13 a. L. Sub-Carb. [558]	
204	Battle Ground. "	
211	Brookston. "	
215	Chalmers. "	707
221	Reynolds. { 13 a. L. Sub-Carb., & 9-12. Devonian[692] }	
229	Bradford. 9-12. Devonian.	
237	Francesville. "	
244	Medarysville. "	
252	San Pierre. "	689
260	La Crosse. "	673
267	Wanatah. 5 c. Niagara.	731
271	Haskell's... "	
273	Lake Huron Cros "	
276	Westville. "	789
279	Otis. "	765
281	Beatty's. "	
288	Michigan City. "	601

Chicago & Eastern Illinois Railroad.

Ms.		Alt.
0	Terre Haute. 14 c. Mid. Cl. Meas.[492]	
5	Ellsworth. "	488
11	Atherton. "	522
15	Clinton.[80][65] "	494
20	Summit Grove. "	520
23	Hillsdale. "	452
25	Highland. "	
28	Opedec. "	510
31	Newport.[81] "	494
37	Eugene.[81] "	507
55	Danville, Ill. (See Illinois.)	

Evansville & Terra Haute Railroad.

Ms.			Alt.
0	Evansville.[80]	14 c. U. Coal Mrs.	875
3	Fair Ground.	"	
5	Erskine.	"	
10	Ingle's.	"	
13	Stacer's.	"	
15	St. James.	"	
17	Haubstadt.	"	
20	Fort Branch.[80]	"	
24	King's.	"	
27	Princeton.	"	483
31	Patoka.	"	
38	Hazelton.	"	
40	Decker's.	"	
45	Purcell's.	"	
51	Vincennes.	"	417
57	John Smith's.	"	
62	Emison's.	"	
64	Busseron.	"	
66	Oak Town.	"	
68	Griswold.	"	
70	Ehrman.	"	
73	Carlisle.	"	
77	Paxton's.	{ 14 c. Middle Coal Measures.	
83	Sullivan.[83]	"	536
88	Shelburn.[33] }[66]	"	
93	Farmersbu'g }	"	
97	Hartford.	"	
101	Young's.	"	
109	Terre Haute.	"	495

St. Louis & Southeastern Railroad.
(Louisville & Nashville.)

Ms.			Alt.
....	St. Louis.	(See Illinois.)	
136	Upton.	14 c. U. Coal Mrs.	869
142	Mount Vernon.	"	407
154	Belknap.	"	456
161	Evansville.	"	378
	(Continued in Kentucky.)		

Chicago & Atlantic Railway.

Ms.			Alt.
0	Marion, O.		965
92	Rivare, Ind.	5 c. Niagara.	847
96	Decatur.	"	820
101	Preble.	"	833
103	Kirtland.	"	846
106	Tocsin.	9-12. Devonian.	849
109	Kingsland.	"	873
113	Union.	"	853
118	Markle.	5 c. Niagara.	829
122	Simpson.	"	827
127	Huntington.	"	762
131	Clear Creek.	9-12. Devonian.	829

31. Coal measures fossils.
32. Caves.
33. Roof of coal frescoed with plant remains.
34. Ancient outlet of Lake Erie.
35. Lower Silurian fossils and glacial marks.
36. Beaver dams.
37. Prehistoric mounds.
38. Oolitic amistone.
39. Coal measures and L.
40. Coal K. and fossils.
41. Ancient outlet of Lake Erie.
42. Choice lime.
43. Sandrock quarries.
44. Elevated plateau.
45. Glacial marks.
46. Coal plants; Lower Devonian fossils.

Chicago & Atlantic Railway.			Chicago & Grand Trunk Railroad.		
Ms.		Alt.	Ms.		Alt
186 West Point.	"	268	0 Chicago. Ill.	5 c. Niagara.	589
138 Willis.	"	854	8 Elsdon.	"	
142 New Madison.	"	834	13 Sherman.	"	609
144 Bolivar.	"	810	19 Blue Island.	"	
146 Newton.	"	769	23 South Lawn.	"	
147 Laketon.	"	769	25 Thornton.	"	813
153 Harrisburgh.	"	842	36 Griffith's.	"	
158 Akron.	"	875	39 Redesdale.	"	
163 Hoover's.	"	824	45 Ainsworth.	"	
168 Rochester.	"	789	55 Valparaiso.	"	738
174 Germany.	"	767	64 Haskell's.	"	
178 Leiter's.	"	762	71 Wellsboro.	9-12. Devonian.	
180 Marshland.	"	757	75 Kingsbury.	"	742
184 Monterey.	"	739	80 Stillwell.	"	
187 Ora.	"	737	84 Fish Lake.	"	
194 Aldine.	"	728	91 Crum's Point.	"	
199 N. Judson.	"	705	99 Oliver's,	"	
205 Mallard.	"	680	100 South Bend.	"	788
208 Wilder's.	"	677	104 Mishawaka.	"	722
214 Kouts.	5 c. Niagara.	691	110 Granger's.	"	
220 Boone Grove.	"	727			
222 Hulburt's.	"	726			
226 Palmer.	"	749	**Indiana, Bloomington & Western R. R.**		
229 Winfield.	"	711			
233 Crown Point.	"	710			
240 Griffith.	"	645	0 Indianapolis.	9-12. Devonian.	709
243 Highlands.	"	626	2 Mass. Avenue.	"	
245 Calumet.	"	600	4 Belt Road.	"	722
249 Hammond.	"	595	9 Hunter's.	"	
261 Auburn, Ill.	"	866	14 Mount Comfort.	"	870
263 Englewood.	"	604	18 Mohawk.	"	
264 51st Street.	"		22 Maxwell.	"	920
268 Archer Avenue.	"		26 Willow Branch.	"	950
269 Chicago.	"	589	31 Wilkinson.	"	
			36 Kennard.	"	1057
Bedford & Bloomfield Railroad.			41 Nixon.	"	1015
			44 New Castle.	4 c. Cincinnati.	1075
0 Bedford.	13 b. L. Carb l. s.	679	49 Messick.	4 c. Cincinnati.	1090
7 Avoca.	"		52 Moorland.	"	
12 Springville.	"		56 Losantville.	"	1140
20 Owensburg.	"		60 Modoc.	"	
22 Dresden.	14 a. L. Coal Meas.		66 Bloomingport.	"	1225
24 Robinson's.	"		71 Lynn.	"	1174
26 Koline.	"		75 Arba.	"	
28 Rockwood.	"		79 Hollandsburg.	"	
30 Mineral City.	{ 14 b. Middle Coal Measure.		84 Clark's.	"	
35 Bloomfield.	"		87 P. C. & St. L. Crossing.	"	
41 Switz City.	"	526			

47. Large perfect earthworks and mounds.
48. St. Louis fossils plants, also Keokuk.
49. Block coal.
50. Bituminous coal.
51. Niagara.
52. Goniatite bed.
54. Devonian quarries.
55. Geodes and Geodized fossils.
56. Kaolin.
57. Good Bituminous coal.
58. Pre-historic mounds.
59. Black slate and knobstone.
60. Knobs and white glass sand.

61. St. Louis limestone; very rich in fossils.
62. Choice oolitic limestone quarries.
63. Hindoostan whetstones.
64. Sandrock quarries.
65. Good Bituminous coal.
66. Roof of coal rich in plants.
67. Black slate.
68. Keokuk fossils.
69. Wyandotte and other caves.
70. Pentemites.
71. Rock houses.
72. Coals, K. L. and M.

Louisville, Evansville & St. Louis R. R. Ms.	Alt.
0 Louisville.	
6 New Albany.[67]	13 a. L. Carb. k. s.[438]
12 Edwardsville.[66]	13 b. L. Carbon. L s.
15 Georgetown.	"
21 Crandall.	"
27 Ramsey's.	"
34 Milltown.	"
39 Marengo.[68]	"
46 English.[69]	14 a. L. Coal Meas.
53 Taswell.	"
56 Boston.[70]	"
60 Birdseye.[71]	14 b. Middle Cl. Meas.
66 Kyana.	"
75 Huntingburg.	"
123 Evansville.	14 c. Up. Cl. Meas.[376]

84 Velpen.[70]	14 a. L. Coal Meas.
91 Winslow.	14 b. Middle Cl. Mers.
99 Oakland.[72]	" [346]
105 Francisco.	14 c. U. Coal Meas.
113 Princeton.	" [483]
114 E. & T. H. Junc.	"
118 Lyles.	"
124 Mt. Carmel.	"

(Evansville Division.)

0 Evansville.[80]	14 c. U. Cl. Meas. [378]
4 Smythe.	" [379]
5 Garvin.	" [378]
8 Stevenson.	"
10 King's Station.	"
12 Chandler.	14 b. Mid Cl. Meas.[406]
14 De Forrest.	" [406]
17 Booneville.	" [391]
26 Tenneson.	"
30 Pigeon.	14 b. Middle Cl. Meas.
32 Centryville.	"
33 Junction.	"
34 Lincoln.	" .
38 Dale.	"
42 Ferdinand.	"
48 Huntingburg.	"
52 Rose Bank.	14 a. L. Coal Meas.
55 Jasper.	14 b. Mid. Coal Meas.

Louisville, Evansville & St. Louis R. R. Ms. (Rockport Branch.)	Alt.
0 Centryville.	14 b. Middle Cl. Meas.
2 Junction.	"
5 Bradley's.	"
9 Chrisney.	"
10 Miller's.	" 625
12 Ritchie's.	"
18 Rockport.	"

Chicago & Great Southern R. R.

0 Fair Oaks.	5 c. Niagara.
9 Mt. Ayr.	"
19 Percy.	9-12. Devonian.
22 Goodland.	{ 13 a. Lower Carbon. Knob Stone. [718]
26 Wadena.	14 a. L. Coal Meas.
32 Orthland.	"
34 Wyndham.	"
40 Oxford.	" 702
45 Pine Village.	" 699
54 Attica.	" 522
63 Rob Roy.[64]	"
68 Stone Bluff.	14 b. Mid. Cl. Meas.
73 Veedersburg.	"
80 Yeddo.	"

Ohio & Mississippi Railroad.
(Louisville Division.)

0 North Vernon.	1-12. Devonian.
25 Lexington.	"
40 Charleston.	"
53 Jeffersonville.	"
55 Louisville.	"

New York, Chicago & St. Louis Railroad.
(Nickel Plate Railroad.)

0 Buffalo.	
364 New Haven, Ind.	9-12. Devonian. 783
371 Fort Wayne.	" 775
397 South Whitley.	" 805
406 Packerton.	"
410 Claypool.	" 902
415 Burkett.	"
419 Mentone,	"
424 Tippecanoe.	" 865

73. *Martinsville.* Glacial bound'y. Glacial deposits to the north, east and west; none to the south.

74. *Edwardsport.* This road runs nearly parallel with the glacial boundary from Martinsville to Edwardsport. Glacial striæ 10 miles west of Spencer, pointing southeast.

75. *Valley Junction.* Tunnel between North Bend and Valley Junction is through a glacial deposit full of finely striated stones.

76. *Aurora.* Split rock, on Woolper Creek in Kentucky, three miles below Aurora, belongs to a post glacial conglomerate, rising more than 200 feet above the river, and marks very nearly the southern boundary of the glaciated area. Gold is found in glacial deposits on Laughery's Creek, five miles southwest of Aurora. See note 62 in Ohio, and No. 62 in West Virginia.

77. *Brownstown.* The glacial boundary running nearly north by south from Charlestown to the northeast corner of Brown County, passes a little east of Brownstown.

78. *Wheatland.* The railroad re-enters the glaciated area at Wheatland.

79. *Quincy.* This railroad from New Albany to Gosport passes through an unglatiated area. The glacial boundary is about three miles south of Quincy.

80. *Fort Branch* and *Evansville.* From Evansville to Fort Branch the country is unglaciated, though covered with Loess. The glacial boundary runs from here nearly parallel with this road to the neighborhood of Vincennes. The above eight glacial notes are by Rev. G. F. Wright.

New York Chicago & St. Louis R. R.— (Nickel Plate Railroad.)			Terre Haute & Indianapolis Railroad. (Vandalia Line.)		
Ms.		Alt.	Ms.		Alt.
431	Argos.	"	0	Terre Haute.	13 c. U. Cl. Meas. 493
438	Hibbard.	"	6	Otter Cr'k Junc.	"
440	Burr Oak.	"	23	Rockville.	"
451	Knox.	"	31	Judson.	14 a. L. Coal Meas.
462	Thomaston.	"	38	Waveland.	9-12. Devonian.
467	Wanatah.	5 c. Niagara. 751	46	New Market.	"
477	Valparaiso.	" 758	53	Crawfordsville.	"
480	Spriggsboro.	"	61	Darlington.	13 a. L. Carb. Knob s.
484	Wheeler.	" 666	69	Colfax.	" 825
488	Hobart,	" 628	79	Frankfort.	9-12 Devonian. 841
493	Joliet Pit.	"	88	Sedalia.	"
503	Hammond.	"	98	Flora.	"
510	Cummings, Ill.	"	102	Camden.	"
512	Stony Island.	"	110	Clymer.	"
514	Grand Crossing.	"	116	Logansport.	" 606
516	Englewood.	" 604	135	Kewanna.	"
521	22d Street.	"	143	Marshland.	"
523	Chicago.	" 589	160	Plymouth.	" 781
			173	Lakeville.	"
			183	South Bend.	" 738

Indiana, Bloomington & Western R. R.			Lake Erie & Western Railroad.		
0	Indianapolis.	9-12. Devonian. 709			
2	Moorfield.	" 705	138	Fort Recovery.	5 c. Niagara.
5	Johnsonville.	{ 13 a. L. Carb. Knob Stone.	149	Portland, Ind.	" 904
			160	Red Key.	"
15	Oakley.	" 898	165	Albany.	"
19	Maplewood.	" 842	176	Muncie.	" 948
23	Montclair.	" 759	176	Muncie.	" 948
27	North Salem.	" 888	192	Alexandria.	" 857
30	Barnard.	" 902	201	Ellwood.	" 858
35	Rochedale.	13 b. L. Carb. l. s. 889	212	Tipton.	" 868
40	Raccoon.	" 745	225	Circlerville.	9-12. Devonian.
45	Russellville.	" 825	237	Frankfort.	" 841
48	S. Waveland.	" 789	246	Mulberry.	13 a. L. C. Knob s. 754
52	Guion.	14 a. L. Cl. Meas. 680	252	Dayton.	" 648
56	Marshall.	{ 14 b. Middle Coal Measures. 700	260	Lafayette Junc.	" 595
			261	Lafayette.	" 595
60	Bloomingdale.	" 662	270	Montmorency.	" 672
67	Montezuma.	" 494	280	Templeton.	14 a. L. Cl. Meas. 675
68	Hillsdale.	" 452	282	Oxford.	" 702
75	Dana.	" 643	289	Boswell.	" 784
78	Illiana, Ill.	"	296	Ambia, Ind.	" 710
81	Scotland, Ill.	"	305	Hoopeston, Ill.	" 718
85	Chrisman.	"	312	East Lynn.	"

81. By the excellent Geological Map of Indiana, published by Professor Collett, with his report for 1884, the following appears to be the full section of the exposed strata of the State, with the thickness of each:

FORMATIONS.	THICKNESS IN FT.	FORMATIONS.	THICKNESS IN FT.
20 c. Alluvium.	0–50	9–12 *Devonian.*	
20 b. Loess.	0–30	Genesee Black Slate.	60–120
20 a. Glacial Drift.	0–311	Corniferous.	5–70
14 c. { Permo Carboniferous or Upper Coal Measures.	50–196	*Upper Silurian.*	
14 b. Middle Coal Measures.	600–888	5 c. Niagara.	20–60
14 a. { Lower Coal Measures, and Conglomerate.	60–210	5 c. Clinton.	0–10
Sub-Carboniferous.		*Lower Silurian.*	
13 b. Chester l. s.	0–74	4 c. Hudson River or Cincinnati.	50–320
13 b. St. Louis l s.	0–330		
13 b. Keokuk l. s.	6–106	The sub-divisions of the Devonian are too narrow to be separately noticed in the Guide.	
13 a. Knobstone s. s.	12–532		

This blank space is intended for additional geological notes in pencil by the traveler.

Illinois.[1]

List of the Geological Formations on the Illinois Railroads.

18 and 19. Cretaceous or Tertiary.	5 c. Niagara Group.
14 c. Upper Coal Measures.	4 c. Cincinnati Group.
14 b.{ Lower Coal Measures and Con-	4 a. Trenton and Galena Limestone.
14 a.{ glomerate.	3 c. St. Peter's Sandstone.
13 a. Low. Carboniferous Limestone.*	3 a. Calciferous and Lower Magnesian
9-12. Devonian.	Limestone.

Baltimore, Pittsburg and Chicago Railroad.

Ms.		(B. & O.)	Alt.
0	Chicago.[74]	5 c. Niagara.	569
12	Kingston.	"	568
14	South Chicago.	"	591
21	Edgemoor.	"	
30	Miller's.	"	625
34	Mich. Cent. Jun.	"	

Illinois Central Railroad.

Ms.			Alt.
0	Chicago.[74]	5 c. Niag. 88 ms.	569
14	Kensington.	"	596
24	Homewood.	"	
27	Matteson.[75]	"	699
34	Monee.	"	796
40	Peotone.	"	
47	Manteno.	"	711
56	Kankakee.[3]	"	626
65	Chebanse.	"	
69	Clifton.	"	644
81	Gilman.	"	652
85	Onarga.	"	
93	Bulkley.	4 c. Cincinnati, 16 ms.	
99	Loda.	"	777
103	Paxton.	"	804
105	Ludlow.	767 {14 a. & b. L. Cl. Mrs & Conglom.	
114	Rantoul.[76]	"	821
119	Thomasboro.	"	
128	Champaign.[782]	14 a. & b. L. Cl. M.	
137	Tolono.	"	729
143	Pesotum.	"	
150	Tuscola.	"	657
158	Arcola.	"	674
173	Mattoon.	14 c. U. Cl. Mrs.	785
185	Neoga.	"	
199	Effingham.	"	588

Ms.	Illinois Central Railroad.—*Continued.*		Alt.
215	Edgewood.	14 c. Upr. Coal Mrs.[872]	
230	Kinmundy.	"	
244	Odin.	"	525
252	Central City.[3]	"	
253	Centralia.	"	494
263	Richview.	"	
267	Ashley.	"	549
274	Dubois.	14 b. L. Cl. Mrs.	
280	Tamaroa.	"	
289	Du Quoin.[459]	{14 a & b. L. Cl. Mrs. & Conglom., 43 ms.	
302	De Soto.	"	
308	Carbondale.[69]	"	494
316	Makanda.	"	
323	Cobden.[5]	"	
328	Anna.[6]	4 a. Trenton, 20 miles.	
339	Dongola,	"	
344	Ullin.[888]	{18 & 19 Cretaceous or Tertiary 21 miles.	
365	Cairo.	"	322

Dubuque to Cairo.

Ms.			Alt.
0	Dubuque.	4 a. Trenton, 71 miles.	
2	Dunleith.[7]	"	
19	Galena.[7]	"	601
26	Council Hill.[7]	"	
31	Scales Mound.[8]	"	
40	Apple River.	"	
46	Warren.	"	1006
49	Nora.	"	
57	Lena.	"	959
70	Freeport.	"	759
74	Baileysville.	5 c. Niagara, 3 miles.	
82	Forreston.[77]	4 a. Trenton, 42 m.[941]	
87	Haldane.	"	
92	Polo.	"	849
105	Dixon.[9]	"	718

* Consisting of the 1. Kinderhook Shale, limestone and sandstone, 2. Burlington limestone, 3. Keokuk limestone, 4. St. Louis limestone and 5. Chester limestone and sandstone.
(In many localities there are no outcrops and the formations are given only in a general way.)
1. The notes are by Prof. A. H. Worthen, State Geologist of Illinois.
2. Rich in Niagara corals.
3. Shelly limestone of Upper Coal Measures filled with fossil shells, bryozoa, &c.
4. Roof shales of coal rich in fossil plants.
5. Upper Chester shales beneath conglomerate with a few fossil shells, corals, &c.
6. Quarries of St. Louis limestone with some small shells, corals, &c.
7. A few fossils characteristic of the Galena limestone.
8. Rich fossiliferous band near the base of the Cincinnati group, and crystals of barite, pyrite and dolomite in pockets of the Galena limestone.
9. Lower Trenton or Blue limestone two miles northeast of Dixon full of charactesintic fossils.

Illinois Central Railroad.
Dubuque to Cairo.—*Continued.*

Ms.		Alt.
117	Amboy.[733]	4 c. Cincinnati, 3 miles.
125	Sublette.	4 a. Trenton, 20 miles.
133	Mendota.	" 749
141	Dimmick.	"
149	La Salle.[10] 510	{ 14 a. Conglo. & 14 b. / L. Coal Mres. 8 ms.
158	Tonica.	"
169	Wenona.	14 b. L. Cl. Mrs.
180	Minonk.	"
188	Panola.	" .
191	El Paso.	" 742
200	Hudson.	"
207	Normal.	"
209	Bloomington.[11]	" 823
227	Wapella.	"
231	Clinton.	" 727
240	Maroa.	14 c. U. Cl. Mrs.
253	Decatur.	" 666
258	Wheatland.	"
263	Macon.	" 716
269	Moawequa.	"
276	Assumption.	"
285	Pana.	" 676
303	Ramsey.[12]	"
315	Vandalia.	" 500
330	Patoka.	"
339	Sandoval.	" 494
344	Central City.	"
345	Centralia.	" 494
358	Cairo.[322]	18. & 19. Creta. & Ter'y

Middle Division.

Ms.		Alt.
0	Kankakee.	5 c. Niagara 626
5	Otto.	No exposure.
29	Kempton Jn.	"
35	Griswold.	"
50	Pontiac.	14 a. & b. Low Cl.M.[666]
71	Kankakee Jn.	" "
73	Minonk.	" "
83	Cullom.	No exposures.
38	Charlotte.	"
42	Chatsworth.	14 a. & b. LowCl. M.[733]
46	Crumpton.	" "
50	Risk.	" "
64	Colfax.	" "
79	Barnes.	" "
85	Bloomington.[50]	14 c. U. Cl. Mres. 323

Illinois Central Railroad.—*Continued.*
Springfield Division.

Ms.		Alt.
0	Springfield.	14 c. Up. Coal Mrs. 589
24	Mount Pulaski.	"
44	Clinton.	" 727
62	Farmer City.	
82	Gibson.	14 a. L. Cl. Mrs. 15 ms.
97	Roberts.	4 b. Cincinnati, 14 ms.
111	Gilman.	5. Niagara, 5 ms. 632

Chicago, Burlington and Quincy Railroad.

Ms.		Alt.
0	Chicago.[74]	5 c. Niagara. 589
30	Naperville.	"
38	Aurora.	" 649
43	Oswego.[13]	4 c. Cincinnati,
47	Bristol.	"
53	Plano.	"
57	Sandwich.	4 a. Trenton, 45 miles.
61	Somonauk.	"
67	Leland.	"
74	Earl.	"
84	Mendota.	" 749
100	Malden.	"
105	Princeton.	{ 14 a. Congl. and 14 b. / Low. Cl. Mrs. 92 ms.
112	Wyanet.	"
118	Buda.	" 768
124	Neponsett.	"
132	Kewanee.[14]	"
140	Galva.[351]	{ 14 a. Cong. and 14 b. / Low. Coal Measures.
148	Altona.	"
152	Oneida.	"
156	Wataga.	"
164	Galesburg.	" 788
179	Monmouth.[15]	"
186	Kirkwood.	"
198	Sagetown.[16]	{ 13 a. Lower Carbon's / Limestone, 15 miles.
207	Burlington.	"
164	Galesburg.[788]	{ 14 a. Con. and 14 b. / L. Coal Mrs. 54 ms.
173	Abingdon.	"
183	Avon.	"
186	Prairie City.	"
192	Bushnell.	" 664
203	Macomb.	"
210	Colchester.[17]	"
212	Tennesee.	"

10. Limestone of the Upper Coal Measures full of fossils.
11. Minute shells in roof of coal seam, probably No. 3.
12. Upper Coal Measure limestone with fossil shells near Ramsey.
13. Cincinnati group, rich in fossils.
14. Fossils in roof shales of coal seam, probably coal No. 5 or 6.
15. Outcrop of Burlington limestone 2 miles north of Monmouth.
16. Burlington limestone rich in fossils.
17. Roof shales of coal rich in fossil plants, coal No. 2.
18. Burlington limestone rich in fossils.
19. Fossils abundant in roof shales of coal No. 5.
20. Fossils in roof shales of coals No. 2. and 3.
21. Fossils in roof shales of coal No. 5.

Chicago, Burlington and Quincy Railroad.

Ms.	*Continued.*	Alt.
223	Plymouth.	13 a. L. Carb. l. s. 5 ms.
227	Augusta.	{ 14 a. Cong. and 14 b.
		{ L. Coal Mrs. 27 ms. " 740
242	Camp Point.	
252	Fowler.	13 a. L. Carb. l.s.13 ms.
263	Quincy.15	" 455

Galesburg and Peoria Division.

164	Galesburg.	14 a. L. Coal Mrs. 755
169	Knoxville.	" 777
180	Maquon.	" 630
188	Yates City.	" 675
190	Elmwood.14	" 631
209	Kickapoo.	"
217	Peoria.	" 455

Galena Junction.

0	Galena Junction.	5. Niagara. 601
6	East Batavia.	"
13	Aurora.	" 649

Aurora and Streator Branch.

0	Aurora.	5. Niagara, 649
6	Oswego. 18	4 c. Cincinnati.
13	Yorkville.	"
23	Millington.	4 a. Trenton, 21 miles.
28	Sheridan.	"
32	Serena.	13 a. Lower Coal Mres.
36	Wedron.	" [3 a Calcif. in
40	Dayton.	" bed of river.]
44	Ottawa.	3 a. Calcif., 2 ms. 455
60	Streator.	13 a. Low. Cl. Mrs. 620

Buda and Rushville Branch.

0	Buda.788	14 b. Lower Coal Mrs.
20	Wyoming.	"
38	Brimfield.	"
45	Elmwood.14	" 621
47	Yates City.	" 673
53	Farmington.	"
64	Canton19	" 656
78	Lewiston.20	"
95	Vermont.	"
110	Rushville.21	" 675

Aurora and Geneva Branch.

0	Aurora.	5. Niagara. 649
9	Batavia.	"
13	Geneva.	"

Chicago, Burlington and Quincy Railroad.
—Continued.

Mendota and Clinton Branch.

Ms.		Alt.
0	Mendota.	4 a. Trenton. 749
9	La Moille.	"
19	Ohio.	"
26	Walnut.	"
32	Deer Grove.	4 c. Cincinnati.
45	Prophetstown.	5. Niagara.
62	Fulton.	"
66	Clinton.	" 727

Galva and Keithsburg Branch.

0	Galva.	551 13 a. Lower Coal Mres.
14	Woodhull.	"
37	Aledo.	"
51	New Boston.	" 573
54	Keithsburg.	" 543
66	Oquawka,	13 a. Burlington l. s.
71	Gladstone.	"

Burlington and Quincy Branch.

0	Burlington.	13 a. L. Carb. Limest.
10	Lomax.	"
24	Adrian.	"
32	Carthage.	" 656
44	West Point.	"
58	Mendon.	"
62	Ursa.	"
72	Quincy.15	" 455

Rock River Division.

0	Shabbona.	4 c. Cincinnati, 3 ms.
8	Paw Paw.	4 a. Trenton.
16	Brooklyn.	4 c. Cincinnati, 5 ms.
26	Amboy.	4 a. Trenton. 733
37	Harmon.	4 c. Cincinnati.
47	Rock Falls.	"

Chicago and Iowa Railroad. (C. B. & Q.)

0	Chicago.74	5 c. Niagara. 559
39	Aurora.	" 649
57	Hinckley.	" 740
64	Waterman.	"
69	Shabbona.	4 c. Cincinnati.
79	Steward.	"
86	Rochelle.	" 507
89	Flag Center.	4 a. Trenton.
94	King's.	"
98	Holcomb.	"
100	Davis Junction.	"
113	Rockford.	"

22. Fossil plants abundant in roof shales of coal No. 2.
23. Limestone of Upper Coal Measures full of fossils.
24. Fossils in roof shales and limestone over coal No. 5.
25. Fine outcrop of Devonian shale and limestone between these points full of fossils.
26. Niagara fossils occur sparingly at each of these points.
27. Fossils abundant in Cincinnati group.
28. Fossil plants in roof shales of coal No. 2.
29. Fossils abundant in roof shales of coal No. 8, and also in that of No. 5. In the shafts opened in this vicinity.
30. Upper Coal Measure limestone with a few fossils.

212   AN AMERICAN GEOLOGICAL RAILWAY GUIDE. (ILL.)

Chicago, Burlington and Quincy Railroad. —Continued.

Quincy, Hannibal and Louisiana Branch.

Ms.		Alt.
0	Quincy [18] [488]	13 a. Low Carbon. l.s.
10	Fall Creek.	"
17	Hannibal.[88]	"
19	Hulls.	" [488]
36	Rockport.[84]	"
41	Pike.	5. U. Silu. Niag. group.
43	Louisiana.[88]	"

St. Louis and Rock Island Division.

Ms.		Alt.
	St. Louis.	13 a. Low Carb. l.s.[416]
	East St. Louis.	" [418]
0	Alton.[86]	" [470]
20	Upper Alton.	14 a. & b. L. Coal Mrs.
25	Brighton.	" [694]
38	Medora.	"
42	Kemper.	"
55	Greenfield.	"
67	Whitehall.	18 a. Low Carbon l. s.
82	Winchester.[88]	"
87	Riggston.	14 a. & b. L. Coal Mrs.
92	Chapin.	"
101	Arenzville.	"
111	Beardstown.	"
115	Frederick.	"
120	Browning.	"
135	Vermont.	"
154	Bushnell.	" [664]
170	Roseville.	"
182	Monmouth.	"
203	Rio.	"
220	Orion.	" [751]
227	Port Byron.[87]	5 c. Niagara.
239	Rock Island.	9–11. Devonian. [684]
242	Moline.	"
246	Port Byron Jun.	"
255	Rock River Jun.	5. Niagara.
268	Erie.	"
278	Lyndon.	"
280	R. I. Junction.	"
291	Sterling.	"

Sheridan and Paw Paw Branch.

Ms.		Alt.
0	Paw Paw.	No outcrop.
20	Sheridan Jun.	"
51	Streator.	13 a. Low. Coal. [620]

Chicago, Rock Island and Pacific Railroad.

Ms.			Alt.
0	Chicago.[74]	5. Niag., 48 miles.	[589]
16	Blue Island.	"	
30	Mokena.	"	
40	Joliet.[36] [78]	"	[541]
51	Minooka.	{ 14 a. Cong. and 14 b. L. Coal Mrs. 41 ms.	
61	Morris.[22]	"	
71	Seneca.	"	
76	Marseillies.	"	
84	Ottawa.	8 a. Cal., 9 ms.	[486]
94	Utica.	"	
99	La Salle.[22]	{ 14 b. L. Cl. Mrs.[510] and Conglomerate.	
100	Peru.[24]	"	
114	Bureau.	"	[485]
0	Bureau.	"	
13	Henry.	"	
20	Sparland.	"	
28	Chillicothe.	"	
46	Peoria.[24]	"	[458]
	Pokin.	"	[475]
	Jacksonville.	"	[619]
114	Bureau.	14 L.C. Mrs.&Cong.	[455]
122	Tishilwa.	"	
126	Sheffield.	"	
146	Annawan.	"	
152	Atkinson.[20]	"	
159	Geneseo.	"	
170	Colona,	"	
179	Moline.[25]	9–12. Devonian.	
188	Rock Island.[25]	"	[584]

Chicago and Alton Railroad.[79]

Ms.			Alt.
0	Chicago.[74]	5. Niagara.	[589]
26	Lemont.[36] [78]	"	
33	Lockport.[26] [78]	"	
38	Joliet.[36]	"	[541]
53	Wilmington.[27]	4 c. Cincinnati.	[551]
58	Braidwood.[28]	{ 14 a. & 14 b. Conglo. and Lower Coal Mrs.	
61	Braceville.[28]	"	[603]
65	Gardner.	"	[605]
74	Dwight.	"	[609]
82	Odell.	"	[726]
92	Pontiac.	"	[668]
103	Chenoa.	"	[724]

31. Outcrop of coal No. 5. 1½ m. west of the station with numerous fossils in the roof shales.
32. St. Louis Limestone with numerous fossils.
33. Coal Measure fossils abundant in this vicinity.
34. Outcrop of Keokuk limestone with characteristic fossils 3 miles northeast of the town.
35. Keokuk limestone 1½ miles south of town with a few characteristic fossils.
36. Outcrop of St. Louis limestone 4½ miles east of the station with numerous fossils.
37. St. Louis limestone in heavy outcrops on Fountain creek 2 miles west of the station, and of Chester limestone 2½ miles southwest, both formations abounding in characteristic fossils.
38. Outcrops of Chester limestone on Prairie du Long creek 2½ miles north of the station with numerous fossils.
39. Fossils abundant in the limestone over the coal No. 6?
40. Fossil plants in roof shales and iron concretions of coal No. 2.
41. St. Louis limestone fossils scarce, 3 miles west of the town outcrops of Hamilton and Corniferous limestone with fossils.
42. Band of ferruginous shale abounding in Upper Coal Measure fossils.

Ms.	Chicago and Alton Railroad.—Cont.	Alt.
111	Lexington.	14 L. C. Ms. 751
119	Towanda.	" 810
124	Normal.	"
126	Bloomington.⁶⁰	" 825
146	Atlanta.	" 744
157	Lincoln.	" 613
164	Broadwell. 611	14 o. Upper Coal Mres.
185	Springfield.²⁹	" 892
194	Chatham.	" 642
206	Virden.	" 691
210	Girard.	" 687
214	Nilwood.	"
223	Carlinville.³⁰ 660	{ 14 a. & b. Low. Coal Mrs. & Congl. 22 ms.
238	Shipman.	" 662
245	Brighton.³¹	" 694
257	Alton.³² 470	13 a. L. Carb. l.s. 2 ms.
258	Upper Alton.	{ 14 a. & b. Lower Coal Mres. and Conglom.
261	Milton.	"
269	Mitchell.	13 a. Lower Carb. l. s.
276	Venice.	"
280	East St. Louis.	" 415
126	Bloomington.⁶⁰	14 a. L. Cl. Mrs. 825
149	Hopedale.	"
157	Delavan.	{ 14 a. & b. Low. Coal Mres. and Conglom.
171	Mason City.	"
187	Petersburg.³³	"
215	Jacksonville.	619
242	Drake.	13 a. Lower Carb. l. s.
265	Pleasant Hill.	"
274	Quincy Junction.	5. Niagara. 408

Jacksonville Division.

Ms.		Alt.
0	East St. Louis.	13 a. Low. Carb. Ls.⁴¹⁵
3	Venice.	"
16	Edwardsville Jn.	14 a. and b.
23	Alton. 470	13 a. Low. Carb. l. s.
28	Godfrey.	14 a. and b. 635
36	Delhi.	"
43	Jerseyville.	"
48	Kane.	13 a. Lower Carb. l. s.
56	Carrolton.³⁴	"
65	Whitehall.³⁵	"
49	Roodhouse.	14 a. and b. L. Cl. Mrs.
91	Jacksonville.	14 a. & b. L. Cl. Mrs.⁶¹⁹
106	Ashland.	" 635
119	Petersburg.	"
135	Mason City.	"
149	Delavan.	"
157	Hopedale.	14 c. Lower Coal Mrs.
180	Bloomington.³⁰	" 825

Ms.	Chicago and Alton Railroad.—Cont. Dwight Branch.	Alt.
0	Chicago.⁷⁴	5 c. Niagara. 589
74	Dwight.	14 a. & b. L. C. Mrs.⁶⁰⁹
96	Streator.	" 620
109	Wenona.	"
118	Varna.	"
128	Lacon.	14 a. & b. L. Coal Mres.
118	Varna.	14 a. Lower Coal Mres.
122	La Rose.	"
128	Washburn.	"
133	Cazenovia.	"
137	Metamora.	"
144	Washington.	" 745

Chicago, St. Louis and Western Railroad.

Ms.		Alt.
0	Chicago.⁷⁴	5 Niagara. 589
37	Joliet.	" 541
89	Streator. 620	14 a. & b. L. Cl Mres..
93	Reading.	"
98	Long Point.	"
108	Minonk.	14 a. Lower Coal Mres.
124	Roanoke.	"
126	Eureka.	"
133	Washington.	" 745
141	Morton.	"
145	Groveland.	14 a. & b. L. Cl. Ms. 823
153	Pekin.	" 705
161	Peoria.	" 463

St. Louis and Cairo Railroad.

Ms.		Alt.
0	East St. Louis.	13 a. Low. Carb. l.s.⁴¹⁵
13	East Carondelet.	"
14	Columbia.³⁶	"
19	Attica.	"
28	Waterloo.³⁷	" 664
32	Cambria.	"
37	Red Bud.³⁸	" 457
45	Baldwin.	"
54	Sparta.³⁹	14 a. & b. L. C. Mrs.⁵⁴⁹
75	Ava.	"
90	Murphysboro.⁷⁰	14 a, Low. Car. l. s. 425
116	Jonesboro.⁴¹	"
135	Hodge's Park.	19 Tertiary.
147	Cairo.	" 312

Ms.	Cairo, Vincennes & Chicago R. R.	Alt.
0	Vincinnes.	
10	St. Francisville.	14 c. Upper Coal Mres.
25	Mount Carmel.	"
41	Grayville.⁴²	" 393
56	Carmi.	" 401
81	Eldorado.⁷¹	" 384
89	Harrisburg. 667	14 a. & b. L. Coal Mrs.
102	Stonefort.	
126	Vienna.	13 a. Low. Carbon l. s.
151	Mound City. 823	18 & 19 Creta. & Ter'y.
157	Cairo.	" 322

43. Numerous fossil shells replaced with yellow pyrite occur in the roof shales of coal No. 7.
44. Fine outcrop of Upper Silurian and Devonian strata with characteristic fossils. .

Chicago & Eastern Illinois Railroad.

Ms.			Alt.
0	Chicago. [589]	5 Niagara, 86 miles.	
20	Blue Island.	"	
34	Bloom.	"	695
38	Crete.	"	733
52	Grant.	"	706
58	Momence.	"	735
69	St. Anne.	"	667
86	Watseka. [645]	14 a. & b. L. Coal Mres.	
108	Hoopston. [735]	" 46 miles.	
132	Danville.[48]	"	615
140	Gessie.	14 c. Upper Cl. Mres.	

Grape Creek Division.

Ms.			
0	Danville Jn.	14 a. & b. L. Cl. M.[615]	
5	Grape Creek.	"	
22	Sidells.	"	

Chicago and Northwestern Railroad.
Council Bluffs and Omaha Line.

Ms.			
0	Chicago.[74]	5. Niagara.	589
6	Austin.	"	
9	Oak Park.[81]	"	
25	Wheaton.[83]	"	
36	Geneva.	"	
38	St. Charles.	"	
44	Blackberry.	"	
55	Cortland.	"	
58	De Kalb.	"	
64	Malta.	"	
75	Rochelle.	4 c. Cincinnati.	607
84	Ashton.	"	
88	Franklin.	4 a. Trenton.	698
98	Dixon.[64]	"	718
110	Sterling.[87]	" & 5. Niagara.	
124	Morrison.	5. Niagara.	
136	Fulton.		
138	Clinton.	4 c. Cincinnati.	727

(Continued in Iowa.)

Chicago, St. Paul and Minneapolis Line.

Ms.			
0	Chicago.[74]	5 Niagara.	589
22	Arlington Heights.[85]	"	
26	Palatine.	"	
38	Cary.[84]	"	
43	Crystal Lake.[84]	"	
51	Woodstock.[84]	"	
63	Harvard Jn.	"	
71	Sharon.	4 c. Cincinnati.	
78	Clinton Jn.	"	727
91	Janesville.[85]	"	

Milwaukee, Green Bay and Marquette Line.

Ms.			
0	Chicago.[74]	5. Niagara.	589
12	Evanston.[86]	"	
21	Highland Park.	"	527
35	Waukegan.[86]	"	
45	State Line.	"	

Chicago and Northwestern Railroad.

Ms.	Rockford, Freeport and Dubuque Line.		Alt.
0	Chicago.[74]	5. Niagara, 66 ms.	589
6	Austin.	"	
9	Oak Park.[81]	"	
25	Wheaton.[83]	"	
30	Junction.	"	
39	Clintonville.	"	727
42	Elgin.	"	700
66	Marengo.	"	
78	Belvidere.	4 c. Cincinnati.	
93	Rockford.	4 a. Trenton.	
100	Winnebago.	"	
107	Pecatonica.	"	
121	Freeport.	"	759

Kenosha and Rockford Line.

Ms.			
0	Rockford.	4 a. Trenton, 18 miles.	
16	Poplar Grove.	"	
21	Capron.	4 c. Cincinnati.	
28	Harvard Jn.	5. Niagara.	
34	Alden.	"	

(See Wisconsin.)

Chicago, St. Paul and Minneapolis Line.

Ms.			
77	Caledonia Jn.	4 a. Trenton.	
78	Caledonia.	"	353
85	Roscoe.	"	
90	Beloit.[87]	"	

Sycamore Branch.

Ms.			
0	Cortland.	5. Niagara.	
5	Sycamore.	"	

Lake Geneva Line.

Ms.			
0	Chicago.[74]	5 c. Niagara.	590
39	Clintonville.	"	727
55	Crystal Lake.	"	

Crystal Lake Short Line.

Ms.			
0	Chicago.[74]	5 c. Niagara.	
43	Crystal Lake.	"	
50	McHenry.	"	
54	Ringwood.	"	
60	Richmond.	"	
61	Genoa Jn.	"	
70	Lake Geneva.	"	

Wabash, St. Louis and Pacific R. R.

Ms.			
93	Pontiac.	14 a. & b. L. Cl. Mr.[668]	
104	Fairbury.	"	
126	Gibson.	"	
134	Foosland.	14 c. Upper Coal Mres.	
145	Mansfield.	"	
158	Monticello.	"	
180	Lovington.	"	
188	Sullivan.	"	698
290	Windsor.	"	
229	Altamont.	"	616

45. Fine outcrop of the Kinderhook division of the Lower Carboniferous, with characteristic fossils, and Burlington limestone capping the bluffs.

Wabash, St. Louis and Pacific R. R.

Ms.	Continued.	Alt.
0	Streator.	14 a. & b. L. Cl. Mrs. 520
6	Manville.	"
11	Cornell.	"
16	Rowe.	"
19	Chicago Jun.	"

Toledo, Kansas City and St. Joseph Division.

Ms.		Alt.
0	Bluffs.	13 a. L. Sub-Carb. l. s.
4	Naples.	" 415
13	Griggsville.	14 a. L. Coal Mres. 555
17	Maysville.	"
6	Pittsfield.	"
20	New Salem.	" 775
27	Hadley.	13 a L. Carb. l. s. 753
37	Kinderhook.45	" 475
40	Hulls.	" 465
50	Hannibal, Mo.	" 470

Cairo, Vincennes and Chicago Line.

Ms.		Alt.
0	Danville. 508	14 a. Low. Coal Mres.
10	Georgetown.	"
16	Ridge Farm.	" 585
23	Chrisman.	"
30	Paris.	" 705
52	Marshall. 519	14 c. Upper Coal Mrs.
81	Robinson.	" 508
90	Flat Rock.	"
97	Pinkstaff.	"
102	Lawrenceville.56	" 424
103	O. & M. Jun.	" 424
108	Beman.	"
112	Vincennes.	"

Chicago, Kansas City and St. Joseph.

Ms.		Alt.
0	Peoria. 468	14 a. & b. L. Coal Mrs.
10	Pekin.	" 475
22	Manito.	"
27	Forest City.	" 676
41	Havana.	" 472
49	Bath.	"
59	Chandlerville.	" .
68	Virginia.	" 606
83	Jacksonville.	" 619

Havana and Springfield Line.

Ms.		Alt.
0	Springfield. 669	14 c. Upper Coal Mres.
13	Athens.	14 b. Lower Coal Mres.
22	Petersburg.66	"
31	Oakford.	"
47	Havana.	" 472

Wabash, St. Louis and Pacific R. R.

Ms.	Continued.	Alt.

Detroit, Toledo, Quincy and Keokuk Line.

Ms.		Alt.
0	Toledo. (see Indiana.)	14 c. U. Cl. Mrs.
242	State Line.	14 a. Lower Coal Mres.
250	Danville.	" 611
262	Fairmount.	" 593
269	Homer.	" 913
275	Sidney.	"
280	Philo.	14 c. Upper Coal Mres.
286	Tolono.	"
303	Bement.	"
311	Cerro Gordo.	" 507
323	Decatur.	"
339	Illiopolis.	"
348	Buffalo.	"
362	Springfield.	" 594
378	Berlin.	"
385	Alexander.	14 a. Lower Coal Mres.
395	Jacksonville.	"
413	Bluffs.	13 a. Low. Carb. l. s.
426	Versailles.	13 a. Low. Carbon. l. s.
436	Mount Sterling.	"
446	Clayton.	" 559
446	Clayton.	" 559
453	Labuda.	"
462	Bowen.	"
467	Denver.	"
476	Carthage.	13 a. Low. Carbon. l.s.
481	Elvaston.	"
488	Hamilton.53	"
452	Camp Point.	14 b. Lower Coal Mres.
457	Coatsburg.	"
463	Fowler.	13 a. Low. Carbon. l. s.
474	Quincy.	" 749

St. Louis and Chicago Line.

Ms.		Alt.
0	St. Louis, Mo.	
3	East St. Louis.	13 a. L. Sub-C. l. s.418
6	Venice.	"
22	Edwardsville.	14 b. Lower Coal Mres.
38	Staunton.55	14 c. Upper Coal Mres.
52	Litchfield.	"
85	Taylorville.	" 656
105	Boody.	"
113	Decatur. 507	"
133	Bement.	14. Coal Mres.
141	Monticello.	14 c. Upper Coal Mres.
146	Lodge.	14. Coal Mrs.
149	Galesville.	"
154	Mansfield.	14 c. Upper Coal Mres.
166	Osman.	"

46. Roof shale and limestone of No. 6 coal full of fossils.
47. Another outcrop of the same.
48. Fossils in the limestone over the coal.
49. Outcrop of nearly 250 feet of Chester limestone and shale abounding in the characteristic fossils of this group.
50. Fossils in limestone and shale over coal No. 6.
51. Fossils of Upper Coal Measures abundant in shale below the mill dam and two miles east of town at the bridge on the wagon road.
52. Fossils in shale and limestone over coal No. 5.

Wabash, St. Louis and Pacific R. R.
Ms. St. Louis and Chicago Line.—*Continued.* Alt.

Ms.			Alt.
162	Howard.	14 a. & b. Low Coal.	
174	Gibson.	"	
182	Sibley.	"	
186	Strawn.	"	
193	Forrest.	"	678
198	Wing.	No exposures.	
209	Emington.	4 c. Cincinnati group.?	
214	Campus.	"	
220	Reddick.	"	
226	Essex.	Upper Silurian.	
233	Ritchie.	"	
239	Manhattan.	"	
262	Alpine.	"	
269	Worth.	"	
272	Oak Lawn.	"	
286	Chicago.74	5 c. Niagara.	589

St. Louis and Jacksonville.

52	Litchfield.	14 Coal Mres.	464
72	Girard.	14 c. Up. Coal Mrs.	687
75	Virden.	"	691
88	Waverly.	14 a. &. b. L. Cl. Mres.	
106	Jacksonville.	"	619

Jerseyville Branch.

0	Springfield.29	14 c. U. Coal Mres.592	
13	Bates.	"	
25	Waverly.	14 a.& b. L. Cl. Mrs.691	
36	Palmyra.	"	
50	Chesterfield.	"	
59	Fidelity.	"	
68	Jerseyville.	"	
81	Jersey Landing.	13 a. Burlington l. s.	
85	Grafton.	5 c. Niagara.	

St. Louis Coal Railroad.

0	Marion.	14 a. & b. L. Coal Mrs.	
3	Bainbridge.	"	
11	Fredonia.	"	
18	Carbondale.	"	394
23	Glenahl.	"	
27	Harrison.	"	
29	Murphysboro.	"	425
29	Grange Hall.	"	
35	Vergennes.	"	
43	Pyatts.	"	
48	Pickneyville.	"	444

Peoria, Decatur & Evansville Railroad.

0	Peoria.68	14 a.& b. L. Cl. Mr.488	
10	Pekin.	"	475
27	Delavan.	"	
37	Hartsburg.	"	615
45	Lincoln.	"	
56	Mount Pulaski.	"	
69	Warrensburg.	"	
78	Decatur.666	14 c. Upper Coal Mrs.	
88	Hervey City.	"	707
96	Dalton.	"	604

Peoria, Decatur & Evansville Railroad.
Ms. —*Continued.* Alt.

Ms.			Alt.
98	Bethany.	14 c. U. Coal Mres.	665
103	Hampton.	"	665
110	Nelson.	"	637
120	Mattoon.	"	733
131	Janesville.	"	
144	Greenup.	"	351
157	Falmouth.		
174	Dundas.	"	
181	Olney.	"	480
191	Parkersburg.	"	
207	Brown's.	"	
227	Stuartsville.	"	
233	New Harmony.	"	
230	Poseyville.	" (?)	
248	Evansville.	14 a., & b. L. Cl. Mres.	

Chicago, Milwaukee and St. Paul R. R.

0	Chicago.74	5 c. Niagara.	589
6	Pacific Jun.	"	
14	Montrose.	"	
24	Deerfield.	"	
32	Libertyville.	"	
39	Gurnee.	"	
47	Russell.	"	
0	Chicago.74	5. Niagara.	589
8	Galewood.	"	
19	Salt Creek.	"	
24	Roselle.	"	607
35	Elgin.	"	706
50	Hampshire.	"	
59	Genoa.	"	
62	Kingston.	4 c. Cincinnati.	
74	Monroe.	4 a. Trenton.	
88	Byron.	"	

Racine and S. W. Division.

0	Racine.	(See Wisconsin.)	
69	Beloit.87	4 a. Trenton.	
90	Davis'.	"	
103	Freeport.	"	759
111	Florence.	5. Niagara.	
117	Shannon.	"	
124	Lanark.	"	
131	Mt. Carroll.64	4. a. Trenton.	
142	Savanna.64	4 c. Cincinnati.	
159	Fulton.	"	
166	Albany.	14 b. Niagara.	
181	Port Byron.66	"	
187	Hampton.	14 b. Low. Cl. Mrs. 665	
194	Moline.67	"	
197	Rock Island.67	Devonian.	534
85	Stillman Valley.	Lower Silurian.	
89	Byron.	"	
97	Leaf River.	"	
101	Adeline.	"	
117	Lanark Jn.	"	
120	Lanark.	"	
138	Savanna.	"	

Cincinnati, Indianapolis, St. Louis and Chicago R. R.

Ms.			Alt.
0	Lafayette, Ind.		595
46	Sheldon, Ill.	5 c. Niagara.	704
49	Iroquois.	"	
59	St. Mary.	"	
65	St. Anne.	"	659
75	Kankakee.	"	626
131	Chicago.74	"	589

Grand Tower and Carbondale Railroad.

Ms.			Alt.
0	Grand Tower.44	{ 9-11. Devonian,	352
		13 a. L. Carbon. l. s.	
10	Sand Ridge.72	14 a. & b. L. C. Mr.	351
15	Mount Pleasant.	"	
19	Mount Carbon.	"	372
24	Carbondale.	"	394

Illinois Midland Railroad.

Ms.			Alt.
0	Terre Haute.	14 a. & b. L. C. M.	493
22	Paris.	" [27 ms.	705
27	May's.	"	
31	Redmon.	14 c. Upper Coal Mres.	
57	Arcola.	"	674
71	Williamsburg.	"	
87	Hervey City.	"	707
96	Decatur.	"	666
128	Waynesville.	"	
142	Armington.	"	
166	Morton.	"	
166	Morton.	"	
170	Groveland.	14 a. & b. L. Coal Mres.	
178	Pekin.	"	476
170	Farmdale.	"	568
176	Peoria.	"	463

Indianapolis, Bloomington & Western R. R.

Ms.			Alt.
74	Mound City.		577
85	Danville.	14 a. & b. L. Cl. Mr.	632
107	St. Joseph.	"	
116	Urbana.	"	
118	Champaign.	"	732
128	Mahomet.	"	
141	Farmer City.	"	
151	Le Roy.	"	
166	Bloomington.	"	828
177	Danver's.	"	
186	Mackinaw.	"	
193	Tremont.	"	
202	Pekin.	"	475
211	Peoria.	"	463
116	Urbana.	"	
118	Champaign.	"	732
128	Mahomet.	"	
139	Monticello.	"	
164	Decatur.	14 c. Up. Cl. Mrs.	666

Indianapolis, Bloomington and Western Railroad.—Continued.

Ms.			Alt.
141	Deland.	14 a. & b. L. Coal Mrs.	
158	Clinton.	"	727
180	Lincoln.	"	613
187	Burtonview.	"	
198	Mason City.	"	
219	Havana.	"	472

Illinois and St. Louis Railroad.

Ms.			Alt.
1	East St. Louis.418	13 a. Low. Carbon. l.s.	
5	Centreville.	"	379
7	Pittsburgh.46	14 a. & b. L. Coal Mrs.	
11	Lenz.	"	
15	Bellville.47	"	479

Indianapolis, Decatur & Springfield R. R.

Ms.			Alt.
0	Decatur.	666 14 c. Upper Coal Mrs.	
20	Hammond.	"	672
36	Tuscola.	14 a. & b. L. C. Mrs.	657
42	Camargo.	"	
52	Newman.	"	641
68	Chrisman.	14 a. & b. L. Coal Mrs.	
76	Illiana.	"	

Wabash, Chester & Western Railroad.

Ms.			Alt.
0	Tamaroa.	14 a. & C. L. Coal Mrs.	
10	Pinckneyville.48	"	444
20	Cutler.	"	
26	Steel's Mills.	"	667
31	Bremen.	13 a. Low. Carbon. l.s.	
41	Chester.49	"	

Jacksonville South-Eastern Railroad.

Ms.			Alt.
0	Jacksonville.	519 14 a. & b. L. Coal Mrs.	
12	Franklin.	"	696
18	Waverly.	"	691
25	Lowder.	"	713
31	Virden.	691 14 c. Upper Coal Mrs.	
34	Girard.	14 c. Up. Cl. Mrs.	687
38	McVey.	14 Coal Mrs.	666
48	Barnett.	"	672
54	Litchfield.	"	464
68	Sorrento.	"	
78	Betterton.	"	
94	Kevesport.	"	
105	Shattuck.	"	
112	Centralia.	14 c. U. Cl. Mres.	494

Lake Shore and Michigan Southern R. R.

Ms.			Alt.
0	Chicago.74	5 c. Niagara.	589
7	Englewood.	"	604
12	South Chicago.	"	591

Michigan Central Railroad.

Ms.			Alt.
0	Chicago. 74	5. Cincinnati.	589
14	Kensington.	"	596
35	Lake.	"	666

53. Burlington limestone and Kinderhook group.
54. Kinderhook group with a few feet of Devonian and Upper Silurian at the base of the bluff.
55. Kinderhook, Devonian and Upper Silurian, the highest bluffs capped with Burlington limestone.

Michigan Central Railroad.—*Continued.*		
Ms.	Joliet Division.	Alt.
0 Lake.	5. Cincinnati.	466
15 Dyer.	"	
24 Matteson.	"	
32 Frankfort.	"	755
37 Spencer.	"	712
45 Joliet.[78]	5 c. Niagara.	541

Ohio and Mississippi Railroad.		
0 St. Louis.	(See Missouri.)	416
2 East St. Louis.	13 a. L. Car. l. s. 5 ms.	
10 Caseyville.[50]	14 a. & b. L. Coal Mrs.	
18 O'Fallon.	"	545
24 Lebannon.	"	441
27 Summerfield.	"	
31 Trenton.	"	500
39 Breese.	14 c. Upper Coal Mrs.	
48 Carlyle.	"	450
61 Sandoval.	"	494
65 Odin.	"	525
70 Salem.	"	536
87 Xenia.	"	
96 Flora.	"	495
103 Clay City,	"	
118 Olney.	"	460
130 Sumner.	"	457
139 Lawrenceville[51]	"	
149 Vincennes.	"	

Springfield Division.		
0 Beardstown.	14 a. & b. L. C. Mrs.[436]	
18 Virginia.	"	608
29 Pleasant Pl'ns[52]	"	606
40 Bradford.	"	581
44 Coal Shaft.	14 c. Upper Coal Mres.	
45 Springfield.	"	582
53 Rochester.	"	569
63 Edinburg.	"	
72 Taylorsville.	"	
88 Pana.	"	
121 Altamont.	"	616
132 Edgewood.	"	
146 Louis.	"	460
153 Flora.	"	
174 Fairfield.	"	535
181 Barnhill.	. "	585
194 Enfield.	"	468
199 Sacramento.	"	418
209 Omaha.	14 a. Low. Cl. Mrs.[369]	
216 Ridgeway.	"	879
225 St. L. & S. E. Jun.	"	
228 Shawneetown.	"	363

Ms.	**Rock Island and Peoria Railway.**	**Alt.**
0 Peoria.	468	14 a. & b. L. Coal Mrs.
15 Dunlap.		"
22 Princeville.		" 719
31 Wyoming.		"
36 Toulon.		" 723
42 Lafayette.		"
48 Galva.		" 851
58 Bishop Hill.		"
62 Cambridge.		" 778
68 Osco.		"
80 Coal Valley.		"
86 Milan.		"
91 Rock Island.		" 9-11 Dev. 584

Pittsburg, Fort Wayne & Chicago R. R.		
0 Chicago.[74]	5 c. Niagara.	
13 Hobart.	"	

St. Louis, Aton & Terre Haute R. R.		
0 East St. Louis.[418]	13 a. Low. Carbon. l.s.	
6 Centreville.	"	379
10 Ogles.	14 a. & b. L. Coal Mres.	
13 West Bellville.	"	
14 Bellville.	"	479
22 Freeburg.[58]	"	514
29 New Athens.	"	404
47 Coulterville.	"	542
61 Pinckneyville.	"	444
71 Du Quoin.	"	459

Louisville & Nashville Railroad.		
St. Louis, Evansville and Nashville Line.		
0 East St. Louis.[418]	13 a. Low. Carbon. l. s.	
14 Bellville.	479	14 a. & b. L. Coal Mres.
0 Bellville.	"	
6 O'Fallon.	"	545
20 Rentchler's.	"	
25 Mascoutah.	"	425
32 New Memphis.	"	411
35 Venedy.	"	412
49 Nashville.	508	14 c. Upper Coal Mrs.
60 Ashley.	"	549
69 Woodlawn.	"	495
87 Belle River.	"	
100 Shawnee Jun.	"	488
0 Shawnee Jun.	"	488
1 McLeansboro.	"	500
13 Broughton.	"	
22 Eldorado.[78]	"	364
30 Equality.	14 b. Lower Coal Mrs.	
36 Cypress Jun.	"	840
42 Shawneetown.	"	363

56. St. Louis limestone and Lower Coal Measures with characteristic fossils.
57. Niagara limestone with numerous fossils.
58. Coal shale 1¼ miles northeast of station full of fossil shells.
59. Limestone over No. 9 coal with fossils.
60. Upper Coal Measure limestone full of fossils.
61. Fossils in roof shales and limestone of coals No. 5 and 6.
62. Coal Measure limestone with fossil corals and shells.

Louisville & Nashville Railroad.—Con.	
Ms. St. Louis, Evansville and Nashville Line. Alt.	
101 McLeansboro.	14 c. Up. Coal Mrs.[500]
113 Enfield.	" [455]
123 Carmi.	" [401]
131 Wabash.	"

St. Louis, Vandalia & Terre Haute R. R.

0 East St. Louis.	13 a. L. Carb. l. s. [416]
11 Collinsville.	14 a. Low. Cl. Mrs.[465]
19 Troy.	" [559]
30 Highland.[59]	14 b. Up. Cl. Mres.[527]
40 Pocahontas.	" [498]
49 Greenville.	" [555]
67 Vandalia.	" [500]
81 St. Elmo.	"
86 Altamont.	" [616]
98 Effingham.	" [588]
102 Teutopolis.	"
122 Greenup.	" [551]
130 Casey.[60]	"
137 Martinsville.	" [573]
148 Marshall.[61]	" [619]
151 Griffiths.	"
155 Dennison.	13 a. Low. Carbon. l. s.
158 Farrington.	"
166 Terre Haute.	" [498]

Toledo, Peoria & Western Railroad.

0 State Line.	5. Niagara.
2 Sheldon.	" [705]
11 Watseka.[62]	" [627]
25 Gilman.	" [652]
29 La Hogue.	4 c. Cincinnati.
40 Chatsworth.	" [733]
47 Forrest.	" [678]
52 Fairbury.	14 a. & b. L. C. Mrs.[697]
63 Chenoa.	" [724]
67 Meadows.	14 c. Up. Coal Mrs.[764]
78 El Paso.	" [742]
92 Eureka.	"
99 Washington.	" [745]
109 Hilton.	14 a. Lower Coal Mrs.
111 Peoria.	" [463]
139 Canton.	" [656]
149 Cuba.	" [674]
171 Bushnell.	" [664]
189 Blandinsville.	" [730]
195 La Harpe.	13 a. L. Carb. l. s. [687]
215 Burlington.	"

195 La Harpe.	" [687]
200 La Crosse.	'
210 Ferris.	" [677]
216 Elvaston.	" [665]
222 Hamilton.[63]	"
227 Warsaw.[63]	"

Lake Erie & Western Railroad.

305 Hoopstown, Ill.	14 a. L. C. M. & Cgl.[718]
312 East Lynn.	"
317 Rankin.	"
318 Pellsville.	"
327 Paxton.	4 c. Cincinnati.
341 Gibson.	14 a. L. C. Ms. & Congl.
351 Saybrook.	"
357 Arrowsmith	"
361 Ellsworth. .	"
364 Padua.	"
367 Holder.	"
377 Blooming.[60]	" [823]

Louisville, Evansville & St. Louis R. R.

0 Mt. Vernon, Ind.	[407]
8 Blueford. "	
20 Wayne, Ill.	14 c. Upper Coal Mres.
30 Fairfield.	" [538]
34 Meriam.	"
47 Albion. ·	"
51 Brown's Cross.	"
56 Bellmont.	"
65 Mt. Carmel.	"
74 E. & T. H. Jun.	"
75 Princeton.	" [455]
88 Francisco.	"
90 Oakland.	" [546]

Chicago and Iowa Railroad.

89 Flag Centre.	4 a. Trenton.
95 Chana.	"
98 Honey Creek.	3 c. St. Peters s. s.
101 Oregon.	" [704]
108 Mt. Morris.	4 a. Trenton. l. s. [906]
114 Maryland.	" [941]
120 Forreston.	"
132 Freeport.	"

Rock Island & Mercer County Railroad.

0 Rock Island.	9-12 Devonian. [584]
4 Milan.	
12 Taylor Ridge.	14 a. & b. L. Cl. Mres.
26 Cable.	"

Chicago & Evanston Railroad.

0 Chicago.[74]	5 c. Niagara. [539]
7 Flaxton.	"
10 Calvary.	"

Kankakee & Seneca Railroad.

0 Kankakee.	5 c. Niagara. [626]
5 Hawkins.	"
11 Bonfield.	4 c. Cincinnati gr.
18 Essex.	"
24 Gardner.	14 a. & b. L. C. Mr.[605]
31 Mazon.	"
36 Hill Park.	"
43 Seneca.	"

63. Fine outcrops of Keokuk limestone with numerous fossils, and geodes containing crystallized quartz, chalcedony, calcite, dolomite, arragonite, blende and pyrite.

Ms.	Indianapolis & St. Louis R. R.	Alt.
72	Terre Haute, Ind.	
84	Vermillion.	14. Coal Measures.	
91	Paris.	"	705
100	Dudley.	"	
105	Kansas.	"	
118	Charleston.	"	
129	Matoon.	"	733
141	Windsor.	"	
152	Shelbyville	"	
168	Pana.	"	
181	Nokomis.	"	
190	Irving.	"	
200	Butler.	"	767
207	Litchfield.	"	
217	Gillespie.	"	
226	Bunker Hill.	"	
232	Dorseys.	"	
237	Bethalto.	"	
242	Wann.	13 a. St. Louis l. s.	
245	Edwardsville Crossing.	"	
262	East St. Louis.	73 a. L. Carb. l. s.	415
265	St. Louis.	"	415

	Danville, Olney & Ohio River R. R.	
0	Danville Jn.	14. Coal Mres.	510
31	Hume.	"	549
49	Kansas.	"	
68	Casey.	"	549
89	Willow Hill.	"	
100	West Liberty.	"	
109	Olney.	"	

	Toledo, Cincinnati & St. Louis R. R.	
272	Humerick.	14 b. Low. Cl. Mres.	
278	Ridge Farm.	"	615
288	Metcalf.	"	618
297	Brocton.	"	
311	Bushton.	"	
332	Trilla.	"	
349	Stewardson.	"	
357	Fancher.	14 c. Upper Cl. Mres.	
370	Herrick.	"	
382	Boyle.	"	
401	Donnellson.	"	
407	Sorrento.	14 b. Lower Cl. Mres.	
418	Alhambra.	"	
431	Edwardsville.	"	
450	East St. Louis.	13 a. L. Sub. Ca. l.s.[415]	

Ms.	Indiana, Illinois & Southern R. R.	Alt.
0	Effingham.	14 c. Up. Coal Mrs.[588]	
14	Wheeler.	"	
23	Newton.	"	
31	Willow Hill.	"	
37	Oblong.	"	
47	Robinson.	"	505
53	Palestine.	"	

	Central Iowa Railway.	
0	Peoria.[68]	14 a. & b. L. Cl. Mr.[465]
13	Hanna.	"
18	Trivoli.	"
24	Farmington.	"
29	Claire.	"
38	London Mills.	"
43	Hermon.	"
49	Abingdon.	"
57	Berwick.	"
61	Phelps.	"
66	Monmouth.	18 a. Low. Carb. l. s.
78	Eleanor.	"
77	Little York.	"
84	Seaton.	"
92	Keithsburg.	"	543

	Champaign and Havana Line.	
0	Urbana.	14 a. L. Coal Mres.	
2	Champaign.	"	732
10	Seymour.	"	
15	White Heat.	"	
21	Monticello.	"	
34	Argenta.	"	
45	Decatur.	14 c. Up. Coal Mrs. 666	
18	Lodge.	"	
28	Weldon.	"	
40	Clinton.	"	737
50	Midland City.	"	
52	Beason.	"	
56	Skelton.	14 b. Low. Coal Mres.	
62	Lincoln.	"	
74	New Holland.	"	
80	Mason City.	"	
88	Easton.	"	
93	Poplar City.	"	
100	Havana.	"	

	Litchfield, Carrolton & Western R. R.	
1	Columbiana.	13 a. Low. Carbon. l. s.
11	Carrollton.	" •
22	Greenfield.	14 b. Low. Coal Mres.

	Fulton County Narrow Gauge Railway.	
0	Galesburg. 766	14 a. Cg. & 14 b, L. C. M.
19	London Mills.	"
30	Fairview.	"
35	Fiatt.	"
40	Cuba.	"
50	Lewiston.	"
61	Havana.	"

64. Cincinnati group with characteristic fossils, and near Savanna the Niagara limestone caps the hills and affords silicified corals in abundance.

Ms.	Havana, Rantoul & Eastern R. R.	Alt.	Ms.	Indiana, Illinois & Iowa Railroad.	Alt.
0	West Lebanon.	Indiana.	0	Streator.	14 a. & b. L. Cl. Mr.[630]
12	Alvan.	14 b. Low. Coal Mres.	6	Missal.	"
17	Henning.	"	12	Budd.	"
26	Armstrong.	"	22	Dwight.	" [609]
34	Gifford.	"	29	Wilson.	"
42	Rantoul.[76]	[331] 14 a. & b. L. C. M. & Cg.	32	Reddick.	4 c. Cincinnati Group?
45	Prospect.	14 b. Low. Coal Mres.	37	Union Hill.	"
52	Fisher.	"	42	Goodrich.	"
56	Dickerson.	"	44	Cagwin.	5 c. Niagara.
58	Howard.	"	52	Kankakee.[2]	" [626]
66	Deians.	"	58	Exline.	"
71	Crumbaught.	"	63	Momence.	" [625]
76	Le Roy.	"	68	Castleton.	"

65. Fossils in limestones over No. 9 coal.
66. Upper Silurian limestone with numerous fossils.
67. Devonian limestone and shale with fossils.
68. Coal Measures fossils.

Glacial Notes by Rev. G. Frederick Wright.

69. Carbondale.—The Glacial boundry is between Carbondale and Mankanda. Fine Glacial striæ are found 2½ miles southwest of Carbondale and 5 miles southeast.
70. Murphysboro.—Glacial boundary about 5 miles south of Murphysboro turning thence to run parallel with the Mississipi to the neighborhood of St. Louis.
71. Eldorado.—The railroad crosses the southern boundary of the glaciated area at Eldorado and runs nearly parallel with it to Carnie. The boundary runs northeast by southwest.
72. Sand Ridge.—The western boundary of the glaciated area passes a mile or two west from Sand Ridge and runs northwest, following the course of the Mississippi River.
73. Eldorado.—The southeastern boundary of the glaciated loop of Illinois, passes through Eldorado, crossing the Wabash near New Harmony.

Glacial Notes by Prof. T. C. Chamberlin.

74. Chicago.—Subaqueous till. Lacustrine plain. Beach line. B. & O. to Michigan Central Junction, and Illinois Central to Desoto, drift plain.
75. Matteson.—Obscure moraine.
76. Rantoul.—Moraine.
77. Forreston.—Osar.
78. Joliet, Lemont, Lockport.—Ancient outlet of Lake Michigan.
79. From Wilmington to Quincy Junction, deep drift plain.
80. Bloomington.—Two vegetal beds in drift.
81. Oak Park.—Beach ridge.
82. Wheaton.—Moraine?
83. Arlington Heights.—Beach ridge.
84. Cary, Crystal Lake and Woodstock.—Moraine.
85. Janesville.—Glacial flood deposit.
86. Evanston, Higland Park and Waukegan. Subaqueous drift, beach formations.
87. Beloit.—Glacial flood deposits; terraces, Trenton, St. Peters.
88. St. Louis and R. I. Division.—Upper Alton to Winchester. Loess.

This blank space is intended for additional geological notes in pencil by the traveler.

Wisconsin.[1] [29]

LIST OF THE GEOLOGICAL FORMATIONS IN WISCONSIN.

20. Quaternary. { Post Glacial.[2] Glacial.	**4 a. Trenton Limestone.**[6]
10. Hamilton (Milwaukee Cement Rock).	**3 c. St. Peter's Sandstone.**
7. Lower Helderberg.	**3 a. Lower Magnesian (Calciferous).**[6]
5 c. Niagara Limestone.[3]	**2 b. Potsdam Sandstone.**[7]
5 b. Clinton.[4]	**Keweenawan or Copper-bearing series.**
4 c. Cincinnati Shale.	**1 b. Huronian.**
4 b. Galena Limestone.	**1 a. Laurentian.**

Chicago & North-Western Railroad.		**Chicago & North-Western Railroad.**	
Ms. Chicago, St. Paul & Minneapolis Line. Alt.		Ms. Chicago, St. Paul & Minneapolis Line. Alt.	
0	Chicago.	(As before.)	153	Dane.	{ 3 a. Lower Magn.l.s. (on top of high dividing ridge.)[1055]
90	Beloit.	{ 4 b. Galena l. s. 4 a. Trenton l.s. 745 3 c. St. Peter's s. s.			
98	Afton.	{ 4 a. Trenton l.s. 756 3 c. St. Peter's s. s.	158	Lodi. 848	{ 3 a. Lower Magn.l.s. capping bluffs. 2 b. Mad.s.s. } bluff 2 b.Mend.ss. } sides.
104	Hanover.	4 a. Trenton l. s. 780			2 b. Potsdam s. s. valley bottom.
107	Footville.	{ 4 a. Trenton l.s. 816 3 c. St. Peter's s. s.	164	Merrimac.	2 b. Potsdam s. s. 796
111	Magnolia.	Junc. Tren. and St. P.	172	Devil's Lake.	{ 1. Archæan q'rtzite. 2 b. Potsdam s. s. and conglom.
116	Evansville.	4 a. Trenton l. s. 893			
122	Brooklyn.	20. Moraine Drift.	175	Baraboo.	{ 1. Archæan q'rtzite. 2 b. Potsdam s.s. 861
128	Oregon.	{ 4 a. Trenton l.s. 952 3 c. St. Peter's s. s.	181	North Freedom.	2 b. Potsdam s. s.
133	Syene. 905	{ 3 c. St. Peter's s. s. 3 a. Lower Magn.l.s.	184	Ableman's.	{ 1. Archæan q'rtzite. 2 b. Potsdam s.s.(in gorge 200 ft. deep, unconformability & exact junc.)878
138	Madison.	Moraines, Drumlins. 3 a. Lower Magn.l.s. 2 b. Madison s.s. 845 Mendota limestone. Potsdam sandstone.			
			191	Reedsburg.	2 b. Potsdam s. s. 877
143	Mendota.	In cut, { 3 a. L. Magn. 2 b. Mad.s.s.	198	Lavalle.	" 897
			205	Wonowoc.	" 911
148	Waunakee.	{ 3 a. Lower Magn.l.s. on bluffs. 922 2 b. Potsdam s. s.	208	Union Centre.	" 944
			212	Elroy.	" 955

1. Prepared by Professor T. C. Chamberlin, of Madison, the State Geologist, and Professors R. D. Irving and M. Strong, Assistant Geologists.
2. Including the Champlain and Terrace epochs.
3. Including four sub-divisions in the southern part of the State and six in the northern, among which are the Racine and Guelph limestones.
4. The Clinton produces the Iron Ridge iron ore, the fossil ore of other States.
5. Including two sub-divisions in the lead region and four in southeastern Wisconsin.
6. The Calciferous may include more than the Lower Magnesian.
7. Including several sub-divisions, among them the Madison sandstone and the Mendota limestone.

Chicago & North-Western Railroad.—Con.

Ms.	Chicago, St. Paul and Min. Line.	Alt.
212	Elroy.	2 b. Potsdam s. s. 955
226	Camp Douglas.[1]	" 929
227	Wis. Val. Junc.	" 925
242	Lowery's.	" 959
244	Warren's.	" 1019
249	Rudd's.	" 974
265	Bl'k River Falls.	{ 2 b. Potsdam s. $\cdot$s., resting on 1 Archæan gneiss. 802
277	Merrillan.	2 b. Potsdam s. s. 988
282	Humbird.	" 1018
289	Fairchild.	" 1066
299	Augusta.	" 965
309	Fall Creek.	" 929
321	Eau Claire.[2]	" 886
323	West Eau Claire.	" 877
332	Elk Mound.	" 926
339	Rusk.	Pots. s.s. { Glacial 901
344	Menomonee.	Pots. s.s. { flood pl. 878
353	Knapp.	3 a. Lower Magn. 919
358	Wilson.	20. Quaternary. 1147
361	Hersey.	" 1168
369	Baldwin.	" 1132
372	Hammond.	{ 20. Quat. & 3 c. St. Peter's. 1100
378	Roberts.	Moraine West. 1086
390	Hudson.[3]	2 b. Potsdam. 700
401	River Falls.	{ 3 a. Lower Magn. 2 b. Potsdam, Glacial flood drift, Moraine.
394	Stillwater Junc.	Moraine hills.
410	St. Paul.	(See Minnesota.)

Kenosha and Rockford Division.

Ms.		Alt.
0	Kenosha.	20. Quaternary. 618
6	Pleasant Prairie.	" 697
10	Woodworth.	" 748
12	Bristol.	" 769
16	Salem.	" 776
19	Fox River.	" 778
22	Bassett.	"
27	Genoa Junction.	" 842
44	Harvard Junc.	(See Illinois.)
72	Rockford.	"

Moraines.

Minnesota Division.

Ms.		Alt.
0	Chicago.	(As before.)
212	Elroy.	2 b. Potsdam s. s. 955
217	Glendale.	"
227	Wilton.	" 986
233	Norwalk.[4]	" 1020
246	Sparta.[4]	" 756
255	Bangor.	2 b. Pots. s. s. Ter. 752

Chicago & North-Western Railroad.—Con.

Ms.	Minnesota Division.—Continued.	Alt.
260	Salem.	2 b. Pots. s. s. Ter. 749
267	Winona Junc.	2 b. Pots. s. s. Ter, 655
276	La Crosse.	{ 2 b. Pots. s. s. L.Mag. Valley drift. 696
267	Winona Junc.	2 b. Potsdam s, s. 655
269	Onalaska.	{ 2 b. Potsdam s. s. Valley drift.
273	Midway.	{ 2 b. Potsdam s. s. Valley drift.
278	Lytles.	{ 2 b. Potsdam s. s. Valley drift.
284	Trempealeau.	{ 2 b. Pots. s. s. Loess. drift. 630
292	Marshland.	{ Potsdam s.s. 3 a. Low. Magn. 659
297	Winona.	(See Minnesota.)

Milwaukee, Green Bay and Marquette Line.

Ms.		Alt.
0	Chicago.	(As before.)
45	State Line.	20. Quaternary.
51	Kenosha.	" 618
60	Racine Junc.[5]	{ 5 c. Niag. (Racine) limestone. 621
62	Racine.[5]	{ 5 c. Niag. (Racine) limestone.
70	County Line.	20. Quaternary. 695
75	Oak Creek.	" 664
81	St. Francis.	" 643
83	Elizabeth St.	"
85	Milwaukee.[6]	{ 10. Hamilton cement rock. 584 5 c. Niagara.
90	Lake Shore Junc.	20. Quaternary. 642
91	Lindivern.	" 632
100	Granville.	5 c. Niagara, Drift. 738
107	Germantown.	" 863
112	Jackson.	" 897
119	West Bend.	{ 20. Moraine, and fluvial drift. 906
120	Barton.	{ 20. Moraine, and fluvial drift.
126	Kewaskum.	{ 20. Mor. and fluvial d'ft. 5c. Niag. 959
133	New Cassel.	{ 20. Mor. and fluvial d'ft. 5 c. Niag.
140	Eden.	{ 20. Mor. and fluvial dft. 5 c. Niag.
148	Fond du Lac.	{ 4 b. Gal. red clay drift. 769 4 b. Galena.
165	Oshkosh.	{ 4 b. Galena. 4 a.Tren. Striæ, Till and Red Clay. 755
178	Menasha and Neenah.	{ 4 a. Tren. Striæ, Till and Red Clay. 756

1. *Camp Douglas.* Remarkable castellated outliers.
2. *Eau Claire.* Glacial valley drift carved into fine terraces.
3. *Hudson.* Potsdam, glacial flood deposits and terraces.
4. *Sparta.* Terraces, artesian wells. Tunnels in or below Lower Magnesian limestone.
5. *Racine.* Glacial and lacustrine drift. Ancient beach lines.
6. *Milwaukee.* Glacial and lacustrine drifts.

Chicago & North-Western Railroad.
Ms. Mil., Green Bay & Marq. Line.—Con. Alt.

Ms.	Station	Alt.
180	West Menasha.	{ 4 a. Tren. Striæ, Till and Red Clay.
185	Appleton.	{ 4 b. Galena. 715 / Tren., Red Clay.
190	Little Chute.	{ 4 b. Galena, red clay drift. 707
192	Kaukauna.	{ 4 b. Galena, red clay drift. 655
198	Wrightstown.	{ 4 b. Galena, red clay drift. Striæ. 626
208	De Pere.	{ 4 b. Galena, red clay drift. Striæ. 591
214	Ft. Howard and Green Bay.	{ 4 c. Cin. shale. 588 / 4. b. Gal., red clay.
218	Duck Creek.	4 b. Galena, Striæ.
222	Big Suamico.	"
228	Little Suamico.	"
233	Brookside.	20. Quaternary.
237	Pensaukee.	{ 4 b. Gal. limestone. / 4 a. Tren. limestone.
242	Oconto.	20. Quaternary.
252	Cavoits.	
256	Peshtigo.	4 a. Trenton l. s.
263	Marinette.	4 b. Galena l. s. Striæ.
264	Monominee.	"
382	Escanaba, Mich.	(See Michigan.)
	(Continued in Michigan.)	

(Lancaster and Woodman Line.)

Ms.	Station	Alt.
0	Galena, Ill.	4 b. Galena limestone.
7	Bell's.	"
15	Benton.	"
20	St. Rose.	"
32	Platteville.	{ 4 b. Galena l. s. / 4 a. Trenton l. s.

(Sheboygan and Western Railroad.)

Ms.	Station	Alt.
0	Sheboygan.	{ 5 c. Niagara. Sub-aqueous drift. 585
5	Sheboygan Falls.	{ 5 c. Niagara. Sub-aqueous drift. 663
10	Town Line.	20. Drift.
14	Plymouth.	20. Red clay. 660
20	Glenbeulah.	{ Kettle Range. 667 / Moraine drift.
26	St. Cloud.	5 c. Niag. l. s. 827
30	Calvary.	Niag. drumlins. 940
43	Fond du Lac.	4 b. Galena l. s. ·746
44	Fond du Lac Jc.	"
47	Woodhull.	20. Quaternary.
52	Eldorado.	" 875
55	Rosendale.	" 891
57	West Rosendale.	" 882
63	Ripon.	{ 4 b. Galena l. s. / 4 a. Trenton l. s. 990 / 3 c. St. Peter's s. s. / 3 a. Lower Magn. l. s.

Chicago & North-Western Railroad.
Ms. (Sheboygan and Western R. R.)—Con. Alt.

Ms.	Station	Alt.
69	Green Lake.	{ 4 a. Trenton l. s. / 3 c. St. Peters s.s. 813 / 3 a. Low. Magn. l. s.
72	St. Marie.	3 a. Lower Magn. l. s.
78	Princeton.	" 766

(Madison and Montford Division.)

Ms.	Station	Alt.
165	Madison.	{ Moraines, drumlins. / 3 a. Low. Magn. 848 / 2 b. Pots. & Mad. s.s.
176	Verona.	Moraines.
182	Riley's.	{ 4 a. Trenton. / 3 c. St. Peter's. / 3 a. Lower Magn.
184	Pine Bluff.	{ 4 a. Trenton. / 3 c. St. Peter's.
188	Mount Horeb,	4 b. Galena.
193	Blue Mounds.	{ 5 c. Niagara. / 4 c. Hudson River. / 4 b. Galena.
197	Barnevel'd.	4 b. Galena.
203	Ridgeway.	"
212	Dodgeville.	{ 4 b. Galena. / 4 a. Trenton. / 3 c. St. Peter.
220	Edmund.	4 b. Galena.
223	Cobb.	"
227	Montford Junc.	"
228	Montford.	"
237	Preston.	"
239	Lancaster Junc.	"
241	Fennimore.	"
248	Werley.	{ 4 a. Trenton. / 3 c. St. Peter.
251	Anderson Mills.	{ 3 a Lower Magn. / 2 b. Potsdam.
257	Woodman.	2 b. Potsdam. 651
243	Stitzer.	4 b. Galena.
246	Liberty.	"
251	Lancaster.	"
234	Livingston.	"
238	Rewey.	"
245	Leslie.	"
247	Mineral Point Jc.	" 935
249	Platteville Jc.	"
253	Platteville.	4 a. Trenton and Ga.
254	Elmo.	4 b. Galena.
256	St. Rose.	"
257	Cuba City.	"
260	Benton.	"
262	Strawbridge.	"
264	Buncomb.	"
268	Millbrig.	"
275	Galena.	{ Loess, Terraces. / 4 b. Galena.

Chicago & North-Western Railroad.—Con.
Ms. (Milwaukee to Madison and Montford.) Alt.

Ms.	Station	Alt.
0	Chicago.	(As before.)
85	Milwaukee.⁶	{ 10. Ham'n cem. rock. / 5 c. Niagara. 584
96	North Greenfield.	20. Drift.
97	Calhoun.	"
102	Waukesha.	5 c. Niagara. 808
110	Wales.	20. Kettle Moraine.
115	Dousman.	"
121	Sullivan.	20. Drift, Kames near.
132	Jefferson Junc.	20. D'ft, Drumlins. 799
139	Lake Mills.	20. Drift Kames.,
144	London.	20. Drift, Drumlins.
154	Cottage Grove.	20. Drift.
165	Madison.	{ 20. Morainic Drift. / 3 a. Low. Magn. 848 / 2 b. Pots. & Mad.s. s.

(Janesville, Watertown & Fond du Lac.)

Ms.	Station	Alt.
0	Chicago.	(As before.)
70	Sharon.	20. Drift.
78	Clinton Junc.	" 941
82	Shopiere.	20. D'ft. 4 b. Gal.l.s.944
91	Janesville.	{ 4 a. Tren. 3 c. St.P'r's / Glacial flood plain.
99	Milton Junction.	20. Quaternary. 877
104	Koshkonong.	20. Drift. 827
110	Ft. Atkinson.	4 b. Gal., Drift. 798
116	Jefferson.	20. Drift. 799
119	Jefferson Junc.	20. Drift, Drumlins.
121	Johnson's Creek.	" 771
129	Watertown Jc.	4 b.Gal., Drumlins.821
130	Watertown.	"
138	Clyman.	Drumlins. 905
145	Juneau.	Drumlins. 912
148	Minnesota Junc.	20. Drift. Galena.
151	Burnett Junc.	" 877
160	Chester.	"
168	Oakfield.	" 888
176	Fond du Lac.	{ 4 b. Galena l. s. / Red Clay. 746
184	Van Dyne.	Lacustrine deposit.
193	Oshkosh.	{ 4 b. Galena l. s. / 4 a Trenton l. s. 759

Chicago, St. Paul, Min. & Omaha R. R,
Ms. (St. Paul and Lake Superior Division.) Alt.

Ms.	Station	Alt.
0	Minneapolis.	{ 4 a. Trenton. / 3 c. St. Peter.
10	St. Paul.	{ Moraine, Glacial / flood deposits.
30	Hudson.	3 b. Potsdam. " 706
33	N. Wisconsin Jo.	20. Quaternary. 872
41	Boardman.	{ 2 b. Potsdam, / Moraine drift. 957
46	New Richmond.	3 a. Lower Magn. 949
55	Deer Park.	20. Moraine.

Chicago, St. Paul, Min. & Omaha R. R.
Ms. (St. Paul and Lake Superior Div.)—Con. Alt.

Ms.	Station	Alt.
63	Clear Lake.	20. Moraine, west.
71	Clayton.	"
75	Turtle Lake.	20. Morainic drift.
79	Perley.	"
88	Cumberland.	"
95	Barronett.	"
104	Shell Lake.	20. Moraine summit.
110	Spooner.	20. Gravel drift,
118	Veazie.	20. Glacial fl'd deposit.
130	Stinnett.	"
136	Hayward.	"
153	Cable.	20. Moraine.
163	Drummond.	"
177	Mason.	20. Red clay drift.
190	Ashland Junc.	"
194	Ashland.	"
190	Ashland Junc.	"
198	Washburne.	2 b. Potsdam, Drift.
211	Bayfield.	"

(Eau Claire and Lake Superior Division.)

Ms.	Station	Alt.
0	Eau Claire.²	Pots. and Val. d'ft.888
10	Chippewa F'lls.²³	{ 2 b. Potsdam. / 1. Archæan granite.
25	Bloomer.	2 b. Potsdam, Drift.
33	Cartwright.	
42	Chetek.	2 b. Pots., gravel hills.
49	Cameron.	2 b. Potsdam. } Gravel
56	Rice Lake.	" } plain.
81	Spooner.	Moraine.
113	Gordon.	{ 20. Ancient outlet of / Lake Superior.
139	Douglass.	{ 2 b. Potsdam. / Keweenawan.
150	Superior.	20. Red clay drift

(Neilsville Branch.)

Ms.	Station	Alt.
0	Neilsville.	2 b. Potsdam s. s.
14	Merillan.	" 935

Chicago, Milwaukee & St. Paul Railroad.
Ms. (Chicago, St. Paul & Minneapolis Line.) Alt.

Ms.	Station	Alt.
0	Chicago.	(As before.)
43	Wadsworth.	20. Quaternary.
52	Kenosha Junc.	" 679
53	Truesdell.	" 879
62	W. U. Junction.	" 722
85	Milwaukee.⁶	{ 10. Hamilton, Mil. / Cement Rock. 884 / 5 c. Niagara l. s.
98	Brookfield.	20. Quaternary. 824
109	Pewaukee.	{ 5 c. Niag., Striæ, / Drumlins east. 842
109	Hartland.	{ 20. Moraine / fluvial drift. 889

Chicago, Milwaukee & St. Paul Railroad.
Ms. (Chicago, St. Paul and Min. Line.)—Con. Alt.

Ms.	Station	Geology	Alt.
111	Nashotah.	{ 20. Moraine, fluvial drift.	
116	Oconomowoc.	"	861
129	Watertown.	{ 4 b. Galena l. s., drumlins.	
130	Watertown Jc.	"	821
139	Reeseville.	20. Drumlins.	
144	Elba.		
148	Columbus.	{ L.. Magn. l. s. drift.	884
152	Fall River.	"	936
158	Doylestown.	"	938
163	Rio.	"	
168	Wyocena.	{ 2 b. Madison s. s. / 2 b. Mendota. s. s. / 2 b. Pots. s. s.	827
176	Portage City.[7]	2 b. Potsdam s. s.	
193	Kilbourn.[8]	{ 2 b. Pots. s. s. finely exposed in dalles of Wisconsin.	893
202	Lyndon.	2 b. Potsdam s. s.	
209	Lemonweir.	"	894
212	Mauston.[887]	" { fine cas-	
220	Lisbon.[893]	" tellated	
225	Camp D'glas.[929]	" outliers.	
238	Tomah.	"	967
242	Greenfield.	"	
249	Lafayette.	"	
255	Sparta.[4]	" "	788
265	Bangor.	2 b. Pots. s. s. ter.	752
270	West Salem.	"	
277	Winona Junc.	"	655
280	La Crosse.	{ 2 b.Pots. s.s.,3 a.Low. Magn.val.d'ft.	699
410	St. Paul.	(See Minnesota.)	
420	Minneapolis.	"	

(Prairie du Chien Division.)

Ms.	Station	Geology	Alt.
0	Milwaukee.[6]	{ 10. Ham. cement r'ck / 5 c. Niagara l. s.	634
6	Wauwatosa.	{ 5 c. Niagara. Striæ, Drift.	631
10	Elm Grove.	20. Quaternary.	746
14	Brookfield Jc.	"	824
17	Forest House.	"	818
21	Waukesha.	{ 5 c. Niagara. Striæ, Drift.	802
28	Genesee.[9]	"	903
31	North Prairie.[10]	20. Quaternary.	941
37	Eagle.[11]	{ Kettle Moraine 948 / Glacial gravel plain.	

Chicago, Milwaukee & St. Paul Railroad.
Ms. (Prairie du Chien Division.)—Con. Alt.

Ms.	Station	Geology	Alt.
		{ Inner border of Ket-tle Moraine.	888
42	Palmyra.		
51	Whitewater.[12]	4 b. Galena l. s.	819
56	Lima.	{ 20. Quat., feeble moraine, E.	888
62	Milton.[13]	Quaternary.	871
64	Milton Junction.	"	877
71	Edgerton.	{ 4 a. Trenton. 820 / 3 c. St.P.s.s. d'ft hills	
81	Stoughton.	20. Quat.heavy d'ft.	887
89	McFarland.	{ 20. Heavy drift. 867 / 3 a. Low. Magn. l. s.	
96	Madison.	{ 20. Mor. drift. 848 / 3 a. Low. Magn. l. s.	
102	Middleton.	{ 2 b. Madison s. s. / 2 b. Mendota l. s. / 2 b. Pots. s. s. 935 / 3 a. Low. Magn. l. s.	
110	Cross Plains.	(Kettle Moraine.) { 2 b. Mad. s. s. { bluff / 2 b. Men. l. s. { sides / 2 b. Pots. s. s. valley bottom.	855
115	Black Earth.	"	810
119	Mazomanie.	"	773
125	Arena.	2 b. Potsdam s. s.	732
132	Spring Green.	{ 3 a. Low. Magn. on bluffs. 722 / 2 b. Potsdam s. s. on low ground.	704
139	Lone Rock.	{ 2. b. Pots. in the valley. Ad-698	
145	Avoca.	jacent bluffs 627	
151	Muscoda.	capped with 3 667	
166	Boscobel.	a. Low. Magn.	688
176	Wauzeka.	limestone.	
183	Wright's Ferry.		
186	Bridgeport.	3 a. Lower Magn.	625
194	P'rie du Chien.[14]		619

Ms.	Station	Geology	Alt.
64	Milton Junction.	29. Quaternary.	877
71	Janesville,	{ 4 a. Trenton. 818 / 3 c. St. Peter's, glacial flood plain.	
78	Hanover.	{ 4 a. Tren. l. s. glacial b'kwater pl'n.	740
83	Orford.	{ 4 a. Tren. l. s. 895 / 3 c. St.P. s. s., Drift.	
80	Brodhead.[15]	St. Peter's s. s.	798
105	Monroe.[16]	4 b. Galena l. s.	870
113	Browntown.	4 b. Galena l. s.	
127	Gratiot.	"	768
138	Shulsburg.	"	

7. *Portage City.* Fluvial drift, moraine between Portage and Kilbourn.
8. *Kilbourn.* Beautiful exhibitions of fluvial erosion in Dalles of the Wisconsin.
9. *Genesee.* Drumlins east and moraines and kames west of Genesee.
10. *North Prairie.* Till, fluvial drift; moraines and kames east and west of this place.
11. *Eagle.* Glacial flood plains.
12. *Whitewater.* Drumlins; striæ. Kettle moraine south of this place.
13. *Milton.* Moraines north and south, glacial flood drift.
14. *Prairie du Chien.* Potsdam; valley drift; artesian wells.
15. *Brodhead.* Trenton (capping bluffs east). Glacial flood plain.
16. *Monroe.* Border of drift. Glacial gravel capped with till.

Chicago, Milwaukee & St. Paul Railroad.

Ms.	Madison Division.	Alt.
0	Madison.	3 a. Lower Magn. 846
12	Sun Prairie.	4 a. Trenton, Drift.
18	Deanville.	{ 4 a. Trent. Drift. Drumlins. 873
20	Marshall.	{ 20. Quat. 4 a. Trent. Drift; Drumlins. 864
23	Waterloo. 17	{ 4 a. Trenton l. s. 819 / 3 a. Lower Magn. l.s. / 1 a. Arch. Quartzite.
27	Hubbleton.	Subaqueous drift.
37	Watertown Junc.	4 b. Galena l. s. 821

Northern Division.

Ms.	Northern Division.	Alt.
0	Milwaukee. 6	{ 10. Hamilton, Milwaukee Cem.Rock / 5 c. Niagara l. s. 854
9	Schwartzburg.	" 845
15	Granville.	" 736
20	Germantown.	" 863
25	Richfield. 18	20. Quaternary. 959
33	Schleisingville.	{ Kettle Moraine. / Glac'l flood d'ft. 1052
37	Hartford.	{ 5 c. Niag. l. s. / 5 b. Clin. iron ore. 986 / 4 c. Cin. shale.
41	Rubicon.	20. Quaternary. 1018
46	Woodland.	" 951
47	Iron Ridge.	{ 5 c. Niagara l. s. / 5 b. Clin. iron ore. 923 / 4 c. Cin. Shale.
76	Fond du Lac.	{ 4 b. Galena. 769 / Red drift clay.
54	Horicon Junc.	20. Quaternary. 854
59	Burnett Junc.	" 877
68	Waupun.	4 b. Gal., Striæ. 892
76	Brandon.	20. Quaternary. 1000
83	Ripon.	{ 4 b. Galena l. s. / 4 a. Trenton l. s. 930 / 3 c. St. Peter's s. s. / 3 a. Lower Magn. l.s.
96	Berlin. 19	{ 3 a. Lower Magn. l.s. / 2 b. Potsdam s. s. 762 / 1 Arch. Porphyry.
90	Picket's.	4 a. Trenton limestone.
102	Oshkosh.	{ 4 b. Galena l. s. 758 / 4 a. Trenton l. s.
90	Rush Lake.	3 a. L. Magn. Striæ. 841
95	Waukau.	L. Magn. Red d'ft clay.
99	Omro.	{ 20. Quat., Red drift clay.
104	Winneconne. 20	3 a. L. Magn. l. s.

Chicago, Milwaukee & St. Paul Railroad.

Ms.	Northern Division.—Continued.	Alt.
54	Horicon Junc.	20. Quaternary. 854
57	Minnesota Junc.	" 926
59	Rolling Prairie.	" 941
68	Beaver Dam.	{ 4 b. Galena l s 918 / Tren. l.s., drumlins.
69	Fox Lake Junc.	4 a. Trenton l. s. 883
74	Randolph.	{ 4 a Trenton l. s. 956 / 3 c. St. Peter's. s. s. / 3 a. Lower Magn. l.s.
80	Cambria.	{ 3 a.Lower Magn. l.s. / 2 b. Madison s.s. 862 / 2 b. Mendota l. s. / 2 b. Potsdam s. s.
90	Pardeeville.	2 b. Potsdam s. s. 810
98	Portage City. 7	"

Madison and Portage Division.

Ms.		Alt.
0	Madison.	(As before.) 845
1	East Madison.	" 845
12	Windsor.	{ 3 a. Lower Magn.l.s. / 2 b. Potsdam s.s. 882
16	Morrison.	3 a. L. Magn. l. s. 965
21	Arlington.	{ 3 c. St. Peter's s. s. / 3 a. L. Mag.l.s. 1004
25	Poynette.	2 b. Potsdam s. s.
39	Portage.	" 792

Racine and Southwestern Division.

Ms.		Alt.
0	Racine. 5	Niag. (Racine) l.s. 618
2	Junction.	" 621
8	W. U. Junc.	Deep drift, (Till) 583
10	Windsor.	" 882
15	Union Grove.	" 760
18	Kansasville.	" 815
27	Burlington.	5 c. Niag., Moraine 781
31	Lyons.	{ Niag. ls. Moraine 800 / Till & gravel hills.
34	Springfield.	{ 20. Till and gravel hills. 848
41	Elkhorn.	20. Heavy drift. 991
46	Delavan.	{ 20. Till & gravel. 984
50	Darien.	20. Moraine. 945
54	Allen's Grove.	Heavy drift. 871
59	Clinton.	" 941
69	Beloit.	{ Galena & Trenton ls. / St. Peter's s. s. / Glac'l flood grav. 740
	(Continued	in Illinois.)
0	Eagle.	Kettle Moraine. 948
6	Troy Center. 21	Heavy drift. 678

17. *Waterloo.* Drumlins; heavy drift; boulder train.
18. *Richfield.* Heavy drift; kettle moraine west.
19. *Berlin.* Red clay drift; boulder train.
20. *Winneconne.* Lower magnesian limestone domes east; heavy drift.
21. *Troy Centre.* Till and glacial flood deposits.
22. *Amherst.* Moraine east; glacial flood plain west of this place.

Chicago, Milwaukee & St. Paul Railroad.
Ms. Racine and Southwestern Div.—Con. Alt.

Ms.	Station	Description	Alt.
9	Mayhew's.	20. Heavy drift.	
11	Fayette.	" "	861
17	Elkhorn.	" "	991

Wisconsin Valley Division.

Ms.	Station	Description	Alt.
0	Tomah.	2 b. Potsdam s. s.	967
7	Valley Junction.	"	934
10	Norway.	"	935
18	Beaver.	"	965
29	Remington.	"	981
42	Port Edwards.	{ 2 b. Potsdam s. s. on / 1. Arc'n Gneiss.	972
46½	Centralia.	"	1015
54	Rudolph.	1. Archæan, Drift.	1146
60	Junction City.	"	1145
70	Knowlton.	"	1131
76	Mosinee.	"	
89	Wausau.	"	1227
08	Trap City.	"	
102	Pine River.	"	
107	Merrill.	"	

Mineral Point Division.

Ms.	Station	Description	Alt.
0	Mineral Point.	{ 4 b. Gal. l. s. / 4 a. Trent. l. s. / 3 c. St. Peter's s. s.	935
10	Calamine.	{ 4 b. Gal. l. s. / 4 a. Trent. l. s. / 3 c. St. Peter's s. s.	812
20	Belmont.	4 b. Galena limestone.	
28	Platteville.	{ 4 b. Galena l. s. / 4 a. Trenton l. s.	
0	Mineral Point.	(As before.)	935
10	Calamine.	"	812
16	Darlington.	4 a. Trent. l. s.	802
26	Gratiot.	{ 4 b. Gal. l. s. / 4 a. Trent. l. s.	783
33	Warren.	(See Illinois.)	

Prairie du Chien Division.—Con.

Ms.	Station	Description	Alt.
119	Mazomanie.	Pots. s.s., Val. drift.	773
127	Sauk City.	{ 3 a. L. Mag.l.s. / 2 b. Pots.	788
129	Prarie du Sac. [25]	{ 3 a. L. Mag. l. s. / 2 b. Pots.	
139	Lone Rock.	2 b. Pots. in val.	704
145	Richland City.	Adjacent bluffs cap'd	
149	Twin Bluffs.	with 3 a. L. Mag.l. s.	
155	Richland Cent.	3 a. L. Mag. l. s.	

Chippewa Valley Division.

Ms.	Station	Description	Alt.
0	Wabasha, Minn.	2 b. Potsdam s. s.	
1	Reads Junc.	Alluvial bottoms.	

Chicago, Milwaukee & St. Paul Railroad.
Ms. Chippewa Valley Division.—Con. Alt.

Ms.	Station	Description	Alt.
19	Durand.	{ 2 b.Pots. Bluffs cap'd / with 3 a. L. Mag.l.s.	
25	Red Cedar.	Valley d'ft, terraces.	
26	Red Cedar Junc.	{ 2 b. Pots. & 3 a. L. / Mag.l.s.in adj.hills.	
32	Meridean.	{ 2 b. Pots. & 3 a. L. / Mag.l.s.in adj.hills.	
43	Porterville.	{ 2 b. Pots. & 3 a. L. / Mag.l.s.in adj.hills.	
47	Shawtown.	{ 2 b. Pots. & 3 a. L. / Mag.l.s.in adj.hills.	
48	Eau Claire. [2]	20. Glac. val. d'ft.	830
54	Lafayette Mills.	{ Terraces, 2 b. Pots. / s. s.	836
56	Badger Mills.	Terraces, 2 b. Pots.s.s.	
62	Chip'ewa Falls.[23]	{ 1. Archæan granite. / 2 b. Potsdam s. s.	

Menomonee Branch.

Ms.	Station	Description	Alt.
26	Red Cedar Junc.	{ Val. d'ft. terraces; / 2 b. Pots. & 3 a. L. / Mag. in hills.	
28	Dunnville.	{ Val. d'ft. terraces; / 2 b. Pots. & 3 a. L. / Mag. in hills.	
41	Menomonee.	{ 2 b Pots., Glac. flood / plain, terraces.	876

Green Bay, Winona & St. Paul Railroad.

Ms.	Station	Description	Alt.
0	Green Bay.	{ 5 c. Niag. l. s. / 4 c. Cin. shale. / 4 b. Galena l. s.	588
10	Oneida.	"	
17	Seymour.	{ 4 a. Trenton l. s. / 3 c. St. Peter's s. s.	
23	Black Creek.	3 a. Lower Magn. l. s.	
31	Shiocton.	20. Quaternary.	
39	New London.	{ 3 a. L. Magnesian l. s. / 2 b. Potsdam s. s., Red clay drift.	
46	Royalton.	20. Quaternary.	822
50	Manawa.	"	824
55	Ogdensburg.	"	870
61	Scandinavia,	Kettle Mor. W. of	883
78	Amherst. [22]	{ Kettle Moraine. / 2 b. Potsdam s. s.	1044
82	Plover.	Glacial flood plain.	
96	Grand Rapids.	{ 1. Archæan Gneiss / overlaid by / 2 b. Potsdam s. s. and / altering into Kaolin.	1024
111	Dexterville.	2 b. Pots. s. s.	1001
119	Scranton.	"	962

23. *Chippewa Falls.* Glacial flood deposit; terraces.
24. *Sauk City.* Drift Margin. Border of the driftless area.
25. *Prairie Du Sac.* Kettle moraine and valley overwash.
26. *Wabasha.* Bluffs capped with Lower Magnesian limestone. Valley drift terraces.

Green Bay, Winona & St. Paul Railroad.—Continued.

Ms.		Alt.
142	Hatfield.	2 b. Potsdam s. s.
149	Merrillan.	" 943
153	Alma Center.	"
159	Hixton.	"
166	Taylor.	"
172	Blair.	"
179	Whitehall.	"
193	Arcadia.	" Val. d't Ter.
210	Marshland.	{ 2 b. Pots. s. s. 859 / 8 a. L. Magn. l.s.
214	Winona.	(See Minnesota.) 655

Milwaukee, Lake Shore & Western R. R.

Ms.		Alt.
0	Milwaukee. 6	{ 10.Hamilton Cement Rock. 584 / 5 c. Niagara l. s.
4	Lake Shore Junc.	20. Quaternary. 642
6	White Fish Bay.	{ 10. Hamilton, Red clay drift. 654
10	Dillman's.	{ 20. Quat., Red clay drift. 653 / "
13	Mequon.	"
20	Ulao.	" 697
25	Port Washington.	{ 5 c. Niag., Red drift clay. 669
31	Decker's.	" 756
33	Belgium.	{ 20. Quat. Red drift clay. 755
38	Cedar Grove.	" 697
42	Oostburg.	" 693
46	Wilson.	"
48	Weeden's.	" 700
52	Sheboygan.	{ 5 c. Niag. l. s., Red clay drift, Striæ. 588
58	Mosel.	{ 20. Quat. Red drift. clay. 689
64	Centreville.	" 687
69	Newton.	" 657
77	Manitowoc.	{ 5 c. Niag. l. s. Red drift clay. 593
84	Branch.	20. Moraine west. 729
89	Cato.	5 c. Niagara. 844
91	Grimms.	" 845
94	Reedville.	"
100	Brillion.	
104	Forest Junction.	20. Quaternary. 880
108	Dundas.	" 832
113	Kaukauna.	4 b. Galena. 655
116	Little Chute.	" 707
120	Appleton.	{ 4 b. Galena l. s. 715 / 4 a. Trenton l. s.

77	Manitowoc.	20. Quaternary. 595
84	Two Rivers.	" 586

78	Manitowoc.	{ 5 c. Niag., Red drift clay. 593
89	Cato.	" 824
94	Reedville.	" 820

Milwaukee, Lake Shore & Western Railroad.—Continued.

Ms.		Alt.
100	Brillion.	{ 5 c. Niag. Red drift clay.
104	Forest Junc.	20. Quaternary. 828
113	Kaukauna.	"
116	Little Chute.	" 722
120	Appleton.	" 706
122	Appleton Junc.	4 a. Trent., Red Clay.
134	Hortonsville Jun.	3 a. L. Magn., drift.
140	New London.	"
141	New London Jun.	"
150	Bear Creek.	20. Drift.
157	Clintonville.	"
164	Marion.	"
176	Tigerton.	1. Archæan granite.
188	Eland Junc.	1. Archæan, Drift.
192	Birnamwood.	"
198	Aniwa.	"
202	Elmhurst.	"
208	Antigo.	Archæan, Glac.gravel.
209	Wolf River Junc.	" "
217	Bryant.	" "
220	Malcom.	" Moraine.
225	Summit Lake.	" "
235	Pelican.	" Heavy d'ft.
241	Monico.	" "
267	Eagle River.	" "
293	Watersmeet.	
310	Gogebic.	{ 1 b. Potsdam. / Keweenawan. / 1 b. Huronian.

0	Eland Junc.	1 Archæan Gran. d'ft.
2	Norris.	20. Drift.
22	Wausau.	1 Archæan.

Milwaukee & Northern Railroad.
Milwaukee Division.

Ms.		Alt.
0	Milwaukee. 6	{ 10.Hamilton Cement Rock Drift. 584 / 5 c. Niagara l. s.
	Schwartzburg.	5 c. Niagara. 648
18	Thienville.	20. Quaternary.
23	Cedarburg.	5 c. Niagara l. s. 773
25	Grafton.	" 752
29	Saukville.	" 763
36	Fredonia.	" 788
41	Random.	20. Quaternary. 877
46	Sherman.	" 835
50	Waldo.	" 836
55	Plymouth.	" 844
		" 944
62	Elkhart Lake.	{ 20. Moraine. / Kettle Range.
68	Kiel.	5 c. Niag., Mor. E. 913
72	Holstein.	20. Quaternary.
	Hayton.	" 822
79	Chilton.	" 815
86	Hilbert.	"

Milwaukee & Northern Railroad.—Con.
Milwaukee Division.

Ms.			Alt.
86	Hilbert.	20. Quaternary.	
91	Forest Junction.	"	830
	Holland.	"	
99	Greenleaf.	"	
	Ledgeville.	5 c. Niagara.	
109	De Pere.	4 b. Gal., R. C. d'ft.	591
113	Green Bay.	{ 5 c. Niagara l. s. { 4 c Cin. shale. { 4 b. Gal. l. s.	833
114	Ft. Howard.	{ 4 b. Gal., l. s., La- { custrine clay.	684
119	Cormier	4 b. Gal., drift.	
124	Tremble.	20 Drift.	
128	Gardner.	"	
141	Grand Trunk Jc.	"	
146	Maple Valley.	3. L. Magn., Drift.	
153	Coleman.[27]	"	
156	Pound.	"	
159	Beaver.	2 b. Pots. s. s., Drift.	
165	Ellis Junc.	2 b. Pots., sand plains.	
177	Porterfield.	1. Archæan, Drift.	
185	Marinette.	4 b. Gal., drift, Striæ.	
187	Menominee.	"	
168	Noquebay.	1. Archæan, Drift.	
176	Wausaukee.	"	
185	Pike.	"	

Appleton Branch.

0	Hilbert.	20. Quaternary.	828
6	Sherwood.	5 c. Niagara l. s.	835
11	Lake Park.	Lacustrine drift.	
15	Menasha. 832	4 b. Gal. l.s 4 a.Tren. l.s.	
16	Neenah.	"	748
21	Appleton.	"	715

Wisconsin Central Line.

0	Milwaukee.6	{ 10. Hamilton Cem't { Rock. { 5 c. Niagara l. s.	884
32	Schleisingerville	{ 20. Kettle Moraine, { Glac. flood deposit.	
39	Allentown.	5 c. Niagara ls., Drift.	
48	Theresa.	"	
57	Hamilton.	"	
66	Fond du Lac.	4 b. Gal. l. s.	746
74	Van Dyne.	Lacustrine drift.	
83	South Oshkosh.	"	
84	Oshkosh.	{ Galena & Trenton ls. { Lacustrine d'ft.	753
88	State Hospital.	Lacustrine drift.	748
93	Snells.	"	
97	Neenah.	{ Galena & Trenton ls. { Striæ, Drift.	
86	Hilbert.	20. Quaternary.	
92	Sherwood.	5 c. Niag. l. s.	835
98	Menasha.	{ 4 Gal. l. s. { 4 a. Trent. l. s.	832

Wisconsin Central Line.—Con.

Ms.			Alt.
107	Medina.	3 a. L. Mag. ls.	813
110	Dale.	"	
124	Weyauwega.	2 b. Pots. ss.	825
131	Waupaca.	1. Archæan.	899
138	Sheridan. 1017	Kettle Moraine.	
144	Amherst.	"	1059
160	Stevens' Point.	{ Pots. ss. and Arch. { Gneiss. Gl. flood { plain.	1090
171	Junction City.	1. Archæan.	1145
175	Milladore.		
183	Auburndale.	{ 1. Arch. overl'd by { heavy d'ft.	1217
192	Marshfield.	"	1259
195	Mannville.	"	1292
200	Spencer.	"	1307
207	Unity.	"	1338
211	Colby.	"	1315
213	Abbotsford.	Drift.	
219	Curtiss.	"	
226	Withee.	"	
236	Thorpe.	2 b. Potsdam, Drift.	
247	Boyd.	"	
254	Cadott.	"	
267	Chip'wa Falls. 33	{ 1. Arch. Granite. { 2 b. Potsdam ss.	
268	St. Croix Junc.	{ 1. Arch. Granite. { 2 b. Potsdam.	
278	Morris.	2 b. Potsdam ss.	
285	Wiswell.	"	
288	Colfax.	{ Pots. ss., Glacial { flood dep. Terraces.	
293	Lochiel.	{ 20. Glacial fi'd dep. { Terraces.	
307	Barker.[27]	2 b. Potsdam, Drift.	
310	Downing.	"	
313	Emerald.	"	
324	Cylon.	3 a. L. Mag., Drift.	
333	New Richmond.	"	
338	Clarendon.	20. Drift. L. Magn.	
346	St. Croix.	" Pots. & L. Mag.	
349	Arcola.	20. Drift.	
363	Castle.	"	
367	Lake Phalen Jc.	"	
372	St. Paul.	(See Minnesota.)	

Northern Division.

0	Abbotsford.	{ 1. Archæan, overl'd { by heavy d'ft.	
4	Dorchester.	"	1466
14	Medford.	"	1413
25	Chelsea.	"	
29	Westboro.	"	
47	Worcester.	"	1605
55	Phillips.	"	1454
62	Wauboo.	"	
68	Fifield.	"	1458
79	Butternut.	"	
..	Chippewa.	"	

27. The formations given for this station and the following four, occur in the vicinity.

Ms.	Wisconsin Central Line.—*Con.* Northern Division.	Alt.	Ms.	Wisconsin Central Line.—*Con.* Southern Division.	Alt.
104	Penokee.²⁸ { 1. Hur'n, with iron ore.	1385	71	Portage. { 2 b. Pots., overlaid by drift.	792
126	White River. 20. Red clay drift.		55	Packwaukee. 20. Drift.	
133	Ashland. { 20. Red clay drift.	678	62	Montello. 20. Drift, Granite.	

Ms.	Southern Division.	Alt.		Minneapolis, Sault Ste. Marie & Atlantic.	
0	Stevens' Point. (As before.)	1090	0	Turtle Lake. Morainic drift.	
5	Plover. { 2 b. Pots., overlaid by drift.	1078	5	Scott's Siding. "	
11	Buena Vista. "		15	Barron. 20. Glac. flood drift.	
22	Plainfield. ¹¹¹⁸ Moraine east.		20	Cameron Junc. "	
28	Hancock. ¹¹⁰² Kettle Moraine.		25	Canton. 20. D'ft., Q'rtzite near.	
46	Westfield. ⁸⁶⁰ " "		31	Hawkins. "	
55	Packwaukee. "	784	42	Tibbets Siding. "	
			45	Bruce. "	

28. Unconformability between Huronian and Laurentian finely shown at Penokee.

29. NOTE.—Where several formations are given it is to be understood that they occur in the vicinity, not necessarily immediately at the station. Also, that where the drift effectually conceals the underlying formations they are not usually given, though in almost all cases definitely known.

Iowa.[1]

LIST OF GEOLOGIC FORMATIONS FOUND IN IOWA.

20 b. Loess, (concealing stratified rocks.	13 b. Burlington.
20 a. Glacial Drift " " "	13 a. Kinderhook.
18 Inoceramus.	10. Hamilton.
18 Woodbury,	5 c. Niagara.
18 Nishnabotna.	4 c. Maquoketa.
18 Fort Dodge.[2]	4 b. Galena Limestone.
14 c. Upper Coal.	4 a. Trenton.
14 b. Middle Coal.	3 b. St. Peter.
14 a. Lower Coal.	3 a. Lower Magnesian.
13 d. St. Louis.	2 b. Potsdam.
13 c. Keokuk.	2 a. Sioux.

Brief Sketch of the Geology of Iowa.

The general geologic structure of Iowa is simple: The prevailing dip of the strata is low, rarely reaching 5°, and south-westerly in direction. In consequence the outcrops of the greater rock series, from the oldest to the newest, form successive zones trending N. W.—S. E., each overlapped on the south-west by the attenuated margin of the next higher series. In detail this structure is modified and complicated by slight diversity in strike and dip and variations in thickness of the several formations, and the regularity of the zones of outcrop is destroyed through erosion by which the north-easterly (and basal) margins of the successive formations are channelled, deeply crenulated, and sometimes cut off in insulated outliers; and some of the major as well as many of the minor features of the stratified rocks are obscured by a mantle of superficial deposits.

The Potsdam is exposed by erosion only in the valley-bottoms of the extreme northeastern corner of the State, where it forms the gently-sloping bases of bluffs 300 to 500 feet high. The steeper medial portion of these bluffs is Lower Magnesian limestone, which, by reason of its firm texture, has well resisted the degradation of the rivers and forms nearly continuous mural or castellated precipices. Both formations disappear on the Oneota (or Upper Iowa) river about the west line of Allamakee county, and on the Mississippi, a few miles south of McGregor. The gentle slopes toward the summits of the bluffs in this region represent the friable St. Peter sandstone, sometimes white as snow, again brown, red or yellow, and elsewhere curiously variegated, as at McGregor, where it forms the "pictured rocks" of Iowa. The generally abrupt escarpment of the Trenton limestone overlooks the easy slopes of the sandstone, and forms a secondary line of bluffs along the Mississippi, Oneota and Yellow rivers in the north, which merges into the immediate river bluffs toward the mouth of Turkey river. The Trenton is the first of the formations to occupy a considerable area. It extends along the Iowa-Minnesota line from a few miles west of the Mississippi to several miles west of Decorah; but by reason of rapid attenuation southward and its confinement to the precipitous Mississippi bluffs below the mouth of the Turkey, the terrane contracts greatly toward Dubuque, where it passes beneath the surface. Almost everywhere the Trenton is richly fossiliferous. The precipitous bluffs at Dubuque represent the Galena limestone, which there has a thickness of 200 or 250 feet, but which rapidly dwindles northwestward. It is the plumbiferous formation of Illinois, Wisconsin, and Iowa, and takes its name from the prevalent form of the ore. From its caverns are brought forth the superb stalactites and crystalline masses of various minerals adorning the lawns and verandas of Dubuque. A narrow belt of soft-contoured hills cleft by spring-born streamlets, or a single gentle slope, rises from the precipices of the Galena and is overlooked by the bold Niagara escarpment. It represents the easily weathered shales and clays of the fossiliferous Maquoketa—a formation typally exposed along the Little Maquoketa river in Dubuque county. The type section is at Lattner's, on the D. & N. W. R. R., and 4 miles north of Peosta, on the I. C. R. R. The most prominent topographic feature in the State is the deeply crenulated escarpment of the western equivalent of the New York Niagara, stretching from the Minnesota line north of Cresco by West Union, Elkport, "Sherrill's Mound" (Dubuque county), Lattner's, and Peosta to the Mississippi at Bellevue, and forming the river-bluffs thence to Lyons. To the north the formation (generally a poorly fossiliferous dolomite abounding in cherty nodules) is thin, and its outcrop but a few miles in width; but toward the south it thickens to 350 feet or more, and its terrane widens greatly. It forms the "rapids" at Le Claire, but passes beneath the Mississippi between that town and Davenport. It is economically important by reason of its building-stone. Each of these formations (Niagara to Potsdam) is clearly differentiated, and conjointly they constitute a topographically distinct section of the State—a section in which the relief is the product of sculpture by rain and rivers during a vast period. Elsewhere the monotonous topography of the State is glacic in origin, with some post-glacial modification by hydric agencies: Here it is exclusively hydric.

To the southwestward the firm dolomites of the Niagara pass beneath the argillaceous limestones and shales of Devonian age which are usually referred conjunctively to the epoch of the New York

1. By W. J. McGee, U. S. Geologist.
2. The Fort Dodge is referred to the Cretaceous with doubt.

Chicago, Milwaukee & St. Paul Railroad.

Prairie du Chien, & Ia. and Minn. Div.

Ms.	Station	Formation	Alt.
0	No. McGregor.[1]	3 b. St. Peter, 625 / 3 a. L. Magnesian in hills, 2 b. Potsdam.	
6	Giard.	3 b. St. Peter.	
15	Monona.	4 a. Trenton.	1221
19	Luana.	"	1122
26	Postville.[2] 1207	4 c. Maq. & 4 b. Galena.	
32	Castalia.	" "	1257
37	Ossian.	"	1251
43	Calmar.	"	1269
46	Conover. 1247	4 c. Maq. & 4. b. Gal.	
53	Ridgeway.	5 c. Niagara.	
62	Cresco.	"	1312
73	Lime Springs.	"	1235
78	Chester.	"	1244
85	Leroy.	"	1298

(See Minnesota.)

Iowa and Dakota Division.

Ms.	Station	Formation	Alt.
0	Calmar.	4 a. Trenton.	1269
6	Fort Atkinson.	"	1025
18	Lawler.	10 b. Hamilton.	
27	New Hampton.	"	1166
35	Chicasaw.	"	1148
38	Bassett.	"	
47	Charles City.	"	1012
50	Floyd.	"	1107
59	Rudd.	"	
65	Nora Springs.	"	
74	Mason City.	"	1130
84	Clear Lake.[3]	20 a. Glacial Dft.	1237
95	Garner.	"	1237
105	Britt.	"	1330
115	Wesley.	"	1254
126	Algona.	"	1300
150	Emmetsburg.	"	
165	High Lake.	"	
173	Estherville.	"	
162	Ruthven.	"	
175	Spencer.	"	
187	Milford.	"	
192	Lakes Okoboji.	"	
196	Spirit Lake.	"	
200	Sanborn.	"	
211	Sheldon.	"	
225	Patterson.	"	
252	Canton.[3]	"	

Chicago, Milwaukee & St. Paul Railroad.

Mason City and Austin Division.

Ms.	Station	Formation	Alt.
0	Mason City.	10 b. Hamilton.	1130
8	Plymouth.	"	1114
21	Carpenter.	"	
28	Lyle.	"	
40	Austin, Minn.	18. Cretaceous.	1197

Dubuque and South-Western Railroad.

Ms.	Station	Formation	Alt.
0	Farley.	5 c. Niagara.	1111
7	Worthington.	"	
14	Sand Spring.	"	938
20	Monticello.	"	800
24	Langworthy.	"	
31	Anamosa.	"	
88	Viola.	"	
45	Paralta.	"	
50	Marion.	10 b. Hamilton.	
56	Cedar Rapids.	"	710

Chicago, Council Bluffs and Omaha Line.

Ms.	Station	Formation	Alt.
0	Sabula.[4]	Maquoketa, 5 c. Niag.	
6	Elk River.	" "	
15	Miles.	" "	
20	Preston.	5 c. Niagara.	
28	Riggs.	"	
83	Delmar Junct'n.	"	
40	Elwood.	"	
52	Oxford Junct'n.	"	730
62	Olin.	"	
74	Martelle.	"	
79	Paralta.	"	
87	Marion.	10 b. Hamilton.	

Sioux City and Dakota Division.[5]

Ms.	Station	Formation	Alt.
0	Sioux City. 1122	20 b. Loess & 18 Woodb.	
8	McCook, Dak.[5]	"	1123
13	Jefferson.	18 b. Mid. Creta's.	1130
14	Davis Jc. "	"	1130
21	Elk Point, "	"	1142
30	Burbank, "	"	1153
84	Vermillion, "	"	1161
44	Meckling, "	"	1167
50	Gayville, "	"	1176
55	James Riv., "	"	
61	Yankton.[6] "	"	1196
14	Davis Jc., "	"	1130
19	Joy.	"	
24	Westfield.	"	1145
29	Portlandville.	"	1163

Hamilton, the precise contact being everywhere concealed by drift save at Fayette and a point on the Wapsipinicon river a few miles above Central City, Linn county. The basal member of the Hamilton is a black shale which does not extend so far eastward as the medial calcareous member, but is exposed by excavations at Independence; while the uppermost member, also a dark shale or clay (typically exposed at Rockford) rarely appears along the Drift-buried western margin of the terrane. The Sub-Carboniferous formations (Burlington, Keokuk, Kinderhook, and St. Louis) cannot be discriminated geographically by reason of their deep burial beneath Drift and Loess; but all have important local exposures;—the type sections of the first two being within the State. The Burlington is noted for its crinoids which have made famous alike the city from which the formation derives its name and their local investigator, Dr. Wachsmuth; the Keokuk is equally noted for the magnificent geodes which have enriched so many collections; and both form the "Lower Rapids" which have so long vexed the spirits of Mississippi pilots and engineers. The Kinderhook yields a valuable oölitic limestone at Le Grand and elsewhere, and the St. Louis is still more important as a source of building material.

Chicago, Milwaukee & St. Paul R. R.—Cont. Ms.	Davenport Line.	Alt.
0 Davenport.* ***	10 b. Ham., 20 a. Gl. Dft.	
5 Mount Joy.	" "	
8 Eldridge.	20 a. Glacial Drift.	
17 Donahue.	5 c. Niagara.	
23 Dixon.	"	
32 Wheatland.	"	***
37 Toronto.	"	
40 Massillon.	"	
46 Oxford Mills.	"	
53 Wyoming.	"	
69 Monticello.	"	***
77 Hopkinton.	"	
85 Delhi.	"	
89 Delaware	"	***
94 Greeley.	"	
99 Edgewood.	"	
106 Enfield.	"	
115 Brush Creek.	"	
125 Fayette.7 ****	" & 10 Hamil.	
140 Hawkeye.	20 a. Drift, "	
149 Waucoma.	{ 5 c. Niag., 10 Ham-ilton in highlands.	
153 Jackson Junc.	20 a. Drift, 10 Ham.	
165 Calmar.	" 4 a. Tren. ***	

Racine and South-Western Division.

Ms.		Alt.
11 Eldridge.	20 a. Glacial Drift.	
14 Long Grove.	5 c. Niagara.	
C. & N.W. Cros'g.	"	
24 De Witt.	"	
31 Wilton.	"	
37 Delmar Junct'n.	"	
44 Maquoketa.	"	

Chicago, Milwaukee & St. Paul R. R.—Cont. Ms.	Dubuque Division.	Alt.
78 LaCrosse.	(See Wisconsin.)	
153 New Albin.	{ 2 b. Potsdam & 3 a. L. Magnesian	
141 Lansing.9	2 b. Pots. & L. Magn.	
126 Harper's F'ry.10	"	
118 Yellow River.11	"	
115 No. McGregor.1	"	***
104 Clayton.12	{ 3 a. L. Magnesian & 3 b. St. Peter.	
95 Guttenberg. ***	{ 4 a. Trenton & 4 b. Galena limestone.	
88 Turkey River.	4 a. Tren., 4 b. Galena.	
84 Buena Vista.	"	
80 Waupeton.13	"	
72 Specht's Ferry.14	"	
Peru.15	"	
60 Dubuque.16	4 a. Trenton. ***	
54 Massey.	4 b. Galena limestone.	
46 Gordon's Ferry.	{ 4 b. Galena Maquo-keta & 5 c. Niag.	
36 Bellevue.	Maq. & 5 c. Niagara.	
28 Green Island.	" " in hills.	
18 Sabula.4	" " "	
2 Lyons.17	5 c. Niagara. ***	
0 Clinton.	" ***	

Volga Branch.

Ms.		Alt.
88 Turkey River.	4 a. Tren. & 4 b. Galena	
103 Elkport.18	"	
111 Littleport.	"	
125 Volga City.	4 b. Gal., 5 c. Nia., Maq.	
138 Lima.	"	

The southwestern third of the State is mainly occupied by the Coal Measures (generally divided into Upper, Middle, and Lower) which, notwithstanding their economic importance, have not yet been adequately studied. It is known, however, that Coal Measure outliers, containing "pockets" of coal, and of such petrographic character as to indicate that they were deposited in bays or estuaries of the coal-period sea, repose unconformably upon the Sub-Carboniferous, the Devonian, and even the Silurian formation, far beyond the normal limits of the terrane; that workable beds of coal (under existing commercial conditions) are confined in the lower member; and that the three members reach a total thickness of not less than 800 or 1,000 feet. The Carboniferous outliers find homologues in the Cretaceous sandstones designated Nishnabotna by Dr. White, after one of the rivers along which they occur; but only slight remnants of the formation they represent (unless it be the Inoceramus, the Woodbury, or both) are preserved in Iowa. It is a good working hypothesis, but nothing more, that the bedded gypsum, of which the Ft. Dodge is composed, was precipitated in one of these Cretaceous estuaries so situated as to receive little drainage and suffer rapid desiccation after the first influx of the Mesozoic ocean. The Inoceramus (named from its characteristic fossil) and the Woodbury (named from the county in which it occurs, and well exposed about Sioux City) represent regularly bedded off-shore deposits not yet finally correlated with the well-developed Cretaceous deposits of Dakota and Nebraska. So far as certainly known they occupy a limited area in extreme western Iowa.

Over the five-sixths of the State lying west and south of the Niagara escarpment the lithified sedimentary strata are over-spread by a sheet of Glacial Drift, which, in the northern-central and northwestern counties reaches a depth of 100 to 200 feet and effectually conceals the subterrane, but which attenuates eastward, southward, and westward to such a degree that stream-corrasion and artificial excavation occasionally expose the subjacent rocks. In the northern part of the State Drift-bowlders frequently lie upon the surface; and within an area of 4,000 or 5,000 square miles centering in Bremer county, these superficial bowlders of northern crystalline rocks reach maxima in dimensions and abundance. Diameters of fifteen to twenty feet are common; and a dozen examples sometimes occur within a radius of half a mile. In eastern, and at least parts of central, Iowa the Drift is bipartite, and the "Upper Till" and "Lower Till" constituting it are frequently separated by a "Forest Bed"; and one of the loops of the great Kettle Moraine of northern United States extends far into the northwestern portion, reaching almost or quite to Des Moines; but tripartition of the Drift inside the loop has not yet been proven stratigraphically. Inside the moraine post-glacial drainage is not yet fully developed, lakes, ponds and sloughs abound, and the topography is the acme of monotony. In extreme southern Iowa the Upper Till disappears, and is replaced by a compact, tenacious, dark clay of aqueous origin, locally known as "hard-pan;" and both (as well as

Ms. Chicago, Milwaukee & St. Paul R. R.	Alt.		Chicago, Milwaukee & St. Pai
194 Rock Island, Ill.	10 Hamilton.	584	Ms. Waukon Branch.³⁴

Ms. Chicago, Milwaukee & St. Paul R. R.	Alt.		Ms. Waukon Branch.
138 Savannah, Ill.	Maquoketa, 5 c. Niag.		0 Waukon Junc. — { 3 b. St. / 3 a. L.
141 Sabula, Ia.⁴	" "		
147 Elk River.	" "		
157 Miles.¹⁹	5 c. Niagara.		9 Waterville. — { 4 a. Tr / 3 a. L. / Peter i
167 Browns.	"		
174 Delmar Junction.	"		23 Waukon. — 4 a. Tren
181 Elwood.	"		
185 Lost Nation.	"		**Cascade Branch.**
193 Oxford Junction.	"	720	
203 Olin.	"		0 Bellevue. — { 5 c. N / Maquo / bottom
215 Martelle.	{ About Junction of / Niag. and Hamilton.		
228 Marion.	10 Hamilton.		11 La Motte. — 20 b. Loe
228 Marion.	"		16 Zwingle.²⁵ — { 20 b, / Drift,
233 Cedar Rapids.	"	719	22 Wash'n Mills.²⁶ — 20 b. Loes
253 Amana.²⁰	"		25 Bernard.²⁷ — 5 c. Niag
295 Sigourney.²¹	13 d. St. Louis.		30 Fillmore.²⁷ — "
310 Hedrick.	"		36 Cascade. — 5 c. Niag
324 Ottumwa.²² 680	13 c. Keok. & 13 d. St. L		**Illinois Central Rail**
228 Marion.	10 Hamilton.		**Iowa Division.**
232 Louisa.	"		0 Dubuque.¹⁶ — 4 a. Tren
238 Covington.	"		10 Julien. — Maquoke
243 Atkins,	"		15 Peosta. — 5 c. Niag
255 Van Horne.	"		23 Farley. — "
260 Keystone.	"		29 Dyersville. — "
267 Elberon.	20 a. Glacial Drift.		37 Earlville.
277 Gladstone.			41 Delaware. — "
282 Tama City.	13 a. Kinderhook. 882		47 Manchester. — "
295 Pickering.	20 a. Glacial Drift.		54 Masonville. — "
310 Melbourne.	"		61 Winthrop. — "
354 Des Moines.²³		8 7	69 Independence. — 10 Hami
333 Cambridge.	14 Lower Coal, etc.		78 Jesup. — "
348 Madrid.	"		86 Raymond. — "
366 Perry.	"	977	93 Waterloo. — "
382 Bagley.	"		98 Jn.C.F.& M.R.R. — "
395 Coon Rapids.	"		99 Cedar Falls. — "
411 Templeton.	20 a. Glacial Drift.		109 New Hartford. — "
421 Aspinwall.	"		118 Parkersburg. — "
435 Defiance.	"		123 Aplington. — 20 a. Gla
446 Panama.	"		132 Ackley. — 13 a. Kin
458 Persia.	"		143 Iowa Falls. — "
468 Neola.	20 b. Loess.		149 Alden. — "
478 Weston.	"		158 Williams. — Gl. Drift.
487 Council Bluffs, Ia	"	989	172 Webster City. — " 1
490 Omaha, Neb.	"		192 Fort Dodge.²⁸ — 13 d. St.

the Lower Till when they are absent) are commonly overlain by Loess, which is g(
formable to all older deposits, but in southern Iowa often merges by impercept
into the Upper Till. The Loess in the south and west is often attenuated or absent
frequently eroded from valleys, and thus forms only the brows of the hills. The cor
the Loess attains its best development along the Missouri River. In north-eastern I
below the Niagara escarpment and overlapping the Drift margin for some miles, is a
the Loess, peculiar in its attitude;—it sometimes descends into valleys, but generally se
and caps the highest ridges and divides in the region. The rivers occasionally exh
behavior in the same region, in that they have manifestly avoided and deserted lowl
sought and corraded their channels in plateaus and in the axes of ridges. (See note 5
portion of the Wisconsin "Driftless Region" extending into Iowa, which is bounded
escarpment, Glacial Drift is absent, and the prevailing superficial covering is a residu
through secular decomposition of the subjacent strata, together with a sheet of l
debris. Alluvium occurs along all the streams of the State, and its amount varies wit

1. *North McGregor.* St. Peter in hills.

Illinois Central Railroad.

Ms.	Iowa Division—Continued.	Alt.
210	Manson.	20. Glacial Drift. 1245
218	Pomeroy.	20 a. Glacial Dft. 1244
226	Fonda.	"
235	Newell.	"
245	Storm Lake.	"
258	Aurelia.	"
268	Cherokee.	" 20 b. Loess. 1211
283	Marcus.	" " 1469
291	Remsen.	" " 1335
302	Le Mars.	" " 1221
319	James'	20 b. Loess & Woodb'y.
327	Sioux City.	" " 1122

Cedar Falls and Minnesota Branch.

Ms.		Alt.
0	Waterloo.	10 b. Hamilton. 862
12	Janesville.	" 892
18	Waverly.	" 942
27	Plainfield.	" 926
35	Nashua.	" 975
46	Charles City.	" 1012
52	Floyd.	" 1107
63	Osage.	" 1178
67	West Mitchell.	"
72	St. Ansgar.	" 1179
80	Mona.	" 1203

Chicago and North-Western Railroad.
Clinton and Anamosa Line.

Ms.		Alt.
0	Clinton.	5 c. Niagara. 617
8	Lyons.[17]	" 617
10	Almont.[29]	" Maquoketa. 692
17	Bryant.	5 c. Niagara. 802
25	Charlotte.	" 711
33	Delmar Junct'n.	" 837
38	Maquoketa.	" 718
44	Nashville.	" 739
47	Baldwin.	" 744
50	Monmouth.	" 791
57	Onslow.	" 936
64	Amber.	" 936
71	Anamosa.	" 844

Council Bluffs and Omaha Line.

Ms.		Alt.
0	Chicago.	(As before.)
138	Clinton.	5 c. Niagara. 609
143	Camanche.	"
147	Low Moor.	" 657
152	Malone.	"
157	De Witt.	" 699

Chicago and North-Western R. R.

Ms.	Council Bluffs and Omaha Line—Cont.	Alt.
163	Grand Mound.	5 c. Niagara. 736
169	Calamus.	" 721
173	Wheatland.	" 695
178	Loudon.	" 733
185	Clarence.	" 641
190	Stanwood.	" 663
195	Mechanicsville.	" 612
202	Lisbon.	" 668
203	Mount Vernon.	10 b. Hamilton. 658
210	Bertram.	" 733
219	Cedar Rapids.	" 744
227	Fairfax.	" 784
234	Norway.	" 809
244	Blairstown.	" 855
240	Luzerne.	"
254	Belle Plaine.	" 840
260	Chelsea.	20 a. Glacial Drift.
270	Tama.	13 a. Kinderhook. 832
277	Montour.	" 868
280	Le Grand.	" 953
283	Quarry.	" 899
288	Marshall.[30]	13 c. Keokuk. 893
296	Lamoille.	14 a. Low. Coal Mres.
303	State Centre.	" 1086 ·
310	Colo.	" 1059
317	Nevada.	" 1017
326	Ames.	13 d. St. Louis. 936
330	Ontario.	14 a. Lower Coal.
335	Midway.	"
340	Boone.	" 1155
346	Moingona.	" 907
352	Ogden.	" 1109
357	Beaver.	" 1041
363	Grand Junction.	" 1055
370	New Jefferson.	" 1071
379	Scranton.	20 a. Glacial Drift.
388	Glidden.	"
396	Carroll.	" 1240
406	Arcadia.	" 1439
408	West Side.	"
415	Vail.	20 b.Loess, 20 a.Gl. Dft.
424	Denison.	" " 1192
433	Dowville.	" "
441	Dunlap.	" "
450	Woodbine.	" "
458	Logan.	928 14 c.Up. or 14 b.Mid.Cl.
467	Mo. Valley Jc.[31]	" " 1022
482	Crescent.[31]	" " 1209
488	Council Bluffs.[31]	" " 939

2. *Postville.* Galena and Maquoketa, with Niagara outlier to south and Trenton exposures to north.

3. *Clear Lake to Canton.* The road traverses a plain of Glacial Drift, characterized by the lakes, marshes and nascent drainage system of the region circumscribed by the Terminal Moraine. The drift is of great thickness and the subterrane wholly unknown.

4. *Sabula.* Maquoketa in slopes, Niagara in hill-tops.

5. *McCook.* One of the finest exposures of Loess in the Missouri basin extends along this Railway from Sioux City to McCook.

6. There are no rock exposures on this division, and the author of this chapter is not responsible for the formations here given.

7. *Fayette.* The contact between Devonian and Silurian rocks, seen only at one other locality in the State (near Central City, Linn Co.), is well exhibited here in a natural exposure in the northwestern part of the town.

Chicago and North-Western R. R.—Cont.
Ms.   St. Paul and Minneapolis Lines.   Alt.

Ms.	Station	Geology	Alt.
0	Des Moines.[23]	14 a. Lower Coal.	824
7	Saylor.	"	934
8	Trent.	"	
11	Ankeny.	"	1034
14	Pelton.	"	
18	Polk City.	"	
21	Ulm.	"	
25	Sheldahl.	"	1060
31	Kelley.	"	
37	Ames.	13 d. St. Louis.	943
44	Gilbert.	" 20 a. Dft.	1154
60	Story.	" "	1199
53	Randall.	" "	1207
59	Jewell.	20 a. Drift.	1078
66	Kamrar.	" 14 c. Low. Coal.	
78	Webster City.	" 13 d. St. L.	1066
81	Woolstock.	20 a. Drift.	1109
88	Eagle Grove.	"	1139
94	Thrall.	"	1163
100	Renwick.	"	
108	Whitman.	"	1169
117	Irvington.	"	1175
121	Algona.	"	1225
131	Burt.	"	1173
187	Bancroft.	"	1139

Maple River R. R. Branch.[22]

Ms.	Station	Geology	Alt.
0	Maple River Jc.	20 a. Glacial Dft.	1059
7	Breda.	"	1193
17	Wall Lake.[33]	"	1059
27	Odebolt. 1188	" and 20 b. Loess.	
38	Ida Grove.	Dft. in valley "	1050
45	Battle Creek.	" "	1023
54	Danbury.	" "	984
60	Mapleton.	" "	939

Sac City Branch[22]

Ms.	Station	Geology	Alt.
0	Wall Lake.[33]	20 a. Glacial Dft.	1059
13	Sac City.	"	1104
21	Early.	"	1144
29	Schaller. 1207	" and 20 b. Loess.	
36	Galva.	" "	1099
44	Holstein.	" "	1254
52	Cushing.	" "	1212
57	Correctionville.	" "	844
70	Kingsley.	" "	1047

Tipton Branch.

Ms.	Station	Geology	Alt.
190	Stanwood.	{ 5 c. Niag. over-	868
		{ lain by Dft.	
194	Walden.	" "	
198	Tipton.	" " & Loess.	

Chicago and North-Western R. R.—Cont.
Ms.   Eagle Grove and Hawarden Line.   Alt.

Ms.	Station	Geology	Alt.
368	Eagle Grove.	20 a. Drift.	1159
377	Thor.	"	1171
386	Dakota City.[34]	13a.Kind'k. Drift.	1144
391	Rutland.	" ? "	1147
398	Bradgate.	20 a. Drift.	1144
404	Rolfe Junction.	"	
413	Havelock.	20 a. Glacial Dft.	1251
421	Lawrence.	"	1333
428	Marathon.	"	1414
437	Sioux Rapids.	"	1363
443	Lime Grove.	"	1276
450	Peterson.	"	1257
455	Waterman Sdg.	"	
459	Sutherland. 1449	" and 20 b. Loess.	
479	Granville.	" "	1489
488	Alton.	- "	1324
499	Maurice.	"	1329
514	Hawarden.		1206

(Continued in Dakota.)

Iowa and South-Western Railway.

Ms.	Station	Geology	Alt.
0	Carroll. 1247	Drift. 14 c. Low. Coal.	
17	Manning.	" "	1149
25	Gray.	" "	1175
35	Audubon.	" "	1122

Ms.	Station	Geology	Alt.
17	Manning.	" "	1149
29	Irwin.	Loess, Drift. "	1089
35	Kirkman.	" "	1054

Iowa, Dakota and Minnesota Division.

Ms.	Station	Alt.	Geology
270	Tama.	839	Loess in plateau to N.W., 13 a. Kinderhook, Drift.
273	Toledo.	873	
281	Garwin.	919	Loess in plateau to the West, Drift, 14 c. Low. Coal in vicinity, 13 a. Kinderh'k.
298	Conrad.	1029	
306	Whitten.	1061	
	Eldora Junc.[57]		20 Alluvium.
310	Gifford.		" 941
314	Lawn Hill.		20 a. Drift, 14 c. L. Cl.
329	Radcliffe.		" " 1209
336	Ellsworth.		20 a. Drift. 1104
339	Jewell Junction.		" 1078
	Stanhope.		" 14 c. L. Cl. 1141
354	Stratford.		" "
364	Dayton.		" " 1109
375	Gowrie.		" " 1158
380	Franklinville.		" "
397	Lake City.		" " 1269

8. *Davenport.* Hamilton in valleys and hillsides, and feruginous sandstone of the Lower Coal on eminences, overlain by Glacial Drift, Forest Bed and Loess. The brown sandstone occurs also at Muscatine, Iowa City, Eldora, and elsewhere. It is referred to Lower Coal with doubt. It occurs in isolated outliers and was probably deposited in independent basins, as indicated by Hall in 1858.
9. *Lansing.* St. Peter in hills.
10. *Harper's Ferry.* St. Peter in hills.
11. *Yellow River.* St. Peter in hills.
12. *Clayton.* St. Peter, with Trenton on hills.
13. *Waupeton.* Trenton and Galena, with Maquoketa and Niagara in hills.

Ms.	Chicago, Rock Isl'd and Pac. R. R.	Alt.
0	Chicago.	(As before.)
183	Davenport.[5] 575	{ 20 a. Gl. Dft., 20 b. Loess, 14 a. Low. Cl. 10 Hamilton.
195	Wolcott.	20 a. Glacial Drift.[783]
199	Fulton.	{ 5 c. Niagara. 755 20 a. Glacial Drift.
208	Wilton.	5 c. Niagara. 672
211	Moscow.	10 Hamilton. 652
216	Atalissa.	"
221	West Liberty.	" 665
227	Downey.	" 655
237	Iowa City.[55]	" 671
252	Oxford.[56]	" 720
257	Homestead.[57]	" 665
267	Marengo.[58]	"
277	Victor. 505	20 a. Gl.Dft., 20 b. Loess
287	Brooklyn.[59]	20 a. Gl. Drift. 556
293	Malcolm.	"
302	Grinnell.[40]	" 1011
313	Kellogg.	14 a. Lower Coal. 559
322	Newton.	" 955
334	Colfax.	13 d. St. Louis. 755
340	Mitchellsville.	14 a. Lower Coal. 656
357	Des Moines.[28]	" 500
372	Booneville.	"
379	De Soto.	"
385	Earlham.	14 c. Upper Coal.
392	Dexter.	" 1146
397	Stuart.	20 a. Glacial Drift.
403	Guthrie.	" 1269
408	Casey.	" 1226
415	Adair.	"
422	Anita.	"
436	Atlantic.	"
455	Avoca.	20 b. Loess, 20 a. Gl. Dft
463	Shelby.	" "
474	Neola.	" "
490	Council Bluffs.	" " 989

South-Western Division.

Ms.		Alt.
208	Wilton.	5 c. Niagara. 672
220	Muscatine.[41]	" 644
233	Onowa.	13 a. Kinderhook.
240	Fredonia.	"
242	Columbus Junc.	" 655
252	Ainsworth.	18 d. St. Louis.
258	Washington.	" 755
271	Brighton.[42]	"
286	Fairfield.[61]	" 767
292	Libertyville.	20 a. Glacial Drift.
304	Eldon.	13 c. Keokuk.
317	Belknap.	14 a. Lower Coal. 557
333	Unionville.	"
345	Centreville.	" 1013
360	Seymour.	14 c. Up. or 14 b. M. Cl.

(Continued in Missouri.)

Indianola and Winterset Branch.

Ms.	Indianola and Winterset Branch.	Alt.
0	Des Moines.[28]	14 a. Lower Coal. 500
8	Avon.	"
10	Carlisle.	"
15	Somerset Junc.	14 b. Middle Coal.
18	Somerset.	"
21	Indianola.	"
15	Somerset Junc.	"
21	Spring Hill.	"
25	Lathrop.	14 c. Upper Coal Mrs.
30	Bevington.	"
34	Patterson.	"
42	Winterset.[43]	"

Oskaloosa Branch.

0	Washington.	13 d. St. Louis. 735
15	Keota.	14 a. Lower Coal.
20	Harper.	"
28	Sigourney.[21]	" 13 d. St. L.
36	Delta.[44]	" "
43	Rose Hill.[45]	" "
52	Oskaloosa. 550	{ 14 a. Lower Coal. Loess. Drift.
58	Knoxville Junc.	Drift, 14 a. L. Cl.
63	Olivet.	"
68	Harvey.	" " 13 d. St.L.
78	Knoxville.	" " "

Keokuk and Des Moines Division.

0	Des Moines.[28]	14 a. Lower Coal. 799
24	Prairie City.	"
35	Monroe.	"
47	Pella.[47]	14 a. Lower Coal.
62	Oskaloosa. 550	" [St.L.
71	Eddyville.[48] 672	" 13c. Keo. 13 d.
86	Ottumwa.[22] 650	" " "
98	Eldon.	" ' "
116	Summit.	13 c. Keokuk. 1054
123	Bentonsport.	"
126	Bonaparte.	"
132	Farmington.	" and 14 b.
137	Croton.	" "
147	Sand Prairie.	" "
162	Keokuk.	13c.Keok.&13a.Kind.

Audubon Branch.

0	Atlantic.	{ Drift, Loess in valleysides, Subterrane probably 14 c. U. CL 18 Nishnabotna near to South-east.
1	Audubon Junc.	
12	Brayton.	
16	Exira.[46]	Drift, Loess. [Cl.
26	Audubon.	" " ov. 14 b. Mid.

Carson and Harlan Branch.

1	Carson.	{ Loess and Drift over 14 c. Upper Coal.
18	Avoca.	
1	Harlan Junction.	
13	Harlan.	Loess and Drift.

14. *Specht's Ferry.* Trenton and Galena, with Maquoketa and Niagara in hills.
15. *Peru.* Trenton and Galena, with Maquoketa and Niagara in hills.
16. *Dubuque.* Trenton in river bed, Galena in hills, Maquoketa on eminences, overlaid by Loess.

Chicago, Rock Island and Pac. R. R.—Cont.

Ms.	Monroe Branch.	Alt.
0 Newton.	14 a. Low Coal.	
10 Reasnor.[49]	"	
17 Monroe.	"	

Guthrie Branch.

Ms.		Alt.
0 Menlo.	D'ft over 14 c. Up. Cl. ?	
6 Glendon.	" Nish'botna.	
15 Guthrie Centre.	" "	

South-Western Division.

Ms.		Alt.
183 Davenport.[8]	As before.	
192 Buffalo.		
197 Montpelier.	{ Fossilifer's 10 Hamilton in valley, 14 c.	
208 Fairport.	Lower Coal in hills.	
211 Muscatine.[41]	{ Loess, D'ft, 10 Hamilton, 14 c. L. Coal.	

Chicago, Burlington and Quincy R. R.
Iowa Division.

Ms.		Alt.
0 Burlington.[50]	13 b. Burlington.	526
9 Middletown. [725]	20a. Gl. Dft., 20b. Loess	
13 Danville. [715]	"	
19 New London.	"	
28 Mt. Pleasant. [735]	13 c. Keok. & 13 d. St. L.	
35 Rome.	13 b. Burl. & 13 c. Keok.	
42 Glendale.	14 b. Lower Coal.	745
50 Fairfield.[51]	13 d. St. Louis.	767
55 Whitfield.	"	677
62 Batavia.	14 a. Lower Coal.	640
69 Agency.	"	801
75 Ottumwa.[22]	13 c. Keokuk.	830
83 Chillicothe.	"	845
88 Dudley.	" & 13 d. St. L.	
91 Frederic. [725]	20 Gl. Dft. & 14 a. L. Cl.	
100 Albia.	" "	945
108 Tyrone.	" "	819
114 Melrose.	" "	853
122 Russell.	" "	1017
130 Chariton.	" "	1080
139 Lucas.	14 c. U. or 14 b. Mid. C.	
146 Woodburn.	14 c. Up. & Mid. Coal.	
156 Osceola.	"	1122
166 Murray.	"	1188
180 Afton.	"	
190 Creston.	"	
195 Cromwell.	"	1220
211 Corning.	"	1127
215 Brooks'.	"	
225 Villisca.	"	
233 Stanton.	"	1004

Chicago, Burlington and Quincy R. R.

Ms.	Iowa Division—Continued.	Alt.
241 Red Oak. 1088	{ 14 c. U. or 14 b. M. C. Nish. & 20 b. Loess.	
255 Hastings.	20 b. Loess.	
261 Malvern.	{ 14 b. or c. U. or Mid. Coal & 20 b. Loess.	
271 Glenwood.	{ 14 c. Up. or Mid. 979 Coal & 20 b. Loess.	
275 Pacific Junc.[52]	14 c. U. or Mid. Cl. 960	
279 E. Plattsmouth.	River mud. 924	

Des Moines, Chariton and St. Joseph Branch.

Ms.		Alt.
0 Indianola.	14 a. L. & 14 b. Mid. Cl.	
5 Ackworth.	" "	
11 Milo.	" "	
19 Lacona.	" "	
26 Oakley.	", "	
30 Indianola Junct.	" "	
30 Chariton. 1080	14 a. Lower Coal Mrcs.	
44 Derby.	" [Mrs.	
50 Humeston.	14 b. U. or 14 c. Mid. Cl.	
56 Garden Grove.	" "	
69 Leon.[54]	" "	1025
190 Creston.	" "	
207 Lenox.	" "	
225 Bedford.	" "	
234 Hopkins.	" "	
241 Red Oak. 1088	{ 14 c. U. or 14 b. Mid. Coal. Nishnabotna.	
254 Essex.	20 b. Loess.	996
259 Shenandoah.	"	979
266 Farragut.	"	968
271 Riverton.	"	981
280 Hamburg.	"	912
291 Nebraska City.	River mud.	

Albia and Des Moines Branch.

Ms.		Alt.
0 Albia.	Drift over 14 a. L. Cl.	
9 Lovilla.	" "	
14 Bussey.	" 13 d. St. L.	
19 Tracey.[55]	" "	
25 Durham.	Loess & Dft. over "	
28 Flaglers.	Drift over 14 a. L. Cl.	
33 Knoxville.	{ Loess, Drift, 14 a. L. Coal, 13 d. St. Louis.	
37 Donnelly.	Drift over 14 a. L. Cl.	
43 Pleasantville.	" "	
49 Swan.	" "	
68 Des Moines.[23]	14 a. Lower Coal. 800	

17. *Lyons.* The Maquoketa passes beneath the Niagara a mile north of Lyons, where the contact is well exhibited in an artificial cutting.

18. *Elkport.* Treaton in valley, Galena [in first bluff, Maquoketa in terrace, and Niagara in second bluff.

19. *Miles.* Maquoketa in slopes, Niagara in hills.

20. *Amana.* Hamilton, locally overlain by Lower Coal ferruginous sandstones.

21. *Sigourney.* St. Louis, with Lower Coal in hills.

22. *Ottumwa.* Keokuk, with St. Louis and Lower Coal on hills to north and south.

23. *Des Moines.* The Loess of Des Moines reposes on Drift in normal relation, but is in turn overlain by a newer sheet of Drift. Such superposition is unknown elsewhere. *Vide Am. Jour. Sci.* 3d, XXIV., 1882. 202-23.

Chicago, Burl. and Quincy R. R.—Continued.

Ms.	Branches.	Alt.
0	Villisca.	14 b. U. Cl., Loess, Drift.
1	Clarinda Junct.	" " "
16	Clarinda.	" " "
36	Burl'ton Jc., Mo.	" " "
0	Creston.	Drift, 14 b. Mid. Coal.
15	Orient.	" "
30	Fontanelle.	" "
0	Bethany Junct.	Loess (sometimes absent). Drift, 14 c. Upper Coal.
11	Kellerton.	
22	Mt. Ayr.	
29	Delphos.	
44	Grant City.	
0	Red Oak.	14 c. Up. Coal, Nishnab'na & 20 b. Loess.
7	Stennet.	Loess, Drift, (sometimes absent), 14 c. Upper Coal.
12	Elliot.	
18	Griswold.	
0	Hastings.	20 b. Loess over 14 c. Upper Coal.
9	Henderson.	
13	Macedonia.	Loess, Drift, (s'times absent), 14c. Up. Cl.
16	Carson City.	
0	Hastings.	20 b. Loess over 14 c. Upper Coal, Drift sometimes exposed at base of Loess.
12	Randolph.	
18	Anderson.	
2?	Sidney.	
0	Clarinda.	Loess, Drift, 14 c. Upper Coal
18	Northboro.	
0	Burlington.	Loess, Drift, 13c.[52?] Keok., 13 b. Burl.
11	Wever.	
19	Ft. Madison.	Loess, D'ft, 13c. Keo. in hills, Allu. in val.
25	Viele.	Loess, Drift.
32	Montrose.	Alluvium, Loess, Drift & 13 c. Keok.
37	Ballinger.	
38	Sandusky.	
43	Keokuk.	Loess, Drift, 13d.[501] St. L., 13c. Keokuk.

Chicago, Burl. and Kansas City R. R.

Ms.	Station	Alt.
0	Burlington.[50]	13 b. Bur. 13c. Keo. [52?]
19	Fort Madison.	" [51?]
25	Viele.	13 c. Keokuk. [54?]
31	Franklin.	" 702
33	Donaldson.	20 a. Glacial Drift. 707
36	Warren.	13 c. Keokuk. 709
44	Farmington.	" 13 d. St. L.[571]
50	Willits.	14 a. Lower Coal. 604
55	Mount Sterling.	" 649
63	Cantril.	" 776
69	Milton.	" 806
75	Pulaski.	" 840
85	Bloomfield.	" 884
99	Moulton.	" 994
108	Caldwell.	" 887
113	Cincinnati.	" 1087
118	Mendota, Mo.	" 885
122	Howland, " [983]	14c. Up. or 14 b. Mid. Cl.
128	Unionville,"	(Con. in Mo.) 1082

Wabash, St. Louis and Pacific Railroad,
St. Louis and Des Moines Branch.

Ms.	Station	Alt.
0	St. Louis.	(See Missouri.)
229	Glenwood, Mo.	979
230	Glenwood Junc.	979
252	Centreville.	14 a. Lower Coal. 1012
266	Moravia.	" overlain by Gl. Drift.
279	Albia.	" " 845
293	Bussey	13 d. St. Louis "
298	Tracy.[56]	"
317	Dunreath.	14 a. Lower Coal.
328	Runnells.	"
343	East Des Moines.	"
344	Des Moines.[28]	" overlain by 20 b. Loess & 20 a. Glacial Drift. 807
0	Centreville.	14 b. Lower Coal. 1012
7	Sedan.	" 827
11	Dean.	" 835
15	Hamilton.	" 947

24 *Waukon Branch.* Entirely in the "Driftless Area." The superficial detritus is residuary clays, sands, and alluvium.

25 *Zwingle.* Attenuated eastern margin of Glacial Drift.

26 *Washington Mills.* Maquoketa a few feet below level of creek.

27 *Bernard, Filmore.* Between these stations lies an insulated basin of Drift, completely surrounded by Loess.

28 *Fort Dodge.* St. Louis overlain by Fort Dodge resting on Lower Coal in hills.

29 *Almont.* Maquoketa in slopes, Niagara in hills.

30 *Marshall.* St. Louis? Lower Coal in eminences. Remarkable crinoid bed near here.

31 *Mo. Valley Junction.* Glacial Drift in valleys. Loess on uplands.

32 *Maple River and Sac City Branches* traverse an area over which the Glacial Drift is of considerable thickness and overlain by Loess, gradually thickening westward from an irregular eastern margin generally coinciding approximately with the Mississippi-Missouri watershed.

33 *Wall Lake* is named from the adjacent lake, which is in part surrounded by a natural wall of rock, formed by the long continued pushing shoreward of the boulders lying upon its shallow bottom by the expansion of the ice in which they become bedded each winter.

34 *Dakota City.* From near Dakota City to the Big Sioux River this railway traverses a heavily drift-mantled area, and the subterrane is wholly unknown empirically. The Sub-Carboniferous probably extends many miles. Northwest of Dakota there may be remnants of the Coal Measures. The Inoceramus and Woodbury are probably developed towards the state line where, too, the red quartzites of the Sioux doubtless lie beneath the Drift and Loess.

35 *Iowa City.* Hamilton in city, and Lower Coal sandstones in hills to northward, overlain by Glacial Drift and Loess. Locality of "Iowa City Marble."

36 *Oxford.* Hamilton with Lower Coal sandstones in hills.

Ms. Wabash, St. L. and Pac. R. R.—Con. Alt.

269	Keokuk. 449	{13 c. Keok. overlain by 20 b. Loess.
274	Alexandria, Mo.	
281	Wayland, "	
287	Clark City, "	
298	Luray, "	
306	Arbela, "	
314	Memphis, "	
325	Downing, "	
335	Lancaster, "	
338	Glenwood Junc.	
352	Sedan.	14 a. Lower Coal. 627
359	Centreville.	" 1018
387	Corydon. 1092	14c. Up. or 14b. Mid. Cl.
400	Humeston.	" "
414	Weldon.	" "
428	Grand River.	" "
453	Goshen.	" "
484	New Market.	" "
492	Clarinda. 1069	" "
500	Yorktown.	" "
513	Shenandoah.	a " 979
535	Malvern.	" "
563	Council Bluffs.	" " 969
	Omaha, Neb.	20 b. Loess.

(513–563 marked: overlain by 20 b. Loess.)

Des Moines Division (Narrow Guage).

0	Des Moines. 23	(As before.) 807
15	Waukee.	14 a. Lower Coal. 1049
22	Adel.	" 901
31	Redfield. 968	" and 18 Nish.
43	Panora.	" " 1074
53	Herndon.	20 a. Glacial Drift.
66	Jefferson.	"
79	Churdan.	"
87	Eads.	"
98	Rockwell City.	"
115	Fonda.	"

Chic., St. Paul, Minneap. and Omaha R'y.
St. Paul, Omaha and Kansas City.

0	Sioux City. 1122	20b. Loess & 18 Woodb.
8	James.	20 b. Loess.
25	LeMars.	" 1221
30	Seney. 1231	" & 20 a. Gl.Dft.
42	East Orange.	20 a. Gl. Drift. 1202
50	Hospers.	" 1222
58	Sheldon.	" 1406
67	St. Gilman.	" 1442
74	Sibley.	" 1509
92	Worthington.	(See Minnesota.

Sioux City and Pacific Railroad.

0	Sioux City. 1122	20 b. Loess & 18 Woodb
9	Sergeant's Bluffs.	" " 1008
22	Sloan.	" " 1089
38	Onawa.	Alluvium & Loess. 1064
58	River Sioux.	" " 1053
60	Mondamin.	" " 1038
66	Modale.	" " 1029
71	California Junc.	" " 1024
77	Missouri Valley.	" " 1022

Kansas City, St. Joseph and Council Bluffs.

1	Council Bluffs.	20 b. Loess. 989
6	Traders' Point.	20. Alluvium. 974
14	Pacific.	" 961
17	Pacific Junct. 52	" 960
20	Haney's. 56	" 955
25	Bartlett.	" 949
30	McPaul.	" 940
34	Percival.	" 933
40	E. Nebraska City.	" 926
51	Hamburg.	" 912
	(Continued in Missouri.)	

Des Moines and Fort Dodge Railroad.

0	Des Moines. 23	14 a. Low. Cl. Mres. 807
8	Ashewa.	" 905
15	Waukee.	" 1049
21	Dallas Centre.	" 1085
27	Minburn.	" 1062
34	Perry.	" 977
42	Rippey.	" 1080
50	Grand Junction.	" 1055
59	Paton.	" 1116
67	Gowrie.	" 1154
73	Callender.	"
82	Tara.	" 1159
88	Fort Dodge. 24	{13 d. St. Louis, 1018 18 d. Fort Dodge.

(34–82 marked: Overlain by Drift.)

82	Tara.	20 a. Drift, 14 a. L. Cl.?
89	Clare.	20 a. Drift.
100	Gilmore.	"
108	Rolfe.	" 13 a. Kind'h'k?
114	Plover.	" "
119	Mallard.	"
130	Ayrshire.	"
137	Ruthven.	"

37. *Homestead.* Hamilton with Lower Coal sandstones in hills.
38. *Marengo.* Hamilton with Lower Coal sandstones in hills.
39. *Brooklyn.* Glacial Drift with St. Louis? in artificial exposures.
40. *Grinnell.* About the undetermined eastern margin of the Lower Coal.
41. *Muscatine.* Hamilton with Lower Coal sandstones on hills, overlain by Glacial Drift and Loess. From Davenport to Muscatine the Mississippi has corraded its channel through one of the Carboniferous outliers (ferruginous sandstone, with pockets of coal) characteristic of eastern Iowa (cf. Hall, Geol. Ia., 1858. Pt. 1, 44, 120 *et seq.*) and into Hamilton strata which decline from perhaps 100 feet above the river at Davenport to its level just below Muscatine. The stratified rocks are overlain by Drift, generally capped by Loess, which is typical in Muscatine.
42. *Brighton.* St. Louis, with Lower Coal to southward in hills.
43. *Winterset.* Lower Coal in river, Upper and Middle Coal generally.
44. *Delta.* St. Louis, with Lower Coal in hills.

Ms.	Central Iowa Railway.[37]	Alt.
0	St. Louis.	(See Missouri.)
176	Keokuk.	13 c. Keokuk. 501
253	Ottumwa.[33]	" 630
269	Eddyville.[44]	" 13 d. St. L. 672
273	Givin.	" "
278	Oskaloosa.	14 a. Lower Coal. 850
291	New Sharon.	" 877
299	Searsboro.	13 d. St. Louis. 810
311	Grinnell.[40]	20 a. Glacial Dft. 1011
322	Gilman.	13 c. Keokuk. 1035
327	Dillon.	" 893
336	Marshalltown.[48]	13c. Keo. & 13 d. St. L.?
343	Albion.	20 a. Glacial Drift. 963
349	Liscomb.	" 1075
354	Union. 1012	14a. L.Cl., ferugin's ss.
363	Eldora.	" " 1153
		" " 1061
367	St'mboat Rock.[57] {	13c. Keo., 13a. Kind.
374	Abbott.	20 a. Glacial Drift. 1176
379	Ackley.	" 1177
384	Franklin.	13 a. Kinderhook. 1193
389	Geneva.	20 a. Glacial Dft. 1181
395	Hampton.	" 1240
404	Chapin.	" 1246
412	Rockwell.	10 Hamilton. 1219
424	Mason City.	10 b. " 1130
93	W. Keithsburg.[57]	20 Alluvium.
100	Elrick.	"
108	Morning Sun.	Gl.Dft. "H'd-pan."
119	Winfield.	" "
126	Olds.	" "
132	Wayland.	" "
135	Coppack.	" " ?
142	Brighton.[42]	13d. St. L., 14a. in hills
147	Clay.	20 Gl. Drift.
151	Richland.	" over 14 a. L. Cl.
169	Hedrick.	" "
176	Fremont.	" " ?
182	Wright.	" " ?
189	Oskaloosa.	14 a. Lower Coal. 850
19	Hickory.	{ Loess, Drift ov. 13 c. Keok. & 13 d. St. L.
23	Maxon.	Loess, D'ft ov. 14a. L.C.
24	Albia.	" "

(vertical note: over undetermined Sub-carb. strata.)

Belmond Branch.[55]

0	Belmond.	{ Drift over undeter-
14	Lattimer.	{ mined Sub-Carbon-
22	Hampton.	{ iferous strata.

Central Iowa Railway—Continued.

Story City Branch.

Ms.		Alt.
0	Marshalltown.	{ 13 c. Keok., 13 d. St. Louis, partly overlain by D'ft & Loess.
4	Minerva Junct.	Drift and Alluvium.
11	Minerva.	Drift over 14 a. L. Cl. ?
13	Bromley.	" "
17	St. Anthony.	" "
22	Zearing.[59]	" "
33	Roland.	" "
39	Story City.	" 13 d. St. Louis.

State Center Branch.

0	Grinnell.[40]	20 a. Glacial Drift.
6	Newburg.	20a. D'ft ov. 14a. L. C.?
24	State Center Jct.	" "
33	State Center.	" "

Newton Branch.

0	Newsharon.	{ 14 a. Low. Coal, 850
14	Lynnville.	generally concealed
30	Newton.	by Drift. 852

Montezuma Branch.

0	Grinnell.[40]	20 a. Drift. 1011
10	Ewart.	"
17	Montezuma.	{ " Loess ov. Eastern margin 14a. L.C.

Burlington, Cedar Rapids and Northern Railroad.

0	Burl'gton.[50] 536	13 b. Burl. & 13 c. Keo.
9	Latty.	20 a. Glacial Drift.
12	Sperry.	" 747
15	Kossuth.	" 769
20	Linton.	" 761
23	Morning Sun.	" 835
29	Wapello.	13 a. Kinderhook. 573
35	Long Creek.	20 a. Glacial Drift.
41	Columbus Junc.	" 535
44	Port Allen.	"
47	Cone.	" 603
55	Nichols.	10 Hamilton. 628
61	West Liberty.	" 666
67	Centredale.	" 715
70	West Branch.	" 703
73	Oasis.	" 790
77	Morse.	" 755
82	Solon.	" 754
89	Ely.	" 751

45. *Rose Hill.* St. Louis, with Lower Coal in hills.
46. *Exira.* About the northern margin of Upper Coal.
47. *Pella.* St. Louis in valleys and south of town.
48. *Eddyville.* Keokuk, with St. Louis and Lower Coal in adjacent hills.
49. *Reasnor.* It is probable that the Chicaqua (Skunk) River, crossed between Reasnor and Monroe, cuts down to the Sub-Carboniferous.
50. *Burlington.* Burlington, with Keokuk in hills overlain by Glacial Drift and Loess.
51. *Fairfield.* St. Louis, with Lower Coal in hills to northward.
52. *Pacific Junction.* Upper or Middle Coal capped by Loess in hills to eastward.
53. *Des Moines, Chariton and St. Joseph Branch of C. B. & Q.* probably passes a short distance east of and parallel with the eastern limit of the Upper Coal, sometimes on the Middle and sometimes on the Lower, sometimes, possibly, over salients or outliers of Upper Coal. The stratified rocks are generally deeply covered by Drift, sometimes overlain by Loess.
54. *Leon.* Streams have rarely cut down to Middle Coal.

Burlington, Cedar Rapids and Northern Railroad—Continued.

Ms.		Alt.
97	Cedar Rapids.	10 b. Hamilton.	744
101	Linn.	"	
107	Palo.	"	741
111	Shellsburg.	"	764
120	Vinton.	"	800
128	Mount Auburn.	"	853
134	La Porte.	"	802
150	Waterloo.	"	862
156	Cedar Falls.	"	844
160	Norris.	"	
164	Finchford.	"	
171	Shell Rock.	"	911
178	Clarksville.	"	916
189	Greene.	"	948
195	Marble Rock.	"	992
202	Rockford.	"	1011
210	Nora Junction.	"	1052
215	Rock Falls.	"	1096
219	Plymouth.	"	1114
250	Lyle.	"	1105
261	Austin.	"	

Decorah Division.

Ms.		Alt.
0	Cedar Rapids.	10 b. Hamilton.	766
4	Linn.	"	
18	Center Point.	"	809
25	Walker.	"	860
39	Independence.	"	1111
53	Oelwein.	"	1089
60	Maynard.	"	1096
69	Donnan.	"	
74	West Union.	"	866
78	Brainard.	5 c. Niag. & Maq'keta.	
81	Elgin.[60]	4 a. Trenton.	833
89	Clermont.	"	856
98	Postville.[2] 1207	4 c. Maq. & 4 b. Gal.	

Muscatine Division.

Ms.		Alt.
0	Muscatine.[41]	10 b. Hamilton.	544
11	Cedar River.	"	
13	Adams.	"	608
16	Nichols.	"	628
23	Lone Tree.	"	718
26	River Junction.	"	
31	Riverside.[61]	"	631
37	Kalona.	Loess, D'ft, 13a. Kind.?	
53	Kinross.	" " "	
66	Keswick.	" " "	
70	Thornburg Junc.	" " 14a. L. Coal.	
76	What Cheer.	" " "	
79	Barnes City.	" " "	
88	Montezuma.	" " ?	

Burl., Cedar Rapids and North. R. R.—Con.
Pacific Division.

Ms.		Alt.
0	Cedar Rapids.	10 Hamilton.	719
10	Palo.	"	741
14	Shellsburg.	"	764
23	Vinton.	"	800
30	Garrison.	"	849
39	Dysart.	"	958
47	Traer.	"	906
59	Reinbeck.	"	916
69	Grundy Centre.	"	966
78	Wellsburg.	20 a. Glacial Drift.	
85	Cleves.	"	
87	Abbott Crossing.	"	
97	Iowa Falls.	13 a. Kinderhook.	444
107	Carleton.	20 a. Glacial Drift.	
119	Galtville.	"	
126	Clarion.	"	
136	Goldfield.	"	
144	Hardy.	"	
153	Livermore.	"	1184
158	Bode.	"	
169	West Bend.	"	
185	Emmetsburg.	"	
195	Graetinger.	"	
201	Wallingford.	"	
207	Estherville.	"	
214	Superior.	"	
223	Spirit Lake,Minn	"	
235	Lake Park.	"	
244	Round Lake.	"	
253	Worthington.	"	

Belmond Division.

Ms.		
0	Dows.	20 a. Heavy Drift, over Sub-Carboniferous.
15	Belmond.	
41	Madison.	

Clinton Division.

Ms.		
0	Elmira.	Drift, 10 Hamilton.
6	Plato.	Loess, Drift, 5 c. Niag.
16	Tipton.	" " "
25	Bennett.	" " "
37	Dixon.	Loess, " "
45	Noels.	" " "
53	McCausland.	All., Loess, Drift, Nia.
58	Folletts.	Alluvium, 5 c. Niag.
69	Clinton.	Loess in hills, Alluvium in valley, 5 c. Niagara.

Iowa City Division.

Ms.		
0	Elmira.	Drift, 10 Hamilton.
3	Graham.	" "
9	Iowa City.[35]	Loess, Drift, "
18	Iowa Junction.	" " "

55. *Tracey.* St. Louis, with Lower Coal on hills to westward.

56. *Haneys.* Upper or Middle Coal capped by Loess in bluffs one mile east.

57. *Steamboat Rock.* At and about this place the Iowa River flows in a gorge 50 to 150 feet deep, which it has eroded in friable ferruginous sandstone and firm limestones. To reach the plateau in which the gorge is excavated the nascent river left a low-lying valley in its direct course, going some miles out of its way. This is one of the finest examples of the anamalous behavior of several Iowa rivers in avoiding valleys and seeking ridges and plateaus for their courses. (cf. Burl. Phil. Soc. Wash., VI, 1884, 93; Science II., 1883, 762; Trans. Iowa Hort. Soc. XVIII., 1883, 526.)

Ms.	Dubuque and Dakota Railroad.	Alt.
0	Hampton.	Drift ov. 13 S.-C. strata
12	Dumont.	" " ?
16	Bristow.	" 10 Hamilton.
22	Allison.	" "
29	Clarksville.	10 b. Hamilton. 9 14
36	Shell Rock.	"
41	Waverly.	" 9 42
64	Sumner.	Drift over 10 Hamilton

	St. Louis, Keokuk and North-West. R. R.	
0	Keokuk.	{ Loess, Drift, 13d. St. / Louis, 13 c. Keokuk.
15	Boston.[64]	20 a. Drift.
17	Charleston.	"
32	Houghton.	"
37	Salem.	" [Keok.
43	Oakland Mills.	Loess, " 13d. St. L., 13c.
49	Mt. Pleasant.[64]	" " " "

	Minneapolis and St. Louis Railway.	
121	Norman.	20 a. Glacial Drift.
127	Lake Mills.	"
142	Forest City.	"
156	Britt.	"
167	Corwith.	"
176	Luverne.	"
182	Livermore.	"
192	Humbolt.	Drift, 13 a. Kinderh'k.
210	Ft.Dodge.[24] 1015	{ Drift, 18 Ft. Dodge, / 14a. L. C., 13d. St. L.
216	Kalo Junction.	20 a. Drift.
230	Dayton.	" 14 a. Low. Cl.
246	Ogden.	" "
259	Angus.	

	Wisconsin, Iowa and Nebraska Railroad.	
0	Des Moines. 799	Drift, Loess, 14 a. L. C.
9	Berwick.	" 14 a. Low. Coal.
26	Mingo.	" "
45	Melbourne.	" "
51	Luray.[65]	
58	Marshalltown.[66]	{ 13 c. Keokuk. 991 / 13 d. St. Louis?
63	Rockton.	Drift.
74	Gladbrook.	{ Loess to S.-W., Drift, / 14a. L. C., 13a. Kind.
80	Berlin.	Drift.
87	Reinbeck.	" over 10 Ham.
95	Hudson.	" "
105	Waterloo.	" "
110	Cedar Falls.	Drift ov. 10 Hamilton.

	St. Louis, Des Moines and Northern R. R.	
0	Des Moines.[23]	14 a. Lower Coal. 800
21	Kelsey.	Drift over 14 a. L. Cl.
43	Boone.	14 a. Low. Coal. 1185

	Des Moines, Osceola and Southern R. R.	
0	Des Moines.[23]	14 a. Lower Coal. 800
11	Norwalk.	{ Drift & Loess ov. 14a. / L. C. & 14 b. Mid. C.?
18	Poole.	Drift, Loess over 14 a.
20	R. I. Crossing.	Drift, Loess.
29	St. Charles.	" "
50	Jamison.[62]	
58	Osceola.	Dft., Loess ov. 14c. U.C.
72	Van Wert.[63]	Drift over "
81	Decatur.	" "
87	Leon.[64]	" "
100	Harding.	" "
111	Cainsville.	" "

	Fort Madison and North-Western R. R.	
0	Fort Madison.	{ Drift, 13 c. Keokuk, / 13 b. Burlington.?
3	Bluff Siding.	Drift.
6	Benbon.	" 14 a. Low. Coal.
32	McVeigh.	" "
41	Birmingham.	

	Burlington and North-Western and Burlington and Western Railroads.	
0	Burlington.[50]	As before. 826
20	Roscoe.	20 a. Drift.
34	Winfield.	"
39	Wyman.	" 20 b. Loess.
42	Crawfords.	20 a. Drift.
47	Havre.	Drift, 13 d. St. Louis?
52	Washington.	" 13 d. St. Louis.
40	Wayne.	"
56	Brighton.[42]	" Loess, 14 a., 13 d.
66	Woolson.	" 14 a. Low. Coal.
84	Hedrick.	" "
95	Cedar.	" "
104	Oskaloosa.	" 8.50

58. *Belmond Branch* traverses the eastern side of the Iowa loop of the Great Terminal Moraine of the Upper Mississippi Valley.

59. *Zearing.* The Terminal Moraine crosses the railway from north to south in this vicinity.

60. *Elgin.* Galena, Maquoketa and Niagara in eminences.

61. *Riverside.* Hamilton, with Kinderhook on south side of river.

62. *Jamison.* Drift—concealed eastern margin of Upper Coal probably near here.

63. *Van Wert.* Drift along valley sides generally overlain by Loess. The phase of Drift known as "hard pan" (a dense, tenacious blue or gray clay, weathering white) occurs in vicinity of this and succeeding stations.

64. *Boston to Mt. Pleasant.* Subterrane includes eastern salients of Lower Coal, the St. Louis and Keokuk, and, possibly, the Burlington.

65. *Luray.* About eastern margin of Lower Coal.

66. *Marshalltown.* Keokuk and St. Louis? with Lower Coal on adjacent hills.

67. *West Keithsburg to Oskaloosa.* Formations only approximately located.

Minnesota.*

LIST OF THE GEOLOGICAL FORMATIONS FOUND IN MINNESOTA.[10]

FORMATIONS PER GENERAL LIST.	MINNESOTA SUB-DIVISIONS.	FORMATIONS PER GENERAL LIST.	MINNESOTA SUB-DIVISIONS.
20. QUATERNARY.	20. Quater. or drift.	4 a. TRENTON.	4 b. Galena l. s.
18 CRETACEOUS.	18 b. Benton.	"	4 a. Trenton l. s.
"	18 a. Dakota.	3 a. CALCIFEROUS.	3 b. St. Peter s. s.
10. HAMILTON.	10 a. Hamilton l. s.	"	3 a. L. Magnesian.†
9 c. CORNIFEROUS.	9 c Corniferous.	2 b. POTSDAM.	3 c. St. Croix s. s.
5 c. NIAGARA.	5 c. Niagara l. s.	"	{ 2 b. Potsdam s. s.
4 c. HUDSON RIVER.	4 c. Maquoketa sh.		of Wisconsin.
			2 a. Potsdam of Min.
	1. ARCHÆAN.	1. Archæan.

Potsdam sandstone of the Wisconsin geologists; 3 c. of this scheme for Minnesota (the St. Croix sandstone), and the Potsdam sandstone of New York is regarded as the equivalent of 2 a. by Prof. Winchell. Under the New York Calciferous are included the St. Peter sandstone, the Lower Magnesian (Shakopee, Jordan and St. Lawrence), and the St. Croix sandstone.   N. H. W.

The course of glacial striæ, and of transportation of the drift in eastern Minnesota, is southwest from Lake Superior to the Mississippi River; but in the west part of the State it is to the south and southeast, from Lake Winnipeg to Big Stone Lake, and into Iowa, excepting the southwest corner of the State, where the course is deflected to the southwest.

A tract adjoining the Mississippi River, from Lake Pepin to the southeast corner of the State, lies in a driftless area, which has a large extent toward the east and south in Wisconsin.   W. U.

The four most notable features of the glacial drift in Minnesota are the following:

a. Its great depth, averaging 100 feet, and sometimes exceeding 200 feet, upon the western two-thirds of the State, where it generally covers all the surface of the older bed rocks.   W. U.

b. The terminal moraines of the last glacial epoch. These belts of hilly and knolly drift reach from St. Paul and Minneapolis, north and northwest, to the Leaf hills and Itasca Lake. A great loop of the same formation also extends from Lake Minnetonka, by Albert Lea, into Iowa, to Pilot Mound, Mineral Ridge, and the vicinity of Des Moines, where it curves like the letter U, thence passing northwest by Storm Lake and Spirit Lake in Iowa, and along the elevated *Coteau des Prairies* through southwestern Minnesota into Dakota.   W. U.

c. Lake Agassiz, which occupied the basin of the Red River of the North and Lake Winnipeg during the recession of the ice sheet, that being a barrier to prevent the water on this area from flowing to Hudson Bay as now. The beach of Lake Agassiz is well exhibited on the Northern Pacific Railroad close east of Muskoda.   W. U.

d. The channel or valley in which lakes Traverse and Big Stone and the Minnesota River lie, excavated 100 to 225 feet in depth and about a mile in width. It was eroded by the outflow from Lake Agassiz; and the river thus formed has been named the River Warren, in honor of Gen'l George K. Warren, who first described this channel and showed its origin from the glacial lake in the Red River Valley.   W. U.

Chicago, Milwaukee & St. Paul R. R. Ms. (Southern Minnesota Division)		Alt.	Chicago, Milwaukee & St. Paul R. R. Ms. (Southern Minnesota Division.)—*Con.*		Alt.
0	Milwaukee.	3 c. St. Croix.	564	86	Grand Meadow.	{ 18. Creta. (proba- bly)	1338
0	La Crescent.	3 a L. Mag. Bluffs.	647	101	Brownsdale.	"	1271
1	Grand Crossing.	"		106	Ramsay.	"	1214
32	Rushford.	"	722	113	Oakland.	"	1265
37	Peterson.	"	756	122	Hayward.	"	1345
46	Whalan.	"	786	128	Albert Lea.	over " Dev.	1221
51	Lanesboro.[1]	"	841	138	Alden.	"	1261
57	Isinours.[2]	"	899	147	Wells.	"	1183
62	Fountain.	{ 3 b. St. Peter.	1302	162	Delavan.	"	1057
		{ 4 a. under village.		171	Winnebago City.	18 "	1096
70	Wykoff.	{ 4 a. Tren. Frequent sink-holes.	1310	174	Winnebago.	20. Heavy drift.	
77	Spring Valley.[6]	{ 10 a. Ham. uncon. on 4 c.Hud. River.	1266	191	Fairmount.	"	
				216	Jackson.	"	

* Prepared expressly for this work by Prof. N. H. Winchell, of Minneapolis, the State Geologist of Minnesota; with elevations and notes on glacial drift by Mr. Warren Upham, Assistant Geologist.
† Sub-divided into 3 Shakopee l. s., 2 Jordan s. s., and 1 St. Lawrence l. s.

1. The three sub-divisions of the Lower Magnesian: 1, St. Lawrence limestone; 2, Jordan sandstone; and 3, Shakopee limestone are here seen.

2. In the immediate river bluffs are the Jordan and Shakopee. Further back are the St. Peter and Trenton.

Chicago, Milwaukee & St. Paul R. R.
Southern Minnesota Division.—Con.

Ms.			Alt.
240	St. P & S.C. Junc.	Heavy Drift.[3]	
254	Fulde.	"	
263	Iona.	"	1705
282	Edgerton.	"	
296	Pipestone.[13]	Quartzite & Catlinite. Dakota Line.	1744

Chicago & North-Western Railroad.

Ms.			Alt.
297	Winona.	{ 3 c. St. Croix & 3 a. L. Mag. in bluffs.	
303	Minnesota City.	"	
308	Stockton.	{ 3 c. St. Croix, 3 a. L. Mag.	755
316	Lewiston.	"	1211
319	Utica.	"	1170
825	St. Charles.	{ 4 a. Tren. in bluffs. 3 b. St. Peter. " 3 a. Low. Mag.	1189
329	Dover.	3 b. and 4 a.	1188
334	Eyota.[6]	4 a. Trenton.	1287
347	Rochester.	(Same as St. Chas.)	991
356	Byron.	4 b. Galena l. s.	1250
362	Kasson.	"	1252
368	Dodge Centre.	18. Cret. probably	1288
375	Claremont.	"	1280
882	Havana.	"	1246
887	Owatonna.	{ 4 a. Trenton. Heavy drift.	1144
896	Meriden.	18. Cretaceous.	1149
402	Waseca.	{ 18. Cretac. Heavy drift.	1155
413	Janesville.	"	1083
428	Mankato Junc.	"	906
428	St.Paul& Sioux City Junction. }	3 a. Low. Magnesian.	
428	Mankato[3]	18. Cretace's clays.	781
437	St. Peter.	"	812
446	Oshawa.	"	932
467	New Ulm.	{ 2 a. Potsdam (conglomerate and red quartzite.) Granite.	887
479	Sleepy Eye.	1. Archæan.	1054
490	Springfield.	18. Cretaceous.	1025
498	Sanborn.	Prob. "	1089
506	Lamberton.	" "	1144
516	Walnut Grove.	" "	1228
526	Tracy.[11] 1405	{ 20. H'vy drift of the Coteau des Prairies	
539	Balaton.	"	1525
545	Redwood.	"	1025
553	Tyler.	"	1750
561	Lake Benton.	"	1759
567	Verdi.	"	1771

Chicago and North-Western Railroad.
Continued.

Ms.			Alt.
574	Elkton.	{ 20. H'vy drift of the Coteau des Prairies	
552	Marshall.	{ 20. H'vy drift, probably underlain by gneiss and schists.	1174
565	Minnesota.	"	1179
576	Canby.	"	1243
593	Gary.[11] (Dakota Line.)	"	1484

Minnesota Valley Railway Division.

479	Sleepy Eye.	Archæan.	1027
481	Redwood Jc.	Heavy drift of the Coteau des Prairies	1008
493	Morgan.	Heavy drift.	1043
499	Paxton.	"	1082
505	Redwood Falls.	{ 1. Archæan and 18. Cret.	1026

Chatfield R. R. Branch.

334	Eyota.[3]	Heavy d'ft 4 a. Tren.	1237
335	Chatfield Junc.	Drift over Tren.	1275
346	Chatfield.	{ 4 a. Trenton, 3 b. St. Peter.	967

Plainview R. R. Branch.

334	Eyota.[6]	As before.	1237
335	Plainview Junc.	20. Drift.	1275
337	Doty.	"	1310
340	Viola Centre.	"	1129
345	Elgin.	{ 4 a. Tren. 3 a. Shakopee.	1069
350	Plainview.	Drift.	1157

Rochester & Northern Minnesota R'y Branch.

347	Rochester.	See main line.	991
348	Zumbrota Junc.	4 a. Trenton.	990
355	Douglass.	"	1091
360	Oronoco.	3 a. Shakopee.	1041
364	Pine Island.	3 a. and 4 a. Tren.	998
368	Lena.	Drift.	1078
373	Zumbrota.	{ 3 a. Shak., 3 b. St. Pet., 4 a. Tren.	971

Chicago, St. Paul, Minneapolis & Omaha Railway.

0	St. Paul.	{ 8 b. St Peter and 4 a. Trenton.	704
6	Mendota Junc.	"	713
11	Nicols.	"	706
19	Hamilton.	{ 20. Quaternary, drift bluffs.	714
22	Bloomington.	"	736
28	Shakopee.	3 a. Low. Magnesian, Shakopee l.	741
34	Merriam.	"	758

3. Overlying 3 a. Lower Magnesian, i. e., its two upper members, the 2. Jordan sandstone and the 3. Shakopee limestone, seen in the bluffs. Artesian well 2,000 feet in sandstone.

4. The cascade at Minneopa Falls, 80 feet high, is caused by the Jordan sandstone. This railroad crosses the gorge one-quarter mile below the fall.

Chicago, St. Paul, Minneapolis & Omaha Railway.—Continued.		
Ms.	Railway.—Continued.	Alt.
39 Jordan.	749 Shakopee l. and Jordan s. s.	
43 St. Lawrence.	{ 3 a Low.Magnesian { St Lawrence.	
47 Belle Plaine.	{ 18. Cretaceous over { 3 a. Low. Mag.725	
51 Blakely.	"	728
58 E. Henderson.	"	734
62 Le Sueur.	{ 3 a L. Mag., Shakopee limestone, Jordan sandstone. 753	
69 Ottawa.	"	790
75 St. Peter.	"	747
77 Kasota.	"	800
86 Mankato.3	791 " 18 a. Creta.	
89 South Bend.	"	808
91 Minneopa.4	"	871
99 Lake Crystal.	18.Cret. H'vy drift.994	
109 Madelia.	"	1021
116 Lincoln.	"	1042
122 St. James.	"	1073
137 Mountain Lake.	"	1300
148 Windom.	"	1353
154 Wilder.	"	1448
160 Heron Lake	"	1417
170 Hersey.	"	1485
178 Worthington.7	"	1582

Blue Earth Branch.		
0 Lake Crystal.	18 Cret. h'vy dr'ft.	994
5 Garden City.	3 a. Shakopee.	966
11 Vernon Center.	Drift.	1028
16 Amboy.	"	1048
24 Winnebago City.13	"	1101
34 Blue Earth City.	"	1085
44 Elmore.	"	1131

Pipestone Branch.		
0 Heron Lake.	18.Cret.,h'vy d'f't.1425	
8 Dundee.	20. Drift.	1453
20 Avoca.	"	1542
31 Hadley.	"	1599
44 Woodstock.	"	1852
65 Pipestone.	Quartzite & catlinite.	
63 Dakota Line.	"	1722 / 1724

Rock River Branch.		
0 Lu Verne.	Drift & Potsdam.	1408
8 Ash Creek.	"	1405
16 Rock Rapids.	"	1464
28 Doon.	"	1294

Minneapolis & St. Louis Railway.		
0 Minneapolis.8	{ 4 a. Trent. 3 c. St { Peter s. s.	825
21 Chaska.	3 a. Calciferous.	725
23 Carver.	"	719
26 Sioux City Jc.	"	758

Minneapolis & St. Louis Railway.		
Ms.	Continued.	Alt.
27 Merriam Jc.	3 a. Shakopee.	753
32 Jordan.	3 a. Jordan s.s.	753
42 New Prague.	Morainic Drift.	973
50 Montgomery.	"	1088
58 Kilkenny.	"	1066
65 Waterville.	Flat Drift.	1008
76 Waseca.	"	1151
88 Richland.	"	1176
94 Hartland.	"	1237
108 Albert Lea.	{ 18. Cret. (prob. over { Devonian) and H'vy { Drift.	1221

Cannon Valley Division.		
0 Waterville.	Flat Drift.	1008
6 Morristown.	Rolling Drift.	1008
9 Warsaw.	"	1007
17 Faribault.	4 a.Tren. 3 b. St.P.	971
27 Dundas.	4 a. Tren. in bluffs.	926
30 Northfield.	3 a. Shakopee.	910
32 Waterford.	"	903
38 Cascade.	"	893
45 Cannon Falls.	{ 4 a. Tren., 3 b.St.Pet. { 3 a. Shak.	814
55 Belle Creek.	Low. Mag.in bluffs.	707
66 Redwing.	{ 3 a. Low. Mag., 3 c., { St. Croix.	706

Pacific Division.		
0 Minneapolis.	{ 4 a.Trenton, 3 b. { Peter s. s.	825
8 Hopkins.	Morainic Drift.	922
12 MinnetonkaMills	"	936
19 Excelsior.	"	947
25 Victoria.	"	936
31 Waconia.	"	936
39 Young America.	"	993
40 Norwood.	"	976
48 Green Isle.	"	999
54 Arlington.	Flat Drift.	995
62 Gaylord.	"	998
69 Winthrop.	"	1018
77 Gibbon.	Flat d'ft on Arch.	1048
86 Fairfax.	"	1041
100 Franklin.	"	1005
Morton.	"	841
107 Redwood.	Archæan.	
123 Echo.	Undulating Drift.	
130 Wood Lake.	"	
135 Hanley.	"	
146 Clarkfield.	"	
162 Dawson.	"	
171 Madison.	"	
189 Revillo.	"	
206 Troy.	"	
223 Watertown.	"	

St. Paul & Duluth Railroad.		
1 St. Paul.	{ 4 a. Trenton. 704 { 3 b. St. Peter s. s.	
3 Post's.	4 a. Trenton.	847

St. Paul & Duluth Railroad.
Continued.

Ms.	Continued.	Alt.
....	W. D. Junction.	4 a. Trenton.	
12	W. Bear Lake.	3 b. St. Peter s. s.	
....	Stillwater Junc.	3 a. Calciferous.	984
17	Centreville.	"	931
25	Forest Lake.	"	909
30	Wyoming.	2. Primordial.(?)	896
42	North Branch.	"	894
47	Harris.	"	895
54	Rush City.	"	916
64	Pine City.	"	949
77	Hinckley.	"	1031
87	Miller.	"	1136
95	Kettle River.	"	1030
110	Moose Lake.	Taconic.	1064
115	Barnum.	"	1097
121	Black Hoof.	"	
132	N. P. Junction.	"	1061
123	Thompson.	"	1052
141	Fond du Lac.	Potsdam.	608
155	Duluth.	Cupriferous.	608

Stillwater Branch.

0	White Bear.	Drift.	935
13	Stillwater.	3 a. Calciferous.	897

Minneapolis Branch.

0	Minneapolis.8	Trent. and St. Peter's.	
16	White Bear.	Drift.	935

Taylor's Falls Branch.

0	Wyoming.	2. Primordial.(?)	896
21	Taylor's Falls. Passenger Dep't.	St. Croix. s. s.	741

Knife Falls R. R. Branch.

0	N. P. Junction.	Huronian Slates.	1082
6	Cloquet.	"	1178

Northern Pacific Railroad.
Fergus Falls and Black Hills R. R.

Ms.			Alt.
0	Wadena.12		1349
1	Wadena Junc.		1350
10	Deer Creek.		1394
14	Parkton.		1394
18	Henning.	20. Heavy drift with many glacial lakes and morainic hills.	1436
24	Vining.		1449
29	Clitheral.		1346
33	Battle Lake.		1354
39	Maplewood.		1360
41	Southwick.		1342
42	Underwood.		1152
53	Fergus Falls.12		1063
60	Ames.		998
68	Everdell.		
77	Breckenridge.		960
	Dakota Line.		

Northern Pacific Railroad.—Continued.
Little Falls & Dakota R. R.

Ms.	Little Falls & Dakota R. R.		Alt.
0	Little Falls.	Staurolitic& garnet-iferous mica schists.	1118
8	La Fond.	Drift.	1184
16	Swanville.	"	1178
25	Gray Eagle.	"	1228
29	Birch Lake.	"	1226
31	Spaulding.	"	1292
38	Sauk Center.	Archæan.	1232
48	Westport.	"	1332
53	Villard.	Drift on Archæan	1358
60	Glenwood.	"	1401
59	Starbuck.	Drift. -	1159
79	Cyrus.	"	1185
88	Morris.	"	1134

Chicago, Milwaukee & St. Paul Railway.
Southern Minnesota Division.

0	Wells.	Heavy Drift.	1153
9	Minn Lake.	"	1036
19	Mapleton.	"	1031
25	Good Thunder.	"	974
37	St. P. & S.C. Jc.5	3 a. Low. Mag.Shak. l. s. 18 Cret.	798
33	Mankato.5	18.Cret. L. M. Shak. l.s Jordan. s. s.	770

Wabasha Division.

0	Wabasha.	3 a. L. Mag. 3 c. St. Croix in bluffs.	712
13	Glasgow.	"	716
20	Theilman.	"	743
29	Millville.	"	787
34	Hammond.	3 a.L. Mag. in bl'fs.	792
42	Zumbro Falls.	"	837
52	Mazeppa.	"	935
53	Forest Mills.	"	970
60	Zumbrota.	" Shak. l.s.	860

Hastings & Dakota Division.

0	Minneapolis.8	4 a. Tren., 3 c.St. Pet.	
9	Hopkins.	Heavy Drift.	912
18	Chanhassen.	"	966
22	Hazeltine.	"	934
27	Augusta.	"	974
31	Benton Jc.	Heavy drift.	943
33	Cologne.	"	943
0	Hastings.	3 a. Low.Mag. & St. Croix bluffs.	707
8	Vermillion.		
12	Auburn.	3 a. Low. Mag.	861
18	Farmington.	3 b. St. Peter s. s.	904
22	Fairfield.	" or 4 a. Tren.	943

5. *Castle Rock.* The outlier of the St. Peter sandstone, 70 feet high, visible from the station toward the east gives the name to the place.

Chicago, Milwaukee & St. Paul R. R.

Ms.	(Hastings & Dakota Div.)—Con.	Alt.
38	Prior Lake	{ 3 a. St. Peter s. s. or 4 a. Trenton. 949
41	Shakopee.	3 a. Shakopee l. s. 756
45	Chaska.	3 a. Cal. heavy drift 728
48	Carver.	" 815
54	Glencoe.	{ 20. Heavy drift, underlain by 1. Archæan rocks.
89	Bird Island.	
114	Granite Falls.9	{ Alternating beds of gneiss and schists.
137	Montevideo.	Red and gray gneiss.
167	Appleton.	20. Drift.
173	Odessa.	{ Heavy exposures of gneiss & granitoid gneiss, with conspicuous glaciation parallel with the Minnesota River Valley.
178	Junc. Switch.	
182	Ortonville.	

(1. Archæan Rock.)

(Dakota Line.)

Chicago, Milwukee & St. Paul R. R.—Con.

Ms.	(La Crosse & St. Paul Division.)	Alt.
306	Winona.	{ 3 a. Low. Mag.& 3 c. St. Croix s. s. compose the bluffs. 642
318	Minnesota City.	" 677
323	Minneiska.	" 673
326	Weaver.	" 674
333	Kellogg.	" 702
340	Wabasha.	" 712
342	Reed's Landing.	" 662
352	Lake City.	" 705
359	Frontenac.	" 720
369	Red Wing.	" 687
390	Hastings.	" 709
396	Langdon.	" 613
401	Newport.	" 761
409	St. Paul.	{ 4 a. Trenton. 704 / 3 b. St. Peter.
.....	Fort Snelling.	"
.....	Minnehaha.	"
424	Minneapolis.8	"

(Iowa & Minnesota Division.)

Ms.		Alt.
0	N. McGregor.	(See Iowa.) 633
85	Le Roy.	10. Hamilton. 1260
96	Adams.	" 1276
111	Austin.	{ 18 a. Cretaceous on Marcellus. 1197
114	Ramsey.	" 1215
117	Lansing.	Heavy drift. 1224
126	Blooming Prairie	" 1256
135	Aurora.	" 1253
144	Owatonna.	{ 4 a. Tren. on river banks. 1144
150	Medford.	3 a. River Terr's. 1098
159	Faribault.	{ 4 a. Trenton. 1002 / 3 a. St. Peter.
170	Dundas.	3 a. L. Mag. (Shak.)955
173	Northfield.	{ 3 a. Cal. & 4 a. Tren. on high bluffs. 915
179	Castle Rock.6	{ 3 b. St. Peter s. s. & 4 a. Tren. near 935
186	Farmington.	4 a. Trenton. Heavy904
193	Rosemount.	" drift.959
199	Westcott.	" 662
206	St. Paul Junc.	" 759
212	St. Paul.	704 " & 3 b. St. Pet.

Minneapolis & St. Louis Railway.

Ms.		Alt.
0	Minneapolis.8	{ 4 a. Trenton. 825 / 3 c. St. Peter s. s.
21	Chaska.	3 a. Calciferous. 725
23	Carver.	" 719
26	Sioux City Junc.	" 753

St. Paul, Minneapolis & Manitoba Ry.9

Ms.		Alt.
0	St. Paul.	{ 4 a. Trenton. 704 / 3 c. St. Peter s. s.
10	E. Minneapolis.	" 842
11	Minneapolis.	" 834
25	Wayzata.	18. Cretaceous.? 956
28	Long Lake.	" 954
33	Maple Plain.	" 1025
35	Armstrong.	"
43	Delano.	2. Primordial.? 925
49	Waverly.	" 999
54	Howard Lake.	" 1010
57	Smith Lake.	" 1064
61	Cokato.	{ 1. Metamorphic probably 1050

(Heavy Drift.)

6. *Spring Valley.* At four miles east is the best exposure of *Rhyaconella, Orthis* and *Strophomena* I have seen. At Spring Grove, on the Preston Branch of the Chicago, Milwaukee & St. Paul, have been found the largest *Trilobites* known of their kind (*Isoteles*). Similar ones have been seen three or four miles northwest of Eyota, on Chicago & Northwestern Railroad. Two miles north Kasson building stone of Galena formation (Upper Magnesian) are quarried of any size, 2½ inches thick. At Stockton and Lewiston, the lower Magnesian of similar dimensions are quarried by the Railroad Co., Same beds are wrought at Mankato somewhat thinner—supply unlimited. Orthoceratidæ, 10 inches in diameter,8 or 10 inches long, have been found in lower Trenton about Rochester. W. D. HURLBUT.

Some persons prefer to call this the Upper Magnesian limestone. In going from Spring Valley east, we ascend over 183 feet of layers of this rock in four miles on the railroad.

7. *Worthington.* The drift here is supposed to be 700 ft. elevation above tide; near town is over 1,800 ft.

8. The Falls of St. Anthony, at Minneapolis, are caused by the rapid wearing out of the very friable St. Peter sandstone under the Trenton limestone, leaving a projecting shelf of the latter.

9. *Granite Falls* is a reef or bar of quartzite (probably metamorphic). It is expected that the most of our quartzites will prove to have been Potsdam. They appear in proper horizon as do those at Devils Lake, Wis., and Sioux Falls, Dakota. Boulders from these quartzite rocks are widely distributed in Minnesota.   W. D. H.

St. Paul, Minneapolis & Manitoba Ry.—Continued.

Ms.		Alt.
67	Dassel.	1. Metamorph. 1069
72	Darwin.	" Probably 1162
78	Litchfield.	" 1129
86	Swede Grove.	" 1192
91	Atwater.	" 1211
98	Kandiyohi.	" 1222
104	Willmar.	" 1129
111	St. John's.	" 1121
118	Kerkhoven.	" 1106
127	De Graff.	" 1061
134	Benson.	" 1047
140	Clontarf.	" 1044
150	Hancock.	" 1155
159	Morris.	" 1129
168	Donnelly.	" 1124
178	Herman.	1. Archæan. 1070
185	Gorton.	" 1022
194	Tintah.	" 995
201	Campbell.	" 982
209	Doran.	" 971
217	Breckenridge.	" *959

(bracketed annotations: Heavy Drift. — Perhaps Cretac's.)

(Branch Line St. Paul, Min. & Man. Railway.)

Ms.		
0	St. Paul.	{ 4 a. Trenton. 704 / 3 a. St. Peter s. s.
10	St. Anthony.	4 a. Trenton. 842
17	Manomin.	3 b. St. Peter s. s. 848
27	Anoka.	3 a. Calciferous. 878
34	Itasca.	" 891
39	Elk River.	2. Primordial. 896
48	Big Lake.	" 940
56	Becker.	" 977
63	Clear Lake.	1. Archæan. 997
75	St. Cloud.	" 1012
76	Sauk Rapids.	" 1004
108	Melrose.	" 1195

(bracketed annotation: Heavy drift.)

0	St. Paul.	{ 2–4. Low. Silur. and / Cam. l. s. and s. s.
.....	Minneapolis.	"

St. Paul, Minneapolis & Manitoba Ry.—Continued.

Ms.		Alt.
11	Parker.	
22	Osseo.	
34	Hassan.	
39	Crow River.	
44	St. Michaels.	
48	Monticello.	
56	Silver Creek.	
63	Clearwater.	
69	Augusta.	
75	St. Cloud.	
82	St. Joseph.	
85	Collegeville.	
90	Avon.	
96	Albany.	
103	Freeport.	
109	Melrose.	
117	Sauk Centre.	
125	West Union.	
130	Osakis.	
142	Alexandria.	
148	Garfield.	
154	Brandon	
166	Interlaken.	
176	Dalton.	
186	Fergus Falls.	
196	Carlisle.	
204	Rothsay.	
212	Lawndale.	
218	Barnesville.	
232	Sabin.	
241	Moorhead.	

(bracketed annotations: Probably Cambrian and 1. Archæan and covered with drift and water deposits. — 1. Archæan, with exposures of granite and syenite (at and near St. Cloud), and gneiss and diorite. (Sauk Centre.) — 20. Heavy drift. The Lake Park Region, with high morainic hills and numerous lakes.)

(The beaches of the glacial lake Agassiz are crossed. — Modified Drift.)

St. Paul, Stillwater & Taylor's Falls R. R.

Ms.		
0	St. Paul.	{ 4 a. Trenton, 704 / 3 a. St. Peter s. s.
3	Post's.	4 a. Trenton. 847
12	St. Elmo.	" 938
16	Stillwater Junc.	3 a. Calciferous. 857
20	Stillwater.	" 697

* The main line of the Northern Pacific Railroad is given in a separate chapter.

10. The standard thickness of the formations in Minnesota of the palæozoic rock is: downward, Galena, or Upper Magnesian, 183 feet; Upper Trenton, gray limestone, 120 feet; a green shale, 15 feet; Lower (blue) Trenton, 17 feet; St. Peter sandstone, 115 feet; Lower Magnesian, 250 feet; Potsdam, perhaps, 1,000 feet. The upper measures are greatly corroded and show but a small part of the several measures, except the Lower Trenton and its invariable associate the St. Peter sandstone, giving such uniformity of escarpment as will be found in no other formations. The Upper Trenton is usually corroded well back from the front of any bluff and shows light slopes. W. D. H.

11. From Tracy to Gary, on the southwest, are to be seen the foothills of the *Coteau des Prairies.* Going west from Tracy the railroad passes into a valley between two morainic hills, and near Canby the ascent of the *Coteau* is begun, the summit of which is reached at Goodwin, Dak., at 1,996 feet above the sea. C. W. H.

12. From Wadena to Fergus Falls the railway passes through the beautiful "Lake Park Region," with the abrupt morainic mounds of the Leaf Hills and numerous glacial lakes. Near Ames and Everdill are the beaches of the glacial lake Agassiz (Upham.) C. W. H.

13. Winnebago City is on the deposits of a glacial lake (Upham.) After crossing the Des Moines River the *Coteau des Prairies* is ascended. The three highest points between the Des Moines and the James Rivers are: Four miles west of Iona, 1,705 feet; four miles east of Pipestone City 1,744 feet; west of Lake Herman, Dak., 1,825 feet. At Pipestone City occur the beds of quartzite and Catlinite (Indian Pipestone), of either Cambrian (Winchell), or Huronian (Chamberlin and Irving). O. W. H.

St. Paul, Minneapolis & Manitoba Railway.			St. Paul, Minneapolis &
Alt.		**Ms.**	**Ms.** **way.—Contin**

Alt.			Ms.
.....	Breckenridge.	(See No. Pacific.)	959
.....	Manston.	20. Drift.	976
.....	Atherton.	20. Drift.	979
218	Barnesville.	Drift.	1007
225	Downer.	"	958
235	Glyndon.	{ Flat drift in the bed of the ancient lake Agassiz.	932
241	Averill.	"	927
249	Felton.	"	925
254	Borup.	"	921
264	Ada.	"	907
275	Rolette.	"	895
280	Beltrami.	"	905
285	Russia.	"	895
290	Kittson.	"	888
297	Carman.	"	885
298	Crookston.	"	868
304	Shirley.	"	905
311	Euclid.	"	895
319	Angus.	"	875
827	Warren.	"	858
837	Argyle.	"	850
346	Stephen.	"	832
357	Donaldson.	"	831
361	Kennedy.	"	830
370	Hallock.	"	820
375	Northcote.	"	807
382	Humbolt.	"	797
389	St. Vincent.	"	792
391	Boundary Line.	"	793

Sauk Centre & Northern Branch.			
0	Sauk Centre.	Sauk Centre.	1252
10	Little Sauk.		1240
19	Long Prairie.	{ Covered with drift.	1286
26	Browerville.		1269
32	Clarissa.		1819
37	Eagle Bend.		1871

Brown's Valley

Ms.		
0	Morris.	Drift
13	Chokio.	
26	Graceville.	

St. Cloud & Hinckle

Ms.		
0	Hinckley.	2 a. 1
7	Pokegama.	
22	Mora.	
26	Ground House.	
39	Millaca.	
41	Bridgman.	
47	Oak Park.	
50	St. Francis.	
53	Foley.	
67	St. Cloud.	See N

Pelican Rapids

Ms.		
0	Pelican Rapids.	
6	Ehrhardt.	
14	Elizabeth.	
21	N. P. Junction.	
23	Fergus Falls.	See N

Duluth & Iron Rang

Ms.		
0	Duluth.	
26	Two Harbors.	Trap
32	Sibwissa.	20.
38	Gakadina.	
49	Wissakode.	
62	St. Louis River.	River
70	Okwanim.[15]	Gabb
75	Mesaba Heights.	Gran
80	Embarrass R.	20. D
93	Tower.	Slate with

Notes signed C. W. H. are by Prof. C. W. Hall.

14. *Taylor's Falls*. The primordeal is here very fossiliferous and lies und rock, supposed to be *Cupriferous*.

15. The great Mesabi range of Gabbro is crossed between St. Louis river at *Mesaba Heights*, as here named, is on a range of granitic rocks, the apparent Giant's range known further northeast and in Canada

Errata: Page 246, after Wisconsin geologists, read, is equivalent to 3 c., etc.
Note 6. For "of *Rhyaconella*," read, for *Rhynchonella*.
Note 7. For "700," read 1,700.
Note 9. For "is a reef or bar of quartzite," read, are caused by a grey gneiss.

North and South Dakota.[1]

Chicago, Milwaukee & St. Paul Railroad, Iowa and Dakota Division.

Ms.	(Mitchell to Chamberlin.)		Alt.
332	Mitchell.[2]	{ 18 a. & b. Cretaceous. / 2d Moraine.	1294
347	Letcher.	{ 18 b. Cretaceous, / Deep Till.	1800
361	Woonsocket.	"	1808
388	Woolsey.	"3d Mor.	1353
420	Redfield.	18 b. Cretaceous.	1295
429	Ashton.	1296 "Lacust'l Alluv.	
461	Aberdeen.	1301 " " "&Till.	
355	Plankington.	Deep Till.	1521
367	Yorkton.	"	1639
379	Kimball.	1731 1st or Principal Mora.	
390	Puckwana.	{ Lacustral Alluvium, / and Till.	1639
399	Chamberlain.[3]	{ 18 b. Cret (Berg)1856 / Till on Uplands.	

	(Canton to Mitchell.)		
252	Canton.	18 b. Cret., Till.	1241
262	Worthing.	"	1357
268	Lennox.	"	1347
381	Parker.[4]	{ 1 b. Red Quartzite, / and 2d Mor.	1341
287	Marion Ju.	"	1440
287	Marion Ju.	"	1440
298	Freeman.	Till and 2d Mor.	1504
309	Menno.	Till.	1517
319	Scotland.	18 b. Creta., Till.	1340
343	Springfield.	"	1327
350	Running Water.	"	1218
287	Marion Ju.		1440
303	Bridgewater.	{ 1 b. Red Quartzite, / Till.	1418
318	Alexandria.	"	1345
332	Mitchell.	{ 1 b. Red Quartzite, / 18 a. and b. Creta-/ceous, 2d Mor.	1294

Sioux City and Dakota Division.

Ms.			Alt.
0	Sioux City.	{ 18 a. Cretaceous, / Drift and Loess.	1097
8	McCook.	Alluvium.	1105
13	Jefferson.	"	1111
21	Elk Point.	"	1124
21	Elk Point.	"	1124
38	Westfield.	"	1124
33	Akron.	{ 18 a. and b. Cretac., / Drift and Loess.	1148
47	Calliope.	18 b. " "	1175
55	Eden.	" "	1215
65	Rock Valley.	" "	1246
58	Austin.	" "	1197

Chicago, Milwaukee & St. Paul.—Con.

Ms.	Sioux City and Dakota Div.—Con.		Alt.
62	Fairview.	{ 18 b. Cretaceous, / Drift & Loess.	1207
68	Beloit.	"	1233
71	Canton.	18 b. Cret. Till.	1241
91	Sioux Falls. 1336	1 b. R. Quartz. 1st Mor,	
21	Elk Point.	Alluvium.	1124
29	Burbank.	"	1133
35	Vermillion.	{ 18 b. Cretaceous, / Drift and Loess.	1143
44	Meckling.	"	1149
50	Gayville.	Alluvium.	1150
61	Yankton.	{ 18 b. Cretaceous, / Drift and Loess	1156
70	Utica.	Drift.	1360
78	Lesterville.	1st Moraine.	1373
90	Scotland.	18 b. Cret., Till.	1340

South Minnesota Division.

Ms.			Alt.
0	Woonsocket.	18 b. Cret., Till.	1308
9	Forestburg.	"	1230
19	Diana.	"	1311
30	Roswell.	"	1398
38	Howard.	"	1561
	Winfred.	" 2d Mor.	1704
	Russell.	" 1st "	
60	Madison.	Drift.	1669
75	Coleman.	Drift Plain.	1687
0	Sioux Falls.	1 b.R.Quartz.,Dft.	1386
20	Dell Rapids.	"	1455
85	Egan.	Drift.	1522
89	Flandreau	"	1562
	Airlie.	"	1641
104	Pipestone.	"	1705

Hastings and Dakota Division.

Ms.			Alt.
0	Ipswich.	18 b. Cret., Till.	1531
13	Mina.	" 3d Mor.	1433
26	Aberdeen.	"Lac'l Silt.	1301
34	Bath.	" "	1301
45	Groton.	" "	1304
55	Andover.	" 3d Mor.	1476
65	Bristol.	" 2d "	1773
77	Webster.	Till.	1842
87	Waubay.	Till and 1st Mor.	1813
	Wilmot.	" 3d "	1196
123	Millbank.	" "	1145
134	Big Stone City.	1 a.Gran.,Till &All.	979
135	Ortonville.	" " "	997

James River Line.

Ms.			Alt.
9	Aberdeen.	Till & Lacust'l Silt.	1301
12	Westport.	18 b. Cretac., Till.	1333
37	Ellendale.	"	1455
64	Edgeley.	" " 3d Mor.	1514

Chicago, Milwaukee & St. Paul R. R.—Con.

Ms.	Fargo Southern Line.		Alt.
0	Ortonville, Minn.	Till. Archæan 997 granites extensively exposed in valley of Minnesota River. 1109	
22	Graceville, "		
49	White Rock.	Lacustrine deposits of Lake 971	
66	Tyler.	Agassiz overly- 967	
		ing till.	
88	Abercrombie.	"	988
120	Fargo.	"	908

Hastings and Dakota Line.—Con.

Ms.			Alt.
0	Ipswick.	Till.	1581
16	Roscoe.	"	1827
31	Bowdle.7	1st & 2d Moraine.1996	

Roscoe and Orient Branch. 6.

Ms.			Alt.
0	Eureka.	Till & 2d Moraine.1885	
8	Hillsview.	"	1850
26	Roscoe.	"	1827
49	Millard.	‿"	1641
58	Faulkton.8	" 2d Moraine.	1574
68	Orient.	"	1600

Chicago and North Western R'y. Eagle Grove and Hawarden Line.

Ms.			Alt.
514	Hawarden.		1181
522	Alcester.	Till and Loess.	1346
531	Beresford.	1st Moraine.	1505
541	Centreville.	18 b. Cret., Till.	1239
554	Hurley.	" "	1268
563	Parker.	1b. Red Quartzite.1340	
579	Canistota.	18 b. Cret.2d Mor.1455	
590	Salem.	" Till.	1517
602	Canova.	" "	1527
612	Vilas.	" "	1480
624	Carthage.	" "	1438
631	Esmond.	" "	1438
640	Iroquois.	" "	1401
	Cavour.	3d Moraine.	1311
658	Huron.	Till.	1285

Minnesota and Central Dakota Line.

Ms.			Alt.
593	Gary.	2d Moraine.	1484
	Altamont.	1st "	1834
	Goodwin.	Old Till.	1996
	Kransburg.	"	1982
631	Watertown.	1st Moraine.	1735
649	Henry.	Till.	1812
662	Clark Centre.	2d Moraine.	1789
	Raymond.	Till.	1458
681	Doland.	3d Moraine.	1355
691	Frankfort.	Alluvium & Till.	1296
702	Redfield.	18 b. Cret.,3d Mor.1800	
713	Athol.	" Lact'l Allu.1296	
723	Northville.	" "	1299
736	Rudolph.	" "& Till.1501	
744	Aberdeen.	" " 1800	
753	Ordway.	" "	1814
759	Columbia.	" "	1815

Chicago & North Western R'y.—Con.

Ms.	(Elkton to Redfield.)		Alt.
574	Elkton.	Drift Plain.	1751
584	Aurora.	"	1680
590	Brookings.	"	1636
597	Volga.	"	1636
608	Nordland.	1st Moraine.	1646
619	Preston.	Till.	1696
644	De Smet.	2d Moraine.	1726
653	Iroquois.	Till.	1401
653	Cavour.	2d Moraine.	1811
662	Huron.	Till.	1285
675	Woolsey.	3d Moraine.	1848
687	Wessington.	"	1419
699	St. Lawrence.	Till.	1580
713	Ree Heights.	{ 18 b. Cretaceous, 1st & 2d Mora.1731	
725	Highmore.	2d Moraine.	1880
739	Harold.	Till.	1601
752	Blunt.	{ 18 b. Cretaceous, 1st Moraine.	1621
761	Canning.	"	1563
781	Pierre. (Missouri River.)	"	1440
662	Huron.	Till.	1285
675	Broadland.	"	1308
684	Hitchcock.	3d Moraine.1839	
703	Redfield.	18 b. Cret., "	1300

(Watertown Junction to Watertown.)

Ms.			Alt.
0	Watertown Ju.		1604
8	Bruce.	Drift.	1640
18	Estelline.	"	1659
30	Castlewood.	"	1686
44	Watertown.	"	1735

St. Paul, Minneapolis & Manitoba R. R.

Ms.			Alt.
241	Morehead, Minn.	{ Plain of Lake Agassiz. Lacus'l Dep.903	
242	Fargo, Dak.	"	901
251	Harwood.	"	886
	Argusville.	"	884
263	Gardner.	"	886
269	Grandin.	"	891
275	Kelso.	"	897
281	Hillsboro.	"	901
289	Cummings.	"	925
295	Buxton.	"	930
300	Reynolds.	"	910
307	Thompson.	"	885
320	Grand Forks.	"	830
333	Manvoel.	"	819
345	Ardock.	"	824
351	Minto.	"	820
360	Grafton.	"	827
374	St. Thomas.	"	840
387	Hamilton.	"	824
392	Bathgate.	"	821
400	Neche.	"	831
402	Gretna, Canada Line.	"	

1. By Profs. T. C. Chamberlin and J. E. Todd, U. S. Geologists, with elevations by Mr. Warren Upham, Assistant on the Geological Survey of Minnesota and the U. S. Survey. The geology of the two States is given in one chapter without reference to the division recently made.

St. Paul, Minneapolis and Manitoba Railroad.—Con.

Breckenridge Extension.

Ms.			Alt.
0	Breckenridge.	{ Lacustrial Champlain.	959
18	Dwight.	"	952
21	Colfax.	"	958
53	Everest.	"	958
80	Greenfield.	Drift.	945
99	Mayville.	"	975
131	Larimore.	"	1134
145	Orr.	"	1098
155	Conway.	"	988
167	Park River.	"	998

Devils Lake Extension.

0	Crookston.	868	Lacustrine Champlain
28	Grand Forks.	"	830
57	Larimore.	Drift & 18. Creta.	1134
83	Michigan City.	"	1617
118	Devils Lake, Sta.	"	1464
	Devils Lake, Water.	"	1432

Hope Branch.

0	Ripon.	1042	Drift, Beach——near.
4	Ayr.	1202	" 18 Cretaceous.?
16	Page City.	"	" ? 1177
23	Colgate.	"	" ? 1179
29	Hope.	"	" ? 1243

Aberdeen Branch. 6

Ms.			Alt.
0	Tintah Jc.	{ Lake Agassiz deposits.	968
25	Hankinson.	Herman Beach.	1068
37	Lidgerwood.[9]	Till.	1123
55	Rutland.	"	1225
58	Sprague Lake.[10]	"	1219

St. Paul, Minn. & Manitoba R. R.—Con.
Aberdeen Branch.—Con.

Ms.			Alt.
64	Havana.	{ Till, Lacustrine plain Lake Dakota.	1294
71	Kidder.	"	1295
78	Burch.	"	1296
84	Amherst.	Till. 4th Mor.(?)	1312
91	Clarmont.	" Lake Dakota.	1302
96	Huffton.	"	1307
102	Putney.	"	1306
110	Hadley.	"	1302
119	Aberdeen.	"	1300

Northern Pacific Railroad. 5
Ms. Jamestown and Northern Railroad. Alt.

0	Jamestown.	1406	18.Cret.,Till & Vy Drift.	
6	Parkhurst.		" "	1500
13	Buchanan.		" "	1546
21	Pingree.		" "	1548
34	Melville.		" "	1601
43	Carrington.		" "	1583
60	New Rockford.		" "	1528
56	Sykeston.		" "	1630

Fargo and Southwestern.6—Con.11

88	La Moure.	{ 18 b. Cretaceous Till.	1505
	Glover.	" "	1370
	Oakes.	{ " Beach of Lake Dakota.	1510
	Berlin.	18 b.Cret. Till.	1468
	Medbury.	"	1520
110	Edgeley.	" 3d Mor.	1516

Chicago, St. Paul, Minneap. & Omaha R. R.
(Sioux Falls Branch.)

0	Sioux Falls.	{ 1. Red Quartzite, Drift Alluvium.	1394
14	Hartford.	Drift.	1561
28	Montrose.	1 & 2d Moraines.	1471
39	Salem.	Till.	1517

2. *Mitchell*. Dakota s. s. (18 a.) finely exposed along Enemy Creek five miles east of south. Also on the Firesteel at and near the crossing of the Letcher Branch. *Niobrara* (?) (Chalkstone) 18 b. along the railroad one mile east, and along the Firesteel a mile northeast and further up. This with the clays of probably the Ft. Benton frequently struck in deep wells.

3. *Chamberlain*. *Niobrara* and Fort Pierre clays (18 b.) exposed over 350 feet in the sides of the bluffs, 40 to 50 feet of Till, probably of glacio-natant origin, cap the bluffs and several feet of Loess frequently covers that.

4. *Parker*. Red Quartzite of Dakota which is 1 b. Huronian, is exposed along the Vermillion near the level of the water two miles east.

5. The main line of the Northern Pacific is given in a separate chapter.

6. Elevations, as well as geology, on this line by Prof. J. E. Todd.

7. *Bowdle*. Unusually fine exhibition of gravel plains and ridges, in a broad re-entrant angle of the first and second moraines which are here united. They are crossed two to three miles east of the town.

8. *Faulkton*. The hills southwest are the eastern head of a re-entrant angle or interlobular portion of the second moraine.

9. *Lidgerwood*. An interlobular portion of the fourth and fifth moraines is well developed a few miles south. The latter is crossed near Geneseo.

10. *Sprague Lake*. Near the head of Coteau des Prairies, third and fourth moraines at its base, the second at its summit.

11. The Fargo and Southwestern is continued from the Northern Pacific chapter.

Ms.	St. Paul, Minneapolis and Manitoba. Continued.	Alt.
352	Shawnee.	{ Drift and 18 c. Ft. Pierre. }
405	Devil's Lake.	"　　" 1444
413	Grand Harbor.13	" 1454
424	Church's Ferry.	" 1458
436	Leeds.	" 1514
442	York.	" 1512
448	Knox.	" 1605
453	Pleasant Lake.	" 1609
468	Rugby Junc.	" 1561
474	Berwick.	" 1482
481	Towner.	" 1475
487	Denbigh.	" 1485
500	Granville.	" 1503
508	Norwich.	" 1528
503	Minot.14	{ 18 d. Laramie 1557 Lignite Mines. }
535	Des Lacs.	18 d. Laramie. 1597
541	Lone Tree.	" 1995
546	Berthold.	" 2052
556	Wallace.15	" 2132
562	Delta.	" 2255
569	Elton.15	" 2195
577	Stanley.	" 2252
584	Ross.	" 2257
589	Manitou.	" 2275
597	White Earth.	" 2057
606	Tioga.	" 2273
615	Ray.	" 2271
622	Wheelock.	" 2374
631	Spring Brook.	" 2113
638	Avoca.	Lignite Mines. 1956
645	Williston.	18 d. Laramie. 1854
656	Trenton.	" 1896
665	Buford,	" 1944
	Montana Line.	

Cando and St. John Line.

Ms.		Alt.
424	Church's Ferry.	D'ft.18c.Ft.Pierre.1455
439	Cando.	"　　" 1416
452	Bisbee.	"　　" 1600

Ms.	St. Paul, Minneapolis and Manitoba. Cando and St. John Line.—Con.	Alt.
459	Perth.	D'ft.18c.Ft.Pierre.1731
471	Rolla.	"　　" 1518
479	St. John.16	"　　" 1945

Bottineau Branch.

Ms.		
463	Rugby Junc.	D'ft.18c.Ft.Pierre.1561
	Barton.	"　　" 1505
484	Willow City.	"　　" 1471
504	Bottineau.16	"　　" 1688

Aberdeen, Bismark and N. Western R'y. 6

Aberdeen.	1295	Till. Lacustral Silt.
Foster.	1331	18 b. Cretaceous, Till.
Leola.		" 1557
Ashley.17	2001	Till (?) Lacustral Silt.
Beaver Creek.		18 c. Cret. Drift. 1987
Red Lake.		" 1970
Lowry.		" 2057
Napoleon.		" 1955
Merriam.		" 1862
Bismark.		" 1672

Fremont, Elkhorn and Missouri Valley.
Elkhorn Valley Line.—Con.18

Ms.		
444	Chadron, Neb.	19 b. Miocene. 3360
449	Dakota Jc.	" 3245
461	Wayside.	"
476	Oelrich, Dak.19	18 Cretaceous.
485	Smithwick.	18 a. "
500	Buffalo Gap.20	"　" 3252
516	Fairburn.	"　"
528	Hermosa.	"　" 3295
540	Brennen.	"　"
548	Rapid City.21	"　" 3192
555	Black Hawk.	Jura-Trias.
562	Sacora.	"
568	Tilford.	"
577	Sturgis.22	" 3467
584	Whitewood.23	" 3640
593	Deadwood.	Surveyed. 4545
597	Pennington.	" 4972

12. Geology, notes, and elevations on this line and branches from Shawnee west by Mr. Warren Upham, Assistant Geologist, U. S. Geological survey.

13. The country is all more or less drift-covered to Great Falls, Montana, but is destitute of drift thence to Helena and Butte.

14. The Laramie formation, extending from Minot to Kintyre, contains occasional beds of Lignite.

15.—Terminal moraine drift hills, marking a stage of halt or re-advance of the ice-sheet, are well displayed along the distance of thirteen miles by Wallace, Delta and Elton, a S. E.-N. W. belt of these deposits being there crossed by the railway.

16. Between St. John and Bottineau, the Turtle Mountain area, elevated about 500 feet above the general level, is an extensive outlying tract of the Laramie formation, overspread with irregularly hilly deposits of glacial drift.

17. Ashley. The first and second moraines are crossed separately seven to twenty miles N. W. of Leola, where they turn sharply from a south-south-westerly direction to nearly due west. Ashley is on a level pebbless plain, covering perhaps twenty square miles. The road between Ashley and Napoleon runs mostly in a valley just outside of the first moraine, which is unusually heavily developed.　J. E. T.

18. By Prof. G. E. Bailey of the Dakota School of Mines, Rapid City, S. Dakota.

19. Oelrich. Cretaceous, with here and there outliers of Miocene.　G. E. B.

20. Buffalo Gap. Bad Lands twenty miles east, the great collecting ground of Prof. Cope and Marsh. Fossil horses, shells with pearl preserved, turtles, etc. Two miles west handsome variegated sandstones, whetstones, fifteen miles west hot springs, tufa.　G. E. B.

21. Rapid City. Black Hills, tin mines, twenty miles S. W. Gold, silver, copper, lead, mica and graphite mines; marble, gypsum, brick, fire and potter's clays.　G. E. B.

22. Sturgis. Homestake mines, ten miles, Galena Smelters, ten miles.　G. E. B.

23. Whitewood. Carbonate and Nigger Hill mining districts. The coal, oil and salt districts of Dakota.　G. E. B.

General Note on the Geology of the Western part of the North American Continent.

It may be useful to those not familiar with the local geology of America, to insert a general account of the well-marked difference between the eastern and western parts of the Continent. Adopting the line of Central Texas, Indian Territory, Kansas, and Eastern Nebraska and Dakota, and extending it in the same general course to the Arctic Circle, we will have North America divided into two great divisions, in each of which the geology of the country has the same general character and each widely different from the other.

The eastern division shows a sub-division into a number of great basins, representing all the older geological formations in their regular stratified order, and each with a carboniferous coal field on its summit, and then the whole area framed on the outside by two or three irregular bands of the Cretaceous, Tertiary and Quaternary formations, and showing also several intermediate lines of Triassic and probably Jurassic.

But on crossing the line above described, we pass from the old to the new geological world, in which the Upper Silurian* and Devonian formations are unknown, and even the Carboniferous appears in so changed an aspect as to be unworthy of the name, inasmuch as it is no longer coal bearing. As our geological table is now numbered, much more than half of it has here become useless in this western district, as none of those formations are there to be seen, and we come into a new geological continent of magnificent distances, covered for thousands of miles chiefly by the Cretaceous and Tertiary, with smaller areas of Triassic and Jurassic formations, with other vast areas of mountains and plains of eruptive and metamorphic rocks, with the minerals peculiar to them, affording but little material for geological notes, and sometimes greatly disturbing and subverting the order of stratification and rendering Metamorphic the Cretaceous and Tertiary. Some of the ranges no doubt contain a central axis of granite and crystalline formations of the older rocks, and in time some small portions of the metamorphic rocks, like those of New England, may prove to have been changed from Palæozoic and other formations well known in the eastern division. A few fossils here and there may show traces of what they once were, but as yet they may be classed under the comprehensive name of Metamorphic.

But the most remarkable point in this description is the vast extent and great persistence and uniformity of these formations of the Far West, so limited in number and spreading from near the Mississippi and Missouri Rivers to the Pacific Ocean, and from the North Pole to the Isthmus of Tehuantepec. This statement gives a correct general impression of the geology of more than half of North America. An examination of this "Geological Railway Guide," along all the lines as yet constructed, and of all the geological maps of the United States and of the Dominion of Canada, and the reports of all travelers, will serve to confirm what has here been stated, and to impress on the mind of the student the important transition he makes in passing west of the Mississippi Valley.

One of the most unfortunate facts in connection with the geology of this western district is, that throughout a large portion of it, especially its central and southern parts, the soil is "alkaline," the rain-fall being less than the evaporation by which soluble salts are brought to the surface, rendering the land unfit for cultivation without irrigation, although portions of it afford pasturage, and there are many lakes and rivers whose waters contain a greater or less per centage of soda salts. The areas, however, are relatively small in which the soil is not able to yield crops, if only water can be supplied to it.

Another point may be worthy of mention, namely, that the study of the formations of the Far West has only been begun, and they are so much more expanded and sub-divided that, for aught we now know, a new geological world may yet be opened, which may greatly enrich the science of geology, modifying our present series of the newer formations, giving us new views of structural and dynamic geology and discovering new forms of ancient life.

It is as true now, as it was when written by Prof. James Hall, thirty years ago, that "our knowledge of the geological formations of the West is so rapidly progressing, and the materials are accumulating in such abundance, that whatever may be presented to-day as new and in advance of previous knowledge, will to-morrow be regarded only as a historical record of our progress."   J. M.

TABLE OF THE TERTIARY AND CRETACEOUS FORMATIONS.

From Dr. Edward D. Cope's Report on the Vertebrata of the Tertiary Formations of the West, United States Geological Survey, 1883.

19. TERTIARY.			18. POST-CRETACE'S.		18. CRETACE'S.	
19 c. Pliocene.	{	Magalonyx Beds.		? Puerco. †		Puerco.
		Equus Beds.		18 d. Laramie.		Fort Union.
		Procamelus Beds.				Bear River.
19 b. Miocene.	{	Ticholeptus Beds.				
		John Day.				
		White River.		18 c. Fox Hills.		Fox Hills.
		Uinta.				Fort Pierre.
19 a. Eocene.	{	Amyzon Beds.		18 b. Colorado.		Niobrara.
		Bridger.				Fort Benton.
		Green River.		18 a. Dakota.		Dakota.
		Wasatch.				

* The Lower Silurian is known in Idaho, Montana, Wyoming, Colorado, New Mexico, Utah, Nevada and Arizona, most largely in the two last named.

† Professor Cope insists there is plenty of evidence, since the publication of his report, that the Puerco is distinct from the Laramie.

Northern Pacific Railroad.[1]

Ms.	MINNESOTA.	Alt.
.....	St. Paul.	{ 4 a. Trenton, 3 a. St. Peter sandstone. 701
11	Minneapolis.	" 882
13	N. Minneapolis.	"
15	Northtown Junc.	3 a. St. Peter sand s.
18	Fridley.	" 845
25	Coon Creek.	" 860
29	Anoka.	3 a. Calciferous. 853
36	Itaska.	" 891
41	Elk River.	2. Primordial. 901
45	Bailey's.	" 915
50	Big Lake.	" 940
57	Becker.	" 976
64	Clear Lake.	1. Archæan. 997
71	Haven.	" 1016
76	E. St. Cloud.	" 1030
77	Sauk Rapids.	" 1004
83	Watab.	" 1053
90	Rice's.	" 1059
97	Royalton.	" 1080
103	Gregory.	" 1095
107	Little Falls.	" 1115
112	Belle Prairie.	Taconic. 1130
116	Topeka.	" 1144
121	Fort Ripley.	" 1153
126	Albion.	" 1173
130	Crow Wing.	" 1186
138	Brainerd.	" 1205
	Miss. River Low Water.	" 1152
146	Gull River.	" 1189
148	Sylvan Lake.	" 1203
151	Pillager.	" 1200
156	Bath.	" 1212
160	Motley.	1. Archæan. 1223
168	Staples Mill.	" 1250
170	Dower Lake.	" 1290
174	Aldrich.	1327 "heavy drift.
178	Verndale.	" 1347
185	Wadena.	" 1349
187	Wadena Junc.	" 1350
190	Bluffton.	" 1310
193	Amboy.	" 1376
197	New York Mills.	" 1409
203	Richmond.	" 1394
209	Perham.	" 1367

Ms.	MINNESOTA.—Con.	Alt.
214	Luce.	1. Arch. h'vy drift 1370
220	Frazee.	" 1384
225	Johnson.	" 1393
230	Detroit.	" 1362
237	Audubon.	" 1308
242	Lake Park.	" 1334
248	Hillsdale.	" 1399
254	Hawley.	" 1150
258	Muskoda.	" 1090
267	Glyndon.	" 924
269	Tenny.	" 920
275	Moorhead.	" 903
	Red River Low Water.	867

DAKOTA.

Ms.	DAKOTA	Alt.
276	Fargo.	1. Arch. h'vy drift. 903
281	Haggart.	" 903
285	Canfield.	" 903
289	Mapleton.	9-12. Up. Devonian 903
292	Greene.	" 913
294	Dalrymple.	" 920
297	Casselton.	" 930
303	Wheatland.	" 965
313	Buffalo.	" 1206
319	Tower City.	" 1170
324	Oriska.	" 1240
329	Alta.	" 1423
333	Valley City.	18. Cretaceous. 1216
.....	Cheyenne River Low Water.	1200
342	Hobart.	18. Cretaceous. 1417
346	Sanborn.	" 1460
349	Eckelson.	" 1444
359	Spiritwood.	" 1477
364	Bloom.	" 1485
369	Jamestown.	" 1395
.....	James River Low Water.	1380
376	Eldridge.	18. Cretaceous. 1540
386	Windsor.	" 1888
390	Cleveland.	" 1840
398	Medina.	" 1790
406	Crystal Springs.	" 1790
415	Tappen.	" 1760
420	Dawson.	" 1746

1. The geology here given of the Northern Pacific Railroad, east of Bismarck, is by Prof. N. H. Winchell, of Minnesota, and that west of Bismarck, through Dakota and Montana, is by Prof. Raphael Pumpelly, whose work, however, was devoted almost wholly to coal explorations, and his journeys were made on horse trails, often off from the route of the railroad, before most of the stations in Montana and Idaho were located. His foot notes are marked R. P., those marked B. T. P. are by his assistant, B. T. Putnam, and those signed G. W. D. are by Dr. George M. Dawson, giving the observations of a passing geological traveler well versed in the geology of the adjoining territory of Canada.
J. M.

Ms.	Northern Pacific R. R.—*Con.*		Alt.
428	Steele.	18. Cretaceous.	1857
435	Geneva.	"	1833
439	Driscoll.	"	1835
446	Sterling.	"	1865
453	McKensie.	"	1696
458	Menoken.	"	1716
467	Apple Creek.	"	1642
471	Bismarck.[2]	18 d. Laramie, Cretaceous.	1668
.....	Missouri River	Low Water.	1616
476	Mandan.	18 c. Pierre & Fox Hill.	1644
484	Marmot.[2]	"	1729
490	Sweet Briar.	"	1683
500	Sedalia.	"	2030
.....	Summit.	"	2165
504	New Salem.	"	2161
507	Blue Grass.[3]	18 d. Ft. Union. "	2042
511	Sims.[4]	"	1960
516	Almont.	"	1918
521	Curlew.	"	1955
528	Kurtz.	"	2023
533	Glenullen.	"	2070
538	Eagle's Nest.	"	2096
547	Knife River.	"	2160
555	Antelope.[5]	18 d. Ft. Union Laramie.	2412
561	Richardton.[5]	"	2464
566	Taylor.	"	2486
574	Gladstone.[6]	"	2346
.....	Green River low	water. "	2275
585	Dickinson.	"	2403
591	Eland.	"	2434
597	South Heart.	"	2470
606	Belfield.[7]	18 d. Fort Union Laramie, Cretaceous.	2577

Ms.	Northern Pacific R. R.—*Con.*		Alt.
611	Fryburg.	18 d. Fort Union Laramie, Cretaceous.	2767
617	Sully Springs.	Pyramid Park Wonderful Bad Land Scenery. "	2647
620	Scoria.[8]	"	2505
625	Medora.	"	2265
.....	Little Mo. River.[7]	"	2246
626	Little Missouri.[9]	" Lignite Mines	2255
633	Andrews.	"	2476
641	Sentinel Butte.	"	2707

MONTANA.

Ms.	Station		Alt.
650	Beach.	18 d. Fort Union Laramie, Cretaceous.	2754
.....	Summit.	"	2819
659	McClellan.	"	2685
661	Mingusville.	"	2639
.....	Summit.	"	
671	Hodges.	"	2535
681	Allard.	"	2299
691	Glendive.[10]	"	2067
701	Iron Bluff.	"	2097
706	Milton.	"	2114
721	Fallon.	"	2206
.....	O. Fallon Creek.	"	2145
731	Terry.	"	2240
.....	Powder River.	"	2199
741	Morgan.	"	2245
751	Ainslie.	"	2272
761	Dixon.	"	2320
770	Miles City.	"	2358
.....	Tongue River.	"	2342
772	Fort Keogh.	"	2365
777	Lignite.	18 d. Laramie, Cretaceous, Lignite Mines.	2375

2. From *Bismarck*, at Missouri Crossing, to a few miles beyond Marmot Station, numerous exposures in cuttings, and banks of Knife River of Pierre shales, capped in places by Fox Hill sandstones. G. M. D.

3. Near *Blue Grass*, detached portions of edge of plateau formed of Fort Union Laramie appear, rocks showing in some places. At Sims, same rocks. G. M. D.

4. *Sims* (Bly's Mine). Several seams of lignite, of which two, 4 feet and 7 feet thick, are opened. R. P.

5. Line runs on up Valley of Knife River, and gradually attains to level of plateau above referred to. This, about Antelope and Richardson, forms a rolling and hilly prairie, which is based directly on Fort Union Laramie, the soil consisting of disintegrated rocks of this formation. No erratics or glacial drift appear anywhere on this plateau, so far as observed. G. M. D.

6. At *Gladstone*, descend into Valley of Heart River continued exposures of Fort Union. G. M. D.

7. From *Belfield Station* to the Little Missouri, pass through fine "bad land" scenery. Fine display of rocks of Fort Union Laramie. Thin seams of lignite, which in many places have been burnt out, reddening the surrounding rocks. Large masses of silicified wood in some places. G. M. D.

In entering the Bad Lands of the Little Missouri, the change in the scene is startling, and the appearance of the landscape wholly novel and singularly grotesque. There are thousands of these buttes, and you ride in a fast train for an hour in the midst of red, gray, black, brown and blue towers, pyramids, peaks, ridges, domes and castellated heights, turrets, battlements, sharp spires, grotesque gargoyles and huge projecting buttresses—an amazing jumble of weird architectural effects, that startle the eye with suggestions of intelligent design. It is a region of extraordinary interest to the tourist and artist. E. V. Smalley.

8. *Scoria*. In Bad lands or Pyramid Park. Near here are extensive burning seams of lignite. R. P.

9. *Little Missouri*. Several seams of lignite, of which one, 7 feet thick is opened. R. P.
At *Little Missouri*, high banks with good exposures of Fort Union Laramie rocks.

10. Beyond *Glendive*, following the Valley of the Yellowstone, numerous banks showing Fort Union, thin lignite seams and much massive soft sandstone. G. M. D.

Ms.	Northern Pacific R. R.—Con.	Alt.	Ms.	Northern Pacific R. R.—Con.	Alt.
782	Horton. { 18 d. Laramie, Cretaceous, Lignite Mines.	2390	.....	Summit of Mt. over Tunnel.	5835
			1046	West End. 18 U. Cre. Juras. &	5540
790	Hathway. "	2426	1046	Timber Line,[18] " [Trias.	5500
802	Rosebud. "	2460	1048	Mountain Side. "	5275
815	Forsyth. "	2512	1049	{ Rock Cañon[19] Chestnut.[20] { 17. Jurassic, 16 Carboniferous.	5225
825	Howard.[11] 18 c. Fox Hill.	2559	1051	Gordon. "	4905
836	Sanders.[11] "	2593	1054	Fort Ellis. 20. Quaternary.	4860
847	Myers.[12] "	2651	1057	Bozeman.[16] "	4752
857	Big Horn. "	2688	1067	Belgrade. "	4435
863	Custer. "	2725	1072	Central Park. "	4295
872	Riverside. "	2777	1076	Gallatin River, "	4280
880	Bull Mountain. "	2840	1085	Gallatin. "	4030
888	Pompey's Pillar.[13] "	2869	1096	Magpie. { 14. Carboniferous. 2. Cambrian.	3980
896	Clermont. "	2951			
904	Huntley. 18 c. Fox Hill.	3012	1103	Painted Rock. "	3958
.....	1st Cross'g Yel. River. "	3077	1112	Toston. "	3919
917	Billings.[13] { 18 c. Fort Pierre, with Bluffs of Fox Hill Group.	3115	1122	Townsend. { 20. Quaternary, Lake Basin.	3809
930	Laurel. "	3258	.....	Missouri River. "	3791
940	Park City. 18. Cretaceous.	3385	1125	Bedford. "	3882
953	Rapids. "	3515	1137	Placer. "	4290
957	Stillwater. "	3570	.....	Summit. "	4345
965	Merrill. "	3655	1144	Clasoil. "	4123
968	Reedpoint. "	3685	1149	Jefferson Junc. "	3887
.....	2d Crossing Yel. River. "	3674	.....	Prickly Pear Ck. "	3865
984	Greycliff. "	3845	1151	Prickly Pear. "	3878
998	Big Timber. "	4070	1155	Helena.[21] "	3930
1012	Springdale.[14] "	4188	.....	10-Mile Creek. 2. Cambrian.	3875
1019	Elton. "	4250	1163	Birdseye. "	4025
1024	Mission. "	4855	1168	Butler. "	4725
.....	3d Crossing Yel. River. "	4435	1176	Mullan (Tun.) { 14. L. Carbon. Limestone & Granite	5548
1032	Livings'n.[15*40] 18. Up. Cretaceous	4485			
1037	Coal Spur.[16] Juras. & Trias.?	4735	.....	Summit.[22] { 18. Cretaceous, with Coking Coal.	5873
1041	Hopper's.[17] "	5175	1184	Elliston. { 14. Carboniferous, 18. Cretaceous.	5036
1044	Muir. "	5500			
.....	Belt Range Tunnel. "	5565			

11. Before reaching *Howard*, and between that station and Saunders, almost continuous exposures of massive yellowish soft sandstone, evidently Fox Hill, and nearly horizontal.   G. M. D.

12. In a cut at *Meyer's*, and just beyond that station, a slight undulation brings the top of the Pierre into view. The base of the sandstone becomes interbedded with dark shales.   G. M. D.

13. Similar sandstones, with top of Pierre occasionally showing below them, extend all along the Yellowstone Valley to *Billing's*, and beyond. At Billing's they form bold cliffs behind the town. The so-called Pompey's Pillar, near station of same name, is an isolated mass of these sandstones.   G. M. D.

14. Near *Springdale*, the rocks become disturbed for the first time, and dip at high angles. Jurassic-Triassic, according to Hayden's map. (? ?)

Beyond Springdale, fine views of Little Belt Mountains to north, and north end of Yellowstone range to south, the former composed (by map) of volcanic rocks, with a belt of Carboniferous tilted up around them, the latter of Metamorphic rocks, surrounded by Silurian, Carboniferous and Jurassic-Triassic.   G. M. D.

15. *Livingston*. Branch railroad to Yellowstone National Park, Lower cañon of the Yellowstone in sight. It is cut across the arch of a pitching anticlinal giving a fine section of Carboniferous, Jurassic, Triassic (?) and Cretaceous fossiliferous beds.   R. P.

16. From *Livingston* to *Bozeman Tunnel*. Cretaceous and possibly Jurassic-Triassic rocks, much disturbed, and at all angles to vertical.   G. M. D.

17. *Hoppers*. Seams of Cretaceous coking coal are worked a mile or so south of the tunnel. R. P.

18. At *Timber Line*, just west of Bozeman Tunnel, spur track to coal mine, which I am informed yields most of coal now used on line.   G. M. D.

19. *Rock Cañon*, just beyond Timber Line, seems to show Carboniferous limestones and other old rocks nearly on edge.   G. M. D.

20. *Chestnut*. Several seams of coking coal, much crushed. Carboniferous, Jurassic and Dakota exposed in a cañon across the end of an anticlinal arch.   R. P.

21. *Helena* is built in a gulch, which has been washed with great profit for gold.   R. P.

22. *Summit*. Cretaceous seams of coking coal.   R. P.

Ms.	Northern Pacific R. R.—Con.	Alt.
1193	Avon. { 14. Carboniferous. / 18. Cretaceous.	4675
1206	Garrison.² { 18. Cretaceous. / 14. Carboniferous.	4815
1207	Lloyd.²⁴ "	4295
1214	Gold Creek.²⁵ "	4203
1227	Drummond.²⁶ { 14. Carboniferous. / Cañon in Carbonif. / limestone.	3943
1239	Bearmouth. "	3787
1247	Carlan. { Deposit of Travertine. / 2. Cambrian, with / eruptive-dykes.	3653
1255	Bonita.²⁷ "	3564
1262	Wallace. "	3435
1269	Turah. "	3305
1279	Missoula.²⁸ { 18. Cretaceous basin / with seams of lignite.	3195
1286	De Smet. "	3213
1296	Evaro.²⁹ 2. Cambrian.	3946
1307	Arlee. { Lake bas. prob-	3057
......	Jocko Creek. ably 19 f. Pli-	2953
1316	Ravalli.³⁰ ocene or Quat-	2690
1323	Jocko. ernary.	2507
1330	Duncan.³¹ { 2.Cambrian containing Plioc. or Quat. / Lake Basin.	3497
1338	Perma. "	3493
......	3d Crossing Clark's F'k. "	2462

Ms.	Northern Pacific R. R.—Con.	Alt.
1344	Victor. { 2.Cambrian containing Plioc. or Quat. Lake Basin.	2453
1350	Paradise. "	2460
1357	Horse Plains. "	2463
1364	Weeksville.³¹ "	2440
1371	Eddy. "	2413
1378	Woodlin. "	2455
1381	Thompson Fs.³² "	2434
1382	Allen.³³ "	2410
......	{ 2d Crossing / Clark's Fork. "	2395
1387	Belknap. "	2405
1394	White Pine. "	2372
1404	Trout Creek. "	2275
1410	Tuscor "	2235
1419	Noxon. "	2186
1429	Heron. "	2261
1435	Cabinet.³⁴ "	2187
1442	Clark's Fork. "	2086
......	1st Crossing Clark's Fork "	2065
1452	Hope. "	2105

The Valley of Clark's Fork is chiefly between Cambrian walls, and contains old lake basins of Quaternary and perhaps also of Tertiary age.

IDAHO TERRITORY.

Ms.	Northern Pacific R. R.—Con.	Alt.
......	{ Lake Pend / d'Oreielle.³⁵ { Clay, Slate and / Trap.	2059
1457	Kootenai. "	2080
1467	Sand Point.³⁶ { Granite & Gneissic / area.	2100
1473	Algoma. "	2214
1480	Cocolalla. "	2224

23. *Powell's peak* on the south occasionally visible between *Garrisons* and *Drummond*, has a granite core, overlaid by Cambrian slates, Carboniferous limestone, and Cretaceous strata. B. F. P.

24. *Lloyd.* Cretaceous, with eruptive; Carboniferous limestone in mountains to the north. B. F. P

25. *Gold Creek.* First discovery of gold in Montana is said to have been made near here. B. F. P

26. *Drummond.* Lower (?) Cretaceous fossils in Colerley's hollow, 5 miles southeast of Drummond. B. F. P.

27. *Bonita.* Bitter Root Mountains seen towards the south are granite; Cambrian slates in foot hills. B. F. P.

28. Near *Missoula* (Evaro), the rocks evidently "Cambrian." These continue in a series of undulations, but often for long distances at low angles, to Sand Point. "Cambrian" rocks, consisting of hard quartzites, shales, slate, etc. G. M. D.

29. *Evaro.* Probably Pliocene or Quaternary, or 2. Cambrian. R. P.

30. *Revalli.* A ride of about 12 miles to MacDonald's Peak, one of the grandest and wildest mountain masses on the continent, remarkable for its great amphitheatres and lakes and high cascades. Here is exposed a great thickness of Cambrian overlaid by lower Carboniferous. The ascent is along the crest of a fine moraine, on a horse trail of the Northern Transcontinental Survey. R. P.

31. *Duncan to Weeksville.* Valley of Clark's Fork is between Cambrian walls, and contains Pliocene or Quaternary lake basins. R. P.

32. *Thompson's Falls.* I have seen no drift in Montana, Idaho and Washington Territory, east of the Cascades, that appeared to me to be truly glacial drift. Moraines occur along the great ranges as remnants of local glaciation; and erratics which may have been brought by icebergs, agreeably to Dr. G. M. Dawson's theory, occur at many points on the high plains at the eastern base of the Rocky Mountains, south of the boundary. R. P.

33. *Allen.* Glaciers exist on a moderate scale in the Wind River Mountains, and others were discovered by the writer in 1883, on the headwaters of the Flathead River in the main range of the Rocky Mountains, just south of the British boundary. Very large glaciers exist on Mount Rainier, in the Cascades, and are accessible by the horse trail of the Northern Transcontinental Survey from Wilkeson. R. P.

34. *Cabinet.* The valley of Clark's Fork is chiefly between Cambrian walls, and contains old lake basins of Quaternary, and perhaps also of Tertiary age. R. P.

35. *Lake Pend de Oreielle.* The islands in south end of Lake Pend de Oreielle are finely glaciated. R. P.

36. Shortly after passing *Sand Point*, enter a granitic or gneissic area. These rocks continue, apparently at least in the hills, to near Spokan Falls, where basaltic rocks set in, and characterize the whole Columbia plain. G. M. D.

Northern Pacific Railroad—Continued.

Ms.			Alt.
1490	Granite.	{ Granite & Gneissic area.	2290
1495	Athol.	"	2210
1499	Chilco.	"	2450
1509	Rathdrum.	"	2210
1519	Idaho Line.	"	2128

WASHINGTON TERRITORY.

Ms.			Alt.
......	Spokane River.	{ Granite & Gneissic area.	1925
1528	Trent.	"	1989
1537	Spokane Fa's.[36]	{ Volcanic basaltic rocks.	1910
......	Hangman Cr'k.	"	1793
1545	Marshall.[39]	{ Volcanic basaltic rocks over the whole Columbia plain.	2134
1553	Cheney.	"	2340
1564	Stevens.	"	2282
1577	Sprague.	"	1906
1587	Harriston.	"	1950
1601	Ritzville.	"	1825
1618	Lind.	"	1868
1628	Providence.	"	1530
1638	Twin Wells.	"	1075
1646	Palouse Junc.	"	858
1656	Lake	"	677
1665	Eltopia.	"	600
1675	Glade.	"	500
1685	Ainsworth.	"	351
... ...	Snake River.	"	328
1686	S. Ainsworth.	"	356
1698	Wallula Junctio n, Ore.	"	326

OREGON.

Ms.	Oregon, R. W. & Navig. Co.'s R. R.		Alt.
1715	Cold Springs.	{ Vol. bas. rocks over the whole Columbia plain.	367
1726	Umatilla Junc.	"	302
1733	Stokes.	"	308
1751	Castle Rock.	"	248
1762	Willows.	"	334
1771	Alkali.	"	
1779	Blalock.	"	220
1794	John Day's.	"	190
1801	Grant's.	"	180
1811	Celilo.	"	160
1824	The Dalles.[37]	"	106
1833	Rowena.	"	140
1847	Hood River.	"	100
1867	Cascade L'ks.[38]	"	108
1871	Bonneville.	"	60
1879	Oneonta.	"	47
1880	Multnomah Fal[39]	"	45
1884	Bridal Veil.	"	46
1887	Rooster Rock.	"	45
1895	Troutdale.	"	60
1910	E. Portland.	"	35
1911	Albina.	"	35
1912	Portland.[40]	"	42

Rocky Mountain R. R. of Montana.
Yellowstone Park Line.[40]

0	Livingston.	18. Cretaceous.	4485
10	Brisbin.[41]	{ 19. Post Tertiary, (Lake Deposit)	4680
20	Chicory.	"	4845
31	Dailey's.	"	4915
41	Sphinx.	"	5070
51	Cinnabar.	"	5179

37. At *Dalls*, basaltic lava in numerous supposed flows forms the hills.
38. At *Cascades*, tufaceous and agglomerate beds appear, and beds of rounded gravels underlie the volcanic materials. Basalts of hills in light, broad undulations.   G. M. D.
39   *The Volcanic Region*   The portion of the Northern Pacific Railroad through the vast volcanic region in Washington and Oregon, affords but little material for interesting geological notes. A recent report of Mr. J. C. Russel, in the 4th Annual Report of the U. S. Geolog'l Survey, gives some descriptions of the little known part of Southern Oregon, south of the railroad. Its rocks are almost wholly volcanic, and spread out in great sheets of lava that once formed a broad, smooth table-land; but in later times it has been broken by faults, so characteristic of the Great Basin region, and thus divided into long, narrow blocks, stretching north and south, and tilted by very recent displacements so as to expose fresh precipitous scarps that have not yet sensibly worn back from the fault lines. In the Warner Valley, for example, the orographic blocks of the dark volcanic rock, miles in length, are literally tossed about like the cakes of ice in a crowded floe, their upturned edges forming bold palisades that render the region almost impassable, which, with the branching fault cracks, combine to make a region of the wildest and roughest description. At present the waters have retreated from the terraces and benches that marked their former level, some, like Summer and Albert Lakes, are permanent sheets of very saline water, but the more numerous are fresh. Mr. Russel finds no evidence of either local or general glaciation in the region he examined. The volcanic history of Oregon and Washington is far from being understood. The points that may be claimed as centres of eruption are rare, so far as has yet been observed, and in only a few instances can the overflows of lava be traced to their sources. Captain C. E. Dutton reports immense flows of lava in the Sandwich Islands, from surprisingly small openings. But those were down the sides of a steep mountain. Neither is there definite and satisfactory evidence obtained that these immense lava fields originated from fissure eruptions. With the exception of very recent deposits of lacustrine origin, nothing is to be seen but volcanic rocks in sections or regularly stratified layers, which from a distance resemble sedimentary beds, but on examination one finds them to be wholly of igneous origin. These black volcanic rocks are composed of rhyolite, together with large quantities of obsidian or volcanic glass. No evidence of volcanic craters were observed, and no basaltic overflows were seen to indicate centres of recent volcanic action. Major Powell reports this region as containing the grandest and most extensive display of volcanic phenomena now known in any part of the world, and the investigation of it promises to supply matter of great importance and instruction to geologic science. We do not yet know even

Ms.	Duluth & Brainerd Line.		Alt.
0	Duluth, Minn.	1. Cupriferous.	608
23	N. P. Junction.	Potsdam Taconic.	1080
28	Pine Grove.	"	1235
33	Norman.	"	1315
39	Corona.	"	1301
45	Cromwell.	Taconic.	1304
51	Wright.	"	1307
57	Tamarack.	"	1269
66	McGregor.	"	1226
75	Kimberly.	"	1235
87	Aitken.	"	1207
92	Cedar Lake.	"	1220
97	Deerwood.	"	1275
108	Jonesville.	"	1286
114	Brainerd.	"	1208

Pacific & Cascade Divisions.

Ms.			Alt.
0	Portland, Ore.	Volcanic.	
38	Kalama, Wash.	"	33
59	Castle Rock.	"	82
75	Winlock.	"	328
88	Chehalis.	"	204
92	Centralia.	"	207
104	Tenino.	"	315
118	Yelm Prairie.	"	387
134	Lake View.	"	324
143	Tacoma.	"	31
152	Puyallup.	"	51
153	Puyallup Junc.	"	67
155	Sumner.	"	80
159	Struck Junc.	"	110
156	Alderton.	"	95
175	Wilkeson.	"	855
177	Carbonado, Wash.	"	1152

Wisconsin Division.

Ms.			Alt.
0	Lake Superior.	20. Red Clay Drift.	602
2	Ashland, Wis.	"	669
6	Omaha Junc.	"	642
24	Summit.	"	1178
64	Superior.	"	608
76	Walbridge.	"	813
79	Carlton.	"	938
88	N. P. Junction.	"	1080

Ms.	N. P. Fergus & Black Hills R R.		Alt.
0	Wadena.	20. Heavy drift with many glacial lakes and moranic hills.	1349
1	Wadena Junc.		1380
10	Deer Creek.		1394
14	Parkton.		1394
18	Henning.		1436
24	Vining.	"	1389
29	Clitheral.	"	1346
33	Battle Lake.	"	1354
39	Maplewood.	"	1360
41	Southwick.	"	1342
52	Fergus Falls.	"	1182
59	French.	"	1085
60	Ames.	"	1063
68	Everdell.	"	993
77	Breckenridge,	"	960
78	Wahpeton.	"	962
86	Ellsworth.	"	960
92	Mooreton.	"	967
98	Barney.	"	1031
105	Wyndmere.	"	1060
120	Milnor.	"	1093

Fargo & Southwestern Division.

Ms.			Alt.
0	Fargo.	{ 20. Lacustrine silt of Lake Agassiz,	903
4	Cotters.	"	909
10	Horace.	"	917
19	Davenport.	"	921
28	Leonard.	"	1045
41	Sheldon.	20. Till.	1078
50	Buttzville.		1171
56	Lisbon.		1089
68	Marshall.	{ 20. Till and 4th Moraine.	1341
76	Verona.	"	1384
88	La Moure.	18. Cret. & Till.	1305

Sanborn, Cooperstown & Turtle Mountain Railroad.

Ms.			Alt.
0	Sanborn.	{ 18. Cret., under very heavy drift.	1460
9	Odell.	"	1441
18	Dazey.	"	1448
27	Hannaford.	"	1437
36	Cooperstown.	"	1447

the extent of this vast volcanic region in Idaho, Washington, Oregon, Nevada and California, but it has been estimated by Prof. Joseph LeConte, at from 200,000 to 300,000 square miles, and its age, he thinks, is Tertiary and probably Miocene. After these vast fields of lava had cooled and consolidated, then came another revolution that affected a region equally great, but situated mostly to the south of it, a force or series of forces, the power and extent of which are utterly beyond the limits of our conception, which broke the earth's crust into thousands of fragments, which were depressed and buried or upheaved into mountain ridges. It will be, when fully explored, one of the wonders of geology for its extent, its remarkable structure, and the mystery of its origin.

40. *Yellowstone Park Line* of Rocky Mountain Railroad of Montana; by Professor Wm. M. Davis, of Harvard College.

41. *Brisbin.* In passing up lower Cañon of Yellowstone, Jurassic (fossils just outside and west of entrance), Carboniferous limestone (very heavy, poor in fossils), and Lower Silurian (Potsdam), are crossed east of river above cañon, contact of Lower Silurian and Archæan. (Hayden.)

The altitudes on the Northern Pacific Railroad were furnished by A. Anderson, Engineer in Chief. They differ slightly from those in Gannett's Dictionary of Altitudes, in Minnesota, but agree with them in Montana, and all west of that. The original datum point was obtained by taking the assumed low water of Lake Superior at 602, as determined by Captain Bayfield, of the Royal Navy, in 1825, by barometrical observations, which have been confirmed by the United States Engineers. From the west, the datum is mean low water of Puget Sound. J. M.

Montana.[1]				Ms.	Montana Central Railroad.	Alt.
St. Paul, Minn. and Manitoba Ry.[2]				0	Great Falls.	3313
Ms.	Continued from North Dakota.		Alt.	14	Ulm.	
673	Willows.[4]	18 d. Laramie	1889	28	Cascade.	
682	Kila.	"	1955	36	Hardy.	
689	Lanark.	"	1976	44	Mid Cañon.	
697	Culbertson.	"	1918	51	Craig.	
703	Blair.	"	1920	59	Wolf Creek.	
711	Calais.	"	1934	66	Wilder.	
720	Brockton.	"	1945	68	Mitchells.	
730	Poplar.	"	1955	80	Silver.	
739	Chelsea.	"	1980		Marysville.	
745	Macon.	"	1976	89	Iron.	1 a. Laur. 1 b. Huron.
751	Wolf Point.	"	1995	97	Helena.	" "
762	Oswego.	"	2018	108	Montana City.	
769	Lenox.	"	2072	113	Clancy.	
775	Kintyre.[3]	"	2082	114	Alhambra.	
181	Milk River.	18 c. Ft. Pierre.	2048	115	Winslow.	
786	Nashua.	"	2060	119	Jefferson.	
794	Whately.	"	2086	121	Corbin.	
801	Glascow.	"	2087	125	Wickes.	
805	Stockholm.	"	2093	133	Boulder.	
811	Tampico.	"	2105	141	Basin.	
818	Vandalia.	"	2120	145	Bernice.	
825	Hinsdale.	"	2162	153	Elk Park.	
834	Beaverton.	"	2167	162	Woodville.	
839	Saco.	"	2175	171	Butte.	
849	Ashfield.	"	2205			
857	Bowdoin.	"	2209			
866	Malta.	"	2242			
871	Exeter.	"	2254		**Washington.**	
877	Wagner.	"	2256			
884	Dodson.	"	2279			
889	Eureka.	"	2301		Northern Pacific Railroad[6].—(Con.)	
897	Savoy.	"	2324		Cascade Division.	
902	Wayne.	"	2332			
911	Harlem.	"	2339	0	Pasco Jc.[7]	See Notes.
919	Zurich.	"	2368	3	Kennewick	"
926	North Fork.	"	2381	41	Prosser.	'
932	Chinook.	"	2401	53	Mabton.	"
940	Yantic.	"	2431	71	Toppenish.	"
947	Toledo.	"	2455	90	Yakima.	" 990
954	Havre.	"	2472	127	Ellensburg.	" 1510
961	Assinniboine.	"	2576	152	Clealum.[8]	"
968	Laredo.	"	2627	158	Nelson's.	"
978	Box Elder.	"	2669	165	Easton.	See Note 9.
989	Big Sandy.	"	2690	173	Martin.	"
994	Verona.	"	2705	175	Stampede.[10]	"
1001	Cairo.	"	2687	183	Weston.	"
1008	Dry Fork.	"	2914	190	Hot Springs.	"
1018	Marias.	"	2561	203	Eagle Gorge.	"
1023	Teton.	"	2626	211	Palmer.	See Note 11.
1030	Benton.	"	2550	220	Enumclaw.	"
1036	Tunis.	See Note 5.	2957	223	Buckley.	"
1043	Sidney.	"	3095	227	Cascade.	"
1048	Flowerree.	"	3203	228	South Prairie	"
1056	Portage.	"	3413	241	Alderton.	"
1065	Watson.	"	3470	243	Meeker.	"
1073	Great Falls.[4]	"	3312	245	Puyallup.	" 67
				254	Tacoma.[14]	" 81

Ms.	Spokane and Palouse Ry.	Alt.
0	Spokane Falls. 910 { Ter. Erup., whose limit on the S. E. is undetermined.	
9	Marshall Jc.	"
20	Spangle.	"
35	Rosalia.	"
46	Oakesdale.	"
52	Belmont.	"
68	Palouse.	"
79	Whelan.	"
84	Pullman.	"
103	Uniontown.	"
112	Genesee.	"

Central Washington.

Ms.		Alt.
0	Cheney. { Tertiary Eruptives, Great Plain of the Columbia.	
10	Medical Lake.	"
15	Deep Creek.	"
25	Fairweather.	"
34	Mondovi.	"
41	Davenport.	"

Seattle, Lake Shore & Eastern.

Ms.		Alt.
0	Seattle.	See Note 12.
5	Ross.	"
6	Fremont.	"
11	Yesler.	"
18	Terence.	"
21	Winsor.	"
28	Snohomish Jc.	"
29	Earle.	"
36	Snohomish.	"
27	York.	"
33	Adelaide.	"
42	Gilman.	"
49	Preston.	"
53	Falls City.	"

Ms.	Olympia and Chehalis Valley Railroad.	Alt.
0	Olympia. .	Drift.
2	Turnwater.	"
6	Bush Prairie.	"
8	Plum.	"
10	Shurlock.	"
12	Gillmore.	"
15	Tenino.	"

Puget Sound Shore Railroad.

Ms.		Alt.
0	Seattle.	Drift.
10	Black River Jc.	"
16	Kent.	"
20	Slaughter.	"
23	Stuck Jc.	"

Columbia & Puget Sound Railroad.

Ms.		Alt.
0	Seattle.	Drift.
10	Black River Jc.	"
18	Renton.	{ Upper Cretaceous. Lignite.
21	Coal Creek.	"
19	Cedar Mt.	?
23	Maple Valley.	?
31	Black Diamond.	{ Upper Cretaceous. Bituminous Coal.
34	Franklin.	"

Oregon Railway and Navigation Co.

Ms.		Alt.
230	Pendleton, Or.	See Note 13. 1070
241	Eastland.	" 1425
244	Adams.	" 1520
248	Athena.	"
252	Weston.	" 1855
258	Blue Mt.	"
267	Milton.	"
271	Spofford.	"
278	Walla Walla, W.	" 925
284	Valley Groove.	" 875

1. The large number of railroads constructed in the "North West" since the preparation of the chapter on the Northern Pacific, has necessitated the addition, out of the proper order, of some lines properly belonging in that chapter. Other new lines are also added.

2. By Mr. Warren Upham, Assistant Geologist U. S. Geological Survey.

3. *Kintyre.* See note 14, N. & S. Dakota.

4. See note 13, N. & S. Dakota.

5. The formations are older than the Cretaceous, including probably Jurassic or Triassic and Carboniferous.

6. The remainder of the chapter is by Mr. Bailey Willis, Assistant U. S. Geologist. The elevations, so far as given, are furnished by Mr. Henry Gannett, Chief Geographer, U. S. Survey. Much of the region traversed by these railroads has not been carefully surveyed, and the assignments of formations and the notes are necessarily of a general character. See note 39 Northern Pacific R. R.

7. Twenty miles west of Pasco, the road leaves the volcanic flows of the Great Plain of the Columbia and enters Yakima Prairie. Thence to ten miles beyond Ellenburg the route is through Ahtanam, Wenass, and Kittittass Prairies and through the cañons of the Yakima, which separate the valleys; the Prairies are Tertiary (?) lake beds, drained through the cañons which the river has cut in volcanic rocks, also Tertiary.
B. W.

8. Branch from Clealum to Rosyln coal mine. Coals of Puget group, (Upper Cretaceous.)
B. W.

9. The road runs across the main range of the Cascades, which consists of granite, Palæozoic crystallines and Cretaceous strata, folded and afterwards cut through and overflowed by Tertiary eruptives. The Cretaceous rocks are sandstone and shale, resting on a basal conglomerate. The volcanic rocks preponderate in this section, but give way to granite northward beyond Snoqualmie.
B. W.

10. The pass is 3,980: the tunnel 2,885 above tide.

Ms.	Oregon Railway and Navigation Co. Continued.	Alt.
287	Hadley, Wash.	See Note 13. 846
291	Berryman.	" 1011
294	Highland.	" 1181
298	Prescott.	" 1036
302	Bolles Jo.	" 1165
306	Menoken.	" 1298
314	Alto.	" 1907
320	Relief.	" 1096
325	Starbuck.	" 645
329	Grange City.	" 522
333	Ripasia.	" 530
346	Hay.	" 1100
353	Meeker.	" 1603
358	La Crosse Jc.	" 1478
361	Sutton.	" 1505
368	Winona Jc.	" 1492
374	Endicott.	" 1700
385	Diamonds.	" 2045
389	Mockonema.	" 2130
391	Crest.	" 2278
394	Colfax.	" 1981
400	Glenwood.	" 2075
406	Elberton.	" 2185
412	Garfield.	" 2470
421	Farmington.	" 2614
427	Seltice.	" 2525
432	Tekoa.	" 2490
439	Latah.	" 2442

Ms.	Oregon Railway and Navigation Co. Continued.	Alt.
448	Truax.	See Note 13.
455	Rockford.	" 2560
0	Bolles Jc.	" 2390
3	Waitsburg.	" 1165
6	Huntsville.	" 127.8
10	Long's.	" 1336
13	Dayton.	" 1472
0	Starbuck.	" 1606
7	Delaney.	" 645
14	Chard.	" 885
24	Zumwalt.	" 1154
29	Pomeroy.	" 1398
0	Connell.	" 1900
9	Sulphur.	" 889
18	Kahlotus.	" 757
29	Washtuona.	" 896
39	Hooper.	" 1012
48	Pampa.	" 1084
53	La Crosse Jc.	" 1850
0	Colfax.	" 1478
7	Riverside.	" 1974
9	Shawnee.	" 2178
12	Guy.	" 2194
18	Pullman.	" 2244
24	Garrison.	" 2345
28	Moscow.	" 2500
		2589

11. Drift Plain, with occasional outcrops of Tertiary eruptives and river cañons cut down into Upper Cretaceous (Puget Group) coal measures.   B. W.

12. This road is probably all on drift (glacial) with occasional outcrops of sandstones of Puget group, coal measures.   B. W.

13. The line lies chiefly through regions of volcanic flows, and the conditions were favorable for the formation of lake deposits during both Tertiary and Quarternary time. It is probable, though not known to be true, that the agricultural lands of this region are very largely dried lake beds. Specific information as to localities is not at present obtainable. The same statement is also applicable to the other line of the O. R. & N. Co., east of Umatilla.   B. W.

14. The following note is on the branch of the Northern Pacific to Carbonado. (See page 263). At South Prairie, Wilkeson, and Carbonado, bituminous coking coal is mined. This is the only producing field of coking coal on the coast; the Strata are Upper Cretaceous, "Puget Group." Similar trip south of Alaska.   B. W.

Wilkeson is the starting point for parties visiting the glaciers of Mt. Tacoma, distance 25 miles over a good horse trail; time required for trip, including ascent over snow fields to 9,500 feet above sea, in three days; the route is through the great forests of the region in their most typical development, and the glacial phenomena are of more striking interest and beauty than those afforded by any.

Some suggestions as to geology on the Oregon and Washington Railway, in Washington, may be gathered by the traveler from the foregoing notes. Nothing more definite can be obtained.   J. R. M.

The following altitudes, taken from Mr. Gannett's Dictionary of Altitudes, are of interest. Mt. Baker, 10,827 feet; Mt. Hood, 11,225; Mt. Jefferson, 15,500; Mt. Olympus, 8,138; Ranier, (Tacoma) 14,444; Mt. Skomegan, 8,400; Mt. Tchopshk, 7,200; Mt. St. Helena, 9,750.   J. R. M.

Missouri.[1]

GEOLOGICAL FORMATIONS OF MISSOURI.

20. Quaternary, Alluvium, Bluff or Loess, and Drift.
19. Tertiary, in Southeast Missouri.
18. Cretaceous, "
14. Coal Measures, 14 c. Upper.
"   "   14 b. Middle.
"   "   14 a. Lower.
13. L. Carboniferous or Sub-Carb., 13 e. Chestergroup.
"   "   13 d. St. Louis.
"   "   13 c. Keokuk.
"   "   13 b. Burlington.
"   "   13 a. Kinderhook or Chouteau.
10. Devonian, 10c. Black Slate (Genesee?)
5-7. Upper Silurian, 8 Oriskany.

5-7. Upper Silurian, 7. L. Helderberg.
"   "   5. Niagara.
2-4. Lower Silurian, 4. c. Hudson River.
"   "   4. b. Galena or Receptaculite l.s.
"   "   4. a. Trenton and Black River.
"   "   1st Magnesian. Saccharoidal s.s.
"   "   2d Magnesian l. s.
"   "   2d Sandstone.
"   "   3d Magnesian l. s.
"   "   Lower Magnesian l. s. and s. s.
"   "   2 b. Potsdam.
1 b. Huronian.
1 a. Laurentian.

(3 a. Calcifer's.)

Ms. Hannibal and St. Joseph Railroad.		Alt.	Ms. Hannibal and St. Joseph R.R.—Cont.		Alt.
0	Hannibal.	470	13 a. & b. Sub-Carb.		
6	Bear Creek.	669	" & 20. Quat.		
10	Barkley.	667	" Lime made.		
15	Palmyra Jc.	649	"		
19	Woodland.	679	"		
30	Monroe.	704	14 a. Coal Mres.		
42	Lakenan.	729	"		
53	Lentner.	790	"		
59	Clarence.	624	20. overlies 13 c.		
70	Macon.	667	14 b. Coal Mres.		
79	Callao.	612	" 4 ft. coal.		
90	Lingo.	609	" "		
104	Brookfield.	767	"		
109	Laclede.	767	"		
121	Wheeling.	740	14 b. Mid. Coal Mres.		
130	Chillicothe.	764	"		
140	Mooresville.	921	14 c. Up. Coal Mres.		
150	Nettleton.	968	"		
156	Hamilton.	967	"		
163	Kidder.	1017	"		
172	Cameron.	1026	"		
177	Osborn.	1044	"		
185	Stewartsv'le.	968	"		
200	Saxton.	661	"		
206	St. Joseph.	663	{ " and hills covered with Bluff clay.		

Ms. Hannibal and St. Joseph R.R.—Cont.		Alt.
0	Quincy.	13 a. Sub-Carb.
9	North River.	13 b. " 470
15	Palmyra.	" 604
206	St. Joseph. 663	14 c. Up. Coal Mres.
211	Lake. 629	20. Alluvial
217	Halls. 604	"
222	Rushville. 768	" & 14 c. U.C.M.
226	Winthrop. 601	"
172	Cameron.	14 c. Up. Cl. Mrs. 1026
187	Lathrop.	" 948
201	Kearney.	" 655
211	Liberty.	" 646
218	Arnold.	" 739
226	Kansas City.	" & 20 746

Wabash, St. Louis and Pacific R. R.[2]

0	St. Louis.	669	13 d. St. Louis group.
6	Bartmer.		14 b. Mid. Coal Mrs.
14	Graham's.		[by 20.
22	St. Charles.	604	13 d. St.Lo. group, cov'd
30	Dardenne.		20. Quaternary.
38	Perruque.		13 c. and d.
48	Foristell.		13 a. & b. rests on 10 c.
58	Warrenton.	663	" on 4 a. & 4 b.
68	Jonesburg.	606	13 a. and 4 a. Trenton.
77	New Florence.		13 a.

1. By Professor G. C. Broadhead, late State Geologist of Missouri.
2. On W., St. L. & P. R. R., in Warren and Montgomery Counties, we pass within a few miles from Carboniferous, chiefly Lower part of Sub-Carboniferous through thin outliers of Devonian to the Receptaculite (Galena Limestone) and Trenton and Black River to the 1st Magnesian limestone and Saccharoidal sandstone; the latter well developed and very suitable for glass-making purposes—thick deposits and easy to crush. It is the equivalent of the St. Peter's sandstone.

Wabash, St. Louis and Pacific Railroad.

Ms.	Continued.	Alt.
0	Wellsville.	14 a. Lower Coal Mrs.
103	Benton City.	"
108	Mexico.	" 825
114	Thompson.	"
122	Centralia.	" 873
130	Sturgeon.	" 847
140	Renick.	" 4 ft. coal.
146	Moberly.	" 882
153	Huntsville. 771	" 4 ft. coal.
160	Clifton.	" 722
167	Salisbury.	. " 721
178	Dalton.	" 637
185	Brunswick.	" 631
192	Dewitt. 644	" [quarry.
195	Miami.	" white s. s.
202	Wakenda.	20. Quaternary.
209	Carrollton.687	14. b. Mid. Coal Mrs.
219	Norborne.	20. Quaternary.
228	Hardin.	"
234	Lexington Junc.	14 b. Coal, middle ser.
239	Camden.724	" 2 ft. coal.
245	Orrick.	20. Quaternary.
254	MissouriCity.722	14 c. base of U. Cl. Ms.
265	N. Missouri Junc.747	"
273	Harlem.	20. Quaternary. 746
275	Kansas City.8	{ 14 c. Up. Cl. Mrs.748 / Good Mollusca of / Up. Carb.

St. Louis and Des Moines.

Ms.		Alt.
146	Moberly.	14 a. Lower Cl. Ms.882
153	Cairo.	" 880
162	Emerson.	" 866
169	Macon.	" 900
180	Atlanta.	" 906
189	LaPlata.	" 940
196	Millard.	" 970
203	Kirksville.	14 a. & b. " 975
211	Sublett's.	"
218	Queen City.	14 a. " 1004
227	Glenwood.	" 990
234	Coatesville.	"

(Continued in Iowa.)

St. Joseph Division.

Ms.		Alt.
0	Lexington Junc.	14 b. Mid. Coal Mres.
9	Swanwick.	14 c. Base of up. Coal.
19	Vibbard	14 c. Up. Coal Mres.
25	Lawson.	"
36	Lathrop.	" 948
44	Plattsburg.	" 948
53	Gower.	" 935
62	Agency Ford.	. "
73	St. Joseph.	" 827

Columbia Branch.

Ms.		Alt.
0	Centralia.879	14 a. Lower Coal Mrs.
22	Columbia.	14 a. and 13 b. & c.

Wabash, St. Louis and Pacific R. R.—*Cont.*

Ms.	Glasgow Branch.	Alt.
0	Salisbury. 721	14 a. Lower Coal Mrs.
15	Glasgow. 680	" base.

St. Louis and Omaha Line.

Ms.		Alt.
	St. Louis.	
0	Brunswick. 644	14 a. Lower Coal Mrs.
38	Chillicothe.	14 b. Mid. Cl. Mrs. 764
64	Gallatin.	"
80	Pattonsb'gh. 772	14 c. Up. Coal Mres.
107	Stanbury.	" 876
131	Marysville.	" 1087
143	Roseberry.	" 977
	Burlington Junc.	"
223	Council Bluffs,Ia.	" 989

Quincy, Missouri and Pacific Railroad.

Ms.		Alt.
2	West Quincy.	20. Quaternary.
11	Maywood.	13 a. Sub-Carb. 524
22	Tolona.	" 697
32	La Belle.	" 741
47	Edina.	13 d. Overlaid by drift
54	Hurdland.	Deep drift. [738
70	Kirksville.	14 a. Lower Cl.Mrs.975
	Cooksville.	14 b. 995
	Milan.	14 b. & 14 c. 840
137	Trenton.	"

Missouri, Iowa and Nebraska Railroad.

Ms.		Alt.	
0	Alexandria.	20. Alluvium. 466	
7	Wayland.	13 d. St. Louis l. s. 551	
15	Kahoka.	14 a. Coal Mres.	
24	Luray. 737	"	Deep drift deposits overlie formations.
32	Arbela. 655	"	
40	Memphis. 787	"	
51	Downing. 869	"	
61	Lancaster. 972	"	
64	Glenwood. 990	"	
70	Hamilton. 987	"	

Missouri Pacific Railroad.4

Ms.		Alt.	
0	St. Louis.5 431	{ 13 d. St. Louis l. s. & / 14 a. Coal Measures.	
7	Benton. 470	13 d. St. Louis l. s.	
13	Kirkwood. 628	"	
34	Carondelet.	13 d. & 13 c. Keok.	
19	Meramec. 420	13 b. Sub-Carbonifer's.	
26	Glencoe.	4 a. Trenton.	
30	Eureka.	"	
37	Pacific. 456	3 a. Calcif. & 4 a. Tren.	
41	Gray's Sum't.630	" 1st sandstone.	
52	South Point. 510	" 2d Magn. l. s.	
54	Washington. 487	" "	
67	Miller's L'd'g.505	" "	cap. with s. s.
75	Berger. 515	" "	
81	Hermann. 511	" "	
88	Gasconade. 468	" "	
92	Morrison. 522	"	

Ms.	Missouri Pacific Railroad—Cont.	Alt.
100	Chamois.	**521**
105	St. Aubert.	**527**
125	JeffersonCity.	**524**
140	Centretown.	**556** lead " 2d sandstone.
150	California.	**556** " 2d Magnes'n.
		" On hills some-
162	Tipton.	**911** lead " times find 13 b.
175	Otterville.	**919** " Bur'n l.s. & 3 a.
188	Sedalia.	**557** 13 a.& b.Burlington l.s.
195	Dresden.	{ " Potter clay { & 13 a. & 14 a.
200	Lamonte.	**546** 14 a. Lower Coal Mres.
208	Knobnoster.	" iron ore & coal Ms.
218	Warrensburg.	**597** " fine s. s. quarries.
230	Holden.	**750** 14 b. Coal Mres.
237	Kingsville.	**594** 14 b. & c. U. Coal Mres.
248	Pleasant Hill.	" **526**
259	Lee's Summit.	" **1026**
272	Independence.	" **995**
282	Kansas City.	" **751**

Lexington Branch.

0	Sedalia.	**559** 13 a. Sub-Carbonifer's.
4	Georgetown.	13 a., b. & c. "
22	Sweet Spgs.	**547** 13 b. Upper Sub-Carb.
38	Aullville.	**709** 14 b. Coal Mres.
55	Lexington.	**755** 2 ft. coal. " coal mines
63	Wellington.	14 b. "
75	Buckner.	
87	Independence.	14 c. Up. Coal Mrs. **995**
97	Kansas City.	" **748**

Versailles and Boonville Branches.

0	Versailles.	{ 3 a. 3d. Magn. l.s. **911** { lead ms. near, beau- { tiful cave 12 mi. so.
19	Tipton.	13 b. Sub-Carb. on 3 a.
33	Palestine.	13 a. Sub-Carb.
44	Boonville.	13 c. " **507**

Lebanon Branch.

0	JeffersonCity.	**418** 3 a. Calcif. 2d Magn. ls.
11	Moreau.	"
19	Russelville.	" **750**
28	Olean.	" Lead mines near
33	Eldon.	"
37	Aurora Sp's.	**1257** 8 a. Calcf. 3d Magn. l.s.
40	Cooper.	"
45	Bagnell.	" Osage River.

Lexington and Southern Branch.

0	Pleasant Hill.	14 c. U. Cl. Mres. **526**
10	Harrisonville.	"
23	Archie.	{ 14 c. Upper & 14 b. { Mid. Coal Mres.
29	Adrian.	14 b. Mid. Coal Mres.
38	Butler.	" **514**
50	Rich Hill.	**756** { 14 a. L. C. Mrs., coal { mines, beds 3 to 5ft.

Ms.	Missouri Pacific Railroad.	Alt.

Lexington and Southern Branch—*Continued*.

54	Bedford.	14 a. Lower Coal Mres.
56	Arthur.	" **710**
69	Nevada.	' **870**
82	Sheldon.	.'
93	Lamar.	" coal and s. s.
99	Carleton.	"
105	Jasper.	13 c. Keokuk.
110	Cary.	"
116	Carthage.	" Lime quar. **1249**
119	Edwin.	" Zinc and lead.
126	Webb City.	" "
133	Joplin.	" " **1010**

Warsaw Section.

0	Sedalia.	{ 13 a. Kinderhook **907** { 13 b. Burlington.
20	Cole Camp.	3 a. Calcif., lead mines.
42	Warsaw.	" on Osage River.

Creve Cœur Lake Branch.

0	Laclede.	13 d. St. Louis. **756**
12	Creve Cœur.	Lower Carb.

St. Louis, Iron Mountain and Southern Division.[6]

0	St. Louis.	13 d. St. Louis l. s. **411**
10	Jefferson Bar'ks.	13 d. Warsaw l. s. **419**
13	Cliff Cave.	13 c. Keokuk l. s.
21	Kimmswick.	**415** 13 b. Burl. l. s., lime.
24	Sulphur Springs.	" **411**
26	Pevely.	4 a. Trenton. **441**
29	Horine.[7]	{ 3 a.Calc., Sandy lead { mine 6 miles north.
35	Hematite.	3 a. Calciferous. **475**
39	Victoria.	"
43	De Soto.	**497** { " Valle lead ms. { 10 miles so., Frumet { lead ms. 10 miles no. { Good building stone.
51	Blackwell.	3 a. Calciferous. **492**
57	Cadet.	" lead mine. **505**
61	Mineral Pt.	**565** " many lead ms.
65	Potosi	" "
66	Hopewell.	" " **985**
70	Irondale.	**756** "
75	Bismarck.	**1024**
83	Loughborough.	2 b. Potsd. & 1 b. Hur.
87	De Lassus.	**559** " [quarry.
95	Knob Lick.	**926** " & granite
102	Mine La Motte.	**547** { " lead, nickel, { cobalt, manganese, { copper, iron and { porphyry.
105	Frederickt'n.	**721** 2 b. Potsd. & 1 b. Hur.
112	Cornwall.	{ 2 b., 1 b. & 3 a. Calc. { Iron and granite.
118	Marquand.	3 a. Calcif's, iron. **570**
125	Bessville.	**551** "
134	Lutesville.	" Lime. **552**

3. Loess is well developed at Kansas City.

Ms.	Missouri Pacific Railroad.	Alt.
	St. Louis, Iron Mount. and South. Div.—*Cont.*	
148	Allenville.	3 a. Calcif's, iron. ⁸³⁹
164	Jackson.	4 a. Trenton & Black riv
158	Sylvania.	3 a. Calciferous.
162	Morley. ³⁴⁵	{ 20. Quaternary, with probably 19. Tert'ry.
174	Dieblstadt.	" ⁸²¹
178	Charleston.	" ⁸²⁶
195	Belmont.	" ⁸¹⁸

Arkansas Division.

76	Bismarck.	3 a. Calciferous. ¹⁰²⁴
81	Iron Mountain.⁸	{ 2 b. Pots. & 1 b. Hur. Specular iron ore in vast quantities.¹⁰⁷⁷
86	Pilot Knob.⁹	" ⁸⁵⁸
88	Ironton.¹⁰ ⁹¹⁹	2 b. Potsd. & 1 b. Hur.
89	Arcadia.	"
96	Hogan. ⁸⁹²	"
104	Ozark. ⁸³⁵	"
108	Annapolis.	"
116	Des Arc. ⁸⁴⁷	{ granite quarry."
127	Piedmont. ⁵⁰⁸	"
134	Mill Spring. ⁴⁴³	3 a. Calciferous.
145	Williamsville⁴⁰¹	"
148	Blums. ⁸⁴⁸	"
166	Poplar Bluff.	" & 20. Quat.
181	Neelyville. ⁸⁰⁶	20. Quat. Swamp.
201	Domphau.	3 a. Calciferous.
186	Moark. ²⁸⁷	20. Quaternary.

Cairo Branch.

0	Cairo. ⁸⁶⁰	{ Low lands. 20. Quat. and probably 19. Tertiary.
10	Hough's.	"
15	Charleston.	" ⁸²⁶
28	Sikeston.	" ⁸³⁰
74	Poplar Bluff.	" ⁸⁴³

St. Joseph and Desloge Railroad.

0	Summit.	{ 3 a. Calciferous and probably 2 b. Potsd.
13	Bonne Terre.	{ 2 b. Pots. with mines of lead with copper, nickel, cobalt and purple calcite.

Ms.	Missouri Pacific Railroad—*Continued.*	Alt.
	Missouri, Kansas and Texas Division.	
0	Hannibal. ⁴⁶⁹	13 a. & b. Sub-Carb's.
12	Rensalier. ⁷⁸⁸	"
22	Monroe. ⁷²⁸	14 a. Lower Coal Mres.
34	Stoutsville. ¹¹⁶⁶	13 b. Sub-Carbonifer's.
44	Paris. ⁶⁵¹	13 c. "
57	Madison. ⁷⁷²	13 c. & d. & 14 a.
70	Moberly. ⁸⁶⁵	"
80	Higbee. ⁸⁷⁷	" 4 ft. coal.
88	Burton. ⁶⁷²	14 a. Coal Mres.
95	Fayette. ⁶⁵⁷	"
99	Talbott. ⁶²⁰	"
108	Boonville. ⁶⁰⁷	"& 13 c. U. S.-C.
122	Harris. ⁸⁵³	13 b. Upper Sub-Carb.
131	Clifton. ⁷²²	13 a. Sub-Carbonifer's.
143	Sedalia. ⁹⁰⁷	"
155	Green Ridge. ⁹⁰³	13 b. Upper Sub-Carb.
164	Windsor. ⁸⁷⁵	14 a. Coal Mrs. 4 ft. cl.
172	Calhoun. ⁷⁷⁴	potter " clay & iron ore
183	Clinton. ⁸⁰⁷	{ " coal mines, fossil ferns, &c.
196	Montrose. ⁸²⁴	" "
202	Appleton C'y. ⁸⁶⁸	" " 4 ft. cl.
215	Schell City. ⁷⁵⁴	"
226	Walker. ⁸⁵⁶	"
233	Nevada. ⁸⁷⁰	"

Kansas and Arizona Division.

0	Holden.	14 b. Mid. Coal Mres.
8	Benton. ⁴⁷⁰	"
16	East Lynn.	14 b. Coal Mres.
22	Harrisonville.⁸¹²	14 c. Upper Coal Mres.

Chicago, Rock Island and Pacific R. R.

South-Western Division.

0	Atchison.	14 c. Upper Coal Mres.
30	Atchison Junc.	"
0	Leavenworth.	"
5	Beverly.⁷⁶⁹	"
11	Platte City.	"
21	Atchison Junc.	"
29	Grayson.	"
36	Plattsburg.⁹⁴⁸	"
47	Perrin.	"
55	Cameron.¹⁰³⁸	"
76	Gallatin.	{ 14 c. Up. Coal Mres. base of. Mollusca.

4. On Missouri Pacific R. R., from St. Louis west, we pass St. Louis group, Lower Coal Measures, St. Louis group Warsaw limestone, Burlington and Chouteau group to the Trenton, but no Devonian. At Hermann we have 2d Magnesian limestone capped in hills back with 1st or Saccharoidal sandstone, and at Jefferson we have 2d Magnesian limestone rising in a few miles south exposing in succession 2d sandstone and 3d Magnesian limestone. West of Tipton the same limestone (2d) is capped by Burlington limestone. The latter west of Sedalia having reposing on it the sandstone at top of Sub-Carboniferous (Millstone Grit?) and underlaid by Chouteau group. Then the Coal Measures appear.
    5. At Cheltenham, four miles from St. Louis, are vast deposits of good fire clay.

Chicago, Rock Island and Pacific R. R. Ms. South-Western Division—*Continued.*	Alt.		St. Louis and San Francisco, formerly Atlantic and Pacific, Railroad.[11] Ms.		Alt.
86 Jamesport.		14 c. Upper Coal Mres.	0 St. Louis. 451		20. & 13 d. St. L. l. s.
102 Trenton.		"	37 Pacific. 658		4 a. Tren. & 3 a. Calcif.
127 Princeton.		"	44 Calvey.		3 a. Calciferous.
143 Lineville.		" Middle	49 Moselle. 923		" Iron.
156 Allerton.		" series in	56 St. Clair. 759		"
169 Seymour.		" valleys.	66 Stanton. 887	Copper.	"
			78 Bourbon. 941		"
Chicago and Alton Railroad.			91 Cuba. 1010		"
Chicago, Kansas City and Denver Line.			104 St. James. 1117		" iron.
275 Louisiana. 460		13 a. & b. & 10 c. & 4 c.	114 Rolla. 1201		" iron.
282 Watson. 904		" Hud. Riv.	124 Ozark.	3d Magnes'n limestone capped with 2d sandstone.	"
286 Bowling Green.	881	{ good building stone.	138 Dixon. 1146		"
293 Curryville.		13 c. Sub-Carbonif's.	144 Hancock. 1109		" iron.
302 Vandalia.		"	150 Crocker. 1132		"
311 Laddonia.		14 a. Low. Coal Mres.	163 Richland. 1143		"
320 Littleby.		"	171 Stoutland. 1166		"
325 Mexico. 798		14 a. Low. Cl. Mrs.	178 Sleeper. 1209		"
339 Centralia. 879		"	185 Lebanon. 1269		"
361 Higbee. 877		" coal mines	217 Marshfield. 1498		{ " Highest pt. in Mo. Good bldg. s.
381 Glasgow.	680	{ " 13 c. and 13 c. Keokuk.	241 Springfield. 1360		13 b. Sub-Carbonifer's.
393 Slater.		14 a. Low. Coal Mres.	266 Logan's.		"
404 Marshall.	678	{ 13 c. Keokuk and 13 e. Chester.	278 Verona. 1282		" and c.
415 Mt. Leonard.		{ 14 a. Low. Coal Mrs. salt springs near.	291 Peirce City. 1235		Lime and 13 c. Sub-C.
434 Higginsville. 647		14 a. Low. Cl. Mrs.	306 Granby C'y. 1080		{ 13 c. Keokuk l. s. (Lead abounds.)
448 Odessa.		14 b. Mid. Coal Mres.	314 Neosho.		13 c. Keokuk l. s 1018
459 Oak Grove.		"	325 Dayton.		" 947
478 Independence. 995		14 c. Up. Cl. Mres.	330 Seneca.		Polishing " stone. 851
489 Kansas City. 746		"	(State Line.)		(See Kansas.) 846
South Branch.			**Arkansas Division.**		
0 Chicago.			0 Peirce City.		{ 13 c. Keo. group.1176 good lime qrs.
325 Mexico. 798		14 b. Mid. Cl. Mrs.	4 Plymouth.		" 1326
345 Callaway.		"	29 Washburn.		"
350 Fulton. 843		14 a., 13 b. & 10 c.	35 Seligman.		" 1525
357 Carrington.		"	**White River Branch.**		
364 New Bloomfield.		"	0 Springfield.		13 c. Keok. group. 1353
370 Hibernia. 880		10 c. and 3 a.	20 Ozark.		{ 13 a. Kinderhook, & 13 b. Burlington.
876 Jefferson City. 414		3 a. Calciferous.	85 Chadwick.		13 a. Kinderhook.

(Along the San Francisco R.R. middle columns: "Occasional lead & iron mines.")

6. Down the St. Louis & Iron Mountain R. R. we have St. Louis limestone then Warsaw limestone, Keokuk limestone, and Burlington limestone within 20 miles. Crossing the Merrimac River, we find the last for a while, then the Receptaculite, Trenton and Black River limestone, 1st Magnesian limestone, and at Horine Station the Saccharoidal sandstone, very soft, used for glass-making, and is very white and pure. Afterwards we have 2d Magnesian limestone. Crossing Big River, the 3d Magnesian limestone near Iron Mountain. De Lassus, Mine la Motte, Fredericktown, Pilot Knob, Des Arc and Annapolis are porphyry hills of Huronian age, and the adjacent limestones and lower sandstones and conglomerates are probably Potsdam. At Mine la Motte and Fredericktown are certainly Potsdam fossils, but the absolute line (if any) has not been determined between the Potsdam and Calciferous beds. Near Iron Mountain, Knob Lick and Cornwall are superior granite quarries, which may be of age of Laurentian.

7. Four miles southeast is Crystal City on the Mississippi River, where glass is made. The Saccharoidal or St. Peter's sandstone is here forty or fifty feet thick, and over one hundred feet thick in Warren County. It is very valuable for glass-making.

8. Iron Mountain is 228 feet high, and its base covers 500 acres.

9. Pilot Knob is a conical hill, nearly circular, 581 feet high, with a north and south diameter of about one mile at its base, which covers 360 acres. Elevation 1,500 feet above sea.

10. Sheppard Mountain magnetic iron ore.

Ms. St. Louis & San Francisco R. R.—Con. Alt.

Ms.	Station	Geology	Alt.
0	Springfield.	13 c. Keokuk.	1360
21	Buckley.	"	
24	Graydon.	L. Carb. probably 13 b.	
39	Bolivar.	"	

Joplin Branch.

Ms.	Station	Geology	Alt.
0	Oronogo.	13 c. Keokuk mines.	
4	Webb City.	" Handsome crystals of Blende, Calcite & Galena Zinc mines.	
10	Joplin.	13 c. Rich in lead & zinc	1018
20	Galena.	"	

Kansas Division.

Ms.	Station	Geology	Alt.
0	Peirce City.	13 c. Keok. lime.	1225
27	Carthage.	" Lime kilns.	
36	Oronogo.	" Zinc & lead.	
44	Smithfield.	"	
	(Continued in Kansas.)		

Girard Branch.

Ms.	Station	Geology	Alt.
	Opolis.	13 c. Keok.	
20	Joplin.	" Lead & zinc.	1018

Kansas City, St. Joseph and Council Bluffs Railroad.

Ms.	Station	Geology	Alt.
0	Kansas City.	{ 14 Upper Carbon. Good fossil mollusca	748
10	Parkville.	14 c. Upper Carbon.	758
17	Waldron.	"	757
25	E. Leavenworth.	"	764
34	Weston.	"	778
54	Winthrop.	"	801
55	Rushville.	"	798
66	Lake Station.	20. Quaternary.	826
70	St.Joseph.	14 c. Upper Carbon.	824
80	Amazonia.	" fusulina abounds.	
99	Forest City.	" " & mollusca.	
109	Bigelow.	20. Quaternary.	861
116	Craig.	" over 14 c.	871
122	Corning.	"	876
135	Phelps.	"	695
149	Hamburg.	" & 14 c. U. C.	
200	Council Bluffs.	"	959
	(Continued in Iowa.)		

Hopkins Branch.

Ms.	Station	Geology	Alt.
70	St. Joseph.	14 c. Up. Carbon.	824
79	Amazonia.	" Fusulina.	833
85	Savannah.	Good " fossil molusca	1100
91	Rosendale.	"	795
101	Barnard.	"	943
108	Bridgewater.	"	
115	Maryville.	"	1087
123	Pickering.	"	1023
131	Hopkins.	"	1046

Kansas City, St. Jos. & Council Bluffs R. R.

Ms.	Nodaway Valley Branch.		Alt.
0	Mound City.	Quaternary.	861
11	Maitland.	14 c. Up. Coal. Mres.	
17	Skidmore.	"	
23	Quitman.	"	
29	Burlington Junc.	{ " Coal and highest Upper Carbonif's rocks in Mo.	526

Tarkio Valley Branch.

Ms.	Station	Geology	Alt.
0	Corning.	Quaternary.	876
	Fairfax.	" on 14 c. U. C. M.	
	Tarkio.	" "	
28	Northborough.	" "	

Chicago, Burlington & Kansas City R. R.
Burlington & South-Western R. R.

Ms.	Station	Geology	Alt.
0	Laclede.	14 b. Mid. Coal Ms.	787
7	Linneus.	Iron. " Clays.	425
20	Browning.	"	780
32	Milan.	14 c. Upper Carb.	840
37	Boynton.	14 b. Mid. Coal Ms.	879
45	Pollock.	"	943
53	Unionville.	14 a. Low. Cl. Ms.	1068
181	Burlington.		505
	(Continued in Iowa.)		

St. Louis, Keokuk & North-Western R. R.

Ms.	Station	Geology	Alt.
0	Keokuk.	13 c. Keokuk l. s.	
5	Alexandria.	"	465
22	Canton.	"	
28	La Grange.	20. Quaternary.	
40	Quincy.	13 b. & c. Keok. ls.	484
53	Helton.	"	
59	Hannibal.	13 b. Sub-Carb.	469
65	Saverton.	13 a. & b. " & 4 c. Cinn.	
74	Ashburn.	4 c. Hudson River.	
84	Louisiana.	{ 4 c., 10 c. and 13 a. & b. Sulphur Sp'gs.	460
94	Clarksville.	{ 13 a. Kinderhook. 13 b. Burlington & 10 Devonian.	
100	Kissenger.	13 a. and 13 b.	
110	Elsberry.	{ 10 Dev'n, 4 a. Tren. and 4 b. Galena.	
	Winfield.	13 d. St. L. Fault near.	
	Monroe.	13 c. Keokuk.	728
138	St. Peters.	20. Quaternary.	

St. Louis, Salem & Little Rock Railroad.

Ms.	Station	Geology	Alt.
0	Cuba.	3 a. Calcif.	1010
9	Steelville.	"	
24	Cook's.	"	
40	Salem.	"	1162
46	Orchard Bank.	"	

(Lead & iron)

11. On St. Louis & San Francisco R. R., going southwest, after leaving Pacific (or Franklin) the 2d Magnesian limestone gradually rises, showing some 2d sandstone, and through Crawford, Phelps, and Pulaski counties the latter is the highest rock, resting on 3d Magnesian limestone, the latter well exposed along the Gasconade River. Crossing it, we are upon the highest lands in Missouri. Descending towards Springfield, we find the Lower members of the Sub-Carboniferous

Kansas City, Fort Scott & Gulf Railroad.
Ms. Kansas City, Sp'gfield & Memphis Line. Alt.

0	Fort Scott, Kan.	{ 14 b. Mid. Coal Mrs. / Coal near.
15	Arcadia.	{ 14 a. Low. Coal Mrs. / Coal mines.
38	Lamar.	" coal and sandst.
50	Golden City.	13 c. Keokuk.
65	Greenfield.	" lead near.
83	Ash Grove.	" lead and lime.
101	Springfield.	" 1552
136	Seymour.	" 1650
		{ Highest land in Mo.
143	Cedar Gap.	3 a. Calciferous. 1700
193	Willow Springs.	" 1270
214	West Plains.	" 950
	Augusta.	" 3d Magn. l. 780
242	Mammoth S'pg.	"
	Spring City.	" Big spring.

Pleasant Hill & De Soto R. R.

0	Pleasant Hill.	14 c. Upper Coal Mrs.
12	Raymore.	"
17	Belton.	"
25	Stanley.	(See Kansas.)
	(Continued in Arkansas.)	

Rich Hill Branch.

0	Miami.	14 a. Lower Coal Mres.
13	Rich Hill.	" coal mines.784
19	Carbon Centre.	" " 772

St. Louis & Emporia Railway.

0	Blue Mound.	14 a. Lower Coal Mres.
20	Pleasonton.	" & 14 b. Mid. Cl. "

Kansas City, Clinton & Springfield R. R.

0	Kansas City.	14 c. Upper Coal Mrs.
21	Olathe, Kan.	" 1030
38	Belton, Mo.	"
43	Raymore.	"
56	Harrisonville	"
62	Dougherty.	14 b. Middle Coal Mrs.
95	Clinton.	14 a. Up. Coal Mrs.807
119	Osceola.	3 a. Calc. & 13 a. & 13 b.
139	Humansville.	13 b. Burlington.
175	Ashgrove.	13 c. Keokuk.

Kansas City and Southern.

0	Osceola.	
13	Otter Creek.	14 a. & 13 b.
16	Browning	14 a. Lower Coal Mrs.
17	Grand River.	13 b. Burlington.
21	Vickers.	14 a. Lower Coal Mrs.
26	Clinton.	807 14 a. Good fossil plants
	Urich.	14 a. & 14 b.
	Index.	14 b. Mid. Coal Mres.
67	East Lynne.	"

Ms. Cape Girardeau Southwestern R. R. Alt

0	Cape Girardeau.	{ 4 a. Trenton. 888 / and 4 b. Galena.
15	Delta.	{ 20. Quaternary with / heavy timber
	Lakeville.	" " 881
40	Idlewild.	" "
52	Wappapello.	" "

St. Louis, Hannibal & Keokuk Railroad.

	St. Louis.	460
0	Gilmore Springs.	13 c. Keok. & L. Carb.
13	Moscow Mills.	" Archimedes fos.
18	Troy.	13 c. Keokuk.
30	Silex.	13 a. and 13 b.
45	Edgewood.	"
53	Bowling Green.	13 b. & Up. Silurian.
60	McCunes.	4 a. Trenton group.
67	Frankfort.	"
	Jones.	{ 3 a. 1st Magnes. l. s. / & Saccharoidal s. s.
76	New London.	4 a. Tren. & Black Riv.
86	Hannibal.	440 13 a. & b. good lime qrs.

Chicago, Burlington & Quincy Railroad.
Des Moines Charlton & St. Joseph Branch.

0	St. Joseph.	14 c. Up. Coal Ms. 792
49	Albany.	"
65	Bethany.	"
90	Andover.	"
93	Bethany Jc., Ia.	
	Grant City.	14 c. Upper Coal Mres.
	Clarinda Jc., Ia.	"
	Burlington Jc.	"

Quincy Hannibal & Louisiana Branch.

0	Quincy.	13 b. & c. Keok. l. s.444
7	Marble Head.	20. Quaternary.
13	Fall Creek.	"
19	Hannibal.	13 b. Sub-Carb. 449
23	Kinderhook, Ill.	{ 10 c. bl. sl. 13 a. Kin- / derh. & 13 b. Burl.
44	Louisiana.	4 c. 10 c. & 13 a. & b.460

Texas & St. Louis Railroad.
Missouri & Arkansas Division.

0	Birds Point.	20. Quat., Swamp dist.
37	Paw Paw Junc.	" } Low,
43	New Madrid.	" } swampy,
58	Malden.	" } Heavy 397
70	St. Francis, Ark.	" } timber.388

St. Louis, Creve Cœur & St. Charles R. R.

0	St. Louis.	13 d. St. Louis.
5	Rinkleville.	14 a. Lower Coal Mres.
16	Florrisant.	20 on 14 a. Rich Valley

limestone resting on the 2d Magnesian limestone or Calciferous. In southern parts of Lawrence County we find a coarse ferruginous sandstone, probably equivalent to Millstone Grit, but more probably a member of the Chester group, resting on Lower Carboniferous limestone. Throughout Newton and Jasper, the Sub-Carboniferous limestone, with much chert is of great development, and is galeniferous. The celebrated lead mines of Joplin and Granby occur in this.

Kansas.[1]

LIST OF GEOLOGICAL FORMATIONS IN KANSAS.

20. Quarternary.	20 d. Alluvium. / 20 c. Loess. / 20 b. Modified Drift. / 20 a. Glacial Drift.	*Mesozoic. 16-18* — 18 Cretaceous.	18 c. Niobrara, Including the "Colorado" above. / 18 b. Ft. Benton. / 18 a. Dakota.
		16-17 Jura-Trias, or Red Beds.	
19. Tertiary.	19 c. Pliocene, including deposits of Volcanic ash—possibly of Quarternary age. / 19 c. Miocene.	*Carbonifer's. 16-18* — Upper Carboniferous.	15. Permian or Permo-Carboniferous. / 14 c. Upp. Cl. Meas. / 14 b. Low.Cl. Meas.
		Lower Carboniferous.	13c. Keokuk, limest. & chert, bearing of Lead and Zinc.

Ms.	Union Pacific Railway. Kansas Division.		Alt.	Ms.	Union Pacific Railway. Kansas Division.		Alt.
0	Kansas City. (Union Depot.)	14 c. Upper Coal Measures.	748		Menoken.	14c.Upp.CoalMres.	902
1	Kansas City, Kansas.	"	748	78	Silver Lake.	"	915
2	Armstrong.	"	755		Kingsville.	"	920
9	Muncie.	"	767	83	Rossville.	"	933
13	Edwardsville.	"	783	91	St. Marys.	"	955
17	Bonner Springs.	"	789	97	Bellvue.	"	965
	Loring.	"	789	104	Wamego.	"	1000
23	Lenape.	"	781	111	St. George.	"	1000
28	Linwood.	"	789	119	Manhattan.[7]	"	1000
32	Fall Leaf.	"	809		Eureka Lake.	15. Permo-Carbonif.	
39	Lawrence.	"	822	130	Odgensburg.	"	1060
45	Buck Creek.	"	846	135	Ft. Riley.	"	1070
48	Williamstown.	"	851	139	Junction City.[8]	"	1082
51	Perryville.	"	852	146	Kansas Falls.	"	1106
53	Medina.	"	858	152	Chapman.	"	1114
55	Newman.	"	861	158	Detroit.	"	1135
61	Grantville.	"	877	163	Abilene.	"	1155
67	Topeka.[3]	"	880	172	Solomon.[9]	"& 18 a. Dak.	1175
				180	New Cambria.	"	1189
				186	Salina.	"	1225

1. By Mr. Orestes St. John of Topeka, Kansas.
2. *Leavenworth.* In the vicinity of Leavenworth and at the State Penitentiary at Lansing, a 21-inch seam of coal is mined by means of shafts at a depth of between 700 and 800 feet. The limestones crossing the bluffs that hem the Missouri are richly stored with characteristic upper coal measure fossils. The Loess heavily covers the bluffs, and in the bed of the Missouri Valley the glacial drift occurs beneath the alluvial deposits. Deposits of modified drift or stratified gravels locally intervene between the Loess and the basis rocks of the region.
3. *Topeka.* The Osage coal crops in the western suburbs of the city, where it is mined to limited extent. An experimental diamond drill boring, authorized by the local government, has penetrated the coal measure series to the depth of between 1,600 and 1,700 feet at this writing, encountering several thin deposits of coal.

Union Pacific Railway. Kansas Division.—Con.

Ms.	Station	Geology	Alt.
194	Bavaria.[10]	18 a. Dakota.	1271
201	Brookville.	"	1348
	Arcola.	"	1423
	Terra Cotta.	"	1470
211	Carneiro.[4]	"	1570
	Mt Zion.		
218	Kanopolis.		1550
223	Ellsworth.	18 b. Benton.	1538
	Black Wolf.	"	1585
	Cow Creek.	"	
239	Wilson.	"	1654
	Dorrance.	"	1730
253	Bunker Hill.	"	1864
	Homer.	"	1874
263	Russell.	"	1882
	Gorham.	"	1912
	Walker.	"	1944
279	Victoria.	"	1928
	Toulon.	"	
289	Hays.	"Up. l. s.1991	
	Hogback.	"	
303	Ellis.	"	2117
313	Ogallah.	18b.Niob.&19.T'r2367	
321	Wakeeney.[5]	" " '2456	
	Colono.	19. Tert'ry in uplands.	
335	Collyer.	" 2586	
	Quinter.	"	
350	Buffalo Park.	"	2755
356	Grainfield.	"	2811
365	Grinnell.	"	2904
377	Oakley.	"	3042
385	Monument.	"	3181
	Boaz.	"	
398	Winona.	"	3364
406	Lisbon.[6]	"& 18 c. Colora.3140	
	McAllaster.	" "	
	Turkey Creek.	" "	
420	Wallace.	" "	3301
429	Sharon Springs.	" "	3460
437	Monotony.	"	3774
	Montero.	"	

Leavenworth and Lawrence Branch.

Ms.	Station	Geology	Alt.
0	Leavenworth.[2]	14 c. Up. Cl. Mres.765	
5	Lansing.	"	751
11	Fairmount.	"	955
15	Hoge.	"	854
18	Big Strainger.	"	834
19	Moores.	"	915
21	Tonganoxie.	"	851
26	Reno.	"	885
34	Lawrence.	"	822

Union Pacific Railway. Leavenworth, Topeka & South Western Line.

Ms.	Station	Geology	Alt.
0	Leavenworth.[2]	14 c. Upper Coal Measures.	765
9	Bolings.	"	905
16	Springdale.	"	1032
21	McLouth.	"	1157
	McIntosh.	"	1125
28	Oskaloosa.	"	959
	Osawkee.	"	876
45	Meriden.	"	964
56	Topeka.[3]	"	886

Blue Valley Line.

Ms.	Station	Geology	Alt.
0	Manhattan.[7]	14 c. Upper Coal Measures, and 15. Permo-Carbon.	1000
	Stockdale.	"	
17	Garrison Cross'g.	"	1081
	Winkl'r'sMills St.	"	
22	Randolph.	"	1088
	Cleburne.	"	
	Florena.	"	
39	Irving.	"	1127
43	Blue Rapids.	"	1141
	Schroyer.	"	
56	Marysville.	"	1179
	Hull.	"	
65	Oketo.	"	1200

Solomon Valley Line.

Ms.	Station	Geology	Alt.
0	Solomon.[9]	15. Permo-Carboniferous and 18 a. Dakota.	1172
	Niles.	"	
9	Verdi.	"	1202
15	Bennington.	"	1223
21	Lindsay.	"	1242
23	Minneapolis.	"	1256
29	Sumnerville.	"	1285
35	Delphos.	"	1310
42	Glasco.	"	1319
47	Brittsville.	"	1334
50	Asherville.	"	1346
57	Beloit.	"	1388

Salina and Upper Solomon Line, or Lincoln and Colorado Branch.

Ms.	Station	Geology	Alt.
0	Salina.	18 a. Dakota, and 15. Permo-Carboniferous.	1172
	Trenton.	"	
	York.	"	
12	Culver.	"	1265

4 *Carneiro.* The Dakota sandstone weathered into picturesque monumental shapes.

5. *Wakeeney.* In the ravine cutting the upland slopes, the chalky limestones of the Niobrara outcrop, affording characteristic vertebrate and molluscan fossils. The manufacture of the chalk into whiting is here successfully engaged in. Copious springs of delicious water issue from the gravel deposit at the base of the Tertiary.

6. *Lisbon.* The Colorado shales appear in the valley sides and outlying buttes, capped by Tertiary conglomerate in places, containing beautifully dendritic marked chalcedony. The Colorado shales abound in selenite crystals, septaria concretions and fossils.

7. *Manhattan.* The light gray limestone in the bluffs, and which form a convenient lithological

Union Pacific Railway.
Salina and Upper Solomon Line, or Lincoln and Colorado Branch.—*Con.*

Ms.	Station	Formation	Alt.
19	Tescot.	{ 18 a. Dakota and 15 Permo-Carb.	1297
24	Beverly.		1324
35	Lincoln.		1373
	Vesper.		
	Sylvan.		
56	Lucas.		1716
66	Luray.		
72	Waldo.		
	Ivamar.		
88	Natoma.		
	Codell.		
104	Plainville.		
111	Zurich.		
	Palco.		
	Daman.		
130	Bogue.		
138	Hill City.		
	Redford.		
	Kalula.		
	Carll.		
	Tasco.		
171	Hoxie.	19. Tertiary.	
	Gerona.	"	
	Zillah.	"	
	Verner.	"	
204	Colby.	"	
225	Oakley.	"	3042

Salina and Southwestern Railway.

Ms.	Station	Formation	Alt.
0	Salina.	{ 15. Permo-Carb. and 18 a. Dak.	1225
	Mentor.	"	
12	Assaria.	"	1282
16	Bridgeport.	"	1300
21	Lindsburg.	"	1350
	Johnstown.	18 a. Dakota.	
	Hilton.	"	
36	McPherson.	"	1490

Junction City and Ft. Kearney Branch.

Ms.	Station	Formation	Alt.
0	Junction City.	15. Permo-Carbo.	1082
8	Alida.	"	1109
14	Milford.	"	1102
19	Wakefield.	"	1152
28	Broughton.	"	1188
33	Clay Centre.	"	1203
41	Morganville.	"	1288
49	Clifton.	18 a. Dakota.	1277
50	Vining.	"	1277
56	Clyde.	"	1299
63	Lawrenceburg.	"	1329
71	Concordia.	"	1366
63	Lawrenceburg.	"	1329
65	Christie.	"	1341
70	Talmo.	"	1365
80	Belleville,	"	1551

Union Pacific Railway,
Kansas Central Line.

Ms.	Station	Formation	Alt.
0	Leavenworth.[2]	{ 14 c. Upper Coal Measures.	765
7	Hund.	"	830
11	Pleasant Ridge.	"	1081
15	Easton.	"	903
20	Lee.	"	1035
25	Winchester.	"	1158
	Boyle.	"	1165
36	Valley Falls.	"	911
	Arrington.	"	
46	Larkin.	"	936
51	Elk.	"	971
55	Holton.	"	1012
63	Circleville.	"	1096
70	Soldier.	"	1184
76	Havensville.	"	1165
79	Savannah.	"	1104
82	Onago.	"	1093
96	Blaine.	15. Permo-Carb.	1503
110	Olsburg.	"	1427
117	Garrison.	"	1058
	Leonardville.	"	
189	Green.	"	1287
147	Clay Centre.	"	(1193 1303)
	Idane.	"	
166	Miltonvale.	18 a. Dakota?	1372

St. Joseph & Grand Island R. R.

Ms.	Station	Formation	Alt.
0	St. Joseph, Mo.	{ 14 c. Upper Coal Measures.	825
1	Elwood.	"	817
6	Wathena.	"	815
9	Blairs.	"	897
14	Troy.	"	1093
19	Norway.	"	1042
23	Ryans.	"	892
25	Severance.	"	903
29	Leona.	"	918
34	Robinson.	"	950
38	Mannville.	"	973
43	Hiawatha.	"	1095
50	Hamlin.	"	984
54	Morrill.	"	1098
61	Sabetha.	"	1303
69	Oneida.	"	1219
77	Seneca.	{ 15. Permo-Carboniferous.	1152
84	Baileyville.	"	1204
89	Axtel.	"	1363
99	Beattie.	"	1293
105	Home.	"	1339
113	Marysville.	"	1155
118	Herkimer.	"	1238
128	Hanover.	18 a. Dakota?	1225
137	Hollenberg.	"	1238

series, are extensively quarried for building purposes. Underlying the quarry ledges is a heavy stratum of soft buff earthy limestone, possessing the properties of an hydraulic limestone, and preparations for the manufacture of cement have been made on quite an extensive scale.

St. Louis and San Francisco Railway. Ms. Monett (Mo.) to Halstead and Ellsworth. Alt		
0 Carthage, Mo.	{ Lower Carbon.: Keokuk limest.	956
23 Crestline.	{ 14 b. Lower Coal Measures.	856
31 Columbus.	"	913
35 Welland, or Wilson.	"	889
37 Sherwin.	" ·	875
39 Hallowell.	"	861
47 Oswego.[14]	{ 14 c. Upper and 14 b. Low. Cl. Mres.	914
Stover.	{ 14 c. Upper Coal Measures.	
58 Altamont.	"	924
64 Mound Valley.	"	839
69 Big Hill.	"	836
74 Cherryvale.	"	853
83 Brooks.	"	897
88 Neodesha.[15]	"	816
Dun.	"	
101 Fredonia.	"	975
107 New Albany.	"	913
113 Fall River.	"	940
119 Greenwood.	"	1011
125 Severy.	{ 15. Permo-Carbon-iferous.?	1124
134 Piedmont.	"	1216
140 Derry.	"	1470
145 Beaumont.[16]	"	1604
152 Keighley.	"	1542
160 Leon.	"	1549
165 Haverhill.	"	1540
171 Augusta.	"	1246
177 Lorena.	"	1356
181 Andover.	"	1370
186 Manchester.	"	1402
192 Wichita.[17]	"	1315
195 Davidson.	"	
197 Wichita Heights.	"	
201 Valley Centre.	"	1339
210 Bentley.	"	
219 Paterson.	"	

St. Louis and San Francisco Railway. Ms. Monett to Halstead and Ellsworth. Alt.		
225 Burrton.	15. Permo-Carb.	
234 Buhler, or Hamburg.	" ?	
238 Medora.	?	
252 Wherry.	?	
264 Lyons.	18 a. Dakota.?	1691
271 Clarence, or Pollard.	"	
275 Dacey.	"	
281 Lorraine.	" ?	
288 Phipps.	18 b. Benton. ?	
295 Ellsworth.	"	1535

Arkansas City and Anthony Line.		
0 Beaumont.	15. Permo-Carb.	1604
7 Burgess.	"	
13 Latham.	"	
19 Wingate.	"	
23 Atlanta.	"	
31 Wilmot.	"	
34 Floral.	"	
40 Younts.	"	
43 Winfield.[18]	"	1112
50 Tresham.	"	
57 Arkansas City. Cale.	"	1064
64 Geuda Springs.	"	
69 Ashton.	"	
73 Portland.	"	
79 South Haven.	"	1134
81 Hunnewell Ju.	"	1102
84 Drury.	"	
86 Falls.	"	
91 Caldwell.	"	
101 Blackstone.	"	
106 Bluff.	"	
Blackburn.		
Anthony.	16 Triassic.	

Wichita and Halstead.		
0 Wichita.[17]	15. Permo-Carb.	1315
10 Valley Centre.	"	1355
17 Sedgwick.	"	1383
25 Halstead.	"	1402

8. *Junction City.* Extensive quarries in heavy ledges of light buff limestone, used in the construction of the east wing of the Capitol at Topeka.

9. *Solomon.* Strong brine wells in gypsiferous shales of the Permo-carboniferous, from which salt has been manufactured quite extensively.

10. *Bavaria.* The Dakota sandstone near this place affords numerous characteristic fossils. Near Brookville Dicotyledonous leaves abundant in the sandstone.

11. *Pittsburgh.* Centre extensive coal mining interests and zinc smelting furnaces. The ores are brought from Galena and adjacent mining districts in Missouri, in the lower carboniferous rocks.

12. *Weir City.* Centre of coal mining district, zinc smelting establishments.

13. *Galena.* Extensive lead and zinc mines in lower carboniferous Keokuk formation.

14. *Oswego.* The Neosho river is excavated into the lower coal measures, the upper coal horizons of which appear at various localities in the vicinity. The plateau upon which the town is located, is formed by the basal limestones of the upper coal measures, including the horizon of the Ft. Scott coal, which is here a bituminous shale and the cement rock. Interesting localities for both upper and lower coal measures fossils.

15. *Neodesha.* Along the Verdigris and Elk rivers a heavy ledge of sandstone occurs, which belongs well up in the upper coal series, and affords remains of large trees peculiar to the coal measures period. Although the Verdigris has cut its bed more deeply, geologically it is more than a thousand feet above the Neosho at Oswego, or on the line of greatest depression between the Ozark region of S. W. Missouri and the first great highland belt traversing Central Kansas from near the south border to the Nebraska line on the north.

St. Louis and San Francisco Railway.

Ms.	Girard Branch.		Alt.
0	Carl Junction.	13. L. Carb. and / 14b.L.Coal Mres.	911
12	Opolis.	14 b. Lower Coal / Measures.	937
18	Litchfield Jc.	"	925
19	Pittsburgh.[11]	"	954
22	Lone Oak.	"	966
29	Girard.	Upper and Lower / Coal Measures.	1003

Weir City Branch.

Ms.			Alt.
0	Pittsburgh.	14 b.Low. Cl. Mres.	954
10	Weir City.[12]	"	934

Joplin and Galena.

Ms.			Alt.
0	Joplin.	LowerCarbonif. / 13 c. Keokuk	1026
9	Galena.[13]	"	893

Missouri, Kansas and Texas Ry.
In Kansas.

Ms.			Alt.
0	Nevada, Mo.	14 b. Lower Coal / Measures.	870
21	Ft. Scott.	Low. and Upper / Coal Measures.	802
28	Ronald.	14 c. Upper Coal / Measures.	
34	Hiattville.	"	1003
41	Hepler.	"	1002
48	Walnut.	"	981
56	Osage Mission.	"	890
62	South Mound.	"	993
69	Parsons.	"	902
78	Labette.	"	864
83	Oswego.	14 c. Upp. and 14 b. / Low. Cl. Mres. ·	895
93	Chetopa.	14 b. Lower Cl. / Measures.	832

Missouri, Kansas and Texas Ry.

Ms.	Neosho Valley Section.		Alt.
0	Parsons.	14 c. Upper Coal / Measures.	902
5	Ladore.	"	909
11	Galesburg.	"	979
17	Urbana.	"	981
26	Chanute.	"	910
35	Humboldt Stat'n, So. K.	"	952
44	Piqua.	"	
50	Neosho Falls.	"	980
56	Moody.	"	
59	LeRoy.	"	994
64	Bristol.	"	
67	Burlington.	"	1037
75	Rockeby.	"	
82	Hartford.	"	1087
88	Wyckoff.	"	
95	Emporia.	"	1132
104	Americus.	"	1158
111	Dunlap.	"	
120	Council Grove.	15. Permo-Car- / boniferous.	1238
127	Downing Station.	"	
132	Parkersville.	"	1337
137	White City.	"	1476
144	Skiddy.	"	1226
152	Wreford.	"	
157	Junction City.	"	1082

Lawrence and Southwestern R. R.

Ms.			Alt.
0	Lawrence.	14 c. U. Coal Mres.	822
10	Clinton.	"	
13	Belvoir.	"	871
19	Richland.	"	901
	Ridgeway.	"	
27	Kinneys.	"	
31	Carbon Hill.	"	1122
32	Carbondale.	"	1072

16. *Beaumont.* Summit of the " Flint Hills," composed of a cherty member and the light buff limestones of the Permo-Carboniferous, forming a highland bench of the type of a monocline, presenting a somewhat abrupt eastern scarp and long gentle westerly slope. A conspicuous topographic feature at intervals across the central portion of the State to the Nebraska line.

17. *Wichita* lies within the area occupied by the heavy series of shaly deposits, to which the great salines and salt beds, occurring in central Kansas, belong. These deposits underlie the "red beds" presumably of Triassic age, and are in conformable sequence with the underlying porous limestones and shales of the so-called Permo-Carboniferous.

18. *Winfield.* Extensive quarries of even, thick, and thin-bedded limestone, affording fine building material and flagging in the vicinity.

19. *Scott City.* Basin receives considerable drainage from the west.

20. The line from La Cross follows the water-shed south of the Smoky Hill, an elevated plain steadily increasing in altitude to nearly 4,000 feet on the west boundary of the State, and blanketed by Tertiary deposits. The Niobrara appears along the more deeply eroded drainage channels flowing to the Smoky Hill, the exposures affording characteristic fossils.

21. *Louisburg.* Natural gas wells, also near Somerset.

22. The highlands west of Mankato are blanketed by Tertiary deposits, the Cretaceous, Niobrara, appearing at intervals in the more deeply cut drainage channels. The latter deposits abound in characteristic fossils, vertebrates and mollusks.

23. *Paola.* Natural gas found in drilled wells in vicinity, in considerable volume.

24. *La Cygne.* Coal shaft, to workable vein in lower portion of Upper Coal measures.

25. *Pleasanton.* Coal shaft, same coal mined at La Cygne. On mine creek, S. E. of the town, the ores of lead and zinc occur in Upper Coal measures strata. Near the town a bituminous sandstone affords flagging layers.

26. *Ft. Scott.* Gas and mineral water developed in drilled wells. Associated with a thin coal which has been extensively worked by surface stripping in the vicinity and south to Arcadia and Mulberry, occurs an hydraulic limestone, which furnishes material for the manufacture of cement, which is extensively engaged in at Ft. Scott.

27. *Farlington.* In the vicinity, extensive quarries have been opened in a flagging sandstone.

Missouri Pacific Railway.
Ms. Omaha, St. Joseph & Kansas City Line. — Alt.

Ms.	Station		Alt.
0	Kansas City.	14 c. Up. C'l Mres.	748
3	Wyandotte.	"	
	Ramapo.	"	
10	Nearman.	"	
13	Pomeroy.	"	
15	Connors.	"	
19	Ross.	"	
	Lansing.	"	
26	Leavenworth.[2]	"	765
29	Ft. Leavenworth.	"	
	Wade.	"	
	Kickapoo City.	"	
37	Oak Mills.	"	
38	Port Williams.	"	
	Dalbey.	"	
47	Atchison.	"	793
55	Shannon.	"	
58	Lancaster.	"	
63	Huron.	"	
67	Pierce Junction.	"	1161
68	Everest.	"	
75	Willis.	"	
79	Baker.	"	
87	Hiawatha.	"	1094
92	Pandona.	"	
96	Reserve.	"	

Denver and Kansas City Line.

Ms.	Station		Alt.
0	Kansas City.	14 c. Up. Cl. Mres.	748
	Martin City.	"	
	Stillwell.	"	
38	Bucyrus.	"	
45	Wagstaff.	"	
53	Paola.	"	
60	Ossawatomie.	"	
65	Obrien.	"	
69	Rantoul.	"	
73	Imes.	"	
80	Ottawa. { Maria s des Cygne s Riv.	"	895
	Pomona.	"	
94	Lomax.	"	
101	Vassar.	"	
104	Lyndon.	"	
112	Osage City.	"	1075
117	Rapp.	"	
121	Miller.	"	
128	Admire.	"	
132	Allen.	"	
137	Bushong.	"	
143	Comiskey.	"	
151	Council Grove.	{ 15. Permo-Carboniferous.	1233

Missouri Pacific Railway.
Ms. Denver & Kansas City Line.—Con. — Alt.

Ms.	Station		Alt.
158	Helmick.	{ 15. Permo-Carboniferous.	
163	Wilsey.	"	
170	Delavan.	"	
177	Herington.	"	1388
185	Hope.	"	
190	Swrayne.	"	
194	Banner City.	"	
197	Carlos.	"	
205	Gypsum City.	"	
207	Chico.	"	
221	Salina.	{ 15. Permo-Car- and Dakota.	1225
230	Smolan.	"	
237	Falun.	"	
246	Marquette.	"	
224	Hallville.	15. Permo-Carbonif.	
230	Bridgeport.	"	
235	Lindsborg.	"	1382
	Smoky Hill.	"	
246	Marquette.	{ 15. Permo-Carb. and Dakota.	
254	Langley.	18 a. Dakota.	
259	Crawford.	"	
265	Geneseo.	"	
272	Frederick.	"	
278	Bushton.	"	
286	Claflin.	"	
299	Hoisington.	"	
309	Great Bend.	"	1841
303	Boyd.	18 a. Dakota.	
309	Olmutz.	" ?	
316	Otis.	18 b. Benton.	
331	La Cross.[20]	"	
346	McCracken.	19. Tertiary.	
349	Holbrook.	"	
357	Brownell.	"	
368	Ransom.	"	
381	Utica.	"	
390	Pen-Dennis.	"	
396	Shields.	"	
406	Healey.	"	
412	Manning.	"	
423	Scott City.[19]	"	
433	Modoc.	"	
	Halcyon.	"	
444	Coronado.	"	
447	Leoti.	"	
457	Tuell.	"	
465	Whitelaw.	"	
471	Horace.	"	
	Reid.	"	

28. *Cherokee.* Extensive mining operations carried on in the main coal of the Lower coal measures, to the south and east as far as Stilson and Weir City.

29. *Galena.* Centre of an important mining district. The ores of lead and zinc occurring abundantly, extensive works for the smelting of the former are located here, the zinc ore being shipped to furnaces located on the coal belt, chiefly to Pittsburgh and Weir City and Rich Hill.

30. *Pittsburgh.* Centre of extensive coal mining operations and zinc smelting establishments. The coal is sought by means of shafts, 40 to above 100 feet in depth; the coal is fairly good, coking

Missouri Pacific Railway. Central Branch Line.

Ms.	Station		Alt.
0	Atchison.	{ 14 c. Upper Coal Measures.	798
13	Farmington.	"	
15	Monrovia.	"	1054
18	Effingham.	"	1144
25	Muscotah.	"	973
31	Whiting.	"	1126
37	Netawaka.	"	1140
42	Wetmore.	"	1153
49	Goffs.	"	1200
55	Corning.	"	1369
62	Centrailia.	{ 15. Permo-Carboniferous.	1270
70	Vermillion.	"	1195
74	Vleits.	"	
78	Frankfort.	"	1155
81	Barrett.	"	1142
85	Bigelow.	"	
91	Irving.	"	1152
95	Blue Rapids.	"	1198
100	Waterville.	"	1183
107	Barnes.	"	1356
113	Greenleaf.	18 a. Dakota.	1462
	Washington.	"	1316
120	Linn.	"	
125	Palmer.	"	
129	Day.	"	
134	Clifton.	"	1281
140	Clyde.	"	1310
155	Concordia.	"	1366
160	Yuma.	"	
167	Norway.	" ?	
174	Scandia.	18 b. Benton.	
	Sherdall.	"	
183	Republic.	"	
190	Warwick.	"	
160	Yuma.	18 a. Dakota.?	
166	Jamestown.	" ?	
176	Randall.	18 b. Benton.	
183	Jewell City.	"	
191	Mankato.	"	
199	Burr Oak.	18 c. Niobrara.?	
166	Jamestown.	18 a. Dakota. ?	
172	Scottaville.	18 b. Benton.	
179	Danville.	"	
184	Beloit.	"	1858
189	Solomon Rapids.	"	
195	Glen Elder.	"	
102	Cawker City.	"	
108	Downs.	"	
	Osborne.	18 c. Niobrara.?	
	Bloomington.	"	
232	Alton.	"	
	Woodston.	"	
250	Stockton.	" .	
208	Downs.	"	

Missouri Pacific Railway. Central Branch Line.—Con.

Ms.	Station		Alt.
217	Portis.	18 c. Niobrara.	
	Harlan.	"	
227	Gaylord.	"	
232	Cedarville.	"	
242	Kirwin.	"	
253	Marvin.	"	
	Big Bend.	"	
268	Logan.	"	
278	Densmore.	"	
282	Edmond.	"	
293	Lenora.	"	

Kansas City and Paola Line.

Ms.	Station		Alt.
0	Holden, Mo.	14 c. Up. Coal Mres.	
22	Harrisonville.	"	
41	Louisburg.[21]	"	
46	Sommerset.	"	
54	Paola.	"	854

Kansas, Nebraska and Dakota Division.

Ms.	Station		Alt.
0	Topeka.[3]	14 c. Upp. Cl. Mre.	892
11	Tevis.	"	
15	Richland.	"	901
21	Swissvale.	"	
26	Overbrook.	"	
33	Michigan.	"	
41	Quenemo.	"	
48	Rosemont.	"	
56	Waverly.	"	
	Amiet.	"	
66	Dickey.	"	
72	Glenlock.	"	
80	Garnett.	"	1056
88	Bush City.	"	
93	Selma.	"	
101	Blue Mound.	"	
106	Yoro.	"	
111	Mapleton.	"	
	Harding.	"	
120	Devon.	"	
125	Azua.	"	
130	Ft. Scott.	14b.L&14c.U.C.M.	802

Denver, Memphis and Atlantic Division.

Ms.	Station		Alt.
	Pittsburgh.[11]	14 b. LowerCl.Ms.	954
	Cherokee.	"	988
	Folsom.	"	
	Sherwood.	"	
	Faulkner.	"	
371	Chetopa.	"	832
	Bartlett.	14 c. Up. Coal Mres	
	Elm City.	"	
386	Edna.	"	
	Valeda.	"	
	Kings.	"	
401	Coffeeville.	"	728
407	Deering.	"	

and averages about 40 inches in thickness. Several thinner overlying coals occur in this region with which are associated fossiliferous shales and limestone. The town is supplied with water from a drilled well .. feet deep, which penetrates to Lower Silurian formations

21. *Weir City.* Coal mines and zinc smelting furnaces.

Missouri Pacific Railway.
Ms. Denver, Memphis & Atlantic Div.—Con. Alt.

Ms.	Station	Formation	Alt.
413	Tyro.	14 c. Upper Coal Mres.	
520	Caney.	"	
431	Peru.	"	
437	Sedan.	"	
	Rogers.	"	
450	Wauneta.	"	
459	Cedarvale.	"	
469	Hoosier.	15. Permo-Carbon.	
476	Dexter.	"	

Arkansas City & Dexter.

Ms.	Station	Formation	Alt.
	Vinton.	15. Permo-Carbon.	
	Cameron City.	"	
	Silverdale.	"	
501	Arkansas City.	"	1064
476	Dexter.	"	
482	Eaton.	"	
	Tisdale.	"	
495	Winfield.	"	1112
	Kellogg.	"	
505	Oxford.	"	
516	Belle Plaine.	"	1209
	Riverdale.	"	1330
	Arson.	"	
536	Conway Springs.	"	
	Milton.	"	
548	Norwich.	"	
558	Belmont.	"	
	Alameda.	"	
570	Kingman.	"	
583	Penalosa.	"	
587	Olcott.	"	

Iuka and Olcott.

Ms.	Station	Alt.
596	Preston or Silverton.	1553
601	Carmi.	
607	Iuka.	
587	Olcott.	
591	Turon.	
	Neola.	
607	Stafford.	
	Bedford.	
	Hudson.	
626	Seward.	
635	Ray.	
643	Larned.	1993

Winfield, Independ'ce & Kan. City Line.

Ms.	Station	Formation	Alt.
0	Kansas City.	14 c. Upper Coal Measures.	745
60	Ossawatomie.	"	

Missouri Pacific Railway.
Ms. Winfield, Indep.& Kan. City Line.—Con. Alt.

Ms.	Station	Formation	Alt.
	Belle Grade.	14 c. Up.Cl. Mres.	
111	Le Roy.	"	994
115	Moody.	'	
121	Vernon.	"	
129	Yates Centre.	"	
	Rose.	"	
142	Buffalo.	"	
145	Roper.	"	
148	Benedict.	"	
151	Guilford.	"	
158	Altoona.	"	
165	Neodesha.	"	
	Sycamore.	"	
174	Larimer.	"	
179	Independence.	"	794
187	Winton.	"	
193	Deering.	"	
198	Coffeeville.	"	728

Roper and Peru.

Ms.	Station	Formation	Alt.
146	Roper.	14 c. Up. Coal Mres.	
	Cordley.	"	
	Sexton.	"	
	Dill.	"	
	Fredonia.	"	
	La Fontaine.	"	
	Costello.	"	
	Elk City.	"	
	Colfax.	"	
	Hale.	"	
	Monett.	"	
	Peru.	"	

Ft. Scott, Wichita and Western Railway.

Ms.	Station	Formation	Alt.
0	Ft. Scott.	{ 14 b. Lower Coal Measures.	802
7	Marmaton.	{ 14 c. Upper Coal Measures.	?917
10	Redfield.	"	
15	Uniontown.	"	
22	Bronson.	"	
28	Moran.	"	
35	La Harpe.	"	
41	Iola.	"	985
48	Piqua.	"	
60	Yates Centre.	"	
68	Batesville.	"	
73	Toronto.	"	
81	Neal.	"	
87	Tonovay.	"	1073

Missouri Pacific Railway.
Ms. Ft. Scott, Wichita & West'rn R'y.—*Con.* Alt.

Ms.			Alt.
147	Greenwich.	15. Permo-Carb.	
152	Tolerville.	"	
158	Wichita.	"	1291
164	Oatville.	"	
169	Bayneville.	"	
174	Clearwater.	"	
179	Millerton.	"	
186	Conway Springs.	"	
190	Ewell.	"	
196	Argonia.	"	
203	Freeport.	{ 16. Triassic Red Beds.	
214	Anthony.	"	
221	Goss.	"	
224	Ruella.	"	
231	Corwin.	"	
236	Hazelton.	"	
242	Kiowa.	"	
0	Pleasanton.	{ 14 c. Upper. Coal Measures.	860
7	Mound City.	"	
12	Critzer.	"	
19	Blue Mound.	"	
27	Kincaid.	"	
	Lone Elm.	"	
39	Colony.	"	1121
46	Northcott.	"	
54	LeRoy.	"	994
	Crandall.	"	
70	Gridley.	"	
	Dunaway.	"	
78	Wilbur.	"	
84	Madison.	"	1068

Chicago, Kansas and Nebraska Railway.
Southwest Line: St. Joseph to Liberal.

Ms.			Alt.
0	St. Joseph, Mo.	{ 14 c. Upper Coal Measures.	840
1	Elwood, Kansas.	{ 20 d. Valley Alluvium.	831
5	Wathena.	"	833
13	Troy.	{ 14 c. Upper Coal Measures.	1112
19	Bendena.	"	1124
24	Dentonville.	"	1088
29	Purcell.	"	1171
34	Pierce Junction.	"	1161
41	Horton Junction.	"	1029
49	Whiting.	"	1118
54	Straight Creek.	"	1007
60	Holton.	"	1057
69	Mayette.	"	1210
76	Hoyt.	"	1180
82	Elmont.	"	960
89	North Topeka.	"	892
90	Topeka.	"	892
101	Valencia.	"	918
105	Willard.	"	927
110	Maple Hill.	"	972
118	Paxico.	"	1006

Chicago, Kansas & Nebraska R'y.
Ms. So'west Line: St. Joseph to Liberal.—*Con.* Alt.

Ms.			Alt.
122	McFarland.	14 c.Up.Cl. Mres.	1035
126	Alma.	"	1071
134	Volland.	"	1191
142	Alta Vista.	{ 15. Permo-Car-boniferous.	1442
148	Dwight.	"	1510
157	White City.	{ Up. Coal Measures. (Permo-Carboniferous.)	1479
164	Latimer.	"	1421
171	Horington.	"	1338
179	Ramona.	"	1446
186	Tampa.	"	1438
192	Durham.	"	1388
198	Waldeck.	"	1578
205	Canton.	"	1602
211	Galva.	"	1564
218	McPherson.	"	1508
224	Groveland.	"	1498
229	Aiken.	"	1535
235	Medora.	"	1494
245	Hutchison.	"	1544
256	Partridge.	"	1525
263	Arlington.	"?	1609
271	Langdon.	"?	1707
278	Turon.	"?	1784
285	Preston.	?	1853
292	Natrona.	?	1890
298	Pratt.	Probably Triassic	1920
307	Cullison.	"red beds," with	2053
314	Wellsford.	remnants of Ter-	2135
319	Haviland.	tiary forming the	2173
324	Brenham.	superficial depos-	2214
329	Greensburg.	its.	2245
339	Mullinville.		2349
348	Bucklin.		2428

Dodge City Branch.

Ms.			Alt.
356	Ford.		2425
366	Wilroads.		
378	Dodge City.	19. Tertiary.	2494
355	Kingsdown.	"	2525
363	Bloom.	"	2600
370	Mineola.	"	2568
381	Fowler.	"	2495
392	Meade.	"	2515
398	Jasper.	"	2713
406	West Plains.	"	2776
412	Kismet.	"	2759
421	Arkalon.	"	2625
435	Liberal.	"	2853

South Line.

Ms.			Alt.
171	Herington.	15. Permo-Carb.	1388
178	Lost Springs.	"	1487
183	Lincolnville.	"	1442
194	Marion.	"	1320
200	Aulne.	"	1414
208	Peabody.	"	1376
216	Elbing.	"	1451
223	Whitewater.	"	1396

Chicago, Kansas and Nebraska R'y.
South Line.—Con

Ms.			Alt.
229	Furley.	15. Permo-Carb.	1424
236	Kechi.	"	1383
245	Wichita.	"	1310
250	Gladys.	"	1285
259	Peck.	"	1280
262	Zyba.	"	1242
267	Riverdale.	"	1330
274	Wellington.	"	1208
283	Perth.	"	1223
287	Corbin.	"	1171
295	Caldwell.	"	1128

Clay Centre Line.

Ms.			Alt.
100	McFarland.	14 c. Up.Cl. Mres.	1035
109	Wabaunsee.	"	1059
114	Zeandale.	"	1007
122	Manhattan.	"	1027
130	Keats.	15. Permo-Carb.	1199
139	Riley.	"	1289
146	Bala.	"	1281
152	Rosevale.	"	1195
158	Clay Centre.	"	1218
165	Morganville.	"	1346
173	Clifton.	18 a. Dakota.	1281
180	Clyde.	"	1310
188	Agenda.	"	1424
195	Cuba.	"	1603
204	Belleville.	"	1522

Salina Line.

Ms.			Alt.
171	Herington.	{ 15.Permo-Car-boniferous.	1338
180	Woodbine.	"	1265
193	Enterprise.	"	1154
198	Abilene.	"	1160
207	Solomon.	{ 18 a. Dakota & 15. Permo-Car.	1161
215	New Cambria.	"	1211
220	Salina.	"	1234

Colorado Line. (In Kansas.)

Ms.			Alt.
41	Horton Junction.	{ 14 c. Upper Coal Measures.	1029
51	Powhattan.	"	1220
59	Fairview.	"	1229
65	Sabetha.	"	1315
68	Berwick.	"	1373
76	Birn, Neb.	"	1295
170	Mahasba, Kan.	18 a. Dakota.	1613
175	Narka.	"	1593
182	Munden.	"	1636

Chicago, Kansas and Nebraska R'y.
Colorado Line. In Kansas.—Con.

Ms.			Alt.
254	Smith Center.	See Note 22.	1610
261	Athol.	"	1792
268	Kensingto .	"	1779
273	Agra.	"	1862
278	Dana.	"	1870
284	Phillipsburg.	"	1945
291	Stuttgart.	"	2010
298	Prairie View.	"	2182
307	Almena.	"	2161
311	Calvert.	"	2203
318	Norton.	{ Tertiary, overlying Niobrara extends thence into Col.	2278
327	South Oronoque.	"	2342
335	Clayton.	"	2424
342	Jennings.	"	2498
351	Dresden.	"	2737
360	Selden.	"	2844
371	Rexford.	"	2937
380	Gem.	"	3099
388	Colby.	"	3145
396	Levant.	"	3317
406	Brewster.	"	3421
415	Edson.	"	3576
424	Goodland.	"	3693
433	Ruleton.	"	3794
441	Kanorado.	"	3913

Kansas City, Wyandotte and Northwestern Railway.

Ms.			Alt.
0	Kansas City.	{ 14 c. Upper Coal Measures.	748
2	Wyandotte.	"	766
4	Quindaro.	"	880
6	Welborn.	"	938
8	Calorific.	"	1002
9	Vance.	"	1007
11	Bethel.	"	1004
12	White Church.	"	1008
13	Horanif.	"	1004
15	Maywood.	"	1015
17	Roper.	"	969
19	Menager Jc.	"	909
22	Baschor.	"	842
28	Edminster.	"	830
31	Tonganoxie.	"	846
36	Neely.	"	933
41	McLouth.	"	1166

Kansas City, Wyandotte and Northwestern Railway.—Con.

Ms.			Alt.
117	Seneca.	15. Permo-Carb.	1121
128	Axtel.	"	1309
134	Mina.	'	1430
139	Summerfield.	, "	1490

Leavenworth Branch.

20	Usher.	14 c. Up.Cl. Mres.	956
21	Wallula.	"	964
26	Lansing.	"	788
28	Soldier's Home.	"	844
30	So. Leavenworth.	"	768
31	Leavenworth.[2]	"	786
34	Ft. Leavenworth.	"	838

Burlington and Missouri River R. R.
(In Kansas.)
Atchison and Nebraska R. R.

0	Atchison.	793	14 c. Upp. Coal Mres.
7	Doniphan.		"
12	Brenner.		"
16	Troy.		" 1112
22	Fanning.		"
24	Highland.		"
30	Iowa Point.		"
35	White Cloud.		"

Nebraska Railway.
Hasting, Republican and Oberlin.

0	Republican, Neb.		1944
10	Woodruff.	18 c. Niobrara in	
17	Long Island.	the deeper valleys;	
27	Almena.	19. Tertiary in	2161
31	Seth.	the uplands.	2203
38	Norton.	"	2278
47	Oronoque.	19. Tertiary.	2342
57	Norcatur.	"	
68	Kanona.	"	
78	Oberlin.	"	

Orleans and St. Francis.

0	Orleans, Neb.	19. Tertiary.
62	Cedar Bluffs.	"
69	Traer.	"
76	Herndon.	"
86	Ludell.	"
91	Atwood.	"
95	Blakeman.	"
102	Beardsley.	"
110	McDonald.	"
118	Bird City.	"
128	Wheeler.	"
134	St. Francis.	"

Lincoln, Wymore and Concordia.

0	Odell, Neb.		1281
7	Lanham.	18 a. Dakota.	
14	Hanover.	"	
23	Emmons.	"	

Burlington and Missouri River R. R.
In Kansas.

Ms.	Lincoln, Wymore and Concordia.—Con.		Alt.
26	Washington.	18 a. Dakota.	
33	Morrow.	"	
40	Haddam.	"	
50	Cuba.	"	1603
58	Wayne.	"	
64	Hollis.	"	
72	Concordia.	"	1366

Kansas City, Ft. Scott and Memphis Railroad.

0	Kansas City.	14 c. Upper Coal Measures.	765
4	Rosedale.	"	825
8	Merriam.	"	920
14	Lenexa.	"	1040
21	Olathe.	"	1060
26	Bonita.	"	1105
29	Ocheltree.	"	1080
30	Spring Hill.	"	1020
36	Hillsdale.	"	900
43	Paola.[23]	"	860
48	Pendleton.	"	855
54	Fontana.	"	920
62	LaCygne.[24]	"	820
68	Barnard.	"	800
74	Pleasanton.[25]	"	850
79	Miami.	"	910
82	Prescott.	"	880
86	Fulton.	"	805
92	Hammond.	"	880
99	Ft. Scott.[26]	Low. & Up. Cl. M.	802
103	Southeastern Jc.	14 c.Upp. Cl. Mres.	930
106	Clarksburg.	" & Low. "	890
110	Garland.	14 b. Low. Cl. "	863
116	Arcadia.	"	850

Baxter and Joplin Line.

99	Ft. Scott.[26]	Lower and Upper Coal Measures.	802
103	Southeastern Jc.	14 c. Upper Coal Measures.	930
105	Godfry.	"	962
111	Pawnee.	"	988
117	Farlington.[27]	"	988
125	Girard.	"	990
130	Beulah.	"	977
136	Cherokee.[28]	14 b. Lower Coal Measures.	988
142	Stilson.	"	909
148	Columbus.	"	905
154	Neutral.	"	862
160	Baxter.	L.Carboniferous. 13 c. Keokuk.	881
163	Lowell Station.	"	823
167	Galena.[29]	"	898
175	Joplin, Mo	"	

Kansas City, Ft. Scott and Memphis Railroad.—Con.

Ms. Cherryvale Line, via Pittsb'gh & Parsons. Alt.

Ms.			Alt.
116	Arcadia.	{ 14 b. Lower Coal Measures.	850
118	Coalvale.	"	888
123	Mulberry.	"	930
130	Minden.	"	967
132	Midway.	"	925
137	Pittsburg.[30]	"	932
143	Weir City.[31]	.,	923
146	Cherokee.[26]	"	933
153	Monmouth.	"	900
157	McCune.[32]	{ 14 c. Upper Coal Measures—base of.	910
161	Mathewson.	"	853
164	Laneville.	"	870
171	Parsons.	"	902
180	Dennis.	"	925
184	Mortimer.	"	895
190	Cherryvale.	"	886

Atchison. Topeka and Santa Fe Railr'd.[35]
Atchison Branch.

0	Atchison.	{ 14 c. Upper Coal Measures.	795
6	Parnell.	"	1039
9	Hawthorne.	"	
11	Cummings.	"	981
17	Nortonville.	"	1153
20	Nichols.	"	1001
26	Valley Falls.	"	907
35	Rock Creek.	"	1037
39	Meriden.	"	964
40	Meriden Junct.	"	945
43	Kilmer.	"	
49	North Topeka.	"	872
50	Topeka.	"	884

Leavenworth Extension.

0	Kansas City.	14 c. Up. Cl. Mres.	748
17	Wilder.	"	770
18	Bonner.	"	
	Jaggard.	"	
29	Fairmount.	"	955
34	Lansing.	"	
36	Home.	"	
39	Leavenworth.	"	765
44	Miocene.	"	
50	Lowement.	"	
56	Potter.	"	
62	Hawthorne.	"	
71	Atchison.	"	793

Atchison, Topeka and Santa Fe R. R.
Ms. Emporia Branch. Alt.

0	Kansas City.	{ 14 c. Upper Coal Measures.	748
13	Holliday.	"	758
57	Ottawa North.	"	
68	Pomona.	"	
72	Quenemo.	"	
80	Melvern.	"	
86	Olivet.	"	
94	Lebo.	"	
102	Neosho Rapids.	"	
112	Emporia Jc.	"	
113	Emporia.	"	

Howard Branch.

0	Emporia.	{ 14 c. Upper Coal Measures.	1182
11	Olpe.	"	
20	Madison.	"	1068
24	Madison Jc.	"	
35	Hamilton.	"	
40	Utopia.	"	
47	Eureka.	"	1078
56	Climax.	"	1018
63	Severy.	"	1098
69	Fiat.	"	
76	Howard.	"	1006
84	Moline.	"	1050

Manhattan, Alma and Burlingame R'y.

0	Burlingame.	{ 14 c. Upper Coal Measures.	1048
8	Harveyville.	"	
18	Eskridge.	{ 15. Permo-Carboniferous.	1403
25	Halifax.	"	
34	Alma.	{ 14 c. Upper Coal Measures.	1051
37	Fairfield.	"	1060
42	Pavillion.	"	1096
45	Wabaunsee.	"	1011
49	Zeandale.	"	
56	Manhattan.	"	1000

Strong City and Ellinor Extensions.

	Bazar.	{ 15. Permo-Carboniferous.	
	Gladstone.	"	
	CottonwoodFalls.	"	
0	Strong City.	"	1172
2	Evans.	"	

32. *McCune.* Coal shaft, sunk to one of the upper workable coals, overlying the main coal of the Lower coal measures of the region.

33. Fine flagging and building sandstone along the Neosho to the northeast.

34. Almost every locality within the Upper coal measures area afford deposits charged with fossils peculiar to the epoch.

35. The Kansas chapter properly ends at the Colorado line on the Atchison, Topeka and Santa Fe, but for convenience, the branches of that road are given first, the main line following and continued through Colorado into New Mexico.

Atchison, Topeka and Santa Fe Railroad.
Ms. Strong City & Ellinor Extensions.—*Con.* Alt.

Ms.			Alt.
7	Rockland.	15. Permo-Carbon.	
11	Hilton.	"	
17	DiamondSprings.	"	
23	Burdick.	"	
29	Lost Springs.	"	
41	Hope.	"	
48	Navarre.	"	
56	Enterprise.	"	1135 C.P.
62	Abilene.	"	1155
71	Talmage.	"	
75	Manchester.	{ 15. Permo-Carb. or 18 a. Dakota.	
82	Longford.	"	
87	Oak Hill.	"	
97	Miltonville.	"	
106	Aurora.	"	
117	Concordia.	"	1366 C.P.
131	Hackley.	{ 18 a. Dakota, or 18 b. Benton.	
138	Courtland.	"	
145	Lovewell.	"	
151	Webber.	"	
155	State Line.	"	
157	Superior, Neb.	"	
0	Abileno.	{ 15. Permo-Carboniferous.	1155
8	Solomon.	{ 15. Permo-Carbonif. & 18 a. Dakota.	1175
17	New Cambria.	"	1189
22	Salina.	"	1225
0	Manchester.	18 a. Dakota.	
7	Vine Creek.	"	
16	Wells.	"	
26	Minneapolis.	"	1257
30	Brewer.		
36	Ada.		
40	Milo.		
45	Barnard.		

McPherson Branch.

Ms.			Alt.
0	Florence.	{ 15. Permo-Carboniferous	1260
4	Owesler.	"	
10	Marion.	"	1299
15	Canada.	"	
20	Hillsboro.	"	1424
26	Lehigh.	"	1520
34	Canton.	"	1582
40	Galva.	?	
47	McPherson.	?	
53	Conway.	"	1527
60	Windom.	"	
66	Little River.	"	1572
72	Mitchell.		1781
78	Lyons.		1691
86	Chase.		1708
98	Ellinwood.	18 a. Dakota.	1780

Atchison, Topeka and Santa Fe R. R.
Ms.   Little River Extension.   Alt.

Ms.			Alt.
0	Little River.	15. Permo-Carb.?	1572
6	Galt.	"	
10	Geneseo.	18 a. Dakota.	
14	Thomas.	"	
21	Lorraine.	?	
26	Holyrood.	"	
29	West line of Ellsworth County.	"	

Great Bend Extension.

Ms.			Alt.
0	Great Bend.	18 a. Dakota.	1841
8	Heizer.	"	
15	Albert.	"	
24	Timken.	"	
32	Rush Centre.	? or Benton.	
39	Nekoma.	{ 18 a. Dakota ? or Benton.	
45	Alexander.	"	
52	Bazine.	"	
64	Ness City.	"	
72	Laird.	"	
80	Beeler.	"	
87	Alamota.	" ?	
95	Dighton.	19 Tertiary.	
103	Ellen.	"	
109	Grigsby.	"	
120	Scott City.	"	
129	Modoc.	"	
133	Halcyon.	"	
141	Coronado.	"	
144	Leoti.	"	
154	Crosby.	"	
159	West Line Wichita County.	"	

Larned Extension.

Ms.			Alt.
0	Larned.	{ 18 a. Dakota, Tertiary ?	1993
6	Sage.	"	
17	Rozel.	"	
24	Burdett.	" ? or Benton.	
30	Gray.	"	
35	Hanston.	"	
46	Jetmore.	"	

Augusta Extension.

Ms.			Alt.
0	Augusta.	{ 15. Permo-Carboniferous.	1212
12	Rose Hill.	"	
21	Mulvane.	"	1085
29	Hukle.	"	1230
35	Clearwater.	"	
42	Viola.	"	
47	Anness.	"	
54	Norwich.	"	
67	Rago.	16. Triassic ?	
71	Spivey.	"	
78	Rochester.	"	
86	Nashville.	"	

Atchison, Topeka & Santa Fe R. R.
Augusta Extension.—Con.

Ms.		Alt.
93	Isabel.	{ Tertiary uplands, Triassic in Valleys.
100	Sawyer.	19. Tertiary.
108	Coats.	"
115	Springvale.	"
124	Belvidere.	18 a. Dakota. ?
135	Wilmore.	" or Tertiary.
144	Coldwater.	19. Tertiary.
154	Protection.	16. Triassic.
164	Sitka.	"
170	Ashland.	"
178	Manning.	"
185	Englewood.	"

Osage City Extension.

0	Quenemo.	{ 14 c. Upper Coal Measures.
5	Deavers.	"
11	Lyndon.	"
20	Osage City.	" 1075

Wichita and Western and Kingman, Pratt and Western Railroad.

0	Wichita.	{ 15. Permo-Car- 1291 boniferous.
3	College Green.	"
14	Goddard.	"
20	Garden Plain.	"
26	Cheney.	"
34	Murdock.	"
45	Kingman.	"
56	Calista.	
63	Ninnescah.	
69	Cairo.	
77	Saratoga.	
80	Pratt.	1920
89	Cullison.	2053
96	Wellsford.	2135
100	Haviland.	2172
106	Brenham.	2214
110	Greensburg.	2245
120	Mullinville.	2349
125	W. Li'e, Kiowa Co.	

Hutchison and Kinsley Line.
(South of the Arkansas River.)

0	Hutchison.	{ 15. Permo-Car- 1524 boniferous.
11	Partridge.	"
17	Abbyville.	"
23	Plevna.	
28	Sylvia.	
39	Safford.	18 a. Dakota ?
48	St. John.	"
55	Dillwyn.	"
60	Macksville.	"
67	Belpre.	
75	Lewis.	
84	Kinsley.	* 2153

Atchison, Topeka & Santa Fe R. R.
Southern Kansas Division.
Lawrence and Burlington Branches.

0	Lawrence.	{ 14 c. Upper Coal 849 Measures.
6	Sibley.	" 817
9	Vinland.	" 881
15	Baldwin.	' 1046
20	Norwood.	" 838
26	North Ottawa.	"
27	Ottawa. (Marais des Cygnes R.)	" 896
0	Ottawa.	" 896
4	Burlington Juct.	"
11	Homewood.	"
14	Ransomville.	"
17	Williamsburg.	"
23	Agricola.	"
27	Waverly.	"
33	Hall's Summit.	"
38	Sharpe.	"
46	Burlington.	" 1037
56	Gridley.	"

Southern Kansas Division.

0	Kansas City.	{ 14 c. Upper Coal 746 Measures.
13	Holliday.	" 758
16	Zarah.	"
22	Elizabeth.	"
26	Olathe.	" 1030
35	Gardner.	"
40	Edgerton.	" 962
45	Wellsville.	" 1041
50	LeLoup.	" 949
57	North Ottawa.	"
58	Ottawa. (Marais des Cygnes R.)	" 896
62	Burlington Jc.	"
67	Princeton.	" 966
74	Richmond.	" 1017
78	Scipio.	"
83	Garnett.	" 1056
91	Welda.	" 1098
99	Colony.	" 1121
105	Carlyle.	" 984
110	Iola.	" 955
118	Humboldt.	" 952
127	Chanute.	" 910
128	Eastern Juct.	"
133	Earlton.	" 960
140	Thayer.	" 1445
148	Morehead.	" 900
156	Cherryvale.	" 836
166	Independence.	" 794
172	Crane.	" 783
178	Elk-City.	"
185	Oak Valley.	"
190	Longton.	" 819
196	Elk Falls.	"
203	Moline.	" 1059

Atchison, Topeka and Santa Fe R. R.
Southern Kansas Division.

Ms.			Alt.
211	Grenola.	15. Permo-Carb.	1112
218	Grand Summit.	"	
226	Cambridge.	"	1246
227	Torrance.	"	
231	Burden.	"	1380
239	New Salem.	"	1242
247	Winfield.	"	1112
248	Winfield Junct.	"	
254	Kellogg.	"	
257	Oxford.	"	
263	Dalton.	"	
269	Wellington.	"	1219
127	Chanute.	14 c.Up.Coal Mres.	910
128	Eastern Junct.	"	
135	Vilas.	"	
144	Benedict.	"	
146	Benedict Junct.	"	
155	Coyville.	"	
163	Toronto.	"	
170	Quincy	"	
176	Virgil.	"	
182	Hilltop.	"	
187	Madison.	"	
146	Benedict Junct.	"	
152	Fredonia.	"	
160	Buxton.	"	
166	Upola.	"	
171	Longton.	"	919
269	Wellington.	15. Permo-Carb.	1219
277	Rome.	"	1216
284	South Haven.	"	1124
287	Hunnewell.	"	1102

Independence Extension.

Ms.			Alt.
166	Independence.	{ 14 c.Upper Coal Measures.	794
173	Bolton.	"	
182	Havanna.	"	
187	Niota.	"	
191	Peru.	"	
199	Chautauqua.	"	
205	Elgin.	"	
206	New Elgin.	"	
214	Hewins.	"	
220	Cedarvale.	"	

Pan Handle Extension.

Ms.			Alt.
261	Wellington.	{ 15. Permo-Car- boniferous.	1219
262	Wellington Junc.	"	
270	Mayfield.	"	
277	Milan.	"	
282	Argonia.	"	
284	Albion.	"	
289	Danville.	16. Triassic.	
297	Harper.	"	
303	Crystal.	"	
308	Attica.	"	
308	Attica.	"	
319	Sharon.	"	
329	Medicine Lodge.	"	

Atchison, Topeka and Santa Fe R. R.
Southern Kansas Division.
Pan Handle Extension.

Ms.			Alt.
308	Attica.	16. Triassic.	
315	Crisfield.	"	
323	Hazelton.	"	
330	Kiowa.	"	

Girard Branch.

0	Chanute.	{ 14 c. Upper Coal Measures.	910
1	Eastern Junct.	"	
10	Shaw.	"	
15	Erie.	"	
25	Walnut.	"	931
33	Brazilton.	"	
41	Girard.	{ 14 c. Upper and 14 b. Lower Coal Measures.	v90
50	Frontenac.	{ 14 b. Lower Coal Measures.	
54	Pittsburgh.	"	
57	Chicopee.	"	

Douglass Branch.

0	Florence.	{ 15. Permo-Car- boniferous.	1260
11	Burns.	"	1488
23	DeGraff.	"	
30	Eldorado.	"	1232
38	White.	"	
42	Augusta.	"	1212
49	Gordon.	"	
54	Douglass.	"	1192
59	Rock.	"	
65	Akron.	"	
74	S. Winfield.	"	1112
	Hackney Sta.	"	
81	Arkansas City.	"	1064

Arkansas City Branch.

0	Newton.	{ 15. Permo-Car- boniferous.	1438
9	Sedgwick Junct.	"	1369
10	Sedgwick.	"	1365
18	Halstead.	"	1366
10	Sedgwick.	"	1366
17	Valley Center.	"	1339
22	North Wichita.	"	1394
27	Wichita.	"	1291
32	Green.	"	
38	Derby.	"	1271
43	Mulvane.	"	1085
53	Udall.	"	1272
58	Seeley.	"	1162
66	S. Winfield.	"	1112
71	Hackney Sta.	"	
78	Arkansas City.	"	1064

Atchison, Topeka and Santa Fe R. R. Southern Kansas Division. Caldwell Branch.

Ms.	Caldwell Branch.		Alt.
0	Mulvane.	{ 15. Permo-Car- boniferous.	1065
6	Belle Plaine	"	1209
11	Cicero	"	1306
17	Wellington.	"	1219
27	Perth.	"	1201
31	Corbin.	"	
39	Caldwell.	"	1102

Atchison, Topeka & Santa Fe Railroad. Main Line.

Ms.	Main Line.		Alt.
0	Kansas City.	{ 14 c. Upper Coal Measures.	745
5	Argentine.	"	745
7	Turner.	"	752
10	Morris.	"	
13	Holliday.	"	755
15	Choteau.	"	764
17	Wilder.	"	770
23	Cedar Junct.	"	772
25	De Soto.	"	790
33	Endora.	"	811
40	Lawrence.	"	849
46	Lake View.	"	828
51	Le Compton.	"	844
54	Glendale.	"	849
56	Grover.	"	
59	Spencer.	"	859
62	Tecumseh.	"	860
66	Topeka.	"	884
73	Pauline.	"	1027
79	Wakarusa.	"	946
84	Carbondale.	"	1072
87	Scranton.	"	1099
93	Burlingame.	"	1043
98	Peterton.	"	1065
101	Osage City.	"	1075
106	Barclay.	"	1169
112	Reading.	"	1073
120	Lang.	"	
127	Emporia Junct.	"	1132
128	Emporia.	"	1132
134	Phillips.	"	1123
137	Plymouth.	"	1135
139	Staffordville.	"	1140
143	Ellinor.	"	1156

Atchison, Topeka and Santa Fe Railroad.

Ms.	Railroad.		Alt.
148	Strong City.	{ 14 c. 15. Per- mo-Carbonifer.	1172
152	Evans.	"	
154	Elmdale.	"	1193
162	Clements.	"	
166	Cedar Grove.	"	1237
173	Florence.	"	1264
180	Horner's.	"	1314
184	Peabody.	"	1349
188	Braddock.	"	
194	Walton.	"	1327
201	Newton.	"	1433
211	Halstead.	"	1366
220	Burrton.	"	
227	Kent.	"	1491
234	Hutchison.	"	1534
239	Bath.	"	
245	Nickerson.	"	1592
253	Sterling.	"	1635
259	Alden.	"	1675
265	Raymond.	18 a. Dakota.	1721
269	Clarendon.	"	
275	Ellinwood.	"	1780
280	Dartmouth.	"	
286	Great Bend.	"	1841
293	Dundee.	"	1895
299	Pawnee Rock.	"	1939
308	Larned.	"	1993
313	Hamburg.	"	
319	Garfield.	"	2066
325	Nettleton.		2112
332	Kinsley.		2163
341	Offerle.	19. Tertiary.	2261
346	Bellefonte.	"	2669
352	Spearville.	"	2440
361	Wright.	"	
368	Dodge City.	"	2473
377	Howell.	"	2535
387	Cimarron.	"	2616
393	Ingalls.	"	
400	Charle. town.	"	
406	Pierceville.	"	2750
412	Mansfield.	"	
418	Garden City.	"	2827
425	Sherlock.	"	2924
433	Deerfield.	"	2933
440	Lakin.	"	2989
449	Hartland.	"	3047

36. The portion of the line in Colorado is by Mr. S. F. Emmons, (see Colorado chapter), and that from Trinidad to the end of the chapter, with the notes, was prepared by James Macfarlane, but from what authority compiled, his notes do not in all cases indicate.

J. R. M

37. The road follows the valley bottom of the Arkansas river; underlying rocks are Cretaceous.
S. F. E.

38. *Pueblo.* Niobrara limestone in R. R. cut north of town. Casts of Inoceramus.
S. F. E.

39. *Trinidad.* Coal mines in Laramie. Sandstones capped by basalt.
S. F. E.

40. *Santa Fe.* New Mexico is a very mountainous country with a large valley in the middle, in which is located the At. Top. and Santa Fe Railroad. The valley is formed by the Rio del Norte, which follows a generally southern direction, at least 2,000 miles from the region of eternal snow to the almost tropical climate of the gulf; and only the lower end of it, about 700 miles from Laredo to the mouth, is navigable. The valley is generally about twenty miles wide, and bordered on the east and west by mountain chains six or eight thousand feet high, and north of Santa Fe ten or twelve

Ms.	Atchison, Topeka and Santa Fe Railroad.	Alt.
458	Kendall. 18 b.Ft. Benton.	
465	Mayline. "	
470	Syracuse. "	3218
477	Medway. "	3234
485	Cooledge. " ?	3389
487	State Line.[35] " ?	
	Colorado.[36]	
491	Holley's.[37] { 20. Quat. River bottom.	
501	Granada. "	3436
515	Blackwell. "	3573
526	Prowers. "	
537	Caddoa. "	3756
546	Hilton. "	3877
552	Las Animas. "	3854
562	Robinson. "	3977
571	La Junta. "	4044
590	Catlin. "	4234
606	Nepesta. "	4356
615	Boone. "	4458
628	Baxter. "	
634	Pueblo.[38] 18 b. Colorado.	4639
579	Benton. "	
588	Tempas. "	4407
599	Iron Springs. "	4674
607	Delhi. "	
616	Thatcher. "	5399
625	Tyrone. "	5515
643	Holhne's. "	5704
652	Trinidad.[39] 18 d. Laramie.	5965
658	Starkville.[36] { 18. Lignitic Group.	6831
663	Morley. "	6746
	New Mexico.	
662	Lansing. "	7053
675	Raton. 18. Cretaceous.	6620
679	Dillon. "	6454
681	Otero. "	6877

Ms.	Atchison, Topeka and Santa Fe Railroad.	Alt.
	Maxwell. 18. Cretaceous.	6061
692	Dorsey. "	5883
716	Springer. "	5766
736	Levy. "	6238
758	Shoemaker. { 18 Cretaceous No. 1.	6254
766	Watrous. "	6396
775	Onava. { 18. Cretaceous.	6728
780	Azul. "	6670
786	Las Vegas. "	6381
792	Hot Springs. "	6709
805	Bernal. { 14. Carboniferous.	6056
815	San Miguel. "	6019
837	Pecos. "	
841	Glorieta. "	7415
846	Canoncito. { 18. Cretaceous No. 1.	6858
849	Manzanares. { 14. Carboniferous.	6869
851	Lamy. { 18. Cretaceous No. 1.	6453
869	Santa Fe.[40] "	6987
863	Ortez. { Lignitic Group.	5819
868	Los Cerrillos. "	
870	Waldo. "	5604
891	Wallace. "	5246
893	Algodones. "	5087
902	Bernalillo. "	5031
910	Alameda. "	4919
918	Albuquerque.[41] { Base 18. Cret. Summits of 16. & 17. Jura Triass. alter'g.	4933
928	Isleta. "	4881
931	A. & P. Junct.[42] "	4874
938	Los Lunas. "	4831
948	Belen.[43] "	4784
958	Sabinal.[44] "	4741

thousand, composed of igneous rocks, granite, sienite, diorite, basalt, etc. On the higher mountains excellent pine timber grows; on the lower, cedars and sometimes oak; in the valleys of the Rio Grande, mezquite. The general dryness of the climate and the aridity of the soil will always confine agriculture to the valleys, by well-managed systems of irrigation; but water courses which contain running water throughout the year are very rare. There are, however, large tracts of land, too distant from water or too mountainous to be cultivated, which afford excellent pasture for millions of stock during the whole year, as horses, mules, cattle, sheep and goats, and no feeding in stables in the winter is necessary.

41. *Albuquerque.* On the east are rugged granite mountains. The country about the place is well cultivated by means of irrigation. It is astonishing how soon this apparently sterile soil is changed into the more fertile by affluence of water.

42. *Atlantic and Pacific Junction.* For the sake of continuity, the railroad from this point by the Needles to Mojave, is given in the chapter on California.

43. *Belen.* Mountain bluffs reach the Rio del Norte, and consist of black amygdaloidal basalt.

44. *Sabinal.* This book is strictly a geological work and not botanical, but it is well to note the beginning here in going south of two of the prevailing plants. The so-called *mezquite*, now first makes its appearance. It is thorny like a locust, bears yellow flowers and long pods, with a pleasant sour taste, and the wood is compact and heavy. The *mezquite* is the most common tree on the high plains of Mexico, and the pest of the country for travelers and forms the endless chaparral. Here it is but five or ten feet high, but in Mexico it is some times forty or fifty feet.

The other new plant is the *yucca*, resembling the palm tree with very fibrous, straight, pointed leaves. It is often the only tree growth visible in the desert, with its awkward branches terminated by tufts of its rigid lance-shaped leaves imparting a weird aspect to the landscape. It bears a cluster of white, bell-shaped, numerous flowers hanging down from their weight, one to two feet in length.

Ms.	Atchison, Topeka and Santa Fe Railroad.	Alt.	Ms.	Atchison, Topeka and Santa Fe Railroad.	Alt.
981	Alamillo. 4634		1128	Las Cruces.3871	
994	Socorro.45 4565			The plains are chiefly 18.Cret. The Mts. in partPaleozoic,etc.	
1004	San Antonio. 4517	The plains are chiefly 18. Cretaceous. The Mountains in partPaleozoic probably Carboniferous limestones and in part eruptive.	1140	Mesquite.	3512
1011	Arny. 4512		1148	Lyndon.	7512
			1152	Anthony.	3773
			1161	Montoya.47	
1021	San Marcial. "	4487	1172	El Paso, Tex.48	3713
1028	Pope. "	4557	1096	Rincon, N. M. "	4014
1037	Lava. "	4708	1101	Hatch, N. M. "	4433
1047	Crocker. "	4707	1110	Sellers. "	4495
1059	Engle. "		1134	Florida. "	4484
1067	Cutler. "	4888	1142	Coleman. "	4356
1079	Upham. "	4587	1149	Deming.36 "	4337
1090	Grama. "	4525	1166	Crawford.	
1096	Rincon, N.M.46 "	4014	1173	Hudson.	
	Tonuco. "		1180	White Water.	
1123	Dona Ana. "	3899	1197	SilverCity,N.M.	3771

Near Santa Fe it is from two to three feet high, but the larger species in Northern Mexico grow as trees of several feet in diameter and forty or fifty feet in height.   W.

Mesquit or Prosopis glandulosa of Gray and Torrey, is a shrub or tree with thorny branches and desiduous foliage, which is composed of thin and scattered leaflets, affording no protection from the heat. Its flowers are greenish white at first, and later yellow. The ripe pods are yellowish white, mottled with red, and the ripe beans are used for food by the Mexicans, and are eaten by animals. As fuel, the wood, both root and stem, is unsurpassed. The roots often afford much fuel when there is hardly any stalk, branches, or foliage. Of roots there are two kinds, some of them spreading laterally, while others are very long top roots. Large mesquite trees indicate the presence of water beneath. The mesquit flourishes in Arizona, New Mexico, Texas, and Mexico, its northern limit being the 37th parallel or the southern boundary of Colorado and Utah.

DR. V. HARVARD. U. S. A. in AM. NAT.

45. *Socorro.* The mountains consist principally of porphyritic rocks, with green trachyte. At *Lopez*, six miles beyond Socorro, the mountains which have generally been ten to twenty miles distant now approach, and the bluffs consist of brown, nodular sandstone; south of this the hills are black basalt.

46. *Rincon. The Jornada del Muerto*, literally the day's journey of the dead man, which refers to an old tradition that the first traveler who attempted to cross it in one day perished on the way, was a part of the old Santa Fe road, 90 miles in length without any water in the dry season. The circuitous course of the river, with rough mountains along side of it, rendered it necessary to resort to this awful Jornada. As to the Colorado Desert, see in the California chapter notes Nos. 24, 25, 29, 30 and 31.

47. *Montoya, Organ Mountain.* The eastern mountain chain has a very broken pointed basaltic appearance, and is called the Organ Mountain, from the resemblance of the basaltic columns of its terminus to the pipes of that instrument.

48. *El Paso.* Note 13 on Texas.

THE DESERT FORMATION. To the traveler from the East, the desert country of the West and Southwest is surprising. The valley of the Mississippi, so called, lying between the Appalachian chain and the desert border of the Rocky Mountains, consists of each an expanse of fertile country, as can be found in one body, nowhere else on the face of the globe, producing all the fruits of the earth, including those found in every zone from the boreal regions to the tropics. The region west of the Mississippi Valley, and extending to the Coast Range of California on the contrary, is widely different, owing to the dryness of the climate and the presence of "alkalies" injurious to vegetation in extensive districts, and the physical structure of the surface formations often consisting of stratified pebbles and coarse sandy layers of great thickness. In these deep porus layers, rapidly absorbing the rain-fall, which is very small, leaving the surface an arid waste under a burning sun we see one important cause, in many places, of the desert character of this region, covering a vast extent of the great Southwest. Except on the borders of streams scarcely anything exists deserving the name of vegetation, in the absence of irrigation. But there seems to be hope for most of these deserts, as in other arid localities population and the cultivation of the soil increases the amount of rain-fall, while irrigation from the streams and artesian wells develop wonderful fertility from the soils of deserts.

This blank space is intended for additional geological notes in pencil by the traveler.

Nebraska.*

GENERAL NOTES ON THE GEOLOGY OF NEBRASKA.

1. A large number of the localities have been personally visited. For lines not traversed, careful consideration of published statements by Hayden, Meek, Aughey, and others, has been employed.

2. The quaternary deposits may be grouped, in the order of formation, as follows: (a) Till or typical Boulder Clay, with numerous striated pebbles and boulders from the north. It is usually yellow or blue and "jointed." (b) Red Clay, showing commonly a red color and always more or less pitratified but otherwise resembling till, into which it passes below. It sometimes shows few, if any aebbles in its upper portion. (c) Loess, a homogeneous straticulate silt usually dull yellow or drab and commonly containing calcareous concretions, always cracked within. (d) A Red Loam, containing sometimes white, water-worn quartz pebbles. This deposit is found beyond the western limits of the till and red clay, underneath the Loess. It is frequently capped, as is also the Red Clay at some points, with a dark chocolate-colored earth, two to four feet thick, commonly called "the old soil." Beds of gravel and sand occur irregularly in all quaternary deposits, except, perhaps, the Loess. In Knox county it is the prevailing drift deposit. The term drift is here used to indicate any deposit containing northern erratics referable to glacial origin.

A volcanic ash stratum, evidently deposited in Quaternary times, is widely deposited in Knox, Cuming, Lancaster, Seward, and Furnas counties, and along the Republican further west.

3. The Tertiary Deposits are not satisfactorily determined, especially in portions of the State most traversed by railroads. Hayden, Aughey, and others agree that the later Miocene, White River Group, and the Pliocene, Loup Fork Group, are both represented. But as they are conformable, quite variable in composition, imperfectly exposed, and fossils are rare, they are easily confounded. Hence the formations given in the table are largely provisional.

4. Another question in several cases is whether certain beds are Quaternary or Tertiary. Certain beds of silt or "silicious marl" do not clearly show whether they were deposited in Lake Cheyenne of the Pliocene age or in Lake Missouri, as we may call its successor or continuation in Quaternary times.

Ms. Burlington & Missouri River R. R.		Alt.
0 Plattsmouth.	Loess, 14 c. Up. Carb.	
4 Oreapolis.	" "	974
9 Concord.	"	
19 Louisville.	"	1040
31 Ashland.[5]	18 a. Dak., "	1101
43 Waverly.	"	1136
55 Lincoln. 1155	18 a. Cret. Dakota Gr.	
65 Denton.	"	1247
71 Berks.	{ Deep till over 1428	
	{ 19 c. Pliocene? sand.	
75 Crete.	18 b. Niobrara. 1868	
83 Dorchester.[6]	"	
92 Friendville.	"	1572
108 Fairmont.		1656
115 Grafton.		1699
123 Sutton.		1689
136 Harvard.		1812
151 Hastings.		1947
166 Kenesaw.		2068
176 Lowell.		2076
182 Fort Kearney.		
191 Kearney Junc.		2150

(Middle column, vertical: Loess. 19 c. Loup Fork ov. 19 b. White River Tertiary?)

Ms. Atchison and Nebraska Division.		Alt.
0 Lincoln.[8] 1155	Loess, 18 a. Dakota Gr.	
9 Saltillo.	" " ?	1178
11 Roca.[9] 1219	" 14 c. Up. Carb.	
15 Hickman.	" "	1247
22 Firth.	" "	1319
36 Sterling.	" "	1185
49 Tecumseh.	" "	1113
63 Table Rock.	" "	1028
72 Humboldt.	" "	985
86 Salem.	" "	915
92 Falls City.	, " 14 b. Cl. Mres.	904
111 White Cloud.	" "	856
(Continued in Kansas.)		

Nebraska Railway Division.

		Alt.
0 Nebraska City.	Till, Loess, 14 Cl.M.	941
11 Dunbar.[10]	" " "	1051
22 Syracuse.[10]	" " "	1056
34 Palmyra.	" " "	1151
41 Bennet.[11]	" " "	
47 Cheney's.		
57 Lincoln. 1164	Loess, 18a. Dak. Group	
75 Germant'n. 1584	Till, Loess, 18 Cret.	
82 Seward.	" " "	1443

5. *Ashland.* Fine exposure of Dakota sandstone a little east along the Platte.

6. *Dorchester.* Six miles northwest, in bank of West Blue, a stratum of volcanic ashes 1 to 5 feet thick with drift above and below. (See Note 2.)

7. *Sutton.* (See General Note 3.)

8. *Lincoln.* Loess and Till found overlying all, the latter not conspicuous throughout this line.

9. *Roca.* Fine quarries near station.

* By Prof. J. E. Todd, of Tabor College, Tabor, Iowa, Assistant Geologist, Glacial Division, U. S. Geological Survey.

Ms.	Nebraska Railway Div.—Cont.	Alt.
89	Tamora.	1559
95	Utica.	1389
102	Waco.	1627
109	York.	1642
117	Bradshaw.	1725
124	Hampton.	1770
131	Aurora.	1603

(19 c. Pliocene. (Loup Fork?) 20 Loess.)

Ms.	Nebraska Railway Div.—Cont.	Alt.
142	Marquett. 1525	19 b. W. River, Loess.
150	Central City.1705	" Alluv.
142	Phillips.	Alluv., 19 b. White Riv.
149	Grand Island.	" " 1871
164	Hastings.	1947
178	Kenesaw.	20 Loess, 19 c. 2056
186	Hartwell.	
195	Minden.	Pliocene Sand over
205	Axtell.	19 b. White River.
219	Holdrege.	
285	Rouse.	20 Loess, 19 c. Plio-
240	Oxford Junc.	cene Sand ? over
242	Oxford.	19 b. W. Riv. ? 2079

Salem Branch.

Ms.		Alt.
0	Falls City.	Loess and Drift. 904
11	Verdon.	14 b. Coal Mres. ?
17	Shubert.	Loess and Drift, 14 c.
25	Nemaha.	Up. Coal Mres.? 885

De Witt Line.

Ms.		
0	De Witt.	20 Drift and 1399
15	Western.	Loess, 18 b. Niobrara
23	Tobias.	Chalkstone.

Hebron Branch.

Ms.		
0	Chester. 1621	20 Loess, 18 a. Niob. ?
5	Stoddard.	" "
11	Hebron.	" "

Nemaha Line.

Ms.		
0	Beatrice.	Drift and Loess. / 18 a. Dakota. 1278 / 14 c. Upper Carb.
21	Crab Orchard.	
35	Tecumseh.	
48	Johnson.	1120
57	Auburn.	1230 / 1052
67	Nemaha City.	865
72	Brownville.	894
79	Peru.	905
85	Barney.	
94	Nebraska City.	941

(20 Drift and Loess. 14 c. Upper Carb.)

Northern Division.

Ms.		
0	Lincoln. 1155	Dft., Loess, 18 a. Dak.
7	Emerald.	" " " 1206
13	Pleasant Dale.	" " " 1511
19	Milford. 1614	" " {18b. Ft.Ben- / ton & Niob.
24	Ruby. 1423	" " {18b. Ft.Ben- / ton? & Niob.

Nebraska Railway Division.

Ms.	Northern Division—Cont.	Alt.
29	Seward.	Dft., Loess, Niob.? 1445
42	Ulysses.	Loess, 19c. W. Riv. 1524
50	Garrison.	" " 1603
56	David City.	" " 1619
64	Bellwood.	Alluv. " 1451
74	Columbus.	" " 1458

Eastern Division.

Ms.		Alt.
0	Table Rock.	1025
7	Pawnee.	1180
19	Birchard.	
28	Liberty.	1272
39	Wymore.	1233
48	Odell.	
57	Diller.	1349
66	Endicott.	1291
72	Kesterson.	" Loess.
80	Reynolds.	" "
90	Hubbell.	" " 1460
97	Chester.	"? " 1621
105	Harbine.	"? " 1678
114	Hardy.	18 b. Niobrara?" 1513
122	Superior.	" ? " 1574
135	Guide Rock.	" ? " 1650
142	Amboy.	" ? " 1693
146	Red Cloud.	" " 1690

(20 Loess and Drift. 14 c. Upper Coal. / Loess and Drift, 1281 / 18 a. Dakota Group.)

Republican Valley Branch.

Ms.		Alt.
0	Hastings.	20 Loess, 19 c. 1947 / Pliocene ? ss.
12	Ayr.	" 1647
19	Blue Hill.	" 1978
31	Cowles.	" 1801
37	Amboy.	" 1693
41	Red Cloud.	1690
49	Inavale.	Loess. 19 c. 1729
54	Riverton.	Pliocene ? over 1820
65	Franklin.	
69	Bloomington.	18 b. Niobrara. 1848
74	Naponee.	Chalkstone. 1878
81	Republican.	
87	Alma.	19 c. Pliocene (Loup 1944
93	Orleans.	" [Fork)?
105	Oxford.	" 2079
120	Arapahoe.	" 2177
134	Cambridge.	" 2262
148	Indianola.	" 2380
160	McCook.	" 2511
171	Culbertson.	" 2572
193	Stratton, Neb.	" 2800
211	Benkleman.	" 2975
233	Haigler.	" 3265
242	Laird.	"
249	Wray, Col.	" 3519
257	Robb.	"
264	Eckley.	" 3579

10. *Dunbar, Syracuse.* Quarries within two miles.
11. *Bennet.* Quarries near, and Striæ.

Ms.	St. Joseph and Western Railroad.	Alt.
0	Kearney Junc.	19b.W.Riv.Tert'y 2050
40	Hastings.	" 1947
48	Glenville.	" ?
58	Fairfield.	" ? 1760
66	Edgar.	"
75	Davenport.	18 b. Niobrara. ? 1660
83	Carleton.	" ? 1554
90	Belvidere.	" ? 1501
99	Alexandria.	" 1505
114	Fairbury.	18 a. Dakota. 1316
124	Steele City.	" 1269

Union Pacific Railroad.

Ms.			Alt.
0	Omaha.	14 c. Upper Carb.	1059
10	Gilmore.	"	998
21	Millard.	"	1078
31	Waterloo.	"	
47	Fremont.[12]	18a. Cret. Dak. Gr.	1208
54	Timberly.	"	
69	Rogers.[13] 1859	18b. Ft. Benton & Niob.	
	Schuyler.	"	
84	Richland.	"	1850
	Columbus.	19 c. White River.	
99	Jackson.	"	
109	Silver Creek.	"	1555
121	Clark's.	19 b.W. Riv.Tert'y	1636
132	Central City.	"	
142	Chapman's.	"	1775
154	Grand Island.	"	1871
162	Alda.	"	1922
170	Wood River.	"	1996
183	Gibbon.	"	2067
195	Kearney Junc.	"	2157
204	Stevenson.	"	
212	Elm Creek.	"	2273
221	Overton.	"	2326
231	Plum Creek.	"	2394
239	Cayote.	"	
250	Willow Island.	"	2529
260	Warren.	"	
268	Brady Island.	"	2657
277	McPherson.	"	2895
291	North Platte.	"	2806
299	Nichols.	"	2920
315	Dexter.	"	3000
332	Roscoe.	"	
342	Ogalalla.	"	3216
357	Brule.	"	
361	Big Spring.	"	3371
387	Chappel.	"	
396	Lodge Pole.	"	3533
406	Colton.	"	
414	Sidney.	"	4095
423	Brownson.	"	4200
433	Potter.	"	4366

Ms.	Union Pacific Railroad—Continued.	Alt.
443	Bennett.	19 b. White Riv. Tert'y
451	Antelope.	" 4712
463	Bushnell.	"
473	Pine Bluffs.	" 5047
479	Tracy.	"
484	Egbert.	"
496	Hillsdale.	"
503	Atkins.	"
508	Archer.	"
516	Cheyenne.	(See Wyoming.) 6059

Omaha and Republican Valley Branch. Nebraska Division.

0	Valley. 1149	Alluv., 18a. Dak. ss.	
7	Clear Creek.	Loess, " ?	1155
19	Wahoo.[14]	" " ?	1153
27	Weston.	" " ?	1261
38	Valparaiso.	{ Drift, Loess, 1316 { 18 b. Niob. Chalkst.	
47	Raymond.	{ Loess, 19c. Plio-1155 { cene sand and clay.	
58	Lincoln.	Dft., Loess, 18a. Dak. ss	
66	Jamaica.	" " ?	
69	Hanlon.	" " ?	
80	Cortland.	" " ?	
90	Pickrell.	" " ?	
98	Beatrice.	{ Dft., Loess, 18a.1261 { Dak. ov. 14c.U. Carb.	
112	Blue Springs.	"	
119	Otoe Agency.	"	
125	Oketo.	" ?	
136	Marysville, Kan.		

38	Valparaiso.	{ Drift, Loess, 1316 { 18b. Niob. Ch'kstone	
51	Brainard.	Drift, ? Loess. 1687	
61	David City.	" 1619	
71	Risings. 1597	Loess, 19c. Plioc. sand.	
78	Shelby.	" "	
85	Osceola.	" " 1642	
90	Stromsburg.	" " 1636	

Omaha, Niobrara and Black Hills Branch.

0	Norfolk.	Till, Loess,19 Tert.1532	
5	Munson.	Loess, 19 c. Plioc. 1595	
15	Madison.	" " 1585	
24	Humphreys.	" " 1650	
36	Platte Center.	" " 1537	
41	Lost Creek.	Alluvium, " 1500	
50	Columbus.	" " 1453	
9	Lost Creek.	" " 1500	
20	Genoa.	" " ? 1584	
31	St. Edwards.	"Loess" ? 1666	
43	Albion.	Loess,19b.W.R.?1736	
0	Genoa.	" 19c. Plioc.?1584	
13	Fullerton.	" ? "	
30	Cedar Rapids.	" ? "	

12. *Fremont.* Very fine exposures of Till, Red Clay, Old Soil and Loess in bluff south of the Platte, 2 to 5 miles southwest. A high terrace extends along north of the Platte from Kearney to Fremont.

13. *Rogers.* Fort Benton exposed 5 to 8 miles south near Linwood and Skull Creek.

14. *Wahoo.* On west bank of an old valley of the Platte.

Union Pacific Railroad—Continued.

Ms.	Grand Island and North Loup Br.	Alt.
0	Grand Island.	20 Alluvium.	1871
47	Scotia.	{ Loess, 19 c. Pliocene
49	North Loup.	{ over 19 b.White Riv.

Sioux City and Pacific Railroad.

Ms.	Elkhorn Valley Line, Nebraska Div.		Alt.
0	Mo. Valley, Ia.		
12	S.C.&P.Bridge[15]	20 Alluvium.	
13	Blair.	20 Dft. and Loess.	1100
20	Kennard.	" "	1157
29	Arlington	{ 20 Drift and	1175
38	Fremont.	{ Loess,	1203
46	Nickerson.	{ 18 a. Dakota.	1211
53	Hooper.	{ 20 Alluv. and	1237
61	Scribner.	{ Loess, 18a. Dak.	1256
73	West Point.[16]	{ 20 Till and	1326
89	Wisner.	{ Loess,	1393
98	Pilger.	{ 18 b. Niob-	1423
106	Stanton.	{ rara.	1486
117	Norfolk Junc.	{ Till, Loess, 19	1582
		{ Tertiary. ?	
117	Norfolk Junc.	" ?	1582
119	Norfolk.	" ?	1532
124	Hadar.	"	
132	Pierce.	"	
140	Morehouse.	"	
149	Plainview.	"	
159	Creighton.	{ Drift and Loess, 19 c. Pliocene (Loup) over 19 b. White River.	
128	Battle Creek.[17]		1603
140	Burnett.		1691
147	Oakdale.		1722
152	Neligh.		1761
171	Ewing.		1875
192	O'Neill.	20 Loess, 19 c. Plio-	1992
200	Emmett.	cene ? sands over 19	2039
210	Atkinson.	b. White Riv. beds.	2125
219	Stuart.		2171
229	Newport.		2249
240	Bassett.	19 b.White Riv. ?	2340
250	Long Pine.	"	2416
259	Ainsworth.	"	2538
269	Johnstown.	"	2618
280	Woodlake.	"	2704
287	Arabia.	"	2785
299	Thatcher.	"	2669
806	Valentine.	"	2598

Ms.	Missouri Pacific Railroad.	Alt.
379	Reserve, Kan.		
384	Falls City, Neb.	{ 20 Drift & Loess, 14 c. Up. Carb.	904
394	Verdon,		
401	Stella.		
408	Howe.		
414	Auburn.	{ Drift, Loess,	1053
418	Glen Rock.	{ 14 c. Upper Carb.	
423	Brock.	"	
427	Talmadge.	"	
432	Delta.	"	
337	Dunbar.	"	1051
444	Berlin.	"	
449	Avoca.	"	
455	Weeping Water.	"	
465	Louisville.	"	1040
471	Springfield.	"	
481	Papillon.	"	1003
486	Gilmore.	"	998
496	Omaha.	"	1039

Chic., St. Paul, Minneapolis & Omaha R. R. Nebraska Division.

Ms.			Alt.
0	Sioux City. 1122	Till, Loess, 18 a. Dak.	
2	Covington.	Alluvium, "	1124
7	Dakota City.	" "	1121
12	Coburn Junc.	" Loess, "	1124
16	Hubbard.	Loess, 18 b. Niob.	1161
29	Emerson.	" "	1450
51	Bancroft.	" " ?	1316
58	Lyons.	" " ?	1306
65	Oakland.	" " ?	1300
81	Tekamah.	Till, 18 a. Dakota.	1075
98	Blair.	{ Drift, Loess, 14 Carb. Coal Mres.	1100
102	De Soto.	" "?	1100
104	Mills.	" "?	
107	Calhoun.	" "?	1327
122	Florence.	" "	
128	Omaha.	" "	1039
12	Coburn Junc.	See above.	1124
15	Jackson. 1141	Drift, Loess, 18 a. Dak.	
28	Ponca.[18]	{ " " " 18 b. Niobrara.	1162

Hartington Branch.

29	Emerson.	Loess, 18 b. Niob.	1450
39	Wakefield.	" Drift, "	1404
49	Concord.	" "	1455
63	Coleridge. 1672	" 19 c. Plioc. sands.	
73	Hartington.	{ Dft., Loess, 19 c. Pliocene sands, 19 b. W. Riv., 18 b. Niob.	1434

Norfolk Branch.

48	Wayne.	20 Loess.	1469
67	Hoskins.	"	1684
75	Norfolk.	Drift, 20 Loess.	1542

15. *S. C. & P. Bridge.* 14 c. Upper Carboniferous limestone 50 feet below low water.

16. *West Point.* A fine exposure of more than 100 feet vertical 5 miles northwest, showing Loess, Red Clay, Volcanic Ash (6 feet) and Till. Chalkstone struck in wells at West Point.

17. *Battle Creek.* "Yellow Banks," a cliff of 60 to 70 feet of sand above as much bluish clay, both without fossils, 3 miles northwest.

18. *Ponca.* A seam of lignite at the ferry landing

Colorado.

BY S. F. EMMONS, UNITED STATES GEOLOGIST.

GEOLOGICAL FORMATIONS IN COLORADO.

20. Quaternary.	**17. Jurassic.**	
19. Tertiary.	**16. Triassic.**	
	14. Carboniferous.	{ 14 c. Upp. Cl. Mres. { 14 b. Weber Grits. { 14 a. Low. Carbon-iferous.
18. Cretaceous. { 18 d. Laramie (Lignitic of Hayden.) 18 c. Fox Hills. 18 b. Colorado. { Fort Pierre. Niobrara. Fort Benton. 18 a. Dakota.	**5-7 Silurian.**	
	2. Cambrian.	
	1. Archæan.	

GEOLOGY OF COLORADO.

Certain broad general features of the geology of Colorado are comparatively simple and, owing to the climatic conditions of the region which leave the rock exposures relatively unobscured, can be easily recognized by the geological tourist. The details of structure for any particular region are, on the other hand, as a rule extremely complicated and have only been worked out over limited areas. Even were they fully known it would not be practicable to explain them in the restricted space of the present guide. The notes given above, therefore, must be understood as only indicating these broad and easily recognizable features. In some few cases, moreover, the country has not been visited since the respective railroads have been built, and in such cases the geological indications given may not be strictly applicable to the actual location of the given railroad station; in other cases there may still be some doubt as to the exact subdivision of a geological formation which is exposed at a given point. It is believed, however, that such cases are sufficiently explained by the accompanying notes to avoid leading the observer into any serious error. The Hayden atlas of Colorado gives a most excellent idea of the general distribution of geological formations throughout the state whenever these notes differ therefrom it is because later and more detailed studies have enabled the writer to make such later corrections, as would naturally be called for in a work of so general a character as that necessarily was.

GENERAL STRUCTURE.

In physical structure this region may be divided into a mountain area and plain areas which border it both on the east and west sides. The plain areas and many of the broad valleys, included within the mountain area proper, show as a rule only exposures of Mesozoic, generally Cretaceous, strata, or of overlying Tertiary beds, either of which may be completely obscured by later Quaternary deposits. In the mountain area on the other hand are found the original Archæan rocks, which form the base of all the deposits, and some considerable areas of upturned Palæozoic beds, and of eruptive rocks. Along the immediate flanks of the mountains, especially on the east flank of the Colorado or Front Range, the upturned Mesozoic strata often form fringing reefs, popularly called "Hogback" ridges, approximately parallel with the shore line of the sea in which they were originally deposited. Large areas of Archæan rocks have undoubtedly never been entirely submerged since Archæan times, and everywhere, where erosion has gone deep enough, they are exposed as the base rock.

While the view of earlier geologists that the time of principal uplift in this region was at the close of the Cretaceous still holds good, evidence has recently been found in local nonconformities, of subsidence and elevation both previous and subsequent to this period.

ARCHÆAN FORMATIONS.

These consist of granite, granite-gneiss, micaceous and hornblendic gneisses and amphibolites. The granite is sometimes found as an immense central mass upon which the more distinctly stratified members of the formation are apparently resting; again as distinctly eruptive or intrusive masses penetrating these members, and still again as a constituent part of them, sharing in their bedded structure. Granite has never yet been found in Colorado penetrating later formations than the Archæan, although some later eruptives have so crystalline a structure that they might on hasty examination be considered to be granite. Granite-gneiss is the name given to a very common development among these rocks in which, while the component minerals are foliated, the rocks have still the massive structure of granite. The true gneisses vary from the extreme micaceous to the extreme hornblendic type, and the amphibolites are massive rocks composed almost exclusively of hornblende. Less crystalline rocks than the above, if present, are very rare, and as yet no limestones whatever have been found among these rocks. For one who wishes to make a study of this oldest known geological formation, which presumably represents the first rock crust of the globe, no better field can be found than is afforded by the many deep cañon exposures of Colorado.

PALÆOZOIC FORMATIONS.

These are much thinner in Colorado than in Nevada or in the Eastern states. The *Cambrian* which is the lowest formation found in contact with the Archæan consists of a few hundred feet of saccharodial quartzites, generally white, and passing up into shaly and more or less calcareous beds carrying fossils of the Upper Cambrian. A still lower unconformable series of beds, about ten thousand feet in thickness and later than the Archæan, has been observed by the writer at a single locality in the state but not on the line of any railroad. Above the Cambrian are a few hundred feet of light colored siliceous limestones, often magnesian, sometimes greenish or pinkish in color, whose fauna corresponds to that of the Pogonip, or *Silurian* limestone of Nevada.

The *Devonian* is apparently wanting in Colorado, as the beds found immediately overlying the above, generally a blue gray limestone or dolomite, carry lower Carboniferous fossils. There is some evidence of a nonconformity by erosion in the upper part of the Silurian which would explain the local absence of Devonian formations. The *Carboniferous* formation has a greater aggregate thickness than all the other Palæozoic formations combined. The lower Blue limestone above mentioned is generally succeeded by black shales and these by a very considerable thickness, amounting to two or three thousand feet, of sandstones and conglomerates with subordinate beds of black shale and limestone, locally known as the Weber Grits. Thin beds of impure anthracite are sometimes found in the lower part of this formation. Its prevailing colors are gray or red. The upper part of the Carboniferous formation is of similar constitution, generally with an increasing proportion of calcareous beds and of coarse red sandstones, which are often difficult to distinguish from the immediately overlying red sandstones of the Trias. Gypsum is found in these upper beds. No unquestionably Permian fauna has yet been found in Colorado.

MESOZOIC FORMATIONS.

The *Trias* is represented by a series of coarse red sandstones and conglomerates, the former often strikingly crossbedded, which are everywhere prominent by their brilliant coloring. Organic remains are apparently almost entirely wanting in these beds, for which reason it is impossible to draw a definite dividing line between this and the preceeding or succeeding formation.

The *Jura* consists of a gray or buff sandstone at base, often crossbedded, succeeded by shales of variegated colors, with lenticular secretions of limestone which sometimes form a distinct and prominent bed. This formation is locally well defined by both molluscan and vertebrate remains.

The *Cretaceous* is the most important of the Mesozoic formations and is subdivided into four members. The *Dakota* at the base is characteristically a heavy bedded sandstone or quartzite, carrying a peculiar conglomerate bed at its base. The formation also includes some beds of shale, and on the eastern slopes of the mountains carries beds of remarkable pure fire clay. The *Colorado* next above is essentially a clay formation, its clays being black when freshly opened and bleaching upon exposure; its topography hence is quite characteristic. It generally carries a bed of light colored limestone, which is known as the Niobrara limestone, being characteristic of the sub-division of that name formerly made by Dr. Hayden. The *Fox Hills* and *Laramie* sub-divisions which succeed consist of alternating friable sandstones and clays, and are only distinguishable from each other by their molluscan remains, which in the former are marine, in the latter brackish, or fresh water. The *Laramie* formation has been formerly considered Tertiary by some geologists on account of its fauna, but later investigations have shown it to be more properly the closing member of the Cretaceous from a paleontological point of view, while its stratigraphical relations have always associated it with the Cretaceous. It is the coal-bearing formation of the West, most all the known coal deposits whose horizon has been accurately determined having been found to belong to it, while those not yet thoroughly studied some have been provisorily assigned to the Fox Hills.

TERTIARY FORMATIONS.

There are many detached remnants of fresh water Tertiary formations in Colorado, the relations of which to each other have not yet been thoroughly worked out, nor in most cases have their ages been satisfactorily determined. In the above notes therefore they have not been assigned to any definite subdivision, and the local names are given only when they are sufficiently known to justify it.

QUATERNARY FORMATIONS.

These have likewise not been subdivided, though it is evident that there were several distinct periods of deposit. They have been indicated in the notes only where they so obscure the underlying formations that the latter can be determined either not at all or only with considerable uncertainty.

ERUPTIVE ROCKS.

These form a most important feature in the geology of Colorado. In the Archæan rocks they occur as narrow dikes of porphyry, diorite and diabase. In the Palæozoic and Mesozoic formations are laccolitic masses and immense intrusive sheets of prophyry, porphyrite and diorite whose principal time of eruption was just preceding and subsequent to the Post Cretaceous upheaval. Among later Tertiary and recent eruptive rocks are found hornblende and hypersthene andesites, basalts, rhyolites and less frequently trachytes. The larger areas of recent surface flows are found in the southwestern part of the State. Here are extensive bedded masses of breccia, formerly considered trachytic but probably in large part, if not entirely, andesitic.

MINERALS.

Colorado is exceptionally rich in rare and precious minerals. The best known locality is in the Archæan area around Pike's Peak, extending west as far as Florissant and north to Platte Mountain. Here are found very fine topaz, amazon-stone, zircon and phenacite crystals and a very complete series of cryolite minerals, hitherto known only in Greenland. Boulder county is famous for its great variety of Telluride minerals, many new to science. Topaz is also found in the Arkansas valley, in druses in the rhyolite of Nathrop and Chalk Mountain, associated in the former locality with fine clear garnets. A great variety of silver, copper and bismuth minerals have been obtained from various mining districts. The San Juan and Elk Mountains offer a most attractive field for the mineralogical explorer and have already yielded many new and rare mineral species.

PRECIOUS METALS.

In the value of its product of precious metals Colorado ranks first among the States. Its average annual product may be estimated in round numbers at four million dollars in gold and sixteen millions in silver (coining value). Of this value the single district of Leadville produces more than half. In other metals its most important products have been lead and copper, amounting in a single year to 70,000 tons of the former metal and a thousand tons of the latter. Its ores present every variety of mineralogical composition, but that which produces the greatest aggregate value is argentiferous galena and its secondary products.

In geological distribution the ores are as diversified as in their mineralogical constitution. In the Archæan are found the Telluride ores of Boulder County, the auriferous pyrites of Gilpin County, the argentiferous galena and other silver minerals of Clear Creek and Hall's Valley, and deposits in in the Wet Mountain valley, the Mosquito, Sawatch and other ranges. Ores have been extracted from the Cambrian and Silurian in the Mosquito Range, at Red Cliff, at Ouray and possibly at other localities. From the Lower Carboniferous limestone is derived most of the ore of Leadville, of Red Cliff, Aspen, Monarch, Ouray and other mining districts. At the Ten Mile district and in various parts of the Elk Mountains and San Juan Mountains ores are obtained from the upper horizons of the Carboniferous. Some of the ores from the vicinity of Breckenridge and of the San Juan region come from Triassic horizons, while those in the vicinity of Irwin, Gunnison County, and probably of several other regions not yet examined, are found in Cretaceous rocks. While eruptive bodies in some form are an almost invariable accompaniment of the valuable concentrations of ore in Colorado, the ore itself is rather more frequently found in the associated sedimentary rocks, especially when the latter are calcareous. Important deposits are found, however, in the eruptive rocks them-, selves, notably in the San Juan region, in Summit District, Rio Grande County and in Wet Mountain Valley, (Rosita and Silver Cliff); moreover the so-called fissure veins in the Archæan are sometimes only mineralized dikes of eruptive rock.

COAL AND IRON.

Although the development of these more useful minerals is still in its infancy, amounting to a million and a quarter tons of the former, and 25,000 tons of the latter, the natural resources of the State are most extensive. The coal horizons surround the mountains on every side and penetrate many of the interior valleys, while many deposits of iron ore have already been discovered, although the industrial conditions h ve not yet developed a very active search.

Scenery. Colorado presents several types of scenery, each in its way of great interest. On the east are the great treeless plains, sloping imperceptibly towards the Mississippi valley. Their soil is naturally rich, but, owing to the slight rainfall, only that portion which can be irrigated is available for agriculture, the balance being utilized as pasturage for cattle and sheep. Facing the plains is the Colorado or Front Range, whose trend is nearly north and south and which is cut by the deep cañons of draining mountain streams, utilized by the various railroads which reach the interior. Back of this are a series of mountain valleys, the principal of which are the Wet Mountain Valley, San Luis Park, South Park, Middle Park and North Park; all but the last of these are penetrated or traversed by railroads. West of these is a second series of mountain ranges forming the general line of elevation known as the Park Range, but which is less regular in structure than the Colorado Range. Opposite the South Park it is split into two ranges, the Mosquito and the Sawatch, by the deep

longitudinal valley of the Upper Arkansas River. West of these two system
the Mesa region of the basin of the Colorado river, characterize(
work of deep, narrow cañons cut through soft horizontal strata, which
development beyond the boundaries of the state, in Utah and Ari
tain masses stretch out on the western flanks of the ranges above n
teau region.   Of these the most important are the San Juan Mountai
tains, on the south and north of the Gunnison River respectively, whicl
of eruptive rocks, and some smaller masses such as the Sierra La Sal
ly owe their elevation entirely to eruptive action.   Types of the vari(
rious regions can be seen from the railroad itself, but a far better kn
short excursions which can be readily made from various central poin
    From Denver excursions may be made 1st to Estes Park, 75 mil(
rail and four hours by stage) a most beautiful mountain valley in the
the only one to which the name "Park," as it is understood outside (
applicable. A good hotel and various ranche boarding houses afford
tourist and a great variety of excursions may be made on horseback
Peak, the most precipitous in the Colorado Range, can be easily ascended
nerves are sufficiently steady.   The air is dry, cool, yet mild, and pe
elevation is about 8,000 feet.
    2nd.   By rail to Boulder and thence by wagon or on horseback t
mines of Boulder County.
    3rd.   By rail past the volcanic mesas of Golden, up Clear Creek (
Central City and by Idaho Springs (thermal baths) to Georgetown; fr
minus, it is an easy two-hours' walk or ride to the summit of Gray's
    4th.   By rail to Morrison—upturned Mesozoic strata, carrying gyps
lanta saurus.
    5th.   By the Denver and South Park Railroad up the Platte can
Thence either across Mount Guyot to Breckenridge, and up the Ten-M
or southwest across South Park to Buena Vista in the Arkansas V(
watch Range, by the Alpine Pass, to Pitkin and Gunnison.
    6th.   By the Denver and Rio Grande to Palmer Lake (summ
grounds) on the divide between the South Platte and the Arkansas
foot hills of the Colorado Range.
    The metallurgist will be repaid by a visit to the Argo (copper) a
ing works on the outskirts of Denver.
    From Colorado Springs (excellent hotel—"The Antlers"). By carria
to Manitou, the fashionable summer resort of Colorado. Many hotels.
vescent) springs.   Caverns in the Silurian limestone. Ute Falls (granit(
(upturned red sandstones). Glen Eyrie (residence of General Palmer),
in Archæan and Cambrian just back of the house. Ascent of Pike
U. S. Signal Service on the summit) can be made in a day either or
Drive across Ute Pass to Manitou Park, a pretty mountain valley co
Cambrian and Silurian strata, deposited in a bay of the original Arcl
have escaped erosion.   Near Cheyenne Mountain are found 'the rare
south of Manitou near Florissant amazon stone, topaz and phenacite.
    The projected Midland Railroad (broad gauge) starting from Colors
the Ute Pass, traverse the lower part of South Park, crossing the)
zoic and Archæan) to Leadville, and thence across the Sawatch Range (Ar
ores in lower Carboniferous limestone) on the Roaring Fork of Grand
    Pueblo is of more importance as an industrial centre, than from a pictur
it are tributary the Cañon City coal fields, and those worked by the Atchi
R. R., and the Denver & Rio Grande Railway in the vicinity of Trinidad
various interior railroad lines centering here communicate with the princip
state.   Two large lead smelting works and one Bessemer plant are already
diate vicinity.
    From Pueblo railroad lines run south, southwest, west, north and eas
Topeka & Santa Fe leads to New Mexico, and the southern overland route.
G. Railway crosses the La Veta pass, just north of the Spanish Peaks and s(
the broad alluvial valley of San Luis Park. From Alamosa a branch follows
to Wagon Wheel Gap, now a favorite summer resort; another branch runs s(
into New Mexico; while the main line crosses a low range of eruptive r
past the Toltec gorge, and then crossing the Cretaceous and Tertiary plain
Juan River to Durango (coal mines and smelting works), penetrates the San
the magnificent gorge of the Amimas, having its present terminus at Sil
This is the centre of the boldest and most precipitous mountain mass in Col
important mining districts.   The Alpine climber will here find many u:
prowess; the geologist many problems to solve, and the mineralogist an er
species to be determined.
    Westward.   The main artery of the D. & R. G. Railway reaches the r
(State Penitentiary, Hot Springs and bath, Soda Springs, Lead smelting wo
and petroleum wells in the country around).   From here a branch runs sor
row gorge of Grape Creek to Wet Mountain valley and the mines of Silver (
lows up the Arkansas river through the magnificent cañon, known as the R(
minor valleys cutting across the north end of the Sangre de Cristo range a
Mosquito Range to Salida at the junction of the South Arkansas with the m(
the original line follows the fine north and south valley of the Upper Arka
of Archæan granite, to Leadville, the great silver mining centre. From Lea(
Lakes, formed by the damming up by terminal moraines of a mount
a deep gorge in the Sawatch Range, can be reached in a drive of 16 miles. A
leads across the Arkansas valley (six miles) to Soda Springs, at the foot o
feet).   Beyond Leadville, branches of the D. & R. G. Railway cross the C

Union Pacific Railway.

Ms.	Denver and South Park Division.	Alt.
0	Denver.[1]	20. Quaternary.	5175
1	West Denver.	"	5179
3	Auraria.	"	
7	Mooreville.	"	
7	Bear Creek.	"	5547
11	Littleton.	"	5350
17	Wheatland.[1]	"	
21	Platte Cañon.[2]		
27	Deansbury.[3]	1. Archæan.	
29	South Platte.[4]	" Granite.	5049
32	Dome Rock.	" "	
35	Dawson's.	" "	
40	Buffalo.	" "	
42	Pine Grove.	" "	
48	Crosson's.[4]	" "	
52	Estabrook.[5]	"	
55	Bailey's.	"	
59	Slagkt's.	"	
62	Meadows.	"	
66	Grant.[8]	"	8491
69	Webster.[6]	"	
74	Hoosier.[7]	"	9905
76	Kenosha.[7]	"	
81	Jefferson.	{ 20. Quaternary over Laramie.	9865
88	Como.[8]	"	
94	Halfway.	Quartz-porphyry.	
97	Selkirk.	"	
99	Boreas.[9]	"	
101	Dwyer.	16. Red Sandstone.	
104	Argentine.[10]	18. }	
106	Mayo.[10]	18. }	
110	Breckenridge.[11]	Quaternary.	
114	Broncho.	"	
116	Dickey.[11]	"	
120	Frisco.	{ 20. Quaternary over Archæan.	
122	Curtin.[12]	"	
126	Wheeler.	"	
133	Kokomo.	14 c.& porphyry.	10609
134	Robinson.	" "	10549
137	Climax.	14 b. Webber Grits.	
139	Alicants.[13]	1°. Archæan.	11146
144	Bird's Eye.[13]	14b.& porphyry.	10161
151	Leadville.[14]	{ 20. Quaternary Lake beds.	10178

Union Pacific Railway.

Ms.	Denver and South Park Division.—Con.	Alt.
88	Como.[8]	{ 20. Quater. over Laramie Cretaceous.	
94	Red Hill.	18 b. Colorado.	
103	Arthur's.	"	
104	Garos.	"	
105	Garo's.	"	
115	Fairplay.[15]	16. Trias.	9941
120	London.	1.Archæan.	
113	Platte River.[16]	20. River Bottom.	
120	Hill Top.	{ 14. Carboniferous Limestones.	
127	McGee's.	1. Granite.	
132	Charcoal.	"	
133	Schwanders.	"	
137	Buena Vista.	{ 20. Quaternary over Archæan.	
133	Schwanders.	1. Archæan.	
137	Nathrop.[17]	{ 20.Quaternary over Archæan.	
142	Hortense.	1. Granite.	
149	Alpine.	"	
153	St. Elmo's.	"	
155	Murphy's.	1. Archæan.	
175	Pitkin.[18]	"	
190	Parlins.	20. Quaternary.	
202	Gunnison.		
216	Baldwin.	18 d. Laramie.	
219	Baldwin Mines.	"	

Colorado Central Branch—Colorado Division. Broad Gauge.

Ms.		Alt.
0	Cheyenne.		
6	Colorado Junct.	{ 19. Niobrara Pliocene.	5514
13	Lone Tree.	"	
24	Taylor's.	18 c. Fox Hills.	
32	Bristol.	"	
40	Fort Collins.	"	
63	Loveland.	18 b. Colorado.	
71	Berthoud.	"	
80	Longmort.	"	
85	Niwat.	"	
92	Boulder.	18 c. Fox Hills.	5508
100	Louisville.[19]	18 d. Laramie.	
110	Church's.	"	

north, one descending Eagle River to the mining town of Red Cliff, the other the Ten-Mile river to the Middle Park, each valley being extremely precipitous and picturesque.

From *Salida* again, the present main line goes westward, past Poncho Springs (Thermal baths), sending off a short branch to the northwest to the Monarch mining district, and southward across Poncho Pass into the San Luis Valley and the iron mines at Hot Springs. The main line crosses the south end of the Sawatch range by the Marshall Pass and follows the Gunnison river down to the Utah boundary line. From Gunnison City (LaVeta Hotel) a branch runs north to Crested Butte, a good centre for visiting the wild and beautiful scenery of the Elk Mountains, and the mines of anthracite and bituminous coal, of silver, copper and lead. The forest growth and vegetation is generally more luxuriant on these western slopes than on the east flanks of the mountains. Below Gunnison the railroad passes part way through the cañon of the Gunnison (known as the Black cañon) and then diverges to the south into the Uncompaghre valley. From Montrose in this valley the San Juan mountains may be reached by stage by way of Ouray, probably the most picturesquely situated town in the state. Further westward the country assumes the somewhat monotonous but striking appearance characteristic of the Colorado plateau region.

Union Pacific Railway.
Colorado Central Branch—Colorado Division.

Ms.	Broad Gauge—Con.		Alt.
118	Ralston.[20]	18 d. Laramie.	
121	Jones' Siding.	{ 19. Monument Creek Tertiary.	
122	Golden.[21]	18 d. Laramie.	5684
130	Arvada.	20. Quaternary	5322
136	Argo.[36]	over Denver	
138	Denver.	Tertiary.	5175

Narrow Gauge.

0	Denver.[22]	20.	
16	Golden.[22]	18 d. Laramie.	5684
19	Chimney Gulch.[23]	1. Archæan.[23]	5909
22	Guy Gulch.	"	6212
24	Beaver Brook.	"	6391
28	Big Hill.	"	6523
29	Forks Creek.	"	6875
31	Cottonwood.	"	7178
34	Smith Hill.	"	7526
36	Black Hawk [24]	"	8031
40	Central City.[24]	"	8484

Georgetown Branch.

29	Forks Creek.	1. Archæan.	
33	Floyd Hill.	"	7201
38	Idaho Springs.	"	7541
45	Lawsons.	"	8111
51	Georgetown.[25]	"	8474
56	Silver Plume.	"	9074
60	Graymont.[26]	"	

Omaha and Denver Short Line.

(Continued from Nebraska.)

361	Big Springs.[27]	20. Quaternary.	
369	Barton.	"	
371	Denver Jc. (formerly Julesberg.)	"	5184
386	Sedgewick.	"	
400	Crook.	"	
417	Iliff.	"	
429	Sterling.	"	
441	Merino.	"	
458	Snyder.	"	
471	Denel.	"	
480	Orchard.	"	
506	Hardin.	"	
522	La Salle.[27]	"	
533	Platteville[28]	"	4612
541	Lupton.	"	4896
549	Brighton.	"	4979
554	Henderson.	"	
556	Jersey.	"	
569	Denver.[28]	"	5175

Union Pacific Railway.
Denver Pacific Branch

Ms.	Colorado Division.		Alt.
0	Denver.	{ 20. Quaternary[5175] over Denver Tertiary.	
2	Jersey.	"	
7	Hatchery.	"	
14	Henderson.	"	5028
19	Brighton.	18 d. Laramie.	
26	Lupton.	"	
35	Platteville.[29]	"	
41	Hautes.	"	
46	La Salle.	"	
48	Evans.	{ 20. Quaternary[4646] River Bottom.	
52	Greeley.	"	4642
60	Eaton.	18 d. Laramie.	
67	Pierce.	"	
76	Dover.	"	
86	Carr.	"	5696
96	Athol.	{ 19. Niobrara Pliocene.	

Boulder Branch.

0	Denver.	{ 20. Quaternary over Denver [5175] Tertiary.	
2	Jersey.	"	
7	Hatchery.	"	
14	Henderson.	"	
19	Brighton.	18 d. Laramie.	5024
26	Dick.	"	
30	St. Vrains.	"	
34	Erie.[30]	"	
35	Northrop.[30]	"	
36	Canfield.[30]	"	
40	Clifton.	18 c. Fox Hills.	
43	Vochmont.	18 c. Ridge of Solerite.	
46	Boulder.	18 c. Fox Hills.	

Boulder and Carbon Branch.

0	Boulder.	18 c. Fox Hills.	5308
6	Marshall.[30]	18. Laramie.	5329

Morrison Branch.

0	Denver.	{ 20. Quaternary[5175] over Denver Tertiary.	
1	West Denver.	"	
7	Mooreville.	"	
8	Bear Creek.	"	
10	Gilman.	"	
13	Mt. Carbon.	18 d. Laramie.	
16	Morrison.[31]	18 a. Dak. 17. Jurass.	

1. *Denver to Wheatland.* The road follows Platte Valley bottom, and edges of benches formed of Denver Tertiary underlain by Laramie Cretaceous.
2. *Platte Canon.* 16, 17, 18 a., 18 b. Hog back ridges of Cretaceous sandstones and Jurassic limestones. Sections from Ft. Benton to Trias, inclusive, from a point one mile east to a point one half mile west of station.
3. *Deansbury.* Granite gneiss and amphibolites.
4. *South Platte to Crosson's.* Massive red granite throughout this distance. In part disintegrating

Ms.	Union Pacific Railway. Greeley, Salt Lake and Pacific Branch.	Alt.
0	Denver.	{ 20. Quaternary over Denver Tertiary.
2	Jersey.	"
7	Hatchery.	"
14	Henderson.	• "
19	Brighton.	18 d. Laramie.
26	Lupton.	"
35	Platteville.	"
41	Hautes.	"
46	La Salle.	"
48	Evans.	{ 20. Quaternary⁴⁶⁴² River Bottom.
52	Greeley.	"
64	Windsor.	"
76	Fort Collins.	18 c. Fox Hills. 4815
80	La Porte.	18 d. Colorado. 5065
91	Stout.³²	1°(?)

	Boulder Cañon Branch.	
0	Boulder.	18 c. Fox Hills. 5505
4	Oredel.	1. Archæan.³³
6	Crisman.	"
7	Gold Hill.	"
9	Sugar Loaf.	"
13	Sunset Branch.	"

Ms.	Union Pacific Railway. Kansas Division.		Alt.
	Continued from Kansas.		
420	Wallace, Kansas.		
429	Eagle Tail, "	18 d. Laramie.	1484
440	Monotony, "	"	3774
452	Arapahoe.³⁴	"	4006
462	Cheyenne Wells.	"	4277
472	First View.	"	
487	Kit Carson.	"	4259
499	Wild Horse.	"	4436
510	Aroya.	"	4645
523	Mirage.	"	4841
534	Hugo.	"	5050
546	Lake.	"	
562	Cedar Point.	"	5712
566	Godfrey.	"	5603
577	Agate.	"	5453
595	Byers.	"	5203
607	Bennett.	"	
617	Box Elder.	"	5525
629	Magnolia.³⁴	"	
637	Jersey.³⁵	{ 20. Quaternary Gravels.	
639	Denver.	"	5175

readily on exposure to the atmosphere, in part resisting disintegration and making handsome building stone. Quarries near Buffalo Station.

5. *Estabrook—Grant.* Granite gnéiss, schists (some amphibolites) and gray granite.

6. *Webster.* Branch Valley leads to Geneva district and Hall Valley mines. Bismuth silver ores.

7. *Hooser—Kenosha.* Gray granite and some eruptives.

8. *Como.* Coal mines west of town. At Hamilton, higher up Tarryall Creek, are abandoned gold placers. Here was the first discovery of gold in Colorado west of the Colorado range.

9. *Boreas.* Mt. Guyot to the east, almost entirely made up of eruptive rocks, with a few caught up fragments of sedimentary beds.

10. *Argentine—Mayo.* The beds are much disturbed and probably faulted on the slopes of the range toward Blue River valley, and the horizons have not been determined with certainty. The sandstones on the lower slopes probably belong to the Dakota, and the black clays higher up may be Colorado.

11. *Breckenridge—Dickey.* Road follows valley of Blue River. Rich gold placers have been washed in this and tributary valleys.

12. *Curtain—Birds Eye.* On the east side of the narrow valley of Ten Mile Creek which the R. R. ascends, the steep slopes of the Mosquito Range furnish excellent exposures of Archæan rocks. White veins of pegmatite and dark bands of hornblendic schists stand out prominently in the generally light-colored mass of granite-gneiss. About three miles above Wheeler the R. R. crosses the Mosquito fault, and passes from Archæan into Upper Carboniferous and intrusive porphyry.

13. *Alicante.* The Mosquito fault crosses the Arkansas valley in a north and south direction about tangent to the curve or loop of the railroad. By its displacement the Archæan rocks forming the high mountains to the east have been lifted up and brought into juxtaposition with Upper Carboniferous and Triassic strata on the west.

14. *Leadville.* Silver mines in Carboniferous limestone. Gold placers in gulches.

15. *Fairplay.* Quaternary gravels which have been washed for gold.

16. *Platte River.* Salt Springs and gypsum deposits west of here.

17. *Nathrop.* Ridge east of station, rhyolite carrying topaz.

18. *Pitkin.* Ridge of Palæozoic limestones to the northwest.

19. *Louisville.* Fault in R. R. cut one half mile south. In opposition are seen the coal a.s. at base of Laramie, and the shales and iron-stones above the sandstone.

20. *Ralston.* Basalt breaking through the Cretaceous formations in hill to the west.

21. *Golden.* Table topped ridges to south and east formed of Denver Tertiary beds, capped and protected from erosion by flow of basaltic lava. Hogback ridges of Dakota sandstone, carrying fire clay to the west. Coal mines in vertical beds of Laramie sandstone. See 22.

22. The road crosses vertical outcrop of Laramie and Dakota Cretaceous and of Triassic Red beds before entering the Archæan. Excellent fire clay found in the Dakota, north of Golden.

23. Granite, granite-gneiss and schists.

24. Gold mines in granite-gneiss often associated with porphyry dikes. Main ore is auriferous pyrites. Treated in amalgamating mills.

25. *Georgetown.* Silver mines mainly in granite-gneiss and intrusive porphyry. Main ore argentiferous galena, pyrite and sulphides of silver. Ore mostly treated in smelting works, after being dressed and concentrated here.

Union Pacific Railway.
Ms. Denver, Marshall and Boulder Branch. Alt.

For dist'es see Col. C. Br., B'd G'ge.

Ms.	Station	Formation	Alt.
	Denver.	{ 20. Quaternary over Denver Tertiary.	
	Argo.[36]	20. Quaternary.	
	Argo Junction.	"	
	Semper.	"	
	C. C. Junction.	"	
	Louisville.[00]	18 d. Laramie.	
	Boulder.	18 c. Fox Hills.	5303
	Ni Wot.	18 b. Colorado.	
	Longmont.[119]	"	
	Highland.	"	
	Berthoud.	"	
	Loveland.	"	
	Fort Collins.	18 c. Fox Hills.	

Denver and Rio Grande Railway.
Denver and Leadville Line.

Ms.	Station	Formation	Alt.
0	Denver.	{ 20. Quaternary over Denver Tertiary	5173
2	Burnham.	"	
4	N. O. Crossing.	"	
8	Petersburg.	"	
11	Littleton.	"	5350
17	Acequia.[37]	{ 19. Monument Creek Tertiary.	5508
25	Sedalia.[38]	"	
29	Plateau.	"	
33	Castle Rock.[39]	"	6195
35	Douglas.	"	
39	Glade.[40]	"	6515
43	Larkspur.	"	6649
47	Greenland.[41]	"	6599
52	Palmer Lake.[42]	"	
56	Monument.	"	6953
58	Borst's.	"	6611
62	Husted's.[43]	"	
67	Edgerton.	"	
71	Pike View.[44]	"	
75	Colorado Springs.[45]	{ 18 d. Laramie	5970
84	Widefield.	{ 20. Valley Quaternary over Colorado Cretaceous.	5697
89	Fountain.	"	5508
94	Butte.[46]	"	3846
96	Wigwam.	"	
106	Pinon.	"	6016
112	Cactus.	"	

Denver and Rio Grande Railway.
Denver and Leadville Line.—Con:

Ms.	Station	Formation	Alt.
120	Pueblo.[47]	18 b. Colorado.	4669
124	Goodnight.	{ 18 b. Colorado Cretaceous.	4703
130	Meadow.[48]	"	4798
135	Swallow.	"	
140	Carlisle.	"	
143	Beaver.[49]	"	
144	Thompson.	"	
153	Florence.[50]	"	
157	Reno.[51]	18 c. Fox Hills.	
161	Cañon City.[52]	{ 18 b. Colorado Limestone.	5332
162	Cañon Junction.	1. Archæan.	5313
165	Gorge.[53]	"	
171	Parkdale.[54]	{ 17. and 18 a. Jura and Dakota Cretaceous.	5715
176	Spike Buck.[55]	1. Archæan.	
186	Texas Creek.[56]	1. Gneiss.	6196
193	Cotopaxie.[57]	1. Red Granite.	6364
199	Vallio.	{ 20. Quaternary and Tertiary beds over Archæan.	6513
205	Howards.[58]	{ 20. Quaternary over Archæan.	6693
207	Badger.[59]	{ 14 a. Upper Carboniferous.	6743
215	Cleora.	{ 20. Quaternary over Archæan.	6993
217	Salida.[60]	"	7028
224	Brown's Cañon.	"	
225	Harp.	1. Archæan.	
226	Hecla Junction.	"	
234	Nathrop.[61]	{ 20. Qaternary over Archæan.	7673
239	Midway.	1. Archæan.	7380
242	Buena Vista.[62]	{ 20. Quaternary over Archæan.	7943
243	Dornick.	"	
246	Americus.	"	8113
250	Riverside.	{ 1. Archæan Granite.	8350
255	Pine Creek.	"	8785
259	Granite.[63]	"	8923
261	Twin Lakes.	"	9005
265	Hayden.	{ 20. Arkansas Valley Quaternary.	9136
270	Crystal Lake.	"	9389
273	Malta.	"	9588
274	Eilers.[64]	20. Quaternary.	9888
277	Leadville.[64]	"	10178

26. *Graymont.* Ascent of Gray's Peak easily made in a few hours.

27. *Big Springs—La Salle.* The railroad follows the bottom of the South Platte River. The country adjoining is formed of Upper Cretaceous beds overlaid on the north by Miocene Tertiary.

28. *Platteville—Denver.* The plain country traversed is underlaid by Laramie Cretaceous covered by quaternary gravels and loess, and in some parts by remnants of Denver Tertiary.

29. *Platteville.* Directly west is Long's Peak (14, 271 ft.), at the southern end of the beautiful valley of Estes Park; it is the highest and finest mountain in this portion of Colorado.

30. Coal mines.

Ms.	Denver and Rio Grande Railway. Denver and Ogden Line.	Alt.	Ms.	Denver and Rio Grande Railway. Denver and Ogden Line —Con.	Alt.
217	Salida.⁶⁰ — 20. Quaternary 7023 over Archæan.		364	Colorow.⁷⁹ — 20. Quaternary.	
221	Poncha Junct.⁶⁵ — 19. Tertiary 7468 Lake beds.		374	Delta. — "	4947
226	Otto.⁶⁶ — 1. Archæan.		376	Escalante. — "	4814
228	Mears Junction. — Andesite.		392	Dominguez. — "	4771
230	Shirley. — "	6654	399	Bridgeport. — "	4727
235	Gray's.⁶⁷ — 1. Archæan Granite.		409	Kahnab. — "	4649
242	Marshall's.⁶⁸ — Andesite.		412	White Water. — "	4635
245	Hillden. — 1. Gneiss.		425	Grand Junct. — "	4561
246	Shamans.⁶⁹ — "		433	Roan.⁸⁰ — "	4509
250	Chester. — Eruptive Rocks.		439	Fruitvale. — "	
254	Buxton. — "		446	Crevasse. — "	
259	Sargent. — 1. Archæan.	8486	452	Shale. — "	4575
264	Elks. — "		457	Excelsior. — "	4595
267	Crookton. — Eruptive Rocks.		463	Acheron.⁷⁹ — "	
271	Doyle. — "	8035	474	West Water.¹²¹ — "	
272	Bonita.⁷⁰		479	Cottonwood. — "	
278	Parlin.⁷¹ — 1. Archæan.	7935		Continued in Utah.	
284	Mounds. — "			**Denver and Silverton Line.**	
290	Gunnison.⁷² — 20. Quaternary.	7655	121	Bessemer.⁸¹ — 18 b. Colorado.	4751
296	Ridgeway. — 1. Archæan.		129	San Carlos. — "	4912
302	Kezar. — "	7409	134	Greenhorn. — "	5076
309	Cebolla.⁷³ — "	7550	141	Salt Creek. — "	5442
316	Sapinero.⁷⁴ — "	7235	147	Granero's. — "	
322	Curecante. — "	7082	151	Huerfano. — "	5667
329	Crystal Creek.⁷⁵ — "	6869	164	Apache. — "	5917
331	Cimarron.⁷⁶ — Fox Hills Sandstone.	6674	176	Walsen's.³⁰ — 18 d. Laramie.	6187
336	Cerro Summit.⁷⁷ — "		181	Wahatoya. — 18 a. Dakota.	6482
343	Cedar Creek. — 18 b. Colorado Clays.	6723	191	La Veta. — 14. Carboniferous Beds,	7002
353	Montrose.⁷⁸ — "	5771	199	Ojo. — "	8167
			202	Mule Shoe.⁸² — "	8762
			206	Veta Pass.⁸³ — "	

31. *Morrison.* Remains of Atlanosaurus found in Jura—Trias (red beds) just above town resting on Archaean Gypsum deposits.

32. *Stout.* Gypsum deposits found in Triassic rocks.

33. Numerous dikes of porphyry and diorite traversing the granite and schists. Mines of gold and silver. In the former a most interesting series of telluride minerals.

34. *Arapahoe—Magnolia.* The outlines of the formations on this plain area are still somewhat uncertain; they are undoubtedly Cretaceous, however, with a varying cover of Quaternary.

35. Underlaid by Denver Tertiary.

36. *Argo.* Large smelting works using the Augustine Ziervogel process for the separation of silver from copper.

37. *Acequia.* High line canal crosses Plum Creek.

38. *Sedalia.* Wild Cat Buttes to the west show folding of Monument Creek beds. Plateau capped by Monument Creek Tertiary.

39. *Castle Rock.* Table topped hills to the east, capped by pink rhyolitic tufa, extensively used as building stone in Denver.

40. *Glade.* Dawson's Butte to west.

41. *Greenland.* White knoll of Tertiary to west, known as Casa Blanca.

42. *Palmer Lake.* Tertiary covers upturned edges of Mesozoic and Palæozoic strata and abuts against Archæan foot-hills.

43. *Husted.* In the distance to the west are some tall monuments, characteristic of the formation.

44. *Pike View.* On the line between Monument Creek and Laramie formations.

45. *Colorado Springs.* Fine view of Pike's Peak. Manitou, a summer resort where the actual springs are situated, lies four miles west, in a recess at the foot of the mountains.

46. *Butte.* Road follows the bottom of the Fontaine-qui-bouille, or Fountain Creek, named by the Canadian trappers from the effervescent springs at its source.

47. *Pueblo.* Niobrara limestone carrying casts of Inoceramus in railroad cut north of town.

48. *Meadow.* Bluffs capped by limestone.

49. *Beaver.* Prominent outcrops of Niobrara limestone along Bluffs on either side of railroad.

50. *Florence.* Oil Wells. Branch to Cañon City coal fields to south.

51. *Reno.* Laramie beds capping cliffs to north.

52. *Cañon City.* Road crosses upturned edges of Dakota sandstone, Jura and Trias—latter capped by later horizontal beds. Effervescent spring in Dakota hog back north of road, and Hot Spring on south near contact of Archæan.

Denver and Rio Grande Railway. Denver and Silverton Line.—Con.				Denver and Rio Grande Railway. Denver and Silverton Line.—Con.		
Ms.			Alt.	Ms.		Alt.
208	Blanca.[54]	14. Carboniferous Beds.		394	Carracas.[94]	18 c. Fox Hills. 6151
213	Placer.[55]	20. Quaternary. 8888		402	Arboles.[95]	19. Tertiary 5991 Sandstones and Shales.
219	Trinchera.[56]	20. Quaternary 8082 over Archæan.		405	Siding No. 22.[96]	"
226	Garland.		7914	409	Vallego.	" 6200
238	Baldy.	20. Alluvial 7592 deposits in the San Luis Valley.		412	Solidad.	" 6355
247	Hayes.	"		415	Serape.	" 6210
250	Alamosa.	"	7524	417	La Boca.[97]	20. Quaternary.
265	La Jara.	"	7587	424	Ignacio.	19. Tertiary 6415 Sandstones and Shales.
279	Artonito.[57]	20. Quaternary 7866 Gravels.		430	Silla.	" 6650
289	Lava.	"	8446	433	Colina.	" 6712
298	Big Horn.	Basaltic Tufa. 9000		436	Florida.	18 d. Laramie. 6695 Fox Hills.
303	Sublette.	Andesitic Creccia. 9215		444	Bocea.	"
309	Toltec.[58]	" 9443		447	Carbon.[50]	"
317	Osier.	" 9615		450	Durango.[98]	18 b. Colorado 6498 Clays.
321	Los Pinos.	" 9615		452	Animas.[99]	18 d. Dakota 6552 Sandstones.
329	Cumbres.	" 9998		457	Home Ranch.	14 c. Upper Carboniferous.
331	Coxo.	" 9781		459	Trimble.[100]	"
334	Cresco.	"		461	Hermosa.[99]	14 b. Weber Grits.[6628]
338	Lobato.	"		468	Rockwood.[101]	1. Archæan Red Granite.
343	Chama.	"	7841	477	Cascade.	1. Granite Gneiss and Schists. 7768
348	Willow Creek.	"	7720	481	Needleton.	" 8118
352	Azotea.	"	7701	489	Elk Park.[102]	" 8761
362	Monero.[59]	18 c. Fox Hills.	7256	495	Silverton.	20. Quaternary 9202 Valley.
365	Amargo.[90]	"	6987			
372	Dulce.[91]	"	6757			
376	Navajo.[92]	"	6566			
385	Juanita.[93]	"	6819			

53. *Gorge.* The Archæan in the Royal Gorge consists of gneiss and schists with intrusive masses of red granite and small dikes of diabase.

54. *Parkdale.* This valley was one of the ancient bays in the original Archæan land mass.

55. Gneiss and amphibolite traversed by red granite.

56. *Texas Creek.* At head of valley to north are horizontal beds of eruptive rocks (andesite?).

57. *Cotopaxi.* Eruptive rock on high hill to north. Carboniferous to the south of Vallio.

58. *Howards.* High peaks of the Sangre de Christo range to the south.

59. *Badger.* A continuous descending series of upturned Palæozoic beds, somewhat faulted, and resting on Archæan is crossed from here to Cleora.

60. *Salida.* Tertiary beds on west side of valley. Andesite hills east of town.

61. *Northrop.* Ridges of Rhyolite just above station. Rock carries Crystals of garnet and topaz.

62. *Buena Vista.* Fine view of the high peaks of the Sawatch Range. Mt. Harvard (14,375 ft.) the northermost, then Mt. Yale (14,187); to south of west, Mts. Princeton (14,196), Mt. Antero (14,246), and Mt. Shavano (14,239).

63. *Granite.* On the west side of the valley are many important gold placers. Twin Lakes, beautiful sheets of water held by terminal moraines, at the north of Lake Creek, a few miles west of railroad. (Good mountain hotel, trout fishing, etc.) Remarkably well defined moraines on either side of lakes.

64. *Eilers—Leadville.* Road rises from Arkansas valley over mesa of lake beds covered by re-arranged moraine material. Above Leadville are argentiferous lead deposits in Carboniferous limestone.

65. *Poncha Junction.* Line of Archæan opposite Spring hotel.

66. *Otto.* Some Andesite on the east side.

67. *Gray's.* Andesite at mile post 237.

68. *Marshall's.* Hills around are largely Archæan.

69. *Shaman's.* Eruptive on the south and at sign of station.

70. *Bonita.* At Bonita are Cretaceous rocks resting on Archæan—eroded. At 273.5 to 274.5 an eroded anticlinal gives a wider outcrop to the Archæan.

71. *Parlin.* Cretaceous on hills to north. Probably eruptives to south capped by Cretaceous beds and eruptives.

72. *Gunnison.* Eruptive cliffs (Andesite) on west and northwest.

Ms.	Denver and Rio Grande Railway. Manitou Branch.		Alt.
75	Colorado Spr'gs.	18 d. Laramie.	5970
78	Colorado City.	18. Colorado.	6092
81	Manitou.[103]	{ 14. Carbonifer-[6302] ous Limestones.	

Silver Cliff Branch.

161	Cañon City.	{ 18 a. & b. Col-[5322] orado Limestone & Dakota Sandstone.	
163	Cañon Junct.	1. Archæan.	
172	Marsh.[104]	"	6325
177	Soda Springs.	"	6323
194	West Cliff.[105]	{ 20. Quaternary[7342] over Archæan.	

San Luis Branch.

217	Salida.	20. Quaternary.	7025
228	Mears Junct.	Andesite.	8417
231	Poncha Pass.	1. Archæan.	8945
247	Villa Grove.	{ 20. Quaternary[7735] of San Luis Valley.	
255	Hot Springs.[106]	{ 14. Carboniferous(?) Limestone.	

Crested Butte Branch.

217	Salida.	{ 20. Quaternary[7025] over Archæan.	
290	Gunnison.[73]	"	7655
301	Almont.[107]	1. Archæan.	
312	Jack's Cabin.	18 c. Fox Hills.	8284
318	Crested Butte.[108]	18 c. Laramie.	8858

Del Norte Branch.[109]

250	Alamosa.	20. Quaternary	7524
268	Henry.	"	
281	Del Norte.	"	7855
297	South Park.[110]	"	8166
311	Wagon Wheel Gap.	{ Eruptive [8427] Cliffs.	

Ms.	Denver and Rio Grande Railway. Monarch Branch.		Alt.
217	Salida.[60]	20. Quaternary.	7025
221	Poncha.	"	7455
228	Maysville.	{ 19. Tertiary [8290] Lake Beds.	
235	Garfield.[111]	1. Archæan	
238	Monarch.	"	

Eagle River Branch.

277	Leadville.	{ 20. Quaternary Lake Beds.	
273	Malta.	{ 20. Arkansas [9558] Valley Quaternary.	
279	Keildar.	"	9943
282	Crane's Park.[112]	{ 1. Archæan [10097] Granite.	
283	Tennessee Pass.	"	
294	Eagle Park.[113]	{ 20. Quaternary[9205] Valley Bottom.	
300	Red Cliff.[114]	{ 2 b. Cambrian [9449] Quartzita	

Blue River Branch.

277	Leadville.	{ 20.Quaternary[10175] Lake Beds.	
282	Birds Eye.	14 b.&Porphyry.[10161]	
290	Fremont Pass.[115]	14 b. Weber Grits.	
294	Robinson.	14 c.& Porphyry.[10849]	
296	Kokomo.	14 c. &Porphyry.[10609]	
302	Wheelers.	{ 20. Quaternary[9759] over Archæan.	
309	Frisco.	"	9064
318	Dillon.	"	9852

El Moro Branch.

120	Pueblo.	18 b. Colorado.	4469
170	Cuchara.	"	5921
180	Santa Clara.	"	
190	Apishapa.	"	6187
199	Chicosa.	"	6095
206	El Moro.[116]	18 d. Laramie.	5857

73. *Cebolla.* Large deposits of magnetite occur in the valley of Cebollo Creek. Capping of Cretaceous sandstone and andesite to north.
74. *Sapinero.* Archæan capped by Cretaceous and eruptive rocks. Cliffs of granite and gneiss.
75. *Crystal Creek.* Archæan capped by Dakota sandstone.
76. *Cimarron.* At contact of Archæan fault line.
77. *Cerro Summit.* Archæan traversed by eruptive dike to north.
78. *Montrose.* Stage line from here south to Ouray (35 ms.), which is beautifully situated in an amphitheatre at the head of the Uncompaghre, almost entirely surrounded by high peaks of the San Juan Mountains. Panoramic view of these mountains seen from higher points on the railroad.
79. *Colerow—Acheron.* Road follows in general valley bottom, ridges around formed of Cretaceous beds, sometimes capped by lavas.
80. *Roan.* Roan or Book Cliffs to the north.
81. *Bessemer.* Steel works of Colorado Coal and Iron Company.
82. *Mule Shoe.* Spanish Peaks to south, porphyry breaking through Carboniferous strata.
83. *Veta Pass.* Red sandstone shales.
84. *Blanca.* Gray sandstones.
85. Quaternary rests on Carboniferous strata. Archæan exposed on railroad cut below. Magnetite mines five miles north of station.
86. *Trinchera.* Blanca Peak to the south is the highest peak in Colorado, (14,464 ft.)
87. Mainly the debris of eruptive rocks, basalt and andesite.
88. *Toltec.* Toltec gorge is cut through Archæan rocks which underlie the eruptives.
89. *Monero.* Coal mines in sandstones.
90. *Amargo.* Stage to Pagosa Springs (Hot Sulphur), beautiful natural pools in a bend of the San Juan river, formerly held in high repute among the Indians for their curative powers.
91. *Dulce.* Narrow vertical dikes of basalt, crossing sandstone strata and standing out like stone walls on the surface.

Burlington and Missouri River Railroad.

Ms.	Railroad.	Alt.
400	Eckley.	20. Quaternary.	8879
439	Akron.	"	4656
452	Pinneo.	"	
463	Brush.117	"	4235
472	Fort Morgan.	"	4500
487	Corona.	"	4547
504	Roggen.	"	
521	Hudsen.	"	4998
544	Derby.	{ 20. Quaternary5159 / over Denver / Tertiary.	
551	Denver.	"	5175

Denver, Utah and Pacific Railroad. 118
Narrow Gauge.

Ms.	Railroad.	Alt.
0	Denver.	{ 20. Quaternary over / Denver Tertiary.	
		"	
1	Argo.	"	
17	Baker.	18 d. Laramie.	
21	Erie.	"	
23	Mitchell.	"	
34	Longmont.119	18 b. Colorado.	
45	Lyons.120	16. Trias.	

Denver, Texas and Gulf Railroad.
Formerly Denver & New Orleans.

Ms.		Alt.
	Denver.	{ 20. Quaternary over Denver Tertiary. 19. Monument Creek Tertiary.
.4	Melvin.	"
23	Parkers.	"
30	Bellevue.	"
39	Elizabeth.	"
47	Cameron.	"
52	Elbert.	"
58	Sidney.	"
64	Easton.	"
72	Granger.	"
78	Bierstadt.	"
81	Manitou Junc.	" 6302
90	Colorado Sp'gs.	18 d. 5970
87	Franceville Juc.	18 d. Laramie.
94	Fountain.	As on D. & R. G. 5503
99	Little Buttes.	" 5346
105	Wigwam.	" 5211
112	Pinon.	" 5016
118	Cactus.	" 4559
112	Pueblo.	" 4669

92. *Navajo.* Quarry of building stone used in new capitol at Denver.
93. *Juanita.* Junction of San Juan River.
94. *Carracas.* Cretaceous rocks dip down to west and are succeeded horizontal.
95. *Arboles.* Tertiary beds.
96. *Siding No.* 22. Junction of Piedra River.
97. *La Boca.* Valley of Los Pinos River.
98. *Durango.* Coal mines and smelting works. Colorado Cretaceous clays, capped by Fox Hill sandstones.
99. From Animas to Hermosa the cliffs on either side of the valley show an excellent section from the Cretaceous down to the Middle Carboniferous.
100. *Trimble.* Thermal bath establishment.
101. *Rockwood.* In the gorge of the Animas river is some of the boldest Alpine scenery in the Rocky Mountains. Especially fine are the Needle peaks to the east.
102. *Elk Park.* At entrance to gorge below are Cambrian quartzites and Silurian limestones resting on Archæan. Mountains around capped by great thickness of andesitic Breccia, often highly altered and mineralyzed.
103. *Manitou.* Good section of Carboniferous and Silurian limestones and Cambrian quartzites resting on Archæan seen in Williams Cañon. Cave is in Silurian limestone. Ute Falls are in the Archæan just below the Palæozoic beds. In Glen Eyrie the red sandstone (Trias), by faulting or non-conformity, comes in contact with the Cambrian quartzite which rests directly on the Archæan. Garden of the Gods—Trias.
104. *Marsh.* Some dark eruptive dikes seen traversing the Archæan schists.
105. Flat hills of Rhyolite at Silver Cliff.
106. Brown hematite mines of the Colorado Coal and Iron Co.
107. *Almont.* Archæan capped by Sandstones of Jura and Dakota Cretaceous.
108. *Crested Butte.* Mines of bituminous coal in hills southwest of town. Anthracite on either side State Creek valley.
109. Road follows alluvial deposits of Rio Grande river.
110. *Wagon Wheel Gap.* Andesitic breccia.
111. *Garfield.* Archæan on west, Carboniferous and Silurian on east.
112. *Crane's Park.* Cambrian quartzite resting on Archæan.
113. *Eagle Park.* Valley cut partly in Archæan, partly in overlying Palæozoic rocks.
114. *Red Cliff.* Archæan cut just below town. On either side cliffs of Cambrian, Silurian and Carboniferous beds.
115. *Fremont Pass.* Archæan forms mountains east of Mosquito fault.
116. *El Moro.* Coal mines and coke ovens.
117. Plains country underlain by Cretaceous beds, either Laramie or Fox Hills.
118. Distances and stations on this line given approximately.
119. *Longmont.* Red sandstone quarries. Flagging and building stone.
120. *Lyons.* Stage starts from here for Estes Park, twenty-two miles.
121. *Sierra La Sal.* High isolated peak to south.

Wyoming, Utah, Nevada and Idaho.*

LIST OF GEOLOGICAL FORMATIONS IN THESE TERRITORIES,

In the region of the Union Pacific and Central Pacific Railroads.

GENERAL TABLE.	WYOMING.	UTAH.	NEVADA.
20. QUATERNARY.	20. Quaternary.	20. Up. Quatern'y. 20. Lower Quat'y.	20. Up. Quatern'y. "
19 c. PLIOCENE. " 19 b. MIOCENE. " 19 a. EOCENE. " "	19 c. Niobrara. 19 b. White River. 19 a. Bridger. 19 a. Green River. 19 a. Vermill'n Ck.	19 c. Humboldt. 19 a. Bridger. 19 a. Green River. 19 a. Vermill'n Ck.	19 c. Humboldt. 19 b. Truckee. 19 a. Green River.
18. CRETACEOUS. " " "	18 d. Laramie. 18 c. Fox Hill. 18 b. Colorado. 18 a. Dakota.	18 d. Laramie. 18 c. Fox Hill. 18 b. Colorado. 18 a. Dakota.	No Cretaceous in Nevada.
17. JURASSIC.	17. Jurassic.	17. Jurassic.	17. Jurassic.
16. TRIASSIC. "	16. Red Beds.	16. Red Beds.	16. Star Peak. 16. Koipato.
14. CARBONIFEROUS. " " "	14 Coal Measures.	14-15. Perm. Carb. 14 c. Up. Cl. Mres. 14 b. Weber Quart. 14 a. Low. Cl. Mres.	14 c. Up. Cl. Mres. 14 b. Weber Quart. 14 a. Low. Cl. Mres.
13. SUB-CARBONIF's. "		13. Sub-Carbonif's.	13. Sub-Carbonif's. Diamond Pk. Quart.
9-11. DEVONIAN.		9-11 Nevada l. s. Ogden Quartzite.	9-11. White Pine Sh'le. Nevada Limestone.
5-7. SILURIAN. " "		5-7. Ute Limestone.	5-7 Lone Mt. l. s. Eureka Quartzite. Pogonip Limestone.
2-4. CAMBRIAN. " " " "		2-4. Cambrian.	2-4. Hamburg Shale. Hamb'rg Limestone. Secret Canon Sh'le. Prospect Mt. l. s. " " Quart.
1. ARCHÆAN.	1 b. Huronian. 1 a. Laurentian.	1 b. Huronian. 1 a. Laurentian.	1. Archæan.

*The Table of Formations and the main line of the Union and Central Pacific Railroads, the Utah and Northern Division, the Eureka and Palisade, and Virginia and Truckee Railroads are by Mr. Arnold Hague, Geologist, United States Geological Survey. Mr. G. K. Gilbert, U. S. Geologist, furnishes the lines in Utah and Mr. John B. Hastings, M. E., of Ketchum, Idaho, and Prof. G. E. Bailey of Rapid City, S. Dakota, have noted the lines given under their authority.

Wyoming.   Ut:

Ms.	Union Pacific Railroad.		Alt.
463	Bushnell, Neb.	19 c. Niobrara,Pl'c'ne.	
473	Pine Bluffs, Wy.	"	5047
484	Egbert.	"	
496	Hillsdale.	"	
508	Archer.	"	
516	Cheyenne.[1]	"	6059
523	Hazard.	"	
531	Otto.	"	
536	Granite Cañon.[2]	1 a. Lauren'n.	7319
542	Buford.	"	7785
549	Sherman.[3]	"	8256
559	Harney.	"	
564	Red Buttes.[7309]	17 Jurassic & Trias.	
570	Fort Sanders.	18 a. Dak., Cretace's.	
573	Laramie City.	"	7158
581	Howell.	"	7090
589	Wyoming. 7066	18 b. Colo., Cretac's.	
599	Cooper's Lake.	"	7078
608	Lookout.	"	7177
616	Miser.	"	
625	Rock Creek.	"	
640	Aurora.[4]	17 Jurassic.	
648	Medicine Bow.	18 b. Colo., Cret. 6571	
657	Carbon.[5] 6880	18 d. Laramie, Cret.	
668	Percy.[6]	"	6971
682	Edson.	"	
690	Walcott's. 6800	18 c. Fox Hill, Cret.	
696	Fort Steele.	' "	
711	Rawlins.[7] 6755	14 b. Coal Measures.	
724	Separation.	18 d. Laramie, Cret.	
739	Creston.	"	7048
754	Wash-a-kie.	19 a. Ver'n Ck.	
764	Red Desert.	"	6722
779	Table Rock.	"	7551
787	Bitter Creek.	"	6705
791	Black Buttes.	18 d. Laramie, Cret.	
801	Hallville.	"	6590
807	Pt. of Rocks.[8]	"	6517
818	Salt Wells.	20. Quaternary. 6851	
826	Baxter.[9] 6800	18 d. Laramie, Cret.	
832	Rock Springs.[10]	"	6270
847	Green River.[11]	19 a. Green R. 6088	
860	Bryan. 6196	19 a. Bridger, Eocene.	
878	Granger.	"	6289
888	Ch'rch Buttes.[12]	"	6568
905	Carter.	"	
915	Bridger. 6687	19 a. Ver'n Ck. E'ne.	
930	Piedmont. 7082	19 a. Green Riv. E'ne.	
939	Aspen. 7405	18 c. Fox Hill, Cret.	

Ms.	Union Pacifi Conti:
957	Evanston.[18]
968	Wasatch.[14]
977	Castle Rock.
993	Echo.
1009	Weber.[15]
1021	Devil's Gate.[16]
1026	Uinta.[17]
1032	Ogden.[20]

Central Pac

0	Ogden.[20]
10	Bonneville.
24	Corinne.
43	Blue Creek.
53	Promontory.
78	Monument Pt.
94	Kelton.
113	Matlin.[18]
124	Terrace.
134	Bovine.
147	Lucin.

Nev

Central Pacific Rs

167	Montello.
183	Toano.
193	Pequo.
195	Otego.
205	Independence.
210	Moors.
220	Wells.[19]
227	Tulasco.
252	Halleck.
257	Peko.
266	Osino.[20]
275	Elko.[21]
287	Moleen.[22]
299	Carlin.
308	Palisade.[28]
326	Be-o-wa-we.
336	Shoshone.
347	Argenta.
360	Battle Mount'n.
379	Stone House.
394	Iron Point. 4875
403	Golconda.

1. At Chalk Bluffs, 15 miles southeast from Cheyenne, the Niobrara P Miocene are both exposed, the latter resting unconformably upon the Cretaceous.

2. Both to the north and south of Granite Cañon the Palæozoic be: against the Archæan rocks.

3. Sherman, the highest station along the line of the Union Pacific above sea-level, and is on the summit of the Colorado range.

4. The railroad passes through the axis of an anticlinal fold, exposing Jurassic strata.

Central Pacific Railroad.		
Ms.	**Continued.**	**Alt.**
414 Tu.e.	19 c. Humb't, Pliocene.	
419 Winnemucca.	"	4332
430 Rose Creek.	"	4332
410 Raspberry.	"	4327
448 Mill City.[34]	4236 "	[side.
459 Humboldt.[25]	16. Triassic, on the east	
471 Rye Patch.	"	4257
481 Oreana. 4181	19 c. Humb't, Pliocene.	
483 Humbolt Bridge.	"	
493 Lovelocks.	"	3977
502 Granite Point.	20. Quatern'y. [stat'n.	
509 Brown's.[26] 3929	Rhyolite west of the	
521 White Plains.	"	3894
528 Mirage.	19 b. Truckee, Mi'c'ne.	
535 Hot Springs.[27]	Basalt on E. side.4072	
546 Desert.	Basalt on west side.	
555 Wadsworth.[28]	20. Quaternary. 4077	
569 Clark's. 4368	Rhyolite, Andesite.	
581 Vista.	20. Quaternary. 4400	
583 Reno.	"	4497
600 Verdi.	"	4895
616 Boca, Cal.	"	5531
(Continued in California.)		

Utah.

Union Pacific Railroad.—Continued.		
Ms.	**Utah and Northern Division.[31]**	**Alt.**
0 Ogden.[45]	20. Quaternary.	4303
9 Hot Springs.	"	4277
14 Willard.	"	4340
22 Brigham.	"	4315
32 Honeyville.	"	4278
34 Dewy.	"	4320
41 Collinston.	"	4691
51 Mendon. 4450	19 c. Humb't Pliocene.	
58 Logan.	"	4499
63 Hyde Park.	"	
65 Smithfield.	"	4555
71 Richmond.	"	4527
78 Franklin.	"	4505

Idaho.

Union Pacific Railroad.—Continued.		
	Utah and Northern Division.[31]	
90 Battle Creek.	20. Quaternary and 19. Pliocene.	4492
101 Oxford.		4768
115 Calvin.		
125 Arimo.		4654

5. Carbon offers an excellent opportunity for studying the Cretaceous coals of Wyoming.

6. To the south of Percy Station, Elk Mountain, which rises conspicuously above the plain, consists of Archæan crystalline schists, with Palæozoic and Mesozoic strata on the slopes.

7. Rawling's Peak consists of an Archæan mass, surrounded by Palæozoic and Mesozoic beds. In the coal measures is an interesting body of iron ore.

8. Northeast from Point of Rocks is a remarkable outburst of leucite rocks.

9. There is exposed here an interesting section of Laramie coal rocks.

10. Near Rock Springs the coal formations are well shown.

11. Along the bluffs of Green River are seen the best exposures of the Green River Eocene. These beds are celebrated for the fine specimens of fossil fishes preserved in the shales.

12. On the south of the railroad, between Church Buttes and Carter, may be seen distant but good views of the Uinta Range.

13. About three miles north of Evanston are situated the Rocky Mountain and Wyoming coal Company's mines, where there is a good section of the Laramie beds. These mines have supplied immense quantities of coal used by the Union and Central Pacific railroads.

14. From Wahsatch to Echo the railroad passes through Echo Cañon, where are exposed both the Vermillion Creek and Laramie formations, the former lying unconformably upon the latter.

15. Passing through Weber Cañon, from Lost Creek to Weber Station, there is exposed a series of beds from the top of the Jurassic, through the Triassic, Upper Coal measures, Weber Quartzite to the base of the Lower Coal measures.

16. At the Devil's Gate the Archæan rocks of the Wahsatch Range are characteristically shown.

17. The terraces of Lake Bonneville, which stand over 950 feet above the present level of Salt Lake, may be seen from Uinta station. They may be easily traced all the way from Ogden to Lucin.

18. On the north side of the railroad at Matlin the old lake terraces are distinctly cut in basalt.

19. From Wells there is a fine view of the East Humboldt range. Mount Bonpland attains an elevation of 11,341 feet above sea-level.

20. Just east of Osino the railroad passes through Osino Cañon, exposing a good section in the Weber Quartzite.

21. In the neighborhood of Elko may be seen the Green River Eocene, Humboldt Pliocene, characteristic outbursts of rhyolite and "Chicken Soup" hot springs.

22. In Moleen Cañon the Carboniferous formations are well shown. The limestones of Moleen Peak, just south of the railroad, carry large numbers of coal measure fossils.

23. Palisade Cañon cuts through rhyolites. Andesites are also exposed.

24. Mill City is the most convenient place to leave the railroad in order to study the characteristic Triassic formations of the West Humboldt Range.

25. From Humboldt there is a fine view of the West Humboldt Range. In the neighborhood are some interesting outbursts of basalt and a deposit of sulphur.

26. In the Montezuma Range, west of Brown's station, the volcanic rocks are well shown. It is an interesting place to study rhyolites and basalts.

27. The Hot Springs, a short distance east of the station, reach the surface near the base of the basaltic hills.

28. The Truckee Cañon, just east of Wadsworth, offers remarkable outbursts of a great variety of volcanic rocks. There may be seen here basalts, rhyolites and andesites. Tourists leave the railroad here for Pyramid Lake.

29. Propylite is the characteristic volcanic rock, which carries the Comstock Lode. A. H.

30. The last rail completing the Pacific railroads, from Omaha to San Francisco, was laid May 10, 1869.

Idaho.

Union Pacific Railroad.—Continued.
Ms.   Utah and Northern Division.[31]   Alt.

Ms.			Alt.
132	McCammon.		4755
142	Inkone.		
148	Port Neuf.	Cambrian in hills.	
155	Pocatello.	Quat'y on basalt.	4465
166	Ross Fork.	"	4452
179	Blackfoot.	"	4505
191	Basalt.	Basalt.	4579
205	Eagle Rock.	"	4714
215	Payne.		
222	Market Lake.	"	4781
235	Hawgood.		
243	Camas. 4832	B's'lt cov. 19 c. Pl'o'ne.	
	Dry Creek.		
	High Bridge.		
	China Point.		
272	Beaver Canon.	"	6025
	Pleasant Valley.	Drift and Basalt.	
	Monida.		6509
	Williams.		

Montana.

Union Pacific Railroad.—Continued.
Utah and Northern Division.[31]

Ms.			Alt.
300	Spring Hill.		5267
	Dell.		
323	Red Rock. 5605	Carbonifer's in Mts.	
	Grayling.	Pal'z'c and ign's rocks.	
	Barratts.	[and Arch. in hills.	
348	Dillon. 5106	19 c. Pl'o'ne, Palz. l. s.	
378	Melrose.		5191
382	Lowell.		
394	Feely.		
410	Silver Bow.		5344
417	Butte City.	Granite.	5454
421	Stuart.		
443	Deer Lodge.		4529
454	Garrison. 4540	Northern Pacific R. R.	

Wyoming.

Union Pacific Railroad.—Continued.
Oregon Short Line.[32]

Ms.			Alt.
876	Granger. 6231	19 a. Bridg'r (Eocene.)	
891	Nutria.	"	6516
900	Waterfall.	Qu. over Wasatch.	6796
918	Ham's Fork.	"	6955
920	Twin Creek.	"	
925	Fossil.	"	6665
932	Nugget.	Jura. Trias.	
	Sage.	Qu. over 18 d.Lar.	6552
947	Beckwith.		6207
959	Cokeville. 6201	Qu. over Jura. Trias.	

Idaho.

Union Pacific Railroad.—Continued.
Ms.   Oregon Short Line.[32]   Alt.

Ms.			Alt.
968	Border.	16–17 Jura. Trias.	6082
974	Nupher.	20.over "	6041
984	Dingle.	" "	
991	Montpelier.	"	5948
997	Piscadero.	20.over Salt L.Ter.	5928
1002	Oasis.	Salt Lake Ter.	5336
1005	Novene.	"	
1020	Stock Yards.	Basalt.	
1021	Soda Springs.	Basalt.	5782
1026	Crater.	Basalt.	5736
1038	Squaw Creek.	Basalt. Cl.in hills.	5427
1053	Lava.	Cambrian Hills.	
1060	Topaz.	Quat., Basalt.	4934
1067	McCammon.	Quaternary.	4755
1072	Onyx.	"	4648
1078	Inkom.	Quat. Camb. in hills.	
1090	Pocatello.	Quat. on Basalt.	4465
1099	Michaud.		4478
1109	Sunshine.		
1115	American Falls.	{ Late Ter. or Quat. Basalt.[33]	4343
1124	Napata.	"	4467
1182	Wapi.	"	
1148	Minidoka.	"	4287
1156	Oniona.	"	
1165	Kimama.	"	4279
1179	Owinza.	"	4211
1188	Waucanza.	"	4073
1197	Shoshone.[34]	"	3975
1213	Toponis.	"	3581
1226	Bliss.	"	
1232	Ticeska.	"	3089
1241	King Hill.	"	2543
1249	Glenn's Ferry.	"	2566
1261	Medbury.	"	2387
1269	Reverse.	"	
1279	Mt. Home.[35]	"	3147
1290	Cleft.	"	
1298	Nameko.	"	
1305	Bisuka.	"	3139
1312	Owyhee.	"	
1324	Kuna.	"	2656
1334	Nampa.[35]	"	2459
1343	Caldwell.	"	2374
1358	Parma.	"	
1376	Ontario.	"	
1378	Payette.	"	
1387	Crystal Springs.	"	
1391	Weiser.	"	2135
1407	Old's Ferry.	"	
	Oregon Line.		

31. The geology of most of the stations on the Utah and Northern Division is given by Mr. Hague, but the editor has not been able to obtain complete assignments of formations. The geology of some parts of the great West has been necessarily done in something of a reconnoissance way, and often before the railroads were located, so that accurate statements are impossible. The altitudes have been kindly furnished by Mr. Henry Gannett, Chief Geographer, U. S. Geological Survey.

Union Pacific Railroad—Continued.
Oregon Short Line.—Continued.

Ms.	(Wood River Branch.)		Alt.
0	Shoshone.	Quat. Basalt.	3975
14	Pina.	"	
30	Tikura.[36]	"	4681
37	Picabo.	"	4839
52	Bellevue.[37] [5173]	Quat. Stratified Dft.	
57	Hailey.[37]	"	5344
69	Ketchum.[38]	"	5825

Wyoming.

Cheyenne and Northern District.[39]

0	Cheyenne.	19 b. Miocene.
4	Ft. Russell.	"
13	Silver Crown.	Granite to 14 c.
17	Stone Spur.	14 c. Upp. C'l. Meas.
26	Islay.	" & 15 Permian.
33	Horse Creek.	16 Trias., 17. Juras.
39	Altus.	19 c. Plioc., 20. Quat.
46	Iron Mt.	14 a. Upp. C'l. Meas.
51	Shultz Spur.	19 b. Miocene.
60	Kelley.	"
71	Chug Water.	"
84	Bordeaux.	"
96	Wheatland.	"
103	Wendover.	"

Fremont, Elkhorn and Missouri Val.[39]—Elkhorn
Valley Line.—Continued from Nebraska.

307	Valentine, Neb.	19 b. Miocene.		
318	Crookston.		"	3570
329	Georgia.		"	
345	Cody.		"	
358	Eli.		"	
370	Merriman.		"	
383	Irwin.	Sand Dunes and Lacustrine Drift.	"	
397	Gordon.		"	3547
412	Rushville.		"	
424	Hay Springs.		"	
433	Bordeaux.		"	
444	Chadron.		"	3880
449	Dakota Jc.		"	3345
459	Whitney.		"	
470	Crawford.		" &20.Q'ty.	
489	Andrews.		"	
498	Harrison, Neb.		"	

Wyoming.

Fremont, Elkhorn and Missouri Val.[39]—Elkhorn

Ms.	Valley Line.—Continued from Nebraska.		Alt.
509	Van Tassell.	14 c.U.C'l to 18 a.	4727
520	Node Ranch.	"	
529	Lusk.	18 b. Cret.	5007
538	Manville.	"	
545	Keeline.	18 a. and 18 c. Cret.	
554	Lost Spring.	18 c. Cret.	
566	Fisher.	18 d. Cret.	4752
576	Irvine.	18 b. Cret.	
584	Douglass.	"	4810
597	Fetterman.	18 c. Cret.	
604	Wolcott.	18 d. Cret.	
606	Glen Rock.	"	
630	Casper.	Granite. 18 c.	5118

Utah.

Denver and Rio Grande Railroad.[40]
Continued from Colorado.

463	Acheron.	18. Lower Cretaceous.	
479	Cotton Wood.	"	4681
490	Cisco.	"	
507	Sagers.	"	
515	Thompson's.	"	
521	Crescent.	"	
529	Little Grand.	"	
536	Solitude.	"	
545	Green River.	"	4086
558	Desert.	"	
570	Lower Crossing.	"	
591	Sunny Side.	"	
600	Farnham.	"	
610	Price.[41]	"	
623	Castle Gate.	18. Cretaceous.	6062
637	Pleasant Val. Jc.	18 Upp. Cret.	7152
644	Soldier Summit.	Tertiary.	7477
658	Mill Fork.	"	8791
669	Thistle.	18 Cretaceous. (?)	
680	Spanish Fork.[42]	Bonnev'le B.Quat.	4665
684	Springville.	"	4566
689	Provo.[42]	"	4525
699	Battle Creek.	"	
702	American Fork.	"	
705	Lehi.[42]	"	

32. The geology from Granger to Squaw Creek is by Prof. W. B. Scott of Princeton University; thence to Michaud; it is given on the authority of an atlas of the U. S. Survey, which was made before the road was located, and the assignments must, therefore, be taken with allowance.
Geology from American Falls to the Oregon line and on the Wood River Branch is by Mr. John B. Hastings, M. E., F. G. S. A., of Ketchum, Idaho. Altitudes on all this line by Mr. Gannett.
33. These late Tertiary and Quaternary basalts form part of the great Northwestern lava-flood, of Northern California, Northwestern Nevada, Oregon, Washington, Montana and British Columbia. The basalt of the Wood River Branch is of later date than the flow from Glenn's Ferry westward.
J. B. H.

34. *Shoshone.* Shoshone Falls of Snake River, 210 feet vertical altitude in basalt.
35. *Mountain Home, Nampa.* Gold and silver mines in Archæan granite in vicinity.
J. B. H.

36. *Tikura.* From Tikura to Lava Creek may be seen a ropy lava field of seventy-five square miles, almost untouched by the elements, a congealed, black, stormy sea.
J. B. H.
37. *Bellevue, Hailey, Ketchum.*—In vicinity, hot springs and argentiferous galena mines in Silurian limestone and slates and various free milling silver ores in Archæan granites. Tertiary trachytes.
J. B. H.

Denver and Rio Grande Railroad.				Utah Central Railroad.40-48		
Ms.	Continued from Colorado.	Alt.	Ms.		Continued.	Alt.
718	Draper.	Bonnev'le Beds. Quat.	46	Lovendahl's.	20. Quaternary.	4277
724	Bingham Jc.	"	49	Junction.	"	
728	Germania.	4296	50	Sandy.	"	4399
735	Salt Lake. 44	4287	54	Draper.	"	4443
743	Wood's Crossing.	"	68	Lehi Junction.	"	4517
750	Farmington.	"	71	American Fork.	"	4554
754	Kaysville.	"	74	Pleasant Grove.	"	4495
764	Hooper.	"	85	Provo.	"	4456
771	Ogden.45	"	90	Springville.	"	4451

Coal Branch.			95	Spanish Fork.	"	4493
0	Pleasant Val. Jc.	18. Upper Cretaceous.	103	Payson. 4548	20. Bonneville Beds.	
14	Schofield.	"	108	Santaquin.	20 Quaternary.	4813
19	Mud Creek.	"	120	Mona.	"	4859
			128	Nephi.	"	5056
Bingham and Alta Branch.			142	Juab.	"	5019
0	Salt Lake.48	Bonnev'le Beds. Quat.	151	Mills.	"	4852
11	Bingham Jc.		167	Lemmington.	20. Bon'v'le Beds.4674	
27	Bingham.	14. Carboniferous.	185	Riverside.	"	4583
			194	Deseret.	"	4541
13	Sandy.	Bonnev'le Beds. Quat.	213	Neels.	"	4356
21	Wasatch.	Granite.	241	Black Rock.	"	4799
29	Alta.	Devonian. (?)	263	Milford.	"	4908
Utah Central Railroad.40-46			280	Frisco.	Volcanic.	6315

Utah Central Railroad.40-46						
0	Ogden.45	20. Quaternary.	4303	Utah and Nevada Railway.40		
16	Kaysville.	"	4298			
22	Farmington.	"	4281	0	Salt Lake.48	20. Bonneville Beds.
26	Centreville.47	"	4253	12	Chambers.42	14. Carboniferous.
26	Wood's Crossing.	"	4299	18	Garfield.	"
37	Salt Lake City.48	"	4261	20	Lake Point.42	"
43	Francklyn.	"		32	Tooele.	20. Bonneville Beds.
44	Germania.	"	4242	37	Terminus.	" 4991

38. *Ketchum.* Near station at Wood River bridge hornblende-andesite. At head of Wood River valley and vicinity many gulches contain deposits of extinct glaciers, including glacial lakes with Chinook salmon and smaller salmon (*oncorhynchus norka*) locally called redfish from the color. Tertiary trachyte underlies stratified drift. J. B. H.

39. Cheyenne and Northern, and Tremont, Elkhorn and Missouri Valley are by Prof. G. E. Bailey, of the Dakota School of Mines, Rapid City, South Dakota. A portion of the latter road should be in the Nebraska chapter, but was overlooked when that chapter was printed.

40. By Mr. G. K. Gilbert, Geologist, U. S. Geological Survey.

41. From Acheron to Price the road follows a great monoclinal valley overlooked on the north by the Book Cliffs (Cretaceous.) G. K. G.

42. The north end of the Oquirrh Range from Chambers to Lake Point is finely carved by old shore lines of Lake Bonneville. These extend up to 1,000 feet above Great Salt Lake. G. K. G.

43. From Spanish Fork to Lehi the road is in Utah valley and commands a view of the old shore lines of Lake Bonneville. A large delta of the old lake forms the terrace near Provo. G. K. G.

44. There is a profound fault along the western base of the Wasatch range. The hot springs close to the track between Salt Lake City and Wood's Crossing rise on the fault line. G. K. G.

45. *Ogden.* View of Wahsatch Mountains to east, a very fine range, as seen in afternoon light, when eastern train arrives; southeast, Archæan, with Weber Canon cut in it, through which the railroad has come out into valley; east, "Fault Canon," faulted Cambrian lying on Archæan, recognized by color: Ogden Canon; northeast, Eden Pass, another fault; north and north-northeast, Palæozoic rocks on Archæan. Lake terraces show all along base of mountains, by gray horizontal line, very distinct. W. M. Davis, Jr., of Harvard College.

46. *Utah Central Railroad.* Leaving Ogden and rounding long Quaternary slope south of Weber River, a long stretch of Wahsatch range comes into view. From Fault Canon, north; Archæan, at base; Palæozoic, above; between Fault Canon and Centreville station, including Weber Canon, all Archæan. Then begins the great synclinal, as seen from along here. The north end, a little south of east from Centreville (Cambrian to Carboniferous) shows on top of mountains; and the south end. Twin Peaks (Cambrian), and Lone Peak (granite intruded through Archæan), in farthest distance, showing over lower Tertiary hills south of Centreville. The axis of the synclinal (of soft, Mesozoic rocks) being low and hidden. The old lake terrace is very clearly seen. W. M. D.

47. *Centreville to Salt Lake City.* Around west base of hills, formed of Palæozoic rock, dipping south (part of synclinal), overlaid by uncomformable Tertiary rocks. W. M. D.

San Pete Valley Railroad.[40]		
Ms.		Alt.
Nephi.	20. Quaternary.	5056
Fountain Green.	19. Tertiary.	
Moroni.	"	

Union Pacific Railroad.[40]—Continued. Echo and Park City Branch.		
0 Echo. 5460	Wasatch; Tertiary.	
8 Grass Creek Jc.	18. Upp. Creta.	5520
5 Coalville.	"	5596
13 Wanship.	"	5654
20 Atkinson.	14. Carbonifer's.	6462
27 Park City.	"	6851

Nevada.

Eureka and Palisade Railroad.[49]		
0 Palisade.[50]	Rhyolite.	4521
12 Evans.	20. Quaternary.	
28 Box Springs.	"	

Nevada.

Eureka and Palisade Railroad.[49] Continued.		
Ms.		Alt.
37 Mineral.[51]	20. Quaternary.	5442
50 Alpha.	"	5911
60 Garden Pass.	"	
63 Summit.[52]	"	
78 Diamond.	"	5941
90 Eureka.[53]	Pumice and Tufa.[6871]	

Virginia and Truckee Railroad.[49]		
0 Reno.	20. Quaternary.	4497
11 Steamboat.[54]	Hot Springs deposits.	
21 Franktown.	Metamorphic rocks.	
30 Carson[55]	19 c. Humb't Plio.4680	
39 Eureka.	20. Quaternary.	
52 Virginia.[56]	Andesite.	6205

48. *Salt Lake City.* Walk north, one hour, to Ensign Peak, (or better, an hour further northeast, to point whence northeast can be seen also—giving fine view in all directions.) The Wahsatch range fills the east, from north to south. Other mountains are: Northwest, Antelope Island, in lake, Archæan; north-northwest, beyond Antelope Promontory Mountains and Island; west, Lakeside, Stansbury and Cedar Mountains; southwest, Oquirrh Mountain; west-southwest, Aqui Mountain; south, Pelican Mountain, (beyond Traverse)—Carboniferous, all running north and south· south, Traverse Mountains, east and west—Trachyte—cut through in middle of River Jordan, coming from Utah Lake (fresh of course), north to Great Salt Lake. From Ensign Peak can be seen the city, the fertile valley of the Jordan (fertile from irrigation); the lake; Camp Douglas (U. S. troops) on terrace east of and commanding city; Emigration Canon, through which the Mormons first came to the valley. Salt Lake is better than Colorado Springs for excursions. D.

49. By Mr. Hague.
50. *Palisade.* Andesite and basalt near by. A. H.
51. *Mineral.* Devonian limestones in the hills of the Pinon Range. A. H.
52. *Summit.* The railway crosses a low pass of the Pinon Range. A. H.
53. *Eureka.*—All the characteristic types of the volcanic rocks of the Great Basin occur in the immediate neighborhood. A. H.
54. *Steamboat.* Well-known steamboat springs depositing Silica. Andesite near the railway. A. H.
55. *Carson.* Fossil remains in the sandstones near the Prison. A. H.
56. *Virginia.* The famous Comstock Lode is here, an excellent place to study the volcanic rocks of the Great Basin. A. H.

Lake Bonneville is the name given to the great Quaternary lake, whose boundary has been traced by its shore lines and deposits to and into Nevada on the west, Idaho on the north, as far east as Salt Lake City and in bays of which Utah and Sevier Lakes are the remnants, to the south as far as Frisco. The Great Salt Lake is the reduced remnant of this great sheet of water. The highest, or Bonneville, shore line is 1,000 feet above the level of Great Salt Lake, and is one of the most conspicuous water lines. Of the numerous lower lines, marking the heights at which the water lingered, one lying 400 feet below the highest is called the Provo shore line. Between the Bonneville and Provo lines are four or five prominent lines.

The following, from Mr. G. K. Gilbert's report on Lake Bonneville, gives, in a general way, its origin. ' The lowlands of the 'Great Basin' are valleys without drainage to the ocean, and when the climate of the Glacial Epoch gave them a more generous supply of moisture, the surplus was accumulated in their lower parts in quantities which bore a definite relation to the climate. When for centuries the climate became more humid, the lake rose and encroached upon the land, and when the reverse was true and aridity prevailed, they dried away and the land was laid bare.'' The origin and history of the great lakes of former periods is a subject of absorbing interest to the student of geologic science, and none offers a better field than Lake Bonneville.—[Ed.]

Oregon.[1]

Ms.	Oregon & California Railroad. (Up the Willamette Valley.)	Alt.	Ms.	Oregon & California Railroad. Continued.	Alt.
0	Portland.	⎧ Hills on west. Basalt alluvial gravel plain east. 19 b. Miocene fossils in the river bed. **45**	87	Tangent.	**269**	⎧ An extended bed of an ancient inland sea, named by Prof. Condon "The Willamette Sound," with abundance of 19. Tertiary fossils.
7	Milwaukee.	Basalt hills. **117**	98	Halsey.	**307**	
11	Clackamas.	" **134**	106	Harrisburg.	**332**	
			110	Junction.	**345**	
16	Oregon City.	⎧ Bed of river and hills on both sides columnar basalt. **99**	124	Eugene.		⎧ The hills again with abundant 19 b. Miocene fossils.**451**
20	Rock Island.	⎧ A transverse dike of trap, with amygdaloid. Hills of basalt. The bed of the river and the now widening valley of 20 Post Pliocene contain abundant fossil remains of *bos*, *latifrous*, *elephas*, mastodon and horse.	135	Creswell.	**565**	⎧ Volcanic tufas and porphyries.
			145	Latham.	**667**	
			148	Divide.		⎧ Carbonaceous shale, with coal 18. Cret.
25	Canby.	**175**	156	Comstock.		"
			161	Rice Hill.		
			181	Oakland.		Metamorphic. **450**
			200	Roseburg.		" **485**
29	Aurora.	**215**	213	Dillard.		⎧ 20. Quaternary of L. Umpqua Valley.
			231	Riddle's.		Metamorphic & Slate.
			267	Glendale.		Metamorphic.
		⎧ The streams here to right and left expose the 20. Post Pliocene mud.	296	Grant's Pass.		⎧ 18. Cre. in foothills. Slate and l. s. 17. Jur. 16. Tri. age.
33	Hubbard.	**206**				
40	Gervais.	**210**				⎧ 18. Cretaceous along foothill; older in the mountains.
		⎧ The river bed is 20. Post Pliocene. The hills are rich with 19 b. Miocene marine fossils.	320	Gold Hill.[2]		
53	Salem.	**157**				⎧ 20. Quaternary and 19. Pliocene of Rogue River Val'y.
61	Turner.	**210**	335	Medford.		
67	Marion.	**222**				"
72	Jefferson.	**204**	340	Phœnix.		⎧ and distant hills Creta. to J. Trias.
	(Exposure a mile above the town on the Santiana River.)	⎧ A ridge of dark colored 19. Tertiary crosses the line of travel here — rich in fossils.				⎧ End of Rogue River Valley, mountains .n sight. 18. Creta. to 17. Jur. 16 Tri., slates, l. s. & granite. Liskiyon Mts
81	Albany.	⎧ The above rock seen across the river.**233**	349	Ashland.**344**		

1. Furnished for this work by Prof. Thomas Condon, of the Oregon State University, Eugene City, Oregon, the State Geologist.

2. *Gold Hill to Ashland.* Gold mining Auriferous slates.

3. Notes on this stage line are by J. S. Diller, of U. S. Geological Survey Corps.

4. *Ashland.* Liskyon Mountains and hills, west of road, chiefly of granite and Metamorphic rocks; those on east chiefly Cretaceous strata and lavas (basalt and andesite).

5. *Yreka.* Cretaceous fossils (chico group) eight miles northeast of Yreka.
Scott's Mountains, chiefly Metamorphic rocks, serpentines and granites.
Six miles northwest of Gazelle, at Cave rock, coarse conglomerate of Cretaceous shore line against Scott Mountains. Three miles west of Gazelle Carboniferous limestone with fossils.
Shasta Valley. Remarkable for great number of volcanic cones. Grand view of Mount Shasta.

6. Ascent of Mt. Shasta from *Sissons*, by good trail to camp at timber line, three hours; to summit from camp about six hours, partly on horseback. Glaciers and cañons on north and east sides of mountain. One of the finest volcanic cones in the world. Shasta chiefly Hypersthene andesite. Sugar Loaf is of Hornblende andesite. Mt. Shasta, 14,442 feet above tide, or nearly 11,000 above Berryvale. Dr. G. W. Dawson says, in its grand isolation, and the remarkable symmetry of its conical form, it is very impressive.

Southern Pacific Railroad.			Oregon Railway nd Navigation Co.		
Ms.	San Francisco and Portland Line.[10]	Alt.	Ms.	Continued.	Alt.
0	Ashland[4]	See Notes.	1453	Encina.	See Note 9. — 3980
36	Hornbrook.	"	1457	Norton.	" — 3580
54	Montague.	"	1463	Baker City.	" — 3440
	(Yreka.[5])	"	1474	Haines.	" — 3335
76	Sission.[6]	"	1483	North Powder.	" — 3250
98	Dunsmuir.	"	1493	Telocaset.	" — 3449
	(U. Loda Sp's.[7])	"	1503	Union.	" — 2720
125	Gibson.	"	1515	La Grande.	" — 2786
134	Delta, Cal.	"	1522	Hilgard.	" — 3004

	Oregon Central Railroad.		1534	Kamela.	" — 4204
0	Portland.[8]	Hills of basalt, overlying 19 b. Mio. [48] salt.	1540	Meacham.	" — 3631
			1548	Laka.	" — 2909
6	Summit.	salt.	1557	North Fork.	" — 2308
9	Ross Landing.	"	1558	Wilbur.	" — 2252
		To Forest Grove over the bed of the 20. Post Miocene inland sea, connected with the main one of Willamette Valley, through the Twalatin and Chehalem Valley.	1568	Mikecha.	" — 1751
			1578	Cayuse.	" — 1414
			1586	Mission.	" — 1182
11	Beaverton. 212		1589	Pendleton Jc.	" — 1130
16	Readsville. 288		1590	Pendleton.	" — 1070
24	Hillsboro. 1·8		1597	Barnhart.	" — 912
29	Cornelius. 200		1605	Yoakum.	" — 835
	For'st Gr've.[198]		1608	Nolin.	" — 786
			1615	Echo.	" — 839
		Hills of fossil rock right and left, 19 b. Miocene. 306	1618	Foster's	" — 592
32	Gaston.		1627	Maxwell.	" — 452
48	St. Josephs.	" 158	1634	Umatilla Jc.	" — 800

	Oregon Railway and Navigation Co.			Heppner Branch.	
1416	Huntington, Or.	See Note 9. 2110	0	Arlington.	See Note 8.
1428	Weatherby.	" 2393	10	Willows Jc.	" — 241
1436	Durkee.	" 2850	25	Cecils.	" — 625
1443	Unity.	" 3123	30	Douglass.	" — 796
1451	Pleasant Val.	" 3750	39	Ione.	" — 085
			46	Lexington.	" — 1425
			55	Heppner.	" — 1905

7. *Upper Loda Springs.* Near Upper Loda Springs, an ancient Lava stream from Mt. Shasta enters the Cañon of the Sacramento River, which it follows for nearly 50 miles. Lava seen at many places clinging to sides of old Cañon, especially near Delta.

8. Dr. Dawson discovered in Oregon, west of the Cascade Mountains, no traces of general glaciation or deposits like northern drift. There is a remarkable absence of any well marked terraces or benches, although the bottoms of the valleys suggest that the sea may have at one time flowed into them. The almost complete absence of lakes or ponds is very remarkable, and contrasts strongly with the innumerable lake basins of British Columbia. The drift appears at Tacoma and other places in Washington.

9. This line of the Oregon Railway and Navigation Co. traverses a region covered by the great lava sheet, but just what formations are exposed at given stations can not be determined from any sources at the command of the editor. Prof. Condon's notes, the general note 39 on the Northern Pacific, and Mr. Willis' notes on pages 265 and 266 will throw some light on the geology of this section. Other lines of the Oregon Railway and Navigation Co. will be found in the chapter on the Northern Pacific. J. R. M.

10. The notes on this line were prepared before the road was built (see Note 3,) and as they are all that I can obtain for this line I have inserted the old stage stations in parentheses. J. R. M.

California.*

LIST OF THE GEOLOGICAL FORMATIONS IN CALIFORNIA.

TERTIARY.	20. Quaternary.		
	19 c. Pliocene.		
	19 b. Miocene.		
	19 a. Eocene.		
	18. Cretaceous.	W. of Sierra Nevada.	
	17. Jurassic.	W. and E. of Sierra Nevada.
	16. Triassic.	"	" "
	14. Carboniferous.	E. of	" "
	13. Sub-Carboniferous.	W. and E.	" "
	9-11. Devonian. ?	E. of	" "
	5-7. Silurian. ?	"	" "
	2-4. Cambrian. ?	"	" "
	1. Archæan.³	W. and E.	" "

*Explanatory Note. This chapter was prepared by my father just before his death, principally from notes furnished by Dr. J. G. Cooper, whose name is given at note 1 as the authority for most of the chapter. Through some misunderstanding the plates were made before Dr. Cooper had finally corrected the proofs, and in the haste to release the type an unusual number of errors, most of them in orthography, were overlooked. Many of these are apparent and need no further explanation; others are explained in the *errata* at the end of the chapter. While it is thought best to publish the chapter as it stands, it is only just to Dr. Cooper to say that he is in no way responsible for the insertion of, or the statements in, any of the notes or tables, except his own, also that he would make some alterations, based upon recent investigations, if the whole chapter were revised.

J. R. M.

General Note on the Topography of California.

The two prominent features, extending through nearly the entire length of the State of California are the snow-capped range of the Sierra Nevada on the eastern border, and the low Coast Range, or rather belt of ranges, bordering the sea coast on the west. Between the two lies the great valley of California, drained from the northward by the Sacramento, and from the southward by the San Joaquin rivers, and these uniting near the middle of the length of the valley, pass westward through the narrow Strait of Carquines into San Francisco Bay, and thence through the Golden Gate into the Pacific Ocean. These two rivers receive nearly all their waters from the Sierra Nevada, the streams flowing landward from the Coast Range being insignificant. The main drainage of the Coast Range is to seaward, through many small rivers bordered by fertile valleys. The immediate coast is mostly abrupt and rocky and frequently mountainous. The Great Valley, from the Tejon Mountains on the south to Red Bluff on the north where the valley proper terminates, is about four hundred miles in length, and its width varies from over sixty to somewhat less than forty miles. The northern part, or Sacramento Valley, is about 160 miles long, from Red Bluff to the Calaveras River, and is seven miles wide at the head, widening in three miles to fifteen, and then expanding suddenly to about forty miles. The southern or San Joaquin valley is two hundred and forty miles long, and its prominent topographical feature is the Tulare Lake and the basin surrounding it.— *E. W. Hilgard, in Cotton Report of U. S. Census.*

General Note on the Geology of California.—Broadly speaking the *Coast Range* of California consists of Tertiary and Cretaceous, mostly sandstones and calcareous clay slates, almost everywhere greatly disturbed, folded, and frequently highly metamorphosed, and traversed by dikes of eruptive rocks and upheaval axes. In the portion north of San Francisco these are frequently by tufaceous and scoriaceous, or crystalline lava flows, emanating from distinct volcanic vents now extinct.

In contrast to the Coast Range the *Sierra Nevada* has in general a central axis of granite or other rocks, occasionally traversed by volcanic vents, on the flanks of which lie more or less crystalline and metamorphic slates or schists of Palæozoic, Triassic, and Jurassic age, with edges upturned at a high angle or sometimes vertical. Abutting against this, the proverbial "bed rock" of the California miners, there lies on the border of the great valley strata of marine deposits, mostly of the Tertiary, but northward also of the Cretaceous age, which are but slightly disturbed, and into which the rivers flowing from the Cañons of the Sierra have cut their immediate valleys, flanked by bluffs from forty to seventy feet high. From opposite San Francisco northward, on the lower foot hills, appear immense gravel beds, mostly gold bearing, and these are partly over-laid by eruptive or volcanic out-flows and tufaceous rocks, also accounted as belonging to the Tertiary age. In the northern portion of the Sierra region the eruptive rocks become more and more prominent, covering an enormous area called the "lava bed" in the northeastern part of the State, and, as in the Cascade Range, in Oregon, forming the body of the comparatively low range, upon which the volcanic cone of Mount Shasta is superimposed. (See Note 39 on Northern Pacific Railroad.)

fic Railroad.	Alt.	Ms.	Central Pacific Railroad—*Continued.*	Alt.
20. Quaternary.		731	Arcade.	20. Quater. Alluvial.[55]
"	5551	744	Sacramento.	" 50
"	5519	.....	Sacramento.	" 50 / 55
"	6995	.....	Elk Grove.	" 49
"	5954	525	Galt.	" 55
"	5221	607	Stockton.[6]	" 56
"	4695	650	Lathrop.	20. Quaternary.
"	3607			⎧ 19. Tertiary, Plio.,
"	5595	706	Banta.	⎨ 19 b. Miocene & lig-
"	5220			⎩ nite, 19. Eocene(?)[50]
"	2422	713	Tracy.	20. Quaternary.
"	1759	745	Byron.	"
"	1550	815	Antioch.	"
"	956			⎧ 18. Cretaceous and
"		859	Martinez.	⎨ 19. Eocene.
"	249	863	Port Costa.	18. Cretaceous.
19 c. Pliocene, "	165	877	San Pablo.	20. Quaternary.
⎧ Quaternary, above		890	Oakland Pier.[9]	" 14
⎩ Granite (Arch.?)[154]		895	San Francisco.[10]	18. Meta. Cretaceous.

:eous and Tertiary beds on the borders of the great valley, there are within
·ench marks showing the existence in *Quaternary* times of a great fresh-
lbsequently drained by the erosion or breaking, first of the Strait of Car-
that of the Golden Gate. Prior to the latter event, the drainage of the great
a Santa Clara and Pajaro valleys into the Bay of Monterey. The latest sur-
an Joaquin valley, mostly sandy, and in the Sacramento valley more com-
responding to the composition of the Coast Ranges opposite to each district.
· *Cotton Report.*
nearly all constructed in the valleys on the Quaternary formations just
lttle variety in the tabular list of formations passed over and immediately
The notes on adjacent mountains impart some interest to the country for

r, of Hayward's, Cal., late Assistant State Geologist under Professor Whitney,
from Prof. E. W. Hilgard's U. S. Census Cotton Report, and other sources.
iarine and fresh water in the Coast Range and Sierra Nevada Mountains, but
ι of it volcanic.
of the Granite is also eruptive (19. Tertiary), but may be remelted Archæan.
·lcanic and glacial, with 1. Archæan (granite) and metamorphosed rocks of
rous but not rich. Mt. Stanford, northward, is 9,500 feet high.
luburn. Glacial and detrital above 16. Triassic and 17. Jurassic sandstones, con-
nined on the western slopes. A fine iron mine seven miles north of Auburn.
lin. Detrital above 1. Archæan granite, surface mining for gold, platinum,
:kel. Diamonds also occur in small quantities.
ountains to the east produce lime, marble, copper ore and some lignite (19 c.

ιlo, 3,876 feet high, is in full view and easily ascended from near the coal mines.
Francisco. The Golden Gate and Bay of San Francisco. This Bay has been
ι of its first discovery, as among the finest in the world, and is justly entitled
ιnder the seaman's view of a mere harbor. But when all the accessory
; to it are taken into the account, it rises into an importance far above that
iay of San Francisco is separated from the sea by low (Cretaceous) mountain
:he peaks of the Sierra Nevada, the Coast Mountains present an apparently
ly a single gap, resembling a mountain pass. This is the entrance to the
' water communication from the coast to the interior country. Approaching
resents a bold outline. On the south the bordering mountains come down
ken hills, terminating in a precipitous point, against which the sea breaks
·n side the mountains present a bold promontory, rising in a few miles to a
ιusand feet. Between these points is the strait, about one mile broad in the
niles long from the sea to the bay. This passage is called the Golden Gate.
; into the Bay of San Francisco, and its advantages for commerce, suggested
ι discovery of gold in California, and by analogy to the Golden Horn of Con-
·ough this gate, the bay opens to the right and left, extending in each direc-
iles, having a total length of more than seventy, and a coast of about two-
e miles. It is divided by straits and projecting points into three separate
hern is called San Pablo, the middle one Suison, and the southern San
view is that of an interior lake of deep water lying between parallel ranges
ι thousand feet above the water, and behind the rugged peak of Mount
dred and seventy feet high, over-looking the bay and surrounding country.
bold character of the shores, some mere masses of rock, and others origi-
ιg to the height of three and eight hundred feet, break the surface of the
esque beauty. J. C. FREMONT.

Ms.	Central Pacific Railroad— *Continued.*	Alt.	Ms.	Central Pacific Railroa *Continued.*	
.....	Sacramento.[12]	20. Quaternary.	30			
13	Davis.	"	34	86	Banta.	⎧ 19 c. Te⎪ ⎨ 19 b. Mi⎪ ⎩ 19 a. Mi
21	Dixon.[11]	"	65			
29	Elmira.[12]	"	75	94	Lathrop.[15]	20. Quater
40	Suisun.	"		105	Ripon.	"
57	Benicia.	"		108	Salida.[16]	"
58	Port Costa.	18. Cretaceous.		114	Modesto.	"
61	Vallejo Junction.	"		119	Ceres.	"
66	Pinole.	19 b. Miocene, Tertiary		127	Turlock.	"
69	Sobrante.	"		137	Livingston.	"
72	San Pablo.	20. Quaternary.		152	Merced.	"
84	West Oakland.	"		162	Athlone.	"
85	Oakland Pier.	"	14	178	Berenda.	"
90	San Francisco.	18. Met. Cretaceous.		185	Madera.*	"
.....	San Francisco.[10]	"		197	Sycamore.	"
5	Oakland Pier.[9]	20. Quaternary.	14	207	Fresno.	"
7	Oakland (16th St | reet).	"	216	Fowler.	"
10	West Berkely.	"		227	Kingsburg.	"
18	San Pablo.	"		235	Cross Creek.	"
21	Sobrante.	19 b. Miocene Tertiary		241	Goshen.[15]	"
24	Pinole.	"		.....	Tagus.[86]	"
27	Tormay.[13]	18 c. Cretaceous.		251	Tulare.	"
29	Vallejo Junction.	"		262	Tipton.[17]	"
32	Port Costa.	"		.....	Alila.	"
36	Martinez.	18. Cre. & 19 a. Eocene.		282	Delano.	"
39	Avon.	20. Quaternary.		294	Poso.	"
42	Bay Point.	19 c. Pliocene Tertiary		302	Lerdo.	"
50	Cornwall.[14]	20. Quaternary.		314	Sumner.[18]	"
55	Antioch.	"		321	Wade.	"
63	Brentwood.	"		329	Pampa.[19]	"
68	Byron.	"		336	Caliente.[86]	"
77	Bethany.	"		342	Bealeville.	1. Arch. G
83	Tracy.	"		350	Keene.[20]	19 c. Plio. (

* The road to Yosemite Valley is from this place.

10. *San Francisco.* The rock on which the city rests belong entirely to the metam ceous series, and is not the Lignite or Eocene, or Tejon beds which bear the coal, as giv edition.   H. W.

11. The islands in the bay are all like San Francisco in structure.

12. *Elmira to Sacramento.* The coast range westward, 5,000 to 8,000 feet high, is lit but resembles that south of San Francisco Bay, with much more volcanic, and towar auriferous, but only granitic or metamorphic rocks, containing the gold quartz, under ceous, as far as now known.

13. *Tormay.* Fossils of both formations are more plenty and better than elsewh Francisco Bay.

14. *Cornwall.* Good fossils are to be found in Kirker's pass, three miles south of Cc coal mines, five miles south, are not now worked, but a ride to the summit of Mt. Diat is interesting.

15. *Lathrop to Goshen.* The " High Sierra," 14,000 to 15,000 feet, can be seen on cleε mountains eastward have the same general character as on the line from Boca to Sacr the addition of some 18. Cretaceous uplifts near base.

16. *Salida.* Table Mountain, made famous by Bret Harte's humorous poem, risiɪ feet above the Stanislaus river, has a length of about 30 miles, its flat top being from 1,20 wide. A prominent feature in the topography of Amador, Calaveras and Tuolumne cc occurrence of belts of lava-capped hills and mountains, as well as deposits of other voIcɜ the remains of what were once lava flows from the Sierra mountains westward. The Ta is a flow of lava, originating in the lofty volcanic region beyond the " big trees " of Cala

17. *Tipton.* A great bed of magnesite twenty miles east.

18. *Sumner.* A great vein of antimony overlies 40 miles due south near Mt. Pinₑ elevation of mountain being 7,000 feet.

19. *Pampa.* For several miles east the roads pass through hills of 19. Pliocene, Teɪ and clays, with volcanic and other detritus overlying metamorphic shales, etc., thₐ Cretaceous or 19. Eocene.

20. *Keene.* Broken terraces of 19 c. Pliocene, Tertiary age, chiefly of volcanic mat or six miles.

Ms.	Central Pacific R. R.—*Con.*		Alt.	Ms.	Central Pacific R. R.—*Con.*		Alt.
....	"The Loop." *			439	Lang.	17. Jurassic.	1681
355	Girard.[21]	13. Sub Carb. l. s.	3301	452	Newhall.	20. Quaternary.	1388
....	Tyler.	"	3305	....	Andrews.	"	1338
362	Tehachapi.[22]	1. Arch. Granite.	3964	456	S. F. Tunnel.[27]	19 c.Plio. Tertiary	1401
....	Summit Siding.	"	4025	461	San Fernando.	20. Quaternary.	1066
371	Cameron.[23]	13. Sub Carb. l. s.	3737	....	Lulmuga.	"	980
....	Nadean.	"	3357	474	Sepulveda.	"	461
382	Mojave.[24]	20. Quaternary.	2731	482	Los Angeles.[28]	"	293
....	Gloster.	" Desert Region.	2355	484	Shorb.	"	460
396	Rosamond.[25]	"	2315	491	San Gabriel.	"	409
407	Lancaster.	"	2380	494	Savanna.	"	296
417	Alpine.	13. Sub Carb. l. s.	2322	496	Monte.	"	286
...	Vincent.	"	3211	502	Puente.	"	323
427	Acton.[26]	17. Jurassic.	3678	512	Spadra.	"	705
431	Ravena.	"	2350	515	Pomona.	"	856

* The railroad here describes a circle and crosses itself.

21. *Girard.* Beds of 13. Lower Carboniferous limestone on granite hills near by, one crossing the road; good marble, common, some vesicular basalt also.

22. *Tehachapi.* Gold mines in gravel, and quartz veins near by.

23. *Cameron.* The pass through Sierra Nevada here resembles other sections northward; some auriferous slates, 17. Jurassic (?), are worked in vicinity also.

24. *Mojave.* The desert region known as the Mojave Desert, and east of the Sierra Nevada the Colorado Desert or basin, reaches far eastward into Arizona, and affords, by this route, one of the strangest railroad rides in the world. It is a sandy barren waste, interspersed with salt lakes and alkali tracts, destitute of all timber growth, except occasional tracts of yucca, small nut pines and juniper In the south it is subject to very frequent and severe sand storms. Enough of it to satisfy the traveler is seen along the line of this railroad for hundreds of miles. A boiling Mud Lake is only a few hundred yards southwest of the road (See notes 25, 29, 30 and 31.) But probably the culminating point of this fearful desert is found in "Death's Valley," far from any railway station, near the eastern line of California. It is four hundred feet below the level of the sea, while but seventy miles west of it are clustered a number of the highest peaks of the Sierra Nevada, many of which are from 12,000 to 15,000 feet in height. For 45 miles in length and 15 in width along its centre it is a salt marsh with a thin layer of soil, and a large portion of the basin is covered with an incrustation of salt and soda several inches thick, destitute of the slightest vegetation. The heat of the valley is fearful during the summer. Whatever may be the rock formation underlying the desert is of no importance, as its existence is not due to that, but to the aridity of the climate and to the excessive deposits of alkali on the surface and mingled with the superficial formations. For a description of the alkali, see note No. 25.

25. *Rosamond.* The Alkali, so injurious to extensive regions of the southwest, has been carefully studied in California by Prof. E. W. Hilgard. His analyses show the presence of from one to four per cent of these injurious salts in 100 of soil. Of these salts, from 20 to 50, and in some cases 75 per cent., the proportions varying very much in different places, is sulphate of sodium or glauber salt; from 10 to 20, and sometimes 30 per cent. chloride of sodium or common salt, from 15 to 60 per cent. of carbonate of soda or sal-soda, sometimes from five to 20 per cent. of sulphate of potassium, a less quantity of carbonate of potassium or saleratus, and other salts injurious to vegetation in various quantities, phosphates, nitrates, etc.

The remedy for the reclamation of alkali lands is, of course, the leaching out of the injurious salts, by flooding with pure water and underdraining. Unfortunately, in many cases, the alkali returns and again increases on irrigated lands, rising from below through the agency of the water evaporated on the surface, which causes a greater depth of sub-soil to be drawn upon for its alkali, where, too, the soil is more highly charged with it than at the surface. The origin of the alkali is not fully determined. Professor Hilgard thinks much of this salty matter pre-existed in the geological strata, as it is seen to "bloom out" from the rocks, and that from these it was continually washed out in Quaternary times by percolating water, when great lakes covered the valleys of California, for a time held in suspense and then precipitated, or in some cases by the drying-up of the lakes the salts were deposited, which are now found accumulated in the soil. But the very great quantities of the alkali may be said not to be satisfactorily accounted for. The alkali has a corrosive action upon the root crowns and upper roots of plants. It seems that the cotton plants, having long tap roots, it is less injurious to them than to others. Another injurious effect it has in hardening clay soils, producing a tamped condition, instead of the flocculent state which we see in a well tilled and productive soil.

26. *Acton.* Iron and copper mines occur near here.

27. *San Fernando Tunnel.* On west side of pass the sandstones reappear with marine fossils. Tunnel through 18. Cretaceous and 19. Tertiary hills.

28. *Los Angeles.* The hills northward are metamorphic (18.Cretaceous?),with a great 19.Tertiary (19 b. Miocene and 19 c. Pliocene) basin between them and the range north of San Fernando. To the east more metamorphic and granitic,with auriferous quartz,copper,etc. The 19.Tertiary contains much petroleum.

Los Angeles. The traveler from the eastward who has begun to despair of ever seeing anything greener than giant cacti and adamantine vegetation which dispenses with water, is agreeably surprised as he approaches Los Angeles. A drive through the place will enable you to appreciate the reasons which induced the Spanish founders to give the city its name. W. H. R.

Los Angeles to Aanaheim. Alabaster and gypsum occur in low 19. Tertiary hills near here.

Los Angeles to El Careo. About half way the metamorphic and granitic hills approach the road. Much 19 b. Miocene Tertiary, with poor lignite, caps these on the west.

Los Angeles to St. Monica. See note 89.

Ms.	Central Pacific Railroad—*Continued.*		Alt.	
521	Ontario.	20. Quaternary.	981	
525	Cucamonga.	"	952	
......	Sansevain.	"	1074	
540	Cotton.	"	965	
543	Mound City.	"	1055	
547	Brookside.	"	1810	
554	El Casco.	"	1874	
563	San Gorgonio.[29]	"	2560	
569	Banning.	"	2317	
575	Cabazon.	Col. Desert Region	1779	
583	White Water.	"	1126	
591	Seven Palms.	"	584	
......	Dry Camp.	"	168	
612	Indio.[30]	"	20	Below Sea Level (rows Indio–Volcano S'gs)
625	Walters.	"	195	
637	Salton.	"	263	
642	Dos Palmas.[31]	"	253	
653	Frinks.	"	260	
......	L. Point 1 mi. E. of Frinks.	"	263	
......	Volcano.	"	225	
661	Volcano S'gs.	"	220	
671	Flowing Well.[30]	"	5	
676	Tortuga.	"	188	
682	Mammoth Tank.	"	257	
694	Mesquite.	"	294	
708	Cactus.	"	396	
716	Ogilby.	"	355	
722	Pilot Knob.	1. Arch. Gran. & Vol.	285	
......	El Rio.[29]	"	164	
......	Col. River Bdge.	"	139	

ARIZONA.

Ms.			Alt.
731	Yuma.	20. Quaternary.	140
......	Araby.	"	144
745	Gila City.	" Desert Region.	171

Ms.	Central Pacific Railroad—*Continued.*		Alt.
......	Rattlesnake.	Desert Region.	198
761	Abonde.	"	212
771	Tacna.	"	325
......	Mohawk Sum't.	"	541
793	Texas Hill.	"	358
806	Aztec.	"	495
......	Stanwix.	"	515
821	Sentinel.	"	688
834	Painted Rock.	"	726
850	Gila Bend.	"	737
860	Bosque.	"	1080
869	Estrella.	"	1521
878	Montezuma.	"	1330
887	Maricopa.	"	1186
902	Sweet Water.	"	1296
913	Casa Grande.	"	1396
923	Toltec.	"	1507
932	Picacho.	"	1616
946	Red Rock.	"	1865
961	Rillito.	"	2058
......	Jaynes.	"	2241
978	Tucson.	"	2390
......	Wilmot.	"	2667
993	Papago.	"	3009
1007	Pantano.	"	3536
1016	Mescal.	"	4084
1024	Benson.	"	3578
1034	Ochoa.	"	4102
1044	Dragoon Sum't.	"	4614
1054	Cachise.	"	4232
1064	Willcox.	"	4164
1073	Railroad Pass.	"	4394
1088	Bowie.	"	3759
1104	San Simon.	"	3609

29. *San Gorgonio.* Metamorphic auriferous rocks (secondary) overlying granite, chiefly on the west side. San Barnardino Mountain is 11,000 feet high.

San Gorgonio to *El Rio.* The railroad plunges into the most remorseless, cruel waste of sand and rock I every beheld. It spreads out up to the foot of the rugged hills of the Barnardino range, an abomination of desolation, compared with which the Lybrian Desert is the Garden of Hesperides. I cannot describe, nor could I at any time hope to give an adequate conception of this dreadful wilderness. For 107 miles there is not a drop of water to be found, but Nature, as if to take away the reproach of permitting such a vast blotch on her fair face, kindly threw in Fata Morgana. We saw with delight wide spread lakes, with fairy islands in the midst; placid seas washing the base of the distant hills. This baked and dreary expanse extends from near San Gorgonio nearly to El Rio.

Wm. Howard Russel.

30. *Indio* to *Flowing Wells.* For 61 miles the road is below sea level, going down to 263 feet on the border of 19. Pliocene Tertiary lake bed which contains fresh water fossil shells, and below them beds of salt, from being once the head of the Gulf of California; on its west side are 19 b. Miocene Tertiary sandstone strata, with marine fossils, lying against east slope of Coast Mountains. Hot springs and mud volcanoes also occur in the lake bed near its centre; some of our rarest minerals are found in the neighboring mountains.

31. *Dos Palmas.* A few miles southwest of this place is a broad valley in which is the dry bed of a lake forty miles in circumference. Nearly in the centre of this plain, there is a *lake of boiling mud* about half a mile in length by five hundred yards in width. In this curious caldron the thick, grayish mud is constantly in motion, hissing and bubbling, with jets of boiling water and clouds of sulphurous vapor and steam bursting through the tenaceous mud and rising high in the air with reports often heard at a considerable distance. The whole district around the lake trembles under foot, and subterranean noises are heard in all directions.

32. *Deming.* The San Luis Mountains, on the Mexican side of the river, rise abruptly from the plain, as they run south, and assume by far the most formidable appearance of any range west of the Rio Grande. Tombstone mining region is in this mountain. This stupendous range of Mexican mountains drops abruptly a few miles north of the boundary, as if to make room for a railroad to connect the Pacific and Atlantic states. In fact the original boundary line was changed by a second treaty, for the express purpose of securing to the United States this great roadway, for at El Paso

NEW MEXICO.
Central Pacific Railroad—Con.
Southern Pacific Branch.

Ms.	Station		Alt.
1118	Stein Pass.	Desert Region.	4351
.....	Pyramid.	"	4301
1138	Lordsburg.	"	4245
1149	Lisbon.	"	4278
1158	Separ.	"	4508
1169	Wilma.	"	4557
1178	Gage.	"	4468
.....	Lunis.	"	4422
1198	Deming.[32]	"	4834
1209	Zuni.	"	4187
1224	Cambray.	"	4224
1237	Aden.	"	4391
1249	Afton.	"	4307
1259	Lanark.	"	4165
1271	Strauss.	"	4053
1281	Rogers.	"	3723
.....	Bridge over Rio Grande.	"	3748

TEXAS.

Ms.	Station		Alt.
1286	El Paso.[33]	Desert Region.	3718
	Low Water in Rio Grande River about		3712

NEW MEXICO.
Atlantic & Pacific R. R.* (Western Div.)
Albuquerque by The Needles to Mojave.

Ms.	Station		Alt.
0	Albuquerque.	{ Base 18. Cre., Summits of 16. and 17. Jurassic & Triassic alternating.	4983
10	Isleta.	"	4881
13	A. & P. Junction.	"	4933
23	Luna	"	
34	Rio Puerco.	"	5026
47	San Jose.	"	5428
60	El Rito.	"	5683
66	Laguna.	"	5767
72	Cubero.	18. Lower Creta.	5905
83	McCarty's.	"	6141
88	Baca.	"	
96	Grant's.	16. Triassic.	6440
107	Blue Water.	"	6609
122	Chaves.	"	6969
130	Continental Divide.	"	
136	Coolidge.	"	
146	Wingate.	"	6714

* By Capt. C. E. Dutton, U. S. Geologist.

NEW MEXICO.
Atlantic & Pacific Railroad—Con.
(Western Division.)

Ms.	Station		Alt.
158	Gallup.	18. Cretaceous.	6477
166	Defiance.	"	6852
174	Manuelito.[34]	Base of 18. Creta.	6382

ARIZONA.

Ms.	Station		Alt.
187	Allantown.	16-17. Jura.-Tria.	6026
200	Sanders.	"	5507
213	Navajo Springs.	"	5605
226	Billings.	"	5372
238	Carrizo.	"	5199
253	Holbrook.	"	5047
263	St. Joseph.	"	4970
275	Hardy.	"	4910
286	Winslow.	14. Carboniferous	4825
298	Dennison.	"	4979
312	Cañon Diablo.	"	4765
323	Angell.	"	5879
333	Cosnino.	{ 14 Car., overlaid in places with lava	6484
344	Flagstaff.	"	6852
356	Bellemont.	"	7099
368	Chalender.	"	6887
378	Williams.	"	6727
381	Supai.	"	6917
391	Fairview.	"	5909
401	Ash Fork.	"	5105
409	Pineveta.	"	5084
419	Crookton.	"	5657
431	Chino.	"	5234
439	Aubrey.	"	5125
452	Yampai.	"	5552
466	Peach Spring.[35]	"	4759
478	Truxton.	"	4172
489	Hackberry.	"	3522
501	Hualapai.	"	3277
514	Beal.	"	3472
516	Kingman.	"	3303
527	Drake.	"	
540	Yucca.	"	1774
553	Franconia.	"	
566	Powell.	"	418
572	East Bridge.	"	
575	The Needles.	"	477
.....	Colorado River Bridge.	"	468
.....	" " Low Water.	"	440

the great Rocky Mountain Range of the United States also terminates, thus forming what is truly the gate-way of the continent. Between the San Luis Mountains and El Paso are wide plains, bounded by detached mountains of metamorphic and other limestones, associated with igneous rocks.

33. *El Paso.* See notes in Texas chapter on El Paso.

34. *Manuelto.* A natural bridge discovered and reported by Frederick Gardner, Jr., is situated about 20 miles north of the railroad, near the line between New Mexico and Arizona. It is 65 feet long, 15 feet wide, two feet thick in the centre, and 15 feet at the sides, and about 30 feet high. This bridge is formed by a remnant of the over-lying grit, which is continuous with it on both sides. The section cut through beneath it is of light and dark red sandstone (16. Triassic.) A short distance off is a petrified forest. The stone tree trunks lie just beneath the soil or half-exposed, fallen in all directions.—F. G., in *Science* for July, 1885.

Ms.	Atlantic & Pacific Railroad—Con. (Western Division.*)		Alt.
575	The Needles, Nev.	20. Quaternary.	477
582	Java.	Desert Region.	961
589	Ibex, Cal.	"	1448
598	Homer.	"	2118
606	Goff's	"	2577
616	Fenner.	"	2087
628	Edson.[36]	"	1727
632	Danby.[37]	1. Arch. Gran. "	1342
644	Cadiz.	"	819
652	Bristol.	"	705
659	Amboy.	"	611
666	Bagdad.[37]	"	784
673	Siberia.	20. Qua. "	1267
684	Ash Hill.[38]	"	1940
690	Ludlow.[39]	"	1778
699	Lavic.	"	2176
710	Haslett.	"	1663
722	Newberry.	"	1826
734	Daggett.[39]	"	2002
745	Waterman.[40]	"	2118
754	Hinckley.	"	2159
763	Harper.	"	2276
777	Kramer.	"	2462
795	Rogers.	"	2281
815	Mojave, Cal.[24]	"	2751

Ms.	Nev. County (N. G.) Railroad.[41]	Alt.
0	Colfax.	20. Quaternary.
5	You Bet.	16. Trias. & 17. Juras.
9	Storm's.	"
11	Buena Vista.	"
14	Kress'.	"
17	Grass Valley.	"
21	Town Talk.	"
23	Nevada City.	"
	San Francisco & N. P. Railroad.	
.....	San Francisco.	18 c. Met. Cretaceous.
6	Port Tiburon.	"
12	Green Bro.	"
15	San Rafael.[42]	"
20	Miller's.	20. Quaternary.
26	Nevada.	"
35	Junction.	"
40	Pems Grove.[37]	"
46	Cotate.	"
51	Santa Rosa.[43]	"
56	Fulton.	"
.....	Guerneville.	"
57	Mark West.	"
66	Healdsburg,	"
75	Clairville.	"
85	Cloverdale.[44]	"

* By Dr. J. G. Cooper, of California, late Assistant Geologist under Prof. Whitney. Dr. Cooper made a journey over this route specially to obtain the geology given in this table and the notes.

35 *Peach Spring* Best point now known from which to visit the Grand Cañon of the Colorado, and the only accessible point from which the descent can be made, by an easily traveled road, into as majestic and peculiar cañon scenery as is anywhere to be seen. The plates and descriptions by Dr. J. S. Newbury, in Ives' Report of 1858, give a fair idea of what is to be seen. Altogether there is nothing like this cañon. The far-famed Yosemite is more beautiful and more varied, but not more magnificent nor half so strange and weird.—A. G., in *Science.*

36. *The Needles to Edson.* Frequent outcrops of Archæan and Metamorphic rocks near road, also erupted lavas and volcanic cones of 19. Tertiary age, some perhaps 20. Quaternary. "The Needles" themselves are of purple porphyry and trachytic granite worn into sharp peaks.

37. *Danby to Bag-dad.* The road passes through the granite pass of Providence Mountains for many miles; the same rocks occur as eastward and containing ores of various kinds. The mountains northward resemble those of Nevada, being Paleozoic rocks containing lead and silver, with a little gold.

38 *Ash Hill.* The west slope of the mountains descends gradually to Soda Lake, the sink of Mojave River. Death's Valley, described in note No. 24, lies nearly due north from Soda Lake, 75 to 100 miles distant.

39. *Ludlow to Daggett.* 1. Archæan Granite metamorphic and 19. Tertiary volcanic rocks lie at the west side of the sink, then cliffs of 19. Tertiary gravels, 50 to 100 feet high for 20 miles, then metalliferous rocks (Metamorphic). Abundance of soda and salt in the sink of Mojave River, other lake beds also containing borax.

40. *Waterman to Mojave.* After rising about 500 feet in the valley of the Mojave River, the road leaves it, and for 70 miles passes over an apparently level plain with little rock in sight, much of it being barren sand hills or alkaline planes, the rest with low shrubbery or groves of yucca trees 30 feet high. It is probable that this Quaternary desert covers Tertiary strata even as old as Eocene, but fossils are absent. (See Colorado Desert notes, No. 24, 25, 29, 30 and 31.)

41 *Nevada County Narrow Gauge Railroad.* The air line distance is about 16 miles, but the road winds among hills containing Archæan granite, 13 b. Sub-Carboniferous limestone, 16. and 17. Auriferous slates and quartz veins; 19. Tertiary gravels and volcanic strata much intermined. It is the richest quarts mining region in California.

42. *San Rafael.* Mt. Tamalpais, 2,604 feet high, may be ascended here. Gives a magnificent view of the country near than San Francisco Bay.

43. *Santa Rosa.* Mark West Creek, north and northwest of this place, a branch of the Russian River, has along its banks beds of Pliocene or Post Pliocene fossils. (See Palæ. of Cal., by Gabb.)
H. M. T.

The hills north of Santa Rosa are full of fossils, 19 b. Miocene and 19 c. Pliocene, but the highest ridges are more or less 18 c. Lignite and Metamorphic Cretaceous, with some coal, quicksilver, sulphur volcanic dikes frequent.

44. *Cloverdale.* The hills to the east of Cloverdale branch contain many small deposits of quicksilver.
H. M. T.

Northern Pacific Coast R. R.[38]		California Pacific Railroad.—Con.	
Ms.	Alt.	Ms. Main Line.	Alt.
0 San Francisco.	{ 18 c. Metamorphic Cretaceous.	31 Napa Junction.	20. Quaternary.
11 San Quentin.	"	39 Bridgeport.[52]	"
15 San Rafael.[42]	"	44 Fairfield.	"
17 Junction.	"	55 Elmira.[12]	"
		59 Batavia.	"
0 San Francisco.	"	63 Dixon.[11]	"
6 Saucelito.	"	71 Davis.	"
10 Lyford's.	"	84 Sacramento.	"

		Marysville Branch.	
15 Ross.	20. Quaternary.	0 San Francisco.	(As before).
17 Junction.	"	71 Davis.	20. Quaternary.
21 Whitesville.[45]	18. Metam. Cretaceous	81 Woodland.[53]	"
26 Langunitas.	"	85 Curtis.	"
30 Taylorsville.	"	90 Knight's Land'g.	"
37 Point Reyes.	"		
47 Marshalls.	"	California Pacific & Northern Railroad.	
54 Tomales.	"	0 San Francisco.	
61 Valley Ford.	19 b. Miocene Tertiary	32 Port Costo.	{ (Via Oakland and
65 Freestone.[46]	"		San Pablo Bridge
73 Sonoma Mill.[47]	"	to	and ferry across
76 Russian River.	"		Straits of Carquines)
79 Moscow.	"	33 Buricio.[1]	
80 Duncan Mills.	"	39 Goodyear.[2]	19. Tertiary Volcanic.
		49 Suison.[3]	20. Quaternary.
California Pacific Railroad.		55 Vancleu.	19 b. Pliocene.
		90 Sacramento.	20. Quaternary.

0 San Francisco.	{ 18 c. Lign. & Meta. Cretaceous.	Napa Branch.	
25 Vallejo.[48]	20. Quat. & 18. Creta.	0 San Francisco to	Valley Jun., 29 miles.
31 Napa Junction.[49]	20. Quaternary.	South Vallejo.	18. Cretaceous.
39 Napa.	"	38 Napa Junction.	"
45 Oak Knoll.	"	46 Napa.	20. Quaternary.
52 Oakville.	"	46 Cordelia.[4]	19. Tertiary Volcanic.
58 St. Helena.[50]	"	51 Suison.[5]	20. Quaternary.
66 Calistoga.[51]	"		

1. Both sides of the straits are 18. Cretaceous.
2. Near here basalt is quarried for paving blocks.
3. Ten miles across marsh.
4. Paving blocks extensively quarried.
5. The beautiful Travertin or "Suisum Marble" found near by.

45. *White Hills.* Tunnels through these ridges are here capped by 19 b. Miocene tertiary.
46. *Freestone.* The great Red Wood forest commences here and covers most of the hills, with part of the valleys, northward near the coast, chiefly west slopes.
47. *Sonoma.* A low ridge of 18. Metamorphic Cretaceous, much broken by 19. Volcanic Tertiary, separate Sonoma, also Santa Rosa Valley.
48. *Vallejo* No Metamorphic Cretaceous visible along the railroad, only thin bedded, unaltered strata. The fossil forest is on this route.
49. *Napa Jun. to Calistoga.* The hills on both sides are metamorphic (18. Cretaceous?), with volcanic outbursts increasing toward the northeast, and with quicksilver deposits.
50. *St. Helena.* Mt. Helena, the culminating point of the volcanic mountains, to the north and east, is 4,343 feet high.
51. *Calistoga.* Twenty-five miles north is Clear Lake, where sulphur and borax occur in abundance.
52. *Bridgeport.* Tunnel through 18. Cretaceous where fossils are found. Near here is a bed of fine arogonite, called suezaric marble.
53. *Woodland.* A branch road runs 80 miles further up the west side of the Sacramento River to Tehara, over level valley lands over 20. Quaternary.
54. *Ewing to Red Bluff.* The mountains eastward resemble those farther to the south, but with more 18. Cretaceous, some 13. Sub Carboniferous near the middle, and a vast 20. Quaternary volcanic field northward.
55. *Marysville.* Buttes in plain sight from the railway, northwest from the town.
56. *Soto.* Lunen's peak, a volcano, 40 miles east, is over 10,500 feet high; the lava beds here compel the railroad to cross the river.

Oregon Division Central Pacific R.R.

Ms.	Oregon Division Central Pacific R.R.		Alt.
0	Sacramento.	20. Quaternary.	
8	Arcade.	"	
15	Antelope.[7]	"	
18	Junction.	19. Tertiary, Plio., with workable lig'e.	
29	Lincoln.		
33	Ewing's.[54]	20. Quaternary.	
40	Wheatland.	"	
46	Reed's.	"	
50	Yuba.	"	
52	Marysville.[55]	"	
70	Gridley.	"	
83	Nelson.	"	
90	Durham.	"	
96	Chico.	"	
105	Anita.	"	
110	Soto.[56]	"	
122	Sesma.	"	
123	Tehama.	"	
135	Red Bluff.	19. Tertiary hills.	
170	Redding.*[57]	19 b. Pliocene	
173	Middle Creek.†[58]	18 c. Cretaceous.	
180	Copley.	17. Jurassic slates.	
187	Kennett.	19. Tertiary volcanic.	
192	Morley.	17. Jurassic or 16. Triassic slates (?) (auriferous), with 19. Ter. Volcanic.	
196	Elmore.		
203	Smithson.		
208	Delta.		

Central Pacific Railroad.
(Northern Division.)

Ms.	Station	Formation	Alt.
108	Marysville.[55]	20. Quaternary.	
120	Honent.	"	
144	Orville.[59]	19 c. Pliocene Tertiary, 18 c. Creta., 14. Sub-Carbon.	

Ms	Sacramento & Placerville R. R.		Alt.
0	Sacramento.	20. Quaternary.	
10	Mayhew's.	"	
22	Folsom.	1. Arch. Granite.	20
29	White Rock.	13. Sub-Carboniferous.	
37	Latrobe.	16. Trias., 17. Jur.	790
42	Dugan's.	‡ "	
48	Shingle Springs.	[60] "	1459

San Jose Branch.

Ms	Station	Formation	Alt.
0	San Francisco.	18. Metam. Cretaceous	
4	Oakland.	20. Quaternary.	
7	Brooklyn.[61]	20. Qua., 19c. Ter. Plio.	
12	Melrose.	"	
16	San Leandro.	"	
18	Lorenzo.	"	
27	Decoto.	"	
30	Niles.[62]	"	
34	Irvington.[63]	Tertiary, Pliocene.	
37	Warm Springs.	"	
39	Haward's.	20. Quaternary.	
42	Milpetas.	"	
48	San Jose.[64]	"	

Stockton & Visalia and Stockton & Copperopolis Railroads.[66]

Ms	Station	Formation	Alt.
0	Stockton.	20. Quaternary.	
6	Charleston.	"	
11	Holden	"	
15	Peter's.	"	
15	Peter's.	"	
22	Waverly.[65]	19. c. Tertiary Plio.	
30	Milton.	1. Arch. Granite.	
15	Peter's.	20. Quaternary.	
20	Farmington.	"	
28	Clyde.	"	
34	Oakdale.	"	

* The gravelly hills, with clay, slates and sandstone of fresh water formation, are here 200 feet thick or more, and may include the whole Tertiary age.

† This formation crosses the river near here full of marine fossils, and lies flat on edges of the slates below.

‡ Very much changed by 19. Volcanic.

57. *Redding.* Mt. Shaska, 14,440 feet high, is in view and easily ascended in summer from the end of the railroad. Fine Cretaceous fossils are found near here and also beds of fossil wood, and an abundance of excellent iron ore is found on Spring Creek, 12 miles to the northwest. The rocks from here north are much covered with 19. Tertiary volcanic fragments and ashes, but exposed by the deep cuts.

The Lava Beds. A large portion of the northeastern part of California, to the northern state line and spreading over Idaho, Oregon and Washington Territories, is covered to a depth of several hundred feet with great beds of lava and other volcanic material. The country has generally a broken surface, and is interspersed with hills and high volcanic cones, frequently cut into deep chasms by the few streams that occur in this region, and extensive caves have been found under the lava beds. This lava section has no arable lands, and it is fit only for grazing purposes. (See Note 30 on Northern Pacific Railroad.) E. W. H.

58. *Middle Creek.* Much placer mining is done, and quartz veins exist.

59. *Oroville.* Tertiary leaves and Lignite, 18. Cretaceous, 14. Sub-Carboniferous fossils found near by toward the northeast.

60. *Shingle Spring.* Iron, lead and zinc occur near.

61. *Brooklyn.* Redwood Peak, 1,635 feet high, is the highest in the range opposite San Fancisco. Mission Peak, 34 miles southeast, is 2,566 feet high.

62. *Niles* to *Haywards.* Follows the 20. Quaternary (alluvial), nearly after passing through Alameda Cañon 10 miles, traversing 19. Tertiary, 19 c. Pliocene and 19 b. Miocene, then lignitic, with little coal.

63. *Irvington.* Mountains on the east side rise to 4,443 feet, and on the west side to 3,780 feet in height.

64. *San Jose.* Alum Rock Cañon, about seven miles easterly from San Jose, is a pretty place, with Miocene fossils and a good hotel. H. M. T.

South Pacific Coast (N. G.) R. R.

Ms.	Station	Alt.
.....	San Francisco.	18. Meta. Cretaceous.
6	Alameda.	20. Quaternary.
14	W. Sanleandro.	"
24	Alverado.[66]	"
31	Moury's.	"
37	Alviso.	"
46	San Jose.[54]	"
56	Los Gatros.	19. Tertiary Gravels.
58	Alma.	18 c. Lign. & Met. Cre.
62	Wright's.[67]	
66	Glenwood.	19 b. Miocene Tertiary
73	Felton.[68]	"
76	Rincon.[69]	19 c. Pliocene Tertiary
81	Santa Cruz.	20. Quaternary.

Southern Pacific Railroad.

Ms.	Station	Alt.	
0	San Francisco.	{ 18 c. Metamorphic Cretaceous.	
6	San Miguel.	"	
12	Baden.[70]	20. Quaternary.	89
17	Millbrae.[71]	"	8
21	San Mateo.	"	22
25	Belmont.	"	31
28	Redwood City.[72]	"	9
33	Menlo Park.	"	64
38	Mountain View.[73]	"	73
44	Lawrence's.	"	64
50	San Jose.[54]	"	86
63	Coyote.	"	351
73	Tennant.[74]	"	327
80	Gilroy.[75]	"	193
83	Carnadero.	"	164
86	Sargent's.	"	136
96	Vega.[76]	"	57
99	Pajaro.	"	28
110	Castroville.[77]	"	17
118	Salinas.	"	44

Southern Pacific R. R.—Con.

Ms.	Station	Alt.	
128	Chualar.	20. Quaternary.	144
134	Gonzales.	"	127
143	Soledad.	"	182
80	Gilroy.[75]	"	193
94	Hollister.	"	254
100	Tres Pinos.[75]	"	514
99	Pajaro.	"	28
101	Watsonville.	"	23
106	St. Andrew's.	19 c. Pliocene, Tert.	153
112	Aptos.	"	102
116	Soguel.	"	53
120	Santa Cruz.	"	18

Goshen Division S. P. R. R.

Ms.	Station	Alt.	
0	Huron.	20. Quaternary.	167
.....	Heinlen.	"	211
.....	Lemoore.	"	230
.....	Hanford.	"	242
40	Goshen.	"	276
.....	Visalia.	"	

Central Pacific Railroad. (Amador Branch.)

Ms.	Station	Alt.
0	Galt.	20. Quaternary.
9	Cicero.	"
20	Carbondale.	19 b. Pliocene, Terti.
28	Ione.[79]	"

Montrey Branch.[80]

Ms.	Station	Alt.	
110	Castroville.[77]	20. Quaternary.	17
115	Martino.	"	14
124	Del Monta.	19 c. Pliocene, Terti.	8
125	Montrey.	1. Archæan Granite.	8

65. *Peter's to Milton.* Passing into 19. Tertiary, 19 c. Pliocene and 1. Archæan (granite) below it. About 18 miles southeast is Copperopolis, on the copper ledge, not worked on account of the low price of the metal.

66. *Alverado.* The hills on east are the same described on San Jose Branch in note.

67. *Wright's.* The east slope is entirely of this formation when ascended, the west being heavily covered by 19 b. Miocene Tertiary.

68. *Felton.* The hills to the west have a core of 1. Archæan Granite, also much 18 c. Cretaceous metamorphic limestone.

69. *Rincon.* Asphalt is common both east and west, and petroleum is obtained by bored wells.

70. *Baden.* A ridge of marine 19 c. Pliocene Tertiary, full of shells, etc., lies west of the road for five miles.

71. *Millbrae.* Metamorphic Cretaceous hills west of road, and granite (1. Archæan?) below.

72. *Redwood City.* 19 b. Miocene (Tertiary) hills come near on the west.

73. *Mountain View.* 18. Metamorphic Cretaceous hills on the west, mostly capped by 19 c. Miocene Tertiary (marine.)

74. *Tennant.* The celebrated New Almaden Quicksilver Mines are not far west.

75. *Gilroy.* Some Lignitic (19 a. Eocene and later) exists to the west, but has not yet been found workable. Much 19. Tertiary on the slopes of hills around, with very fine marine fossils (19 b. Miocene and 19 c. Pliocene.)

76. *Vega.* Passes through the 18. Cretaceous hills, flanked by 19. Tertiary (19 a. Miocene and 19 b. Pliocene) on the west. Some lignite in it.

77. *Castroville.* The hills to the southward are metamorphic and granitic, with 19. Tertiary on their flanks as before.

78. *Tres Pinos.* The New Idra Quicksilver mine lies 50 mile southeast in the highest part of this range of mountains, near 5,000 feet elevation. Iron, lead, silver and arsenic also occur.

79. *Ione.* Some lignite of very little value is found here.

80. *Montrey Branch.* Passes through a zone of 19 b. Tertiary containing fossils, which lie upon the granite, and shows the effects of change by heat at the junction, from which the granite is supposed by some to be eruptive 19. Tertiary. Tropolite or infusorial polishing sand is common near here.

Ms.	Pacific Coast Railroad. (Near latitude 35°.)	Alt.
0	Port Harford.[81]	19 b. Miocene, Tertia.
10	Ocean Side.	"
15	Steele's.	"
22	Verde.	"
30	Los Berros.	"
35	Nipoma.	20. Quaternary.
42	Santa Maria.	"
46	Lake View.	19 b. Miocene, Tertia.
55	Harris.	20. Quaternary.
64	Los Alamos.	"

Ms.	California Southern Railroad.	Alt.
0	National City.	20. Quaternary.
4	San Diego.	19 c. Pliocene, Tertiary
9	Old Town.	20. Quaternary.
20	Selwyn.[82]	19. Eocene, Tertiary.
26	Cordero.	19 b. Miocene, Tertiary
35	Encinitas.	"
42	Stewart's.	"
47	San Luis Rey.	20. Quaternary.
52	Ysidora.	18 c. Metam. Creta.
60	De Luz.	1. Archæan Granite.
66	Fallbrook.	"
78	Temecula.	20. Quaternary.
86	" Car B."	"
96	Elsinore.	"
104	Pinacate.[83]	"

Ms.	California Southern Railroad—Continued.	Alt.
110	San Jacinto.	20. Quaternary.
122	Riverside.	"
127	Colton.	"
133	San Barnardino.	"

Los Angeles & San Diego Railroad.

Ms.		Alt.	
0	Los Angeles.[28]	20. Quaternary.	393
5	Florence.	"	151
.....	Downey.	"	112
.....	Norwalk.	"	93
.....	Costa.	"	84
27	Arnheim.	"	134
.....	Orange.	"	150
34	Santa Anna.	"	135

Los Angeles Division.

Ms.		Alt.
0	Los Angeles.	20. Quaternary.
18	San Monica.'	"
0	Los Angeles.	"
5	Florence.	"
10	Compton.	"
15	Cerritos.	"
22	Wilmington.[84]	"
25	San Pedro.	"

There are several short lines in different parts of California, which traverse Quaternary strata, but they show nothing beyond what is contained in these notes.

81. *Port Harford.* A branch runs northeast of San Luis Obispo, nine miles over rolling table land by Tertiary and 20. Quaternary; beds of enormous fossil oyster and other shells are common near by; also lignite and petroleum, volcanic and metamorphic hills also lie near, containing quicksilver. Limestone, etc., is further north.

82. *Selwyn.* Fossils are numerous in the nearly level strata near the coast and probably include all the 19. Tertiary divisions. Under these, at Pt. Loma, 18. Cretaceous fossils are found with lignite in up-tilted strata, and the bed near Selwyn was confounded with these and described as Cretaceous, Division B., at first, but agrees better with the Tertiary. The true Cretaceous again occurs on the west slope of the Santa Anna Coast Mountains, five miles north of Fall Brook station. Fine felspar, tourmaline and garnets also occur in this range in granite.

83. *Pinacate.* A few miles north of the Tamesca Mountains are the tin mines, which will probably become of much value, going up to 60 per cent.

84. *Wilmington.* A metamorphic (18. Cretaceous) hill north of this harbor. The islands visible are similar, with some 20. Quaternary sandstone and Paleozoic rocks.

85. *Goshen to Caliente.* The mountains westward are like those from Pleasanton to Niles, with more 19. Tertiary, 19 b. Miocene and 18. Cretaceous. Also 20. Quaternary, volcanic and granite in places. The only coal now worked is north of Mt. Diablo and south of Livermore. The granite, of the coast ranges at least, is eruptive, and belongs rather to the Quaternary than the Archæan.

86. *Stockton & Visalia Railroad.* The most northern group of "Big Trees" is approached by this route.

The Big Trees. One of the greatest curiosities in California consists of the Big Tree Grove, situated on the divide between the middle fork of the Stanislaus and the Calaveras rivers, about 20 miles east of Mokelumne hill, and at an elevation of 4,759 feet above the level of the sea. The trees range in height from 150 to 327 feet, and in diameter from 15 to 30 feet.

87. *Pems Grove to Santa Rosa.* The foothills are full of Tertiary fossils (Miocene and Pliocene). The metamorphic and volcanic mountains contain valuable quicksilver mines.

88. *Northern Pacific Coast Railroad.* The only groves of celebrated " Redwood " tree, accessible by railroad, are on this route and northward.

Errata:—Note 6, for " telburet" read telluret; page 320, at Cornwall and Antioch, read Pliocene; at Brentwood, etc., Quaternary; at Banta, for 19 a. " Miocene" read Eocene; page 321, at Nadean, Quaternary; Note 28, for " El Carco," El Casco; page 324, for " Pem's Grove," Penn's Grove; Note 41, for " intermined," intermixed; for "quarts," quartz; Note 43, after sulphur place a semicolon; page 325, for " Buricio," Benicio; " Vanclen," Vanden; 327, "St. Andrews," San Andreas; Note 80, for "Tropolite," Tripolite; page 328, "San Monica," Santa Monica; throughout the chapter for " Central," read Southern Pacific.

Delaware.*

GEOLOGICAL FORMATIONS OF DELAWARE.[12]

GROUPS.	DELAWARE SUB-DIVISIONS.	
20. QUATERNARY.	Post Glacial. Glacial.	Bog Clay, River Shore, 20 c. Brick Clay, 20 b. Red Gravel and Estuary Sands, 20 a.
19. TERTIARY.	19 c. Pliocene.	Blue Clay, Glass Sand, 19 c.
	19 b. Miocene.	Potters Clay, 19 b.
18. CRETACEOUS.	18 c. Upper Cretaceous. 18 b. Middle Cretaceous. 18 a. Lower Cretaceous.	Green Sand, 18 c. Sand Marl, 18 b. Wealden Clays, 18 a.
	Crystalline Rocks. Age undetermined.	Eruptive Gabbros and Horn-blende Rocks. Philadelphia Gneiss. Magnesian Marble. Quartzite.

Philadelphia, Wilmington, and Baltimore R. R.

Ms.	STATIONS.	GEOLOGICAL FORMATIONS.
0	Philadelphia.	Phila. Gneiss,
19	Claymont.	Gabbros, [20]
22	Bellevue.	" [14]
24	Edge Moor.	18 a. L. Cre. & Gab.
28	Wilmington.[4]	" [7]
32	Newport.	" [21]
34	Stanton.	" [17]
40	Newark.[1]	" [10]

Newark and Delaware City R. R.

Ms.	STATIONS.	GEOLOGICAL FORMATIONS.
0	Newark.[1]	L. Cretaceous, [100] 18 a. (Plastic Clays.)
2	Wilson.[2]	"
3	Cooche.	Plastic Clays & Trap.
4	Keeney.	"
5	Glasgow.	"
6	Porter's.	"

Newark and Delaware City Rail-road—Continued.

Ms.	STATIONS.	GEOLOGICAL FORMATIONS.
8	Corbitt.	Middle Cretaceous. 18 b. (Sand Marl.)
10	Reybold.	"
12	Delaware City.	18 b & c. Middle & Up. Cre. Sand Marl & Green Sand Marl.

Pennsylvania & Delaware R. R.

Ms.	STATIONS.	GEOLOGICAL FORMATIONS.
0	Newark.[1]	18 a. L. Cretaceous Amphibolites and Phila. Gneiss. [100]
8	Landenberg.[11]	Quartzite, Marble, and Philadelphia Gneiss.
11	Avondale.	(See Pennsylvania.)
26	Pomeroy.	"

* By Prof. Fred'k D. Chester, of Delaware State College, Newark, Delaware.

Delaware Railway.			Delaware, Maryland & V: Railroad.		
Ms.	STATIONS.	GEOLOGICAL FORMATIONS.			
			Ms.	STATIONS.	GEOLOGICAL FO
0	Wilmington.⁴	18 a. L. Cre. & Gab.	0	Harrington.	19 c. U. Pli. t
6	New Castle.⁵	18 a. L. Cre. (Pl. Cl.)	9	Milford.	"
16	Kirkwood.	18 b. Cre.(Sand Marl).	12	Lincoln.	"
21	Mt. Pleasant.⁶	18 c. U. C.(Ind Marl).	17	Ellendale.	"
25	Middletown.	18 c. U. C. (Gr. S'd)⁶⁶	25	Georgetown.	"
29	Townsend.	19 b. Mio. (Pot. Cl.)⁷¹	25	Georgetown.	"
37	Clayton.	" "	31	Harbeson.	"
39	Smyrna.⁷	"	33	Cool Spring.	19 c. U. Pli. t
48	Dover.⁸	" ⁹⁹	36	Nassau.	"
51	Wyoming.	"	40	Lewes.	20 c. Modern.
56	Viola.	" ⁸⁹	25	Georgetown.	19 c. U. Pl. to
58	Felton.	"	41	Frankfort.	"
64	Harrington.	19 c. U. Pl. to P. Pl.⁶⁸	54	Berlin.	"
68	Farmington.	"	68	Snow Hill, Md.	"
76	Bridgeville.	" ⁶⁶	77	Stockton, "	"
84	Seaford.⁹	"	81	Franklin, "	"
90	Laurel.	"			
97	Delmar.	" ⁸²			

NOTES ON DELAWARE.

1. *Newark.* On the plane to the south of Newark, red and white (mottled) clays rise ε above the surface, covered by a great thickness of Red Gravel and brick clay of Quaternary ε mottled clays are probably the equivalent of the Wealden, the latter sub-division being re most authors to the Lower Cretaceous, and by a few to the Upper Jurassic. Passing to the the town, you walk for a mile over a belt of Amphibole trap, beyond which are soft mica so granitic gneisses of doubtful Palæozoic age. Hills from the background of the town, along ι of which can be traced the terrace of Quaternary gravel.

2. *Wilson.* Iron Hill is three miles long by one mile wide, the back bone being a mass ε trap and jaspery quartz. The trap is decomposed into a serpentine earth, which is completely nated with masses of limonite. Several iron ore pits are at present wrought. This dike i confined to the area of Wealden clays, but was evidently an island when the latter clays were α or at least of an earlier origin than the clays.

3. *Delaware City.* At this place a yellow sand marl is succeeded by a calcareous Greeı an ashy color. This can be seen well exposed along the level of the canal, particularly near St. ι

4. *Wilmington.* Excellent exposures of Eruptive rocks are obtained along the Brandyw sisting of alternate masses of syenitic gneiss, with a predominance of a coarse feldspathic Hyı Gabbro.

5. *New Castle.* One mile south of New Castle, upon the river, is a bluff of white, sandy This is the only exposure in the State of the lowest member of the Plastic Clay Series, and is ov 50 feet of mottled clays.

6. *Mt. Pleasant.* Two miles to the northwest of this station is the deep cut made by tl For nearly two miles the green sand rises as high banks upon each side, offering the best exp the marl in the State.

7. *Smyrna.* The Miocene clays are well exposed along Duck Creek, and abound in characteristic fossils.

8. *Dover.* The Miocene clays can be seen back of the town on Jones Creek, and a litt south on Murderkill Creek, Miocene fossils are found in abundance.

9. *Seaford.* To the east of Seaford, upon Nanticoke River, a dark blue clay is well exp its junction with the overlying loam are found nests of the modern Oyster. This blue clay is cover all of Sussex County, but is rarely seen, except in the deeper cuttings of the creeks.] ness varies from three to ten feet, beneath which is over forty feet of fine glass sand. The gl is probably the equivalent of the New Jersey glass sand of Pliocene age. The modern shells, found at the junction of the Blue clay with the overlying gravel, are more imbedded in the l therefore regard the gravels as early Quaternary, and the Blue clay as later Pliocene.

10. *Hockessin.* At this place are excellent quarries of pure dolomitic marble. Kaoli worked in abundance. The dolomite beds in Jackson's quarry form a perfect anticlinal, over corresponding anticlinal of Mica schist. This dolomitic area is the extremity of a tongue of rock extending in from Pennsylvania.

11. *Landenberg.* Near this place in the limestone quarries the relation of the Potsdam ε calciferous marbles and mica schists to each other can be well studied; there are seen three aι capping each other, with the mica schists uppermost.

12. The northern part of the State of Delaware is underlaid by Crystalline rocks, whic from the northern curved boundary of the State to a line crossing the State a little north of tł delphia, Wilmington and Baltimore Railroad, and running in the same direction about N. 50° latter area is divided into two belts of about equal extent.

(a) A southern club-shaped area, composed of amphibolite schists, with which is assc bluish gray trap, ranging from a quartz diorite to a true hyperite. This area is a continuatio

Wilmington & Northern R. R.			Wilmington & Western R. R.		
Ms.	STATIONS.	GEOLOGICAL FORMATIONS.	Ms.	STATIONS.	GEOLOGICAL FORMATIONS.
0	Reading, Pa.	See Pennsylvania.	0	Wilmington.[6]	Gabbro. & 18 a. L. C.
57	Chadd's Ford.	Phila. Gneiss.	7	Greenbank.	Phila. Gneiss.
61	Granogue.	"	12	Ashland.	
63	Adams.	Hypersthene Gab. [8a9]	15	Hockessin.[10]	" with Marble.
65	Dupont.	" [8a9]	17	Southwood.	{ Quartzite, Marble, and Mica Schists.
66	Greenville.	"			
68	Lancaster R'd.	"	20	Landenberg.[11]	Same as above.
72	Wilmington.[4]	L. Cre. & Gabbro. [7]			

syenitic areas of southeastern Pennsylvania, referred by Mr. C. E. Hall to the Laurentian, although they may prove to be Huronian, or even later, and probably forms an intrusive mass between the Philadelphia gneiss.

(b) A northern area, the shape of a double convex lens, covered by granitic gneisses and mica schists, the equivalent of the Philadelphia gneiss, which by earlier writers has been referred to the Montalban, and by later to the Palæozoic.

This part of the State has an uneven surface of beautifully rounded hills, with a bold and rounded outline, and is elevated several hundred feet above tide water. Limestone also occurs in this primary region. It is a nearly pure dolomite in a coarse and fine grained crystalline mass of a white color, with at times a bluish tinge. About six miles N. W. of Wilmington is a limited body of serpentine of various shades of green, with a heavy vein of granite passing through it.

South of the Primary or Rocky regions of the State and, indeed, from its lower limit to the southern boundary of Delaware, the general features of the country are widely different. Instead of a constant succession of irregular and boldly rounded hills, is presented a comparatively level country or table land, gently sloping east and west towards either bay from an elevated strip of land several miles in breadth. The streams flow from this east and west through the soft and yielding strata which constitute the geological formations of a very large portion of the State; these formations being composed of clays and sands which are more or less loose in their texture. The surface of the country, originally rather flat and level, has been scooped out by brooks and creeks and rain torrents into an undulating surface, presenting low hills and bowl-like depressions, sometimes gently sloping, at others with abrupt declivities, where the formations offer a sufficient resistance to the agents of denudation. From the lower limit of the primary formation nearly to the southern border of New Castle County, is a series of clays and marls of the Cretaceous and upper Jurassic formations. Between the lower or southern limit of the Cretaceous and the lower part of Kent County exists a series of beds of clay and sand which are of the tertiary (miocene) formation. The surface of the country in the lower part of Kent and the whole of Sussex County is much more level than that farther north. The aggregate thickness of all the formations south of the primary will probably not fall short of five hundred feet, and the general bearing of all the formations, like that of the primary, is nearly N. 50° E.

The little State of Delaware furnishes us with a general description of the Geology of the whole Atlantic Coast, including considerable portions of the States of New Jersey, Maryland, Virginia, North and South Carolina and Georgia, comprising the primitive Archæan backbone or foundation formation, with the Cretaceous, Tertiary and Quaternary extending eastward from it to the Ocean.

Eastern Shore of Maryland and Virginia.[*]

New York, Phila. & Norfolk R. R.			Wicomico and Pocomoke R. R.		
0	Delmar, Del.	19 c. U. Pl. to P. Pl.[53]	0	Salisbury.	19 c. U. Pl. to P. Pl.
6	Salisbury, Md.	"	10	Pittsville.	20 c. P. Pl. & Modern
10	Fruitland.	20 c. Modern.	19	St. Martin's.	" "
19	Princess Anne.	"	23	Berlin.	" "
22	King's Creek.	"	30	Ocean City.	" Ocean Sand.

New York, Phila. & Norfolk (cont.)			Baltimore and Del. Bay R. R.		
28	Kingston.	20 c. Modern.	0	Clayton, Del.	19 b. Miocene.
38	Crisfield.	" Salt Marsh.	20	Kennedyville.	19 a. Eocene.
72	Exmore.	20 c. Modern.	31	Chestertown.	"
95	Cape Charles.	" Ocean Sand.	86	Parsons.	19 a. Eocene & Creta.
119	Old Pt. Comfort	By Steamer.			
131	Norfolk.	"			

Queen Anne's & Kent & Townsend.

0	Townsend.	19 b. Miocene.
18	Sudlersville.	"
35	Centreville.	"

Cambridge and Seaford R. R.

Delaware and Chesapeake R. R.

Cambridge and Seaford R. R.			Delaware and Chesapeake R. R.		
0	Seaford, [9]	19 c. U. Pl. to P. Pl.	0	Clayton, Del.	19 b. Miocene.
14	Williamsburg.	"	14	Marydell.	"
88	Cambridge.	"	32	Queen Anne.	"
			41	Easton.	"
			51	Oxford.	"

* That is the Eastern Shore of Chesapeake Bay in those States.

Maryland.*

Philadelphia, Wilmington and Baltimore Railroad.

Ms.	STATIONS.	GEOLOGICAL FORMATIONS.
0	Philadelphia.	(See Pennsylvania.)
28	Wilmington.	18. Cret. & 17. Juras.'
30	Delaware Junc.	"
32	Newport.	" 91
34	Stanton.	" 17
40	Newark.	" 100
46	Elkton.	" 99
52	Northeast.[1]	1. Azoic " 49
55	Charlestown.	1. Azoic " 99
61	Perryville.	17. Juras.& Archæan[31]
	(Susquehanna River.)	
62	Havre-de-Gr'ce	{ 1. Granite, Gabbro-Diorite, 17. Jur.[10]
67	Aberdeen.	17. Jurassic. 700
74	Bush River.	"
77	Edgewood.	" 90
79	Magnolia.	" 99
89	Stemmer's Run	" 94
94	Bay View.	" 99
98	Baltimore.	"

Phil. and Baltimore Central R. R.

0	Philadelphia.	(See Pennsylvania.)
36	Kennett.	"
52	Oxford.	"
60	Rising Sun.	1 a. Laure'n, Serpent.
67	Rowlandville.	"
71	Port Deposit.[3]	" Granite.
75	Perryville.	17. Jurassic & Archæ.
112	Baltimore.'	"

Baltimore and Potomac Railroad.

0	Baltimore.'	17. Jur. & 1 b. Huro'n
19	Odenton.'	18. Cret. and recent.
21	Patuxent.	"
26	Bowie.	"
34	Wilson's.	" 18. Cret. n'r
41	Navy Yard.'	" "
43	Wash., D. C.	" "

Pope's Creek Branch.

0	Baltimore.'	
26	Bowie.	17. Jurassic.
40	Marlboro.	Upper Eocene.
46	Linden.	19 a. Eocene.
51	Brandywine.	19 b. Miocene.
65	La Plata.	"
69	Cox.	"
75	Pope's Creek.	"

Baltimore and Ohio Railroad.
Washington Branch.

Ms.	STATIONS.	GEOLOGICAL FORMATIONS.
0	{ Baltimore.' Camd'n Stat.	17. Jurassic.
9	Relay House.'	1 b. Hur., Intru. Gran.
19	Annapolis Jun.	17. Jurassic.
22	Laurel.	" & Dior. Hur.
28	Beltsville.	"
34	Alex'ndria Jun.	"
34	Bladensburg.	"
40	Washington.'	" 1 b. Huron'n.

Alexandria Branch.

0	Baltimore.	(As before.)
84	Alexandria Jc.	17. Jurassic.
40	Banning's.	"
42	Uniontown.	"
46	Shepherd.	Cretaceous & Jurass.

Annapolis and Elk Ridge R. R.

0	Annapolis Jc.	19. Cret. & 17. Jurass.
3	Patuxent.	"
6	Odenton.	17. Jurassic.
9	Gambrill's.	"
10	Millersville.	Cretaceous.
12	Waterbury.	"
14	Crownsville.	"
16	Iglehart.	" & 19 a. Eocene
18	Camp Parole.	Eocene.
21	Annapolis.[3]	{ Eocene. "

Northern Central Railroad.

0	Baltimore.	{ 17. Jurassic and 1 b. Huronian. 99
2	Mt. Vernon.	" 121
7	{ Green Spr'gs Junction.[3]	{ 2–4. Siluro-C'mbr'n Serpentine.
12	Timonium.	" 201
15	Cockeysville.	{ " large quarries of white marble
20	Sparks'.	{ 11 c. Montalban. 2–4. Siluro-C'mbr'n Limestones.
23	Monkton.	Hur'n & Mica Schists.
29	Parkton.	{ 1 c. Montalban and Serpentine. 420
85	Freeland's.	1 c. Montalban. 896
42	Glenrock.	"
47	Hanov. Ju., Pa.	2–4. Siluro-Cam. 442
57	York, Pa.	" 366

(Continued in Pa. See page 280.)

* By Prof. P. R. Uhler, of the Peabody Institute, Baltimore, except B. & O. R. R. west.
1. Kaolin occurs near Annapolis, near Northeast, and near the Metropolitan Railroad in Montgomery County.

Ms.	Stations.	Geological Formations.
	Western Maryland Railroad.*	
0	Baltimore.[7] [14]	{ 17. Jurassic & 1 b. Huroaian.
3	Fulton Station.	"
5	Oakland. [300]	"
6	Arlington. [420]	"
9	Ho'rdsville. [445]	"
10	Pikesville. [429]	" Ser. Mo. n'r.
11	Greenwood [422]	"
14	Owing's Ms. [490]	"
19	Reisterstown.	" & Montalb'n.
22	Finksburg.	Montalban. Copper.
31	Tannery. [610]	Huronian.
34	Westm'ster. [700]	" Marble
41	N. Windsor. [440]	" Var. Marble.
45	Un. Bridge.[399]	Trias. & Silur.-Cam.
48	Middleb'rg. [411]	Triassic, Var. Marble.
49	FrederickJc. [418]	16. Triassic.
54	Rocky Ridge.	" Diabase.
61	Emmitsburg.	16. Tri. Diab. dyke.
59	Mech'cst'n. [120]	2 b. Potsd. (Marble.)
69	Blue Ridge.[1372]	"
82	Waynesboro.	Slate "
77	Smithsburg. [700]	4 a. Trent. limestone.
86	Hagersto'n. [620]	"
93	W'msport. [300]	4 c. Hudson River.
106	Martinsburg.	3a. & 4c Cal. & Hud.
	Baltimore and Ohio Railroad.*	
0	Baltimore.[7] [14]	17.
15	Ellicott City. [3]	1 a. Lau., Gran. quar.
20	Elysville. [3]	"
25	Woodstock. [3]	"Gra. & Stea. qu.
27	Marriottsville.	1 b. Huronian?
32	Sykesville. [410]	
43	Mt. Airy. [813]	1 c. Montalban.
50	Monrovia.	" Slate quar.
58	Frederick Junc.	" Trias. near.

Ms.	Stations.	Geological Formations.
	Baltimore & Ohio R. R.—*Continued.*	
62	Frederick.	1 b. Hur. limestone.
69	Point of Rocks	16. Trias. Pot. marb.
0	Washington. *[	17. Up. Jur.? & Azoic.
7	Sil'r Spring. [330]	"
11	Knowles'. [300]	"
16	Rockville. [421]	1 b. Hur. & 1 c. Mont.
22	Gaithersb'g. [310]	" Serpentine.
27	Germant'n. [300]	"
29	Boyds. [410]	" Tal. sc. Mon.
33	Barnesville. [800]	"
36	Dickerson's.	16. Tri. n. Dia. dykes
43	Pt of Rocks.[13*]	" Poto. Marble.
69	Point of Rocks.	16. Trias. Pot. Marb.
75	Berlin.	1 b. Huronian?
79	Weverton. [340]	Montalban.
90	Sandy Hook.	"
81	Harper's F'y?[1*]	Potsdam and Slate.
67	Duffield's, Va.	3a. to 4c.Sil.-Cam. l.s.
92	Kearneysville.	" } See note 11
95	Vanclievesv'le.	"
100	Martinsb'g. [432]	"
107	Nor. Mount. [13]	5-13 Sil. & Devonian.
117	Sleepy Cr'k. [410]	"
122	Hancock. [13*]	10 Ham. & 7 L. Held.
128	Sir John's Run.	8-13 Devon. [488] See note 13
133	Orleans Road.	"
153	Paw Paw.	"
163	Green Spring.	7. L. Hel. & 8 Ori
170	Patterson's Ck.	10. Hamilton.
178	Cumbl'd, Md.[14]	{ 8. Oriskany. 7. Lower Held'g to 13 a. Vespertine.[888]

2. Hartford County, a few miles northwest of the Philadelphia, Wilmington & Baltimore Railroad yields a fine green serpentine in blocks, equal to verd-antique in splendor and polish, besides the common building sort. In the Jurassic beds on the same railroad, also on the Washington branch of the Baltimore and Ohio Railroad, vast beds of nodular carbonates of iron occur, rich in metal.

3. The Woodstock, Ellicott's City and Port Deposit granites are superior of their kind.

4. Bare Hills mineral region. It has chrome and copper ores, asbestos, serpentine and magnesian rocks.

5. The Western Maryland Railroad runs near copper mines, chrome, serpentine, talc, steatite, asbestos, carbonate of iron, and most beautiful marbles of every color, from black, dark red, salmon, etc., to pure white—even statuary marble—besides the breccias of every degree of size in their component pebbles or pieces, both round and angular. P. R. U.

6. By Prof. William M. Fontaine, of Morgantown, West Virginia.

7. *Baltimore* is located upon rocks of 1 b. Huronian and 1 c. Montalban ages and upon clays and sands which rest upon the eroded edges of both of these. The clays approach the neocomian in position, while the sands and drifts belong to various more recent horizons. P. R. U.

8. The rocks of the eastern portion of the Azoic area in Maryland, as in Virginia, are granites, gneisses and hornblendic rocks. This belt extends to near Parr's Ridge, where it is succeeded by Argillites, with some metamorphic limestone, probably of Montalban age.

9. The Azoic area passes some distance to the west of the railroad from Baltimore to Washington, consequently this road runs chiefly in formations similar to those found at Baltimore. Washington has a geological position similar to that of Baltimore, but here the subjacent rocks are plainly similar in age to the Fredericksburg sandstones, and are probably Upper Jurassic.

10. On the west side of the Monocacy River a belt of Mesozoic rocks occurs, extending to near the east base of the Catoctin Range. Along the west margin of this belt occurs the remarkable lime-

Cumberland & Pennsylvania R. R.			Cumberland and Pennsylvania Railroad.—*Continued.*		
Ms.	STATIONS.	GEOLOGICAL FORMATIONS.	Ms.	STATIONS.	GEOLOGICAL FORMATIONS.
0	Cumberland.[14]	⌈ 10. Hamilton.[639]	13	Morantown.	14 c. Up. Coal Mres.
		8. Oriskany.	17	Frostburg. [16]	[1920]
	to	7. Low. Helderb'g	20	Borden Shaft.	
		5 b. Clinton.	22	Ocean Mines.	
		5 a. Medina.	25	Jackson.	
		5 a. Oneida.	29	Barton.	
2	Will's Gap.	4 c. Hudson Riv.	24	Pi'dm't, W. V.	[939]
4	C. & P. Junc.				
7	Patterson's. [18]	4 c. up to 14 b. Low.			
8	Barrelville.	Coal Measures.[706]			
10	Mt. Savage.	⌊			

Geology of the Vicinity of Baltimore.*

Northern Central Railroad.			Western Maryland Railroad.		
Ms.	STATIONS.	GEOLOGICAL FORMATIONS.	Ms.	STATIONS.	GEOLOGICAL FORMATIONS.
0	Baltimore. [17]	Hornbl. sch. Gn. age?	0	Baltimore.	Hornblen. schist age?
3	Woodberry.	Gneiss "	3	Fulton Station	Decomp. Mica sch."
5	Melvale. [18]	" "	4	Highland Park.	Hypersth. Gabbro "
6	Mt. Wash'ton.	" "	5	Oakland.	" "
7	Hollins. [19]	" "	6	Arlington.	" "
14	Texas.	Crys. l. s. Marb. "	8	Mt. Hope.	" "
15	Cockeysville.	" "	9	Howardsv'le.[20]	" "
			10	Pikesville.	Mica schist "
			12	McDonough.	Gneiss "
				etc.,	etc., etc.

stone breccia called the Potomac Marble. This is well exposed near Point of Rocks. This Mesozoic belt is flanked immediately on the northeast and east by a belt of rather impure slaty limestone.

11.   The gorge at Harper's Ferry is cut through metamorphic rocks, of in part probably Huronian age.   One and a half miles west of the station the Calciferous limestone appears.   From this point, 83 miles, to near North Mountain, 107 miles, a wide belt of Lower Silurian limestone occurs, with occasional bands of slate, embracing the rocks from the 3 a. Calciferous to and including the 4 c. Hudson River.   These have never been separated in this region.   The limestone predominates by far, and will be spoken of as the 2-4. Siluro-Cambrian.

12.   On the west side of this limestone belt, a great fault brings down in North Mountain the various Silurian and Devonian formations from the 5 a. Medina to the 13 a. Vespertine or No. X, which are to be seen in North Mountain and its immediate vicinity.

13.   From North Mountain to Cumberland a wide belt of highly disturbed strata occurs.   Owing to the close compression of the folds in which the strata are thrown, many of the formations contained in this belt are always to be seen at any given locality, and hence when any formation is given for a station it must not be inferred that this alone occurs there.

In this belt the following formations are to be found: The 5 a. Oneida, 5 b. Clinton, 7. Lower Helderberg, 8. Oriskany, 10. Hamilton, 11 a. Portage, 11 b. Chemung, 12. Catskill, and 13 a. Vespertine. These have never been clearly separated from each other.   The hard sandstones, such as the 5 a. Oneida and 8. Oriskany, usually form the crests of the ridges, and the softer strata, more commonly the Hamilton, compose the valleys and foot hills.   W. M. F.

14.   *Cumberland, Md.* Beautiful Oriskany sandstone fossils occur at the quarries in and about the city.   Also Lower Helderberg and Clinton group fossils on Wills Creek below the town and Wills Gap. Also Fucoids of the Medina sandstone.   R. P. WHITFELD.

15.   *Patterson Creek.* A short distance south of the road good Hamilton fossils are obtained on the Patterson farm,   R. P. W.

16.   *Frostburg.* Coal plants of various kinds, Hamilton fossils as casts occur in and on the hills on the N. E. of the city, some of them very fine.   R. P. W.

*As it would seem advisable to give with some fullness what is known about the rocks near a large city like Baltimore, the following notes on the crystalline rocks in that neighborhood have been furnished for this book by Dr. George H. Williams, associate in Mineralogy at the Johns Hopkins University, in which he has brought to light some interesting points which are easy of access.   J. M.

Baltimore & Ohio Railroad.			Maryland Central (Delta) R. R.		
Ms.	STATIONS.	GEOLOGICAL FORMATIONS.	Ms.	STATIONS.	GEOLOGICAL FORMATIONS.
9	Relay.	{ Granite & Granitoid Gneiss, age?	0	Baltimore.	Gneiss quarries age?
10	Avalon.	Gn. & Horn. sch. "	2	Guilford.	Gn. & Horn. sch. "
11	Or'ge Grove.²¹	{ Gneiss with Erupt Gran. Dykes age?	7	Towsontown.	Gneiss "
12	Ilchester.	Hornblend. Gn. "	11	Loch Raven.²⁴	{ Mica sch., Quartzite & Crys. limest'ne
14	Grays.	Gneiss "	13	Notch Cliff.	
15	Ellicott City.²²	Granite "	27	Belair.	
20	Elysville.	Gneiss & Granite "	24	Fern Cliff.	
25	Woodstock.²³	Gneiss "	36	The Rocks.	
			44	Delta.	

17. On the outskirts of the city on the right are the large Gneiss quarries of Jones Falls, which furnish Baltimore with much building and paving stone. They also produce many beautiful minerals, including the species Beaumontite (Heulandite) and Haydenite (Chabazite). The Gneiss is intersected by large veins of pegmatite containing fine specimens of microcline and frequently tourmaline, apatite, sphene, garnet, etc.

18. Between Melvale and Woodberry a tongue of the Hypersthene-gabbro is crossed, and a contact between this rock and the gneiss well exposed.

19. Just west of Hollins Station, but not visible from the railroad, is the lenticular mass of serpentine, known as the Bare Hills. It contains considerable chromite, which, however, is now no longer worked. Just south of the Bare Hills is a mine of chalcopyrite, occurring in the hornblende gneiss in connection with octahedral crystals of magnetite, and an interesting monoclinic variety of anthophyllite. G. H. W.

20. This most interesting eruptive rock, locally known as "Niggerhead," covers an area of about fifty square miles west and north-west of Baltimore. It is most admirably exposed at the above-named stations, especially at Mt. Hope, where a long cut reveals a section of it over 1,000 feet in length. In general appearance it strongly resembles the normal triassic trap, but is petrographically altogether different. It weathers to a dark vermilion soil, through which huge blocks of the fresh purple rock may be seen protruding. The most interesting feature of this gabbro is the partial alteration which it has suffered into a hornblendic rock which is generally massive, although sometimes schistose. This may be designated as Gabbro-Diorite, and has been formed by the paramorphosis of the pyroxene to hornblende without chemical change (see Am. Jour. Sci., Oct., 1884). This change may be most advantageously studied at the Mt. Hope cutting. Just south of Highland Park the contact of the Gabbro and Schists may be seen with large dykes of the former rock alternating with the schists before the actual contact is reached. G. H. W.

21. A few hundred yards above Orange Grove, on the Patapsco River, there is a most interesting profile 250 feet in length exposed by the railroad excavations. Hornblende schists, dipping over 70° to the west, are cut by apparently eruptive granite. In the center a huge trunk, nearly 20 feet broad, emerges from the ground parallel to the dip of the schists, and from this two lateral arms are given off on each side which traverse the schists nearly at right angles to their bedding. The lower of these lateral arms on the west side, although only four feet broad at its origin, may be traced over 150 feet in a horizontal direction, and when it disappears is less than five inches in width. On the east side the arms are equally well marked, but are not exposed for so long a distance. Inclusions of the schist in the granite are very numerous; one in the main trunk is over 14 feet long. These dykes exhibit in an admirable manner the effect of the cooling surface on their structure, being always very coarse grained in the center but fine grained at the edge. Smaller dykes of granite are frequently exposed between Orange Grove and Avalon. G. H. W.

22. The granite at Ellicott City is generally porphyritic; on the edges of the mass, however, this structure disappears and the rock seems to pass gradually into Gneiss. G. H. W.

23. The granite extensively quarried at Fox Rock and Granite P. O., a few miles north of Woodstock, is of a very superior quality, closely resembling the "Richmond Granite" of Virginia. G. H. W.

24. Loch Raven is a romantic spot on the Gunpowder River, which has been dammed as part of the Baltimore water supply. A conduit, cut through five miles of solid rock, leads the water to the city. From the station northward along the river the road exposes a fine section of quartzite and mica schist in contact with crystalline limestone. On the railroad are exposed quartz rocks and gneisses, with tourmaline and secondary mica developed on the cleavage planes. These are immediately overlaid by crystalline limestone, which is in turn succeeded by mica schists, often rich in garnet and fibrolite, and resembling the well known Philadelphia mica schists. At many points, however, the rocks on both sides of the limestone appear to be identical. At the upper contact is a huge dyke of very coarse grained granite. This is on the road just opposite the Water-works building on the dam. G. H. W.

This blank space is intended for additional geological notes in pencil by the traveler.

West Virginia.[1]

TABLE OF GEOLOGICAL FORMATIONS IN WEST VIRGINIA.

20. Quaternary, Glacial dam and river deposit			**Devonian.**	**10 c. Genesee**	150–200 VIII.
15. Permian or Permo Carboniferous 1,500	XVI.			**10 b. Hamilton**	600–800 VIII.
				10 a. Marcellus	500–600 VIII.
Carboniferous. **14 c. Upper Coal Measures** 275–374	XV.		**Upper Silurian.**	**8. Oriskany**	75–150 VII.
14 b. Barren Measures 585–800	XIV.			**7. Lower Helderberg**	400–500 VI.
14 b. Lower Coal Measures 250–1,100	XIII.			**6. Salina**	800–900 V.
				5 b. and c. Niagara(?) and Clinton	400–500 V.
14 a. Pottsville Conglomerate and New River Coal Series 150–1,300	XII.			**5 e. Medina and Oneida**	1,400–2,000 IV.
Sub Carboniferous. **13 c. Mauch Chunk Shales** 300–2,000	XI.		**Lower Silurian.**	**4 c. Hudson River**	2,000–3,000 III.
13 b. Mt. or Green Brier L. S. 100–800	XI.			**4 a. Shenandoah L. S.**	4,000–5,000 III. and II.
13 a. Pocono S. S. 500–1,200	X.			**2 b. Potsdam**	2,000–3,000 I.
Devonian. **12. Catskill** 800–1,500	IX.		**Archean.**		
11-12 Chemung-Catskill 800–1,000	VIII.			**1 b. Huronian**	
11 b. Chemung and **11 a. Portage** 2,500	VIII.				

DESCRIPTION OF THE GEOLOGICAL FORMATIONS.

As the descriptions of the formations given in the introductory part of this volume do not give a detailed account of the carboniferous rocks, and as West Virginia can lay claim to greater development of these beds than any other State, Professor I. C. White has kindly furnished the following resumé of their structure and characteristics, and has extended it briefly to the other formations of that State, besides the Carboniferous. As these are the results of Professor White's very recent explorations as United States Geologist, they will be especially valuable to those who have not the time or opportunity to look through the official geological reports, and they may serve to correct many erroneous statements as to the geology of West Virginia which have obtained currency.

J. M.

20. QUATERNARY. Cincinnati Ice Dam and Flooded River epochs.

The only Quaternary deposits found in West Virginia are those made along the Ohio River and its tributaries during the existence of the Glacial dam at Cincinnati, and those made along all the streams which drain the Allegheny Mountains plateau. (See Note 62.) The rounded boulders at high levels along the Potomac, Cheat and other rivers resemble glacial deposits, but no glacier ever existed in West Virginia, the deposits in question having been made during the "Flooded River" epoch which closed the glacial period, when the snows that had doubtless accumulated to a considerable thickness on the Allegheny plateau melting away filled the draining streams with water to a depth probably exceeding 100 feet. The entire area of West Virginia was elevated above sea level during the Appalachian revolution, and has remained above the same ever since, hence none of the formations between the (15) Permian and (20) Quaternary are found in this State.

15. Permian or Permo-Carboniferous, Upper Barrens.[2] [XVI. Seral.][*]

The Permian beds, according to Fontaine and White, include all the stratified rocks in West Virginia above the horizon of the Waynesburg coal. The series has a maximum thickness of 1,500 feet, and consists of red shales, sandstones and limestones, there being three or four thin coal beds in the lower half of the group, but none whatever in the upper. The beds are all apparently of fresh water origin, since the limestones contain no fossils except *Spirorbis*, *Cypris*, *Estheria*, and other bivalve crustaceans. The plant remains are principally Ferns of Permian type, including *Callipteris conferta*, though *Taeniopteris*, *Baiera* and others recall Mesozoic forms. The formation enters the State from the southwest corner of Pennsylvania and stretches across it to the Great Kanawha River in a belt 30–50 miles wide.

1. By Professor I. C. White, United States Geologist, and lately on the Second Geological Survey of Pennsylvania.

2. *Permian.* The evidence of the existence of the Permian or Permo-Carboniferous formation in West Virginia is contained in Vol. P.P. of the Second Geological Survey of Pennsylvania, by Wm. M. Fontaine and I. C. White, 1880. J. M.

* The names and numbers enclosed in square brackets are those given to the formations by Wm. B. Rogers, late State Geologist of Virginia.

14c. Upper Coal Measures, Monongahela Series. [XV. Seral.]
In the northern portions of the State contains four coal beds in descending order, as follows:

Waynesburg bed, merchantable coal	4–6 ft.
Interval limestones, shales and sandstones	250 ft.
Sewickley bed, merchantable	4–5 ft.
Interval limestones and shales	65 ft.
Redstone bed, worthless	3–4 ft.
Interval limestones, shales and sandstones	40 ft.
Pittsburg bed, merchantable coal	6 ft.
Total thickness	374 ft.

In Southern West Virginia, on Great Kanawha River, the group has undergone the following changes: The Sewickley and Redstone coals are absent; the Waynesburg is thin and worthless: the group has lost all its limestones except one thin stratum; it has also lost 100 feet of rock, intervals being reduced to 275 feet; red shales are abundant on the Kanawha River; there are none in these measures on the Monongahela; the Pittsburg coal maintains 5 ft.–6 ft. of merchantable coal, but it is often absent entirely from wide areas, or only 1 ft.–2 ft. thick on others.

14b. Barren Measures. [XIV. Seral.]
Northern West Virginia shows the following structure:

Shales, sandstones and limestones, sometimes including a thin coal	200 ft.
Morgantown sandstone	25 ft.
Elk Lick coal	0–4 ft.
Shales	75 ft.
Green crinoidal limestone, very fossiliferous	2 ft.
Coal	0–1 ft.
Red and variegated marley shales	100 ft.
Bakerstown coal	0–4 ft.
Shales and sandstones	40 ft.
Upper Mahoning sandstone, pebbly	50 ft.
Brush Creek coal	0–3 ft.
Lower Mahoning sandstone	75 ft.
Shales	12 ft.
Total	585 ft.

On the Great Kanawha this group thickens up to 800 feet; the green crinoidal limestone disappears, but is exactly replaced strati-graphically by one of fresh water origin. The Brush Creek coal attains important dimensions, and two new ones are introduced below it, while the series is terminated by the "Black Flint," a marine deposit of dark gray, or blackish flint peculiar to the Kanawha valley, and exhibiting every gradation between sandy shale and compact silex.
The coals of the barrens are everywhere variable and uncertain. A bed may be present in good thickness on one farm, while on the adjoining land it may be absent entirely, or so impure as to prove worthless. The Brush Creek seam is the persistent and important one.

14b. Lower Coal Measures. Allegheny River Series. [XIII. Seral.]
These measures are 250 feet thick at the northern line of the State, and usually contain five coal beds, in the following order:

Upper Freeport Coal—	
Interval	50 ft.
Lower Freeport Coal—	
Interval	75 ft.
Middle Kittaning Coal—	
Interval	35 ft.
Lower Kittaning Coal—	
Interval	60 ft.
Clarion Coal—	
Interval to top of XII	20 ft.

The Upper Kittaning Coal, which is often present in Pennsylvania, seems to be absent in Northern West Virginia, though it comes into the section on the Kanawha River. The Upper Freeport and Lower Kittaning are the only ones of these five that are valuable, since the others are usually too thin and slaty. The first is generally 4 ft.–6 ft. thick and the latter 3 ft.–5 ft. This series gradually expands southwestward, and on the Kanawha River attains a maximum thickness of 1,100 ft., in which its six productive coal beds are disposed somewhat as follows:

Upper Freeport ("Cannelton Lower") bed—	
Interval	100 ft.
Lower Freeport ("Coalburg") bed—	
Interval	75 ft.
Upper Kittaning ("Winnifrede") bed—	
Interval	350 ft.
Middle Kittaning ("Cedar Grove") bed—	
Interval	115 ft.
Lower Kittaning ("Campbell Creek") bed—	
Interval	120 ft.
Clarion (Eagle) bed—	
Interval to top of No. XII. in which two or three thin coal streaks occur	340 ft.

The six coal beds given above are never all workable in the same section; in fact it is rare that more than two of them furnish valuable coal on the same property. The Lower Kittaning is probably the most persistent of the Kanawha coals.

14a. Pottsville conglomerate. New River Coal Series. [XII. Seral.]
 The No. XII. series has the following structure in Northern West Virginia, on Cheat River:
  Massive, pebbly, sandstone, sometimes in two or more beds with intervening
  shales, the whole representing the Homewood and Cannoquenessing sand-
  stones of Pennsylvania...150 ft.
  Coal ....1-2 ft.
  Black Slate... 10 ft.
  Gray Sandstone to base of XII... 25 ft.
 Southwestward across the State this series thickens, even to a greater extent than XIII., and in
the New River (southward continuation of the Kanawha) region, attains a maximum of 1,300 ft., in
which are three important coal beds in the following order, descending from top of XII.:
  Massive sandstones and conglomerate with a thin coal, 175 ft. below top......... 400 ft.
  Nuttall Coal ....
  Shales and massive sandstones.. 250 ft.
  Coal ....
  Shales and sandstones.. 100 ft.
  Coal ....
  Shales and massive sandstones to base of No. XII... 550 ft.

  Total...1,300 ft.
 These three beds are coking coals of the finest quality, and one of the two lower appears to be
identical with the great ten-foot seam of the Flat Top country. These coals are found of workable
thickness only around the southern margin of the coal area, in a belt of country 20-30 miles wide,
north from which they thin away to insignificant streaks. The Nuttall bed would correspond to the
Quakertown coal of Pennsylvania, and the other two would represent the *Sharon* and its "rider."
13. Sub-Carboniferous.
13c. Mauch Chunk Shales. [XI. Umbral Shales.]
 On Cheat River consists of shales, green sandstones, and thin limestones, with iron ore next the
top; total thickness 300 ft., in which are only 10 ft.-15 ft. of red shale. On New River this series is
not less than 2,000 ft. thick, consisting of red shales, green and gray sandstones, with an impure
limestone at the top of the group.
13b. Mountain or Greenbrier Limestone. [XI. Umbral Limestone.]
 100 ft.-150 ft. thick in Monongalia Co., but increases to over 800 ft. in Greenbrier Co. Is absent
entirely over a large portion of the Northern region of the State west from Chestnut Ridge.
13a. Pocono Sandstone. [X. Vespertine Sandstone.]
 Hard gray current bedded sandstone and conglomerate, 500 ft.-600 ft. thick on Cheat River, and
1,000 ft.-1,200 ft. in the Allegheny Mountains along B. & O. R. R. No measurements have been made
in southwestern portion of the State.
9-12. Devonian.
12. Catskill. [IX. Ponent.]
 Red shales, green and red sandstones, and an occasional conglomerate, 800 ft. thick at Rowlesburg, B. & O. R. R., and 1,200 ft.-1,500 ft. in Allegheny Mountains; thins away to almost nothing west
from Chestnut Ridge.
11-12. Chemung-Catskill. [VIII. and IX. Ponent and Vergent in part.]
 Green and gray flaggy sandstones, fossiliferous, also containing occasional red beds, and a conglomerate with flat pebbles, (1st Venango oil sand and gas rock at Washington and Murraysville), thickness near Keyser down to lowest red bed 800 to 1,000 ft. These rocks have sometimes been classed
with the Catskill and again with the Chemung. In Penna. Geol. Report G[7], p. 63, the desirability of
the present classification is fully set forth.
11b. Chemung⎫
  and ⎬[VIII. Vergent.]
11a. Portage.⎭
 A series of hard, flaggy sandstones and shales, with a massive conglomerate (3d Venango oil sand)
100 to 200 ft. below the top; no red beds whatever; sparingly fossiliferous; thickness about 2,500 ft.
10c. Genesee. [VIII Cadent.]
 Black slate and dark shales; thickness 150 to 200 ft. along B. & O. R. R.
10b. Hamilton. [VIII. Cadent.]
 Dark brown sandstones and sandy shales, very fossiliferous; thickness along B. & O. R. R.,
600 to 800 ft.
10a. Marcellus. [VIII. Cadent.]
 Black and gray slates with beds of impure gray limestone at base. The entire group 500 to 600
ft. along the B. & O. R. R.
9. Corniferous. [VIII. Cadent.]
 Wanting in West Virginia.
5-8. Upper Silurian.
8. Oriskany. [VII. Meridian.]
 A coarse, dirty yellow fossiliferous sandstone, 75 to 150 ft. thick.
7. Lower Helderberg. [VI. Pre Meridian.]
 Highly fossiliferous gray and blue limestones, 400 to 500 ft. thick.
6. Salina. [V. Scalent.]
 Greenish magnesian limestones, red and variegated shales, the whole having a thickness of 800
to 900 ft. along B. & O. R. R.
5c. Niagara (?) and⎫
5b. Clinton. ⎬[V. Scalent and Surgent.]
 Hard, flaggy sandstones; thin limestones and shales, in which occur two beds of iron ore, the
thickness of all being 400 to 500 ft. along B. & O. R. R.
5a. Medina and Oneida. [IV. Levant.]
 Hard, white sandstone (White Medina) at top 400 to 500 ft. thick, succeeded by red shales and
sandstones 800 and 1,000 ft.(Red Medina), and followed by gray sandstones and conglomerate (Oneida)
200 to 500 feet thick.

Baltimore & Ohio Railroad,			Baltimore & Ohio Railroad—Con.		
Ms..	From Harper's Ferry West.[3]	Alt.	Ms.	From Harper's Ferry West.[3]	Alt.
81	Harper's Ferry.[4]	Huronian.	272	139	Rockwell's Run.	Devonian.	499
87	Duffield's.	Sil. Cam. L. S.	562	140	Doe Gully Tun'l.[8]	Catskill.	545
92	Kearneysville.	"	589	155	Little Cacapon.	Devonian.	562
95	Vanclieveville.	"	500	161	S. Br. Pot. River.	"	550
100	Martinsburg.[5]	"	485	163	Green Spr. Run.[9]	Hamilton.	553
....	{ Shepardstown	"	467	170	Patterson's C'k.[10]	"	568
	{ Road.			....	N. Br. Potomac.	"	604
107	North Mountain.[6]	Sil. and Dev.	547	178	Cumberland.[11]	L. Helderberg.	639
113	Cherry Run.	Devonian.	396	185	Brady's Mill.	L. Helderberg.	642
117	Sleepy Creek.	"	410	191	Rawling's.	"	693
122	Hancock.	"	428	193	Black Oak Bottom.	"	736
128	Sir John's Run.[7]	Medina.	434	198	Potomac Bridge.	Hamilton.	736
131	Great Cacapon.	Hamilton.	449	201	Keyser.[12]	L. Helderberg.	800
133	Willett's Run.	Devonian.					

2–4. Lower Silurian or Cambrian.
4c. Hudson River Shales. [III. Matinal.]
Dark brown shales and slates usually cleaved, probably 2,000 to 3,000 ft. thick on B. & O. R. R., west from North Mountain; no exact measurements have been made.
4a. Shenandoah Valley Limestone. [II. and III. Matinal and Auroral.]
Limestones of great thickness, and some of it very pure; no trustworthy measurements have been made, but it is probably not less than 4,000 to 5,000 ft. thick along B. & O. R. R.
2b. Potsdam Sandstone. [I. Primal.]
Found only in Blue Ridge at eastern line of State, where it consists of quartzites and slates, whose thickness has not been accurately determined, but it is probably not less than 2,000 to 3,000 ft.
1. Archæan.
1b. Huronian. Rocks of this age supposed to exist in the gap of the Potomac through the Blue Ridge at Harper's Ferry.

3. Professor White thinks the geology of West Virginia can be best studied by beginning at Harper's Ferry, in Maryland, at the bottom of the series of formations. By this means the road between that place and Cumberland is given twice. J. M.
4. The gorge at Harper's Ferry is cut through metamorphic rocks, of probably Huronian age. One and a half miles west of the station, a fault brings down the Potsdam and Calciferous rocks against the Azoic. From this point, 83 miles, to near North Mountain, 107 miles, a wide belt of Lower Silurian limestone occurs, with occasional bands of slate, embracing the rocks from the 3 a. Calciferous to and including the 4 c. Hudson River. These have never been separated in this region. The limestone predominates by far, and will be spoken of as the 2–4. Siluro-Cambrian. (F).
5. *Martinsburg.* Splendid quarries in No. II. limestone here. One mile east from Martinsburg a syncline catches the Hudson River slate and the limestone goes under for two or three miles, then reappears, and again goes under to come up once more near Kearneysville. These crumples near the centre of the valley are the northeastern extension of the great trough which holds Massanutten Mountain, 50 miles south from Martinsburg.
6. *North Mountain.* On the west side of this limestone belt a great fault brings down in North Mountain the various Silurian and Devonian formations, from the 5 a. Medina to the 13 a. Vespertine or No. X., which are to be seen in North Mountain and its immediate vicinity. (F).
7. *Sir John's Run.* From this point westward to Cumberland the rocks are thrown into a series of great arches, whose corresponding troughs catch the *Pocono* beds in the tops of the mountains, and bring up the Lower Helderberg limestone on the anticlinals, so that frequently several formations may be seen near one station. (F).
8. *Doe Gully.* Fine exposures of Catskill rocks in the approaches to the tunnel, which cutting through them parallel to the strike, permits the highly inclined beds to slide down into the cuts from a long distance up the sloping side.
9. *Green Spring Run.* The valley here is a syncline of Genesee, Hamilton and Marcellus rocks, enclosed on either side by anticlinal ridges of Oriskany sandstone, making Mill Creek Mountain on the east and Patterson's Creek Mountain on the west.
10. *Patterson's Creek.* Another synclinal valley of Hamilton beds, bordered east and west by anticlinal ridges of Oriskany. Under the arch of the eastern one the Lower Helderberg limestone is brought above water level and quarried on the Maryland side of Potomac.
11. *Cumberland.* Good geological headquarters. The great Will's Creek Mountain anticlinal just east from the city, brings up the Red Medina, spanned by a splendid arch of White Medina, through which the creek has carved a narrow cañon, in which there is barely room for the two R. R's and the National turnpike. The Clinton, L. Helderberg, Oriskany and Hamilton all exposed near city. The low mountain which begins on the Virginia side at Cumberland, and trends away to the southwest, is made by the massive Oriskany sandstone and called Knobby or "Knobley."
12. *Keyser.* Splendid ground for geologists. The Potomac river turns squarely around to the northeast on leaving Cumberland and the R. R. follows this direction almost parallel to the strike of the rocks, and hence along the crest and sides of the great Will's Creek Arch, which the river has worn down and converted into a valley from Cumberland to Keyser, with Knobley Mountain (Oriskany) on the south, and Dan's Mountain (Pocono and No. XII.) on the north, from the highest peak of which, opposite Brady's Mill, is one of the grandest views in all the Appalachian region. Queen's point, opposite Keyser, is an arch of Oriskany, under which comes fine exposures of L. Helderberg, both

Ms.	Baltimore & Ohio Railroad.	Alt.	Ms.	Baltimore & Ohio Railroad—Continued.	Alt.
0	Baltimore, Md.				50' under the U.
206	Piedmont.	14 a. Pottsville Cg925	.....	E. P. Kingwood T.	Freeport Coal.[18][19]
.....	Potomac Bridge.	" 999			Freeport limestone at
208	Bloomington.	" 1024	261	W. P. " 16	track level. 1779
214	Frankville.	13 b. M. Chunk.1699			U. Freeport Coal at
220	Swanton Water St.	" 2282	264	E. P. Murray's T.[17]	track level. 1554
223	Altamont.	13 a. Pocono. 2620	267	Newburg.[18]	Barrens. (XIV.) 1215
226	Deer Park.[13]	11 b. Chemung. 2442	.....	Hook's Run.	" 1164
229	Mt. Lake Park.	" 2400	268	Indepenence.	" 1158
.....	Little Yough Br.	" 2392	.....	Helvetia.	" 1110
232	Oakland.	13 b. M. Chunk.2372	.....	Raccoon Creek Br.	" 1105
.....	Little Yough Br.	14 a. P'tville Cg2871	274	Thornton.	" 1088
233	Great Yough Br.	" 2372	.....	Water Sta. No. 59.	" 1032
.....	Chisholm Summit.	" 2487	.....	Three Fk. C. Br.[19]	" 1020
238	Hutton's.	" 2477	280	Grafton.	" 987
240	Snowy Creek Br.	12 Catskill. 2469	281	Fetterman.	" 984
242	Terra Alta.	11 b. Chemung. 2549	.....	Plum Run Bridge.	" 975
243	E. P. McGuire's T.	" 2882	287	Valley River F. [20]	Nos. XII., XIII. 969
246	Rodemer's Tunnel.	12 Catskill 2083	.....	Nuzum's Mills.	No. XIII. 936
250	Salt Lake Bridge.	" 1819	294	Texas.	Barrens. (XIV.) 883
253	Cheat River Br.	11 b. Chemung. 1892 / 12. Catskill.	297	Benton's Ferry.	" 883
			.....	Mon. River Br.	" 877
253	Rowlesburg.[14]	" 1892	302	Fairmont.[21]	" 877
254	Buckeye Run Vt.	Base Catskill. 1515	308	Barnesville.	14 c. Up. Coal M.871
255	Tracy Run Vt.	Fine ex. of Cat.1572	.....	Buffalo Creek Br.	" 891
257	Buckhorn R. Vt.[15]	13 b. M. Chunk.1720	307	Barracksville.	" 901
259	Cassidy's Summit.	Tp. 14 b. L. Cl. M1855	.....	Davis Run.	" 916
260	Tunnelton.	14 b. L. Col. M. 1820	.....	Dunkard Mill.	" 922

very fossiliferous. The R. R. cut at Bull Neck, just below Keyser, is through a sharp syncline of Oriskany. The L. H. limestone, Salina, Clinton and White Medina, all finely exposed along Limestone run near town; while the Hamilton, Chemung, Catskill, Pocono, Mauch Chunk and Pottsville conglomerate come down in succession along the R. R. between Keyser and Piedmont.

13. *Deer Park.* West of Altamont the railroad continues on a broad, undulating plateau, the Savage and Allegheny Mountains of Pennsylvania having here coalesced into one. This remarkable flat mountain top, from 2,400 to 2,600 feet in height above tide, has always attracted much attention from the comparative softness of the outlines, giving the park-like character to its topography. (F.)

14. *Rowlesburg.* Here the R. R. starts up another steep grade to the crest of Laurel ridge, and the view to the right (in going west) down the course of Cheat, is the grandest of all the B. & O. R. R. scenery. The geological picture is no less interesting, since the road bed is almost a continuous rock-cut for 5 miles, thus giving a nearly clean exposure of the column of rocks from the top of the Chemung up through 700 ft. of Catskill, 566 ft. of Pocono, 712 ft. of Mauch Chunk, 368 ft. of Pottsville Conglomerate, 310 ft. of Lower Coal Measures, and 200 ft. of the Barrens (No. XIV).

15. *Buck Horn Run.* All of these viaducts cross wild gorges 75 ft.-100 ft. deep, and at the Gray Run gorge the cars are apparently directly over Cheat River, 200 ft. below.

16. *W. Portal Kingwood Tunnel.* Kingwood Tunnel is 4,132 ft. long and passes through Laurel Hill, the anticlinal axis of which crosses the R. R. somewhere near the eastern end of the tunnel, since the U. Freeport coal has there an elevation of 1,805 ft. A. T. and dips eastward, while at the western portal the same coal is 1,805 ft. A. T. and dipping rapidly westward. The summit of the mountain is made by 200 ft. of Mahoming sandstone.

17. *East Portal Murray's Tunnel.* U Freeport coal here 3½ ft.-4½ ft. thick, and extensively coked at Austin mines 20 ft. under R. R. track, just west from Murray's Tunnel.

18. *Newburg.* A small area (300-400 acres) of the Pittsburg coal is caught in the summit of the hills here near the centre of the trough between Laurel Hill and Chestnut Ridge anticlinals. The Pittsburg coal has an elevation of 500 ft. above R. R. and is transported to the latter over a long incline. A shaft has recently been sunk near the foot of the incline which passed through the *U. Freeport coal,* 4 ft. thick at 160 ft., and the *Lower Kittaning bed,* 7 ft. thick at 359 ft.

19. *Three Fork Creek Bridge.* Three miles up Three Fork Creek is Irondale Furnace where native ore (from 150 ft. above U. Freeport coal) is principally used, and the U. Freeport coal furnishes the coke. A branch R. R. connects it with B. & O. at mouth of Three Fork.

20. *Valley River Falls.* The anticlinal axis of Chestnut Ridge crosses the river here and brings up the conglomerate rocks of No. XII. to 150 ft. above water level, over which the stream descends in a series of wild cascades. The hills are capped by the Mahoming sandstone, thus exposing all of No. XIII.

21. *Fairmont.* The Pittsburg coal comes about 75 ft. above the track here and is extensively mined and shipped east for gas and steam purposes.

Ms.	Baltimore & Ohio Railroad—Continued.		Alt.	Ms.	Parkersburg Branch B. & O. Railroad.		Alt.
312	Farmington.[22]	14 Up. Coal M.	927	0	Grafton.	Barrens (XIV.)	967
.....	Wood's Run.	"	957	4	Webster.	"	1019
319	Mannington.[23]	Permian (XVI.)	967	7	Bartlett C'k Sum.	"	1141
326	Glover's Gap.	"	1150	10	Flemington.[29]	"	1030
.....	Glover's Gap Tun.	"	1146	17	Bridgeport.	"	975
330	Burton.[24]	"	1060	20	Carr's Tun., W. E.	"	1102
.....	E. Por. U. Eaton T.	"	993	22	Clarksburg.[30]	"	1030
.....	E. Por. L. Eaton T.	"	962	26	Wilsonburg.[31]	"	979
337	Littleton.	"	936	30	Wolf's Summit.	14 c. Up. Coal M.	1136
340	E. P. B. Tree Tun.	"	1104	36	Salem.	Permian (XVI.)	1042
.....	W. P. B. Tree.[25]	"	1077	46	Smithton.	14 c. Up. Coal M.	790
344	Bellton.[26]	Permian (XVI.)	886	48	West Union.[32]	',	852
.....	E. Por. Welling T.	"	1202	52	Central.	Permian (XVI.)	808
. ...	W. Por. "	"	1193	59	Tollgate.	"	787
351	Cameron.	"	1049	62	Pennsboro.	"	852
356	Easton.	"	967	67	Ellenboro.[33]	"	777
.....	E. P. Shepard's T.	"	838	72	Cornwallis.	"	676
361	Op. Rosby's Rock.	"	787	75	Cairo.	"	667
362	Rosby's Rock.	"	773	82	Petroleum.[34]	"	684
368	Moundsville.[27]	14 c. Up. Coal M.	640	94	Kanawha.	"	599
373	McMechens Cut.	"	664	94	Claysville.	"	599
375	Benwood.	P'burg C. nr. T. L.	646	104	Parkersburg.[35]	"	626
379	Wheeling.[28]	"	645				

22. *Farmington.* The Waynesburg bed is mined here about 150 ft. above track, the Pittsburg being more than 200 ft. under water level.

23. *Mannington.* The Waynesburg coal, or highest number of the Carboniferous proper, goes under the R. R. track 2½ miles east from Mannington, and from there to near the Ohio river the rocks belong to the Permian or Permo-Carboniferous series, the No. XVI. of Rogers. The Washington coal is 75 ft.-100 ft. above track at Mannington.

24. *Burton.* In the region between here and Bellton are to be found the highest rocks of the Permian series, some of the summits attaining an elevation of 1,200 ft.-1,500 ft. above the Waynesburg coal.

25. *West Portal Board Tree Tunnel.* Ninevah coal, the uppermost small bed of the Permian series, 50 ft. over track here.

26. *Belton.* A fine locality for Permian exposures in the steep hills, which rise 600 ft. to 700 ft. above water level. A hole bored for oil a short distance above Bellton, passed through the Waynesburg coal at 400 ft. below creek level.

27. *Moundsville.* The Pittsburg coal underlies the Ohio river about, 90 ft. at Moundsville, and is mined by shafts. The Waynesburg bed is 170 ft. above the river, but impure, and only 2½ ft.-3 ft. thick.

28. *Wheeling.* The Pittsburg coal is about 100 ft. above river here, and fine exposures of the entire Upper Coal Measures (260 ft. thick), and the lower portion of Permian may be seen in the steep hills around Wheeling.

29. *Flemington.* Here the Lower Coals and Lower Barren Measures are shown, with a small remnant of the Pittsburg bed in the tops of the hills, it being the seam worked there. (F).
At this station is the eastern outcrop of the Pittsburg coal bed, west from the anticlinal of Laurel Hill (Chestnut Ridge of Pennsylvania). From this locality the coal and the railroad level constantly approach, until at Wolf's Summit, a little west from Wilsonburg, the coal is under the track. (S. & F.)

30. *Clarksburg.* Pittsburg coal extensively mined here and westward to Wilsonburg. It is also coked and shipped to Chicago and elsewhere for purposes other than the manufacture of iron.

31. *Wilsonburg.* Just before reaching Wolf's Summit, the Pittsburg coal bed is at the railroad level, and is worked near the track at the Summit. The Redstone coal bed is seen two inches thick in the Summit cut. Between the Summit and the Brandy Gap Tunnel the Waynesburg coal bed is seen and is worked just south from the railroad, the opening being visible from the track. At the west end of the tunnel the Washington coal bed is exposed above the track. This is in the Upper Barren Measures. (S),

32. *West Union.* The Waynesburg coal is mined to a small extent here and eastward beyond Smithton, but is thin (2 ft.-4 ft.) and impure. The roof shales contain numerous finely preserved fossil plants at West Union.

33. *Ellenboro.* Prof. Stevenson is now inclined to believe that what he has described in this region as faults are only very sharp anticlinal axes, and that what is known as the "Oil Break" is simply a great anticlinal arch, and in this view Prof. White coincides, though he has made no special investigation of the question. The oil obtained at Volcano and other localities in this region comes from the Pottsville conglomerate, according to Stevenson.

34. *Petroleum.* About one-fifth of a mile east of this station, a fault crosses the railroad, which brings up the Lower Barren Series against the Upper Barren Series. Thence, from Ellenboro to within a short distance of Petroleum station, the rocks are nearly horizontal, and the Upper Freeport coal bed is exposed in several of the cuts. But, near Petroleum, there is a most remarkable upheaval,

Wheeling & Pittsburg Branch B. & O. R. R. Ms.		Alt.	Chesapeake & Ohio Railroad— Continued. Ms.		Alt.
0 Wheeling.[28]	Barrens (XIV.)	645	307 Caldwell.	11 b. Chemung.	1765
2 Mt. DeChantel.	14 c. U. Coal M.	672	312 Ronceverte.[42]	13 b. Mauch Chunk	
4 Carbon.[36]	"	667		(XI.)	1660
9 Roney's Point.[37]	"	829	319 Fort Spring.	"	1625
10 Point Mills.	Permian (XVI.)	896	326 Alderson..	"	1550
16 West Alexander.	"	1043	328 Mohler.	"	1540
21 Claysville.[38]	"	1143	334 GreenbrierSt'kYds	"	1530
28 Chartier.	"		336 Lowell.	"	1510
32 Washington.[39]	"	1049	337 Talcott.	"	1510
			343 Don.[43]	"	1432
Chesapeake & Ohio Railroad. *			348 Hinton.[44]	"	1377
			350 Barksdale.	"	1345
297 Alleghany Tun.[40]	Pocono(X.),Cat.(IX.)		356 New Richmond.[45]	"	1290
298 Tuckahoe.	11 b. Chemung.	2036	360 Meadow Creek.	"	1265
302 White Sulphur.[41]	10 b. Hamilton.	1920	364 Slade.	"	1237
305 Hart Run.	"	1614	369 Quinnimont.[46]	"	1196

* *Chesapeake & Ohio Railroad.* Prof. Wm. B. Rogers' account of the geology of this road in Virginia and in West Virginia, as given in the first edition, is re-produced in the chapter on Virginia; but since its publication the country has been greatly developed and studied, and Prof. White has therefore prepared a more extended and minute description of the portion of that road in West Virginia.

which has brought up the lower coals, the strata suddenly rising within a few yards to an angle of 80 degrees. Just west of Laurel Fork Junction the rocks dip down again, the conditions being here on the west side similar to those at Petroleum on the east. After passing the first cut west from the station, the dip is suddenly reduced from 50 degrees to nearly horizontal. This forms the so-called "Oil Break,"as all the productive oil wells are found along the line of this belt. This belt is about one and a half miles wide, running in a direction a little east of north and gradually flattening out toward each extremity, and forms one of the most remarkable geological features in this State. This curious disturbance is well worth a visit. Near it, a few miles off by a branch road from Cairo, is the vertical chasm, 4 feet wide, which was filled with the mineral Grahamite, now worked out. There is a fault at Kanawha, forming the western boundary of the disturbed region, as that at Ellenboro is the eastern. (S. & F.)

35. *Parkersburg.* The Washington coal, about 100 ft. above the base of the Permian series, is found at low water of the Ohio here, while the horizon of the Pittsburg bed would be about 360 ft. under the river, but it is altogether probable that the Pittsburg has here thinned away, since borings give no trace of it, and at Burning Springs where the "Oil Break" anticlinal brings up its horizon, the coal is absent.

36. *Carbon.* Pittsburg coal mined here by shaft 65 ft. deep.

37. *Roney's Point.* Waynesburg coal mined locally, only 2½ ft.-3 ft. thick, and impure.

38. *Claysville.* Washington coal at track level, 1½ miles west from borough. Claysville anticlinal of Stevenson crosses R. R. one-quarter mile west from station.

39. *Washington.* The Harvey, Hoff and Hess gas wells supply the town with fuel; these three gas wells all on a line along the crest of the Washington anticlinal, were so located on scientific grounds by Prof. I. C. White. The Gantz Well, one mile southeast from the anticlinal obtained oil from the same sand (1st Venango) that the others get gas from. The Gantz Well struck the sand at 2,200 ft., passing through Pittsburg coal at 350 ft.; while the Hess well got gas at 2,068 ft., passing the same coal at 250 ft.

40. *Alleghany Tunnel.* The line between Virginia and West Virginia is crossed near center of tunnel through the Alleghany Mountain, the backbone of which is the Pocono sandstone.

41. *White Sulphur.* A well known summer resort, famed for the curative properties of its mineral water, which issues from the Oriskany sandstone in a large spring, flowing 75 to 100 gallons per minute.

42. *Ronceverte.* The railroad passes through the Pocono sandstone (X.) at Louisa tunnel, between Ronceverte and Caldwell, and then enters a long stretch of No. XI. limestone and shales along the Greenbrier River. The limestone is over 800 ft. thick, and forms the rich belt of blue grass country, which extends through Monroe, Greenbrier and Pocahontas counties. In the Pocono rocks at Louisa tunnel many fossil plants may be found.

43. *Don.* Near Don is the Big Bend tunnel, 6,080 ft. long, through No. XI. red shale, which cuts off several miles of meanders in the Greenbrier river.

44. *Hinton.* Junction of Greenbrier with New River. Here the railroad enters the cañon of the latter stream, a great gorge cut down 1,000 to 1,500 ft. below the tops of the bounding mountains, and in which the railroad runs for nearly 60 miles through some of the wildest scenery on the continent.

45. *New Richmond.* A splendid sandstone for building purposes crops out in the No. XI. sandy beds above the railroad here, and the West Virginia block for the Washington monument was quarried from the same. In the vicinity of Ronceverte and Alderson these sandy beds of XI. seem to be almost unrepresented, for the limestone there extends nearly up to the base of No. XII.; but as we enter the New River region a great mass of red shales, green and gray sandstones, etc., 1,500 to 2,000 ft. thick, wedges in between the main Greenbrier limestone below and 30 to 40 ft. of impure fossiliferous limestone at top, which immediately underlies the Pottsville (XII.) conglomerate. This upper limestone along New River holds the same fossils as an impure limestone in Monongalia County, which is separated from the main sub-carboniferous limestone by 50 ft. of sandstones and red shales,

Ms.	Chesapeake & Ohio Railroad— Continued.	Alt.	Ms.	Chesapeake & Ohio Railroad— Continued.	Alt.
370	Prince.	13 b. Mauch Chunk (XI.) 1192	416	Frederick.	14 b. L. Coal Meas., Clar. (Eagle) and L. Kit. coals. 641
372	McKendree.47	" 1150	417	Crescent.	" 638
379	Stone Cliff.48	Base of (XII.) 1076	418	Cannelton.54	14 b. L. Coal Meas. (Eagle bed.) 636
381	River View.	" 1072	421	Dego.	14 b. L. Coal M., 75' under L. Kit.
382	Dimmock.	" 1045	423	Paint Creek.55	100' under L. Kit.622
385	Fire Creek.49	Top of No. (XI.)1029	425	Blacksburg.56	5' above L. Kit. Cedar Grove (U. Kittan.) mined here. 626
387	E. Sewell.	Base of (XII.) 1008	427	Coalburg.57	14 b. L. Coal M. 625
388	Sewell.50	" 1004	431	Winnifred Junc.58	14 b. L. Coal M. 616
390	Caperton.	" 984	435	Brownstown.	14 b. L. Coal Meas., axis crosses here608
392	Nuttall.51	" 948	438	Malden.59	14 b. L. Coal M., 20' under L.Kit. coal605
394	Fayette.	L. half of (XII.) 908	444	Charleston.60	Base XIV. (Bar.)602
396	Elmo.	" 860	449	Spring Hill.61	Mahoning sands. 600
399	Hawk's Nest.52	Middle of (XII.) 828	455	St. Albans.	Middle of Barrens594
401	Cotton Hill.	Up. half of (XII.)796	459	Scary.62	" 590
406	Gauley.	Base of Homewood sandstone. 708			
408	Kanawha Falls. 53	Top of (XII.) 672			
413	Loup Creek.	Homewood s. s. 647			
413	Mt. Carbon.	14 b. L. Coal Meas., Clar. and Lower coals mined. 639			

and the two are very probably identical, though the intervening rocks have increased 30 fold in thickness on New River.

46. *Quinnimont.* The No. XII., or New River coal series, comes into the tops of the adjoining mountains here, and one of its coal beds, which comes 600 ft. above the base of XII., has been mined and coked for use in the iron furnace situated at Quinnimont. It makes a splendid coke, as does each of the three workable beds in No. XII. The elevation of the Quinnimont bed is 1,050 ft. above railroad.

47. *McKendree.* About half way between this station and Prince, the upper or Chester limestone mentioned in Note 45 comes down to track level, and presents a fine opportunity for collecting sub-carbo niferous (Chester) fossils.

48. *Stone Cliff.* Mines in Fire Creek and Nuttall coals, the former at 650 ft. above river, the latter at 950 ft.

49. *Fire Creek.* The Fire Creek coal here mined at 700 ft. above railroad, steepest incline on river.

50. *Sewell.* All of the three New River coals may be seen here. The Nuttall bed in the tops of the mountains, and the Quinnimont and Fire Creek below. These coals are of excellent coking varieties and very pure.

51. *Nuttall.* Nuttall coal, 400 ft. under top of XII. and 600 ft. above railroad, mined here. Uppermost great cliff rock of XII. seen capping the mountain here, from which the scenery is very grand.

52. *Hawk's Nest.* The Hawk's Nest cliff is on right bank of river, one mile below station, and here the upper members of XII. rise almost vertically from the bed of the river to 500 ft. above the same. The view from it is well worth a visit. The Anstead coal mines are in Gauley Mountain, four miles distant, and 855 ft. above C. & O. R. R. A narrow-gauge railroad leads out to them. The Lower Kittanning coal is the one mined. Nuttall coal is only 75 ft. above track at Hawk's Nest, and 2 ft. 8 in. thick.

53. *Kanawha Falls.* The falls are a series of cascades aggregating about 20 ft. in height over the hard current-bedded upper portion of the Homewood sandstone.

54. *Cannelton.* A good locality to study the lower coal measure series. The Clarion (Eagle) is just below track level. The Lower Kittanning bed is 105 ft. above, and extensively mined for gas coal, while on the north side here the U. Freeport coal may be seen at 750 ft. above river changed to a splendid cannel. From Mt. Carbon to near Charleston the track runs in No. XIII. beds, and coal openings are numerous on both sides of river. A general section of these measures is given in another connection.

55. *Paint Creek.* Paint Creek axis crosses here, and a railroad extends up Paint Creek for 10 miles to coal mines.

56. *Blacksburg.* Splendid example of erosion during coal measure times in cuts just above Blacksburg.

57. *Coalburg.* Splendid geological headquarters for seeing Coalburg, Cedar Grove and Brush Creek coals, and collecting fossil plants in roof of Lower Kittanning and Cedar Grove beds in Watson's Hollow, North Coalburg.

58. *Winnifrede Junction.* A railroad leads up Field's Creek seven miles to Winnifrede coal mines, the typical locality of Winnifrede bed (Upper Kittanning). On the other side of the river directly opposite, and in plain sight from the cars, is the mine of the Macfarlane Coal Company, in the Winnifrede bed, one of the best mines along the Kanawha, furnishing a very pure coal of splint and bituminous mixed, and in quality unsurpassed for domestic and steam purposes.

59. *Malden.* Cross to opposite side and examine extensive mines on Campbell's Creek (Lower Kittanning) coal, also salt works, the water being derived from base of XII.

60. *Charleston.* Good headquarters for studying barrens (XIV.). Three miniature faults in

Chesapeake & Ohio Railroad—Continued.			Ohio River Railroad—Continued.		
Ms.		Alt.	Ms.		Alt.
463 Scott[63].	Barrens XIV., (upper half.)	653	38 New Martinsv'le.	Permian (XVL)	626
469 Hurricane.	Barrens (XIV.)	655	41 Sardis.	"	622
476 Milton.	"	586	43 Paden's Valley.	"	622
479 Thorndyke.	"	640	47 Sisterville.	"	642
480 Ona.[64]	"	622	51 Friendly.	"	617
482 B. Sulphur Spgs.	"	598		{ Permian (XVI.) and	
485 Barboursville.	"	580	54 Long Reach.	14 c. U. Cl. M. (XV.)	
491 Guyandotte.	"	560		{ Waynes Coal 20'	
493 Huntingdon.[65]	"	566		above river.	617
501 Ceredo.	"	501	59 Raven's Rock.	{ Waynes Coal 20'	
502 Big Sandy, Ky.	"	502		above river.	616
			61 Grape Island.	14 c. U.Cl. M.(XV.)[615]	
Ohio River Railroad.			63 St. Mary's.	"	615
0 Wheeling.[28]	Barrens. (XIV.)		65 Vaucluse.	{ Barrens (XIV.)"Oil Break" crosses river here.	617
4 Benwood.	Pitts. Cl. nr. track.[639]		68 Eureka.	Barrens (XIV.)	620
11 Moundsville.[27]	{ 14 c. Upper Coal Meas.(XV.)	635	71 Willow Island.	"	607
	{ 14 c. Up. Coal Meas.		74 Bull Creek.	"	610
19 Powhatan.	300' of XVI. in		81 Williamstown.	14 c. U.Cl.M.(XV.)[602]	
	hills.	638	83 Henderson.	"	
23 Woodland.	14 c.U. Cl. M.(XV.)[638] "		87 Briscoe.	Permian (XVI.)	
26 Clarington.	{ Waynes Coal 75' above river.	631	88 Vienna.	"	
			94 Parkersburg.[25]	"	596
31 Proctor.	{ 70' under Waynes Cl. at river level.	629	**Ohio Central Railroad—Kanawha Division.**		
	{ Permian (XVI.)		0 Charleston.[60]	{ 14 b. Base of (XIV.) Barrens.	600
36 Baresville.	Waynes Coal nr. water level.	626	4 Lock No. 6.	14 b. Barrens.	592
			7 Smith's.	"	588

cuts of railroad, one mile above station, where U. Freeport coal and overlying "Black Flint" may also be examined. Great deposit of rounded pebbles and stones at junction of Elk and Kanawha here, finely exposed along cemetery road and extending to 385 ft. above river, the upper limit of the glacial dam-lake in which the deposit was made. From Charleston to Huntingdon the railroad runs in No. XIV., or the Barren Coal Measures.

61. *Springhill.* Great terrace of rounded boulders extend up over 200 ft. above river, just below mouth of Davis Creek, up which a railroad extends 15 miles to coal and Black Band iron ore mines.

62. *Scary.* Here the railroad leaves the Kanawha River following up Scary Creek, which leads out into an old valley (Teazes), at Scott, four miles distant. This singular valley, one mile wide and 200 ft. above the Kanawha River, bounded on either side by hills 200 feet higher, and extending through to the Guyandot River, which finally debouches into the Ohio, was once occupied by an arm of the Kanawha River, when the great ice dam at Cincinnati during glacial times backed the waters of the Ohio and its tributaries to a height of 500 to 600 ft. above present low water at Cincinnati. This hypothetical dam of Prof. G. F. Wright is demonstrated beyond any doubt by the great beds of clay, gravel, boulders and other trash which cover Teazes Valley to a great depth all along its course, except where subsequent erosion has removed them. When the ice dam melted away at Cincinnati, the water that had previously filled this valley was withdrawn, passing down to the Ohio by its former and present route, the Kanawha, thus leaving the ancient valley high and dry, though littered up with "Black Flint," pieces of cannel coal, quartzite, sandstone and other rocks that testify to their Kanawha and New River origin.

The traveler should also notice the remarkably level character of the Kanawha Valley flats, on which the railroads are built, as shown by the altitudes given from Point Pleasant to Charleston, on the Ohio Central Railroad, and above Charleston, on the Chesapeake & Ohio Railroad. Another important fact is that the deposit which fills this valley is true loess, a lacustrine deposit similar to that on the Mississippi and Missouri River and elsewhere. · J. M.

63. *Scott.* An excellent locality to study the ice dam lake deposits in a deep cut through them just east from station. The rounded boulders extend up to 750 ft. above tide here.

64. *Ona.* Lake deposits abundant.

65. *Huntingdon.* Mahoning sandstone makes cliffs along the hills from here to the State line at Big Sandy River.

66. *Sattes.* An interesting group of mounds, the work of the Mound-builders, occurs in the wide bottoms toward the river, half way between this station and Charleston.

67. *Poca.* The Pittsburg coal is extensively mined in this vicinity by the Marmet Mining Co. The coal is absent in the immediate river hills, but comes in about one mile back. The horizon of this coal emerges from the bed of the Kanawha, between Buffalo and Red House, being mined at

Ohio Central Railroad—Kanawha Division. Ms.		Alt.
10 Ryans.	14 b. Barrens.	588
12 Sattes.[66]	"	586
15 Bowling.	"	584
18 Poca.[67]	"	579
19 Raymond City.	"	586
20 Queen City.	"	579
21 Energetic.	"	576
26 Red House.[68]	14 c. Up. Coal Me.	577
31 Martin's.	"	572
35 Buffalo.·	"	570
38 18-Mile Creek.	"	564
40 Grimm's.[69]	"	563
42 Maupin's.	"	570
45 Leon or 13 m. Ck.	"	567
48 Beech Hill.	"	562
50 Bright's.	"	564
51 Rock Castle.	"	563
56 River Switch.	14 b. Barrens.	557
57 Ohio Riv. Bdge at Pt. Pleasant.[70]	"	597

Pittsburg, Cincinnati & St. Louis R. R. Pittsburg, Wheeling & Kentucky Div.		
0 Steubenville.	Barrens (No. XIV.)[728]	
1 Wheeling Junc.	"	
3 Middle Ferry.	"	
4 Lower Ferry.	"	
6 Cross Creek.	"	
9 Wellsburg.[71]	"	
12 Beech Bottom.	"	
16 Short Creek.	"	
21 Glenns.	"	
25 Wheeling.	"	645

Grafton & Greenbrier Railroad.[72] Ms.		Alt.
0 Grafton.	Barrens(No. XIV.)	985
3 Fresh Ford.	"	988
6 Foreman's.	"	995
8 Sandy Creek.	L. Coal Meas.	1021
11 Cove Run,	"	1072
14 Moatsville.	Cong. No. XII.	1155
17 Arden.	L. Coal Meas.	1260
19 Bryan's Mill.	"	1286
21 Newman's Trest.	"	1289
22 Kelley's.	"	1287
24 Philippi.	"	1288

Clarksburg & Weston R. R.		
0 Clarksburg, (B. & O. Depot.)	16' under Pitts. Coal.	1030
2 West End.	130' "	945
6 Mouth of Brown's Creek.[73]	100' "	946
8 Mt. Clare.	Barrens (XIV.)	1001
11 Bond's Summit.	"	1175
13 Lost Creek.	"	1013
14 Curry's Summit.	"	1196
18 Jane Lew.[74]	"	1006
21 Fisher's Summit.	"	1223
25 Weston.[75]	"	1009

Weston & Buckhannon R. R.		
0 Weston.[76]	Barrens(No.XIV)	1009
5 Gaston.	"	1040
6 Seymour.	"	1035
11 Stone Coal Sum.[77]	Up. Cl. Me. (XV.)	1444
11 Lorenz.	"	1435
15 Buckhannon.[78]	Barrens (XIV.)	1405

Oak Ridge, four miles below Red House, where it is 20 ft. above river level. Its height is 175 ft. at Poca, and on up the river is carried into the air along the valley.

68. *Red House.* The great cliff near the hill top is the Waynesburg sandstone.

69. *Grimm's.* Here the Waynesburg coal has been opened 190 ft. above river level, where it is slaty, worthless, and only 3 ft. thick. A well, bored in search of the Pittsburg coal, found only a trace of that bed at 80 ft. under river.

70. *Point Pleasant.* The Pittsburg coal is here about 75 ft. above the Ohio River, but only 1½ ft. -2 ft. thick. The Waynesburg sandstone at the base of the Permian, or No. XVI. of Rodgers, makes cliffs near the summit of the hills.

71. *Wellsburg.* In this town, and the immediate vicinity, many strong gas wells have been struck at a depth of 1,300 ft. below the Ohio river. The gas is utilized for both heat and light in the town, and also supplies the glass and other manufactories. The geological position of the gas sand is about 1,650 ft. under the Pittsburg coal, and is possibly identical with the Murraysville sand. A shaft has also been sunk to the same coal that is mined at Steubenville, which Prof. Orton identifies with the Lower Freeport, and which is here about 210 ft. under the railroad.

72. The Grafton & Greenbrier is a narrow-gauge railroad, which follows the Tygart's Valley River southward from Grafton to Philippi, its track running for about six miles in the Barrens, No. XIV., then passing down through the Lower Coal Measures and into No. XII. three or four miles in the vicinity of Moatsville, and emerging at the horizon of the Upper Freeport coal at Philippi.

73. Pittsburg coal is mined and shipped from this point.

74. Pittsburg coal in tops of the hills about 300 ft. above track.

75. The Mahoning sandstone crops out along west fork of Monongahela River here, according to Prof. Stevenson. The State Insane Asylum, built of Barren Measures sandstone, is located at Weston.

76. This is a continuation of the Clarksburg & Weston Narrow Gauge Railroad.

77. The Pittsburg coal is 40 to 50 ft. under the track here.

78. The Pittsburg coal is mined in the hills around Buckhannon, probably 100 ft. to 150 ft. above the depot. It is 4 ft. to 4½ ft. thick.

79. By Mr. James Parsons, C. and M. E., Piedmont, W. Va.

80. From *Piedmont* to within one mile of Gorman the road runs at the base of the Piedmont sandstone, the north branch of the Potomac having cut its circuitous course through that stone and bedded itself upon the upper series of the conglomerate. The cliffs and bluffs formed by that stone tower high above the road on both sides, and the scenery becomes grand, beautiful and interesting.

| West Virginia Central & Pittsburg R. R.[79] | | West Virginia Central & Pittsburg R. R.— | |
Ms.	Alt.	Ms. Continued.	Alt.
0 Piedmont.[80]	14 a. Homewood s.s.926	47 Fairfax.[83]	Top 14 b. Bar. Me.3051
1 Junction.	" 949	50 Thomas.[84]	14 b. Freeport. 2958
4 Empire.	" 1045	53 Porter.	{ Between 14 b. Freeport and Kit.3101
6 Warnicks.	" 1054		
7 Barnum.	" 1130	56 Davis.[85]	14 a.Homew'd s. s 3170
9 Windom.	" 1214		
11 Shaw.	" 1287		
14 Chaffee.	" 1455	Branch to Mineville.	
18 Blaine.	" 1605		
25 Schell.	14 a. Potts. Cong.1980		
30 Gorman.[81]	Base of (XIII.) 2295	0 Shaw.[86]	{ 14 a. Homewood sandstone. 1287
33 Elkins.	14 b. L. Coal M. 2313	4 Mineville.[86]	14 b. Kittanning. 1703
35 Bayard.	Top of XIII. 2340	.... Plane.	L. Barren Meas. 2253
37 Camden.[82]	14 b. Barren Me. 2498	5 Elk Garden.	{ Bottom of 14 c. Up. Coal Meas. 2305
39 Dobbins.	" 2579		
41 Hambleton.	" 2672	.... Mine No. 1.	{ 14 c. Pittsburg seam. 2305
44 Kearns.	" 2837		

81. At *Gorman* the road begins, geologically, to rise up through the Lower Coal Measures *in a red shale*, as observed also by Prof. I. C. White, a thing unheard of or unreported in the Lower Coal Measures, and at Bayard it has passed through the Kittanning and Freeport coals to the base of the Lower Barren Measures.

82. From *Camden* to *Fairfax* it still continues to rise, until by the time it reaches the summit at the latter place it rests upon the top of the Lower Barren Measures and at the base of the Upper Coal Measures.

83. From *Fairfax* to *Thomas* it gradually descends through the same barren measures and down until it reaches the bottom of the Freeport.

84. From *Thomas* to *Davis* it still continues to descend through the Lower Coal Measures until it reaches the Piedmont or Homewood sandstone at the latter place.

85. *Davis* is situated in the renowned valley of Canaan on the Black Water, at its junction with Beaver. Here the bottoms are broad, and stand on an elevation of 3,072 feet above tide water, while the plateaus running back both ways rise still higher—to an elevation of 3,170 feet. Davis, standing upon this bottom and plateau, is destined to become the frequent resort, not only of the seeker after pleasure, but of the scientific traveler, for from this point a great and grand panorama presents itself.

The Plane rises about 600 feet, passing up through the Lower Coal Measures and the Lower Barren Measures to the base of the Upper Coal Measures. Here the Pittsburg seam is opened and worked in several places at and near Elk Garden. This seam is 14 feet thick and of the finest quality.

86. The branch road from Shaw to Mineville passes up through the Piedmont or Homewood sandstone to the Kittanning coal, which crops out of the mountains at the foot of the plane.

The notes signed "F." are by Prof. Wm. M. Fontaine, and those signed "S." by Prof. J. J. Stevenson, taken from the first edition.
The altitudes for West Virginia have been all carefully collected, from original sources, by Prof. L. C. White; many of them are here published for the first time.

Fairmount, Morgantown & Pittsburg R.R.* Ms.		Alt.
0 Fairmount.88	Up. p't'n of (XIV.)	888
1 Junction Bridge.	B'r'ns or No.(XIV)	894
Low water, } Monong. Riv. }		850
3 Houltown.	Base of (XV.) or Up. Coal Meas.	869
4 Rievesville.89	No. (XV.)	888
Monong. R. here.		848
7 Pricket's C'k B'g.	Top of (XIV.)	882
River here.		843
7 Catawba.	Top of (XIV.)	880
11 Opekiska. 874	Up. portion (XIV.)	
River here.		839
17 Little Falls.90	Top of (XIII.)	855
M'th Tom's Run.		822
20 J. Kigers.	U. Freeport Coal.	837
22 Offington.91	Base (XIV.)	823
River here.		791
26 Morgantown.92	See note.	816

Monongahela River Railroad.

Ms.		Alt.
0 Fairmount. 879	75' under P'gh Coal.	
6 Camdensburg.93	Pittsburgh Coal.	889
11 Worthington.	P'gh Coal in riv.	898
13 Enterprise.	Pittsburgh Coal.	901
16 Shimston.	"	911
23 Simpsons Creek.	"	928
27 Bartlett.	"	981
32 Clarksburg.	"	1082

West Virginia and Pittsburgh Railroad. Ms.	Braxton Extension.	Alt.
0 Weston.	Pittsburgh Coal.	1018
12 Roanoke.	14 c. in hills.	1058
14 Arnolds.	14 c. Up. Coal M.	1096
25 Burnsville.	Barrens, (XIV.)	758
L. Kanawha Riv.	{ 250' under P.C'l	741
32 Salt Lick B'dges.	Barrens, (XIV.)	788
35 Hecter's.	Barrens.	858
38 Flat Woods.	"(XIV.)	1059
39 Summit.	"	1268
44 Sutton. 838	Barrens, Mah. s. s.	

Buckhannon River Extension.

Ms.		Alt.
0 Buckhannon.	Barrens, (XIV.)	1408
7 Sago.	"	1425
13 Ten Mile.94	14 b. L. C'l M.	1606
17 Alton.	"	1818
25 Newlon.	"	1917

Ohio River Railroad.—Continued.

Ms.		Alt.
94 Parkersburg.	Perm. C'b.,(XVI.)	622
107 Harris' Ferry.	"	596
111 Belleville.	"	591
117 Murraysville.95	Waynesburg s. s.	592
120 Muse's Bottom.	Perm. C'b., (XVI.)	588
123 Portland.	"	592
125 Sherman.	"	587
128 Ravenswood. 885	Waynesburg "A" C'l.	
132 Pleasant View.	Perm. C'b., (XVI.)	581
135 Willow Grove.	"	584
138 Ripley Landing.	"	579

* Since the stereotypes were made of the foregoing pages of this chapter, (which had been edited by my father), Prof. White has furnished these additional lines and surveys.   J. R. M.

87. *Errata in Note 45.* The statement in Note 45 with reference to the thinning away of No. XII. red beds in vicinity of Alderson, etc., was made upon information which I considered reliable at the time, but a subsequent personal examination shows that what was taken for the Pottsville conglomerate is simply a massive, white pebbly sandstone in the No. XI. shales and that instead of having thinned away, these shales are here thicker than anywhere else in the state, approaching 2,500 feet and holding two immense white conglomerates, along with the red beds and impure limestones.   I. C. W.
The casting of the plate in which Note 45 occurs prevented the making of this correction in its proper place.   J. R. M.

88. *Fairmount.* The levels are brought from Fairmount on main line of B. & O. by Major Whiting of the B. & O. engineer corps. The elevation here gives 779 feet for low water at Morgantown, but the river survey from Pittsburgh makes it 786 feet. See Note 21.

89. *Rievesville.* Sewickley coal crops out along railroad cuts.

90. *Little Falls.* Upper Freeport coal in cuts. Rapids in river made by Upper Freeport sandstone.

91. *Offington.* Mahoning s. s. makes great cliffs here known as "Raven Rocks."

92. *Morgantown.* Upper Freeport coal 75 feet under river. Pittsburgh coal 440 feet above same level. Fine show of terrace deposits extending to 275 feet above river. Good locality for fossils in crinoidal limestone. Cheat river gorge nine miles distant. Grand view from crest of Chestnut Ridge. Subcarboniferous fossils under great arch below.

93. *Camdensburg.* The Pittsburgh coal dips under the river about two and a half miles above Fairmount to about 50 feet below the same, but comes up just below Camdensburg and is soon 25 to 30 feet above water. Extensive coking works of ex-Senator Camden and others, 250 ovens. Coal 9 to 10 feet thick. This bed is never less than 8 feet thick between Fairmount and Clarksburg, and is of excellent quality for fuel, gas and coke. This road passes through one of the finest coal fields in the world, which must in the near future replace the Connellsville field.

94. *Ten Mile.* Upper Freeport coal in hills here and at the level of the track four miles below, near mouth of Grassy Run, where it is only 3 to 4 feet thick, but roofed with 12 feet of cannel slate.

95. *Murraysville.* The Waynesburg sandstone is frequently seen between Parkersburg and Letout Falls, sometimes a great cliff as at Murraysville; again its top is just seen in the bed of the Ohio. At Letout it rises from the river to the northwest and makes the rapids in the river. Below here it forms long lines of cliffs near the summits nearly to Guyandotte.

96. *Graham.* Pittsburgh coal mined on the other side of the river by shaft 170 feet deep. Coal about 5 feet thick and dips rapidly southeast toward the center of the Appalachian basin.

97. *Hartford.* Hartford, Mason City, Clifton and the town of Pomeroy on the Ohio side are celebrated for the manufacture of salt and bromine. Salt bearing stratum reached by borings at about 1,150 feet under the Pittsburgh coal. It appears to be the top portion of the Pocono, (No. X.) sandstone and the same as the Mt. Morris oil rock ("Big Injun.")

road.—Continued.	Alt.	West Virginia Central R. R.—Continued.		
		Ms.	Extension from Thomas to Elkins.	Alt.
'erm. C'b., (XVI.)	⁵⁷⁴	74	Fairfax.	Barrens, (XIV.) ³⁰⁵¹
"	⁵⁷⁶	78	Thomas. ³⁹⁵⁰	Top L. Coal M.,(XIII.)
.4 c. Up. C'l Meas.	⁵⁷⁴	79	Davis. ²⁸⁶⁵	Low Kittanning Coal.
'gh Coal in riv.	⁵⁷⁶	80	Globe Falls.	No. (XII.) Congl.²⁷²⁴
Pittsburgh CoaL	⁵⁷³	81	Pt. Lookout.¹⁰²	" ²⁶⁴⁰
"	⁵⁷⁴	82	²⁴⁵⁰	Top Mauch C'k Reds.
"	⁵⁵⁴	84	Big Run.	No. (XI.) beds. ²¹⁵⁰
"	⁵⁶⁶	87	Hendrick's.	12. Catskill. ¹⁷²⁰
"	⁵⁶⁷	90	Black Fork.	11 b. Chemung. ¹⁶⁵⁰
"	⁵⁷⁰	91	Shaver's Fork.	" ¹⁶⁴⁸
"	⁵⁷¹	93	Haddix Run.	" ¹⁶⁸⁰
Barrens,P'gh CoaL	⁵⁷⁷	98	Haddix Summit.	" ²¹⁷⁹
Barrens, (XIV.)	⁵⁵¹	101	Montrose.	10 b. Hamilton. ¹⁹⁸³
"	⁵⁷⁰	106	Kerens.	" ¹⁹³⁸
"	⁵⁵⁰	112	Old Leadsville.	" ¹⁹¹²
"	⁵⁵¹	113	Elkins.¹⁰⁸	" ¹⁹²⁴
"	⁵⁴⁸		Survey, Elkins to Gauley River.	
"	⁵⁷⁹	0	Elkins.	10 b. Hamilton. ¹⁹²⁴
"	⁵⁶⁷	6	Beverly.	" ¹⁹⁵²
"	⁵⁴⁹	8	Burnt Bridge.	" (water.) ¹⁹⁸⁹
"	⁵⁴⁸	13		" ¹⁹⁷⁴
"	⁵⁴⁶	16	Mill Creek.	" ²⁰⁰²
Branch.		17	Huttonsville.	" ²⁰⁶²
{ 15. Permo. Carb.		26	Elk Water.	11 b. Chemung. ²³⁵⁸
{ Wash'gton Coal.⁵⁵⁴		32	Brady's Summit.	No. (XI.) l. s. ²⁹⁹²
.5. Permo. Carb.	⁵⁵⁰	34	Riggles.	No. (XI.) Shales. ²⁷¹⁴
"	⁵⁵²	35	Red Lick Run.	Top (XI.) l. s. ²⁴²⁹
"	⁵⁶⁰	36	Elk River.	No. (XI) Shales. ²³³¹
"	⁵⁹⁰	38	Whitacre's Falls.	" ²¹⁷¹
"	⁵⁷¹	39	Big Run.	" ²¹³⁶
		46	Burgoo.	" ¹⁹⁰⁴
.4 c.Up.Coal Meas.⁹⁰⁵		48	Leatherwood.	" ¹⁶⁴¹
Barrens, (XIV.)	⁷²⁰	56	Elk River.	" ¹⁵⁸³
Railroad.—Continued.		59	Addison.¹⁰⁴	Top (XI.) l. s. ¹⁶⁴⁸
Cumberland.		63	Payn's Summit.	Base of No.(XII.)²⁴⁵⁶
'. Low'r Helderb'g.⁶³⁰		71	Gauley Riv.¹⁰⁵	No. (XII.) Congl. ²³⁰³
; b. Clinton.	⁵⁹⁶	78	Williams Riv.	" ²²¹⁵
"	⁷³⁴		Stony River Survey.	
.0 b. Ham. (Marc'lus.)		0	Mouth of River.	No. (XII.) Congl. ²⁰⁷⁶
'. Low'r Helderb'g.		6	Pike Cross'g. ¹⁰⁶	Barrens, (XIV.) ²⁵⁴⁵
{ 14 a. Pottsv'le Cong.,		10		Low. Coal Meas. ²⁷⁹⁹
{ Top of (XII.)	⁹¹⁵	13	Falls.¹⁰⁷	Clarion Coal. ²⁹⁷⁷
		15		No. (XII.) Congl. ³¹⁰²

irgh coal, 4 to 5 feet thick, mined here. It thins away down the river to 18
sant. Occasionally, as at Mercer's Bottom, it thickens to 4 or five feet.
to a few inches and not mined until near Huntington, where it is 3 to 4 feet.
recently an attempt was made to sell lands as containing tin ore. The
d lime-tone 40 to 60 feet below the Pittsburgh coal and on analysis proved
in. Another "tin syndicate" explored this same stratum for that metal on
) miles above Grantsville.
ly. The "Ridge Limestone" near the summits of the hills over a large
7 is often 10 to 20 feet thick, and is probably the Ninevah Limestone of
Stevenson's Green county series.
urning Springs or Volcano anticlinal passes along the valley of Spring
en Measures to the surface. Pittsburgh coal is absent or but feebly repre-
the state and especially along the line of the Volcano anticlinal everywhere.
Grandest scenery in the Appalachian Mountains. The Black Fork of the
feet deep through the Back Bone Mountain range, which is capped by the
The railroad grade down this gorge is 160 feet to the mile and it runs
400 feet above the river, which has a fall of 100 feet to the mile. The New
along the railroad grade, both the Nuttall (2½ feet thick) and Quinnemont
The Quinnemont and Five Creek beds are split into a half dozen thin
ville Conglomerate series is here over 700 feet thick.

West Virginia Central R. R.—*Continued.*		
Ms.	Survey, Elkins to Buckhannon.—*Con.*	Alt.
7 Roaring C'k.[103]	14 c.Low. Coal M.	1860
10	"	2121
11 Roaring.	Barrens, (XIV.)	2368
12 King's Ridge.	"	2450
17 Toll Gate.	"	1851
18 Burnt Bridge.	Top Low. Coal M.	1840
21 White Oak S'm't.	Barrens, (XIV.)	2031
27 Buck. R. Divide.	"	1743
32 Buckhannon.	"	1418
Elk River.		
0 Charleston.	Base of Barrens.	556
21 Big Sandy.	"	091
24 Queen's Sh'ls.[109]	"	611
60 Big Otter. 726	Top of Low. C'l Meas.	
70 Grove's Creek.	Barrens, (XIV.)	751
80 Birch River.	"	770
93 Little Otter.	"	794
Beall's Mills.	"	798
100 Sutton.[110]	"	806
Gauley River.—C. & O. Survey.		
0 Mouth.	Top of No. (XII.)	650
5 M'th of 20-Mile.	Base of No. (XII.)	667
10 Little Elk.	"	691

Gauley River.—C. & O. Survey.—*Continued.*		
Ms.		Alt.
15 Peters.[111]	Top of (No. XII.)	679
21 Carnifax Ferry.	No.(XII.)N't'l C'l.	1208
25 Hughes Ferry.	No.(XII.)Congl.	1546
29 Brock's.	"	1589
51 Beaver Creek.	"	1694
40 Cherry River.	14 a.Nutall Coal.	1777
43 Cranberry.	"	1915
46 Stroud's Creek.	No. (XII.) Congl.	2009
55 Williams River.	"	2167
75 Laurel Fork.	"	3011
80 Stony Creek.	"	3228
85 Marlin's Bottom.	{ No. (XI.)or Greenb'r l.s.to Cherry R.	2130
Little Kanawha River.		
0 Parkersburg.[112]	No.(XVI.)P'm-C'b.	568
2 Lock One.	"	564
14 Lock Two.	"	574
22 Lock Three.[113]	"	584
32 Lock Four.[114]	No.(XIV.) Bar'ens.	596
43 Spring Creek.	"	612
Buffalo Rock.	(?)	625
L'r Leading C'k.	No. (XVI.)P'm-C'b.	631

103. *Elkins.* The Tygarts valley in which the town is situated, is geologically a great arch, or rather two anticlinal axis which have come nearly together. These are the anticlinals which cross the B. & O. R. R. at Terra Alta and Mountain Lake Park respectively, having there a trough between them deep enough to catch the Lower Coal Measures,but here at Elkins the axes are less than a mile apart and the trough holds only the basal beds of the Chemung. On one side (west) of this double arch at Elkins, the Rich—Big Laurel Mt. rises to 3,500 feet above the sea, and on the other (east) Cheat Mt. attains a greater height, while both are crowned with the Pottsville Conglomerate, thus rendering the wide valley between, one of the most beautiful and picturesque in the country.

104. *Addison.* County seat of Webster county. On the summit of an anticlinal axis, which brings the top of the Greenbriar Limestone 40 feet above water level and exposes 800 feet of the Mauch Chunk Red Shales between the top of the limestone and the base of the Pottsville Conglomerate in the summit of the Mountain above. Near the crest of this arch at Addison a hole was once bored for oil many years ago, but at about 100 feet a strong stream of salt and sulphur water was struck, which still continues to flow and has attained much celebrity as a mineral water for medicinal purposes, especially for kidney troubles Where the Gauley Turnpike crosses McGuires Gap, opposite Addison, a coal bed 2½ to 3 feet thick has been mined only 20 feet above the Mauch Chunk red beds.

105. Near here on Land Run is the out crop of a coal bed 7 feet thick, of poor quality and it would seem to come at the same horizon as the Pocahontas or No III. bed of the Flat Top region.

106. Capt. Joseph Parsons, chief engineer of the W. Va. C. R. R. who has kindly furnished all the elevations on that railroad and its surveys, states that the Lower Kittanning coal passes under Stony river about three and a half miles above its mouth and reappears at nine miles up. The center of the trough is near where the northwestern pike crosses Stony river, and here the Pittsburgh coal is in the summits of the hills just north from the river. This is the northern end of the Elk Garden Pittsburgh coal basin, since northward from here that coal misses the hills by only 50 to 100 feet for twenty miles, till it is caught in the Fairfax summit on the Cheat-Potomac Divide.

107. There is a large area of the lower Kittanning coal from here on down the river for four miles and it has a thickness of eight feet with its customary partings. It is forty feet above water at the Falls.

108. Half way between Roaring creek and Elkins the Tygarts Valley river cuts squarely through the great Rich-Laurel Mt. uplift, and exposes a splendid section from the Hamilton up to the Lower Coal Measures. Along and in the vicinity of Roaring creek is a large field of the Upper Freeport coal where the bed has a thickness of 8 to 10 feet. The Freeport sandstone is very massive and pebbly along the lower part of Roaring creek and makes the numerous falls.

109. *Queen's Shoals.* A few miles above here the river bends southward and the Upper Freeport coal comes above water level, and keeps above the same till the stream turns northwestward above Clay C. H. There is a fine area of this coal on Big and Little Sycamore creeks. With this exception only the Barren Measures crop out along Elk between Sutton and its mouth, a distance of 100 miles, and as these beds have a greater thickness (800') here than anywhere else in the country, I have termed them the Elk River series.

110. *Sutton.* The Mahoning coal (about 100 feet above the base of the Barrens) crops 30 to 40 feet above river level and has been mined to a small extent, while at Frametown 16 miles below, the Pittsburgh coal is in the summits of the hills, 500 feet above the river and 6 to 7 feet thick.

111. From the mouth of the Little Elk up to the Cherry River the Gauley flows in a narrow cañon 300-400 feet deep, excavated out of the top members of No. XII., while the softer Lower Coal Measures occur back in the summits of the hills on the broad plateau at the top of No. XII. The Nutall coal comes up at the mouth of Meadow River, but it thins there. It has a thickness of 5 to 6 feet on the waters of Hommony, Cherry and other streams, which put in from the south, and is a splendid coking coal.

Ms.	Little Kanawha River.—*Continued.*	Alt.
61	Down's Ripple.	No.(XVI.)P'm-C'b.	635
63	Anna Maria C'k.	"	641
68	Big Root.	"	644
76	Pine Creek.	Upp. Coal Meas.	554
78	Grantsville.[115]	"	556
80	Steer Creek.[116]	"	566
85	Acre Island.	"	571
89	Musch Shoals.	"	577
92	Tanner Fork.[117]	"	582
96	Cedar Creek.	No.(XIV.)Barrens.	587
98	3d Run Sh'ls.[118]	"	589
101	Leading Creek.	"	690
103	Glenville.[119]	"	702
105	Stewart's Creek.	"	702
106	Mud Lick Run.	"	710
110	Sand Fork.	Upp. Coal Meas.	711
115	Stout's Mill.	"	723
118	Hyer's Run.	No.(XIV.)Barrens.	735
121	Oil Creek.	"	741
122	Burnsville.(Lumber port.)	"	741
	Bennett's Run.	"	752
131	Bulltown.	"	760

Kentucky.[120]

Chesapeake and Ohio Railroad.—*Continued.*

Ms.	Cincinnati Division.	Alt.
504	Catlettsburg.	Low. Coal. (XIII.)	544
506	Williams.	"	
509	Norton.	"	
510	Ashland.	"	544
511	A. C. & I. Cr's'g.	"	
512	Bellefonte.	14 a. Pottsv., (XII.)	
515	Russell.	"	
519	Wurtland.	"	

Kentucky.[120]

Chesapeake and Ohio Railroad.

Ms.	Cincinnati Division.—*Continued.*	Alt.
522	Riverton Jc. [629]	14 a. Pottsv., (XII.)	
523	Greenup.	13. Sub-Carboniferous.	
528	Gray's Branch.	"	
535	Siloam.	"	
541	S. Portsmouth.	"	
551	Quincy.	"	
553	Kinney.	"	
556	Buena Vista.	Huron Shale.	
560	Fairview.	"	
563	Vanceburg.	9 c. Cornif. l. s. in riv.	
568	Rome.	5 c. Niagara.	
575	Concord.	"	
577	Pence.	4c. Cincinnati.	
586	Springdale.	"	
592	M. & B. S. Junc.	"	
593	Maysville.	"	502
601	S. Ripley.	"	
603	Dover.	"	
610	Augusta.	"	
614	Wellsburg.	"	
617	Bradford.	"	
621	Foster.	4 c. Cincinnati.	
628	Belmont.	"	
630	California.	4 a. Trenton.	
682	New Richmond.	"	494
634	Oneonta.	"	
638	Ross.	4 c. Cincinnati.	
649	Dayton.	"	542
651	Newport.	"	
653	K. C. Jc.	"	515
654	Covington.	"	
655	Cincinnati.	"	

112. *Parkersburg.* Low water here as given by Col. Roberts is 562.804. See Note 35.

113. The elevations given for these locks is the top of the mitre sill below the dams. From Parkersburg for 25 miles up the river the rocks are nearly horizontal and the Upper Meretta sandstone of the Permian Series, which is quarried at Parkersburg, (Jackson quarry,) makes cliffs in the river hills for a long distance. It is extensively quarried at Elizabeth.

114. *Lock Four.* Near here is Burning Springs, the famous oil district, from which oil was collected and marketed as far back as 1841. The Eureka Volcano Anticlinal (called the "Oil Break") passes through this region, and brings up 400 feet of the Barren Measures. The Pittsburgh coal is absent, or only a few inches thick, while the Crinoidal coal is 20 inches thick and mined below the village for local supply. Oil is obtained here in the Mahoning, Conglomerate, "Big Injun" (Pacors) and Maxburg (Gantz) sands.

115. *Grantsville.* Here the Waynesburg is in the summit of the hills.

116. *Steer Creek.* At the mouth of this stream the massive sandstone above the Pittsburgh coal comes above water level, and the base of the great Waynesburg sandstone cliff is 275 feet above the same.

117. *Tanner Fork.* Along this stream the Waynesburg coal is mined for local use. It is only 18 to 24 inches thick and at Tannersville 6 miles up the stream is 135 feet above the latter.

118. *Third Run Shoals.* The Waynesburg Coal shows in summit of hill here 360 feet above the river or 1050 A. T. The horizon of the Pittsburgh coal is about 50 feet above the river, but the coal is absent.

119. *Glenville.* A broad anticlinal, which is probably identical with the Chestnut Ridge axis, crosses the river above Glenville and hoists the Pittsburgh coal 225 feet above the same. This coal makes its first appearance here it being absent or but feebly developed everywhere below until its horizon dips under water near the mouth of Steer Creek; near one and a half miles above Glenville it is 4 to 5 feet thick and 200 feet above the river. It runs along the hills at near this level for a mile or two further and then dips rapidly down below water level, passing under the river 1¼ miles below Land Fork or 109¼ miles from Parkersburg. The sandstone above the coal has an immense development in this region, being 130 feet thick. The horizon of the Pittsburgh coal keeps 50 to 75 feet below river level till we come to Stout's Mills, when the basin is crossed and it begins to rise rapidly appearing 10 feet above river level, one mile above Stout's Mills, and one-half mile further upstream is 75 feet above the same. It is here 7 feet thick and there is a great coal field in this basin between Burnsville and Glenville.

120. This Division of the C. & O., (formations by Prof. I. C. White) belongs in the Kentucky chapter, but for lack of space is inserted here, just before publication. J. R. M.

Virginia.[28]

By Prof. William B. Rogers.

List of the Geological Formations Found in Virginia and West Virginia.

	GENERAL GROUPS.	SUB-DIVISIONS IN VIRGINIA AND WEST VIRGINIA.	Numbers marking the Paleozoic Formations of Penn. and Va., as used in the Annual Reports of W. B. and H. D. Rogers.	Names adopted H. D. and W. B. R. the Paleozoic Form tions of Pennsylvat and Virginia and u in H. D. Rogers' Fi Report of the Geolo of Pennsylvania.
Mesozoic. | Cenozoic	QUATERNARY.	20. Quaternary.		
	TERTIARY.	19 c. Pliocene. 19 b. Miocene. 19 a. Eocene.		
	UPPER AND LOWER MESOZOIC.	(18 & 17.) Jurasso-Cretac's.[1] Upper Secondary s. s. (17, 16.) Jurasso-Triassic.[2] Mid. Secondary Sandstones and Coal Measures.		
Paleozoic.	UPPER CARBONIFEROUS.	14 c. Upper Barren Group. 14 c. Upper Coal Group. 14 b. Lower Barren Group. 14 b. Lower Coal Group. 14 a. Great Conglomerate and Conglo. Coal Group.	XVI. XV. XIV. XIII. XII.	Seral. Seral. Seral. Seral. Seral.
	MID. CARBONIFEROUS. (UPPER SUB-CARB.)	13 b. Greenbriar Shales. 13 b. Greenbriar Limestone. (Carb. Limestone.)	XI. XI.	Umbral Shales Umbral Limesi
	LOWER CARBONIFEROUS. (LOWER SUB-CARB.)	13 a. Montgomery Grits and Coal Measures. (Tuedian ?)	X.	Vespertine Sar stone and Co
	DEVONIAN.	Names of N. Y. Survey chiefly: 12. Catskill. 11 b. Chemung. 11 a. Portage. 10 c. Genesee. 10 b. Hamilton. 10 a. Marcellus.	IX. VIII. VIII. VIII. VIII. VIII.	Ponent. Vergent. Vergent. Cadent. Cadent. Cadent.
	SILURIAN.	8. Oriskany. 7. Lower Helderberg. 6. Salina. 5 c. Niagara. 5 b. Clinton. 5 a. Medina.	VII. VI. V. V. V. IV.	Meridian. Pre-Meridian. Scalent. Scalent. Surgent. Levant.
	SILURO-CAMBRIAN[3] OR UPPER CAMBRIAN.	4 c. Hudson River. 4 b. Utica. 4 a. Trenton.	III. III. III.	Matinal. Matinal. Matinal.
	MIDDLE[4] AND LOWER CAMBRIAN.	3 c. Chazy. 3 b. Levis. 3 a. Calciferous. 2 b. Potsdam Group.[5]	II. II. II. I.	Auroral.[4] Auroral. Auroral. Primal.[5]
	ARCHÆAN.	Archæan. A. B. C. D.[6]		

Virginia.

Baltimore and Ohio Railroad.

Ms.	Harper's Ferry and Valley Branch.	Alt.	
0	Harper's Ferry.	277	Altered Cambri'n(b) or Archæan B, followed west by Cambrian, 2 b., 3 a.
1	Shenandoah.	377	
6	Halltown.	339	Cambrian 3 a., b.
10	Charlestown.	513	" 3 b., c.
14	Cameron.	547	" "
23	Wadesville.	495	Siluro-Cam. 4 a. & 4 b.
27	Stephenson's.	499	Siluro-Cam. & Cam. 4 a. and 3 c.
32	Winchester.	717	The road runs close to boundary of Cambrian 3 c., and Sil.-Cambrian, 4 a., of the belt lying east, composed largely of 4 c.
36	Kernstown.	744	
39	Newtown.	770	
42	Vaucluse.	7	
44	Middletown.	700	
46	Cedar Creek.	695	
50	Capon Road.	740	
51	Strasburg Jc.	703	Siluro-Cam ori'n, 4 a. and 4 b., on switch track.
55	Tom's Brook.		Cambrian, 3 b., c. 745
57	Maurerstown.		" " 733
61	Woodstock.	820	" "
66	Edinburg.	845	" "
74	Mount Jackson.	914	Cam. & Siluro-Cam. 3 c. and 4 a.
81	New Market.		" 971
88	Broadway.		" 1038
94	Linville.		" 1242
00	Harrisonburg.	8	" 1840
105	Pleasant Valley.		Cambrian, 3 b., c. 1245
117	Fort Defiance.	9	" " 1275
126	Staunton.	1366	Cam. & Siluro-Cam. 3 c. and 4 a.

Chesapeake & Ohio Railroad.

Ms.		Alt.	
0	Richmond.	44	W. outcrop of Tert'y and Upper Mesozoic, all resti'g on Arch.C.
9	Atlee's.	202	19. Tertiary.
18	Hanover C. H.	83	"
28	Hanover Junct.		Upper Mesozoic, Jurasso-Cretaceous.
33	Noel's.	257	1. Archæan, C.
40	Beaver Dam.	339	Gneiss & MicaSlates, with veins of Gran.
45	Bampass'	341	1. Archæan, A.
50	Frederick's Hall.		" 351
56	Tolersville. 10	463	Mic.Hornb.& Hydro. Mic.Slat., with Aurif. q'rtz. The gold belt.
62	Lousia C. H.	452	1. Archæan, C.
76	Gordonsville.	500	" B.
81	Lindsay's.	457	Argil.Mic. & Hydro. Mic.Sla., with patches of SlatyLimestone & Steatite Epidotic,
83	Cobham.	595	Chlor. and Sil. Grits
90	Keswick.	489	& Slates of S. W. Mt.
97	Charlottesville.	449	followed west by Gneissoid Sandst'ne.
104	Ivy.	544	1. Archæan, D.
107	Mechum's River		Horn.& Chl Gnei.Syen.
115	Greenwood.		1. Arch.,B. Bl. Ridge Epid. Chlor. Argil. Slates,&c.,flank'd W. by Camb. I,2 b. Pots.
124	Waynesboro. 1301		Cambrian, 3 a., adjoining slates of 2 b.
129	Fishersville. 1321		Sil-Camb., 4 a. & 4 b. Edge of slate belt.
136	Staunton.	1367	Camb. & Sil-Camb., 3 c. and 4 a.
144	Swoope's.	1645	"

1. The term Jurasso-Cretaceous is chosen to designate the Upper Secondary Sandstones of the Virginia reports and the associated sands and clays which in their prolongation, northeast through Maryland, Delaware and New Jersey, are found to underlie the Cretaceous green-sand formation of those States, because the fossils found in the vicinity of Fredericksburg, etc., in Virginia, as well as near Baltimore, suggest the upper stage of the Jurassic period; while it is stated that the sands and clays of this belt in New Jersey are referable to the base of the Cretaceous. The whole group would seem in the main to be one of transition, and it is probably best comparable to the European Wealden.

2 The name Jurasso-Triassic is preferred for the Mid-Secondary rocks of the Virginia reports, as it is thought to correspond best with the fossil indications thus far furnished by the several belts included in it. Of these, the most western area is in part continuous with the so-called Triassic belt of Maryland and Pennsylvania, and in part with the coal bearing rocks of Dan River, North Carolina. The middle belt is in the line of prolongation of the Deep River coal rocks of North Carolina, and the eastern belt, including the Grits and Coal Measures of Chesterfield, Henrico, etc., is topographically without a counterpart. The middle and eastern belts in Virginia, and the western tract in North Carolina, show a close agreement in their fossil flora, which in many particulars has a decidedly Jurassic character, and all three belts are connected by certain species of Estheria, Candona, etc., held in common. Collectively these beds represent most probably a group of deposits ranging through Upper Triassic, and Lower Jurassic time, and are in large measure of a transitional character.

3. In grouping the Lower Paleozoic formations, Sedgewick's classification is used, including as Cambrian and Siluro-Cambrian, all the formations from the base of the Paleozoic to the top of the Trenton period (4 c.), and as Silurian the succeeding formations to the top of the Oriskany (8.); these correspond in limits to the Upper and Lower Silurian periods of the table.

4. The Middle Cambrian, or Auroral group, occupying much of the surface of the great valley west of the Blue Ridge, and exposed in numerous anticlinals and faults in the mountain belt farther west, is marked by a great preponderance of magnesian limestones in the lower two-thirds of its mass, passing below in many cases into Arenaceous and Argillaceous limestones, and followed above by oolitic and by cherty and sandy beds these latter giving place still higher to the

Ms.	Chesapeake & Ohio R. R.—Con.	Alt.	Ms.	Chesapeake & (
150	North Mountain. 20,74	Devonian, 10 a., adjoining Silurian of the Gap, 5 a., 5 b. to 8, inverted.	195	Jackson's River. 13 1133
159	Craigsville. 1516	Silurian, 7.,Encrinal Marble. 8. Oriskany.		
168	Goshen. 1f 1410	Devonian, 10 a. and 10 b., between ridges of Silurian, 5 a. to 8.	205	Covington. 14 1425
175	Millboro. 12 1679	Devonian 10 a., near 8. of Sideling Hill.	221	Alleghany. 2068

more purely Calcareous and Argillo-Calcareous strata appertaining to the
brian, Trenton, or Matinal group. The frequent faults, inversions and r
the great valley, and the rarity of fossils in the Auroral rocks, have i
demarcation of formations, but there can be little doubt, from fossil and o
cover the period of the formations 3 a., 3 b., 3 c., assigned to them in th
indicating the formations *near* as well as *at* the localities, the designatio
these rocks up to the top of the magnesian, without distinguishing b
Quebec (or Levis), and 3 b. c., for the remaining strata up to the well de
Cambrian, Trenton or Matinal group, 4 a. b. and c.

5. The Potsdam, or Primal group, includes in Virginia, where comple
proper, the ferriferous shales next above, and the slates, shaly grits and c
formation. It is exposed in varying mass and completeness on the wester
flanking hills of the Blue Ridge throughout much of its length, often, by i
southeast, in seeming conformity beneath the older rocks of the Blue Rid
uncomformably upon or against these older rocks. These older rocks, comprising m
to Huronian and Laurentian age, include also a group of highly alter
apparently to the copper-bearing or Keweenian series of Northern Mich
lately described Dimetian rocks of Wales.

6. The letters A, B, C, D mark four rather distinct groups of Archæan
of which the first three may probably be referred to the Laurentian, B
periods respectively, and the fourth to an intermediate stage—the Norian

7. This belt of Siluro-Cambrian slates extends continuously from the I
about ten miles south of Staunton, a distance of 140 miles, beyond which
discontinuous. In the tract corresponding to the interval, from Strasb
encloses the complex synclinal of the Massanutten Mountains, consistin
Silurian rocks 5 a. 5 b., with some bands of 7 and a few traces of Devonia
wide undulated trough of the slates. From Strasburg southwest, the ra
distance of from one-half to one mile west of the edge of the slates, but so
it, affording ready access to fossiliferous beds of 4 a., b. and c.

8. About 13 miles west-by-north from this are the Rawley Springs, an
remarkable fissured rocks known as Moravian Town, both in Ponent 12.
miles are the Dora coal mines, in Vespertine 13 a., of Narrowback mour
and crushed. The irregular fault, which, with many interruptions, extend
River along the northwest edge of the Great Valley in the line of the L
about 120 miles, is seen near these localities to bring the Siluro-Cambr
juxtaposition with the Devonian 10, to 12.

9. ' About eight miles east of this are Weyer's and Madison's caves, sit
dipping limestone, 3 a. b., near the South River.

♦10. In this part of the gold belt are situated the old workings, kno
Baker's, Triple Fork and Walton's Mines.

11. This is a good point of departure for examining the rock structure
mostly inverted, and the wild passage of the North River through the sar
ler's Gap, "The Goshen Pass." About 10 miles southwest are the Rock
10. a. b.

12. About three miles north of this, on the Cow Pasture River, is th
County, in an anticlinal of 8. Oriskany; and twelve miles farther north
river, is the noted intermitting stream called the Ebbing Spring, in a ridg
of Tower Hill, east of Warm Spring Axis. Twelve miles southwest to Ba
and thence 5 miles to Warm Springs, 3 c–4 a.

13. Where traversed by the Jackson's River, this anticlinal shows it
up of the successive concentric beds of 5 a. b. c., and flanked by 7. and
having a span, as measured by the highest sandstone bed, of about 3,300 f
Levant, or Medina, white sandstone, is regular and unbroken, but the out
up of the hard members of 5 b. c., are distorted and in part inverted on
where by a slight fault the beds of 7, pass suddenly from a nearly vertica
Towards the southwest, this axis opens to form the Rich Patch Valle
Siluro-Cambrian 4 a, b, c, and still farther southwest becomes the closed
Pott's Creek Mountain. Heavy beds of iron ore (Hematite) have been
this axis, as at Roaring Run, Callie's, Low Moor, and Kayser's near Clifto
formation 8. Oriskany. The fossil ore of 5 b. is also mined at several point

West Virginia.[23]

Ms.	Chesapeake & Ohio R.R.—Con.	Alt.
227	White Sulphur Springs. 1920	Devon., 10 a. & 10 b. Spring issues from 8.
238	Ronceverte. 1660	Lower Sub-Carb., 13 a. Vespertine.
244	Fort Spring. 1625	Upper Sub-Carb.,13 b. Umbral lim'tone.
251	Alderson. 1550	Upper Sub-Carb., 13 b. Umbral shale.
263	Talcott.	" 1510
272	Hinton[15] 1377	Upp. Sub-Car.,overlaid west by Congl. Coal group 14 a.
294	Quinnimont. 1196	Upper Sub-Carbon. shales, overlaid by Conglo. Coal group 14 a. The shales disappear west near Buffalo Creek.
324	Hawk's Nest. 828	Congl. Coal gr'p 14 a.
326	Cotton Hill.	" 796
333	Kanawha Falls. 672	Great Conglo. overlaid by Lower or main Coal group, 14 a. and 14 b.
352	Coalburg. 625	Main Coal group, 14 b.
359	Brownstown.	" 605
368	Charleston.	" 602
381	St. Albans. 594	Low. barren gr'p,14 b.
395	Hurricane.	" 583
401	Milton.	" 586
409	Barboursville.	" 580
416	Guyandotte.	" 580
421	Huntington.	" 565

Virginia.

Washington City, Virginia Midland and Great Southern Railroad, now Virginia Midland.

Ms.		Alt.
0	Alexandria.	20.Quat. drift on denu.
5	Alex. & Fred'b'g Crossing.	Upper Mesozoic, Jurasso-Cretaceous.
9	Springfield.	1. Archæan, C. 340
14	Burke's.	" A. 238
18	Fairfax.	" A. 182
21	Clifton.	" A. 170
27	Manassas Junct.	Mes.,17-16Jur.-Tri. 317
31	Bristoe.	" 190
34	Nokesville.	" 270
39	Catlett's.	" 250
41	Warrenton Junc.	" 265
44	Midland.	" 321
47	Bealton.	" 290
51	Rappahannock.	" 275
56	Brandy.	" 359
62	Culpeper. 403	" W. margin.
69	Mitchell's.	" 330
74	Rapidanne. 806	" S. margin.
79	Orange. 506	1. Archæan, B.
83	Madison. 395	Argil. Mic. & Hydro. Mic.Slates,with patches of Limestone &
89	Gordonsville. 495	
93	Lindslay's. 477	Steaschist E. of S.W.
96	Cobham. 401	Mt.,followed by Epidotic and Chloritic
102	Keswick. 436	
105	Shadwell. 308	Quartzites & Slates
110	Charlottesville.	of S.W.Mt. & thence
	450	W.byGneissoidGr'ts.
111	Lynchburg Junc.	1. Archæan, D.
119	Red Hill.	"

14. The Anticlinal Valley, which includes the group of thermals known as the Warm, Hot, Healing, etc., Springs, closes up about ten miles northeast of this, and its axis subsides towards the southwest in broad spurs which reach the river a few miles below Covington, in low arches of 7. and 8., overlaid by 10. The heated waters issue at numerous points throughout a distance of thirty miles; from Cambrian and Siluro-Cambrian rocks, 3. c., 4 a., usually inverted and often faulted along the west side of the valley, the eastern boundary of which it formed by the massive Warm Spring Mountain, 5 a. 5 b., dipping east, while its western limit consists of a narrow, broken ridge of the same formations in a vertical or inverted position. Stages to Healing, Hot and Warm Springs, severally 15, 19, and 22 miles. Near the first is the Cascade (200 feet) of Falling Spring Creek, which, cutting through the west wall of the anticlinal, flows over a mass of calcareous tufa, deposited from the waters.

The anticlinal of Peter's Mountain, rising a few miles northwest of Covington and exposing at the tunnel 7. and 8., expands towards the southwest, until it opens out into the valley of the Sweet Springs, containing another group of thermals of lower temperature than the preceding. This anticlinal, extending southwest, does not close up, but passes into the great Peter's Mountain and East River Mountain fault, which for a distance of fifty miles brings the Cambrian in contact with the Vespertine and Umbral formation, Sub-Carb., 13 a., 13 b.

15. The Upper Subcarboniferous, or Umbral Shales, here include a considerable thickness of brown and gray flaggy sandstone, the same which forms the hard rock of Swope's Knobs.

16. About 20 miles northwest of this point (by canal or road) we enter the gorge by which the James River traverses the Blue Ridge, where are exposed fine sections of Archæan rocks, A and B, and of the Cambrian, Primal 2 a., resting unconformably on the western slope of the former, and occupying the flanking ridges, which adjoin the valley. The Natural Bridge, the remnant of a former tunnel or cave in 3 a. b., is about 8 miles northwest from the upper end of the gap.

17. A few miles east of this, between Bannister and Dan Rivers, is a small patch of Jurasso-Triassic rocks, 18–17., corresponding to the Farmville or Middle belt, (see note 2), and containing Estheria, etc.

18. This deposit, made up largely of Diatoms, lies near the base, but within the limits, of the Miocene Tertiary. It contains occasional casts of Miocene shells, and is generally overlaid by beds of this formation, and rests either upon or but little above the top of the Eocene. Having formerly traced this deposit from the Patuxent River in Maryland to the Meherrin in Virginia, I have lately found by an examination of the artesian borings at Fortress Monroe, that a similar

Washington City, Virginia Midland and Great Southern R. R.—Con.			Richmond, Frederic| Rail|	
Ms.	Station	Alt.	**Ms.**	Station	
121	North Garden.	⎧ From one and a half		Washington.	
127	Covesville.	miles west of Char-		. (Steamboat.)	
131	Fabers.	lottesville to near	0	Quantico.	
133	Rockfish.	Lynchb'g the prev'l-			
137	Elmington.	ing rocks are Syen-	5	Richland.	
140	Lovingston.	⎨ ite, Granite, Protog-			
145	Arrington.	ine,Mic.Chlo.Gneiss.	12	Brooke's.	
149	Tye River.	Near base of S.W.Mt			
152	New Glasgow.	are belts ofGneiss'id	14	Potomac Run.	
157	Amherst.	sand and steaschist.	21	Fredericksburg.	
163	McIvor's.	⎩ Mic.&Hor.,Sl.&Tr'p.			
166	Burford's.	1. Archæan, C.	33	Guiney's.	
171	Lynchburg.[16]	" B. 529	42	Milford.	
177	Lucado. 838	⎧ Micaceous & Argil.	47	Penola.	
		Slates, includ'g pat-	53	Rutherglen. 203	
182	Lawyer's Road.	⎨ ches of Limestone &	58	Junction.	
	759	Steatite, Epidotic &	60	Taylorsville.	
188	Evington. 724	⎩ Chloritic Quartzites.			
192	Otter River.	1. Archæan, C. 665	65	Ashland. 221	
195	Lynch's.	" 730			
199	Staunton River.	"	82	Richmond.	
205	Sycamore.	" 238	84	Manchester	
209	Ward's Springs.	" 797		Crossing.	
		" 812			
215	Whittle's.		87	Temple's.	
220	Chatham. 624	⎰ Mesozoic, 17-16.Jur. ⎱ asso-Trias'c,W. mar.	90	Drewry's Bluff.	
226	Dry Fork.	" 624	93	Halfway.	
230	Fall Creek.	" 535			
236	Dundee.	1. Archæan, C. 413	95	Chester. 148	
237	Danville.	"			

Manassas Division.					
			98	Port Walthall J.	
0	Alexandria.	(As before.)	105	Petersburg. 70	
27	Manassas Ju.317	Mes.,17-16.Juras-Tria.	115	Ream's. 71	
36	Gainesville.	" 357	127	Stony Creek. 74	
38	Haymarket.	" 337	135	Jarratt's. 154	
40	Thoro'ghfare. 899	⎧ 1. Archæan,B, Slaty Quartzite, Epid. Chl.	147	Bellefield. 107	
		⎨ Argil.&Mic.Slates or	154	Greensville Jun.	
44	Broad Run. 395	Bull Run and Pond	164	Pleasant Hill.	
		⎩ Mountains.	168	Weldon. 105	
49	Plains.	1. Archæan, C. 565	**Piedmont Air**		
54	Salem.	" 633	0	Richmond. 83	
60	Rectortown.	" B. 444	2	R. F. & P. Junct.	
63	Delaplane.	" 455	22	Powhatan. 820	
67	Markham.	" 552	36	Amelia C. H.	
72	Linden.	" 916	58	Burkeville.	
76	Happy Creek.	" 790	73	Keysville.	
79	Front Royal. 540	Cambrian, 3 a. Calcif.	90	Roanoke.	
81	River. 493	Sil.-Camb.4a.&b. Tr. &	101	Scottsburg.	
85	Buckton. 508	Ut. 4 c. Hudson Riv'r.	109	Boston.[17]	
86	Water Lick. 550	⎰ Fort Mt. Synclinal ⎱ (5 a. & b.)ends near.	127	Barksdale.	
90	Strasburg. 487	" 4 a. & b. Tr.& Ut.	135	Ringgold.	
91	Strasburg Juc.	" " 694	141	Danville.	
			156	Ruffin, N. C.	

deposit exists in that region at the depth of 558 feet below the surface,
Pliocene beds, and resting upon an Eocene deposit identical with that wt
mond. We are thus assured of the great extension seaward of this depo
of estimating the thickness of the Tertiary formations as far east as
River.

Ms.	Richmond, York River and Chesapeake Railroad.	Alt.
0	Richmond.[18]	(Same as before.)
7	Fair Oaks. 163	At Richmond tunnel cutsTert'yInfusorial bed, 19 b. Miocene.
13	Dispatch. 67	In this interval both Lower andUpper 19.
15	Summit.	Tertiary are accessible above tide level.
20	Tunstall's. 80	Eocene andMiocene.
24	White House. 18	In this interval,only
26	Fish Haul. 44	Upp. 19. Tertiary is
31	Sweet Hall. 40	acces'ble above tide
38	West Point. 9	level. 19 b. Miocene.

Norfolk and Western R. R.

Ms.		Alt.
0	Norfolk.	20.Quaternary, resting on Upp.Tertiary 19 c. Pliocene.
23	Suffolk. 53	Up.19.Ter.& 19b.Mioc.
34	Windsor.	" 54
41	Zuni.	" 8
45	Ivor.	" 87
52	Wakefield.	" 100
60	Waverley. 114	Lower 19. Tertiary here probably above tide level.
68	Disputanta.	" 117
81	Petersburg. 9	E. marg. of 19. Tertiary & U.17-18Mes. resting on Gneiss, C.
96	Church Road.	1. Archæan, C. 303
101	Ford's.	" 307
108	Wilson's.	" 367
112	Wellville.	1. Archæan, A. 420
118	Blacks & Whites.	" 425
124	Nottoway C. H.	" 421
133	Burkeville.	" 523
141	Rice's.	" 396
149	Farmville. 316	16. Mesozoic, 17-16. Jurasso-Triassic.
161	Prospect.	1. Archæan, A. 575
169	Pamplin's.	" 673
181	Appomattox.	"

Ms.	Norfolk and Western R. R. Continued.	Alt.
191	Concord.	1. Archæan, B. 533
204	Lynchburg.	" 529
215	Forest.	1. Archæan, A. 877
229	Liberty.	" 959
241	Buford. 1014	2-4 Cambrian,3 a. Cal.
246	Blue Ridge. 1298	" 3 a. b.
251	Bonsack's.	" "
254	Gish's.	" " 932
252	Big Lick. 907	"&Sil-Cambr'n.
264	Salem.[19] 633	"3c&4aCh.& Tr.
277	Big Spring.	" " 1762
281	Allehany. 1260	" 3 b. c.
285	Big Tunnel.	" " 1930
290	Christiansb'g.[20]	" " 2012
301	Central.[65]	" " 1755
302	New River.	" " 1757
309	Dublin.	" " 2066
316	Pulaski.[66] 1919	Fault of Draper'sMt. Silurian & Devonian against Sub-Carbon.
329	Max Meadows.	2025	
337	Wytheville.[21]	2-4. Camb. 3 b. c. 2242
350	Rural Retreat.	" " 2575
364	Marion. 2136	"&Sil-Ca.,3c.&4 a.
380	Glade Spring.[22]	" " 2088
393	Abingdon.	" " 2069
408	Bristol, Tenn.	" " 1689

Continued as East Tennessee, Virginia & Georgia Southwestern Railroad.

Seaboard and Roanoke Railroad.

Ms.		Alt.
0	Portsmouth.	20. Quat. on 19. Ter. and 19 c. Pliocene.
17	Suffolk.	20. Quat. on 19. b. Mic.
31	Carrsville.	"
37	Franklin.	"
42	Nottoway.	"
50	Newsom's.	"
55	Boykin's.	"
63	Margaretsville.	"
68	Seaboard.	"
78	Gary's.	"
80	Weldon.	Outcrop of Gneiss.

19. From this point, for many miles towards the southwest, the railroad runs near to and almost parallel with the broken synclinal, (about 25 miles long), of which the lofty Catawba and Fort Lewis Mountains are the principal parts. The former, composed of southeast dipping 4 a. b., etc., forms the farther or northwest rim of the synclinal, and bending abruptly around at its northeast end, becomes the Tinker Mountain, which closes the basin in that direction. A shorter and gentler bend at the southwest end, terminates in a fault. The corresponding rocks of the southeast, or near side of the synclinal, are only partially preserved in a narrow inverted ridge at either end, the remainder of this rim of the synclinal having been engulfed in the prolonged fault, which, for many miles along the margin of the basin, has brought the Siluro-Cambrian rocks (4 a. c.) of the valley to abut against, and over-ride the Devonian 10. to 12. and the Vespertine 13 a., of which the Fort Lewis Mountain, the central mass of the synclinal, is mainly composed.

20. A few miles west-by-north of this is an area of Vespertine rocks, 13 a., including one or more workable beds of coal, mined on Strouble's Run and elsewhere. This area once probably continuous with the Vespertine of Fort Lewis Mountain, is almost encompassed by faults. Farther to the northwest, and separated from the above by a belt of Cambrian and Siluro-Cambrian rocks 3 c., 4 a., etc., the Vespertine beds of the southeast slope of the Brushy Mountain, contain a similar coal, mined on Tom's Creek, etc., all these seams being more or less affected by the neighboring faults. The dislocation which, southeast of Brushy Mountain, brings Vespertine and Umbral in apposition with Siluro-Cambrian Matinal, is part of the great fault which, with some changes of direction and character, extends along the northwest edge of the great valley, from near the James River to the end of the Brushy Mountain, northeast of Abingdon, a distance of about 125 miles.

Washington, Ohio and Western Railroad.				Washington, Ohio and Western Railroad.—Con.		
Ms.			Alt.	Ms.		Alt.
0	Alexandria.	(Same as before.)	17	27	Guilford.	415 { Mesozoic, 17-16 Jurasso-Triassic.
7	Carlin's.	"				
11	Falls Church.	1. Archæan, C.		31	Farmwell.	" 820
15	Vienna.	1. Archæan, A.	395	38	Leesburg.	321
18	Hunter's.	"	345	42	Clark's Gap.	" W. mar. Cong.
21	Thornton.	1. Archæan, B.		45	Hamilton.	1. Archæan, B. 578
23	Herndon.	{ Mesozoic, 17-16. Jurasso-Triassic. 395		49	Purcellville.	" 454
				52	Round Hill.	" 553
						" 558

At a distance of 23 miles, in a northwest direction, is the sheet of water called "Mountain Lake," situated near the top of Salt Pond Mountain, at a height of 4,000 feet above tide. Here the Potts and Johns Creek Mountains and the other ridges of 5 a. b. coalesce at their southwest termination, into a lofty rugged table-land, overlooking the New River, and commanding wide views.

21. A few miles south, the Lick Mountain range divides the valley for some miles into two and in the southern of these belts, on the New River, below the mouth of Cripple Creek, are the Austenville lead mines, in 3 b., near the Primal 2 b. of Popular Camp Mountain, and about 15 miles distant from Wytheville.

22. From this point a short branch railroad leads north into the valley of the north fork of the Holston River, between Walker's Mountain, 5 a, etc., and Poor Valley ridge, Vespertine 13 c., etc., which flanks the Clinch Mountain on the southeast side. Here, near Saltville, are the remarkable salt wells, which penetrate into a thick mass of rock-salt; and in the same vicinity, and at various points higher up the valley, for a distance of 20 miles, beds of gypsum have been opened and extensively wrought. These deposits are found near and in a line of fault, along which the Siluro-Cambrian 3 c. 4 a., of the southeast side of the valley, has been made to abut against and sometimes over-ride the Umbral 13 b., which, with the Vespertine 13 a. of the Poor Valley Mountain, form a belt on the northwest side of the valley. Both deposits are most probably referable to the Subcarboniferous period. The fault here spoken of extends, with some local changes of character and direction, in a west-by-southwest course, from a point in Giles county to the Tennessee line, a distance of 125 miles, and is prolonged many miles into Tennessee.   WILLIAM B. ROGERS.

23. So few details have been published on the geology of Virginia, that no chapter in this volume will be more welcome to geologists than this, which has been wholly and very carefully prepared by Professor William B. Rogers, late State Geologist of Virginia.   J. M.

NOTE TO THE SECOND EDITION:—The first seven pages of this chapter are from the first edition without material change, except the addition of the altitudes. The larger portion of the Baltimore and Ohio is given again in the succeeding pages, with notes by Prof. J. L. and H. D. Campbell, and the portion of the Chesapeake and Ohio in West Virginia, will be found more fully described in the chapter on that state.

Chesapeake & Ohio Railroad.[*] Peninsula Extension.

Ms.			Alt.
0	Richmond[24]	(Same as below.)	44
2	Orleans Street.	{ 20. Quaternary and	
		{ 19. Tertiary.	83
18	Roxbury.	{ 20. Quaternary and	
		{ 19 b. Miocene.	31
24	ProvidenceForge.	"	29
32	Lanexa.	19 b. Miocene.	21
38	Toano.	{ 20. Quaternary and	
		{ 19 b. Miocene.	101
48	Williamsburg.	19 b. Miocene.	66
57	Lee Hall.	20. Quaternary.	35
69	Morrison.	"	35
75	Newport News.	"	5

Baltimore & Potomac Railroad.[*]

Ms.			Alt.
0	Washington.	{ 20. Quaternary, and	
		{ 17. Jurassic,	
		{ 18. Cretaceous.	
2	Long Bridge.	"	
7	Alexandria.	"	35
13	Franconia.	{ 17. Jurassic.	
		{ 18. Cretaceous.	234
17	Long Branch.	"	82
24	Woodbridge.	"	73
30	Cherry Hill.	"	7
34	Quantico.	"	16
116	Richmond.	{ Junction of 1. Lau-	
		{ rentian, 17. Juras.,	
		{ 18. Cretaceous, and	
		{ 19. Tertiary.	64

Brighthope Railway.[*]

Ms.		Alt.
0	Winterpock.	17. Jurassic, 16. Trias.
8	Summit.	{ Margin of 7. Juras.,
		{ Triassic, and 1.
		{ Laurentian.
14	Fendley.	1 a. Laurentian.
22	Chester.	{ 20. Quaternary,
		{ base of Eocene
		{ near by. 143
33	Bermuda.	20. Quaternary.

Richmond & Alleghany Railroad. †

Ms.		Alt.
0	Richmond.[24]	{ W. margin Tertiary,
		{ Mesozoic, 18., 19.[83]
5	Korah.[25]	1 a. Granite. 106
7	Westham.	" 116
12	Lorraine.	17. Jurassic Coal. 142
13	Vinita.	17. Mesozoic. 142
17	Manakin.	" 141
19	Boscobel.[26]	17. Nr. marg. Meso.[143]
20	Dover.	" 143
25	Lee's.	1 a., 1 b. Archæan.
30	Maiden's Ad.[27]	{ 1 a. In River.
		{ 1 b. On Hills. 143
33	Cedar Point.	" 159
34	Irwin.	" 159
40	Rock Castle.	" 175
42	Stokes.	"
47	Pemberton.	" 190
52	Elk Hill.	" 198
54	Elk Island.	" 198

* By Professor William M. Fontaine, of the University of Virginia.
† By Professors J. L. and H. D. Campbell, of Washington and Lee University, Lexington, Va.

24. *Richmond* is on the west margin of the Mesozoic and Tertiary belt. (See Rogers Note 18.) These formations may be seen in railway cut near Tredegar Iron Works, at the York River Railway station, and on the margin of Shocco Creek, near the Medical College. The bed of the river is gneissoid granite at the city, and for several miles above.

25. At *Korah* large quantities of granite, doubtless of Laurentian age, are quarried for shipment. Another large quarry is opened opposite Westham, on south side of the river. Between Westham and Lorraine the road passes from the Archæan to the Mesozoic coal-bearing beds (17, 18), and continues on them for about 10 miles to Dover.

26. *Boscobel*, or Dover, near the west margin of the coal field, is near the old Dover Mines. Fossils in the debris of the coal slates.

27. Between this point and Goochland C. H., a mica mine was formerly worked (in 1 b.), but not exhausted.

[N. B.—In our notes on the Archæan rocks, we recognize only *Laurentian* (1 a.) and *Huronian* (1 b.); and even the horizon between these is uncertain in this part of Virginia.]

28. At *Columbia* a granite quarry is worked in 1 a., overlaid by mica and hydro-mica slates and schists of 1 b. This is the best point from which to visit the several gold mines in the vicinity.

29. *Bremo Bluff* is a good point of departure for examining several objects of interest. (*a*) "The Bluff," near the station, is apparently a closed anticlinal fold of beds of hard gneissoid sandstone and arenaceous slates, nearly vertical in position. A second bluff of the same general structure occurs about 200 yards farther up the river. The syncline between them and outside flanks of both are occupied with argillaceous slates. The same ledges appear on the opposite side of the river. (*b*) At this point a branch (Buckingham Branch) railway crosses the river to extensive slate quarries, about five miles distant, and apparently in the same formation (I b.) as the slates about the "Bluff." Future explorations may modify this view. (*c*) Willis Mountain, about 20 miles east of this station, is an isolated mass of gneissoid rocks, containing numerous crystals of kyenite of different shades of color, and of hornblende and tourmaline, with other minerals. (*d*) This is one of the best portions of the gold belt. Iron ores—limonite, hematite and magnetite—abound here.

30. From Richmond to *Scottsville* the road cuts the strata by a route generally at right angles, or nearly so, to their strike; and for several miles below the town the outcroppings, mostly of 1 b., show frequent changes of dip, and are occasionally nearly horizontal. The route here changes towards the southwest.

Ms.	Richmond & Alleghany Railroad—Continued.	Alt.	Ms.	Richmond & Alleghany Railroad—Continued.	Alt.
57	Columbia.²⁸ { 1 a. Granite, 1 b. / Mica Shists. / Gold Belt.	206	131	Stapleton.³² 1 b. L. S. Spec. Ore.⁴⁴⁷	
63	Boswell. 1 a., 1 b. Archœan. ²¹⁵		133	Galtville. { Mica Schists, Spec. / Ore.	⁴⁵⁵
67	Bremo Bluff.²⁹ { 1 b. Gneissoid Sand / s. and Slates. ²³¹		136	Joshua Falls. { 1 b. Archœan, Lime- / stone and Ores.⁴⁵⁵	
70	Middleton Mills. 1 b. Archœan. ²³¹		147	Lynchburg.³³ { 1 a., 1 b. Gneiss, / Mica, Slate.	⁵²⁹
73	Hardware. { 1 b. Archœan, Schists / and Slate. ²⁶⁶		148	Va. Mid. Junc. "	⁵²⁹
75	Payne. " ²⁶⁶		149	Smith's Lock. "	⁵¹⁶
80	Scottsville.³⁰ " ²⁷⁵		151	Rolling Mill. 1 a. b. Archœan.	⁵³⁰
83	Brown's. 16. Marg.Mesozoic.²⁹³		159	Bethel. "	⁵⁴⁸
86	Warren. 16. Mesozoic. ²⁹⁹		159	Holcomb Rock. "	⁵⁶²
91	Howardsville.³¹ " ³¹⁵		.....	Pedlar's. "	
96	Manteo. 1 b. Archœan. ³²⁴		161	Coleman's Falls. "	⁵⁷⁸
99	Warminster. { 1 b. Archœan, Lime- / stones & Schists.³³²		166	Big Island. "	⁵⁹⁶
102	Wingina. " ³⁵⁰		.....	Jordan. 1 a. and 2. a b. Margin.	
105	Norwood. "		170	Rope Ferry.³⁴ { 2 a. b. Cambrian, / (Potsdam) Sand- / stone, Slate.	⁶⁶⁸
109	Buffalo Springs. "		175	Balcony Falls.³⁵ "	⁷⁰¹
114	Greenway.³² { 1 b. Limestone, Spec. / Ore. ³⁸³		178	{ Glenwood.³⁶ / Nat. Bridge. } 3 b. L. Silurian.	⁷¹⁵
118	Gladstone. " ³⁹⁹		189	Indian Rock.³⁷ { 3 b. L. Silurian, / near 4 a.	⁷⁸⁰
123	Riverville. " ⁴²³				

31. About three miles below *Howardsville* the river and road cut into the lowest beds of a Mesozoic trough, or oval basin, that covers several square miles of area, the larger portion on the north side of the river. The remarkable coarse conglomerate that forms the base of this series of rocks is well exposed in contact with Archæan rocks along the banks of Rockfish River, near the station, and along a little stream running through the neighboring village, while the overlying ferruginous sandstones and slates appear in the surrounding hills. After passing this Mesozoic tract, the route, following the windings of the James River, keeps within the general trend of a belt four or five miles wide, in which are several beds of limestone and ores of iron imbedded in still heavier strata of micaceous, talcose and chloritic slates and schists, all most probably of Huronian age. After following this limestone and ore belt for about 40 miles, the bearing is abruptly changed toward the northwest about six miles below Lynchburg.

32. At points between *Greenway* and *Stapleton* numerous ore mines and limestone quarries have been opened on both sides of the river.

33. At *Lynchburg* the river has cut the beds (1 a. and b.) nearly at right angles, so as to expose a well-defined waving arch on the cliff opposite the city. For about 20 miles above the city the road continues on the gneisses, granites and slates of Archæan age.

34. At about a mile below *Rope Ferry* is the margin of a belt of alternating conglomerates, sandstones and slates about two miles wide, which were formerly classed as Huronian by Rogers and others. This belt flanks the southeast slope of the Blue Ridge, and is cut by the river so as to give fine exposures of its beds both above and below the railway bridge. The discovery we recently made of *scolithus* borings of the kind characteristic of Cambrian (Potsdam) sandstones in its beds, determines its age to be Cambrian. The "Snowdon Slate Quarries" are in this Cambrian belt three miles towards the northeast.

35. At *Balcony Falls*, between one and two miles below the station, the river has cut obliquely through the core of the main Blue Ridge and exposed a fine section of Archæan rocks. These have been formerly spanned by the Cambrian beds, the upper portions of which were doubtless ruptured at the time of the upheaval and swept away. At this point occurs the finest natural section of the whole Cambrian series to be found anywhere in Virginia. The alternations of conglomerates, shales and sandstones present an aggregate thickness of about 1,200 ft. The uppermost sandstone, about 350 ft. thick, is the typical Potsdam, and abounds in borings of *scolithus linearis*, thousands of which may be seen in the broken rocks at the junction of the Lexington branch, 150 yards above the station house. Here the road enters the Great Silurian Valley.

36. *Glenwood* is the station for stage line to Natural Bridge. (See Note 16.) The road here passes through a depression in the Sallings Mountain, an anticlinal ridge of primordial strata, 2 a. b. The *Natural Bridge*, three miles from this station by stage line, is in Lower Silurian limestone; the abutments in Quebec (3 b.); the arch and the adjacent hills in Chazy (3 c.) This great natural curiosity has been supposed by some observers to be the remnant of a natural tunnel, and by others the remains of an extensive cave, the top of which has all fallen in and been washed away except the narrow arch that now spans the chasm. Our belief is that it has resulted from a vertical fissure in the beds of limestone, which, by its opening, failed to rupture the portion of the uppermost beds that now forms the arch, but simply dragged them a few yards toward the west and left them stretched across the deep chasm, which has been subsequently enlarged by erosion. The entire absence of stalactites and stalagmites along the faces of the cañon militates strongly against the cave theory, while the secondary fissures still to be seen just above the bridge, together with the general

Ms.	Richmond & Alleghany Railroad.—Continued.	Alt.
195	Buchanan.[88]	8 b. L. Silurian.	887
200	Jackson.	"	845
203	Glen Allen.	"	855
205	Saltpetre Cave.	3 b. c. "	892
208	Salisbury.	{ 3 b. c. L. Silurian. (Iron Furnace.)[894]
212	Eagle Rock.[39]	4 a., 4 b. Trenton.	936
216	Gala Water.	10 a. Devonian.	936
.....	Ore Siding.	"
.....	Price's Bluff.[40]	Arch of 7 and 8.
.....	Hadons.	10 a. b. Devonian.
221	Baldwin.	"	970
224	Wilton.	{ 10 a. b. Devonian. (Princess Fur.)[99]
226	Lick Run.	10 a. b. Devonian.
228	Iron Gate.	"
230	Clifton Forge[41]	"	1052

Lexington Branch.[*]

Ms.		Alt.
0	Balcony Falls.	(See above.)	701
5	Miller.	2 b., 3a. Nr. Margin[725]
10	Loch Laird.[42]	3 a.,3 b. L. Silurian.[784]
12	Green Forest.[43]	3 b. L. Silurian.
16	South River.	3 b., 3 c. L. Silurian[850]
19	E. Lexington, jun. of Valley Ry.	4 a. Trenton.	910
20	Lexington.[44]	"	1000

Ms.	Richmond & Alleghany R. R.—Con. Henrico R. R. Branch.	Alt.
0	Lorraine.	17. Jurassic Coal.	142
7	Henrico.[45]	{ 17. Jurassic Coal. (Coal mine.)
11	Hungary.	{ Archæan, near margin Tertiary. [214]

Ms.	" Shenandoah Valley Railway.[*]	Alt.
0	Hagerst'n, Md.[47]	{ 4 a. Trenton, dip S. E.	566
6	St. James.	"
9	Grimes.	3 c., 4 a. Nr. Margin[332]
14	Antietam.	3 b. Siluro-Cambrian.
17	Shep'n, Va.[48]	"
23	Shenandoah Jun.	3 b. c. "
29	Charlestown.	"
34	Ripon.	"
37	Fairfield.	"	523
40	Berryville.	"	571
47	Boyce.	"	575
50	White Post.	"	510
54	Ashby.	"	600
57	Cedarville.	"	569
60	Riverton.[49]	"	497
62	Front Royal.	{ 2 b. Cambrian and 3 a. Calcif.	495
67	Manor.	{ 3 b. c. Sil.-Camb., dip changes to N.W.[497]
73	Bentonville.	"	732

* By Professors J. L. and H. D. Campbell, except those notes marked "M," which are by Dr. A. S. McCreath, Chemist of the Second Geological Survey of Pennsylvania.

appearance of the place seem to favor the view here proposed. On the opposite side of the river are the Glenwood Iron Mines of Judge Anderson.

37. *Indian Rock.* Trenton limestone, gray coralline, quarried largely here for lime.

38. Purgatory Mountain terminates abruptly near *Buchanan.* It is a somewhat isolated outlier of North Mountain. Its base is Trenton limestone (4 a.), its main mass Utica and Hudson shales (4 b. and 4 c.), while its cap is Medina (5 a.); and in a synclinal trough held in a position where its top is double, it carries fine beds of limonite and red shale ores.

39. From *Buchanan* to *Eagle Rock* the limestones of 3 b. and 3 c. are exposed to view in several cuts, and at Eagle Rock they disappear beneath the groups of Trenton (4), of Medina (5), Salina? (6), Lower Helderberg (7), Oriskany (8), Marcellus, etc. (Devonian slates, 10 a. and 10 b.) The mountain at this pass is a prolongation of North Mountain, and has its higher members partially inverted, a feature very characteristic of this range throughout the greater portion of Virginia. The road here passes into a synclinal valley with Helderberg (7) and Oriskany (8) for its bottom, and most of its surface covered with Devonian slates, 10 a. b.

40. *Price's Bluff* is an anticlinal arch of 7 and 8, and furnishes good limestone and ore of iron.

41. *Clifton Forge* is a point of great interest to geologists. (See Rogers Note 13.)

42. *Loch Laird.* A small bed (or dike) of trap between two beds of calcareous shale (3 a.) may be seen 100 yards above the Shenandoah Valley Railway junction.

43. *Green Forest* is the station for the extensive Buena Vista Iron Mines, in the primordial (2 b.) shales at the northwestern base of the Blue Ridge.

44. For *Lexington* and its surroundings, see note No. 74.

45. *Henrico* Coal Company's station for shipping coal and coke.

46. This road, throughout its whole length of 240 miles, runs on the Siluro-Cambrian and the Cambrian formations, chiefly on the former.

47. *Hagerstown* stands on what seems to be the eastern portion of a closed and inverted syncline of Trenton age; the axis in the shales farther west. The Trenton limestones crop out near both of the depots, and are quarried for local building purposes. The road continues on this formation for several miles, but soon after passing Grimes it runs obliquely across the margin to 3 b. c.

48. At *Shepherdstown* are extensive exposures of 3 b. on the margins of the Potomac. Hydraulic limestone has been extensively quarried here for the manufacture of cement.

49. Between *Riverton* and *Port Republic* the Massanutten range of mountains is conspicuous on the northwest side of the road. (See Rogers note 7). The Blue Ridge is seen from the train on the southeast at nearly all points along the whole line. Over a large portion of the route the country rocks are very much obscured by the local drift from the adjacent mountains. In the larger boulders from the Blue Ridge, the burrows of the *scholithus linearis* are abundant.

Ms.	Shenandoah Valley Railroad—Continued.	Alt.
76	Overall.[50] 3 a. Near Sil.-Camb., dip ch. to N.W.	662
80	Rileyville. 3 a. Calcif.	726
85	Kimball. "	895
89	Luray.[51] Sta. on 3 b. entrance to cave on '3 c.	822
96	Marksville.[52] 2 b. Spur of Cam.	1066
102	Ingham. 3 b. c. Sil.-Cambrian.	
104	Grove Hill. "	966
107	Milnes.[53] "	
113	Elkton. "	958
128	Port Republic.[49] "	
129	Weyers Cave.[54] 3 a. b. Sil.-Cambrian cave in 3 b. c.	1123
132	Patterson. "	1135
137	Crimora.[55] "	1242
144	Waynesboro Jun. Margin of 2 b., 3 a.	1298

(Much obscured by drift and alluv. — brace spanning rows 104–128)

Ms.	Shenandoah Valley Railroad—Continued.	Alt.
148	Lyndhurst. Obscured by drift, etc.	1340
151	Lipscomb.[56] "	
153	Stuart's Draft. "	1388
160	Greenville. 3 b. c. Sil.-Camb., drift high on hills.	1550
163	Lofton.. "	
168	Vesuvius.[57] 3 a. Sil.-Camb.	1420
173	Marlbrook. 3 a. b. '	1165
175	Midvale. Bed of Tufa., cut by railroad.	
177	Irisk Creek.[58] 3 a. b. Ore in 2 b., 3 a. b. Sil.-Cam.	1010
180	Riverside. "	938
186	Loch Laird.[59] 3 a. near 3 b.	800
189	Thompson.[60] 3 b. Sil.-Camb.	790

50. *Overall.* Half a mile east of Overall station, Umber deposit, which has been partially developed. (M.)

51. At *Luray*, the station, the junction, and the greater part of the village, appear to rest upon the ledges of 3 b., Quebec (Levis), dipping 20° to 30° northwest, and passing beneath a ridge of 3 c. (Chazy), in which is the entrance to the caverns; and most probably the higher chambers are in the same formation, while the lower ones are either within or rest upon beds of 3 b. Everywhere in the great valley of Virginia the limestones of the Quebec, as a rule, are much more ferruginous than those of the Chazy, and consequently produce darker and more fertile soils. The Quebec also carries several thick beds of shale, while the Chazy is characterized in many places by beds of chert that contain characteristic fossils. The lithological peculiarities of these two formations, especially those which determine differences of soils, are well defined at Luray. (See note 75.)

52. *Marksville.* Considerable deposits of light brown ochre worked here by Oxford Ochre Company.

53. *Milnes.* About five miles south southeast of Milnes there is a fine exhibition of the Potsdam ores (in the slates above the Potsdam sandstone), the principal development being on Fox Mountain, a low flat crested ridge, a foot hill of the Blue Ridge. The present working face is 85x300 ft., and the daily output is over 100 tons, shipped over the branch road to the Shenandoah Iron Co.'s furnace, near Milnes. (M.)

54. *Weyers Cave* has the same geological relations as the Luray Cave, except that it is nearer the margin of the Trenton trough, which carries the Massanuttens, and here extends to the southwest beyond the termination of the mountain range.

55. *Vesuvius.* The *Rockbridge* tin mines are in the Archæan core of the Blue Ridge, and may be reached by ordinary road, from either Vesuvius or Irish Creek Station.

55. *Crimora.* Two miles east from Crimora there is a large valuable deposit of Manganese ore, chiefly pyrolusite. The ore is very rich, and is now being mined in quantity for shipment to England and to Pittsburg, Pa., at the latter place for use in the production of a remarkably high grade of ferro manganese. (M.)

56. *Sherando.* Near Sherando (Lipcomb Station), deposits of China Clay and Fire clay are being worked. (M.)

57. *Vesuvius.* Eight miles southeast of Vesuvius Station, and on a bank of Irish Creek, there is quite an interesting exhibition of tin ore. The ore is Cassaterite; and at one point on the Cash property the ore showed remarkably rich, at times being almost pure Cassaterite, and some of the specimens showing one to one and a half inches in thickness of the pure ore. (See page 134 McCreath's Mineral Wealth of Virginia). Occasionally the tin ore has associated with it the mineral *Mispickel*, carrying more or less silver and gold. On the Vesuvius furnace property, and two and a half miles from the railroad, occurs a bed of brown hematite ore, ten feet wide, between nearly vertical walls of Potsdam sandstones. (M.)

58. Near *Irish Creek* a remarkable deposit of Dufrenite (Hydrated Ferric Phosphate), nearly a foot thick, of nodular and radiating structure, was found several years ago in the Potsdam shales, resting on a heavy bed of limonite ore. (See American Journal of Science, July 1881, pp. 65, etc.)

59. At *Loch Laird*, about sixty yards northeast of the crossing of the Richmond & Alleghany Railway, a *trap dike* about six feet thick may be seen thrust up between two beds of calc-shale of 3 a.

59. *Loch Laird.* On the Buena Vista property there is a fine exhibition of the Potsdam ores (in the slates overlying the Potsdam), showing perhaps the finest development of these ores in the Shenandoah Valley. On the same property where Marl Branch crosses the Lexington Turnpike, there is exposed a bed of so called Marl, fully 40 ft. thick. It yields over 95 per cent. carbonate of lime. (M.)

60. At *Thompson* is an old cement quarry.

61. *Arcadia.* Near Buchanan, on the Arcadia furnace property, there are numerous openings made on the so-called specular ore of the Blue Ridge. The ore is a red hematite, more or less intimately mixed with fine grained quartz. Geologically it lies in the slates underlying the Potsdam sandstone. (M)

62. *Lithia* is near the border of the extensive Cloverdale iron property; ore in 2 b. and 3 a.

Ms.	Shenandoah Valley Railroad—Continued.		Alt.
191	Buffalo Forge.	3 b. Sil.-Camb.	753
199	Natur'l Br.[18 & 36]	{ Station 3 a. b., Bridge 3 b. c.	
209	Arcadia.[61]	2 b. nr. 3 a. Camb.	796
215	Buchanan.	3 b. c. Sil-Camb.	837
220	Lithia.[62]	"	963
225	Houston.[63]	{ 3 a. " 1348	
228	Troutville.	Ore of 2 b. near.	
233	Cloverdale.[64]	} See note. 1125	
237	Tinker Creek.	961	
240	Roanoke.	{ 3 b. c. Sil.-Camb., nr. Trenton 4 a. 907	

Norfolk & Western Railroad.

Ms.			
283	Central [65]		
298	Pulaski.[66]		

Baltimore & Ohio Railroad.
Harper's Ferry and Valley Branch. *

Ms.			Alt.
0	Harper's Ferry.	{ 2 b., 3 a. Altered Cambrian (b) or Archæan B, followed west by Cambrian.	277
1	Shenandoah		
6	Halltown.	3 a. b. Cambrian.	339
10	Charlestown.	3 b. c. "	513
14	Cameron.	"	547
23	Wadesville.	4 a. b. Sil.-Camb.	495

Ms.	Baltimore & Ohio Railroad—Con. Harper's Ferry and Valley Branch.		Alt.
27	Stephenson's.	{ 4 a., 3 c. Siluro- Cam., and Cam.	499
32	Winchester.	{ The road runs	717
36	Kernstown.	close to bound-	744
39	Newtown.	ary of Cam., 3 c.,	770
42	Vaucluse.[7]	and Sil.-Cam., 4	
44	Middletown.	a., of belt lying	700
46	Cedar Creek.	east, composed	695
50	Capon Road.	largely of 4 c.	740
51	Strasburg Junc.	{ 4 a. b., Sil.-Camb, on switch track.	703
55	Tom's Brook.	3 b. c. Cambrian.	745
57	Maurertown.	"	768
61	Woodstock.	"	820
66	Edinburg.[67]	"	845
74	Mount Jackson.	{ 3 c., 4 a. Camb., and Sil.-Cambrian.	916
81	New Market.	"	971
88	Broadway.[68]	"	1038
94	Linville.	4 a. Trenton.	1242
100	Harrisonburg.[3]	4 a. and 3 c.	1340
105	Pleasant Valley.	3 b. c.	1245
106	Mt. Crawford.	3 b. c.	1172
112	Weyers Cave.[54]	3 b. c. nr. 4 a. S. E.	1155
115	Mt. Sidney.	4 a. near 3 c.	1257
117	Fort Defiance.	{ 4 a. nr. 3 c. Grapto- lites in Tr. sha.	1275

* From 88 Broadway, South, by Profs. J. L. and H. D. Campbell; north of that by Prof. W. B. Rogers.

63. *Houston.* Near Houston Station are the Houston Mines of the Crozer Steel and Iron Co., extensively worked to supply their furnace at Roanoke. Rich Manganese ore is also mined here and shipped to Johnstown and Pittsburg. (M.)

64. Between *Cloverdale* and *Tinker Creek* the road skirts the northwest base of a Trenton ridge, capped with 5 a. b. sandstones. It is known locally as Mill's mountain; an outlier of Tinker Mt.

65. The New River Division of the Norfolk & Western starts from *Central,* and has its present terminus at Pocahontas, where it strikes the great Flat Top coal field. It passes through a very interesting geological field. At Ripplemead Station there is a promising deposit of Magnetic Iron ore, in the No. 3 Lower Silurian Limestone opened up on the bank of New River. Some 5,000 tones of 63 per cent. ore have been taken out. (M.)

66. The "*Cripple Creek*" extension of the Norfolk & Western Railroad (now being built) starts from *Pulaski,* and will open up the Cripple Creek region (see note 21 on Virginia), with its vast stores of brown hematite ores in 3 b. and c. (and 2 b.), perhaps the finest and richest, and most uniform quality of (3 b. c., Lower Silurian) brown hematite ores in the United States. It will also bring within railroad communication (for the railroad will pass close to it) the 100 year old lead mine at Austinville, and the Bertha Zinc mine near New River, showing rich Zinc ore (Silicate and Carbonate of Zinc) almost free from lead, and now used at the Bertha Zinc Works), at Pulaski (Martins). Near Blue Ridge, and also near Roanoke (about two and a half miles south of it), important and seemingly very large deposits of Potsdam ores are now being mined at the former point, by the Crozer Iron and Steel Company, of Roanoke, and at the latter by Roxer Iron Company.

From eight to ten miles south southeast of Bristol there are interesting deposits of hematite ore in the No. 11 limestones. These were opened, many years ago, to supply stock for the local charcoal furnaces, but the ores were found too refractory for economical use in such furnaces, and the workings were abandoned. The ore is a dense and fine grained hematite, and shows 64 to 66 per cent. iron and .020 and .030 of phosphorus. (M.)

67. *Edinburg* is the depot for the Liberty and Columbia furnaces, a few miles northwest, in the North Mountain range—good geological field.

68. *Broadway* is a good starting point for studying geology, etc., of Brock's Gap, an interesting region in North Mountain range.

69. *Staunton,* a flourishing little city at the junction of the valley railroad with the Chesapeake & Ohio, is situated on a number of somewhat distinct hills, and surrounded by others of still greater height. These are composed chiefly of Quebec (3 b.) magnesian limestones at their bases, especially on the northwest flanks, and Chazy limestones of lighter color above, with interbedded cherty masses, the fragments of which are seen strewn over the surfaces in great profusion. Several species of gasteropod and cephalopod shells have been found fossil in these chert beds. The northeastern margin of the city rests on Trenton, 4 a., adjoining 4 c.; but the line of contact of these formations sweeps around the southeast and south flanks of two very conspicuous hills, known as "Betsy Bell" and "Mary Gray," and appears again on the valley road near Folly Mills Station, and continues near the line of road for several miles. (See Note 75 as to the Quebec group.)

Ms.	Baltimore & Ohio Railroad—*Con.* Harper's Ferry and Valley Branch.	Alt.	Ms.	Baltimore & Ohio Railroad—*Con.* Harper's Ferry and Valley Branch.	Alt.
119	Verona.	4 a. Tr.-Cal. shales[1310]	144	Raphine.[71]	{ 3 b. c. Iron Ore in 3 c. 1855
126	Staunton.[69]	{ 4 a. at N. E. corner, 3 c. Chief Rocks, 3 b. west margin of city. [1366]	149	Fairfield.[72]	{ 3 b. c. Iron Ore in 3 c., Houston's.[1780]
131	Folly Mills.	{ 4 a. near junc. with 3 c. [1490]	154	Timber Ridge. [73]	3 c. [1434]
133	Mint Spring.	" [1565]	160	{ R. & A. Junc., E. Lexington.	{ 4 a. Trenton limestone forms high river cliffs. Drift on hills. [910]
138	Greenville.[70]	{ 3 b. c. Iron Ores in Cambrian of Blue Ridge, S. E. [1600]	162	Lexington.[74]	{ 4 a. b. on south, 3 c. west of town. [1000]

70. Near *Greenville* the Quebec (3 b.) limestones, producing ferruginous clay soils, crop out in the cuts for a mile northeast of the town, and along the banks of the adjacent stream both above and below the crossing; but the Chazy beds form the country rock of the town and region between it and Raphine Station. The Primordial (Cambrian) ridges of the Blue Ridge range extend much farther into the Great Valley opposite Greenville, than they do at any other point seen from the line of this road, and carry some productive beds of limonite ore.

71. About 2½ miles northwest of *Raphine* Station are very extensive beds of limonite ores on the lands of Samuel Carson, Esq., and Messrs. Gibbs & Rawlings. The beds of ore have been partially opened, and, where seen in place, appear to occupy about the same relative position among the Chazy (3 c.) limestones as the chert beds found in such abundance in other parts of the same formation. The Vesuvius Iron Mines are in 2 b., about four or five miles southeast of this station. The tin mines, now in process of development, are in the Archæan core of the Blue Ridge, about 12 miles southeast by turnpike.

72. At *Fairfield* the road crosses to the west side of Timber Ridge, and on the northwest margin of the valley, the elevated outliers of the North Mountain range—the Jump, the Hogback and House Mountains—become conspicuous features of a striking landscape.

73. From *Timber Ridge* Station a line of conveyances extends to Rockbridge Baths, a pleasant summer resort. The thermal water of these baths issues from the Quebec (3 b.) limestones near a fissure or fault where the beds of 4 a. Trenton have dropped down to the level of 3 b., and apparently dip beneath the latter, as may be seen at points northeast and southwest beyond the accumulations of river drift, which is found on hills here more than 100 feet above the bed of the river. About two miles northwest of the baths is the entrance to the famous "Goshen Pass," the deep cañon through which North River finds its way to the Great Valley This cañon gives a complete section of the whole North Mountain range from 4 a. Trenton up to Devonian shales, 10 a. b. Fossils are abundant here. For sketch and geological section, see Am. Jour. of Sci., Vol. XVIII., 1879, p. 119.

74. About one mile southwest of *Timber Ridge* Station the railway passes abruptly from the Chazy (3 c.) to the Trenton (4 a.), entering the irregular synclinal trough in which *Lexington* is situated. In the town, along the cliffs of the adjacent north branch of James River, and over about six miles of area towards the northeast and four miles southeast, the Trenton limestones (4 a.) are the country rocks; but in the Poplar Hills toward the southwest and south, the Utica shales, with very fossiliferous thin beds of limestone, become conspicuous. The Brushy Hills, west of the town, are composed of Chazy limestones and cherts (3 c.), as regards their southeastern slopes, while the northwestern slopes present exposures of 3 b. dipping beneath the hills. As far as measurements can be made here 3 c. is about 300 feet, and 3 b. about 450 feet thick. Along the eastern base of Brushy Hills the outcrop of the lower Trenton limestone, 4 a., is apparently an ancient coral-reef, now a very compact, pure coral limestone, quite largely quarried for local building purposes, and for the manufacture of lime. This coralline bed contains shells as well as coral. It varies from 100 to 150 feet in thickness. The House Mountain (or rather *pair* of mountains), about six miles west northwest from Lexington, is one of the most striking features of the grand scenery in this portion of the Great Valley. This isolated mountain group rests upon Trenton limestone which crops out around the base. Then in nearly horizontal strata other formations, 4 b., as shales and shaly limestones, 4 c., as purplish, ferruginous shales and shaly sandstones, and above all a cap of Medina sandstones, 5 a.; the whole rising 2,000 feet above the limestone valley below. Lexington is a good point of departure for the geological study of either the Blue Ridge range on the S. E. or the North Mountain range on the N. W. Washington and Lee University and the V. M. Institute, both located here, have good mineral and geological cabinets. For fuller details, and geological section across the Great Valley near Lexington, see Am. Jour. of Sci., Vol. XVIII., 1879, p. 16.

75. *Quebec Group.* Dr. A. R. C. Selwyn, the successor of Sir Wm. Logan, as Director of the Geological Survey of Canada, does not recognize the Quebec as a geological formation, and in Professor J. D. Dana's table, as given in this guide, it is omitted, being considered as merged in the Calciferous. Professor Campbell, of Virginia, is not prepared to adopt this view as suitable for that State. He reports that throughout the Great Valley of Virginia, 350 miles in length, with continuous ledges of limestone, there exists what is known as the Canadian group, consisting of three tolerably well defined sub-groups of limestones, with extensive beds of interstratified shales and calcareous sandstones in the lowest 3 a. *Calciferous;* very regular stratified beds of dolomitic limestones more or less ferruginous and producing rich soils in the next higher 3 b. *Levis;* and, in the last, some beds of pure limestone, with a stratum of brown sandstone in the lower portion, abounding in molluscean fossils, not well preserved, but doubtless 3 c. *Chazy;* and still higher, near the Trenton, beds of chert abounding in cephalopods and gastropods of undoubted 3 c. Chazy age. He, therefore, prefers to retain the three divisions, at least until additional palæontological evidence settles the question at issue.   J. M.

North Carolina.[1]

LIST OF GEOLOGICAL FORMATIONS IN NORTH CAROLINA.

20. Quaternary. 19. Tertiary. 18. Cretaceous. 16. Triassic.		1. Archæan. Igneous.	1 b. Huronian. 1 a. Laurentian.

1. Revised and the notes added for the first edition by W. C. Kerr, State Geologist of North Carolina. Enlarged and revised for the second edition by Dr. H. M. Chance, of Philadelphia, geologist in charge of explorations of North Carolina coal fields.

Sketch of the Geology and Topography of North Carolina.

Derived from the State Geological Reports of Prof. W. C. Kerr.

North Carolina is the Mountain State of the Atlantic slope. As a general description, it may be said that the surface of this State is covered by but two of the great formations. The (1) Archæan, sub-divided into the (1 a.) Laurentian and (1 b.) Huronian, the lowest occupies the western and the (20) Quaternary the upper system covers the eastern portion, the oldest and the youngest, with a vast geological blank between them. Some of the railways run for long distances on a single formation. An irregular line drawn on the map of the State, in a northeast and southwest direction, through the City of Raleigh, will show the relative portions of the State covered by each. The (16) Triassic, the only one of the intermediate groups which appears, covers but a comparatively insignificant area in the middle region. It contains the coal beds of Deep River and of Dan River. The (18) Cretaceous and (19) Tertiary, underlie the (20) Quaternary, but they only appear on the surface in a few localities, of small area, on the river bluffs, and in water courses and ravines in the eastern division. The complete geological series of the State is as follows: (20) Quaternary, (19) Tertiary, (18) Cretaceous, (16) Triassic, (1) Huronian, (1 a.) Laurentian and Igneous.

Most of the metamorphic rocks of North Carolina belong to the (1 a.) *Laurentian* system, which prevails so extensively in Canada, Michigan, Wisconsin, Minnesota, etc. The prevalent species are Granite, Gneiss, Syenite and other Hornblendic rocks, Diorite and Crystalline limestone, and these contain graphite and much magnetic and specular iron ore, frequently in very large beds. This formation, besides iron, produces gold, silver, lead, copper, and other minerals. The (1 b.) *Huronian*, the *Taconic* of Emmon's report on this State, occupies several disconnected areas on the Great Smoky Mountain, at the Tennessee line and on the Blue Ridge, and another considerable area west of Raleigh, extending across the State with two smaller exposures. The rocks are quartzyte and clay slates, light colored, drab, and greenish. With these exceptions, and the small area of (16) Triassic, all the remainder of the western part of the State is (1 a.) Laurentian.

The North Carolina Mountains. The great continental system of the Appalachian Mountains, which extends a thousand miles, from near the mouth of the St. Lawrence to the State of Georgia, reaches its greatest elevations and developes its grandest features in the western part of this State. The system is here represented by two great parallel chains, the Smoky Mountains and the Blue Ridge, with a net-work of heavy cross chains connecting them and numerous spurs thrown off to the east and south, some of them as high as the parent chain and some more than fifty miles long. There are also several other disconnected minor chains to the eastward, with the same general trend. These mountains extend across the State, and their entire length from their southwestern termination, the Blue Mountains in Georgia, to their northern, which is prolonged 50 miles into Virginia, is 275 miles, of which two-thirds, or about 5,000 square miles, lie within North Carolina.

The main or western chain, which more to the north borders the great valley in Virginia and is there called the Blue Ridge, gradually deviates towards the southwest. A new chain, detached on the east and curving a little more to the south, takes now the name of the Blue Ridge, and in this State attains gradually to 5,000 and 5,900 feet, composed of many fragments, scarcely connected into a continuous and regular chain. These groups are separated by long intervals of depression, in which are gaps but little above the interior valleys.

West of this, and separated from it by a valley, is the great western chain of mountains, named locally the Iron Mountain in the northern portion, and Unaka in the southern, the whole being known as the Smoky Mountains, and forming the line between Tennessee and North Carolina. This is much more continuous, more elevated and regular in its direction and height, and increases very uniformly from 5,000 to nearly 6,700 feet. The valley comprised between these two main chains, the Smoky Mountain and the Blue Ridge, is divided by transverse chains into many basins of great altitude. The height of these transverse chains is greater than that of the Blue Ridge, being from 5,000 to 6,000 feet and upwards, and the gaps that cross them are as high, and often higher, than those of the Blue Ridge. The whole chain of valleys extends for more than 180 miles, and from 20 to 50 miles wide, with a mean height of more than 2,000 feet, and portions of them 3,500 to 4,000 feet, this being the highest plateau of the same extent east of the Rocky Mountains. These are all valleys of erosion, and they, as well as the mountains and plateaus have, in Prof. Kerr's opinion, no anticlinal or synclinal origin, being in fact wholly independent of geological structure.

The mountains which reach 6,000 feet are more than fifty in number, and the loftiest peaks rise to 6,700 feet. Here, then, in all respects, is the culminating region of the vast Appalachian system. This mountain region, where the most striking natural objects in the State are to be seen, has not yet been penetrated by the railroads, except that the Western North Carolina R. R. crosses the mountains, connecting with the East Tennessee, Virginia & Georgia R. R.

Richmond & Danville Railroad.			Western North Carolina Railroad.		
Ms.		**Alt.**	**Ms.**		**Alt.**
0	Richmond, Va.		0	Salisbury.	1 a. Lauren. 106 m.⁷⁶⁰
141	Danville, Va.	1 a. U. Lauren. 42 m⁴²⁰	25	Statesville.⁴	" 955
156	Ruffin, N. C.	" 707	48	Newton.	" 1070
165	Reidsville.	" 828	58	Hickory.⁴	" 1140
181	Moorehead.	"	78	Morganton.	"
189	Greensboro.	1 a. L. Lauren. 6 m.⁸⁴³	99	Marion.	" 1425
204	High Point.	" 943	114	Henry.	1 b. Huronian. 8 m.
211	Thomasville.	"	126	Black Mountain.	"
222	Lexington.	" 776	139	Ashville Junc.	1 a. Laurentian.
238	Salisbury.	" 760	142	Ashville.	"
261	Concord.	"	143	Ducktown Junc.	"
282	Charlotte.	" 725	165	Marshall.	" 1647
312	State Line.	"	182	Warm Springs.	2 a. Oc., Cg. & Sh.¹⁸³⁵
			.....	Wolf Ck., Tenn.	E. T. V. & Ga. R. R.
			190	Paint Rock.	

Goldsboro Branch.			Ducktown Branch.		
0	Greensboro.	1 a. Lauren. 30 m.⁸⁴³	0	Ashville.	1 a. Laurentian.
21	Company Shops.	"	30	Waynesville.	"
32	Mebanesville.	1 b. Huronian. 20 m⁶⁹⁷		**Raleigh & Gaston Railroad.**	
41	Hillsboro.²	" 539			
46	University.	"	0	Portsmouth, Va.	1 a. Laurentian.
55	Durham.	16. Triassic. 22 m.⁴⁰⁰	0	Weldon.	" 72
69	Morrisville.³	" 308	12	Gaston.	" 152
73	Carey.	1 b. Huronian. 6 m.⁴⁹⁵	58	Henderson.	" 505
81	Raleigh.	" 317	61	Kittrells.	" 417
96	Clayton.	" 347	97	Raleigh.	" 303
106	Neuse River.	20. Quatern. 24 m.¹¹²		**Raleigh & Augusta Railroad.**	
109	Selma.	"			
118	Princeton.	" 160	0	Weldon.	
130	Goldsboro.	" 102	97	Raleigh.	1 a. Lauren. 3 m. ³⁰³
			107	Cary.	1 b. Huron. 10 m.
			114	Appex.	16. Triassic. 20 m.⁵⁰²
	Salem Branch.		140	Sanford.	{ 16. Triassic, and 20 Quater. 11 m.³⁵³
0	Greensboro.	1 a. Laurentian. ⁸⁴³	152	Cameron.	16. Tr., Huron. 13 m³⁰⁹
17	Kernesville.	" 1016	174	Kyser.	20. Quat., princi'ly ²⁸⁶
28	Salem or Winston	" 884	104	Hamlet.	" 331

2. At Hillsboro depot a good exposure of typical North Carolina Huronian slate, hydromicaceous.
3. At Morrisville depot a dike of dolerite visible. One and a half miles east of station beds of very coarse incompacted conglomerate, the bottom beds of the Triassic, and probably glacial.
4. From Statesville west in the numerous deep cuts are seen fine examples of the *frost* drift, characteristic of sub-glacial regions. Also from Hickory to Morgantown many sections of the purple paragonite schists, which are peculiar to this region.

There is very little exposure of solid rock, and that only on the tops of a few high mountains or an occasional cliff. The mountains are covered to their very summits with dense forests, but with a deep and strong soil which is, however, according to Dr. T. Sterry Hunt's description, very unlike the layers of clay and loam with which we in the North are familiar. The rocks themselves, he says, although of gneiss and mica slate, like that which prevails over so great a part of New England, have undergone a process of decay which has rendered them so soft that they may be readily cut by a spade, although retaining all the veins and layers which mark their original stratification. Without having been broken or ground up, these hard rocks have moldered into a soft clayey mass, forming a soil fifty feet and often much more in depth, which from its peculiar structure has a natural drainage, and possesses great fertility. North Carolina, evidently, never was subjected to the action of glaciers like the Northern States. Only the valleys of the streams are covered with alluvium, consisting of sand, gravel and clay, the debris of the rocks of the higher ridges and mountains.
The middle and eastern part of the State is a long slope, extending from the rugged mountain plateau to the Atlantic. Next, however, to the plateau is a *piedmont* or *middle region* of hill country, with an average elevation of about 1,000 feet. This is divided by its rivers into three regions, drained by the Broad, Catawaba and Yadkin rivers, the slope of the first being toward the south, and that of the others a little east of north. These drainage surfaces are separated by two, nearly parallel, easterly chains of mountains, the South and Bushy Mountains, and are from 2,000 to 4,000 feet high. There are other easterly spurs of the Blue Ridge of similar elevation. This middle division or hill

Cape Fear & Yadkin Valley Railroad.

Ms.		Alt.
0	Fayetteville.	{ 20. Quaternary, 1 b. Huron. 33 m. ³²⁰
37	Sandford.	{ 16. Triassic, 20. Quaternary. ²⁵³
44	Egypt.[5]	" 262
47	Gulf.[6]	" 279
54	Richmond.	"
58	Ore Hill.	1 b. Huronian. 496
63	Siler.	"
70	Staley.	"
75	Liberty.	"
82	Julian.	"
90	Pleasant Garden.	"
98	Greensboro.	1 a. Laurentian. 845

Carolina Central Railroad.

Ms.		Alt.
0	Wilmington	20. Quater. 117 m. 10
68	Lumberton.	" 185
111	Hamlet.	" 331
117	Rockingham.	{ 20. Quaternary, and 1 b. Huronian.²¹⁰
123	Pee Dee River.[7]	1 b. Huronian. 6 miles.
128	Lysleville.	1 a. Laurentian. 5 m.
135	Wadesboro.	16. Triassic. 19 miles.
168	Monroe.	1 b. Huron. 25 m. 556
187	Charlotte.	1 a. L. Laurentian.725
199	Catawba River.	"
.....	Lincolnton.	" 866
229	Shelby.	" 575

Wilmington & Weldon, and Wilmington, Columbia & Augusta Railroad.

Ms.		Alt.
0	Weldon.[8]	20. Quaternary.⁷²
8	Halifax.	"
37	Rocky Mount.	"
78	Goldsboro.	" 102
92	Mount Olive.	"
114	Magnolia.	"
148	Rocky Point.	"
162	Wilmington.[8]	" 10
162	Wilmington.[8]	" 10
191	Maxwell's.	"
208	Whiteville.	"
227	Fair Bluff.	"
.....	S. C. Line.[8]	"

Tarboro Branch.

0	Rocky Mount.	20. Quaternary.
17	Tarboro.	"
.....	Bethel.	'
45	Williamston.	" 878

Halifax & Scotland Neck Railroad.

0	Halifax.	20. Quaternary.
20	Scotland Neck.	"

Ashville & Spartansburg Railroad.

0	Spartansb'g, S. C.	
. ...	Flat Rock.	1 a. U. Laurentian.
49	Hendersonville.	" 505

5. *Egypt.* Old coal shaft, 460 feet deep.

6. *Gulf.* Bituminous coal beds 2 ft. and 3½ ft.-4 ft. thick, worked on a small scale during the war. Not now worked. Much troubled by trap dykes.

7. On both sides of the Pedee River are high dikes of dolerite for more than a mile, and 2 miles east a very coarse porphyritic granite, as well as between Lilesville and Wadesboro.

8. *Wilmington & Weldon Railroad,* 162 miles; north and south. This road runs throughout its whole length from Wilmington to Weldon on the (20) Quaternary formation, with occasional small exposures of the Tertiary (19 a.) Eocene and (19 b.) Miocene and of the (18) Cretaceous in the banks of the streams.

9. *Dismal Swamp.* This road skirts around the *Great Dismal Swamp.*

country extends 200 miles from east to west, and 150 miles northeast and southwest, and comprises nearly one-half of the territory of the State. It rises in going west about four feet to the mile, and attains an elevation of 1,000 to 1,500 feet at the foot of the Blue Ridge. The channels of the large rivers, however, are cut 100 to 300 feet below the intervening divides.

Between the swamp country, along the coast, and the hilly region of the interior, is a belt of level, sandy, barren territory, extending from near the line of Virginia across the entire State, and from 30 to 80 miles wide, covered by the long leaved pine. Spirits of turpentine produced in this pine region is the most important branch of manufacturing in the State.

The eastern division of the State extends from the coast, about 100 miles, to the lower falls of the rivers, and constitutes nearly two-fifths of the State. This region is for the most part nearly level or very gently undulating, except along the rivers on the upper reaches of which are bluffs and small hills. Its slope seaward is between one and two feet to a mile and it is covered by the horizontal strata of the quaternary underlaid by the tertiary. They consist of the noncompacted sands, clays, marls and gravels, coarser materials predominating westward, and becoming successively finer towards the coast.

The Coast of North Carolina is remarkable for the shallow sounds and bays that extend along the entire sea front nearly 300 miles, the largest of which are Pamlico and Albermarle Sounds, the former 75 miles long by 15 to 20 miles wide, and the latter 50 by 5 to 15 miles, with a depth of water from a few feet to 20 feet. There are also along the coast 3,000 to 4,000 square miles of swamp lands, of which the Great Dismal Swamp, on the line between this State and Virginia, is well known.

The foregoing description of North Carolina will serve to give a general idea of the geology of *South Carolina,* also where the same formations are found. J. M.

Atlantic, Tennessee & Ohio Railroad.

Ms.			Alt.
0	Charlotte.	1 a. L. Laurentian.[725]	
47	Slatesville.	"	955

Cheraw & Wadesboro Railroad.

Ms.		
0	Wadesboro, N. C.	16. Triassic.
7	Bennett's.	20. Quaternary.
10	Morven.	"
15	Cheraw, S. C.	"

Charlotte, Columbia & Augusta R. R.

Ms.			
0	Charlotte.	1 a. L. Laurentian.	747
10	Pineville.	"	575
14	S. C. State Line.	"	
44	Chester, S. C.	"	543

Chester & Lenoir Railroad.

Ms.			
0	Chester, S. C.		543
23	Yorkville.		
45	Gastonia, N. C.	1 a. U. Laurentian.	832
49	Dallas.	"	944
63	Lincolnton.	1 b. Huronian.	866
79	Newton.	1 a. U. and L. Lau.[1070]	
89	Hickory.	"	1222
109	Lenoir.	1 a. U. Laurentian[1186]	

Atlantic & North Carolina Railroad.

Ms.		
0	Goldsboro.	⎰ 20. Quaternary with
14	La Grange.	⎸ 18. Cretaceous and
50	Newbern.	⎸ 19. Ter. in banks of
		⎱ the streams. 102
85	Newport.	"
95	Moorhead.	"

Danville, Mocksville & Southwestern R. R.

Ms.		
0	Danville, Va.	16. Triassic.
8	Leaksville, N. C.	1 a. U. Laurentian.

E. Tennessee & W. North Carolina R. R.

Ms.		
0	Johnson City, T.	
26	Roan Mt., N. C.	
33	Cranberry.	1 b. Huronian.
34	Mine.	" Iron Mines.

Norfolk South

Ms.	
0	Norfolk.
9	Prince Anne.
42	Camden C. H.
46	Elizabeth City.
62	Hertford.
74	Edenton.

Jamesville & Was

0	Jamesville.
29	Washington.

Midland North C

0	Goldsboro.
22	Smithfield.

Milton & Suthe

0	Sutherlin, Va.
9	Milton, N. C.

Oxford & Hend

0	Henderson.
13	Oxford.

Petersburg

0	Petersburg, Va.
10	Reams.
53	Pleasant Hill.
64	Weldon.

Seaboard & Ro

0	Portsmouth, Va.
70	Seaboard.
78	Garys.
80	Weldon.

University

0	University.
11	Chapel Hill.

South Carolina.[1]

Ashley River Railroad.

Ms.		Alt.
0	Charleston.[9]	Post Plioc. at depth of 90 ft. Eocene 900 ft. Cretaceous. (H.)
4	Northeastern R.R	"

Asheville & Spartanburg Railroad.

Ms.		Alt.
0	Spartanburg.	1 a. U. Laurentian (K.) Gneiss. 787
2	Air Line Junc.	"
10	Campton.	"
12	Inman.	Mica Slate. (L.)
18	Campobello.	1 a. U. Laurentian (K.) Gneiss.
23	Landrums.	"
27	Tryon, N. C.	"

Atlanta & Charlotte Air Line Railroad.

Ms.		Alt.
0	Atlanta, Ga.	
102	Fort Madison.	Hornblende slate. (L.)
107	Harbins.	Gneiss. (L.)
111	Westminster.	Mica slate. (L.) 919
116	Richland.	Hornblende slate. (L.)
121	Seneca.	Gneiss. (L.) 954
127	Keowee.	"
134	Central.	Mica slate. (L.)
142	Liberty.	Steatite. (L.)
148	Eastley's.	Gneiss. (L.)
154	Saluda.	Mica slate. (L.)
160	Greenville.	Gneiss. (L.) 976
168	Tayler's.	Dike aphanitic porphyry. (L.)
173	Greer's.	Mica slate. (L.)
178	Duncan's.	Gneiss. (L.)
181	Wellford.	"
187	Fair Forest.	"
190	A. L. Junction.	"
192	Spartanburg.	" 787
196	Mount Zion.	Mica slate. (L.)
200	Cowpens.[3]	Gneiss. (L.)
206	Thicketty.	Mica. (L.)
212	Gaffney's.	Itacolumite. (L.)
221	Black's.	Blue Lime s. (L.) 774
226	Whitaker's.	Melaphyre Dike (L) 907
234	Kings Mt., N. C.[2]	942

Augusta & Knoxville Railroad.

Ms.		Alt.
0	Augusta, Ga.	
16	Woodlawn.	Gneiss. (L.)
20	Merriwether.	"
24	Clark's Hill.	"
29	Modoc.	Clay Slate. (L.)
32	Parksville.	"
38	Plum Branch.	Talc Slate. (L.)
43	McCormick.[3]	"
49	Troy.	"
54	Bradley.	Dike of Dioritic por'y.
59	Verdery.	Talc slate.
67	Greenwood.	Mica, Slate and Dior.

Central Railroad of South Carolina.

Ms.		Alt.
0	Lanes.	19 c. Plio. Marls.(T.)
4	Heinneman's.	"
8	Greeley's.	"
10	Mt. Hope.	"
13	Forreston.	"
19	Wilson.	19 a. Eocene Marls.(T.)
22	Manning.	"
26	Dudley.	"
28	Harbin's.	"
30	Durant.	"
33	Lawrence.	"
40	Sumter.	"

Charleston & Savannah Railroad.

Ms.		Alt.	
0	Charleston.	Post Pliocene. (S.)	
7	Charleston Junc.[4]	"	Beds of Phosphate Rock.
10	Dorchester.	"	
12	Drayton.	"	
16	John's Island.	"	
19	Rantowles.	"	
25	Ravenal.[4]	"	
35	Adams Run.	19 a. Eocene Marls(T.)	
37	Jacksonboro.	Post Pli. Phosphate.[17]	
42	Ashepoo.	19 a. Eocene Marls (T.)	
46	Greenpond.	"	
51	White Hall.	"	
58	Saltkehatchie.	"	
60	Yemassee.	Post Pliocene. 25	
68	Coosawhatchie.	19 a. Eocene Marls.(T.)	

1. Prepared for this work by Mr. Harry Hammond, of Beech Island, South Carolina. The authorities for the geology are designated as follows: H. stands for Prof. Francis Holmes; K. for W. C. Kerr, of North Carolina; L. for Oscar M. Lieber; T. for M. Tuomey; S. for Charles N. Shepard.

The great group of crystalline rocks which extends from New England to Alabama is Metamorphic without fossils, and hence of doubtful age. In the opinion of some geologists, instead of attempting to classify them, it is better to insert in this guide, as Mr. Hammond has done for South Carolina, the kind of rock along the line of the railroad, e. g.: Gneiss, mica schists, granite, etc., which gives us some positive knowledge. J. M.

2. *Cowpens to King Mountain.* Itacolumite, or Diamond rock, the prevailing rock, with seams of marble, limestone, barytes, hematite, specular and argillaceous schist, with numerous gold and iron mines, and quarries of various rocks.

3. *McCormick.* Ores of gold manganese and copper abound.

Charleston & Savannah Railroad—Continued.

Ms.		Alt.
77	Ridgeland.	19 a. Eocene Marls.(T.)
84	Terribee Switch.	"
91	Hardeeville.	"
96	Savannah River.	"

Charlotte, Columbia & Augusta R. R.

Ms.			Alt.
0	Charlotte, N. C.		
17	Fort Mills.	Steatite. (L.)	
20	Catawba River.	Granite. (L.)	
25	Rock Hill.	Gneiss. (L.)	
31	Warren's.	Known as Black Jack lands. { Dike of Aph. por'y (L.)	
34	Smith's.	"	
37	Lewis.	"	543
44	Chester.	"	
55	Blackstock's.	"	621
58	Woodward's.	Mica Slate.	
63	White Oak.	"	548
66	Adger's.	Gneiss.	
71	Winnsboro.	"	543
74	Robertson's.	"	
77	Simpson's.	"	
82	Ridgeway.	Mica Slate.	626
90	Blythewood.	Clay Slate. (T.)	
93	Sharps.	"	
96	Killian's.	Eocene Buhrstone. (T.)	
100	100-Mile Siding.	"	
106	Columbia.	Granite. (T.)	296
108	W. C. & A. Junc.	"	
120	Lexington.	"	370
125	Barr's.	Eocene Buhrstone. (T.)	
130	Keisler's.	"	
131	Gilbert Hollow,	"	
133	Summit.	"	
138	Leesville.	Granite. (T.)	
140	Batesburg.	"	
149	Ridge Spring.	"	
153	Ward's T. O.	"	
158	Johnson's T. O.	"	
165	Trenton.	"	
170	Miles Mills.	"	
174	Vaucluse.	"	
178	Graniteville.	"	
179	Aiken Junction.	19 a. Eo. Buhrstone(T.)	
182	Langley.	"	
184	Bath.	"	
189	Dead Fall.	"	
191	Augusta, Ga.		185

Cheraw & Che

Ms.	
0	Chester.
6	Orr's.
8	Knox.
10	McDaniels.
12	Richburg.
15	Bascomville.
18	Cedar Springs.
20	Fort Lawn.
22	River.
25	Waxhaw.
27	Miller's Crossing.
29	Lancaster.

Cheraw & Darli

0	Florence.
5	Palmetto.
10	Darlington.
18	Doves.
27	Society Hill.
34	Cash's.
40	Cheraw.

Cheraw & Sali

0	Cheraw.
11	McFarlan's, N. C.

Chester & Le

0	Chester.
8	Lowrysville.
14	McConnellsville.
16	Guthriesville.
23	Yorkville.
33	Clover.
37	Bowling Green.
39	Crowder's C'k.

Columbia & Gr

0	Columbia.
6	Frost's Mill.
9	Swygert's Mill.
11	Montgomery's M.
13	Bookman's.
20	Wallaceville.

4. *Charleston Junction to Revanel.* Beds of phosphate rock. The phos lina, from which large quantities of valuable fertilizers are manufactured, of phosphate of lime, and 5 to 10 per cent. of carbonate of lime, with sma sulphuric acid, etc. It is in the form of nodules, very rough, rounded and perforated with irregular cavities of an olive, blueish, black, yellowish, color, and from a few inches to several feet in diameter. The River Roc sometimes as a continuous sheet 6 to 18 inches thick. It is profitably dredg and a royalty of one dollar per ton is paid to the State for all taken from na' rock is found about the level of meantide in layers 6 to 30 inches thick of loos mined under 7 feet of earth. It is found in various places from Florida to raised in artesian wells from a depth of 300 feet, and brought up from sea miles from shore.—*Harry Hammond, in Hand-Book of South Carolina.*

nville Railroad.
ued.

	Alt.
Clay Slate. (T.)	259
"	
Mica and Talc Slate. (T.)	330
Dike of Feldspathic and Horneblende Rocks.	
"	502
Granite. (T.)	532
Gneiss. (T.)	
"	
"	
Dioritic aphanitic felspathic porphyry with epidtosite.(L.)	
"	570
Gneiss. (L.)	
Mica Slate. (L.)	671
Gneiss. (L.)	714
Crosses Sandstone, Hornestone and Quartzic Schists. Gneiss (L.)	760
"	810
"	896
"	840
"	
"	
"	989

Branch.

	Alt.
Gneiss (L.)	714
"	
Dioritic Por'y (L.)	533

Railroad.

	Alt.
Gneiss. (L.)	896
"	784
"	
"	
"	
Mica Slate. (L.)	
Gneiss. (L.)	
"	
Gneiss and Hornblende Slate.(L.)	985

Railroad.

	Alt.
Granite. (T.)	
Gneiss. (T.)	
"	
"	
"	

Georgetown & Lane's Railroad.

Ms.			Alt.
0	Georgetown.	Post Pliocene. (T.)	
18	Harper's.	18. Cretaceous of secondary. (T.)	
26	Trio.	"	
36	Lane's.	Pliocene Marls. (T.)	

Northeastern Railroad.

Ms.			Alt.
0	Charleston.	Post Pliocene.	16
2	Magnolia.	"	
6	C. & S. Junction.	"	
8	8-Mile Turnout.	Post Pliocene, Phosphate Rock. (S.)	
14	Otranto.	"	
18	Mount Holly.	"	
23	Strawberry.	"	
25	Oakley.	"	
30	Monck's Corners.	"	
35	Macbeths.	19 a. Eocene, Ashley & Cooper Marls.(T.)	
38	Bonneaus.	"	
45	St. Stephens.	19 a. Eocene Santee Marls. (T.)	
49	Santee.	"	
51	Gourdin.	"	
54	Cane's.	19 c. Pliocene Mar.(T.)	
59	Salter's.	"	
64	Kingstree.	18. Cretaceous of secondary. (T.)	
75	Cade's.	"	
79	Graham.	"	
82	Scranton.	"	
86	Coward's.	"	
92	Effingham.	"	
95	Willoughby.	"	
102	Florence.	"	

Port Royal & Augusta Railroad.

Ms.			Alt.
0	Augusta, Ga.		188
6	Beech Island.	19 a. Eocene Buhrstone. (T.)	
10	Brown's Hill.	"	
15	Jackson.	"	
22	Ellenton.	19 a. Eocene Santee Marls(T.)	149
28	Robbins.	"	
32	Hattieville.	"	
37	Millett.	"	
44	Beldoc.	"	
49	Appleton.	"	
53	Allendale.	"	192
58	Campbellton.	"	
62	Brunson.	"	
68	Hampton.	"	
70	Varnville.	"	
72	Almeda.	"	

Port Royal & Augusta Railroad.

Ms.	Continued.	Alt.
75	McNeils.	{ 19 a. Eocene. Santee Marls. (T.)
81	Early Branch.	{ 19 a. Eocene. Cooper & Ashly Marls. (T.)
87	Yemassee.	19 c. Post Pliocene. 25
92	Tomotly.	"
99	Seabrook.	"
103	Island Tank.	{ 19 c. Post Pliocene Marls, Phos. Rock.
108	Beaufort.	" " 20
112	Port Royal.	" " 27

South Carolina Railroad.

Ms.		Alt.
0	Charleston.	Post Pliocene. (T.) 16
1	Magnolia.	"
4	West's.	"
7	Seven Mile.	{ Post Pliocene, Phosphate Rock. (S.)
10	Ten Miles.	"
12	Sineath's.	"
15	Woodstock.	{ 19 a. Eocene, Ashley and Cooper Marl(T.)
17	Ladson's.	"
22	Summerville.	" 68
26	Jadburg.	"
31	Ridgeville.	{ 19 a. Eocene, Santee Marls. (T.)
37	Rosses.	"
38	Whartons's.	"
41	Forty-One.	"
44	Birds.	"
47	George's.	"
52	Reeve's.	"
58	Fifty-Eight.	"
62	Branchville.	" 140
67	Edisto.	"
72	Midway.	"
75	Bamberg.	" Buhrstone. (T.)
81	Grahams.	"
86	Lee's.	"
89	Blackville.	"
93	Reynold's.	"
96	Elko.	"
99	Williston.	"
102	White Pond.	"
107	Windsor.	"
115	Montmorence.	"
120	Aiken.	"
126	Graniteville.	" Kaolin Clay(T.)
128	Langley.	"
131	Bath.	"
132	Horse Creek.	"
136	Hamburg.	"
138	Augusta, Ga.	"

Branchville to Columbia.

Ms.		Alt.
62	Branchville.	{ 19 a. Eocene, Santee Marls. (T.)
66	Sixty-Six.	"
70	Rowesville.	"
75	Felder.	"
79	Orangeburg.	" 265
81	Stilton's.	"
85	Jameson's.	"
88	Riley's.	{ 19 a. Eocene Buhrstone. (T.)
92	St. Mathew's.	"
95	Singleton's.	"
99	Fort Motte.	"
102	Congaree.	"
106	Kingville.	"
110	Gadsden.	"
118	Hopkins.	"
124	Hampton.	"
127	Taylor's.	"
129	Columbia Junc.	Granite.
130	Columbia.	" 233

Kingsville to Camden.

Ms.		Alt.
106	Kingsville.	19 a. Eo. Buhrstone(T.)
110	Wateree.	"
115	Middleton.	"
118	Camden Junc	"
121	Dixie.	"
125	Claremont.	"
131	Sanders.	"
135	Boykin's.	"
138	Stockton.	"
144	Camden.	"

Spartanburg, Union & Columbia Railroad.

Ms.		Alt.
1	Alston.	Clay Slate. (T.) 259
2	Parr's.	Mica "
8	Dawkin's.	"
18	Blairs.	Gneiss.
19	Shelton.	Granite. (T.)
26	Fish Dam.	Gneiss. (L.)
31	Santuc.	Granite. (L.)
39	Union.	" 579
49	Jonesville.	Mica Slate. (L.)
56	Pacolet.	"
59	Rich Hill.	Gneiss. (L.)
63	Glendale.	"
68	Spartanburg.	" 787

Wilmington, Columbia & Augusta Railroad.

Ms.		Alt.
0	Columbia.	Granite. 233
6	Simms.	19 a. Eo. Buhrstone(T.)
16	Congaree.	"
22	Eastover.	"
25	Acton.	"
31	Camden Crossing	"

Ms.	Wilmington, Columbia & Augusta Railroad—*Continued.*	Alt.	Ms.	Barnwell Railway.	Alt.
33	Wedgefield.	19 a. Eo Buhrstone.(T.)	0	Blackville.	19 a. Buhrstone of Eo.
37	Cane Savannah.	"	4	Ashleigh.	"
43	Sumter.	"	6	Woodward's Jun.	"
52	Maysville.	19 c. Plioc. Marl. (T.)	9	Barnwell C. H.	{ 19 a. Santee, or Cor-
57	Atkins.	"			alline Marls of Eo.
61	Lynchburg.	"			
65	Cartersville.	"		**Cape Fear & Yadkin Valley Railroad.**	
71	Timmersville.	{ 18. Cret. Marls of secondary. (T.)	0	Bennetsville.	19 a. Eocene.
77	Ebenezer.	"	6	Tatum.	"
82	Florence.	"	9	McCall.	"
88	Mars Bluff.	"	13	Hasty.	"
95	Pee Dee.	"	15	Johns, N. C.	
99	Laughlins.	19 c. Plioc. Marls. (T.)			
103	Marion.			**Greenwood, Laurens & Spartanburg R. R.**	
112	Mullins.	19 a. Eo.Buhrstone.(T.)	0	Greenwood.	
118	Nichols.	"	7	Coronaco.	Gneiss.
127	Fair Bluff, N. C.		15	Waterloo.	Granite.
			20	High Point.	Gneiss.
			24	Maddens.	Trap Rock.
			28	Lauren's.	Gneiss.

Georgia.[1]

GEOLOGICAL FORMATIONS OF GEORGIA.

The Metamorphic area of the State extends from a line crossing the State from Augusta t Columbus, extending by Milledgeville and Macon, and extending beyond the line of the State on th northeast. The lithological characteristics of the Metamorphic is that of the Archæan in general.
The paleozoic includes the counties of Dade, Walker, Chattooga, Catoosa, Whitfield, Floyd, Murra; Gordon, Barton and Polk, all in the northwest corner of the State.
The Silurian groups represented, beginning with the lowest, are the Potsdam sandstone, Kno Shale and Dolomite, Chazy, Trenton, Cincinnati, Medina, Clinton and Oriskany. The Devonian represented by a black shale of from 10 to 50 feet in thickness. The Sub-Carboniferous by limeston(and shales of 800 feet. The Coal Measures, confined mostly to the counties of Dade, Walker an Chattooga, cover an area of nearly 200 square miles, and contain several beds of coal.

Charleston & Savannah Railroad.		
Ms.		Alt.
0	Savannah.	19 c. Tertiary.	82
24	Fleming.	"
39	Walthourville	"
53	Doctortown.	"
57	Jesup.	"	100
86	Blackshear.	"
122	Homersville.	"
130	Dupont.	"
139	Stockton.	"
157	Valdosta.	"
174	Quitman.	19 a Tertiary.
188	Boston.	"
200	Thomasville.	"
214	Cairo.	"
226	Climax.	"
236	Bainbridge.	"
200	Thomasville.	19 a. Tertiary.
232	Camilla.	"
258	Albany.[2]	"	232
130	Dupont.	19 c. Tertiary.
151	Statensville.	"
163	Jasper, Fla.	"
179	Live Oak, Fla.	"

Brunswick & Albany Railroad.		
0	Brunswick.	19 c. Tertiary.	14
13	Hazlehurst.	"	261
24	Waynesville.	"
60	Waycross.	"	100
67	Waresboro.	"	117
78	Milwood.	"	130
93	Kirkland.	"
101	Willicoochee.	"	220
151	Isabella.	19 a. Tertiary.	340
171	Albany.[2]	"	165

East Tennessee, Virginia & Georgia R. R		
Ms.	Macon & Brunswick Division.	Al
0	Brunswick.	19 c. Tertiary.	1
40	Jesup.	"	10
70	Baxley.	"	21
93	Lumber City.	19 a. Tertiary.	15
100	Town's.	"	18
140	Dubois.	"	39
148	Cochran.	"	84
161	Buzzard Roost,	"	24
171	Bullard's.	"	26
186	Macon.	Met. and Tertiary.	33
148	Cochran.	19 a. Tertiary.	34
159	Hawkinsville.[3]	"	23

Central Railroad of Georgia.		
0	Savannah.	19 c. Tertiary.	8
50	Halcyondale.[2]	19 a. "	11
62	Ogeechee.	"	10
79	Millen.[3]	"	16
134	Tennille.	19 a. Tertiary.
154	Toomsboro.	"
170	Gordon.	"	34
192	Macon.[4]	Met. and Tertiary.	33
79	Millen.	19 a. Tertiary.	16
100	Waynesboro.	"	11
132	Augusta.[4]	Met. and Tertiary.	13
179	Gordon.	19 a. Tertiary.	24
187	Milledgeville.	20. Ter. and Met.	31
208	Eatonton.	Metamorphic.
0	Macon.[4]	Met. and Tertiary.	33
25	Forsyth.	"	78
41	Barnesville.	"	87
59	Griffin.	"	97
67	Fayette.	"
76	Lovejoy's.	"
80	Jonesboro.	"	90
96	East Point.	"	104
103	Atlanta.[5]	"	105

1. Revised and the notes added for the first edition by Dr. George Little, State Geologist (Georgia; and for the second edition by A. R. McCutchen, of the Department of Agriculture (Georgia.
2. Buhrstone groups.
3. Northern limit of the open pine and wire grass section.
4. Located on the line of Metamorphic and Tertiary.
5. Strangers should visit the Geological Collection Room in Capitol Building.

Central Railroad of Georgia—Con.

Southwestern Railroad.

Ms.		Alt.
0	Macon.[4]	Met. and Tertiary. 334
8	Seago.	Tertiary. 362
29	Fort Valley.	19 a. Tertiary. 530
49	Montezuma.	"
60	Andersonville.[6]	" 396
71	Americus.	" 362
83	Smithville.	" 334
96	Leesburg.	"
107	Albany.[2]	19 a.Ter. Buhrstone 232
....	Walker's.	"
....	Ducker.	"
....	Arlington.	"
29	Fort Valley.	19 a. Tertiary. 530
50	Butler.	20. "
70	Geneva.[4]	"
75	Box Spring.	"
78	Upatoi.[4]	Metamorphic.
100	Columbus.[7]	Met. and Creta. 262
29	Fort Valley.	19 a. Tertiary. 530
42	Perry.	"
83	Smithville.	19 a. Tertiary. 334
98	Dawson.	" 354
118	Cuthbert.	" 448
133	Hatchie Station.	18 c. Cretaceous.
142	Georgetown.	"
144	Eufaula, Ala.	" 200
157	White Oak, Ala.	"
165	Clayton, Ala.	"
120	Junction.	19 a. Tertiary.
128	Coleman.	" 393
132	Fort Gaines.	" 166

North and South Railroad.

Ms.		Alt.
100	Columbus.[4]	Met. and Creta. 262
108	Cleghorn.	Metamorphic.
120	Kingsboro.	" 612

Upson County Railroad.

Ms.		Alt.
0	Macon.[4]	Met. and Tertiary. 334
43	Barnesville.	Metamorphic. 375
51	The Rock,	"
59	Thomaston.	"

Georgia Railroad.

Ms.		Alt.
0	Augusta.	134
38	Thomson.	Metamorphic. 517
47	Camak.	" 592
57	Barnett.	" 647
65	Crawfordville.	" 603
76	Union Point.	" 658
84	Greensboro.	" 612

Georgia Railroad.

Continued.

Ms.		Alt.
104	Madison.	Metamorphic. 651
130	Covington.	" 748
141	Conyers.	" 894
147	Lithonia.	" 937
156	Stone Mountain.[8]	"
165	Decatur.	" 1033
171	Atlanta.	Asbestus, 3 miles. 1050
0	Camak.	Metamorphic. 592
....	Warrenton.	" 506
....	Sparta.	" 567
....	Milledgeville.	" 310
78	Macon.	{ 3 miles Artope's quarry, Lyell's Eocene fossils. 384
57	Barnett.	Metamorphic. 647
75	Washington.	"
76	Union Point.	Metamorphic. 658
....	Lexington.	" 770
116	Athens.	{ Metamorphic. State University and Agricult'l College. 694

Atlanta & West Point Railroad.

Ms.		Alt.
0	Atlanta.	Metamorphic. 1050
6	East Point.	" 1043
18	Fairburn.	" 1034
25	Palmetto.	" 1025
40	Newman.	R. R. to Carrollton. 959
52	Grantville.	{ Gold mine, 3 miles. Metamorphic. 869
58	Hogansville.	" 731
72	La Grange.	{ Metamorph. Asbestus and Chromic Iron, 7 miles. 742
87	West Point.	{ Metamorph. Asbestus & Corundum 384

Piedmont Air Line Railroad.

Ms.		Alt.
312	N. C. State Line.	Metamorphic.
337	Gaffney's, S. C.	"
357	Spartanburg.	" 747
387	Greenville.	" 976
454	Tuccoa City, Ga.[9]	"
....	Mt. Airy.[10]	" 1587
....	Bellton.	"
481	Lula City.	{ Met. N. E. R. R. to Athens, 39 ms. 1334
492	New Holl. Spr'gs.	Limestone & Tremolite
494	Gainesville.[11]	{ 3 b. Metamorphic, flexible s. s. 1227

6. View of old Prison stockade and U. S. Cemetery east of railroad.
7. Fine falls, Lover's Leap and rapids, on Chattahoochee River.
8. Stone Mountain—a mass of granite—height, 1,686 feet.
9. Toccoa Falls, 2 miles, 185 feet. Tallulah Falls, 15 miles distant, nearly 400 feet high.
10. From this point a fine view of Yonah Mountain and the Blue Ridge chain. Clarkesville, 8 miles; Nacoochee Valley, 15 miles; Nacoochee gold mines, 20 miles.
11. Point of departure for Dahlonega gold mines and Porter's Springs.

Piedmont Air Line Railroad—Continued.

Ms.			Alt.
.....	Flowery Branch.	3 b. Metamorphic.	
.....	Buford.	"	1207
.....	Suwanee.	"	1027
.....	Duluth.	Metamorphic. Pine tree visible 4 ms. in center R. R. tk.	1107
527	Norcross.	Metamorphic.	1078
540	7-Mile Track.	Met. Granite quarry.	
547	Atlanta.[5]	"	1050

Rome Railroad.

0	Rome.	Knox Shale.	627
20	Kingston.	"	710

Cherokee Railroad.

48	Cartersville.[12]	Knox Shales.	760
.....	Rockmart.	Cal. and Potsdam.	

Selma, Rome & Dalton Railroad.

0	Dalton.	Tren. & K. Dolomite	757
6	Stark's.	"	
.....	Barnett's.	"	647
15	Sugar Valley.		
21	Skelley's.		
39	Rome.	Knox Shale.	627
45	Six Miles.	"	684
56	Cave Springs.	"	672
63	Pryor's.	Potsdam.	819
76	Anderson's, Ala.	4 b. Quebec or Knox	702

Western & Atlantic Railroad.

0	Atlanta.	Metamorphic.	1050
23	Marietta.	"	1133
34	Acworth.	" Gold mines.	926
40	Allatoona.	"	878
48	Cartersville.	Knox Shale, Potsdam s.s., 1 m. east	760
68	Kingston.	Knox Shale.	710
78	Adairsville.	"	710
84	Resaca.	Cal. & K. Shale.	654
90	Tilton.	Tren.& K. Dolomite	665
99	Dalton.[13]	" Red Marble.	757
107	Tunnel Hill.[14]	K. Sh. and K Dol.	852

Western & Atlantic Railroad—Continued.

Ms.			Alt.
115	Ringgold.[15]	Trenton.	785
120	Graysville.	K. Shale and Lime quarry.	706
125	Chickamauga.	"	685
130	Boyce, Tenn.	"	694
137	Chattanoog.,Tenn	5 b. Clin. iron ores & 3 b. Calhoun, K. Sh. & K. Dol., Que.	684

Northeastern Railroad of Georgia.

0	Athens.	Metamorphic.	694
12	Nicholson.	"	893
18	Harmony Grove.	"	954
26	Maysville.	"	1001
39	Lula City.	"	1334

Savannah, Griffin & North Alabama R. R.

0	Macon.	Metamorphic.	334
60	Griffin.	"	975
70	Brooksville.	"	
78	Senoia.	"	
86	Sharpsburg.	"	
96	Newnan.	Meta. Snake Creek. Factory, m.	959
.....	Whitesburg.	Metamorphic.	
123	Carrollton.	"	
0	Tennille.	19 a. Tertiary.	
4	Sandersville.	"	

East Tennessee, Virginia & Georgia R. R.

351	Rome.	2-4. Lower Silurian.	
349	Atlanta Junc.	"	
349	Silver Creek.	"	
339	Brice.	"	
337	Seney.	"	
335	Hamlet.	"	
329	Rockmart.	Primordial & Canadian	
323	Braswell.	Primordial.	
317	McPherson.	1. Archæan.	
312	Dallas.	"	
306	Hiram.	"	
301	Powder Springs.	"	

12. Ladd's lime kiln, 3 miles; Rockmart slate quarries, 20 miles; Ward's ferro manganese furnace, 11 miles; Bear Mountain, fine view, 18 miles; Etowah rolling mill site at Falls, 5 miles. Ocoee Conglomerate here and at Rowland Springs, also 5 miles from Cartersville. Flexible sandstone 13, and manganese 3 and 10, and iron ore beds 3, 5, 7 and 10 miles.

13. *Dalton* is situated upon a synclinal, the ridges on each side being Knox Dolomite, and the intervening valley in which most of the town is built is made up of Chazy and Trenton Strata. The fossils of the last named group may be seen in the limestone exposed on Hamilton Hill, immediately north of the town. The Chattoogata Mountain, four miles west, is Upper Silurian.

14. *Tunnel Hill.* The tunnel here is cut through a ridge of Knox Dolomite. The Calciferous and Potsdam is in close proximity to the town on the western side.

15. *Ringgold.* The Upper Silurian occurs in a high sandstone ridge immediately east of the town. The groups here well represented are Medina and Clinton with red fossiliferous iron ore. Oriskany fossils are found abundantly in a single bed of about one foot in thickness. These beds are followed on the east by Devonian and Sub-Carboniferous strata.

NOTE. The Knox Shale and Knox Dolomite of Prof. Safford extends from Tennessee into Georgia, with all the Tennessee characteristics of the groups.

East Tennessee,Virginia & Georgia R. R.*— Continued. Ms.		Alt.	Northeastern Railroad of Georgia. Ms.		Alt.
296	Austell.	1. Archæan.*	0	Athens.	1. Archæan.
293	Mableton.	"	8	Center.	"
286	Chattahoochee.	"	12	Nicholson.	"
285	Peyton.	"	19	Harmony Grove.	"
279	Atlanta.	"	26	Maysville.	"
272	Constitution.	"	32	Gillsville.	"
268	Moore's Mill.	"	39	Lula.	" Stacolumite.
265	Ellenwood.	"	.....	Bellton.	"
259	Stockbridge.	"	.....	Longview.	"
250	McDonough.	"	51	Rabun Gap.	"
243	Locust Grove.	"	59	Clarksville.	"
232	Jackson.	"	63	Anandale.	"
227	Indian Springs.	"	68	Turnersville.	"
218	Frankville.	"	72	Tallulah Falls.	"
206	Dames' Ferry.	"			
199	Holton.	"			
190	Macon.	19. Tertiary.			

Georgia Pacific Railroad.

The portion of this road in Georgia will be found in the chapter on Alabama.

Elberton Air Line Railroad.

0	Toccoa.	1. Archæan.
12	Martin's.	"
24	Bowersville.	"
26	W. Bowersville.	"
39	Bowman.	"
51	Elberton.	"

* This and the following railroads by Prof. A. R. McCutchen.

Alabama.

Dana's Table of Formations.	Alabama Divisions by Prof. Gesner.	Dana's Table of Formations.	Alabama Divisions by Prof. Gesner.
20. Quaternary.	20 c. Alluvium.	10 c. Genesee.	10 c. Black Shale.
"	20 b. Bluff Loam.	7. L. Helderberg.	7. Lo. Helderberg.
"	20 a. Orange s. or dt.	5. Niagara.	5 d. Niagara l. s.
19. Tertiary.	19 c. Pliocene.	5. Clinton.	5 c. Dyestone Group
	19 b. Miocene.	5. Medina.	5 b. Wh. Oak Mt. s.s.
	19 a. Eocene.	"	5 a. Clinch Mt. s. s.
18. Cretaceous.	18 c. Upper Creta's.	4. Trenton.	4 b. Cincinnati.
"	18 b. Middle Creta's.	"	4 a. Trenton.
"	18 a. Lower Creta's.	3. Canadian.	3 c. Chazy.
17. Jurassic.	17 b. Marlstone.	"	3 b. Quebec Knox
"	17 a. Lower Lias.		dolomite.
14. Carboniferous.	14 c. Upp. Coal Mrs.	2. Primordial or Cambrian	3 a. Calciferous.
"	14 b. Low. Coal Mrs.	"	2 b. Potsdam s. s.
"	14 a. Millstone Grit.		
13. Sub-Carbonif's.	13 b. Mountain l. s.	1. Archæan.	2 a. Acadian.
"	13 b. Coral or St. L. ls	"	1 b. Huronian.
	13 a. Barren Group.		1 a. Laurentian.

South and North Alabama, or Louisville and Great Southern Railroad.

Ms.			Alt
0	Decatur.	13 b.L. Ca.,St.Louis	[577]
7	Flint.	"	[569]
13	Hartsell's.	"	[673]
18	Falkville.	"	[603]
23	Wilhite's.	"	[608]
28	Summit.[2]	14 b. War'r coal field. "	[540]
31	Milner's.	"	[502]
33	Cullman's.	"	[592]
35	Phelan's	" Plateau of Sand Mount'n	
42	Hanceville.	"	
49	Bangor.	"	[468]
52	Blount Springs.[3]	{ 13 b. Up. Sub. Carb. / 13 a. Low. Sub. Carb. / 10 c. Blk. Shale.[434] }	
57	Reid's.[20]	14 b. War'r cl. field[592]	
63	Warrior.[4]	"	[549]
68	Morris.	[408] " Jeffe.Cl.Co.	
74	Cunningham.	" [Co.	
76	New Castle.[21]	[440] " N. C. Cl. & I.	
79	Black Creek.	Coalburg Co's colliery.	
81	Boyle's Gap.[22]	14 b. War'r cl. field[524]	
86	Birmingh'm.[5 4 23]	{ 4 a. Trenton. / 3 c. Chazy[602] / 3 b. Quebec. / 3 a. Calcifer. } An Jones Valley	

South and North Alabama, or Louisville and Great South. Railroad.—Con.

Ms.			Alt
90	Grace's Gap.[5]	(See foot note.)	
93	Oxmoor.*	14. Cahawba c. fld	[652]
95	Shade Creek.	"	[564]
99	Brock's.	"	[400]
102	Cahaba Mines.[6]		
104	Helena.[7]	{ 3 a. Calcifer's fault. / 14 b. Coal Meas. [400] }	
109	Siluria.	{ 3 c. Chazy and [464]	
112	Whiting's.	Tren. Lime Wks.[555] }	
119	Calera Hills.	{ 13. Sub-Carbon., 3 c. / Chazy & 4 a. Tren[502] }	
125	Clear Creek.	1 b. Metamorphic.	[540]
130	Jemison.	"	[706]
135	Strasburg.	"	
139	Lomax.	"	[625]
141	Clanton.	"	[596]
148	Cooper's.	"	[458]
151	Verbena.	"	[450]
155	Mountain Creek.	20. Quaternary.	[542]
164	Deatsville.	"	[300]
170	Elmore.	"	[199]
174	Coosada.	"	[175]
179	Alabama River.	18. Cretaceous.	
....	Commerce St. Ju.	" rotten l. s.	
182	Montgomery.	"	[162]

1. Prepared expressly for this work by Prof. William Gesner, of Birmingham, Ala., Geologist and Analytical Chemist, and by Prof. Eugene A. Smith, the State Geologist.
2. Ascending the mountain from Wilhite's to Summit, Flint Creek shows looming above it cliffs of millstone grit, sandstone and shales, as seen from the car windows. W. G.
3. White and red sulphur and Chalybeate waters of great sanitary value at Blount Springs are much resorted to, particularly in the summer season, from all the States; and the Jackson House, by S. D. Holt, is a well kept hotel. The 10 c. Black Shale gives rise to the sulphur springs. The mountain on west side is 14 a. Carboniferous. W. G.
4. The Pierce Coal Mine Company and Alabama M. & M. Company's mines here. W. G.

* Eureka furnaces and coke ovens.

Ms.	Selma, Rome & Dalton Railroad, or Blue Mountain Route.	Alt.	Ms.	Alabama Great Southern Railroad—Continued.	Alt.
0	Selma.	18. Cretaceous. 147	28	Cloverdale.	4 c. Cin. & 4 a. Trenton
9	Burnsville.	" 207	32	Sulphur Sp'gs.24	13 a.b. L. Sub-Carb.668
22	Plantersville.	20. Quaternary. 266	34	Eureka.	" 960
32	Maplesville.	"	40	Valley Head.	" 1012
40	Randolph.	" 351	46	Hollman's.25	" 918
49	Ashby.	471	51	Fort Payne.	" 864
51	Briarfield.8	3 b. Knox Dolomite413	56	Brandon's.	" 877
55	Montevallo.9	3 a. Calcifer's, 1 m. 494	61	Porterville.	"
		3 b. Quebec, 5 miles.	65	Collinsville.	" 710
62	Calera.	{ 3c.Cha.,Tren & ridge of 13 a. Sub-Car.522	74	Greenwood.	" 672
.....	Gardner's.	14. Coosa coal field.567	82	Reases.	" 560
67	Shelby Spr'gs. 10	" 554	87	Attalla. 26	" 568
72	Columbiana.11	3 b. Quebec or Knox560	95	Steele's.27	" 591
82	Wilsonville.	" 452	102	Whitney or Ashville.	" 594
.....	Coosa River.12	" 445	115	Springville.28	3 b.Quebec or Knox706
.....	Coosa Station.	" 472	131	Trussville.	13 a. b. Sub-Carb. 668
90	Childersburg.	" 441	137	Irondale.	5 b. Clinton.
99	Alpine.13	" 495	143	Birmingham.	{ 4 a.Tren. & 3 c.b.& a. of Can. anti. axis877
109	Talledega.	" 586	155	Jonesboro.	3 c. and 3 b. Cana. 508
	(Alabama Fur.)		167	Tannehill.18	3 b.or 3 a.Canadian495
126	Munford.	" 446	170	Woodstock.30	3 b.Quebec or Knox800
.....	Silver Run.14	" 655	174	Red Gap.29	3 b. Knox Dolomite.
130	Oxford.15	" 678	178	Vances.	" 410
131	Anniston.	" Woodstock	183	Clement's.	14b.War'r coal field269
139	Weaver's.	" Iron Wks.	191	Cottondale.	"
145	Jacksonville.	" 653	198	Tuscaloosa.	20.Quat. over L. Cre162
156	Patona.	" 714	204	Maxwell's.31	" 157
.....	Cross Plains.	" 722	213	Carthage.	"
...	Ladiga.	696 " Tecumseh	.....	Stewart's or Havanna.	"
160	Amberson.	727 " Iron Co.	223	Akron.	18 b. Rotten l. s. 170
164	State Line.16	930 " Stonewall Ir.	233	Eutaw.	"
168	Pryor's, Ga.	5 b. Clinton.844[Works	239	Haysville.	"
170	Cave Springs.	4 a. Trenton. 697	243	Boligee.	"
172	Rome, Ga.	" 652	250	Epps.	"
Alabama Great Southern Railroad.18			259	Livingston.	"
0	Chattanooga, T19	4 a. Trenton 665	263	Hooks.	19 a. Tertiary, 36 miles
6	Wauhatchie, "	4 b. Cincinnati. 671	269	York.	" 159
9	Wildwood, Ga.	4 a. Trenton.	274	Cuba.	" 219
12	Morganville, Ga.	"	279	Kewanee.	"
18	Trenton, Ga.	" 720	283	Toomsuba.	" 276
23	Dademon, Ala.	" 815	290	Russell's.	" 393
26	Rising Fawn.	4 c. Cin. & 4 a. Tren.778	295	Meridian, Miss.	" 319

5. The prosperous city of Birmingham is in Jones' Valley. The railroad then passes through Red Mountain by Grace's Gap. The rocks of the anticlinal axis show, at the junction of the Lower Carboniferous with the 5 c. Clinton, an exposure of Fossiliferous Hematite Iron Ore, 28 feet thick, which is being used in the production of an excellent quality of Iron by the Eureka Company, at Oxmoor, at the next station. This bed of iron ore extends from a few miles below Pratt's Ferry on the Cahaba River, in Bibb County, through St. Clair, Cherokee and De Kalb counties, into Tennessee, a distance of 120 miles.

6. S. D. Holt and Davis and Carr's collieries.   W. G.

7. Eureka Company's colliery and Central Iron Works Company, at Helena.   W. G.

8. Branch railroad to Briarfield Rolling Mills and Furnaces.   W. G.

9. Cahaba coal field on the west, with branch railroad to the Montevallo coal mines of Dr. T. H. Aldrich.   W. G.

10. Shelby Springs, Chalybeate and sulphuretted Hydrogen water of great renown, and much frequented.   W. G.

11. Columbiana branch to Shelby Iron Works.   W. G.

12. From Coosa River to Childersburgh, mountains of 2 b. Potsdam sandstone are seen to the southeast from car windows.   E. A. S.

13. From Alpine to Talladega, 2 b. Potsdam sandstone mountains on the west, and 2 a. Acadian slate hills toward the east.   E. A. S.

14. At Silver River, 2 a. Acadian on the east, and 2 b. Potsdam on the west.   E. A. S.

Ms.	Memphis & Charleston Railroad.	Alt.
0	Memphis.	20. Qua., bluff loam 245
5	Buntyn.	" 303
9	White's.	"
15	Germantown.	" 378
19	Bailey's.	{ 19. Tertiary, Orange Sand, LaGrange group.
23	Collierville.	" 378
31	La Fayette.	" 315
39	Moscow.	" 352
52	Somerville.	"
49	La Grange.	" 531
52	Grand Junction.	" 575
58	Saulsbury.	" 535
64	Mile Siding.	19. Ter., Porter's Ck.
74	Pocahontas.	" 394
79	Big Hill.	18. Cre., green sand.
84	Chewalla.	" 409
93	Corinth, Miss.	18 c. Ripley group. 434
107	Burnsville.	" 463
115	Iuka.	13 b. a. Sub-Carbon 555
124	Margerum, Ala.	"
127	Dickson.	" 488
129	Cherokee.	"
133	Barton.	" 498
139	Pride's.	"
145	Tuscumbia.	13 b. L. Carbonif. 468
156	Leighton.	" 563
163	Town Creek.	" 560
169	Courtland.	" 560
176	Hillsboro.	" 599
182	Trinity.	" 534
188	Decatur.	" 573
195	Mooresville.	" 601
203	Madison.	" 573
212	Huntsville. 32	{ 14 a. b. Coal Meas. 13 c. Sub-Carb. 612 13 b. St. Louis l. s.
223	Brownsboro.	" 631
229	Gurley's.	"
233	Paint Rock.	13 b. Sub-Carbon. 596
237	Woodville.	" 601
248	Larkinsville.	" 620
254	Scottsboro.	" 652
259	Bellefonte.	" 639
265	Fackler's.	
271	Stevenson.	{ 3 b. Quebec or Knox Dolomite, with hills of Sub-Carbon and Coal Meas. 603

Ms.	Nashville & Chattanooga R. R.	Alt.
....	Stevenson Junc.	3 b. Quebec or Knox.
....	Bass Station.	"
49	Anderson.	13 a. Sub-Carbon.
39	Stevenson.	3 b. Quebec or Kn. 602
29	Bridgeport.	3 c. Canadian.
22	Shellmound.	20. Quat., Alluvium.
14	Whiteside.	14 b. Coal Mrs. & 13 c.
	(Etna Coal Mines.)	
6	Wauhatchie.	4 b. Cincinnati. 671
0	Chattanooga. 19	4 a. Tren.&3 c. Can. 665

Nashville & Decatur Railroad.

Ms.		Alt.
0	Decatur.	13 b. L. Sub-Carb. 577
3	Harris Station.	" 564
13	Athens.	" 709
22	Elkmont.	13 a. Sub-Carb. 778
....	Pittensville.	"
27	State Line.	13 a.L. Sub-Car. or bar.

Western Railroad of Alabama.

Ms.		Alt.
0	West Point.	1. Archæan.
11	Cusseta.	"
13	Mt. Jefferson.	"
18	Rough & Ready.	"
22	Opelika.	"
28	Auburn.	" & 20. Quat.
35	Loachapoka.	20. Quaternary.
42	Notasulga.	"

Fisher Branch—(Narrow Gauge to Tuskeg.e.)

Ms.		Alt.
48	Chehaw.	20. Quaternary. 252
	(To Tallahassee F	actory.) 1 b. Huronian.
56	Cowles' Station.	20. Quaternary.
65	Shorter's.	b. Cre., rotten l. s.
75	Mt. Meigs.	"
88	Montgomer	" 162
101	Manack.	"
107	Lowndesboro.	"
113	Whitehall.	"
119	Benton.	"
127	Alabama River	"
138	Selma.	" 121

Columbus Branch.

Ms.		Alt.
0	Columbus.	1 b. Huronian. 262
4	Smith's or Dover.	"
6	Mott's Mill.	20. Quaternar
8	Salem,	"
19	Hollis.	1. Archæan.
25	Yonges.	"
29	Opelika.	" 512

15. At Oxford, the railroad crosses through a gap of 2 b. Potsdam, and thence to Cross Plains the mountains of 2 b. Potsdam are on the east side. Beyond Cross Plains, to the State line, these mountains can be seen from the cars. E. A. S.

16. The railroad is built on 3 b. Quebec or Knox dolomite almost all the way from Montevallo to the State line, crossing 3 c. Chazy and 4 a. Trenton near Calera and the Coosa coal field above Calera. E. A. S.

17. Yongesborough narrow gauge railroad, 2¾ miles to Chewackla Lime Company's kilns, southeast. The limestone of this company's quarries is a highly crystalline dolomite. W. G.

18. The hills on the west of the railroad consist principally of limonite, and their detritus constitutes the bright red banks of the cuts and fills for many miles. The Thomas ore bank is on east

Mobile & Girard Railroad.

Ms.			Alt.
0	Columbus, Ga.	1. Archæan.	262
9	Fort Mitchell.	18. Cretaceous.	
20	Seale.	"	
25	Hatchechubbee.	"	
35	Hurtville.	"	
39	Guerryton.	"	
54	Union Springs.	494. " Ripley Gp.	
63	Thomas Station.	"	
72	Linwood.	"	
77	Jonesville.		
84	Troy.	19. Tertiary.	

Mobile & Montgomery Railroad.

0	Montgomery.	18. Cretaceous.	
10	McGchee's.	" rotten l. s.	
16	Morgansville.	"	
21	Letohatchie.	"	
28	Calhoun.	"	
33	Fort Deposit.	520 " Ripley Gp.	
44	Greenville.	19. Tertiary.	
53	Bolling.	"	
60	Georgiana.	"	
67	Garland.	"	
.....	Madge's Mills.	"	
76	Gravella.	"	
81	Evergreen.	"	
86	Sparta.	"	
91	Castleberry.	"	
106	Brewton.	"	
114	Pollard.	"	
119	Whiting or Pensacola Jun.	19. Tertiary.	
134	Williams.	"	
155	Bay Minette.	"	
163	Tensas River.	"	
178	Mobile.	"	

Selma & Gulf Railroad.

0	Selma.	18. Cretaceous.	147
.....	Pleasant Hill.	" rotten l. s.	
.....	Snow Hill.	" Ripley Gp.	
35	Allenton.	19. Tertiary.	
40	Pine Apple.	"	
.....	Cokerville.	"	

Mobile & Alabama Grand Trunk R. R.

Ms.			Alt.
0	Mobile.	19. Tertiary.	6
9	Cleveland.	"	15
20	Cold Creek.	"	34
29	Mount Vernon.	"	49
39	Leona.	"	54
50	Sunflower.	"	28
59	Jackson.	"	42

Mobile & Ohio Railroad.
Part in Alabama.

0	Mobile.	19. Tertiary.	6
5	Whistler.	"	41
18	Chunchula	"	78
33	Citronelle.	"	317
44	Deer Park.	"	148
51	Escatawpa.	"	
63	State Line.	"	236

Alabama Central Railroad.

0	Selma.	18. Cretaceous.	121
.....	Marion Junction.	253 " rotten l. s. gp.	
23	Brown's.	"	282
30	Uniontown.	"	
35	Fawnsdale.	"	
42	Macon.	"	
.....	Van Buren.	"	
50	Demopolis.	"	
66	Coatopa.	" Ripley Gp.	
81	York.	19. Tertiary.	159
.....	Cuba.	"	219
.....	Toomsuba.	"	
108	Meridian.	"	

Montgomery & Eufaula Railroad.

0	Montgomery.	18. Cretaceous.	162
10	Oak Grove.	236 " rotten l. s.	
13	Perry's Mill.	"	
16	Pike Road.	"	295
21	Matthews'.	"	262
25	Mitchell's.	"	252
28	Fitzpatricks.	"	262
33	Thompson's.	"	289

Crossing of Mobile & Girard Railroad.

side, close to the main track, nearly opposite the station house. The hills seen beyond these belong to the Warrior coal field. W. G.

19. In addition to the 4 a. Trenton, there are, within the limits of the city of Chattanooga the 3 a. Calciferous, 4 b. Cincinnati, 5. Clinton, 10 a. Black shale, and 14. Carboniferous formations.
[J. SAFFORD.]

20. *Reids.* Branch railway, 3 miles, of the Warrior Coal and Coke Company to mines working the Warrior Coal bed. (W. G.) The Pierce Warrior Coal Co. working the Warrior Coal bed. The Watts Coal and Coke Co., working the Watts bed. (W. G.)

21. *Newcastle.* Branch railway of Milner Coal and Railway Company, working the Black Creek beds. Also in the Warrior coal field. (W. G.)

22. At *Boyle's Gap* the railroad passes from the Coal Measures, between almost perpendicular walls of 14 a. Millstone grit, into Jones Valley. E. A. S.

23. *Birmingham.* Branch railway, 12 miles. The Birmingham Mineral Railway Station, between the Alice Furnace and Rolling Mills, following the foot of Red Mountain down Jones Valley, principally on the Knox, with the upper Silurian and Clinton Hematite Ore beds to be seen all the way, as presented on the western brow of the Red Mountains nine miles south of Birmingham. (W. G.) Pratt Coal and Coke Company's railway westerly to Coketon mines on the Warrior coal field Pratt coal mines on the Pratt bed, capacity 500 tons per day. (W. G.)

24. From *Sulphur Springs* down to Attalla, the railroad follows the valley lying between Lookout Mountain, 14 a. b. on the east, and the Red Mountain Ridge (5 c., 10 c. 13 a.) on the west, and all the stations are upon the Lower Sub-Carboniferous, 13 a. and b. E. A. S.

Montgomery & Eufaula Railroad—Continued. Ms.		Alt.
40 Union Springs.	18. Cre., Ripley Gp	494
50 Three-Notch R'd.	"	492
54 Midway.	"	506
62 Spring Hill.	"	312
66 Batesville.	"	280
74 Cochran.	"	
81 Eufaula.	{ 18. Cre., marl bluff of the ChattahoochieR. Ripley Group.	200

Selma, Marion & Memphis Railroad.		
.... Selma.	18. Cre., rotten l. s.	147
0 Marion Junction.	"	
14 Marion.	"	253
21 Grove Cottage.	"	
29 Newbern.	"	
37 Greensboro.	"	
45 Sawyersville.	"	

Savannah & Memphis Railroad.		
0 Opelika.	1. Archæan.	819
10 Gold Hill.	"	770
15 Waverly.	"	805
22 Camp Hill.	"	736
(Dudleyville	gold mines).	
30 Dadeville.	1. Archæan.	760
35 Jackson's Gap.	"	695
40 Sturdevant.	"	502
42 Salisbury.	"	
47 Alexander City.	"	747
53 Kellyton.	"	800
60 Goodwater.	Steatite (soap s.)qr.	872

East Alabama & Cincinnati Railroad.		
0 Opelika.	1 b. Huronian.	819
10 Oak Bowery.	"	
23 Buffalo Wallow.	"	

Vicksburg & Brunswick Railroad. Ms.		Alt.
0 Eufaula.	18. Cre., Ripley Gp.	200
5 White Oak.	"	
25 Clayton.	" or Tertiary	

Anniston & Atlantic R. R. (Narrow Gauge.)		
0 Anniston.	Quebec and Knox.	
..... Jenifer.	"	
..... Munfroid. ·	"	
..... Irona.	"	
..... Talladega.	"	561
23 Sycamore.	"	

The Birmingham Mineral Railroad. Branch of the N. & S. Alabama R. R.	
0 Birmingham.	{ 4 a. Tren., 3 c. Chazy, 3 a. Cal., 3 b. Que.615
8 Magella.	3 c. Chazy.
6 Newton.	"
9 Alice.	{ Hematite ore bk.in 5. Clin.of Alice Fur.Co.
10 Woodward.	{ Hematite ore bk.in 5. Clin. Wood. Iron Co.
12 Sloss Mines.	{ Hematite ore bk.in 5. Clin. Sloss Fur. Co.

Montgomery Southern Railroad. (Narrow Gauge.)		
0 Montgomery.	Cretaceous.	162
6 Catoma.	"	
10 Snowden.	"	
13 Pleasant Grove.	"	
17 Reamer.	"	
20 Ada.	"	

Wetumpka Branch S. & N. Alabama Railroad.		
0 Decatur.		575
170 Elmore.	20. Qu. over 1 b. Hu.	197
184 Wetumpka.	1 b. Huronian.	183

25. *Hillman Station.* Branch railway, southeast, 1½ miles long, leaving Quebec or Knox and entering 5 c. Clinton of Red Mountain terminus at the Alice Furnace Co.'s Hematite Mines. 10½ miles south of Birmingham, *Wheeling,* station No. 1, branch railway leaving Quebec or Knox and entering Coal Measures of the Warrior Coal field terminus, 5½ miles northwest Woodward Iron Co.'s mine on the Pratt coal bed. Also, branch railway, southeasterly, 2½ miles to terminus in 5 c. Clinton Hematite ore mines of The Woodward Iron Company. (W. G.)

26. At *Attalla* Lookout Mountain ends abruptly, and the Red Ore Ridge rises to a considerable height on west. Just south of Attalla, through a gap in Red Mountain, the escarpment of Blount Mountain, 14 a. b., is seen to westward.   E. A. S.

27. From *Steele's* to near Whitney, Chandlers Mountain, 14 a. and b., is seen on the west, and below Steele's to Springville the ridge on the west is Red Mountain (5 c., 10 c., 13 a.) All the stations from Attalla to Springville are on Knox Dolomite or Knox shale, 3 a., 3 b.   E. A. S.

28. A short distance below *Springville* the road enters the valley between a Red Ore Ridge on the west and the Cahaba coal field on the east, and continues thus to Irondale.   E. A. S.

29. At *Red Gap* the railroad passes from 13 b. Sub-Carboniferous at Irondale, through a gap in Red Mountain (made up of 5 c., 10 c. and 13 a.) in Jones Valley. Thence to Vances down Jones Valley. At Vances, road enters Warrior coal field and passes out of it at Tuscaloosa. Below Tuscaloosa to Eutaw the surface material is Quaternary, but it overlies the Lower Cretaceous beds, and perhaps beds still older than Cretaceous. Just below Eutaw the rotten limestone begins and is left at Living-stone, where the road enters Tertiary formation, continuing in it to Meridian.   E. A. S.

30. *Woodstock.* Here is Edward's Furnace and a branch railway, almost due south, nine miles, leaving Quebec or Knox and passing over Sub-Carboniferous into Coal Measures of the Cahaba coal field, having passed over the southwesterly extremity of the Clinton ore bed of Red Mountain in Alabama terminus, at two coal mines about two miles apart, Blocton being the first one said to be on the Montevalle coal bed. All the property of the Cahaba Coal Mining Co. (W. G.)

31. *Marvells,* Carthage and Stewart are on Quaternary, overlying a formation older than Creta-ceous, but whether Jurassic, Triassic or Permian, not yet determined, probably the former. E. A. S.

32. The Mountains about *Huntsville* are outliers of the Cumberland Mountains capped with 14 a. and b. Coal Measures, and showing on their flanks Mountain limestone 13 c. and underlying beds down to 13 b. Saint Louis limestones.   E. A. S.

Georgia and Alabama.

Ms.	Georgia Pacific Railway.***	Alt.	Ms.	Georgia Pacific Railway—Continued.	Alt.
0	Atlanta, Ga.³³ { 1 b. Huronian, Mica, Slates & Schists	1050	18	Austell. { 1 a. Lauren. and 1 b. Huronian.	940
3	Howell. { 1 b. Huro. Gneiss in Mica Slates.	962	21	Salt Springs. "	1055
7	Peyton. "	869	27	Douglasville. 1217 " Granite.	
8	Chattahoochee. 1 b. Hu. Mica Slates	822	32	Winston.³⁴ "	1132
9	" River. { 1 a. Lauren. 1 b. Hu. Granite in bed of River.	809	38	Villa Rica.³⁵ 1160 " Gold Mine.	
12	Concord. { 1 a. Lauren. and 1 b. Huronian.	867	45	Temple.³⁶ { 1 b. Huronian, Hornblende, Slates and Schists.	1180
15	Mableton. "	995	52	Summit. "	1424
17	Sweetwater. "	914	54	Bremen. "	1413
			56	Waco. "	1343
			68	Tallapoosa River. "	962

* The geology of this road is furnished by Professors J. L. & H. D. Campbell, of Washington and Lee University, Lexington, Va., and where not otherwise credited the notes are by them also. Those signed W. G. are by Dr. Wm. Gesner, of Birmingham, Ala.

33. *Atlanta.* The broad belt of METAMORPHIC ROCKS, extending from Maryland to central Alabama, belongs to the Archæan age. It has the Blue Ridge of Virginia, the Unica of Tennessee, and the Blue Mountain of Georgia for its northwestern border. Its southwestern margin is approximately defined by the falls and shoals of the rivers at Washington, D. C., at Richmond and Petersburg, Va., at Raleigh, N. C., at Columbia, S. C., at Augusta, Milledgeville and Columbus, Ga., and at Opelika and Wetumka, Ala. An air line from Milledgeville, passing near Atlanta to the limit of the Blue Ridge rocks, would measure the width of the Archæan belt in Georgia, showing it to be about one hundred miles wide.

The Archæan rocks are recognized in Georgia under only two divisions, 1 a. Laurentian and 1 b. Huronian. They constitute the country rocks from Atlanta westward to the margin of Choccolocco Valley at Davisville Tunnel, Alabama, 88 miles. The 1 a. Laurentian group consists chiefly of granite, gneiss and hard schists; while the 1 b. Huronian group consists of less metamorphosed beds of chlorite micaceous and talcosa schists and slates, and some beds of argillites. Both groups are exposed along the railway cuts, but 1 b. Huronian constitutes by far the greater portion of the surface rock. The hard rocks of the 1 a. Laurentian, however, are exposed to view in the bed of the Chattahoochee River, eight miles west of Atlanta, and are quarried a short distance west of the river. The Laurentian also occurs, as shown by the Guide, in the excellent granite quarried at Douglasville, also at Villa Rica. *Concord* to Douglasville, mica and Hornblende slates and schists with beds of granite and gneiss exposed in cuts along railroad. From this point westward to the limit of the Archæan rocks in Alabama are but little exposed.

34. *Winston.* Corundum has been found in considerable quantities near Powder Springs, in Cobb County; also near Villa Rica, Ga., and in Tallapoosa County, Ala.

35. *Villa Rica.* The granite beds make their appearance near Villa Rica, where they seem to underlie the hornblende schists and slates that carry the copper ores (chalcopyrites) of that region, as well as the mica schists in which the gold-bearing veins of quartz in the same vicinity are found. A belt of copper ore (chalcopyrite) crosses the Georgia Pacific Railway, west of Villa Rica, in Carroll County. This ore has been mined to some extent at several points in Douglas, Carroll and Haralson Counties. It is transported to Atlanta where the copper is extracted and the sulphur utilized in the manufacture of sulphuric acid. The same belt of copper ore continues its southeasterly course into Cleburne County, Ala., where the Wood Copper Mines were worked for some years.

The *gold belt* of the Atlantic Slope extending from the Potomac in Virginia, and across North Carolina passes through the northwestern portion of Georgia and terminates in Alabama. It is intersected by the Georgia Pacific Railway at Villa Rica and other points between that and the State line. At Villa Rica gold was very extensively mined forty or fifty years ago; also at Arbacoochee, Cleburne County, Alabama, and at other points in both States.

36. *Temple.* *Mica, talc* and *asbestos* are found in Cobb, Douglas and Carroll Counties, Georgia, and in Cleburne County, Alabama. *Roofing slates* and *flagging stones* have been quarried in Polk and Haralson Counties, Georgia, and are found in Cleburne County, Alabama.   J. L. & H. D. C.

37. [From *Muscadine* to Heflin, metamorphic slates and schists, chloritic and micaceous with some gneiss. Southwest of *Heflin* Station, 14 miles in Cleburne County, are the celebrated Arbacoochee gold mines, and 26 miles the Goo, Smith's and Wood's copper mines; and in Randolph County, near High Shoals, the tin ores lately discovered by Wm. Gesner, Analytical Chemist, Birmingham, Alabama.]
W. G.

38. *Davisville.* Soon after passing the tunnel near Davisville, the road leaves the Archæan rocks and passes abruptly upon the Lower Silurian sandstones, limestones and slates of the beautiful Choccolocco Valley. These sandstones, slates and limestones, of Cambrian and Lower Silurian age, along the southeast margin of the valley, apparently dip under the older Archæan beds, which seems to be due to a fault by which the Cambrian rocks have slipped downward, while by an inversion the Archæan beds have been thrown upon them, so as to give a reversed order of superposition. From Davisville

ALABAMA.

Georgia Pacific Railway— Ms.		Alt.	Georgia Pacific Railway— Ms.	Continued.	Alt.
70	Muscadine.³⁷	1 b. Huronian. 9 4 1	134	Eden.⁴²	{ 14. Coosa Coal Field, { 12. Sub-Carbon. 5 3 8
72	Main's Gap.	" 1 1 1 3	139	Cane Creek Tun.	14 b. Coosa Cl. Fd.⁶³⁸
78	Edwardsville.	" 9 2 3	140	Cook's Springs.	" 6 1 0
84	Heflin.³⁷	" 9 8 6	143	Bald Rock Mt.	{ 14 b. Coosa Coal Fd. { & Millstone Grit.⁷³⁴
87	Davisville Tun.	{ 1 a. Lauren., 1 b. { Huron., nr. fault.⁹⁴⁸	144	Kerr's Gap.⁴³	" 7 5 4
90	Davisville.³⁸	{ 3 b. Silurian and l. s. { Iron Ores. 7 7 5	146	Brompton.	{ 3 b. c. Queb. & Chazy { Silurian Valley. 7 4 6
93	Choccolocco.	" 6 8 2	147	Summit.	"
97	De Armanville.	6 9 8 " Linamite Ores.	150	Leeds.	{ 14 b. Cahaba Coal { Fields. 6 5 6
101	Oxford. ³⁹	{ 2 b. Potsdam, Sand- { stone and Shale.⁶⁵⁰	151	O'Barr's Gap.⁴⁴	" 7 1 2
103	Junction.	3 b. Alluvium. 6 8 2	153	Cahaba River.	" 5 9 0
104	Anniston.⁴⁰	6 9 5 " ore & drift.	158	Weems' Gap.	8 2 3 " & 13.Sub-Carb.
112	Berclair.	{ 3 b. c. Quebec and { Chazy. 6 4 8	161	Irondale.	13 a. Sub-Carbon. 7 6 0
116	Estaboga.	5 3 2 " lime, ore.	162	Red Gap.⁴⁵	{ 5 b. c. Clinton and { 10 c. Genesee.⁷³⁶
122	Lincoln.	" 5 0 5	167	Birmingham.⁴⁶	3 b.Queb. & 3 c.Chy.⁶¹³
127	Coosa River.	" 4 8 8	177	Coalburg.⁴⁷	{ 14 b. Warrior Coal { Field, Pratt seam.
127	Riverside.	" 4 8 9			
129	Seddon.⁴¹	" 5 0 0			

Tunnel the road runs southwest for 12 miles, along the beautiful Choccolocco Valley, passing frequent cuts through Lower Silurian rocks, the lower portion of which are considerably metamorphosed—some of the beds being partially changed to Hydromica slates. *Limonite* ores are very abundant in this valley, are easily mined, and await only capital and labor to make them profitable.

39. Near *Oxford*, Calhoun County, the road changes its course northward through a gap of Ladiga Mountain, cut by Snow Creek. Here the sandstones and shales of the Potsdam group (2 b.) are exposed in well defined arches. These rocks constitute the main mass of the Ladiga and Cold Water Mountains—the ridges which flank the narrow valley in which Oxford and Anniston are situated. These ridges are two great stone-waves, between which we find a synclinal trough which holds the rich beds of Limonite ores, mined to supply the furnaces at Anniston. Oxford is a good starting point for the geological study of this region.

40. *Anniston.* From Anniston the railway turns westward and crosses the wide Silurian limestone valley of the Coosa River, the country rocks of which belong mostly to the Quebec, Chazy, and Trenton epochs.   J. L. & H. D. C.

41. *Sheddon* station is on the western border of the Coosa Valley, upwards of 25 miles wide, diagonally as the railway crosses it; and a little east of Eden Station it passes abruptly into the Sub-Carboniferous formation of the Coosa, or third or most easterly Alabama coal field. (W. G.) The Coosa Valley is a prolongation of the great Silurian Valley of Virginia and Tennessee, while the Choccolocco and Anniston Valleys on the one side, and the Cahaba and Birmingham Valleys on the other, may be regarded as its branches or outliers. The width of the Coosa Valley by the line of the Georgia Pacific Railway is 25 miles. Many promising beds of iron ore are found near this line. The Coosa Valley is the southern terminus of one of the most interesting and important valleys in the World, in a geological view. Tracing the 4 a. Trenton limestone, and the 4 c. Hudson River slate formations from their classical localities, from which they derive their names, Trenton Falls, N. Y. (see note 62 of that State), and the Hudson River, we find them in the Mohawk Valley of New York, with branches extending far into New England and Canada. Following it southwestward it crosses New Jersey and southeastern Pennsylvania by Easton, Lebanon, Harrisburg, Carlisle and Chambersburg, as the Cumberland or Kittatinny Valley, into Maryland, past Hagerstown and through Virginia as the Shenandoah or Great Valley, by Winchester and Stanton; and, being divided by the Massanutten Mountain, on the east side by Sheperdstown, Luray, to Roanoke, and into Tennessee, where it is the valley of East Tennessee, and finally in Alabama its two divided branches sink and disappear beneath the cretaceous plains of the South. In Alabama the Trenton is much less conspicuous than the Canadian group. (3 a. b. c.)   J. M.

42. *Eden.* [North of this station are the Broken Arrow and Front Creek coal mines, in the Coosa coal field. (W. G.)] A few miles west of Coosa River we find an abrupt transition to the Sub-Carboniferous of the Coosa coal field. Near Eden station the road passes through a ridge of Sub-Carboniferous limestone, directly upon the highest coal-bearing beds of this region, which dip beneath the older Sub-Carboniferous strata. This can be best accounted for on the hypothesis of a fault. Sub-Carboniferous fossils are found in this neighborhood in abundance. Promising seams of coal are found in this field and have been mined to some extent. The Broken Arrow Wells, valued for their mineral waters, are situated in this region.

43. *Kerr's Gap.* At Kerr's Gap, where the road passes from the Coosa field into Cahaba Valley, the Millstone Grit (here a coarse conglomerate, 80 to 100 feet thick) has a high outcrop on the Coosa or Bald Rock Mountain: Dipping beneath this are the Sub-Carboniferous formations, followed by the Silurian limestones, all dipping to the southeast. Valuable iron ores and limestones, with one good vein of Baryte are found here. Along the western margin of this valley the Silurian limestones have been abruptly cut off by a fissure, and the coal-bearing beds (14) of the Cahaba field have dropped down so as to abut against them. The geological structure of this field is very analogus to that of the Coosa field—both apparently *monoclines*, limited by faults along their eastern margins. Valuable coal mines have been opened here.

44. [*O'Barr's Gap* is in the western boundary of the Second or Cahaba coal field of Alabama; and as this railway crosses the Big or West Cahaba River, at Sycamore Ford, and keeps the face of its western bluff a considerable distance, a good view of the strata of shales, sandstone, and some of the Cahaba coal beds can be seen from the cars.] (W. G.)

45. *Red Gap.* The road passes from Sub-Carboniferous of Cahaba field into the Birmingham (or Jones) Valley through *Red Gap*, which presents a section of the Clinton group that carries the great bed, 30 feet thick, of fossil ore so extensively worked in this part of Alabama. Here the road cuts beds that are probably Genesee (10 c.)

46. *Birmingham* is a rapidly growing city, in and around which are several large iron furnaces and other manufacturing enterprises. Here ores, limestones, coal, and building material are found in unusual contiguity and abundance.

47. *Structure of the Alabama Coal Fields.* There is good reason to believe that the Coosa, Cahaba and Warrior coal fields were originally one common field, which, previous to the Appalachian Revolution, stretched across the areas that are now [the Cahaba and Birmingham Valleys. But these valleys and their margins are now only the relics of a monoclinal uplift, in the one case, and of an irregular anticlinal stone-wrinkle in the other, which were thrust up so high and bent so sharply as to fracture, not only the coal-bearing strata on top, but also the underlying Sub-Carboniferous and Clinton beds and many of the Silurian limestones that now form the bottoms of the valleys.

48. When this railway has been extended westward from Coalburg until it meets its western division, now under construction east of Artesia on the Mississippi & Ohio Railway, it will traverse the Great Warrior coal field over its most productive portions. Between this coal field and the Mississippi it will cross a wide belt of timber, cotton and corn lands. The line will intersect every geological formation found in the Southern States, from the Archæan, at Atlanta, up to the Quaternary, and must always be an interesting route for scientific travellers. J. L. & H. D. C.

Mississippi.[1]

LIST OF GEOLOGICAL FORMATIONS IN MISSISSIPPI.

20. QUATERNARY.	20 e. Alluvial. 20 d. Yellow Loam. 20 c. Loess. 20 b. Port Hudson. 20 a. Orange Sand or Stratified Drift.	19. TERTIARY EOCENE.	19 e. Vicksburg. 19 d. Jackson. 19 c. Claiborne. 19 c. Burstone. 19 a. LaGrange.
		18. CRETACEOUS.	18 d. Ripley Group. 18 c. Rotten Lime s. 18 b. Tombigbee S'd
19. LATER TERTIARY.	19 f. Grand Gulf.		18 a. Eutaw.
		13. SUB-CARBON'S.	13 a. Keokuk or St. Louis Lime s.

[1] By Prof. E. W. Hilgard, Berkeley, Cal., late State Geologist of Mississippi, but, owing to the distance, he was unable to correct the proof sheets.

Notes on the Geological Formations of Mississippi.

Brief descriptions of some formations peculiar to the Southern States seem to be required. Mississippi is a Tertiary and Cretaceous State, by far the greater portion of it being occupied by the former, if we leave out of consideration the strata of the Orange Sand, which undoubtedly forms the greater portion of the actual surface. These formations have been well studied and described by Professor Eug. W. Hilgard, from whose reports the following brief descriptions of the several subdivisions have been taken.

20 Quaternary.

20 e. *Alluvial Deposits.* These include all the soils, first bottom deposits, and sand bars now in process of formation, or attributable to causes now in action. The lower bottoms of the Mississippi River, now frequently overflowed, are bordered by level tracts of land sometimes several miles in width, evidently formed in flowing water, but of too high a level to have been formed by the present river, and being probably due to ancient glacial rivers.

20 d. *Yellow Loam.* The yellow, brown, or reddish loam forms the surface and furnishes the soils of the greater portion of the State of Mississippi, and is the source of its wealth as a great cotton-growing State. Professor Hilgard thinks it was an independent acqueous deposit posterior to the Bluff and Orange Sand, and anterior to the alluvial formations of the present epoch. Its prevalent character is that of a yellow clay or loam, without any definite structure or cleavage, variously tinged with iron, and it forms the best upland soils and sub-soils of the State, averaging about three feet in thickness, and sometimes twenty feet.

20 c. The *Bluff*, or *Loess*, of Mississippi, or cane-hills belt, presents the same remarkably uniform features as in other States and in all parts of the world, as described in the introduction to this volume. It consists of a fine silt, almost too silicious to be called a loam, of a grayish or yellowish buff tint. A certain degree of firmness is imparted to the mass, caused as Professor Hilgard thinks, by rough, irregular concretions, varying in size from fine sand grains to the weight of several pounds, (Loess puppets), into which the fine material has been cemented by earthy carbonates. Hence, it is little subject to erosion, maintains itself readily in even vertical cuts, and valleys cut into it have steep slopes, at times almost vertical walls.* Its thickness is sometimes as much as seventy feet, but it shows only obscure marks of stratification. Its fossils are terrestrial snails and quadrupeds.

20 b. *Port Hudson.* This is a formation consisting, in its landward portion chiefly of paludal, mostly dark-tinted and well stratified calcareous clays, often overlaid by brownish ill stratified loams, which intervene between it and the Loess proper. Its chief fossils are a fresh water and land fauna, among many vegetable remains, including cypress stumps. To seaward the beds become more brackish and finally of purely marine character. It underlies the Mississippi alluvium at least as far as Memphis, rises into "Crowley's Ridge," in Arkansas and Southeast Missouri, and also underlies the Red River alluvium to Shreveport. It is most widely developed in Louisiana.

20 a. The *Orange Sand*, or stratified drift, is an important formation. It covers nearly the whole State of Mississippi, except the alluvial bottoms of the river, being, however, itself often covered by the later formations above described. It forms the main body of most of the ridges of the State, and to a great extent their surface. It gives character to the surface conformation, which, contrary to the popular impression, is generally hilly back from the river, though nowhere mountainous. All the sandy hills seen from the railroad, from 30 to 120 feet high, few of them as high as 400 feet, which are conspicuous features in the landscape, are due to the Orange Sand formation, out of which the hills have been formed by denudation of the valleys and lower ground. The sand of which it is chiefly com-

* In *Science*, for August, 1884, I maintained that the steep slopes of the Loess were owing to its laminated structure, like the Genesee, and other shales.   J. M.

Chicago, St. Louis & New Orleans Railroad. Illinois Central Line.		
Ms.		Alt.
0	New Orleans, La.	16
48	Ponchatoula.	
78	Tangiphoa.	
88	Osyka.	{ 20 a. Orange Sand. / 19 f. Grand Gulf.
98	Magnolia.	" 93
108	Summit.	"
118	Bogue Chitto.	"
128	Brookhaven.	"
139	Beauregard.	"
149	Hazlehurst.	"
158	Crystal Springs.	20 d. Yellow Loam.
167	Terry.	{ 20 d. Yellow Loam, / 20 c. Alluvial.
174	Byram.	{ 19. Eocene and / 20 c. Alluvial.
183	Jackson.	{ 20 d. Yellow Loam, / 19 d. Jackson.
195	Madison.	"
206	Canton.	"
220	Vaughan's.	{ 20 c. Alluvial and / 19 d. Jackson.
234	Goodman.	{ 20 c. Alluvial and / 19 c. Claiborne.
242	Durant.	"
251	West's.	{ 20 c. Alluvial and / 19 b. Burstone.
262	Vaiden.	{ 20 d. Yellow Loam, / 19 d. Burstone.
271	Winona.	"
283	Duck Hill.	{ 20 d. Yellow Loam, / 19 a. LaGrange.
295	Grenada.	{ 20 c. Alluvial and / 19 a. LaGrange. 213
310	Coffeeville.	{ 20 d. Yellow Loam. / 19 a. LaGrange.
323	Water Valley.	{ 20 c. Alluvial and / 19 a. LaGrange.
333	Taylor's.	"
340	Oxford.	{ 20 c. Alluvial, / 20 a. Orange Sand. / 19 a. LaGrange.
357	Abbeville.	"
369	Holly Springs.	{ 20 d. Yellow Loam, / 19 a. LaGrange.
378	Hudsonville.	{ 20 c. Alluvial and / 19 a. LaGrange.
382	Lamar.	{ 20 d. Yellow Loam, / 19 a. LaGrange.
394	Grand Jun., Tenn.	575

Mississippi & Tennessee Railroad.		
Ms.		Alt.
0	Grenada.	{ 20 c. Alluvial, 213 / 19 a. LaGrange.
22	Oakland.	{ 20 b. Yellow Loam, / 19 a. LaGrange.
41	Bateville.	
50	Sardis.	{ 20 b. Yellow Loam, / 19 a. LaGrange,
63	Senatobia.	
88	Hernando.	{ 20 c. Loess, / 19 a. LaGrange.
100	Memphis.	" 258

Natchez, Jackson & Columbus Railroad.		
0	Natchez.	{ 20 c. Loess, / 19 f. Grand Gulf.
26	Fayette.	{ 20 d. Yellow Loam, / 19 f. Grand Gulf.
43	Martin.	"
78	Oakley.	"
100	Jackson.	"

Mobile & Ohio Railroad.		
63	State Line.	19. Later Tertinary.
71	Buckatunna.	" 150
82	Waynesboro.	{ 20 d. Yellow Loam, / 19 e. Vicksburg. 191
96	Shubuta.	{ 20 d. Yellow Loam, / 18 d. Ripley Gp. 197
109	Quitman.	{ 20 d. Yellow Loam, / 19 c. Claiborne. 231
120	Enterprise.	{ 20 c. Alluvial, / 19 b. Burstone. 248
135	Meridian.	{ 20 c. Alluvial, / 19 b. Burstone .336
147	Lockhart.	19 b. Burstone. 360
164	Narkeeta.	" 183
176	Scooba.	{ 20 c. Alluvial, 193 / 18 c. Rotten Lime s.
188	Shuqulak.	{ 20 d. Yellow Loam, / 18 c. Rotten l. s. 321
198	Macon.	" 185
211	Crawford.	" 316
219	Artesia.	" 244
232	West Point.	" 243
241	Muldon.	" 304
254	Egypt.	" 306
262	Okolona.	" 311
275	Verona.	" 307
287	Saltillo.	" 318
297	Baldwyn.	" 379

posed is in color of an orange yellow, sometimes very deep and glaring, but more frequently it is a dull rust color; in some places of a delicate rose color, with frequently bright yellow tints, and there are some deposits of white sand. There are, of course, an endless variety of intermediate tints, and sometimes crimson, purple and almost blue tints are observed. It also contains extensive gravel beds, usually forming belts of a general north and south direction; and irregular beds and bands of clayey materials are common where clayey formations underly. Its origin is not yet clearly ascertained, but it appears very much like a glacial river deposit, the materials being mainly derived from places south of the Ohio River on either side of the Mississippi. As the Mississippi must have been the great outlet of the vast glacial rivers of the age of Ice, it is not to be supposed that it would leave no

Mobile & Ohio Railroad—Continued.

Ms.	Continued.	Alt.
309	Booneville.*	20 d. Yellow Loam, / 18 c. Rotten l. s.[511]
318	Rienzi.	20 d. Yellow Lm.[441] / 18 b. Tombigbee Sd.
329	Corinth.	20 d. Yellow Loam, / 18 c. Rotten l. s.[434]

E. Tennessee, Virginia & Georgia R. R.
Memphis & Charleston Division.

Ms.		Alt.
79	Big Hill, Tenn.	20 a. Orange Sand, / 19 a. LaGrange.
84	Chewalla.	18 c. Rotten l. s. [409]
93	Corinth.	20 d. Yellow Loam, / 18 c. Rotten l. s.[434]
107	Burnsville.	20 a. Orange Sand, / 18 a. Eutaw. [463]
115	Iuka, Ala.	20 a. Orange Sd.,[455] / 13 a. Keokuk or St.L.

(See Alabama for this Railroad.)

Cincinnati, New Orleans & Texas Pacific Railroad.
Vicksburg & Meridian Division.

Ms.		Alt.
0	Vicksburg.	20 c. Loess, / 19 e. Vicksburg.[808]
10	Bovinia.	"
18	Edwards.	"
27	Bolton.	"
35	Clinton.	20 d. Yellow Loam, / 19 d. Jackson.
45	Jackson.	"

Cincinnati, New Orleans & Texas Pacific Railroad. Continued.

Ms.		Alt.
59	Brandon.	20 d. Yellow Loam, / 19 f. Grand Gulf. / 19 e. Vicksburg.
70	Pelahatchie.	20 d. Yellow Loam, / 19 a. Vicksburg.
79	Morton.	"
90	Forrest.	"
100	Lake.	20 d. Yellow Loam, / 19 c. Claiborne.
109	Newton.	"
122	Chunky.	20 d. Yellow Loam, / 19 b. Burstone.
140	Meridian.	20 c. Alluvial, / 19 b. Burstone. [336]

New Orleans & Northeastern Railroad.

Ms.		Alt.	
0	Meridian.	19 b. Burstone.	[336]
17	Enterprise.	19 c. Claiborne.	[248]
30	Barnet.	19 f. Grand Gulf.	[306]
47	Sandersville.	" (Generally over-laid by 20 a. Orange Sand.)	
64	Ellisville.	"	[239]
85	Hattiesburg.	"	[144]
101	Purvis.	"	[360]
131	Derby.	"	[168]
147	Mitchell.	"	[69]
160	Pearl River		
167	Slidel, La.	20 c. Loess, / 20 b. Port Hudson.	[8]
191	Lake Shore.	"	
196	New Orleans	"	[16]

* Booneville, highest railroad point in the State.

traces of that period behind in some of the States on its borders. There is no doubt the deposition of the orange sand took place in flowing water, whose current had a general direction from north to south. This formation is 40 to 60 feet thick; 100 feet is not unusual, and even 200 feet. It contains the fossils of the underlying formations, but none of its own. The materials are non-calcareous and peroxidized throughout; highly ferruginous, and in part silicious sandstones form limited deposits, very frequently capping hills and ridges which have thus been preserved from erosion, profoundly influencing the surface conformation.

19. Later Tertiary.
19 f. *The Grand Gulf.* The highest Tertiary formation apppearing on the surface of the State is the Grand Gulf group of blue, green and white, compact clays, and mostly soft whitish sandstones overlying the same. No fossils save a few leaves and small lignite beds have been found in it, although it occupies, in the southern part of the state, the large area covered by the long leaved pine. It is supposed to be of Miocene age.

19. Tertiary.
19 e. *Vicksburg* Miocene, the highest of the marine tertiary formations, occupies a narrow belt of nearly uniform width, extending across the State to the Tombigbee River in Alabama, and it contains a valuable crystalline limestone, associated, however, with blue and white marls and important beds of lignite, but the chief material is a soft white limestone.

19 d. *Jackson.* The territory of this group is characterized by the occurrence of the black prairie soil on its surface, and also of bald prairies, both very similar to those of the Rotten Limestone region. The material is either a soft yellowish limestone or indurated marl or a soft gray or yellowish calcareous clay, in which the large bones of the Zeuglodon are found.

19 c. *Claiborne.* This group of blue and white calcareous marls occupies but a small area in the state, its fossils are poorly preserved, and it imparts no obvious features to the surface of the country underlaid by it.

19 b. *Burstone.* ("Silicious Claiborne," of Hilgard's Mississippi report). This group forms a wide and to northward ill-defined belt, northward of the Claiborne and Jackson area. Its materials are mostly soft yellowish or whitish sandstones and claystones, alternating with dark-tinted lignito-gypseous clays and sands; sometimes unconsolidated fossiliferous sands and silicious sandstone of the "burstone" character; also, highly ferruginous clays. Northward it passes insensibly into

Louisville & Nashville Railroad.			Louisville, New Orleans & Texas R. R.—		
Ms.	New Orleans & Mobile Division.	Alt.	Ms.	*Continued.*	Alt.
0	New Orleans.	16	245	Redwood.	{ 20 d. Alluvium over 20 b. Port Hudson.
52	{ Bay St. Louis, Miss.	{ 20 c. Alluviai, 34 20 b. Post Hudson.	257	Halpin.	"
59	Pass Christian.	" 10	271	Cary.	"
71	Mississippi City.	" 10	278	Rolling Fork.	"
82	Ocean Springs.	" 38	284	Anguilla.	"
101	Scranton.	"	288	Nitta Yama.	"
141	Mobile.	" 6	306	Arcola.	"
			316	Leland.	"
Louisville, New Orleans & Texas R. R.			331	Nicholson.	"
Baton Rouge to Memphis.			342	Coleman.	"
			363	Duncan.	"
89	Baton Rouge.	{ 20 c. Loess over 20 b. Port Huron.	370	Bobo.	"
			378	Clarksdale.	" 87
108	Slaughter.	"	398	Lula.	"
113	Ethel.	"	415	Tunica.	"
122	Wilson.	{ 20 a.Orange Ld. over 19 b. Port Hudson.	426	Robinsonville.	"
			440	Walls.	"
135	Centreville.	"	442	Lakeview.	{ 20 c. Loess over 20 a. Orange Sand and 19 a. Eocene.
144	Gloster City.	"			
152	Day's.	"	455	Memphis.	" 227
160	Knoxville.	'			
175	Hamburg.	"	**Grand Gulf & Port Gibson Railroad.**		
186	Harriston.	"			
193	Hays.	20 c. Loess.	.....	Grand Gult.	{ 20 c. Loess, 19 f. Grand Gulf.
206	Port Gibson.	"			
218	Allens.	"	.....	Port Gibson.	"
222	Yokena.	"			
227	Warrenton.	" over 19 Eocene.			
235	Vicksburg.	" " 308			

19 a. *La Grange or Lignite* ("Northern Lignitic" of Hilgard), which underlies all of the northern part of the state outside of the Cretaceous area, itself mostly covered by the Orange Sand. It consists of mostly dark-tinted shaly clays, interstratified with gray sands and lignite beds of some economic importance; shows a few marine outliers showing near relation to the Burstone, or more probably to the " Woods Bluff" beds of Alabama, the base of the Eocene Tertiary.

18. Cretaceous.

18 d. *Ripley Group* is composed of hard crystalline limestone, the highest strata and bluish micaceous marls more or less sandy below. The country suddenly becomes hilly and broken as you enter this formation. It is a hard, sandy limestone, with strata of blue shale marl between, and one of heavy gray calcareous clay on top.

18 c. *The Rotten Limestone* is an important formation 700 to 1,000 feet thick in the southwest, and thinning down in the northeast to 70 to 100 feet at the Tennessee line. The material is of great uniformity, a soft, chalky rock of a white or pale bluish tint, with a very little sand. When the rotten limestone appears on the surface it appears white or yellowish white, and preserves the same tint from 2 to 18 feet deep. Below that it is often bluish gray, which, when wet, looks quite dark. These white clay marls or soft limestone form a level or gently undulating surface with a heavy calcareous soil in the Prairie Region proper, and comprises some of the best land in the State.

18 b. *Tombigbee sand* has as its prevalent material a fine grained micaceous sand, usually of a greenish tint, but not unfrequently gray, bluish, black, yellow, and sometimes even orange red. The region is hilly and sandy and the soil generally inferior.

18 a. *Eutaw*. The territory occupied by this formation offers no striking characteristics in Mississippi, by far the larger portion of it being covered thickly by the Orange Sand. It consists of unconsolidated sands and dark-tinted clays.

14. The Sub-Carboniferous occupies a very small territory in the northeastern section of the State adjoining Alabama, and its geological relations can hardly be satisfactorily studied in Mississippi.

The Cretaceous and Tertiary formations of Mississippi are rich in fossils and afford favorite localities for the palæontologist. The geology of Mississippi may become important in the study of the vast, almost unknown region between the Mississippi River and the Sierra Nevada, where the same formations seem to prevail. In this connection see Mr. Loughridge's notes on the Indian Territory

The foregoing descriptions of the sub-divisions of the Cretaceous, Tertiary and Quaternary apply to these formations in the adjoining States of Tennessee, Alabama and Louisiana. J. M.

Louisiana. [1]

LIST OF THE GEOLOGICAL FORMATIONS IN LOUISIANA.

General Table.	Louisiana Formations.	General Table.	Louisiana Formations.
20. Quaternary.	20 d. Alluvium. 20 c. Bluff or Loess. 20 b. Port Hudson. 20 a. Orange Sand or Stratified Drift.	19. Tertiary.	19 f. Grand Gulf Miocene. 19 a. Eocene.
		18. Cretaceous.	18. Cretaceous.

General Geological Note on Louisiana.

Louisiana is not wholly alluvial, as is the general impression; only about one-half of the State, in fact, belonging to the alluvium of the Mississippi and Red Rivers and to the marsh region of the coast. A considerable portion of this, too, is older than the present river channels. Such is the case with the greater part of the "buck-shot" soils, where certain strata of dark colored clay come to the surface. These clays underlie the entire plain from the Gulf coast as high as Memphis and Shreveport at depths of from one to forty feet, and are the older portions of the Champlain formation, most definitely exhibited at Port Hudson Bluff, 20 b.

Next above and north of these prairies occur the beds of sand and gravel belonging to the "Stratified Drift," capping the higher ridges all over the upland portion of the State. It is the 20 a. Orange Sand.

The next formation is the 19 f. "Grand Gulf" group of the Tertiary formation, blue, green and white clays, clay stones and clay sandstones, rising into high ridges as we advance northward, and forming a prominent hilly belt across the State.

Northward, again, of this transverse ridge we find a narrow belt of the calcareous marls and limestones of the Marine Tertiary, 19 e. Vicksburg and 19 d. Jackson groups approaching the surface.

In northwestern Louisiana fossiliferous rocks, mostly ferruginous and red, or sometimes calcareous of Upper 19 c. Claiborne or Lower 19 d. Jackson of Tertiary age, are found and known as the Red Lands. The upper portion of the ridges is composed of or capped by the irregularly bedded sands of the 20 b. Stratified Drift.

See the descriptions of the formations in the Mississippi chapter.—*From E. W. Hilgard's Cotton Report.*

Louisville & Nashville Railroad. New Orleans & Mobile Division.		
Ms.		**Alt.**
0	New Orleans.	20 c. Alluvium.
5	Pontchartrain Junc.	"
9	Lee.	"
13	Micheaud.	"
20	Chef Menteur.	"
26	Lake Catherine.	"
31	Rigolets.	"
36	Lookout.	{ 20 c. Alluvium. 20 b. Port Hudson.
40	Claiborne.	"
45	Toulme.	"
48	Waveland.	"
52	Bay St. Louis.	"

(Continued in Mississippi.)

Cincinnati, New Orleans & Texas Pacific Railroad.		
0	New Orleans.	{ 20 d. Alluvium over 20 b. Port Hudson.
5	Lake Shore.	"
18	Pt. Aux Herbra.	"
28	Slidell.	"

Cincinnati, New Orleans & Texas Pacific Railroad—*Continued.*		
Ms.		**Alt.**
36	Pearl River.	{ 20 d. Alluvium over 20 b. Port Hudson.
43	Nicholson.	"
49	Mitchell.	{ 20 a. Orange S'd over 19 f. G'd Gulf Mioc.
53	Highland.	"
64	Derby.	"

Illinois Central Railroad. (Chicago, St. Louis & New Orleans Division.)		
0	New Orleans.	{ 20 c. Alluvium over 20 b. Port Hudson.
10	Kenner.	"
37	Manchac.	"
48	Ponchatoula.	"
53	Hammond.	20 b. Port Hudson.
68	Amite.	{ 20 a. Orange S'd over 19 f. G'd Gulf Mioc.
78	Tangipahoa.	"
88	Osyka.	"

(Continued in Mississippi.)

[1] By Prof. E. W. Hilgard, Berkeley, Cal., late State Geologist of Louisiana; but, owing to the distance, he was unable to correct the proof sheets.

Louisville, New Orleans & Texas R. R.

Ms.		Alt.
0	New Orleans.	20 d. Alluvium.
5	Sauve.	"
10	Kenner.	"
23	Sarpy's.	"
34	St. Peter's.	"
40	Mount Airy.	"
56	Whitehall.	"
71	Southwood.	"
76	St. Gabriel.	"
89	Baton Rouge.	{ 20 c. Loess over 20 b. Port Hudson.
90	Baker.	"
108	Slaughter.	"
113	Kilbourne.	"

Morgan's Louisiana & Texas R. R.

Ms.		Alt.
0	New Orleans.	20 d. Alluvium.
3	Gretna.	"
12	Jefferson.	"
24	Boutte.	"
40	Raceland.	"
52	Lafourche.	"
60	Thibodaux.	"
55	Terrebonne.	"
70	Houma.	"
66	Tigerville.	"
73	Bœuf.	"
80	Morgan City.	"
81	Berwick.	"
100	Franklin.	"
113	Jeannerette.	20 b. Port Hudson.
125	New Iberia.	"
144	Lafayette.	"
157	Grand Coteau.	"
166	Opelousas.	"
172	Washington.	"
179	Garland.	20 d. Alluvium.
186	Whiteville.	"
195	Eola.	"
204	Cheneyville.	"
215	Lamourie.	"
228	Alexandria.	{ 20 d. Alluvium over 20 b. Pt.Hud's & 19 f. G'd Gulf Miocene.

Galveston, Harrisburg & San Antonio R. R.
(New Orleans to Orange.)

Ms.		Alt.
0	New Orleans.	20 d. Alluvium.
....	Algiers.	"
55	Terrebonne.	"
80	Morgan City.	"
101	Franklin.	"
125	New Iberia.	20 b. Port Hudson.
144	LaFayette.	"
172	Estherwood.	"
184	Jennings.	"
206	Pine Grove.	"
228	Sulphur Mine.	{ 20 b. Pt.Hudson over 19 a. & 18 Creta.
235	Edgerly.	"

Galveston, Harrisburg & San Antonio Railroad- *Continued.*

Ms.		Alt.
246	Sabine.	20 d. Alluvium.
256	Orange.	"

Missouri Pacific Railroad.
(New Orleans to Marshall.)

Ms.		Alt.
0	New Orleans.	20 d. Alluvium.
3	Harvey's Canal.	"
19	Davis.	"
39	Johnson.	"
54	Forstall.	"
64	Donaldsonville.	"
85	Plaquemine.	"
89	Baton Rouge Jun.	"
97	W. Baton Rouge.	"
127	Ravenwood.	"
140	Goshen.	"
154	Morrows.	"
172	Cheneyville.	"
188	Moreland.	"
210	Boyce.	19 f. Grand Gulf Mio.
224	Chopin.	"
237	Prudehomme.	"
247	Provencal.	"
260	Marthaville.	19 a. Eocene.
270	Sodus.	"
288	Mansfield.	"
303	Gloster.	"
318	Reisor.	"
328	Shreveport.	20 d. Alluvium.
343	Greenwood,	19 a. Eocene.
352	Jonesville.	"
360	Scottsville.	"
368	Marshall.	"

Cincinnati, New Orleans & Texas Pac. R. R.
(Vicksburg to Shreveport.)

Ms.		Alt.
0	Vicksburg.	19 a. Eocene.
0	Delta.	20 d. Alluvium.
7	Mounds.	"
11	California.	"
18	Tallulah.	"
25	Quebec.	"
32	Waverly.	"
36	Delhi.	20 b. Port Hudson.
41	Carpenter's.	"
48	Bee Bayou.	20 d. Alluvium.
52	Rayville.	"
65	Gordon.	"
73	Monroe.	"
82	Cheniere.	"
87	Forksville.	19 a. Eocene.
89	Calhoun.	"
93	Averitt.	"
97	Choudrant.	"
105	Ruston.	"
110	Allengreene.	"
114	Simsboro.	"
122	New Arcadia.	"
144	Minden Junction.	"
157	Haughton.	"
170	Shreveport.	20 d. Alluvium.

Florida[1]

General Note on the Geology of Florida.

The first intimation given to the scientific world of the true geology of Florida was by Dr. Eugene A. Smith in his report upon the "Soils of the Cotton Region" in Vol. VI. of the U. S. Census of 1880. The western, northern and middle highland regions mostly occupied his attention. To him is due the discovery that the oldest rocks of the Peninsula are of the division of the Eocene, known in Alabama and Mississippi as the Vicksburg Formation. In 1885, the U. S. Geological Survey prosecuted some work in Florida, principally for the collection of Tertiary fossils, and the observations there made, so far as published, (see Article in "The American Journal of Science," October, 1888, by L. C. Johnson,) show that the Eocene Axis is quite narrow, and not manifest by outcrops further south than Sumter County; by some of its effects it is traceable to Polk County. It is the basis of the "Interior Basin." The next and the most extensive development was called the "Waldo," from the place where the most abundant and decisive fossils were found. This has proved to be Miocene. Most of the phosphatic rocks belong to it. It is also the basis of the Lake region and of the "High Hummocks." It reaches the "Trail Ridge" and highlands of the eastern slope, and occupies the western slope to the Gulf as far south as Tampa.

The greater part of the St. John's River country is Pliocene, with much that is even later. The Jacksonville Formation, exposed at the water works, has been assigned to the Pliocene; while the "cochina" of St. Augustine and the marls of Indian River belong, probably all of them, to Post Pliocene times. The phosphatic rocks of Black Creek and of Enterprise—perhaps on insufficient grounds—are supposed to belong to the Jacksonville Formation.

In 1887, Prof. Angelo Heilprin, in a "Report of a Visit to the Southwest of Florida" decided the formations at Tampa to be Miocene, south of that, as far as explored and definitely settled by fossils, Pliocene. The actual coast and coral reefs and islands must be later.

The underlying limestones in many sections of the state have been dissolved in an irregular and often fantastic manner, producing sink holes, underground channels and numerous ponds and lakes.

The soils on the immediate surface of the country consist mainly of such sands as would be left by a receding ocean. In some places these are drifted into dunes, such as the high "Trail Ridge" and its continuations east, and the lower sand dune hills westward, which overlook the Hummock region, and separate it from the "Interior Basin." Probably the clays and "red lands" generally are derived, by disintegration and leaching from Miocene rocks. The interior "High Hummocks" are Miocene, or a few to the north Eocene, and the "Low Hummock" of the coast Pliocene or later.

The elevations of the highest ridges seldom exceed two hundred feet, whilst the Interior Basin and highest of the hills of the western region are not often much over one hundred feet, while the lower part of the state, south of Polk County, has an average elevation of only about thirty to forty feet above low tide.

Ms.	Louisville and Nashville Railroad. Pensacola Railroad.	Alt.	Ms.	Florida Central and Peninsular. Florida Central and Western.	Alt.
0	Flomaton.	19 a. Eocene. (?)	0	Chattahoochee R.	19 a. Eocene. (?)
5	Bluff Springs.	20.Quat.&19 a.Eoc. (?)	2	River Junc.	19 b. Miocene. (?)
12	McDavid.	"	3	Chattahoochee.	"
20	Molino.	"	20	Quincy.	"
28	Cantonment.	"	32	Midway.	19 a. Eocene.
33	Muscogee.	"	44	Tallahassee.	19 b. Miocene.
31	Gonzalez.	"		Ferrello.	"
44	Pensacola.	"	65	St. Marks.	"
	Pensacola and Atlantic.		56	Chaires.	"
			62	Lloyd's.	"
0	Pensacola.	Coast Qu.&19a. Eo. (?)	71	Drifton.	"
9	Escambia.	"	75	Monticello.	"
20	Milton.	"	78	Ancillo.	19 a. Eocene.
60	Deer Land.	"	85	Greenville.	19 b. Miocene. (?)
67	Mossy Head.	19 b. Miocene. (?)	99	Madison.	"
80	De Funiak Sp'gs.	"	106	Lees.	"
91	Ponce de Leon.	"	114	Ellaville.	19 a. Eo. (Vicksburg.)
98	Westville.	"	127	Live Oak.	"
100	Caryville.	19 a. Eocene. (?)	133	Houstown.	"
127	Cottondale.	"	138	Welborn.	19 b. Miocene. 250
136	Marianna.	19 a. Eo. (Vicksburg.)	142	Dowlings.	"
147	Cypress.	"	150	Lake City.	"
156	Sneads.	19 b. Miocene.	162	Olustee.	"
161	River Junc.	"			

L. By Mr. Lawrence C. Johnson of Meridian, Miss., Assistant Geologist U. S. Geological Survey. The survey of the state was not completed by Mr. Johnson when he ceased work in that field, for which reason, or because the superficial deposits render the boundaries of the formations uncertain, he assigns many of the stations with a ?, denoting the probable formation.

Florida Central and Western Railway. Continued.		
Ms.		Alt.
172 Sanderson.	19 b. Miocene. (?)	
181 Darbyville.	"	(?)
190 Baldwin.	19 c. Pliocene. (?)	
192 Clark's Junc.	"	
208 Waycross Junc.	"	
208 Jacksonville.	"	
0 Jacksonville.	"	
12 Hart's Road.	"	
23 Fernandina.	"	10
34 Hart's Road Jc.	"	10
41 Italia.	"	
50 Callahan.	"	80
59 Dutton.	"	45
Brandy Branch.	"	
60 Baldwin.	"	47
Maxville.	"	57
88 Highland.	"	210
89 Lawtey.	19 b. Miocene. (?)	140
Temples.	19 b. Miocene.	
Starke.	"	130
108 Waldo.	"	150
Fairbanks.	" [Vicksb'g.	
122 Gainsville. 128	19 b. Mio. underl'd by	
Arredondo.	19 a. Eocene.	70
Archer.	"	70
Bronson.	19 b. Miocene.	27
Otter Creek.	"	19
Rosewood.	"	12
178 Cedar Key.	"	10
108 Waldo.	"	150
122 Hawthorne.	"	150
Lockloosa.	"	52
134 Citra.	"	
Sparrs.	"	
Anthony.	"	72
147 Silver Spring Jc.	19 a. Eocene.	
151 Silver Spring.	"	89
153 Ocala.	19 b. Miocene.	100
Lake Wier.	"	
Wildwood.	"	
Panasoffkee.	"	
Withlacoo'ee.	19 a. Eocene.	
190 Leesburg.	19 b. Miocene.	
201 Tavares.	"	

Green Cove Springs and Melrose.		
Green Cove Spgs.	19 b. Miocene. (?)	
Sharon.	"	

Jacksonville, St. Augustine and Halifax R.		
0 Jacksonville.	19 c. Pliocene.	
3 Phillips.	"	
16 Bayard.	"	
19 Clarkville.	"	
28 Sampson.	"	
37 St. Augustine.	"	
52 Tocoi.	19 b. Miocene. (?)	
45 Smith's.		
.48 Middleton.	19 c. Pliocene.	

Jacksonville, St. Augustine and Halifax R. Continued.		
Ms.		Alt.
54 Olds.	(?)	
56 Merrifield.	19 b. Miocene. (?)	
59 Pattersonville.	"	
69 Palatka.	"	
81 Velvington.	"	
89 Dinner Isle.	19c Pliocene.	
97 Windemere.	"	
120 Ormond.	"	
123 Holly Hill.	"	
126 Daytona.	"	

Florida Southern Railway.		
0 Palatka.	19 b. Miocene. (?)	
18 Interlaken.	"	
40 Rochelle.	19 b. Miocene.	
50 Gainsville.	"	168
49 Micanopy.	19 a. Eocene.	
49 Boardman.	19 b. Miocene.	
57 Reddick.	"	
72 Ocala.	"	
88 Ocklawaha.	19 a. Eocene.	
East Lake.	19 b. Miocene.	
96 Conant.	"	
106 Leesburg.	"	
Dragen Junc.	"	
135 Pemberton Fe'ry.	19 a. Eocene.	
146 Brooksville.	19 b. Miocene.	
106 Leesburg.	"	
120 Ft. Mason.	"	
122 Eustis.	"	
126 Tavares.	"	
129 Lane Park.	"	
120 Ft. Mason.	"	
124 Umatilla.	"	
127 Altoona.	"	
129 Pittman.	"	
145 Astor.	"	
135 Pemberton Fe'ry.	19 a. Eocene.	
179 Lakeland.	19. b. Miocene.	250
192 Bartow.	"	
204 Ft. Meade.	"	
241 Arcadia.	"	
251 Ft. Ogden.	"	
261 Cleveland.	"	
268 Punta Gorda.	"	

Orange Belt Railway.		
0 St. Petersburg.	19 b. Miocene.	
15 Armour.	"	
18 Clearwater.	"	(?)
25 Yellow Bluff.	"	(?)
31 Tarpon Springs.	"	
51 Drexel,	"	
64 San Antonio.	"	(?)
71 Blanton.	"	(?)
73 Lenard.	"	(?)
76 Macon.	"	(?)
86 Tarrytown.	"	(?)

Orange Belt Railway.—Continued.

Ms.		Alt.
91	Cedar Hammock.	19 b. Miocene. (?)
101	Sheridan.	"
106	Clermont.	"
108	Minneola.	"
115	Killarney.	"
117	Oakland.	"
128	Lakeville.	"
133	Forest City.	"
138	Groveland.	"
144	Paola.	"
145	Sylvan Lake.	"
148	Monroe.	"

Jacksonville, Tampa and Key West.

0	Jacksonville.	19 c. Pliocene.
4	Edgewood.	"
10	Black Point.	" (?)
14	Orange Park.	"
20	Black Creek.	"
28	Magnolia.	"
29	Green Cove Sp's.	19 b. Miocene. (?)
34	Walkill.	"
41	W. Tocoi.	"
46	Bostwick.	"
56	Palatka.	"
63	Buffalo Bluff.	"
64	Satsuma.	19 c. Pliocene. (?)
67	Sisco.	"
72	Como.	"
78	Denver.	"
84	Seville.	"
92	Eldridge.	"
94	Barbersville.	"
108	Deland Jc.	"
113	Orange City Jc.	"
119	Enterprise Jc.	"
125	Sanford.	"
0	Enterprise Jc.	"
4	Enterprise.	19 b. Miocene. (?)
11	Osteen.	"
24	Maytown.	19 c. Pliocene. (?)
40	Titusville.	"
0	Sanford.	19 b. Miocene.
6	Paola.	"
18	Sorrento.	"
29	Tavares.	"

South Florida Railroad.

0	Sanford.	19 b. Miocene.
10	Longwood.	"
22	Orlando.	"
34	McKinnow.	"
40	Kissimmee.	19 c Pliocene. (?)
57	Davenport.	19 b. Miocene. (?)
68	Bartow Jc.	"
72	Auburn Dale.	"
83	Lakeland.	"
115	Tampa.	"
124	Port Tampa.	"

Savannah, Florida & Western Railway.
Gainesville Line.

Ms.		Alt.
130	Dupont, Ga.	19 b. Miocene.
163	Jasper.	"
171	Suwannee.	19 a. Eocene.
179	Live Oak.	"
190	McAlpin.	"
203	New Branford.	"
216	Ft. White.	19 b. Miocene.
249	Gainesville.	"

Pemberton Ferry Branch.

0	Pemberton F'y.	19 a. Eocene. (?)
23	Richland.	19 b. Miocene.
43	Lakeland.	"
56	Bartow.	"

Sanford and Indian River Railroad.

0	Sanford.	19 b. Miocene.
18	Lake Charm.	"
0	Lake City.	"
19	Lake City Jc.	"
22	Ft. White.	"

Jacksonville Division.

211	Waycross, Ga.	19 b. Miocene. (?)
246	Folkston, Ga.	19 c. Pliocene. (?)
251	Borlogne.	"
257	Hilliard.	"
267	Callahan.	"
280	Jacksonville.	"

Jacksonville and Atlantic.

0	Jacksonville.	19 c. Pliocene.
17	Pablo Beach.	20. Quaternary.

Atlantic and Western.

0	Blue Spring.	19 b. Miocene. (?)
1	Orange City Jc.	"
3	Orange City.	19. c. Pliocene (?)
25	Glencoe.	"
28	New Smyrna.	19 c. Plio. or 20. Quat.

Western Railway of Florida.

0	Green Cove Sps.	19 c. Pliocene. (?)
10	Sharon.	"
15	Belmore City.	19 b. Miocene.

Silver Springs, Ocala and Gulf.

0	Ocala.	19 b. Miocene.
25	Dumeelton.	"
48	Homosassa.	" (?)

Tavares, Apopka and Gulf.

0	Tavares.	19 b. Miocene.
23	Waits Jc.	"
29	Clermont.	"

Jacksonville, Mayport and Pablo.

0	Jacksonville.	19 c. Pliocene.
8	Cohasselt.	"
16	Burnside Beach	20. Quaternary.
20	Mayport.	"

Kentucky.[1]

GEOLOGICAL FORMATIONS FOUND IN KENTUCKY.[2]

20 d. Alluvium.	10 c. Black Shale.
20 c. Bluff or Loess.	9 c. Corniferous.
20 b. Port Hudson.	
20 a. Gravel (equivalent of Orange Sand of Tennessee).	5 c. Niagara.
	5 b. Clinton.
19. Tertiary, Lower Eocene.	
18. Cretaceous, Ripley.	4 c. Hudson River. $\begin{cases} \text{4 c.}^3 \text{ Upper.} \\ \text{4 c.}^2 \text{ Middle} \\ \text{4 c.}^1 \text{ Lower.} \end{cases}$
14. c. Upper Coal Measures.	
14 b. Lower Coal Measures.	4 a. Trenton.
14 a. Millstone grit.	
	3 a. Chazy.
13 c. Chester.	
13 b. Upper Sub-Carboniferous.	
13 a. Lower Sub-Carboniferous.	

1. By John R. Proctor, Director of the Kentucky Geological Survey.
2. The geological survey is in progress, and the formations of the State not fully determined.
3. *Louisville*, the metropolis of Kentucky, very interesting to the geologist. At this point the Ohio River falls 23 feet over ledge of Corniferous and Niagara limestone. At low water the limestone is exposed over a wide area, and discloses the finest collecting ground for corals in this country. Several large collections of Devonian and Upper Silurian corals are owned in Louisville.
5. *Cincinnati.* As to ancient glacial dam at Cincinnati, see Note 62 Ohio, 76 Indiana, 62 West Virginia. . G. F. WRIGHT.
6. *Bagdad.* About six miles to the south of this place can be seen an isolated hill capped with Niagara limestone. This hill is about 1,250 feet above the level of the sea, and the Niagara is found here at a greater elevation than elsewhere in the State.
7. *Benson.* In descending the hill to Benson the road passes through the Middle Hudson.
8. *Frankfort.* Hills around Trenton, the Birdseye limestone reaches up the bank of the Kentucky River as high as the tunnel. Good collecting ground for Trenton fossils.
9. *Springs Station.* Near here are some of the most celebrated stock farms. They are on the (4 c.) Lower Hudson River formations.
10. *Payne's.* Stage from here to Georgetown passes through some of the most beautiful lands of the Blue Grass region.
11. *Colesburg.* This place is at the base of Muldrow's Hill, the road ascends this hill between this point and Elizabethtown. This hill extends around central Kentucky, from the mouth of Salt River on the west to Lewis County on the east, retaining for its entire length the same geological formations, viz.: Black shales (10 c.) at base, and Waverly sandstones and shales (13 a.), and Upper Sub-Carboniferous limestone (13 b.) In Madison County the hill attains its greatest height (1,650 feet above sea), where it is capped with the Carboniferous conglomerate, having a workable bed of sub-conglomerate coal. The Chester (13 c.) is also present in this portion of the hill. It is there known as Big Hill. Muldrow's hill represents the retreating escarpment of the rocks formerly extending over central Kentucky. Siliceous remains of these Palæozoic rocks have been found scattered over the uplands of central Kentucky, and have been by some erroneously classed as glacial drift.
12. *Elizabethtown.* County town of Hardin County. St. Louis Group of Sub-Carboniferous limestone.
13. *Mumfordsville.* County town of Hart County. The road crosses Green River at this point. The high hill on south side of river is capped with Chester sandstone, as are also the hills to the left of road between Cave City and Glasgow Junction.
14. *Glasgow Junction.* Branch road to Glasgow. This is the nearest station to Mammoth Cave. Several beautiful caverns in this neighborhood. All of these caverns are in the St. Louis limestone, and some of them reach up to the Chester sandstone which caps the hills seen to the north of the road from this point to Bowling Green, 41 miles, all the drainage being subterranean.
15. *Bowling Green.* County seat of Warren County. Road crosses the Big Barren River at this point. Boats run from here to Evansville, on the Ohio River.
16. *Franklin.* County seat of Simpson County. The division between 13 a. and 13 b. is not far from this place. Geology of county not yet studied in detail.
17. *Hopkinsville.* County Seat of Christian County. Surrounded with very fertile lands. This county produces more wheat and tobacco than any county in the State. The best lands in this and adjoining counties are not excelled by any in America. The superior body of land beginning near Smith's Grove, in Warren County, and comprising a portion of Warren, Simpson, Logan, Todd, Christian, Trigg, Caldwell and Lyon, is the largest body of all good land with which the writer has any acquaintance. The Western State Asylum for the Insane is located near Hopkinsville.

Louisville & Nashville Railroad. Ms. (Louisville, Cincinnati & Lexington Div.) Alt.		
0	Louisville.[3]	⎧ 10 c. Black Slate, 9 ⎨ c. Corniferous, 5 c. ⎩ Niagara, 4. Trenton.
10	Ormsby's.	"
12	Anchorage.	9 c. Corniferous.
16	Pewee Valley.	5 c. Niagara.
27	La Grange.	5 b. Clinton. 860
33	Pendleton.	4 c.[3] Up. Hudson. 838
36	Sulphur.	" 691
41	Campbellsburg.	" 904
54	English.	"
56	Worthville.	" 486
65	Sparta.	" 505
70	Glencoe.	" 550
75	Elliston.	" 593
84	Verona.	" 870
89	Walton.	" 927
98	Independence.	"
106	Wilder's.	"
109	S. Covington.	" 537
109	Newport.	" 523
110	Cincinnati.[5]	"

(Lexington Division.)		
27	La Grange.	5 b. Clinton. 860
32	Jericho.	4 c.[3] Upper Hudson.
35	Smithfield.	
40	Eminence.	"
44	Pleasantville.	"
49	Christianburg.	"
52	Bagdad.[6]	"
59	Benson.[7]	4 c.[1] Lower Hudson.
65	Frankfort.[8]	4 a. Trenton.
76	Spring Station.[9]	4 c.[1] Hudson River.
79	Midway.	
83	Payne's.[10]	"
87	Yamallton.	"
94	Lexington.	" 946

(Shelbyville Division.)		
12	Anchorage.	9 c. Corniferous.
17	Eastwood.	5 c. Niagara.
23	Simpsonville.	4 c.[3] Upper Hudson.
30	Shelbyville.	"
38	Finchville.	"
42	Normandy.	"
47	Taylorsville.	"
57	Bloomfield.	"

Louisville & Nashville Railroad. Ms. (Main Line.) Alt.		
0	Louisville.[3]	⎧ 20 b. Loess, ⎨ 9 c. Corniferous. ⎩ 5 a. Niagara. 438
3	S. Louisville.	10 c. Black Shale.
18	Shepherdsville.	⎰ 9 c. Corniferous.[424] ⎱ 5 c. Niagara,
22	Bardstown Junc.	5 c. Niagara. 415
30	Lebanon Junc.	10 c. Black Shale. 426
34	Colesburg.[11]	13 a. L. Sub-Carb. 423
42	Elizabethtown.[12]	13 b. Up. Sub-Car.[641]
50	Glennale.	" 633
55	Sonora.	" 697
73	Munfordsville.[13]	" 568
81	Horse Cave.	" 601
85	Cave City.	" 611
91	Glasgow Junc.[14]	" 621
96	Rocky Hill.	" 594
100	Smith's Grove.	" 605
114	Bowling Green.[15]	" 466
118	Memphis Junc.	" 531
125	Woodburn.	" 608
134	Franklin.[16]	" 689
141	Mitchellville.	" 748
146	Fountainhead.	" 778
159	Gallatin.	4 c. Hudson River.[494]
.....	Edgefield Junc.	" 414
185	Nashville.	4 a.Tren.,20 b.Loess[430]

(Memphis Division.)		
118	Memphis Junc.	13 b. Up. Sub-Carb.[531]
123	Rockfield.	" 566
132	Auburn.	" 603
143	Russelville.	" 533
148	Cave Spring.	" 586
157	Allensville.	" 552
164	Guthrie.	" 525

(Nashville & St. Louis Division.)		
0	Nashville.	13 b. Up. Sub-Carbon.
47	Guthrie.	" 525
.....	Trenton.	"
.....	Pembroke.	"
71	Hopkinsville.[17]	" 550
84	Crofton.	"
95	Nortonsville.[18]	14 c. Coal Meas. 410
102	Earlington.[19]	" 370
107	Madisonville.	" 435
118	Slaughter's.	"
145	Henderson.[20]	⎰ 20 b. Loess. 402 ⎱ 14 c. Coal Measure.

18. *Nortonville.* Junction Chesapeake, Ohio & Southwestern Railway fault here. Coal No. 9 west, and coals No. 11 and 12 east of station.

19. *Earlington.* St. Bernard Coal Co., one of the largest mines in the State.

20. *Henderson.* Bottom lands Loess (20 b.) resting on Carboniferous.

21. *New Hope.* Prosperous city, large tobacco market, fine bridge over Ohio River; about 1½ miles from New Hope. At Coal Hollow distillery, is a fine collecting ground of the fossils *Beatricha Columnaria Alveolata.*

22. *Lebanon.* County town of Marion County. Junction of Cumberland & Ohio Railroad, southern division. The streams around Lebanon cut down to Upper Hudson rocks. Hills seen to south, continuation of Muldrow's Hill (see Note 11). Fine localities for collecting Sub-Carboniferous fossils in the hills a few miles south from Lebanon.

23. *Riley's.* Fine collecting grounds near Riley's Station of Corniferous fossils.

Louisville & Nashville Railroad—*Con.* (Knoxville Division.) Ms.		Alt.
0	Louisville.[3]	(As before).
30	Lebanon Junc.	10 c. Black Shale. [413]
35	Boston.	" [431]
45	New Haven.	{ 10 c. Black Shale, 9 c. Corniferous, 5 c. Niagara. [441]
50	New Hope.[21]	{ 5 c. Niagara, [444] 4 c. Upper Hudson.
57	Loretto.	10 c. Black Shale.
62	St. Mary's.	5 c. Niagara. [733]
67	Lebanon.[22]	{ 9 c. Corniferous[754] 10 c. Black Shale.
76	Riley's.[23]	{ 9 c. Corniferous, 10 c. Black Shale, 5 c. Niagara.
85	Mitchellsburg.	10 c. Black Shale.
89	Parksville.[24]	{ 10 c. Black Shale, 9 c. Corniferous, 5 c. Niagara. [1052]
95	Junction City.	10 c. Black Shale. [997]
96	Shelby City.	{ " [997] 9 c. Corniferous.
104	Stanford.	4 c. Upper Hudson.[844]
105	Rowland.	"
115	Crab Orchard.[25]	{ 10 c. Black Shale, 9 c. Corniferous, 5 c. Niagara. [929]
129	Mt. Vernon.	13 b. U.Sub-Carb.[1113]
135	Pine Hill.	{ " [964] Hills capped with 14 a. Millstone Grit.

Louisville & Nashville Railroad—*Con.* (Knoxville Division.) Ms.		Alt.
140	Livingston.[26]	14 a. Millstone Grit[856]
152	East Bernstadt.[27]	14 b. Low. Coal Meas.
155	Pittsburg.[28]	"
157	London.	"
165	Lily.	"
174	Woodbine.	"
181	Rockhold.	"
189	Williamsburg.[29]	"
201	Jellico.[30]	"

Chesapeake & Ohio Railroad.

(Lexington Division.)

0	Lexington.	4 a. Trenton. [946]
11	Pine Grove.	" [960]
18	Winchester.	{ 4 c.[1] Lower Hudson River. [964]
.....	Hedges Station.	4 c.[2] Middle Hud. [976]
33	Mt. Sterling.[31]	4 c.[3] Upper Hud. [984]
49	Olympia.[32]	5 c. Niagara. [751]
57	Farmer.[33]	10 c. Black Shale. [668]
65	Morehead.	13 a. Waverly. [712]
83	Olive Hill.[34]	" [752]
99	E. K. Junction.[35]	14 b. Coal Meas. [613]
102	Denton.	" [601]
109	Rush.	" [647]
116	Mean's.	" [622]
122	Ashland.[36]	{ 20 b. Loess, [544] 14 b. Coal Measure.
128	Catlettsburg.[37]	" [544]
138	Huntington.	" [566]

24. *Parkville.* Hills to the south capped with St. Louis limestone; fine collecting ground for *Lithrostotion Canadensis.* A section may be obtained in a distance of four miles on a north and south line from the Trenton limestone to the top of the Sub-Carboniferous. The hills have waste of the Carboniferous conglomerate on top.

25. *Crab Orchard.* Springs of same name located near here. Caudi Galli found beneath the Corniferous near springs.

26. *Livingston.* Crossing of Rock Castle River. Coal mines in Lower or Sub-Conglomerate here. Fine section of St. Louis and Chester rocks on south side of river. Quarries of fine building stone. Hills on south capped with massive conglomerate sandstone.

27. *East Bernstadt.* Mines in the coal above the conglomerate, probably No. 1. The coal from these mines and from Pittsburg Station, a few miles south, takes high rank in the market, and the output is increasing rapidly. It is known as " Laurel Coal."

28. *Pittsburg.* Several extensive coal mines here.

29. *Williamsburg.* County town of Whitley County. Crossing of Cumberland River.

30. *Jellico.* State line. Extensive coal mines in lower measures near here. Coal of excellent quality. The great Pine Mountain fault can be seen a short distance southeast from this station.

31. *Mt. Sterling.* County town of Montgomery County. Junction of the Kentucky & South Atlantic Railway. The hills seen to the east are a continuation of Muldrow's Hill. (See Note 11.)

32. *Olympia.* Near here extensive deposit of iron ore now being mined. Ore supposed to be in Corniferous. Clinton iron ore is also found in Bath County.

33. *Farmer.* Crossing of Licking River.

34. *Olive Hill.* Very thick deposit of superior fire clay near this station; fine clay also near Enterprise. An excellent building stone is obtained from the Waverly sandstone along the line of the road in Rowan County.

35. *Eastern Kentucky Junction.* Crossing of the Eastern Kentucky Railway. The Mt. Savage furnace is one mile east from here, and fine veins of coals No. 3 and 7.

36. *Ashland.* Extensive iron manufactory. Junction of the Chatterol Railway. Bottom lands Loess (20 b.) resting on Carboniferous.

37. *Catlettsburg.* County town of Boyd County. Confluence of the Big Sandy River with the Ohio River.

38. *West Point.* Crossing of Salt River. Road ascends Muldrow's Hill (see Note 11) after crossing river. Fine sections of Sub-Carboniferous rocks exposed.

39. *Grayson Springs.* Celebrated summer resort; good collecting ground for Chester fossils.

40. *Litchfield.* County town of Grayson County. Sandstone seen here; base of Chester Group; same as massive sandstone above St. Louis limestone at Mammoth Cave and elsewhere. A mile south of here thick deposit of marly shale, containing potash.

Chesapeake, Ohio & Southwestern R. R.

Ms.	Station	Formation
0	Louisville.[3]	{ 20 b. Loess, 436 / 10 c. Black Shale, / 9 c. Corniferous.
9	Pleasant Ridge.	{ 10 c. Black Shale, / 13 a. L. Sub-Car.445
21	West Point.[38]	{ 20 b. Loess, 410 / 10 c. Black Shale.
27	Muldraugh.	13 b. Up. Sub-Carb.738
37	Vine Grove.	" 719
47	Cecelia.	13 c. Chester. 688
52	Stephensburg.	13 b. Up. Sub-Carb.662
62	Big Clifty.	13 c. Chester. 733
67	Grayson Sp'gs.[39]	" 709
72	Litchfield.[40]	" 710
78	Milwood.	14 b. L. Coal Meas. 854
84	Caneyville.	" 450
97	Horse Branch.	" 527
100	Rosine.	" 597
109	Beaver Dam.	14 c. U. Coal Meas.482
118	Rockport.[41]	" 485
127	Central City.[42]	"
134	Greenville.[43]	" 537
147	White Plains.	" 477
151	Nortonville.	" 492
157	St. Charles.	" 509
165	Dawson.	14 b. Low. Coal Meas.
180	Princeton.[44]	13 b. Up. Sub-Carb.624
192	Eddyville.	" 487
194	Kuttawa.[45]	13 a. L. Sub-Carb. 487
209	Calvert City.	{ 20 c. Alluvium, 494 / 13 a. Low. Sub-Carb.
226	Paducah.[46]	{ 20 c. Alluvium, bluff, / gravel and loam.484
240	Boaz.	"
244	Hickory.	"
250	Mayfield.	"
255	Pryor's.	"
259	Wingo.	"
266	Water Valley.	"
271	Fulton.	" Bluff loam.

Cincinnati, New Orleans & Texas Pacific Railroad.

Ms.	Station	Formation	Alt.
0	Cincinnati.[5]	4 c. Hudson River.	
5	Kenton Heights.	"	845
7	Erlanger.[47]	"	915
14	Richwood.	"	939
18	Walton.	"	927
21	Bracht.	"	934
25	Crittenden.	"	923
28	Sherman.	"	939
32	Dry Ridge.	"	964
35	Williamstown.	"	958
44	Blanchet.	"	968
46	Corinth.	"	968
49	Hinton.	"	958
54	Sadieville.	"	872
60	Roger's Gap.	"	928
63	Kinkaid.	"	877
67	Georgetown.	"	883
71	Donerail.	"	897
76	Sandersville.	"	961
79	Lexington.	4 a. Trenton.	975
85	Windom.	"	1034
87	Catnip Hill.	"	990
91	Nicholasville.	"	960
96	Wilmore.	"	887
100	High Bridge.[48]	"	777
106	Burgin.	"	902
107	Harrodsburg Jun c.	"	915
114	Danville.	"	970
118	Junction City	10 c. Black Shale. 997	
124	Moreland.	" & 5 c. Niag.1101	
129	McKinney.[49]	5 c. Niagara. 1028	
136	King's Mount.[50]	{ 13 a. Waverly, 1153 / 10 c. Black Shale.	
139	Waynesburg.	13 b. St. Louis. 1230	
143	Eubanks.	" 1187	
148	Pulaski.	" 1135	
151	Science Hill.	" 1230	
152	Norwood.	" 1137	
158	Somerset.	" 882	
163	Cedar Grove.	" 851	

41. *Rockport.* Crossing of Green River. Coal mined here, and at McHenry Station (Coal No. 9).

42. *Central City.* Extensive coal mines. Coals 11 and 12 near level of railway.

43. *Greenville.* County town of Muhlenburg County. Deposits of limonite iron ore in county, in Lower Coal Measures.

44. *Princeton.* County town of Caldwell County. Fine quarries in the oolite bed of St. Louis limestone near here.

45. *Kuttawa.* Near the base of St. Louis Group. Road crosses Cumberland river west of this station. Large deposits of limonite ore near here.

46. *Paducah.* County town of McCracken County. At this point extensive deposit known as the Paducah Gravel Beds, affording one of the best and cheapest road materials to be found in this country. This gravel (20 a.) is composed of waste from the degraded beds to the eastward, and is principally quartz pebbles from the Corniferous conglomerate, and angular fragments of chert from the Lower Sub-Carboniferous rocks, with coarse, angular sand all quite ferruginous. When properly put on streets or roads it soon cements, needs little after repairs, affording a smooth, hard road. It also affords a superior material for concrete.

47. *Erlanger.* Glacial deposits are found on the highlands, 550 feet above the river, both south and west of Greenwood (Erlanger). A noteworthy collection of Jasper conglomerate boulders from Lake Superior occurs on the road to Burlington, three miles west of Florence.   G. F. W.

48. *High Bridge.* Crossing of Kentucky River. Bridge, 275 feet above water. Cliffs composed of Birdseye and Chazy limestones.

49. *McKinney.* The Upper Hudson is crossed between Moreland and McKinney's Station.

50. *King's Mountain.* The tunnel south of King's Mountain 4,000 feet long, is in the Waverly shales. King's Mountain is a continuation of Muldrow's Hill. (See Note No. 11.) The hills here are capped with the St. Louis limestone.

Cincinnati, New Orleans & Texas Pacific Railroad—Con.

Ms.	Railroad—Con.		Alt.
165	Burnside.[51]	13 b. St. Louis.	770
167	Tatesville.	"	874
170	Sloan's Valley.	"	914
176	Greenwood.	14 b. L. Cl. Meas.	1195
179	Cumberland Fall s.[52]	"	1245
182	Flat Rock.	"	1296
187	Whitley.	. "	1340
194	Pine Knot.	"	1415
198	State Line.	"	1345

Chesapeake & Ohio Railroad.
(Kentucky Central Division.)

Ms.		Railroad	Alt.
0	Covington.	4 c. Hudson River.	
14	Visalia.	"	
21	Morning View.	"	
24	Demossville.	"	
28	Butler.	"	
39	Falmouth.	"	540
50	Boyd.	"	
53	Berry.	"	
65	Cynthiana.	"	700
72	Shawhan.	"	
79	Paris.	"	840
86	Hutchinson.	"	
89	Muir.	"	
99	Lexington.	4 a. Trenton.	867
79	Paris.	4 c.[1] L. Hudson R.	840
95	Winchester.	"	964
106	Boone.	4 c.[3] Up. Hudson River	
118	Richmond.	4 c.[2] Mid. Hud. R.	924
122	Argenta.	"	
133	Paint Lick.	4 c.[3] Up. Hudson R.	792
144	Lancaster.	"	997
151	Rowland.	"	842

Kentucky Central Railroad.
(Northern Division.)

Ms.		Railroad	Alt.
....	Lexington.	4 a. Trenton.	867
....	Muir.	4 c. Hudson River.	
79	Paris.	"	840
88	Millersburg.	"	
95	Carlisle.	"	
109	Ewing.	"	
113	Johnson.	"	
128	Maysville.	"	

Kentucky Central Railroad—Con.

Ms.	(Knoxville Division.)		Alt.
0	Paris.	4 c. Hudson River.	
9	Austerlitz.	"	
16	Winchester.	4 c.[1] Lower Hudson.	
25	Riverside.	"	
38	Richmond.	4 c.[3] Upper Hudson.	
48	White's.	"	
51	Berea.	10 c. Black Shale.	
58	Conway.	13 a. Waverly.	
65	Langford.	"	
72	Link's.	"	
75	Livingston.	13 b. St. Louis.	

Kentucky Union Railway.

Ms.		Railroad	Alt.
0	K. U. Junction.	4 c.[2] Middle Hud.	980
6	Kidvills.	5 c. Niagara.	950
9	Abbott's.	{ 10 c. Black Shale, / 5 c. Niagara.	668
12	Wattersville.	10 c. Black Shale.	562
14	Clay City.	"	564

Eastern Kentucky Railroad.[63]

Ms.		Railroad	
0	Riverton.[54]	14 b. Low. Coal Meas.	
3	Three Miles.	"	
5	Worthington.[55]	"	
6	Argillite.[56]	"	
9	Laurel.	"	
10	McAllister.	"	
12	Hunnewell.[57]	"	
15	Denning's.	"	
16	Hopewell.[58]	"	
18	Anglin's.	"	
21	Pactolus.[59]	"	
23	Grayson.[60]	"	
26	Vincent's,	"	
28	Mt. Savage.[61]	"	
29	Reedville.	"	
34	Willard.[62]	"	

Chattorol Railway.

Ms.		Railroad	
0	Ashland.[34]	14 b. Low. Coal Meas.	
6	Catlettsburg.[37]	"	
14	Lockwood's.	"	
19	Rockville.	"	
26	Fuller's.	"	
31	Louisa.	"	
36	Walbridge.	"	
40	Northrup.	"	
46	Peach Orchard.[63]	"	
50	Richardson.	"	

51. *Burnside.* Crossing of Cumberland River.
52. *Cumberland Falls.* A few miles from railway, perpendicular fall of Cumberland River of 63 feet, over the Carboniferous conglomerate. Beautiful scenery and excellent fishing.
53. This railroad runs through the heart of the Kentucky division of the Hanging Rock Iron Region. On the line of the road all of the coals are to be found, from No. 1 to No. 11, and most of the iron ores.
54. *Riverton.* No. 1 Coal near water level.
55. *Worthington.* No. 3 Coal in the hills, about 150 feet above grade of road.

| Illinois Central Railroad. | | | Kentucky & South Atlantic R. R. | | |
Ms.	(New Orleans Division.)	Alt.	Ms.		A
0	Cairo.	{ 20 Alluv. over ³²²	0	Mount Sterling.³¹	4 c.³ Upper Hudson.
2	East Cairo.	{ Port Hudson.	6	Spencer.	"
6	Wickliffe.⁶⁴	{ "	10	Johnson's.	"
16	Bardwell.	{ 20.Quater. loam.³⁵⁰	12	Pollard's.	"
22	Arlington.	{ and gravel over³³⁰	14	Heges.	"
30	Clinton.	{ Eocene T e r t i-³⁵⁰	15	Chamber's.	5 c. Niagara.
44	Fulton.	{ ary. ³⁵⁰	19	Cornwall.	"
			21	Rothwell.	"
	Mobile & Ohio Railroad.		23	Frenchburg Jc.	10 c. Black Shale.
0	Cairo.	{ 20. Alluv. over ³²²		**Evansville, Owensboro & Nashville R. R.**	
2	East Cairo.	{ Port Hudson. ³²²	0	Owensboro.	14.Carboniferous.
6	Wickliffe.⁶⁴	{ " ³²²	7	Sutherland.	"
18	Berkeley.	{ 20. Quater. loam³⁵⁰	15	Riley's.	"
23	Columbus.⁶⁵	{ and gravel over³⁰⁹	21	Livermore.	"
34	Moscow.	{ Eocene T e r t i-³¹³	27	Stroud's.	"
42	Jordon.	{ ary. ⁴⁰⁴	35	Owensboro Junc.	"

56. *Argillite.* Near site of Old Argillite Furnace, probably the oldest furnace in the Hanging Rock Iron Region, erected in 1822. About three miles east of station is the Pennsylvania Furnace, and three miles west the Buffalo Furnace.

57. *Hunnewell.* Hunnewell Furnace located here; also the machine and repair shops of the railroad. Mines of No. 3 and No. 4 Coal, the latter known as the Hunnewell Cannel Coal.

58. *Hopewell.* The former site of an old furnace of that name.

59. *Pactolus.* The former site of an old furnace of that name.

60. *Grayson.* The county seat of Carter County. Coals No. 2 and No. 3 are found here. Iron Hills Furnace, the largest charcoal furnace in this section, is situated about eight miles northwest from Grayson, where also is the celebrated Lambert Ore Bank, a local deposit 14 feet 10 inches thick, of great value. Thirteen miles west of Grayson are the celebrated Carter Caves, situated in the St. Louis group of the Sub-Carboniferous limestone. These caves and the wild scenery of Tigart Valley, surrounding them, are well worth visiting.

61. *Mt. Savage.* Near here is Mt. Savage Furnace, and fine veins of coals No. 3 and No. 7, the latter known as the Coalton Coal.

62. *Willard.* At Willard are the ores and coal mines of the Bellefonte & Etna Company of Ironton, Ohio. Most of the coals are represented in this vicinity.

63. *Peach Orchard.* Extensive mines, Coal No. 3.

64. *Wickliffe.* County seat of Ballard County. The railroad just south of this passes at the foot of an exposure of lignite three feet thick.

65. *Columbus.* The town lies at the foot of river bluffs, 120 feet high, showing Quaternary and Tertiary strata. Port Hudson clays exposed beneath Alluvium in river bank at low water.

The Quaternary gravel and brown loam beds, that cover almost the entire region lying between the Tennessee and Mississippi Rivers, are very generally underlaid by black and blue clays of the lignitic group of Eocene Tertiary. These clays have, in and near Paducah, been penetrated to a depth of 100 feet. Cretaceous sands and clays underlie the Quaternary thirty-five miles southeast of Mayfield.

Errata for Kentucky.

In note 20 and 21. The first line of 21 belongs to 20, *Henderson.*

In note 46, *Paducah.* Corniferous conglomerate should be Carboniferous conglomerate.

In the Chesapeake, Ohio & Southwestern R. R. the geological formation of Calvert City and Paducah should be "20. Quaternary, Port Hudson." That of Boaz, *et al.*, to Fulton, should be "20. Quaternary gravel and loam over Eocene Tertiary."

The elevation of Princeton should be 524; Calvert city, 351; and Paducah, 341 feet. The same error effects the elevations of all stations south of Paducah and east to Elizabethtown.

Tennessee.[1]

LIST OF GEOLOGICAL FORMATIONS FOUND IN TENNESSEE:

DANA'S TABLE OF FORMATIONS.	TENNESSEE DIVISIONS. BY PROF. SAFFORD.	DANA'S TABLE OF FORMATIONS.	TENNESSEE DIVISIONS. BY PROF. SAFFORD.
20. QUATERNARY.	20 c. Alluvium.	7. HELDERBERG.	7. Held. or Linden.
"	20 b. Bluff Loam.	5. NIAGARA.	5 d. Niagara lime s.
"	20 a. Orange sand, or drift.	" CLINTON.	5 c. Dyestone Group
19. TERTIARY EOCENE	19 b. La Grange s.	" MEDINA.	5 b. White Oak Mt. sandstone.
"	19 a. Flatw'ds s. &c.	" "	5 a. Clinch Mt. s. s.
18. CRETACEOUS.	18 c. Ripley Group.	4 b. CINCINNATI.	4 b. Nashville.
"	18 b. Rotten lime s.	4 a. TRENTON.	4 a. Lebanon.
"	18 a. Coffee sand.		3 d. Lenoir or Chazy
14. CARBONIFEROUS.	14. Coal Measures	3. CANADIAN. QUEBEC	3 c. Knox dolomite.
13. SUB-CARBONIFE'S.	13 c. Mountain l. s.	" "	3 b. Knox shale.
"	13 b. Coral or St. Louis l. s.	" CALCIFEROUS.	3 a. Knox sandstone
"	13 a. Barren Group.	2. PRIMORD'L. POTS'M.	2 b. Chilhowee s. s.
10. HAMILTON.	10 c. Black Shale.	" ACADIAN.	2 a. Ocoee Group.
		1. ARCHÆAN.	1. Metamorphic.

Chesapeake, Ohio & Southwestern R. R. Ms.		Alt.	Chesapeake, Ohio & Southwestern R. R.— Continued. Ms.		Alt.
0	Paducah, Ky.	20. Quaternary.	68	Polk's.	20 b. Bluff loam.
5	Bond's.	"	74	Obion.	"
9	Florence.	"	78	Trimble.	"
14	Boaz.	"	85	Newbern.	"
16	Viola.	"	94	Dyersburg.	"
20	Hickory.	"	98	Foulkes.	"
26	Mayfield.	"	107	Gates.	"
32	Pryor's.	"	119	Ripley.	"
37	Wingo.	"	125	Hennings.	"
44	Water Valley.	"	133	Covington.	"
50	Fulton.	"	145	Atoka.	"
53	Pierce, Tenn.	20 b. Bluff loam.	151	Kerrville.	"
56	Harris.	" Resting on 20 a.,	154	Millington.	"
59	Paducah Junct'n.	" and that on 19 b.	158	Lucy.	"
63	Troy.	" La Grange sand.	170	Memphis.[2]	"

(Right column side note: Resting on 20 a. Orange sand (gravel), and that on 19 b. La Grange sand.)

1. Revised, and the notes added by Prof. James M. Safford, the State Geologist of Tennessee, and the portion in Kentucky by Prof. N. S. Shaler, the State Geologist of Kentucky.

2. *Memphis.* The Bluff loam is well displayed in the bluffs at Memphis, no other formations appearing, excepting in very low water.

Vicksburg. The peculiar property of the Loess, or Bluff formation is shown in the following passage from General Grant's article on the Siege of Vicksburg, in the *Century* magazine, for September, 1885: "The ridges upon which Vicksburg is built, and those back to the Big Black, are composed of a deep, yellow clay, of great tenacity. When roads and streets are cut through, perpendicular banks are left, and stand as well as if composed of stone. The magazines of the enemy were made by mining passageways into this clay, at places where there were deep cuts. Many citizens secured places of safety for their families by carving out rooms in these embankments. A door-way, in these cases would be cut in a high bank, starting from the level of the road, or street, and after mining it in a few feet a room of the size required would be carved out of the clay, the dirt being removed by the door-way. In some instances I saw where two rooms were cut out for a single family, with a door-way in the clay wall separating them; some of these were carpeted, and furnished with considerable elaboration. In these the occupants were fully secure from the shells of the enemy, which were dropped into the city night and day, without intermission." A lady who was in the city during the siege, reported the hills as honey-combed with caves, the digging of which became a regular business. They were well propped with thick posts, as in a coal mine.

Mobile & Ohio Railroad.

Ms.			Alt.
0	Columbus, Ky.	{ 20. Quat., 20 b. Bluff loam 10 miles.	309
7	Clinton.	"	521
13	Moscow.	"	513
16	Cayce's.	"	400
20	Jordan, Ky.	"	404
26	Union City, Tenn.	"	546
31	Troy.	"	
45	Crockett.	"	296
43	Kenton.	{ 2 a. Orange sand, resting on La Grange sand.	309
48	Rutherford.	"	521
52	Dyer	"	565
59	Trenton.	"	521
70	Humboldt.	"	529
79	Carroll.	"	575
87	Jackson.	"	425
89	Pinson.	19 a. Flatwoods.	884
103	Henderson.	"	427
114	McNairy.	18 c. Ripley.	454
120	Bethel.	"	483
132	Ramer, Tenn.	18 b. Rotten l. s.	416
143	Corinth, Miss.	"	434

Illinois Central Railroad.
(N. O., Louisville & Chicago Division.)

Ms.			Alt.
0	New Orleans.		
382	Lamar, Tenn.		
394	Grand Junction.	{ 20 a. Orange sand, resting on La Grange sand.	575
413	Bolivar.	"	430
441	Jackson.	"	425
455	Medina.	"	
464	Milan.	" ·	405
475	Bradford.	"	
481	Greenfield.	"	
487	Sharon.	"	
495	Frost.	"	
550	McConnellville.	"	
506	Fulton, Ky.	20 b. Bluff loam.	

Louisville & Nashville Railroad.
(Memphis Division.)

Ms.			Alt.
0	Louisville, Ky.		438
164	Guthrie.	{ 13 b. Sub-Carbon., St. Louis l. s.	525
168	Hampton's, Tenn.	"	513
171	Dudley's.	"	494
177	Clarksville.	"	592

Louisville & Nashville Railroad.—Continued.

Ms.			Alt.
184	Steele's.	{ 13 b. Sub.-Carbon., St. Louis l. s.	565
189	Palmyra.	"	567
190	Carbondale.	"	562
198	Cumberland.³	13 a. Sub.-Carbon.	350
205	Erin.	"	404
210	Tenn. Ridge.	13 b. Sub.-Carbon.	720
214	Stewart's.	"	464
220	Tenn. River.	13 a. Sub.-Carbon.	
230	Big Sandy.	7. Helderberg.	345
235	Springville.	{ 20 a. Orange sand, 18 c. Ripley.	540
241	Porter's.	19 a. Flatwoods.	552
246	Paris.⁴	{ 20 a. Orange sand, 19 a. Flatwoods.	447
256	Henry.	20 a. Orange s.	513
264	McKenzie.	"	470
274	Trezevant.	"	443
284	Milan.	"	408
296	Humboldt.	"	329
301	Gadsden.	"	406
308	Bell's.	"	320
312	Jones's.	"	314
321	Brownsville.	"	533
329	Shephard.	"	279
333	Stanton.	"	303
341	Mason.	"	296
349	Galloway.	"	277
352	Withe.	20 b. Bluff loam.	271
358	Shelby.	"	249
366	Bartlett.	"	263
377	Memphis.²	"	227

(*The column 256–349 is bracketed with the vertical label* "Resting on La Grange sand.")

(Division to Nashville and Montgomery.)

Ms.			Alt.
0	Louisville, Ky.		438
114	Bowling Green.	13 b. Sub.-Carbon.	266
118	Memphis Junct.	"	
122	Rich Pond.	"	
125	Woodburn.	"	
134	Franklin.	"	617
141	Mitchellville, Tn.	13 a. Sub.-Carbon.	748
144	Richland.	"	774
146	Fountain Head.	"	778
149	Buck Lodge.	"	715
153	(Tunnel.)⁵	10 c. Bl. Sh. " 5 d.	
159	Gallatin.	4 b. Cin. or Nash.	494
164	Pilot Knob.	"	447
166	Saundersville.	"	545
170	Hendersonville.	"	446
175	Edgefield Junct.	{ 4 b. Cin. or Nash., and 4 a. Tren.	414
178	Madison.	4 b. Cin. and Nash.	466

3. Very soon after leaving Cumberland, the road traverses one end of the *Wells Creek Basin* and crosses the 10 c. Black Shale, also 7. Helderberg, 5 d. Niagara, 4 a. Lebanon, 4 b. Nashville, and 3 c. Knox Dolomite strata, which have been brought to the surface by an uplift. The only exposure of Knox Dolomite in Tennessee west of the Cumberland Mountains. In the bluff on the river just below Cumberland are good presentations of the 10 c. Black Shale, as well as the 5 Niagara, and 7. Helderberg rocks.

4. *Paris.* At the Paris depot the Orange Sand is well seen in the railroad cuts, and in the washes about the town. In the cuts of the railroad just east of the depot, and also on roads leading to the southeast from the town, the Flatwoods clay can be observed to advantage.

5. At this Tunnel is a good section of the (10 c.) Black Shale, with the strata above and below.

Louisville & Nashville Railroad.—Continued.

Ms.			Alt.
184	Edgefield.	4 b. Cin. or Nash.	414
185	Nashville.	"	409
189	N. and C. Junc.	"	
197	Brentwood.	"	693
206	Franklin.	"	617
215	Thompson's.	"	477
219	Ewell's.	"	767
223	Carter's Creek.	4 a. Lebanon.	692
233	Columbia.	"	644
243	Pleasant Grove.	"	719
246	Campbell's.	"	636
251	Lynnville.	"	734
254	Buford's.	"	702
256	Reynold's.	"	724
261	Wales.	"	668
266	Pulaski.	"	641
272	Harwell.	"	617
273	Aspen Hill.	"	646
275	Lester's.	"	723
278	Prospect.	"	588
280	State Line.	4 b. Cincinnati.	
286	Elkmont, Ala.	13. Sub-Carbon.	796
	(Continued in Alabama.)		

East Tennessee & Western North Carolina Railroad.

Ms.			Alt.
0	Johnson.	3 c. Knox.	
9	Elizabethtown.	"	
15	Hampton.	"	
24	Crab Orchard.	"	
33	Cranberry.	1 b. Huronian.	
34	Mine.	"	

Louisville & Nashville Railroad. (St. Louis Division.)

Ms.			Alt.
0	St. Louis.		
261	Trenton, Ky.		
269	Guthrie.	13. Sub-Carbon.	525
274	Forts, Tenn.	"	
280	Cedar Hill.	"	
287	Springfield.	"	
299	Baker's.	{ 5 a. Niagara, with bl'k shale above. A good section here.	
303	Goodlett's.	4 b. Nashville.	
306	Edgefield Junc.	{ 4 b. Nashville and 4 a. Lebanon.	414
309	Madison.	4 b. Nashville.	466
315	Edgefield.	"	
316	Nashville.	"	409

East Tennessee, Virginia & Georgia R. R.

Ms.			Alt.
0	Memphis, Tenn.[2]		
5	Buntyn.	20 b. Bluff l'm.	244
9	White's.	"	
15	Germantown.	"	375
19	Bailey.	"	664
23	Colliersville.	"	679
31	{ Rossville, or La Fayettte.	20 a. Orange s.	316
39	Moscow.	"	352
52	Somerville.	"	
49	La Grange.	"	551
52	Grand Junc.	"	575
58	Saulsbury.	"	586
64	64 Miles Siding.	19 a. Flatwoods.	
69	Middleton.	18. Cretaceous.	408
74	Pocahontas.	"	394
79	Big Hill	{ 20 a. Orange sand, 19 a. La Grange.	
84	Chewalla.	18 c. Rotten l. s.	409
93	Corinth, Miss.	{ 20 d. Yellow loam, 18 c. Rotten l. s.	434
107	Burnsville, "	{ 20 a. Orange sand, 18 a. Eutaw.	663
115	Iuka, Ala.	{ 20 a. Orange s., 485 13 a. Keokuk or St. L.	
124	Marguren, Ala.	13. Sub-Carboniferous.	
127	Dickson.	"	485
129	Cherokee.	"	
	(Continued in Alabama.)		

(Left sub-column heading, running vertically: "Resting on La Grange sand.")

Nashville, Chattanooga & St. Louis R. R.

Ms.			Alt.
0	Chattanooga.[6]	{ 4 a. Lebanon, and 3 c. Knox dolomite or Quebec.	664
6	Wauhatchie.	4 b. Nashville.	690
13	Ætna Cl. Mines.	{ 13 c. Upper Sub-Carb., 14. Cl. Measures near by..	
14	Whitesides.		
22	Shellmound.	{ Alluvium (Tenn. river bottom.)	
28	Bridgeport.	{ 3 c. Knox dolomite or Quebec.	
39	Stevenson.[7]	3 b. Knox shale.	769
49	Anderson.	13. Sub-Carboniferous.	
62	(Tunnel.)[8]	13 c. Mountain l. s.	
64	Cowen.	13 b. Sub-Carbon.	
69	Decherd.	"	
82	Tullahoma.	13 a. Sub-Carbon.	
89	Normandy.	4 b. Nash. or Cin.	
96	Wartrace.	{ 4 b. Nashville and 4 a. Lebanon.	
101	Belle Buckle.	4 a. Lebanon.	
109	Christiana.	"	
119	Murfreesboro.	"	
126	Florence.	"	
131	Smyrna.	"	
136	Lavergne.	"	
142	Antioch.	"	
150	Nash. & Dec. Jc.	4 b. Nashville.	
151	Nashville.	"	

6. Upper Silurian beds, the Black Shale and the lowest carboniferous strata, may also be seen in the high hill on the west side of the city.

7. *Stevenson.* A fault here bringing Knox Shale and Sub-Carboniferous together.

8. *Tunnel.* Coal measures on the tops of the mountains each side of the tunnel.

Nashville, Chattanooga & St. Louis R. R.— Continued.

Ms.		Alt.
158	{ Bellemeade, or Harding's.	4 b. Nashville.
164	Bellevue.	"
168	Newsom's.[9]	5 a. Niagara.
176	Kingston Spring.	13. Sub-Carboniferous.
189	Burns.	"
198	Dickson.	"
208	McEwen.	"
218	Waverly.	"
229	Johnsonville.	{ 10 c. Bl'k shale, and 13. L. Sub-Carbon.
238	Camden.[10]	13. Helderberg.
258	Huntingdon.	19 a. Flatwoods Terti.
270	McKenzie. 470	20 a. Orange s.
278	Gleason.	"
285	Dresden.	"
303	Paducah Junc.	"
307	Union City. 348	20 b. Bluff loam
314	State Line, Tenn. (Continu'd in Ky)	"
321	Hickman, Ky.	" 301
333	Columbus, "	" 309
499	St. Louis, Mo.	"

(Resting on 19 b. La. Orange sand.)

(Lebanon Branch.)

0	Nashville.	4 b. Nashville. 430
2	Mt. Olivet.	4 b. Nash., 4 a. Tren.
8	Donelson.	"
12	Hermitage.	"
18	Mt. Juliet.	"
24	Leeville.	"
26	Tucker's Gap.	4 b. Nashville.
31	Lebanon.	4 a. Lebanon.

(Shelbyville Branch.)

0	Chattanooga.	684
96	Wartrace.	4 b. Nash., 4 a. Leban.
104	Shelbyville.	4 a. Lebanon.

(Fayetteville Branch.)

0	Decherd.	{ 13 b. Sub-Carbon., St. Louis l. s.
3	Winchester.	"
10	Belvidere.	13 a. Sub-Carbon.
16	Hunt's.	"
26	Cunningham.	4 b. Cin. or Nashville.
28	Brighton.	"
32	Kelso.	"
37	Fayetteville.	"

Nash., Chattanooga & St. Louis R. R.—Con.

Ms.	(McMinnville and Sparta Branch.)	Alt.
0	Tallahoma.	{ 13 a. Sub-Carbon., barren ground.
12	Manchester.	
35	McMinnville.	{ 13 b. Sub-Carbon. St. Louis l. s.
46	Rock Island.	"
61	Sparta.	"

(Jasper Branch.)

0	Bridgeport.	3 c. Knox dolomite.
6	S. Pittsburgh.	"
12	Jasper.	13 b. Sub-Carbon.
19	Victoria.	"
24	Sequatchee.	Silurian.
25	Inman.	Iron ore mines.

(Centerville Branch.)

0	Dickson.	13 b. Sub-Carb.
11	Bon Aqua.	"
17	Warner.	"
24	Graham.	"
34	Centerville.	5 d. Niagara.

Tennessee Coal and Iron Co.'s R. R.

0	Cowan.	{ 13 b. Sub-Carbon., St. Louis l. s.
9	Sewanee.	14. Coal Measures.
15	Monteagle.	"
21	Tracy City.[11]	"

East Tennessee, Virginia & Georgia Railroad.

0	{ Bristol, at Va. Line.	{ 3 c. Knox dolomite, or Quebec.
11	Union.[12]	" 1457
20	Carter's.[12]	"
25	Johnson's.[12]	" 1643
32	Jonesboro.	" 1734
43	Limestone.	"
47	Fuller's.	· "
56	Greeneville.[13]	" 1581
65	Midway.	"
74	Rogersville Jc.	4 b. Nashville.
82	Russellville.	{ 3 c. Knox dolomite, or Quebec.
88	Morristown.	" 1298
97	Talbot's.	"
101	Mossy Creek.[14]	"
105	Newmarket.	" 1057
114	Strawberry Pls.	"

9. At Newsom's a section may be conveniently seen extending from the upper part of the 4 b. Nashville to the 13. sub-carboniferous.

10. *Camden.* Half a mile west of Camden depot the railroad crosses "the old shore line" and passes from the ancient Paleozoic strata on to the Tertiary and Quaternary ones, the limestones, cherts, etc., disappearing, and the softer sands and clays takng their place.

11. At *Tracy City* is a good bed of coal, extensively mined. In this vicinity a good section of the coal measures of this part of Tennessee can be obtained. (See "The Coal Regions of America," pages 351 to 373.

12. Within a few miles of these Stations are ridges and knobs made up of dark shales of Cincinnati or Nashville age. At Johnson's a point of one of these ridges is very near the Station.

13. The high mountains so conspicuous from the depot at Greeneville are made up of 2 b. Chilhowee (Potsdam) sandstone, and of a 2 a. Ocoee slates and conglomerates.

14. Veins of zinc ore are found at this point in the 3 c. Knox dolomite.

Ms.	East Tennessee, Virginia & Georgia Railroad.—Con.	Alt.
120	McMillan's.	3 c. Knox dolomite, or Quebec.
130	Knoxville.[15]	3 c. Knox dolomite and Trenton. 900
135	Erin.	4 a. Tren. & Nash. 404
145	Concord.	3 c. Knox dolomite.
154	Lenoirs.[16]	"
159	Loudon.	" 816
165	Philadelphia.	"
175	Sweetwater.	"
180	Reagan's.	3 b. Knox shale.
186	Athens.	3 c. Knox dolomite 933
193	Riceville.	3 b. Knox shale.
201	Charleston.	3 c. Knox dolomite.
213	Cleveland.	3 c. Knox dolomite and shale. 678
	State Line. (Continued in Georgia.)	
240	Dalton.	3 c. Knox dolomite.
213	Cleveland.	" 678
227	Ooltawah.[17]	4 a. Trenton.
232	Tyner's.	3 b. Knox shale.
242	Chattanooga.	See N. C. & S., and S. R. R. 684

East Tennessee, Virginia & Georgia R. R.
(North Carolina Division.)

Ms.	Station	Alt.
0	Morristown.	3 c. Knox dolomite, or Quebec. 1283
4	Sulphur Springs.	3 b. Knox shale and dolomite.
6	Witt's Foundry.	"
19	Dandridge Road.	
12	Leadville.	4 b. Shales of Cin. or Nashville age.
15	Rankin's.	3 c. Knox dolomite, Nashville shales.
.....	Newport.	
26	Bridgeport.	3 c. Knox dolomite.
33	Big Creek.	3 c. Knox dolomite, and 2 a. Ocoee Conglomerate & shales.
39	Wolf Creek.	2 a. Ocoee Conglomerate and shales.

(Marysville Branch.)

Ms.	Station	Alt.
0	Knoxville.	3 c. Knox dolomite, and 4 a. Trenton. 900
.....	Bruce's.	Unknown.

Ms.	East Tennessee, Virginia & Georgia R. R. (Marysville Branch.)—Con.	Alt.
.....	Little River.	Unknown.
16	Marysville.	3 c. Knox dolomite.
	(Ohio Division.)	
0	Knoxville.[16]	2-4. Lower Silurian.
9	Powell's.	"
14	Heiskell's.	"
21	Clinton.	4 a. Trenton and 3 c. Upper Knox.
27	Cane Creek.[18]	2-4. L. Silurian.
31	Offutt's.	" (?)
38	Careyville.	14. Coal Measures.
47	Buckeye.	"
55	Elk Valley.[19]	" (fault.)
62	Newcomb.	"
66	Jellico.	"

Cincinnati. N. O. & Texas Pacific R. R.
(Late Cincinnati Southern Railroad.)

Ms.	Station	Alt.
0	Cincinnati.	(See Ohio.)
198	State Line of Tn.	11 b. L. Cl. Measures.
201	Winfield.	"
206	Oneida.	"
211	Helenwood.	" 1454
216	New River.	" 1400
219	Robbins.	" 1215
221	Rugby Road.	" 1382
223	Glen Mary.	"
229	Sunbright.	" 1289
234	Annadel.	" 1356
238	Lancing.	" 1243
243	Nemo.	" 1197
251	Oakdale Junc.	" 917
257	Elmore Gap.	" 812
265	Rockwood.[20]	" (?) 840
270	Glen Alice.	L. Silurian Knox. 881
273	Roddy.	" 824
277	Lorraine.	" 784
280	Spring City.	" 811
285	Sheffield.	" 781
291	Darwin.	"
297	Dayton.	" 751
304	Coulterville.	" 711
307	Rock Creek.	" 711
309	Retro.	" 751
314	Rathbun.	" 741
318	Melville.	" 781
326	Hixon's.	" 711
331	Boyce.	" 694
335	Chattanooga.[20]	" 684

15. The high portion of the city on the former, the depot on the latter. Shales of Nashville jus west of depot. On the side of the Holston River opposite Knoxville high knobs covered with dee red soil are conspicuous, which are made up in good part of a dark ferruginous limestone, called Iro Limestone, and which belongs to the 4 b. Nashville (Cincinnati) group.

16. *Lenoirs.* Depot on junction of the Lenoir or Chazy limestone and the Knox dolomite. Th former lies to the southeast, and the latter to the northwest.

17. About one mile east of Ooltawah the railroad passes through a gap of the White Oak Moun tatns, in which is an interesting section embracing 4 b. Nashville, 5 d. Niagara, Devonian (10 c. Blacl Shale) and 13 Sub-Carboniferous rocks.

18. From *Knoxville to Cane Creek* the stations are either on the Knox divisions or the Trenton.

19. *Elk Valley* is on a fault, and in the upper part of the narrow valley the Trenton, the re Clinton ore, the Sub-Carboniferous limestone, and the Coal Measures may be seen and studied.

20. Although Professor Safford knows the geology of the country passed over, he has not travele on this railroad, and therefore the sub-divisions of the Lower Silurian are not given. From Rockwoo to Chattanooga the stations are mostly on his Knox divisions, but in a few cases on Trenton.

Arkansas.

GENERAL GEOLOGY OF THE STATE.—Dividing the State diagonally from northeast to southwest, beginning near the easterly boundary of Randolph county and running thence past Grand Glaise and Little Rock, through to Fulton in Hempstead county on Red River, (consequently nearly in the line of the St. Louis, Iron Mountain & Southern Railroad), almost all the State, *east* of said line, will be found of the 19. Tertiary formation, except along the river bottoms, where it is 20. Quaternary. The northern portion, *west* of said line, is mostly 2–8. Silurian, with some 9–12. Devonian and 14. Carbon-iferous further south; the middle western part of the State being 14. Carboniferous, while the south-west part (namely, from Arkadelphia and Murfreesboro south and west) will be found 18. Cretaceous.

In consequence of the above general arrangement of the geological formations in the State, it will be readily perceived that the St. Louis, Iron Mountain & Southern Railroad runs mainly near the junction between the Silurian, Carboniferous and Cretaceous of the west side, and the 19. Ter-tiary, with some 30. Quaternary, of the east side. Further, that the Arkansas Midland is chiefly in the 19. Tertiary and 20. Quaternary, while the Little Rock & Fort Smith Railroad passes through the 14. Carboniferous formation; also, that the Memphis & Little Rock Railroad runs through 19. Tertiary and 20. Quaternary.

The State affords abundance of manganese, zinc and kaolin.

The expression, "Quaternary over Silurian," is intended to indicate that the superficial deposits of the locality, opposite which the remark is placed, are Quaternary; but that when lower formations are exposed by denudation, &c., they would be found Silurian. A similar interpretation is designed to be given to "Tertiary over Cretaceous," and the like expressions.   R. O.

Arkansas Midland Railroad.

Ms.		Alt.
0	Helena.	20. Quat. over 19. Ter.
10	Bushville.	"
21	Marvell.	"
30	Palmer's.	"
40	Duncan.	"
48	Clarendon.	"
63	Brinkley.	" 200

Little Rock & Fort Smith Railroad.

Ms.		Alt.
0	Argenta.	14. Carboniferous. 301
10	Warren.	" 331
30	Conway.	14 b. Lower Coal 361
44	Plummerville.	Mrs. " 333
63	Atkins.	" 399
83	Georgetown.	"
95	Cabin Creek.	" 449
101	Clarksville.	" 409
125	Ozark.	" 424
150	Alma.	" 477
159	Van Buren.	" 449
168	Cherokee.	"

Memphis & Little Rock Railroad.

Ms.		Alt.
0	Memphis.	20. Quat. over 19. Ter.
17	Edmondson's.	"
33	Black Fish Siding.	"
41	Madison.	" 207
53	Palestine.	"
70	Brinkley.	" 200
87	De Vall's Bluff.	{ 19. Tertiary over Mills. Grit. 181
103	Carlisle.	"
112	Lonoke.	"
125	Galloway.	"
135	Little Rock.[1]	14. Carboniferous. 363

Missouri Pacific Railroad.

Ms.	St. Louis, Iron Mount'n & South'n Div.		Alt
186	Moark.	20. Allu. over Sil.	287
192	Corning.	"	294
203	Peach Orchard.	"	290
214	O'Kean.	"	276
225	Walnut Ridge.	"	275
232	Minturn.	"	251
244	Swifton.	"	253
262	Newport.	"	232
273	Grand Glaise.	14 a. Mills. Grit	226
278	Bradford.	"	246
292	Judsonia.	"	222
305	Garner.	"	211
312	Beebe.	"	250
320	Austin.	"	254
332	Jacksonville.	"	257
345	Little Rock.[1]	14. Carboniferous.	363
355	Mabelvale.	"	
368	Benton.	"	233
388	Malvern.	"	277
410	Arkadelphia.[2]	{ Junc. of 14. Carb., 18. Creta. & 19. Ter.	191
437	Boughton	19. Ter. over 18. Creta.	
449	Emmet.	"	
457	Hope.	"	357
471	Fulton.	"	272
490	Texarkana.	{ 20. Quaternary over 19. Tertiary.	303

Hot Springs Railroad.

Ms.			Alt
388	Malvern.	{ 14 b. Lower Coal Measures.	277
406	Rockport.		
413	Hot Springs.[3]	{ 14 a. Millstone Grit.	716

*This page is by Richard Owen, M. D., LL. D., of New Harmony, Indiana, the rest of the roads were prepared by Professor R. H. Loughridge, now of the Kentucky Geological Survey.

1. *Little Rock.* In Pulaski county, west of Little Rock, excellent granite is quarried. R. O.
2. *Arkadelphia.* In the ridges pervading Montgomery county, which adjoins Clark county on the northwest, there are gorges which furnish the "crystal hunter" vast quantities of rock crystal, sent extensively to mineralogical cabinets.   R. O.

Missouri Pacific Railroad.
St. Louis, Iron Mountain & South'n Div.—Con.

Ms.	(Helena Branch.)	Alt.
0	Knobel.	{ 20. Quaternary over 19. Tertiary. 271
13	Gainesville.	" 500
21	Parmly.	"
34	Brookland.	"
45	Ridge.	"
58	Harrisburg.	"
69	Cherry Valley.	"
76	Vanndale.	"
98	Forrest City.	" 281
114	Marianna.	"
127	Lexa.	"
140	Helena.	"

(White River Branch.)

Ms.		Alt.
0	Newport.	{ 20. Quaternary over 5–7. Silurian.
3	Diaz.	"
9	Paroquet.	5–7. Silurian.
14	Newark.	13. Sub-Carb.
24	Moorefield.	"
29	Batesville.	"

(Camden Branch.)

Ms.		Alt.
0	Gurdon.	{ 20. Quaternary over 19. Tertiary. 213
7	Whelan.	"
18	Chidester.	"
24	Dowling.	"
34	Camden.	"

Texas & St. Louis Railway.
(Missouri and Arkansas Division.)

Ms.		Alt.
0	Birds Point, Mo.	20. Alluvium. 221
58	Malden, Mo.	" 297
70	St. Francis.	" 335
79	Greenway.	20. Quat. over 19. Ter.
86	Rector.	"
104	Paragould.	"
116	Brookland.	"
125	Jonesboro.	"
155	Fisher.	"
179	Bemis.	"
199	Brinkley.	" 200
214	Clarendon.	"
238	Goldman.	20. Alluvium.
251	Wabbaseca.	"
260	Rob Roy.	"
267	Pine Bluff.	20. Quat. over 19. Ter.
284	Big Creek.	"
300	Kingsland.	"
337	Camden.	" 125
348	Senter.	"
368	McNeil.	"
389	Lewisville.	"
397	Garland City.	20. Alluvium.
418	Texarkana.	{ 20. Quaternary over 19. Tertiary. 303

Arkansas Valley Route.
(Little Rock Division.)

M.s		Alt.
0	Little Rock[1]	14. Carboniferous. 283
5	Sweet Home.	20. Quat. over 19. Ter.
12	Wrightsville.	"
22	Redfield.	"
27	Jefferson Springs.	"
42	Pine Bluff.	"
55	Linwood.	"
69	Varner.	"
81	Dumas.	"
94	Tillar.	"
106	Trippe Junc.	"
113	Arkansas City.	20. Alluvium.

(Ouachita Division.)

M.s		
0	Arkansas City.	20. Alluvium.
7	Trippe.	20. Quat. over 19. Ter.
17	Dermott.	"
25	Collins.	"
40	Monticello.	"
56	Warren.	"

Kansas City, Fort Scott & Gulf R. R.
(Thayer to Memphis.)

M.s		
340	Thayer.	5–7 Silurian.
343	Mammoth Spring	"
369	Williford.	"
381	Imboden.	"
390	Black Rock.	20. Quat. over Sil. (?)
399	Hoxie.	{ 20. Quaternary over 19. Tertiary. 290
412	Bonnerville.	"
424	Nettleton.	"
431	Big Bay.	20. Alluvium.
459	Gilmore.	"
474	Marion.	"
484	West Memphis.	"
487	Memphis.	20 c. Quaternary, bluff.

St. Louis & San Francisco R. R.
(Arkansas Division.)

M.s		
0	Fort Smith.	14. Carboniferous. 467
7	Van Buren.	" 449
27	Mountainburg.	"
47	Brentwood.	"
65	Fayetteville.	"
85	Rogers.	"
98	Garfield.	"
104	Seligman, Mo.	13 c. Low. Carbon.

Eureka Springs Railway.

M.s		
0	Eureka Springs.	14. Carboniferous.
9	Walden.	"
19	Seligman, Mo.	13 c. Low. Carbon.

3. *Hot Springs.* Celebrated alkaline hot springs. In the southwestern part of this county is the noted Magnet Cave, in and around which are found many beautiful minerals, especially magnetite, or magnetic iron ore, garnets, actinolite, epidote and crystallized hornblende, also the celebrated novaculite or Ouachita, sometimes spelled "Washita," honestone, also called Arkansas whetstone.

R. O.

Indian Territory.

The list of Formations is at the head of the Texas Chapter.

Geology of Indian Territory.—The eastern part of the Indian Territory is made up almost entirely of the representative sandstones, limestones, etc., of the Coal Measures, the former rock capping the mountains of the east, and becoming the prevailing feature in the lower hills and country westward, while the limestone which appears prominently in the mountain sides and valleys of the east, disappears almost entirely in the west, or is exposed only in the beds of the largest streams. Carboniferous coal mines are extensively worked on the south of the Canadian river, by companies who have leased them from the Nation. The Permian is said to cover an area south of the Wichita Mountains on the southwest, while the remainder of the western part of the Territory is thought to belong to the Triassic and Jurassic, except the regions of the mountains which are of granitic structure, their granites flesh colored, and associated with greenstone, quartz, porphyry, etc.—*Dr. R. H. Loughridge's Cotton Report, Census of 1880.*

Missouri, Kansas & Texas R. R.			Missouri, Kansas & Texas R. R.—Continued.		
Ms.		Alt.	Ms.		Alt.
355	Vinita.	14 b. Coal Meas.	698	556	Durant.	18. Cretaceous.	689
379	Pryor Creek.	"		568	Colbert.	"	658
388	Chouteau.	"	624	576	Denison, Texas.	"	728
410	Gibson.	"	533				
419	Muskogee.	"	599	**Atlantic & Pacific Railroad.**			
449	Eufaula.	"	617				
470	Reams.	"	609	337	Shawnee.	14 b. Coal Measures.	
479	McAllister	"	684	342	Prairie City.	"	
491	Savanna.	"		348	Oseuma.	"	
506	Limestone Gap.	"	645	353	Afton.	"	
525	Atoka.	"	656	358	Albia.	"	
536	Caney.	"	530	364	Vinita.	"	698
544	Caddo.[1]	18. Cretaceous.	705				

1. The white "Rotten limestone," with an abundance of fossils, is the prevailing rock in this black prairie region, extending southward into Texas, and westward to within a few miles of Tishomingo, Chicasaw Nation.   R. H. L.

Texas.[1]

LIST OF GEOLOGICAL FORMATIONS FOUND IN TEXAS AND INDIAN TERRITORY.

20. Quaternary.	20 c. Alluvium.	18. Cretaceous.	18 b. Upper Creta.
"	20 b. Port Hudson.	"	18 a. Lower Creta.
"	20 a. Stratified Drift.	16 Triassic.	16. Triassic.?
	{ 19 b. Miocene or	14. Carboniferous.*	14. Coal Measures.
19. Tertiary.	{ Grand Gulf.	2. Lower Silurian.*	2. Cambrian.
"	19 a. Eocene.		

International & Great Northern R. R. Gulf Division.		Alt.	International & Great Northern R. R. San Antonio Division—Con.		Alt.
Ms.			Ms.		
. ...	Galveston.	20. Quat. Pt. Hudson.³	119	Rockdale.	19. Ter., a. Eoce. 469
0	Houston.	" 53	145	Taylor.	18. Cretaceous.
		{ 19. Tertiary,	162	Round Rock.	" 720
23	Spring.	{ b. Miocene. 126	181	Austin.	" 477
		{ (Grand Gulf.)	212	San Marcos.	"
47	Willis.	" 381	230	New Braunfels.	"
66	Phelps.	" 377	261	San Antonio.⁹	" 688
78	Riverside.²	" 169	274	Medina.	"
85	Trinity.	" 234	315	Pearsall.	19. Ter., a. Eocene.
99	Lovelady.	19. Ter., a. Eoce. 300	331	Frio.	"
114	Crockett.	" 350	376	Encinal.	"
127	Grapeland.	" 480	394	Webb.	"
139	Elkhart.	" 390	415	Laredo.⁴	"
152	Palestine.	" 495	0	Troupe.	" 467
164	Neches.	" 411	19	Tyler.	" 531
180	Jacksonville.³	" 525	44	Mineola.	" 402
198	Throupe.	" 487		Columbia Division.	
211	Overton.	" 507			
223	Kilgore.	" 371	0	Columbia.	{ 20. Quaternary,
235	Longview.	" 336			{ c. Alluvium. 34
259	Marshall.	" 371	18	China Grove.	" b. Pt. Hudson.⁵⁰
275	Jefferson.	" 221	30	Houston.	" 37
334	Texarkana.	" 303		Georgetown Railroad.	

San Antonio Division.			Georgetown Railroad.		
			0	Round Rock.	18. Cretaceous. 720
			10	Georgetown.	" . 753
0	Palestine.	19. Ter., a. Eoce. 495		Henderson & Overton Branch.	
18	Oakwood.	" 280			
44	Jewett.	" 495	0	Overton.	{ 19. Tertiary,
55	Marquez.	" 410			{ a. Eocene. 507
75	Englewood.	" 420	16	Henderson.	"
90	Hearne.	" 305			

* The sub-division of the Carboniferous and Silurian represented here have not been fully ascertained. The Devonian and Upper Silurian seem to be entirely absent.

1. By Professor R. H. Loughridge, now of the Kentucky Geological Survey, the information being derived largely from his personal observations.
2. *Riverside.* Fine exposures of Grand Gulf sandstones.
3. *Jacksonville.* Tertiary iron ore hills a few miles south.
4. *Laredo.* Lignite in heavy beds near here.

Ms.	Texas & Pacific Railroad. Trans-Continental Division.	Alt.
0	Texarkana.	19. Ter, a. Eoce.	303
17	Whaley's.	"	
34	DeKalb.	"	
61	Clarkesville.	18. Cretaceous.	464
68	Bagwells.	"	
91	Paris.	"	592
112	Honey Grove.	"	682
128	Bonham.	"	582
139	Savoy.	"	
142	Bells.	"	675
155	Sherman.	"	747
173	Whitesboro.[5]	"	
209	Denton.	"	
244	Fort Worth.[10]	"	623

Southern & Rio Grande Division.

Ms.		Alt.
0	Texarkana.	{ 19. Tertiary, a. Eocene.	303
16	Sulphur.	"	
44	Kildare.	"	
58	Jefferson.	"	221
74	Marshall.	"	371
98	Long View.	"	336
120	Big Sandy.	"	336
143	Minneola.	"	402
157	Grand Saline.	"	400
174	Will's Point.	"	530

Ms.	Texas & Pacific Railroad. Southern & Rio Grande Division—Con.		Alt.
190	Terrell.	18. Cretaceous.	514
209	Mesquite.	"	494
222	Dallas.	"	466
241	Arlington.[6]	20. Quater, a. drift.	
254	Fort Worth.[10]	18. Cretaceous.	623
284	Weatherford.[7]	20. Quater., a. drift.	864
308	Brazos.	14. Carboniferous.(?)	
358	Eastland.	"	1299
368	Cisco.	"	1611
414	Abilene.	18. Probably Creta.	
455	Sweet Water.	"	
473	Loraine.	"	
492	Westbrook.	"	
512	Signal Mount.	"	
522	Big Springs.[8]	"	
543	Mariefield.	"	
562	Midland.	"	
572	Warfield.	18. Cretaceous.	
592	Douro.	"	
602	Metz.	"	
612	Sand Hills.	"	
623	Aroya.	"	
641	Quito.	"	
654	Pecos River.[11]	"	
664	Hermosa.	{ The plains are chiefly Cret.; the mountains are part Palæozoic (Carbon.) in part eruptive.
684	Gomez.	
705	Kent.	
736	Wild Horse.	
754	Carrizo.	

5. *Whitesboro.* The belt of Lower Cross Timbers is crossed between this and Denton.
6. *Arlington.* Lower Cross Timbers—a belt of sandy land, 10 to 15 miles wide, timbered with post oak, and reaching from within the Indian Territory southward to the Brazos near Waco.
7. *Weatherford.* Upper Cross Timbers—similar in many respects to the lower belt with which it is united on the north of Red River, but is wider, more irregular in outline, and interspersed with high Cretaceous prairie outliers. It reaches southward from Red River along the western border of the Cretaceous, and crosses the Brazos nearly to the Colorado River.
8. *Big Springs.* Llano Estacado, or the Staked Plain, lying north of this road, is a district of 75,000 square miles in Northwestern Texas, besides the portion in New Mexico, and is a vast and level prairie, as smooth and firm as marble, apparently boundless. The soil is chiefly a brown loam, sometimes sandy, and with no vegetation other than gramma and mesquite shrubs, which appear a few inches above the surface. Alkali ponds or lakes occur frequently, and a number of springs whose waters are suitable for use. Day after day in traveling here, the country is almost perfectly level, except in crossing the sand hills, which are really an object of curiosity. Part of the sand is black; then comes the white sand hills, miniature Alps of sand perfectly white and clean, summit after summit in every direction, not a sign of vegetation upon them, nothing but sand piled upon sand.
9. *San Antonio.* About 80 miles northwest of this place and 18 north of Fredericksburg, in Gillespie County, is a granite hill called Enchanted Rock, a huge granite and iron formation about eight hundred feet high, covering at its base several acres of space, its top being about four hundred yards square. Its name is derived from its magnificent appearance, for when the sun shines upon it in the morning and at evening, it resembles a huge mass of burnished gold. The Azoic rocks found in this central part of the State are mostly of the pink feldspathic variety, resist disintegration, and form high and prominent points or hills throughout the region.
10. *Fort Worth* and *Cleburne.* The Lower Cross Timber Belt passes east of town. Professor R. P. Whitfield says, Fort Worth is an excellent locality for Cretaceous fossils.
11. *Pecos.* Dr. R. H. Loughridge, in his U. S. Census Cotton Report, describes the several chains of almost treeless mountains in Western Texas, west of the Pecos River, as largely granite, with accompanying sandstones and limestones. In some of the mountains characteristic eruptive rocks are reported as penetrating the later formations, and rising above them in huge masses or forming vertical columns, as in the Organ Mountains near El Paso.
12. *Sierra Blanca.* The great mountain ranges consist, first, next the Pacific coast, and lying from ten to two hundred miles distant from it, the Cordilleras or Coast range, and second the Sierra Nevada, for which see the California chapter. The third is an irregular ill-defined chain, the Sierra Madre, and at El Paso we encounter the western flank of the fourth great mountain chain, the Rocky Mountains, which terminate in what is called the Organ Mountain. Going east from El Paso,

Texas & Pacific Railroad.

Ms.	Southern & Rio Grande Division—Con.		Alt.
777	Sierra Blanco.[12]	18. Cretaceous,	4512
823	Porter.	Plains, Mts.,	3541
832	Rio Grande.	Palæ. and erup.	5564
857	Ysleta.	"	3664
869	El Paso.[13]	"	3713

Gulf, Western Texas & Pacific Railroad.

Ms.			Alt.
0	Indianola.	{ 20. Quaternary, b. Port Hudson.	26
25	Placedo.	"	
38	Victoria.	"	87
55	Thomaston.	"	
70	Cuero.	"	177

Houston & Texas Central Railroad.

Ms.			Alt.
0	Houston.	{ 20. Quaternary, b. Port Hudson.	37
6	Hockley.	"	225
51	Hemstead.	"	245
71	Navasota.	19. Ter., a. Eoce.	219
100	Bryan.	"	371
121	Hearne.	"	305
130	Calvert.	"	337
143	Bremond.	"	467
162	Thornton.	"	496
170	Groesbuck.	"	481
181	Mexia.	"	537
211	Corsicana.	"	427
239	Palmer.	18. Cretaceous.	471
265	Dallas.	"	466
296	McKinney.	"	615
329	Sherman.	"	747
338	Denison.	"	723

Western Division.

Ms.			Alt.
0	Hempstead.	{ 20. Quaternary, b. Port Hudson.	245
11	Chapel Hill.	{ 19. Ter. b. Miocene, Grand Gulf.	337
21	Brenham.	"	350
34	Burton.	"	436
47	Ledbetter.	" a. Eocene.	464
56	Giddings.	"	536
78	McDade.	"	539
115	Austin.	18. Cretaceous.	513

Houston & Texas Central R. R.—Con.

Ms.	Waco Branch.		Alt.
0	Bremond.	19. Ter., a. Eoce.	467
9	Marlin.[14]	18. Cretaceous.	394
43	Waco.	"	
98	Morgan.	"	734
128	Hico.	"	1007
150	Dublin.	"	1449
197	Cisco.	14. Carboniferous.[16]	1611
229	Albany.	" (?)	1402

New York, Texas & Mexican Railroad.

Ms.			
0	Rosenberg.	{ 20. Quaternary, b. Port Hudson.	109
26	Wharton.	{ 20. Quaternary, c. Alluvium.	
92	Victoria.	{ 20. Quaternary, b. Port Hudson.	87

Galveston, Harrisburg & San Antonio R. R.
Texas & New Orleans Division.

Ms.			Alt.
0	Houston.	{ 20. Quaternary, b. Port Hudson.	37
41	Liberty.	"	43
63	Sour Lake.	"	47
83	Beaumont.	"	
105	Orange.	"	10
0	Houston.	"	37
10	Pierce Junction.	"	63
34	Richmond.	"	73
53	East Bernard.	"	123
70	Eagle Lake.	"	213
86	Columbus.	{ 19. Tertiary, b. Miocene, Grand Gulf.	213
102	Weimar.	"	420
111	Schulenburg.	19. Ter., b. Mioc.	341
148	Harwood.	" a. Eocene.	463
158	Luling.	"	413
180	Seguin.	"	559
185	Marion.	"	566
216	San Antonio.[9]	18. Cretaceous.	683
241	Lacoste.	"	
266	Hondo.	"	
287	Sabinal.	"	
308	Uvalde.	"	891
343	Anacacho.	"	
350	Spofford Junc.	"	

following the river, we encounter two other ranges of mountains at intervals of about eighty miles, called the Eagle Springs or Sacramento Mountains, and the Limpia or Gaudalupe Mountains, in passing through which the river forms a series of cañons (see Note 16). On the Mexican side of the river all these mountains arise again, and expand in width and height and attain a great elevation.

13. *El Paso* is justly considered one of the garden spots of the interior of the continent. The climate is dry, but the settlements are irrigated by water from the river by means of a dam and canal, and are not dependent on rains for their fertility. The place is more than two hundred years old, the settlement having been commenced about 1680, when the Spaniards were driven from New Mexico by the Indians. It is situated in a charming valley, the Rio del Norte having escaped the mountain passes, here runs in an open fertile plain, stretching out along the river to the length of many miles, all the houses surrounded by gardens, orchards and vineyards, and rich settlements, the result of judicious irrigation, with cornfields as far as the eye can trace the stream lining its green banks. Such a scene will always be attractive, but to a traveler who has passed over the lonesome plains it appears like an oasis in the desert. The mountains southwest of the town consist almost entirely of

Galveston, Harrisburg & San Antonio R. R. Ms. Texas & New Orleans Div.—*Continued.*		Alt.
387 Del Rio.	18. Cretaceous.	
..... Pecos River.[15]	"	
450 Shumla.	"	1413
462 Langtry.	"	1304
491 Lozier.	"	1535
..... Thurston.	"	1911
534 Sanderson.[16]	"	2774
559 Rosenfield.	"	3665
566 Maxon Springs.	"	3538
573 Taber [17]	"	3505
579 Haymond.	"	3583
..... Warwick.	"	4071
595 Marathon.	"	4043
626 Murphysville.	"	4485
653 Maria.	"	4692
663 Aragon.	"	4899
689 Valentine.	The Plains are mostly Cretaceous; the Mountains Paleozoic and eruptive.	4424
720 Haskell.		4013
757 Sierra Blanca.[12]		4512
780 Finlay.		3665
795 Camp Rice.		3519
..... Porter.		3541
811 Rio Grande.		3564
836 Ysleta.		3664
848 El Paso.[13]		3713
0 Columbus.	{ 19. Tertiary, b. Mioc. (Grand Gulf.)[213]	
31 La Grange.[13]	"	

Galveston, Harrisburg & San Antonio R. R. Ms. Texas & New Orleans Div.—*Continued.*		Alt.
0 Harwood.	{ 19. Tertiary, a. Eoc. (Grand Gulf.)[463]	
..... Gonzales.	"	276
0 Pierce Junc.	20. Quat., b. Pt.Hud.[63]	
8 Harrisburg.	"	38
8 Spafford Junc.	18. Cretaceous.	
33 Eagle Pass.	19.Ter., a. Eoc. (?) [500]	

Gulf, Colorado & Santa Fe Railroad.

0 Galveston.	20. Quat., b. Pt.Hud.	8
43 Arcola.	"	66
64 Richmond.	"	73
94 Sealy.	"	189
107 Belleville.	{ 19.Tertiary, b. Mioc. (Grand Gulf.) [262]	
126 Brenham.	"	301
141 Somerville.	"	
158 Caldwell.	" a.Eoc. [411]	
174 Milano.	"	500
188 Cameron.	"	407
218 Temple.	18. Cretaceous.	695
242 McGregor.	"	
270 Clifton.	"	670
280 Meridian.	"	791
287 Morgan.	"	734
317 Cleburne.[10]	"	933
345 Fort Worth.	"	623

limestone, below which at the foot of the mountain are horizontal layers of compact quartzose sandstone, such as underlie the basaltic and granitic rock for several hundred miles in the prairie toward Santa Fe, and granitic and porphyritic rock seem to a small extent to have burst through the limestone and overlown it.   A. W.

The Carboniferous limestone is supposed to underlie the whole extent of the country of the southwest, where the Cretaceous and Tertiary appear on the surface. Although of Carboniferous age it is not coal-bearing, being a marine deposit. An ocean existed in the Far West during the Carboniferous period, and the conditions were never such as to admit of the deposit of such materials as form coal beds. All the coal west of Kansas and Indian Territory is Cretaceous.

14. *Marlin.* Cretaceous rotten limestone forms the Brazos Falls, five miles south.

15. *Pecos River.* On the Mexican side, five miles south of the river, is a singular peak called the Picotena, rising abruptly from amid the surrounding limestone ranges, shooting up a sharp conical peak of basaltic structure. This peak, by its height and external features, presents a most striking landmark. It is the most northern outlier of an extensive igneous development of the mountain range, rising in jagged peaks to Alpine heights, and presenting in the forest growth which clothes its sides agreeable features of verdure, contrasting strangely with the river valley and its bare outline of desert hills.

16. *Sanderson.* The river cañons. Although the railroad, to shorten distance and for a better route, diverges from the river far to the northward, cutting off the great bend, yet the traveler may wish to know something of the general character of the river valley forming the Mexican boundary. The Rio Grande, from El Paso to the mouth of the Pecos River, south of Langtry station, is characterised by extensive cañons. The river presents a series of basins, more or less extensive, with descending steps and then a cañon. The scenery is unsurpassed for singularity and grandeur. Seventy miles below El Paso, south of Sierra Blanca, the Eagle Springs Mountains converge, and the river makes its way through them in deeply cut chasms, exposing the geological structure in sectional faces presented by its precipitous walls. At the gigantic cañon of San Carlos, twenty miles long, the river presents unbroken walls of limestone, from 200 to a perpendicular height of 1,500 feet. A faint conception only can be formed of the truly awful character of the chasm, which in ascending begins 85 miles and ends 105 miles above the mouth of the Pecos River, and is far from the railroads. Another, the San Vincente cañon, is below the great bend to the northward of the Rio Grande, and equals the San Carlos in many places in ruggedness and grandeur. These cañons were reported by Lieut. Emory to be among the most remarkable features on the face of the globe, namely, a river traversing at an oblique angle a chain of lofty mountains and making through these on a gigantic scale, what in Spanish-America is called a cañon, that is, a river hemmed in by vertical walls. The river is from 80 to 300 feet wide, and at a few points narrows down to 25 or 30 feet, where of course it is very deep and rapid.—*Rep. Mex. Boundary Com.*

17. *Taber* The igneous rocks. From the commencement of the table land in going westward on this road, broad belts of the Cretaceous formation occur, interrupted here and there by isolated dykes or mounds of trap or other igneous rocks, of modern age, producing a greater or less degree of

Gulf, Colorado & Sante Fe Railroad—Con.

(Dallas Division.)

Ms.		Alt.
0	Cleburne.[10] 18. Cretaceous.	933
13	Alvarado. "	
40	Duncan. "	1460
53	Dallas. "	466

(Lampasas Division.)

0	Temple. 18. Cretaceous.	695
8	Belton. "	620
56	Lampasas.[19] "	

(Montgomery Division.)

0	Somerville. 19.Tertiary, b. Miocene	
28	Navasota. " [219] (G'd Gulf.)	
55	Montgomery. "	

Houston, East & West Texas Railway.

0	Houston. 20. Quat., b. Pt.Hud.[53]	
56	Sheperd. "	
72	Livingston. { 19. Tertiary, b. Mio. (G'd Gulf.)	
88	Moscow. "	
140	Nacogdoches. " a. Eoc.	

Missouri Pacific R. R. (Texas Extension.)
(Fort Worth Section.)

0	Denison. 18. Cretaceous.	722
25	Whitesboro.[5] "	
43	Pilot Point. "	
61	Denton. "	
96	Fort Worth.[10] "	623
123	Alvarado. "	
150	Hillsboro. "	
184	Waco. "	
198	Lorena. "	
219	Temple Junction. "	695
258	Taylor. "	

0	Whitesboro.[5] "	
15	Gainesville. "	

0	Temple Junction. "	695
7	Belton. "	620

0	Denton. "	
15	Lewisville. "	
38	Dallas. "	466

(Mineola Section.)

0	Denison. 18. Cretaceous.	722
52	Greenville. "	
103	Mineola. 19. Ter., a. Eoc.	402

Missouri Pacific R. R. (Texas Extens'n)—Con.
(Jefferson Branch.)

Ms.		Alt.
0	Jefferson. 19. Ter., a. Eoc.	221
34	Dangerfield. "	403
50	Pittsburg. "	402
70	Winnsboro. "	552
93	Sulphur Spring. "	462
123	Greenville. 18. Cretaceous.	
139	Farmersville. "	
155	McKinney. "	615

Texas & St. Louis Railroad.
(Texas Division.)

0	Texarkana. 19. Ter., a. Eoc.	303
61	Mt. Pleasant. "	
72	Pittsburg. "	402
98	Gilmer. "	
106	Big Sandy. "	336
128	Tyler. "	531
165	Athens. "	
202	Corsicana. "	427
258	Waco. 18. Cretaceous.	
278	McGregor. "	
305	Gatesville. "	1000

Mexican National Railroad.

0	Corpus Christi. 20. Quat., b. Pt.Hud.[20]	
53	San Diego. 19. Ter., b. Mio. (?)	
100	Pena.[20] " (?) (G'd Gulf.)	
162	Laredo.[4] " a. Eocene. 506	

Rio Grande Railroad.

0	Brownsville. 20. Quat., b. Pt. Hud.[33]	
22	Point Ysabel. " (?) [5]	

Fort Worth & Denver City Railroad.

0	Fort Worth.[10] 18. Cretaceous.	623
14	Calef. "	
25	Rhone. "	
40	Decatur. "	
51	Alvord. 20.Quat.(?) } Upper Cross Timbers.	21
59	Sunset. "	
68	Bowie. "	
89	Alma. 14. Carboniferous.	
95	Henrietta. "	
114	Witchita Falls. "	915

metamorphism of the Cretaceous strata. Toward the west the igneous rocks, which first appear in small isolated knolls, gradually assume more importance and expand into long belts. In the Limpia range the second east of El Paso, these rocks become a mountain chain, having an elevation of 6,000 feet, and extending hundreds of miles north and south. These igneous protusions are composed of greenstone or basalt.—*Idem.*

18. *Lagrange.* A high bluff of Grand Gulf sandstone on south side of the Colorado River; heavy sand beds of Quaternary drift on the north of town.

19. *Lampasas.* A large sulphur spring here.

20. *Pena.* The Sandy Desert is a broad area of white sand, commencing about 20 miles southwest of Corpus Christi, extending northwesterly nearly to the Colorado, and up that river to near Eagle Pass, in a wedge shape. In many places it forms hills from 50 to 100 feet above the grassy plain, and being of a light yellow color are visible at a great distance.

21. *The Cross Timbers.* The peculiar belt of timbered country in Texas, and extending from the Brazos into the Indian Territory and to the Arkansas River, is of undetermined age; but, whatever may underlie the top material at 20 or 30 feet, or perhaps less, it can hardly be questioned that the ferruginous sandstones, pebble conglomerates, sands, and clays that form the surface material, are Quaternary. Their origin will be a matter of doubt until their extent northward is fully ascertained.

This blank space is intended for additional geological notes in pencil by the traveler.

Mexico.

GENERAL NOTE ON THE GEOLOGY OF MEXICO.

As long ago as 1830, William Maclure, the father of American geology, visited Mexico and reported in the American Journal of Science, that "the regular order of original stratification was so much deranged throughout that country by the intimate and frequent alternations of volcanic rocks, as to have subverted the original order of nature, and to have changed the class every mile. This leaves the geologist in doubt concerning the sub-strata, and would reduce most of his investigations to hypothetical results." In the previous year, probably the same observer reported in the same journal: "Lava, volcanic tufa, trachyte, clay-slate and a little granite, with porphyry, are predominant rocks in Mexico. Volcanic tufs, trachyte and lava form about ninety-nine hundredths of the country. It affords an extensive field of volcanic rocks, none of which appear to be recent, nor is there any volcano in activity." His travels may have only extended from Vera Cruz to the city of Mexico.

Not being able to procure a detailed report of the geology along the lines of the several Mexican railroads, such general information is here given as to some localities as could be collected from the reports of travelers, and in attempting this, some valuable and unexpected contributions have been received from some of the Pennsylvania geologists, rendering important aid in an almost hopeless task. The reader is also referred to the notes on Texas as to the formations found along the United States and Mexican boundary, which, together with what is given in the chapters on New Mexico and California, will throw some light on the great table-land of Mexico, now traversed by the Mexican Central and other railroads. Also, see the General Note on the Geology of the Far West.

In Mexico the altitudes are an interesting study. At the United States and Mexican boundary the lowest depression of the great table-land occurs, but even that is nearly 4,000 feet above the sea. North of this it ascends again even in the valley to 7,000 feet, and near the 49th parallel it is again depressed. South of the boundary line the plateau rises rapidly to the table-land of Mexico, where the mountains assume a loftier and more rugged and diversified appearance than on the Texas side. In the more northern portions of Mexico the deposits in the valleys seem to be Tertiary, and farther south they are probably the same, and from the prevalence of volcanic deposits portions of them may be metamorphosed. We have no reports of the Cretaceous. The mountains show surprising developments of Carboniferous limestone, and of Huronian and Laurentian formations. Probably they are an extension or repetition of the granitic, porphyritic, basaltic and other eruptive rocks, and of the Carboniferous limestone of our far Western States and Territories, and the latter of very great thickness. Any differences which Mexico may discover, will probably be such as the more recent and more extensive volcanic action, and an enlargement of some of the formations would produce. There is a boundless field for geologists in Mexico, the country is being made accessible by railroads, and there is a charm about the unknown which imparts an interest to that which, when known, may perhaps be neither interesting nor very important. At present there is surprisingly little generally known about the geology of Mexico, and this chapter is a first attempt in that direction. It is given as founded on imperfect observations. J. M.

The Great Mountain Table-Land of Mexico.—There is scarcely a point on the globe, says Humboldt, where the mountains exhibit so extraordinary a formation and magnitude as in Mexico. Switzerland is considered a very elevated country, but this opinion is merely founded on the aspect of a great number of summits perpetually covered with snow, and disposed in chains parallel to the great central chain. The summits of the Alps rise to 12,500 and 15,500 feet, while the neighboring plains are not more than 1,300 to 2,000 feet in height. The chain of mountains which forms the vast plain of Mexico is the same with that which, under the name of the Andes, runs through all South America; but the construction of this chain varies to the north and south of the equator. In the Southern Hemisphere the Cordillera is everywhere torn and interrupted by crevices like open furrows or transverse valleys. The elevated plains of Quito are not to be compared in extent with those of Mexico. In Peru the most elevated summits constitute the narrow crest of the Andes; but, in Mexico, as shown by the railroad altitudes, even the lowest valleys are from 4,000 to 6,000 feet high, and the general altitude of the whole country, except a narrow border on the Atlantic and Pacific coasts, is 7,000 to 8,000 feet, and upon this are disposed the high volcanic peaks, less colossal, it is true, than the Andes, but still 16,000 to 17,000 feet and, taken together, there is no such mountain on the globe, taking into view its extension northward into the United States. Peru and New Grenada contain deep transverse valleys, but in Mexico carriages (or in our day railroad cars) roll on from Mexico to Santa Fe, a distance of 1,500 miles, at altitudes of from 4,000 to 8,000 feet. On the whole road there are few difficulties for art to surmount, so little is the table-land of Mexico interrupted by valleys.

The Volcanic Mountains. In the part of the great plain of Mexico between the capital and Vera Cruz, a group of mountains appears which rivals the most elevated summits of the new continent. It is enough to name four of these colossi: Popocatepetl, or Smoke Mountain, 17,716 feet; Iztaccihuatl, or White Woman, 15,700 feet; Citlaltepetl, or Orizaba, the Star Mountain, 17,371 feet, and Nauhcampatepetl, or Perote, the Square Mountain, 13,414 feet high, and so called from the form of a small porphyritic rock at the summit. Besides the four volcanic mountains mentioned, there are the Navado de Toluca, the Volcan de Colima, and a modern one, the new Volcan de Jorullo. As a general statement we may say that the general level of the whole country being some 7,000 feet above the sea, these volcanic cones situated upon it rise 8,000 to 10,000 feet higher.

The few observations that have been made by geologists are not sufficient to found an opinion upon as to the formations composing the core or main body of this vast mountain chain, or whether it is uniform throughout. Carboniferous limestone forms the visible portion at many places, and is no doubt an important element in its structure. There are other mountains of basalt or trap; others are Laurentian and Huronian, and at Mexico and southward are the chains of remarkable extinct volcanoes. J. M.

Ms.	Mexican Railway.	Alt.	Ms.	Mexican Railway.—Continued.	Alt.
	Vera Cruz.			**Puebla.***	
0	Vera Cruz.[1]	19 b. Loup Fork Mio.(?)			{ The great volcano
9	Tejeria.[2]	"	94	Maltrata.[6]	25 miles to N. E.,
19	Purga.	"			17,368 feet. 5550
26	Soledad.	" 305	97	Bota.	{ Orizaba Mt. near
39	Camaron.	"			on the N.
47	Paso del Macho.	"	107	Boca del Monte.[7]	" to N. E. 7924
53	Atoyac.[3]	Volcanic soil. 1512	111	Esperanza.[8]	Orizaba Mt. to E.7941
66	Cordoba.[4]	" 2713	126	San Andres.	"
71	Fortin.	"	139	Ruoconada.	" 7731
82	Orizaba.[5]	{ The great volcano 25 miles to N. E., 17,368 feet. 4028	150	San Marcos.[9]	{ Malinche Mt. in view, 13,470 feet high.

* The road also passes through the States of Tlaycala and Mexico, but the boundary lines on the railroad are not ascertained.

1. *Vera Cruz.* The coast region extending between the beach at Vera Cruz along the Mexican Railway to the entrance into the gorges of the high Cordillera at Atoyac, fifty miles, is a low, sandy and marshy plain.   A. F. BANDELIER.*
The 19 b. Loup Fork Miocene, 2000 feet in thickness, has been proved over a territory six miles by eighteen, in the State of Hidalgo and the adjoining parts of Vera Cruz, north of this railroad, by Professor Edw. D. Cope, who visited the region, and obtained bones and teeth of Tertiary animals. Several thin beds of coal occur in it, with shales between, apparently composed of volcanic ash and beds of excellent clay.—*Am. Nat. Mag.*, 1885. It probably underlies this part of the railroad. (See Note 16, by Dr. H. M. Chance, as to the coal beds at Jimulco.)
2. *Jalapa.* There is a branch railroad from Vera Cruz to Jalapa, and the table land and mountains at that place are reported to be principally limestone, doubtless the same with the Carboniferous limestone on the Mexican Central Railroad. There are many marble quarries, and some sandstone or quartzite.
3. *Atoyac.* The Cordillera presents an abrupt dark-green front of lofty mountains, above which towers the snow-clad Orizaba. The railway enters the highlands through the narrow and very picturesque pass of the Atoyac, and the scenery changes. In appalling curves we wind our way upwards through groves, along fearful chasms and slopes covered with the most luxuriant vegetation of the tropics. It is the landscape of the tropics, resting, as it were, on the Southern Alps, where they descend towards the plains of Lombardy. The summit of Orizaba rises above the glorious landscape of this wonderful region, like a cone of molten silver, in a cloudless sky.   A. F. B.
.4 *Cordoba.* Much of the superficial formations of this part of Mexico must necessarily be of volcanic origin. The plains and valleys in many places owe their present topography and physical basis to the wasting of the high volcanoes, whose ruins and debris constitute the soil, being volcanic detritus or sand. These masses of volcanic debris thin out as they spread eastward to a fertile layer of black volcanic soil of a sandy appearance, reaching nearly to the eastern brow of the tableland at the Rio Atoyac.   A. F. B.
5. *Orizaba.* Here the giant, of which glimpses were before obtained, bursts out into full view. The railroad at this city is 4,028 feet above tide, and the mountain 17,368 feet, and is twenty-five English miles distant to the N. N. E.   A. F. B.
6. *Maltrata.* From Orizaba, the ascent by the road increases in steepness, and the scenery grows correspondingly wilder. The graceful palms gradually disappear, and beyond Maltrata the rise becomes extremely rapid. We are left in doubt as to which should be most admired—the sublime grandeur of nature, or the remarkable efforts of man to improve every chance, every inch almost, for establishing safe, rapid transit.   A. F. B.
7. *Boca del Monte.* We pass through tunnel after tunnel, until at last Boca del Monte is reached. The air blows cool, even chilly; dark pines cover the mountain sides, and on our right towers, in close proximity, the summit of the Volcano of Orizaba. Less than nine hours have carried us one hundred and seven English miles by the railroad, but a horizontal basis of less than fifty miles; and in altitude through three zones, representing a vertical stratum of 8,000 feet. We have passed through a series of changes and contrasts in vegetation and climate of the most striking kind, and perfectly characteristic of Mexico.   A. F. B.
8. *Esperanza.* The region through which the road passes in the vicinity of Esperanza, is a cold, rather barren looking highland, without any of the wildly picturesque scenery of the lower mountains; but the change is so sudden, that its very bleakness, with enormous prickly pears, dwarfish and ill-shapen palms, and tall *maguey* plants as types of vegetation, and the gigantic pyramid of Orizaba towering in full view to the east, has the effect of a successfully performed change in theatrical scenery.   A. F. B.
9. *San Marcos.* A downward grade is struck beyond Esperanza, the highest point is passed at Guadalupe, and then the insensible and gradual decline to the central basin of Mexico begins. More and more the isolated peak of Malinche or Perote becomes prominent above the surrounding landscape. It is 13,470 feet (English) above sea level.
10. *Huamantla.* Beyond Huamantla the traveler is treated to a change in scenery again, and one of a very peculiar nature. Two remarkable sights burst into view almost simultaneously; the two great volcanic peaks of Mexico looming up like immense monuments. The most northerly,

* Archæological Tour in Mexico.

Ms.	Mexican Railway.—Continued.		Alt.	Ms.	Ferrocaril Central Mexicano, or Mexican Central Railroad.		Alt.
		The two greatest volcanoes come in view to E. and continue so to city of Mexico, to E., S. W., S. and S. E.			**Dist. Federal.**		
161	Huamantla.[10]			0	Mexico.[12]	20. Quaternary.	7349
177	Apizaco.	7912		7	Tlalnepantla.	"	7382
186	Guadalupe.	8833		11	Barrientos.	"	7541
193	Soltepec.			13	Lecheria.	"	7392
				17	Cuautitlan.	"	7390
206	Apam.	Vol., and recent. 8226		22	Teoloyucan.	"	7392
215	Irolo.	" 8046		29	Huehuetoca.	"	7410
221	Ometusco.	"		33	Nochistongo.	"	7375
225	La Palma.	"			**Hidalgo.**		
229	Otumba.[10]	"		39	El Salto.	"	7095
236	San Juan Teotihuacan.	" 7531		50	Tula.	"	8860
243	Tepexpan.	20. Quat., and recent.		58	San Antonio.	Lauren. or Huro.	7175
263	Mexico.[11]	" 7347					

(Right side column for Central Railroad stations marked "Valley of Mexico.")

Yzac-tepetl, or White Woman, commonly called the Sierra Nevada, presents a serrated ridge covered with perpetual snow, and resting on a broad platform, which very gradually descends into dark forests. It has three summits; the northern, the highest, is 15,662 feet. While this mountain is lower than Popocatepetl, it is much more massive, its base being twice as long. From the west its long, icy crest appears, strikingly like a woman in her last repose, in a white shroud, lying on her back upon a steep-sided platform. The other, Popocatepetl, or moke Mountain, lies south of the former, and therefore at a greater distance from the railroad. It ppears as a perfect cone, slightly truncated, or rather with a cup-shaped summit. This concavity is the line of the crater here visible lengthwise, this part of the wall having fallen in, in the year 1664, whereas from Puebla it disappears, the top of the mountain rising above it to a sharp point. The height of Popocatepetl is 17,682 feet, being 314 feet higher than Orizaba. It thus appears to be the highest point of Mexico and of North America. The crater of Popocatepetl is a valuable mine of native sulphur. Its vast cup has a diameter of half an English mile, with such precipitous sides that it is considered impossible to descend into it, unless by means of a rope and crane.

The skeleton or frame of the mountain is formed of dark porphyritic and basaltic rocks, while its ribs and protuberances are covered over and smoothed down by an enormous deposit of volcanic scoriæ, to which is due the regular form of the peak. The rock of the other mountain is more compact, lighter colored, sometimes reddish, seldom amygdaloid, or spongy and very uniform. The limits of vegetation reach to about one-half the height of the mountain, a vast forest of pines of various species. Above this for two or three thousand feet the slopes are composed of dark gray or dirty red volcanic sand, with few crags and rocks protruding. Above this begins the ever-varying snow line, above which eternal snows cover the final slopes of the volcano, wherever they are not too steep to permit its lodging. Geologists state that Popocatepetl has had no eruption or emission of lava for centuries, but earthquake shocks occur every year in its vicinity, and the neighboring inhabitants are occasionally startled by dull sounds, like a plaintive moan uttered by a sleeping giant. History records the emission of smoke at various times. It is a tedious, but not in the least degree dangerous, journey to ascend it and stand on the brink of the crater, a yawning caldron in which the smoke of the three solfataras may be seen often mingled with the whirling clouds of a regular snow fall.

The two summits of Popocatepetl and Yztac-cihuatl are connected by an apparently eroded ridge, which presents itself like a deep gap, notwithstanding its mean altitude of 10,000 feet, so that they shoot up in bold relief like perfectly isolated masses. Their bases are hid by lower mountains running northward, and the railroad rounds the outer spur of these ranges in order to descend into the valley of Mexico from the northeast. We, therefore, see the volcanoes in the course of six hours, in going from Vera Cruz to Mexico, successively from the east, northeast, north, and finally upon reaching the city of Mexico from the northwest. It was while Cortéz and his Spaniards were yet in the higher timbered regions of Popocatepetl, they enjoyed that first glorious view of the valley and the lakes which Prescott has so graphically described. A. F. B.

11. *Mexico.* Few countries inspire so varied an interest as the valley of Mexico. It is the site of an ancient civilization of American people, and recollections the most affecting are associated with the city of Mexico and more ancient monuments, such as the Pyramids of Teotihuacan, dedicated to the sun and moon. Those who have studied the history of the conquest, delight to trace the military positions of Cortéz and of the Tlascaltee army. The naturalist contemplates with interest the immense elevation of the Mexican table-land, and the extraordinary form of a chain of porphyritic and basaltic mountains which surround the valley like a circular wall. He perceives that the whole valley is at the bottom of a dried up lake. The basins of fresh and salt water which fill the centre of the plain, and the five marshes, are to the eye of the geologist the small remains of a great mass of water which formerly covered the whole valley. HUMBOLDT.

The valley of Mexico, however beautiful it may appear under certain aspects of light, is in fact the remnant, not of a deep mountain-lake, but of an enormous marsh, formed by the accumulation, without natural outlet, of the waters collected on the tops and running down the slopes of the high ranges surrounding it. In the very centre of the Lake of Tezcoco flat barges or scows sometimes are in danger of grounding. The descriptions furnished by eye witnesses of the conquest by Cortéz, of the beauty and fertility of the Mexican valley, need not surprise us. The effect from a distance, on a clear day, in the limpid and transparent sky of these altitudes, 7,340 English feet above sea-level, is enchanting. To the little band of Spaniards, traveling along the lake shore by the sides of the cultivated patches which the Indians had grouped around their pueblos, near the placid water, the first which they had seen since leaving the coast, the sight must have been charming. And when, through the filling up of the marsh, parts of it became transformed into sober corn fields, we need not wonder at the regret expressed by some respecting the change. It was the feeling which we ourselves experience at seeing the picturesque supplanted by the useful. A. F. B.

Ferrocaril Central Mexicano, or Mexican Central Railroad.—Con.

Ms.		Alt.	
	Mexico.		
70	Angeles.	7913	⎫ The geology, so far as known, is given in the notes.
74	Lena.	8109	⎭
	Hidalgo.		
76	Marquez.	"	7961
81	Nopala.	"	7681
86	Danu.	"	7838
	Mexico.		
94	Polotitlan.14	"	7520
	Hidalgo.		
100	Cazadero.	"	7880
	Queretaro.		
107	Palmillas.	"	7093
118	San Juan del Rio.	"	6251
127	Chintepec.	"	6217
134	Ahorcado.	"	6359
149	Hercules.	"	6049
158	Queretaro.	"	5949
	Guanajuato.		
164	Mariscala.	"	5867
173	Apaseo.	"	5798
181	Celaya.	"	5765
192	Guaje.	"	5708
207	Salamanca.	"	5648
213	Chico.	"	5645
219	Irapuato.	"	5655

(left bracket note: Mountains supposed to be the same as Zacatecas.)

Ferrocaril Central Mexicano, or Mexican Central Railroad.—Con.

Ms.		Alt.	
229	Villalobos.	5728	⎫ The geology, as far as known, is given in the notes.
238	Silao	5828	
249	Trinidad.	5964	⎭
258	Leon.	"	5889
268	Francisco.	"	5790
	Jalisco.		
278	Pedrito.	"	5889
287	Loma.	"	6302
295	Lagos.	"	6188
306	Serrano.	"	6613
308	Los Salas.	"	6676
323	Santa Maria.	"	6051
334	Encarnacion.	"	6078
	Aguascalientes.		
350	Penuelas.	"	6164
364	Aguascalientes26	"	6151
382	Pabellon.	"	6261
388	Rincon de Romois	"	6321
400	Soledad.	"	6492
	Zacatecas.		
423	Summit.	"	7859
432	Guadalupe.14	"	7645
439	Zacatecas.15	Por'y Hu. Schists.	8011
447	Pimienta.	"	7556
457	Calera.	"	7062
474	Fresnillo.21	"	6862
484	Mendoza.19	"	6900

(right bracket note: Mountains supposed to be the same as Zacatecas.)

12. Very interesting human remains were found in January, 1884, some two and a half miles east of the city of Mexico, imbedded in a rock composed of silicified calcareous tufa. They are described and illustrated in the *American Naturalist*, for August, 1885.

12. *Mexico.* The valley of Mexico is eighteen and one-third leagues or fifty-five miles long, and twelve and a half leagues or thirty-seven miles in breadth. The crest of the mountains which surround it like a circular wall, is most elevated on the southeast, where the great volcanoes La Puebla, Popocatepetl, and Iztaccihuatl bound the valley. The city is no longer built in the midst of a lake, connected with the continent merely by three dikes, owing to the diminution of water of the lake Tezcuco. Humboldt pronounced Mexico, undoubtedly one of the finest cities ever built by Europeans in either hemisphere, but much less from the grandeur and beauty of its structures, than from its uniform regularity, its extent and position, leaving a recollection of grandeur which he attributes to the majestic character of its situation and the surrounding scenery. The beautifully cultivated valley forms a singular contrast with the wild appearance of the naked mountains which enclose it, among which the three famous volcanoes above named, with their enormous cones covered with perpetual snow, are the most distinguished.

14. *Guadalupe.* Dr. H. M. Chance, mining engineer, and lately an assistant on the second Geological Survey of Pennsylvania, who has been over this road, describes the plateau on which it is built as resembling to the traveler a flat valley, for mountains are seen on both sides of the railroad. But the chains, upon close examination, are seen to be simply a series of ranges, broken at many points. The flat plateau seems to have been formed by Tertiary (?) deposits, filling in what were formerly deep valleys between these mountain ranges, thus forming a network of level connected valleys, the Tertiary deposits filling them up above the lower connecting ridges, leaving them in the condition of half buried mountains. This description by Dr. Chance is probably as true as it is picturesque.

Between Zacatecas and the City of Mexico, Dr. Chance had less opportunity of examining the geology than at at Zacatecas, but he thought the mountains on this part of the route are Laurentian or Huronian, consisting of granites, porphyry, etc., and that the plateau or apparent valleys are Tertiary or Quaternary. The mountains nearer Mexico are partly volcanic, and at some points north also volcanic deposits are seen. These lava beds generally lie west of the railroad and form "buttes" or flat top mountains, the lava beds protecting the soft Tertiary deposits from erosion. (See Note 15.)

15. *Zacatecas.* In the Zacatecas mining region an entirely different series of rocks from those to the northward is seen, apparently Huronian schists, with porphyry and Laurentian granites. This same series also occurs all along the range extending northwest, and lying, as at Chihuahua, twenty to one hundred miles west of the railroad. It probably also comes up in some of the ranges east of the railroad.   H. M. C.

16. *Jimulco.* The coal at Jimulco occurs in the plateau Tertiary deposits, and is apparently a lignitic bed of fluvio-marine origin. The bed opened in 1885 was too largely mixed with clay, etc. to be of any commercial value. See Note 1. Dr. Chance examined the mountains only at Jimulco, and found them to consist of an enormously thick series of limestone, partly metamorphosed, and probably of Upper Carboniferous age.

Ms.	Ferrocaril Central Mexicano, or Mexican Central Railroad.—Con.	Alt.	Ms.	Ferrocaril Central Mexicano, or Mexican Central Railroad.—Con.	Alt.
493	Gutierrez.	Huronian Schists. 6647	844	Dolores.	Valley 20 ms wide 4521
507	Canitas.	6652	853	Jimenez.	Mt. l. s. to south. 4531
515	Cedro.	6439	865	La Reforma.	" 4423
528	La Colorada.	6421	877	Diaz.	" 4261
544	Pacheco.	6197	889	Bustamante.	" 4127
556	Guzman.	5941	898	Santa Rosalia.	{ Hills of Amigdaloid Basalt. 4022
568	Gonzalez.	5765			
581	Camancho.	5461	908	La Cruz.	{ Same wide val. running N.E. & S.W 3192
	Coahuila.				
595	San Isidoro.	5991	921	Concho.	" 4003
609	Symon.	5147			
624	La Mancha.	5110	931	Saucillo.	{ Limestone instead of the prevailing porphyry. 3971
637	Calvo.	5003			
652	Peralta.	4439	941	Las Delicias.	" 3889
662	Jimulco.16 4151	{ Mountains of enormously thick beds of Up. Carbon.	945	Ortiz.19	" 3797
671	Jalisco. 4042		960	Bachimba.	" 4147
	Durango.		971	Horcasitas.	{ Narrow pass 6 miles long and 1 mile wide. 4453
680	Picardias	" 3953			
	Coahuila.		985	Mapula.	" 4946
695	Matamoros.	" 3758	999	Chihuahua.20	See Note. 4634
	Durango		1014	Sacramento	4956
709	Lerdo.	" 3735	1023	Torreon.	5221
720	Noe.	" 3664	1030	Sauz.	5133
732	Mapimi.17	" 3694	1043	Encinillas.	5032
747	Peronal.18	Not on the valleys 3655	1051	Agua Nueva.	5011
761	Conejos.16	3761	1060	Laguna.	5088
775	Yermo.	3802	1072	Puerto.	5311
787	Saez.	3999	1085	Gallego.	5321
	Chihuahua.		1103	Chivatito.	4557
798	Zavalza.	3942	1112	Montezuma.	4656
807	Escalon.	4144	1120	Las Minas.	4524
819	Rellano.	4368	1129	Ojocaliente.21	4046
832	Corralitos.	4734	1136	Carmen.	Porphyritic rocks 3969
			1150	San Jose.	" 3919

Left bracket notes: "The main chain of the mountains is limestone." — "The main chain of the mountains is limestone. W."

Right bracket note: "Mountains, igneous rocks, porphyritic and trachyte, red, blue, white and grey."

17. *Mapimi*, lies in an eastern corner of the valley, surrounded by high mountains, in which silver are worked. Five miles south of it the ;Bolson de Mapini begins, beyond a cañon, a very large open level valley, like a pouch or pocket, whence the name. A steep high limestone mountain on the east, and another chain to the left. W.

18. *Peronal* and *Conejos.* This whole country is one large network of encasea valleys, connected with each other by good mountain passes and defiles. Some of the mountains are compact limestone. W.

19. *Mendoza.* From the topographical appearance of the mountains and the natural escarpments seen all along the road for three hundred miles from above Chihuahua, to within fifty miles of Zacatecas, Dr. Chance thinks the mountain rocks to be of similar character throughout this distance to those at Jimulco, namely, a very heavy formation of metamorphic Upper Carboniferous limestone.

20. *Chihuahua* was settled in 1691, and has a beautiful site amidst a circle of mountains opening to the south, with its churches and steeples, flat-roofed and commodious houses, its acqueducts and evergreen alameda. The rocks about Chihuahua, and at a point twenty miles northward, are porphyritic and trachytic, red, blue, white and gray. W.

The Mountains West of Chihuahua. Dr. Wislizenus was, during the Mexican war, detained six months a prisoner at Corihunachi, in the Sierra Madre Mountains, about ninety miles west of Chihuahua. The place is 6,275 feet above the sea, and the highest peak of the chain of mountains, directly above the place, called the Bufa, a prominent landmark, is 7,918 feet. This is in the very heart of the Sierra Madre, and there were some renowned silver mines there, all found in the porphyritic rocks, the prevailing formation in this part of the country. He reports the geology of the country as quite uniform, and although he roamed in hunting for months in that vicinity over the Sierra Madre, which occupies the whole western portion of the State of Chihuahua, the connecting link between the Rocky Mountains of the north and the Andes of the south, he observed no other formations than porphyritic, except stratified limestone. These mountains contain old mines of silver, gold, lead, iron and tin, which were celebrated in their day.

21. *Fresnillo. General Aspect of the Country.* From a short distance south of El Paso nearly to Zacatecas, some seven hundred miles, the plateau on which the railroad is built is (in 1885) little better than a desert. The grass is generally scattered and bunched, and there is very little grass to be seen at all, the principal vegetation being cactus and scrubby mesquite, and there is an almost

Ferrocarll Central Mexicano, or Mexican Central Railroad.—Con.

Ms.		Alt.
1165	Rancheria.²²	Amygdaloid basalt, Mt. with l. s. 4205
1176	Candelaria.	Granite and porphyritic Mts. 4397
1183	Los Mendanos.	Chiefly limestone. 4259
1194	Samalayuca.²³	Some granite & 4161
1204	Tierra Blancha.	porphyritic. 4145
1213	Mesa.	Limestone, 50 3960
1224	Paso del Norte.	miles. 3717
	El Paso.²⁴	

Mexican National Railway

(Southern General Division.)

Ms.		Alt.
0	Mexico.	7847
4	Tacuba.	Geology unknown 7897
9	Rio Hondo.	7550
24	Cima.*	(Summit.) 9974
32	Jajalpa.	" 8872
37	Lerma.	" 8486
45	Toluca.	" 8653
69	Ixtlahuaca.	" 8423
98	El Oro.	" 8344
139	Maravatio.	" 6612
178	Acambaro.	" 6084
235	Moretia.	" 6202

Mexican National Railway.
(Northern General Division.)†

Ms.			Alt.
	Nuevo Leon.		
0	Laredo.‡	19 a. Eocene.	508
1	Nuevo Laredo.	"	
23	Jarita.	"	
49	Rodriguez.²⁵	{ 19 c. Pliocene, or	
76	Lampazos.	{ 20. Quaternary.	
109	Bustamante.²⁶	" Mt. granite.	
111	Villaldame.	"	
128	Palo Blanco.	"	
151	Salinas.	"	
163	Topo.	"	
172	Monterey.²⁷	Up. Carb. l. s.	1626
174	Gonozalitos.	"	
176	San Geronimo.	"	
173	Leona.	"	
180	Santa Catarin.	"	
193	Carcia.	"	
	Cohahuila.		
209	Rinconada.	"	8381
215	Los Muertos.	"	
222	Ojo Caliente.	"	
226	Santa Maria.	"	
240	Santillo.	"	8242
246	Buena Vista.	"	
279	Encarnacion.	"	
323	El Salado.	"	6104

* The highest railroad point in Mexico.
† The altitudes of the places on this division are barometrical, taken by Dr. Wislizenus before the railroad was built.
‡ See Note 4 in Texas chapter.

entire absence of trees. But wherever the road approaches one of the principal water courses the scene changes. Irrigating ditches are seen on both sides of the stream, which is fringed as are the ditches by trees. These spots are as oases in a desert, and the land is apparently very fertile.   C.

22. *Rancheria.* A porous, black-looking basaltic rock known as amygdaloidal basalt is very common throughout the whole of Mexico. Below it, in New Mexico and at El Paso, is a compact quartoze furruginous sandstone, appearing as if changed by volcanic action.

23. *Samalayuca.* After leaving El Paso, Texas, or Paso del Norte, Mexico, to the west is a mountain chain, and to the east the receding valley of the Rio del Norte, from which, in going south, a high chain of mountains soon separate you, the road passing over a wide sandy plain covered with mesquite and similar shrubbery, and then runs for many miles through sand hills or "dunes," that are apparently of recent age. These sand hills similar to those in Texas, are an immense field of steep sandy ridges, without shrubs or vegetation of any kind, looking like a piece of Arabian desert transplanted into this plain, or like the bottom of the sea uplifted from the deep.

24. *Paso del Norte and El Paso.* See Notes 12, 13, 16, and 17 in Texas chapter.

25. Dr. Persifor Frazer, who passed over this road says, the valley traversed by it is a calcareous formation, much crushed and altered, which is clearly newer than the Upper Carboniferous mountains between which it lies. It may be 19 c. Pliocene or that and Quaternary, but no fossils have yet been found, and it may be 19 b. Loup Fork Miocene.

26. The Caudela Mountain is granite, also the Panuco, and a spur of the former reaching towards and near Bustamante. They protrude from the Upper Carboniferous. There is a large trap *mesa* about seven miles northeast of Caldera.   P. F.

27. The limestone mountains on this road are reported, by those who have seen them both, to be similar to those on the Mexican Central (See Notes 18 and 19.) It forms steep, often rugged, mountains, rising on an average 2,000 feet above the plain. It is metalliferous, containing silver and lead mines, and has all the appearance of the limestone found at El Paso and Chihuahua, but as yet we have no report of the discovery here of any fossils.

28. *Aguascalientes.* Here are famous hot springs, as indicated by the name. The place is a celebrated resort for invalids, and one of the cleanest provincial towns in Mexico. Population reported 20,000.   H. M. C.

There are several other railroads in Mexico, but as yet I have learned nothing in regard to their geology.'   J. M.

INDEX OF RAILROADS.

N. B.—Branches, or minor roads, will generally be found under the name of the main or controlling line. The latest names, owing to the constant changes, can not always be given, but in some instances roads, given in the body of the book under an old name, will be found indexed under the new, as well as the old. The Guide is in itself an Index, and this Index is only an additional help to the traveler.